Lecture Notes in Computer Science   **12662**

More information about this subseries at http://www.springer.com/series/7412

Alberto Del Bimbo · Rita Cucchiara ·
Stan Sclaroff · Giovanni Maria Farinella ·
Tao Mei · Marco Bertini ·
Hugo Jair Escalante · Roberto Vezzani (Eds.)

Pattern Recognition

ICPR International Workshops and Challenges

Virtual Event, January 10–15, 2021
Proceedings, Part II

 Springer

Editors
Alberto Del Bimbo ⓘ
Dipartimento di Ingegneria
dell'Informazione
University of Firenze
Firenze, Italy

Stan Sclaroff ⓘ
Department of Computer Science
Boston University
Boston, MA, USA

Tao Mei
Cloud & AI, JD.COM
Beijing, China

Hugo Jair Escalante ⓘ
Computational Sciences Department
National Institute of Astrophysics,
Optics and Electronics (INAOE)
Tonantzintla, Puebla, Mexico

Rita Cucchiara ⓘ
Dipartimento di Ingegneria "Enzo Ferrari"
Università di Modena e Reggio Emilia
Modena, Italy

Giovanni Maria Farinella ⓘ
Dipartimento di Matematica e Informatica
University of Catania
Catania, Italy

Marco Bertini ⓘ
Dipartimento di Ingegneria
dell'Informazione
University of Firenze
Firenze, Italy

Roberto Vezzani ⓘ
Dipartimento di Ingegneria "Enzo Ferrari"
Università di Modena e Reggio Emilia
Modena, Italy

ISSN 0302-9743   ISSN 1611-3349   (electronic)
Lecture Notes in Computer Science
ISBN 978-3-030-68789-2   ISBN 978-3-030-68790-8   (eBook)
https://doi.org/10.1007/978-3-030-68790-8

LNCS Sublibrary: SL6 – Image Processing, Computer Vision, Pattern Recognition, and Graphics

This Springer imprint is published by the registered company Springer Nature Switzerland AG
The registered company address is: Gewerbestrasse 11, 6330 Cham, Switzerland

Foreword by General Chairs

It is with great pleasure that we welcome you to the post-proceedings of the 25th International Conference on Pattern Recognition, ICPR2020 Virtual-Milano. ICPR2020 stands on the shoulders of generations of pioneering pattern recognition researchers. The first ICPR (then called IJCPR) convened in 1973 in Washington, DC, USA, under the leadership of Dr. King-Sun Fu as the General Chair. Since that time, the global community of pattern recognition researchers has continued to expand and thrive, growing evermore vibrant and vital. The motto of this year's conference was *Putting Artificial Intelligence to work on patterns*. Indeed, the deep learning revolution has its origins in the pattern recognition community – and the next generations of revolutionary insights and ideas continue with those presented at this 25th ICPR. Thus, it was our honor to help perpetuate this longstanding ICPR tradition to provide a lively meeting place and open exchange for the latest pathbreaking work in pattern recognition.

For the first time, the ICPR main conference employed a two-round review process similar to journal submissions, with new papers allowed to be submitted in either the first or the second round and papers submitted in the first round and not accepted allowed to be revised and re-submitted for second round review. In the first round, 1554 new submissions were received, out of which 554 (35.6%) were accepted and 579 (37.2%) were encouraged to be revised and resubmitted. In the second round, 1696 submissions were received (496 revised and 1200 new), out of which 305 (61.4%) of the revised submissions and 552 (46%) of the new submissions were accepted. Overall, there were 3250 submissions in total, and 1411 were accepted, out of which 144 (4.4%) were included in the main conference program as orals and 1263 (38.8%) as posters (4 papers were withdrawn after acceptance). We had the largest ICPR conference ever, with the most submitted papers and the most selective acceptance rates ever for ICPR, attesting both the increased interest in presenting research results at ICPR and the high scientific quality of work accepted for presentation at the conference.

We were honored to feature seven exceptional Keynotes in the program of the ICPR2020 main conference: David Doermann (Professor at the University at Buffalo), Pietro Perona (Professor at the California Institute of Technology and Amazon Fellow at Amazon Web Services), Mihaela van der Schaar (Professor at the University of Cambridge and a Turing Fellow at The Alan Turing Institute in London), Max Welling (Professor at the University of Amsterdam and VP of Technologies at Qualcomm), Ching Yee Suen (Professor at Concordia University) who was presented with the IAPR 2020 King-Sun Fu Prize, Maja Pantic (Professor at Imperial College UK and AI Scientific Research Lead at Facebook Research) who was presented with the IAPR 2020 Maria Petrou Prize, and Abhinav Gupta (Professor at Carnegie Mellon University and Research Manager at Facebook AI Research) who was presented with the IAPR 2020 J.K. Aggarwal Prize. Several best paper prizes were also announced and awarded, including the Piero Zamperoni Award for the best paper authored by a student, the BIRPA Best Industry Related Paper Award, and Best Paper Awards for each of the five tracks of the ICPR2020 main conference.

The five tracks of the ICPR2020 main conference were: (1) Artificial Intelligence, Machine Learning for Pattern Analysis, (2) Biometrics, Human Analysis and Behavior Understanding, (3) Computer Vision, Robotics and Intelligent Systems, (4) Document and Media Analysis, and (5) Image and Signal Processing. The best papers presented at the main conference had the opportunity for publication in expanded format in journal special issues of *IET Biometrics* (tracks 2 and 3), *Computer Vision and Image Understanding* (tracks 1 and 2), *Machine Vision and Applications* (tracks 2 and 3), *Multimedia Tools and Applications* (tracks 4 and 5), *Pattern Recognition Letters* (tracks 1, 2, 3 and 4), or *IEEE Trans. on Biometrics, Behavior, and Identity Science* (tracks 2 and 3).

In addition to the main conference, the ICPR2020 program offered workshops and tutorials, along with a broad range of cutting-edge industrial demos, challenge sessions, and panels. The virtual ICPR2020 conference was interactive, with real-time live-streamed sessions, including live talks, poster presentations, exhibitions, demos, Q&A, panels, meetups, and discussions – all hosted on the Underline virtual conference platform.

The ICPR2020 conference was originally scheduled to convene in Milano, which is one of the most beautiful cities of Italy for art, culture, lifestyle – and more. The city has so much to offer! With the need to go virtual, ICPR2020 included interactive **virtual tours** of Milano during the conference coffee breaks, which we hoped would introduce attendees to this wonderful city, and perhaps even entice them to visit Milano once international travel becomes possible again.

The success of such a large conference would not have been possible without the help of many people. We deeply appreciate the vision, commitment, and leadership of the ICPR2020 Program Chairs: Kim Boyer, Brian C. Lovell, Marcello Pelillo, Nicu Sebe, René Vidal, and Jingyi Yu. Our heartfelt gratitude also goes to the rest of the main conference organizing team, including the Track and Area Chairs, who all generously devoted their precious time in conducting the review process and in preparing the program, and the reviewers, who carefully evaluated the submitted papers and provided invaluable feedback to the authors. This time their effort was considerably higher given that many of them reviewed for both reviewing rounds. We also want to acknowledge the efforts of the conference committee, including the Challenge Chairs, Demo and Exhibit Chairs, Local Chairs, Financial Chairs, Publication Chair, Tutorial Chairs, Web Chairs, Women in ICPR Chairs, and Workshop Chairs. Many thanks, also, for the efforts of the dedicated staff who performed the crucially important work behind the scenes, including the members of the ICPR2020 Organizing Secretariat. Finally, we are grateful to the conference sponsors for their generous support of the ICPR2020 conference.

We hope everyone had an enjoyable and productive ICPR2020 conference.

Rita Cucchiara
Alberto Del Bimbo
Stan Sclaroff

Preface

The 25th International Conference on Pattern Recognition Workshops (ICPRW 2020) were held virtually in Milan, Italy and rescheduled to January 10 and January 11 of 2021 due to the Covid-19 pandemic. ICPRW 2020 included timely topics and applications of Computer Vision, Image and Sound Analysis, Pattern Recognition and Artificial Intelligence. We received 49 workshop proposals and 46 of them have been accepted, which is three times more than at ICPRW 2018. The workshop proceedings cover a wide range of areas including Machine Learning (8), Pattern Analysis (5), Healthcare (6), Human Behavior (5), Environment (5), Surveillance, Forensics and Biometrics (6), Robotics and Egovision (4), Cultural Heritage and Document Analysis (4), Retrieval (2), and Women at ICPR 2020 (1). Among them, 33 workshops are new to ICPRW. Specifically, the ICPRW 2020 volumes contain the following workshops (please refer to the corresponding workshop proceeding for details):

- CADL2020 – Workshop on Computational Aspects of Deep Learning.
- DLPR – Deep Learning for Pattern Recognition.
- EDL/AI – Explainable Deep Learning/AI.
- (Merged) IADS – Integrated Artificial Intelligence in Data Science, IWCR – IAPR workshop on Cognitive Robotics.
- ManifLearn – Manifold Learning in Machine Learning, From Euclid to Riemann.
- MOI2QDN – Metrification & Optimization of Input Image Quality in Deep Networks.
- IML – International Workshop on Industrial Machine Learning.
- MMDLCA – Multi-Modal Deep Learning: Challenges and Applications.
- IUC 2020 – Human and Vehicle Analysis for Intelligent Urban Computing.
- PATCAST – International Workshop on Pattern Forecasting.
- RRPR – Reproducible Research in Pattern Recognition.
- VAIB 2020 – Visual Observation and Analysis of Vertebrate and Insect Behavior.
- IMTA VII – Image Mining Theory & Applications.
- AIHA 2020 – Artificial Intelligence for Healthcare Applications.
- AIDP – Artificial Intelligence for Digital Pathology.
- (Merged) GOOD – Designing AI in support of Good Mental Health, CAIHA – Computational and Affective Intelligence in Healthcare Applications for Vulnerable Populations.
- CARE2020 – pattern recognition for positive teChnology And eldeRly wEllbeing.
- MADiMa 2020 – Multimedia Assisted Dietary Management.
- 3DHU 2020 – 3D Human Understanding.
- FBE2020 – Facial and Body Expressions, micro-expressions and behavior recognition.
- HCAU 2020 – Deep Learning for Human-Centric Activity Understanding.
- MPRSS - 6th IAPR Workshop on Multimodal Pattern Recognition for Social Signal Processing in Human Computer Interaction.

- CVAUI 2020 – Computer Vision for Analysis of Underwater Imagery.
- MAES – Machine Learning Advances Environmental Science.
- PRAConBE - Pattern Recognition and Automation in Construction & the Built Environment.
- PRRS 2020 – Pattern Recognition in Remote Sensing.
- WAAMI - Workshop on Analysis of Aerial Motion Imagery.
- DEEPRETAIL 2020 - Workshop on Deep Understanding Shopper Behaviours and Interactions in Intelligent Retail Environments 2020.
- MMForWild2020 – MultiMedia FORensics in the WILD 2020.
- FGVRID – Fine-Grained Visual Recognition and re-Identification.
- IWBDAF – Biometric Data Analysis and Forensics.
- RISS – Research & Innovation for Secure Societies.
- WMWB – TC4 Workshop on Mobile and Wearable Biometrics.
- EgoApp – Applications of Egocentric Vision.
- ETTAC 2020 – Eye Tracking Techniques, Applications and Challenges.
- PaMMO – Perception and Modelling for Manipulation of Objects.
- FAPER – Fine Art Pattern Extraction and Recognition.
- MANPU – coMics ANalysis, Processing and Understanding.
- PATRECH2020 – Pattern Recognition for Cultural Heritage.
- (Merged) CBIR – Content-Based Image Retrieval: where have we been, and where are we going, TAILOR – Texture AnalysIs, cLassificatiOn and Retrieval, VIQA – Video and Image Question Answering: building a bridge between visual content analysis and reasoning on textual data.
- W4PR - Women at ICPR.

We would like to thank all members of the workshops' Organizing Committee, the reviewers, and the authors for making this event successful. We also appreciate the support from all the invited speakers and participants. We wish to offer thanks in particular to the ICPR main conference general chairs: Rita Cucchiara, Alberto Del Bimbo, and Stan Sclaroff, and program chairs: Kim Boyer, Brian C. Lovell, Marcello Pelillo, Nicu Sebe, Rene Vidal, and Jingyi Yu. Finally, we are grateful to the publisher, Springer, for their cooperation in publishing the workshop proceedings in the series of Lecture Notes in Computer Science.

December 2020   Giovanni Maria Farinella
                                                              Tao Mei

Challenges

Competitions are effective means for rapidly solving problems and advancing the state of the art. Organizers identify a problem of practical or scientific relevance and release it to the community. In this way the whole community can contribute to the solution of high-impact problems while having fun. This part of the proceedings compiles the best of the competitions track of the *25th International Conference on Pattern Recognition (ICPR)*.

Eight challenges were part of the track, covering a wide variety of fields and applications, all of this within the scope of ICPR. In every challenge organizers released data, and provided a platform for evaluation. The top-ranked participants were invited to submit papers for this volume. Likewise, organizers themselves wrote articles summarizing the design, organization and results of competitions. Submissions were subject to a standard review process carried out by the organizers of each competition. Papers associated with seven out the eight competitions are included in this volume, thus making it a representative compilation of what happened in the ICPR challenges.

We are immensely grateful to the organizers and participants of the ICPR 2020 challenges for their efforts and dedication to make the competition track a success. We hope the readers of this volume enjoy it as much as we have.

November 2020   Marco Bertini
                                                                                             Hugo Jair Escalante

ICPR Organization

General Chairs

Rita Cucchiara	Univ. of Modena and Reggio Emilia, Italy
Alberto Del Bimbo	Univ. of Florence, Italy
Stan Sclaroff	Boston Univ., USA

Program Chairs

Kim Boyer	Univ. at Albany, USA
Brian C. Lovell	Univ. of Queensland, Australia
Marcello Pelillo	Univ. Ca' Foscari Venezia, Italy
Nicu Sebe	Univ. of Trento, Italy
René Vidal	Johns Hopkins Univ., USA
Jingyi Yu	ShanghaiTech Univ., China

Workshop Chairs

Giovanni Maria Farinella	Univ. of Catania, Italy
Tao Mei	JD.COM, China

Challenge Chairs

Marco Bertini	Univ. of Florence, Italy
Hugo Jair Escalante	INAOE and CINVESTAV National Polytechnic Institute of Mexico, Mexico

Publication Chair

Roberto Vezzani	Univ. of Modena and Reggio Emilia, Italy

Tutorial Chairs

Vittorio Murino	Univ. of Verona, Italy
Sudeep Sarkar	Univ. of South Florida, USA

Women in ICPR Chairs

Alexandra Branzan Albu	Univ. of Victoria, Canada
Maria De Marsico	Univ. Roma La Sapienza, Italy

Demo and Exhibit Chairs

Lorenzo Baraldi   Univ. Modena Reggio Emilia, Italy
Bruce A. Maxwell   Colby College, USA
Lorenzo Seidenari   Univ. of Florence, Italy

Special Issue Initiative Chair

Michele Nappi   Univ. of Salerno, Italy

Web Chair

Andrea Ferracani   Univ. of Florence, Italy

Corporate Relations Chairs

Fabio Galasso   Univ. Roma La Sapienza, Italy
Matt Leotta   Kitware, Inc., USA
Zhongchao Shi   Lenovo Group Ltd., China

Local Chairs

Matteo Matteucci   Politecnico di Milano, Italy
Paolo Napoletano   Univ. of Milano-Bicocca, Italy

Financial Chairs

Cristiana Fiandra   The Office srl, Italy
Vittorio Murino   Univ. of Verona, Italy

Contents – Part II

CARE2020 - International Workshop on pattern recognition for positive teChnology And eldeRly wEllbeing

Visual-Textual Image Understanding and Retrieval (VTIUR) - Joint Workshop on Content-Based Image Retrieval (CBIR 2020), Video and Image Question Answering (VIQA 2020), Texture Analysis, Classification and Retrieval (TAILOR 2020)

CVAUI 2020 - 4th Workshop on Computer Vision for Analysis of Underwater Imagery

DLPR - Deep Learning for Pattern Recognition

CAIHA - Computational and Affective Intelligence in Healthcare Applications for Vulnerable Populations

CAIHA: Computational and Affective Intelligence in Healthcare Applications (Vulnerable Populations)

Intelligent monitoring systems and affective computing applications have emerged in recent years to enhance healthcare. Examples of these applications include intelligent monitoring of physiological changes and sedation states as well as assessment of affective states such as pain and depression. The intelligent monitoring of patients' conditions and affective states can: 1) lead to better understanding of their pattern; 2) provide consistent and continuous assessment, and hence prompt intervention; 3) expedite the early hospital discharge; and 4) decrease the family stress and financial burden. Though the intelligent monitoring of affective states has been around for the past several years, integrating contextual, personalized, uncertainty, and multimodal information recorded specifically from vulnerable populations has been less explored. This workshop provides an interdisciplinary forum for the exchange of ideas on novel applications, new datasets, current challenges, and future directions to enhance healthcare of vulnerable populations.

The first edition of the International Workshop on Computational and Affective Intelligence in Healthcare Applications - Vulnerable Populations (CAIHA 2020) will be held in conjunction with the International Conference on Pattern Recognition (ICPR), Milan, Italy, 10 January 2020. The format of the workshop includes a keynote (Dr. Rosalind Picard), then technical presentations followed by another keynote (Dr. Jiebo Luo), and finally poster presentations. We expect that the workshop will be attended by around 25–30 people.

This year, we received 13 submissions for reviews from authors belonging to 6 distinct countries: Brazil, China, France, Tunisia, Turkey, and United States. The papers were reviewed (single blind) by at least two reviewers with Strong background in computational and affective intelligence. The review process focused on the quality of the papers and their applicability to existing healthcare problems of vulnerable populations. The acceptance of the papers was the result of the reviewers' discussion and agreement. After an accurate and thorough peer-review process, we selected 8 papers (62%) for presentation at the workshop. The accepted articles represent an interesting mix of techniques/applications that solve current healthcare problems of vulnerable populations. Examples of these problems include assessing pain in neonates and patients with sickle cell disease, classification of autism spectrum disorder, supporting children with hearing impairments, and mental health diagnosis.

November 2020

Ghada Zamzmi

Organization

General Chair

Ghada Zamzmi   National Institutes of Health, USA

Program Committee Chairs

Shaun Canavan	University of South Florida, USA
Dmitry Goldgof	University of South Florida, USA
Rangachar Kasturi	University of South Florida, USA
Yu Sun	University of South Florida, USA
Michel Valstar	University of Nottingham, UK

Program Committee

Saeed Alahmari	Najran University, Saudi Arabia
Gang Cao	Communication University of China, China
Baishali Chaudhury	Intel corporation, USA
Palak Dave	University of South Florida, USA
Joy Egede	University of Nottingham, UK
Hamid Farhidzadeh	GumGum, USA
Mustafa Hajij	Santa Clara University, USA
Saurabh Hinduja	University of South Florida, USA
Rabiul Islam	Nanyang Technological University, Singapore
Yi Jin	Beijing Jiaotong University, China
Ahmad Jelodar	University of South Florida, USA
Rahul Paul	Harvard Medical School, USA
Sivarama Rajaraman	National Institutes of Health, USA
Mohamed Rahouti	University of South Florida, USA
Md Sirajus Salekin	University of South Florida, USA
Junyi Tu	Salisbury University, USA
Md Taufeeq Uddin	University of South Florida, USA
Ruicog Zhi	University of Science and Technology Beijing, China

Towards Robust Deep Neural Networks for Affect and Depression Recognition from Speech

Alice Othmani[1]([⊠])[iD], Daoud Kadoch[2], Kamil Bentounes[2], Emna Rejaibi[3], Romain Alfred[4], and Abdenour Hadid[5]

[1] University of Paris-Est Créteil, Vitry sur Seine, France
`alice.othmani@u-pec.fr`
[2] Sorbonne University, Paris, France
[3] INSAT, Tunis, Tunisie
[4] ENSIIE, Évry, France
[5] Polytechnic University of Hauts-de-France, Valenciennes, France

Abstract. Intelligent monitoring systems and affective computing applications have emerged in recent years to enhance healthcare. Examples of these applications include assessment of affective states such as Major Depressive Disorder (MDD). MDD describes the constant expression of certain emotions: negative emotions (low Valence) and lack of interest (low Arousal). High-performing intelligent systems would enhance MDD diagnosis in its early stages. In this paper, we present a new deep neural network architecture, called EmoAudioNet, for emotion and depression recognition from speech. Deep EmoAudioNet learns from the time-frequency representation of the audio signal and the visual representation of its spectrum of frequencies. Our model shows very promising results in predicting affect and depression. It works similarly or outperforms the state-of-the-art methods according to several evaluation metrics on RECOLA and on DAIC-WOZ datasets in predicting arousal, valence, and depression. Code of EmoAudioNet is publicly available on GitHub: https://github.com/AliceOTHMANI/EmoAudioNet.

Keywords: Emotional Intelligence · Socio-affective computing · Depression recognition · Speech emotion recognition · Healthcare application · Deep learning.

1 Introduction

Artificial Emotional Intelligence (EI) or affective computing has attracted increasing attention from the scientific community. Affective computing consists of endowing machines with the ability to recognize, interpret, process and simulate human affects. Giving machines skills of emotional intelligence is an important key to enhance healthcare and further boost the medical assessment of several mental disorders.

A. Del Bimbo et al. (Eds.): ICPR 2020 Workshops, LNCS 12662, pp. 5–19, 2021.
https://doi.org/10.1007/978-3-030-68790-8_1

Affect describes the experience of a human's emotion resulting from an inter-action with stimuli. Humans express an affect through facial, vocal, or gestural behaviors. A happy or angry person will typically speak louder and faster, with strong frequencies, while a sad or bored person will speak slower with low fre-quencies. Emotional arousal and valence are the two main dimensional affects used to describe emotions. Valence describes the level of pleasantness, while arousal describes the intensity of excitement. A final method for measuring a user's affective state is to ask questions and to identify emotions during an interaction. Several post-interaction questionnaires exist for measuring affective states like the Patient Health Questionnaire 9 (PHQ-9) for depression recognition and assessment. The PHQ is a self report questionnaire of nine clinical questions where a score ranging from 0 to 23 is assigned to describe Major Depressive Dis-order (MDD) severity level. MDD is a mental disease which affects more than 300 million people in the world [1], *i.e.*, 3% of the worldwide population. The psychiatric taxonomy classifies MDD among the low moods [2], *i.e.*, a condition characterised by a tiredness and a global physical, intellectual, social and emo-tional slow-down. In this way, the speech of depressive subjects is slowed, the pauses between two speakings are lengthened and the tone of the voice (prosody) is more monotonous.

In this paper, a new deep neural networks architecture, called EmoAudioNet, is proposed and evaluated for real-life affect and depression recognition from speech. The remainder of this article is organised as follows. Section 2 intro-duces related works with affect and depression recognition from speech. Section 3 introduces the motivations behind this work. Section 4 describes the details of the overall proposed method. Section 5 describes the entire experiments and the extensive experimental results. Finally, the conclusion and future work are presented in Sect. 6.

2   Related Work

Several approaches are reported in the literature for affect and depression recog-nition from speech. These methods can be generally categorized into two groups: hand-crafted features-based approaches and deep learning-based approaches.

2.1   Handcrafted Features-Based Approaches

In this family of approaches, there are two main steps: feature extraction and classification. An overview of handcrafted features-based approaches for affect and depression assessment from speech is presented in Table 1.

Handcrafted Features. Acoustic Low-Level Descriptors (LDD) are extracted from the audio signal. These LLD are grouped into four main categories: the **spectral LLD** (Harmonic Model and Phase Distortion Mean (HMPDM0-24), etc.), the **cepstral LLD** (Mel-Frequency Cepstral Coefficients (MFCC) [3,13], etc.), the **prosodic LLD** (Formants [21], etc.), and the **voice quality LLD**

(Jitter, and Shimmer [14], etc.). A set of statistical features are also calculated (max, min, variance and standard deviation of LLD [4,12]). Low *et al.* [16] propose the experimentation of the Teager Energy Operator (TEO) based features.

A comparison of the performances of the prosodic, spectral, glottal (voice quality), and TEO features for depression recognition is realized in [16] and it demonstrates that the different features have similar accuracies. The fusion of the prosodic LLD and the glottal LLD based models seems to not significantly improve the results, or decreased them. However, the addition of the TEO features improves the performances up to +31,35% for depressive male.

Classification of Handcrafted Features. Comparative analysis of the performances of several classifiers in depression assessment and prediction indicate that the use of an hybrid classifier using Gaussian Mixture Models (GMM) and Support Vector Machines (SVM) model gave the best overall classification results [6,16]. Different fusion methods, namely feature, score and decision fusion have been also investigated in [6] and it has been demonstrated that: first, amongst the fusion methods, score fusion performed better when combined with GMM, HFS and MLP classifiers. Second, decision fusion worked best for SVM (both for raw data and GMM models) and finally, feature fusion exhibited weak performance compared to other fusion methods.

2.2   Deep Learning-Based Approaches

Recently, approaches based on deep learning have been proposed [8,23–30]. Several handcrafted features are extracted from the audio signals and fed to the deep neural networks, except in Jain [27] where only the MFCC are considered. In other approaches, raw audio signals are fed to deep neural networks [19]. An overview of deep learning-based methods for affect and depression assessment from speech is presented in Table 2.

Several deep neural networks have been proposed. Some deep architectures are based on feed-forward neural networks [11,20,24], some others are based on convolutional neural networks such as [27] and [8] whereas some others are based on recurrent neural networks such as [13] and [23]. A comparative study [25] of some neural networks, BLSTM-MIL, BLSTM-RNN, BLSTM-CNN, CNN, DNN-MIL and DNN, demonstrates that the BLSTM-MIL outperforms the other studied architectures. Whereas, in Jain [27], the Capsule Network is demonstrated as the most efficient architecture, compared to the BLSTM with Attention mechanism, CNN and LSTM-RNN. For the assessment of the level of depression using the Patient Health Questionnaire 8 (PHQ-8), Yang *et al.* [8] exerts a DCNN. To the best of our knowledge, their approach outperforms all the existing approaches on DAIC-WOZ dataset.

3   Motivations and Contributions

Short-time spectral analysis is the most common way to characterize the speech signal using MFCCs. However, audio signals in their time-frequency

Table 1. Overview of shallow learning based methods for affect and depression assessment from speech. (*) Results obtained over a group of females.

Ref	Features	Classification	Dataset	Metrics	Value
Valstar et al. [3]	prosodic + voice quality + spectral	SVM + grid search + random forest	DAIC-WOZ	F1-score	0.410 (0.582)
				Precision	0.267 (0.941)
				Recall	0.889 (0.421)
				RMSE (MAE)	7.78 (5.72)
Dhall et al. [14]	energy + spectral + voicing quality + duration features	non-linear chi-square kernel	AFEW 5.0	*unavailable*	*unavailable*
Ringeval et al. [4]	prosodic LLD + voice quality + spectral	random forest	SEWA	RMSE	7.78
				MAE	5.72
Haq et al. [15]	energy + prosodic + spectral + duration features	Sequential Forward Selection + Sequential Backward Selection + linear discriminant analysis + Gaussian classifier uses Bayes decision theory	Natural speech databases	Accuracy	66.5%
Jiang et al. [5]	MFCC + prosodic +	ensemble logistic regression	hand-crafted	Males accuracy	81.82%(70.19%*)
	spectral LLD + glottal	model for detecting	dataset	Males sensitivity	78.13%(79.25%*)
	features	depression E algorithm		Males specificity	85.29%(70.59%*)
Low et al. [16]	teager energy operator	Gaussian mixture model +	hand-crafted	Males accuracy	86.64%(78.87%*)
	based features	SVM	dataset	Males sensitivity	80.83%(80.64%*)
				Males specificity	92.45%(77.27%*)
Alghowinem et al. [6]	energy + formants + glottal features + intensity + MFCC + prosodic + spectral + voice quality	Gaussian mixture model + SVM + decision fusion	hand-crafted dataset	Accuracy	91.67%
Valstar et al. [7]	duration features+energy	correlation based feature	AViD-	RMSE	14.12
	local min/max related functionals+spectral+voicing quality	selection + SVR + 5-flod cross-validation loop	Corpus	MAE	10.35
Valstar et al. [17]	duration features+energy	SVR	AVEC2014	RMSE	11.521
	local min/max related functionals+spectral+voicing quality			MAE	8.934
Cummins et al. [9]	MFCC + prosodic + spectral centroid	SVM	AVEC2013	Accuracy	82%
Lopez Otero et al. [10]	energy + MFCC + prosodic + spectral	SVR	AVDLC	RMSE (MAE)	8.88 (7,02)
Meng et al. [18]	spectral + energy + MFCC + functionals	PLS regression	AVEC2013	RMSE	11.54
				MAE	9.78
	features + duration features			CORR	0.42

Table 2. Overview of deep learning based methods for affect and depression assessment from speech.

Ref	Features	Classification	Dataset	Metrics	Value
Yang et al. [8]	spectral LLD + cepstral	DCNN	DAIC-WOZ	Depressed female RMSE	4.590
	LLD + prosodic LLD +			Depressed female MAE	3.589
	voice quality LLD +			Not depressed female RMSE	2.864
	statistical functionals +			Not depressed female MAE	2.393
	regression functionals			Depressed male RMSE	1.802
				Depressed male MAE	1.690
				Not depressed male RMSE	2.827
				Not depressed male MAE	2.575
Al Hanai et al. [23]	spectral LLD + cepstral	LSTM-RNN	DAIC	F1-score	0.67
	LLD + prosodic LLD +			Precision	1.00
	voice quality LLD +			Recall	0.50
	functionals			RMSE	10.03
				MAE	7.60
Dham et al. [24]	prosodic LLD + voice quality LLD + functionals + BoTW	FF-NN	AVEC2016	RMSE	7.631
				MAE	6.2766
Salekin et al. [25]	spectral LLD + MFCC + functionals	NN2Vec + BLSTM-MIL	DAIC-WOZ	F1-score	0.8544
				Accuracy	96.7%
Yang et al. [26]	spectral LLD + cepstral	DCNN-DNN	DAIC-WOZ	Female RMSE	5.669
	LLD + prosodic LLD +			Female MAE	4.597
	voice quality LLD +			Male RMSE	5.590
	functionals			Male MAE	5.107
Jain [27]	MFCC	Capsule Network	VCTK corpus	Accuracy	0.925
Chao et al. [28]	spectral LLD + cepstral LLD + prosodic LLD	LSTM-RNN	AVEC2014	*unavailable*	*unavailable*
Gupta et al. [29]	spectral LLD + cepstral LLD + prosodic LLD + voice quality LLD + functionals	DNN	AViD-Corpus	*unavailable*	*unavailable*
Kang et al. [30]	spectral LLD + prosodic LLD + articulatory features	DNN / SRI's submitted system to AVEC2014 / median-way score-level fusion	AVEC2014	RMSE	7.37
				MAE	5.87
				Pearson's Product Moment Correlation coefficient	0.800
Tzirakis et al. [36]	raw signal	CNN and 2-layers LSTM	RECOLA	loss function based on CCC	.440(arousal)
					.787(valence)
Tzirakis et al. [19]	raw signal	CNN and LSTM	RECOLA	CCC	.686(arousal)
					.261(valence)
Tzirakis et al. [22]	raw signal	CNN	RECOLA	CCC	.699(arousal)
					.311(valence)

representations, often present interesting patterns in the visual domain [31]. The visual representation of the spectrum of frequencies of a signal using its spectrogram shows a set of specific repetitive patterns. Surprisingly and to the best of our knowledge, it has not been reported in the literature a deep neural network architecture that combines information from time, frequency and visual domains for emotion recognition.

The first contribution of this work is a new deep neural network architecture, called EmoAudioNet, that aggregate responses from a short-time spectral analysis and from time-frequency audio texture classification and that extract deep features representations in a learned embedding space. In a second contribution, we propose EmoAudioNet-based approach for instantaneous prediction of spontaneous and continuous emotions from speech. In particular, our specific contributions are as follows: (i) an automatic clinical depression recognition and assessment embedding network (ii) a small size two-stream CNNs to map audio data into two types of continuous emotional dimensions namely, arousal and valence and (iii) through experiments, it is shown that EmoAudioNet-based features outperforms the state-of-the art methods for predicting depression on DAIC-WOZ dataset and for predicting valence and arousal dimensions in terms of Pearson's Coefficient Correlation (PCC).

Algorithm 1 EmoAudioNet embedding network.

Given two feature extractors f_Θ and f_ϕ, number of training steps N.

---

for *iteration in range(N)* **do**

     $(\mathbf{X}_{\text{wav}}, \mathbf{y}_{\text{wav}}) \leftarrow$ batch of input wav files and labels

     $\mathbf{e}_{\text{Spec}} \leftarrow f_\Theta(\mathbf{X}_{\text{wav}})$ Spectrogram features

     $\mathbf{e}_{\text{MFCC}} \leftarrow f_\phi(\mathbf{X}_{\text{wav}})$ MFCC features

     $\mathbf{f}_{\text{MFCCSpec}} \leftarrow [\mathbf{e}_{\text{MFCC}}, \mathbf{e}_{\text{Spec}}]$ Feature-level fusion

     $\mathbf{p}_{\text{MFCCSpec}} \leftarrow f_\theta(\mathbf{e}_{\text{MFCCSpec}})$ Predict class probabilities

     $L_{\text{MFCCSpec}} = \texttt{cross\_entropy\_loss}(\mathbf{p}_{\text{MFCCSpec}}, \mathbf{y}_{\text{wav}})$

     Obtain all gradients $\Delta_{\text{all}} = (\frac{\partial L}{\partial \Theta}, \frac{\partial L}{\partial \phi})$

     $(\Theta, \phi, \theta) \leftarrow \texttt{ADAM}(\Delta_{\text{all}})$ Update feature extractor and output heads' parameters simultaneously

end

---

4   Proposed Method

We seek to learn a deep audio representation that is trainable end-to-end for emotion recognition. To achieve that, we propose a novel deep neural network called EmoAudioNet, which performs low-level and high-level features extraction and aggregation function learning jointly (See Algorithm 1). Thus, the input audio signal is fed to a small size two-stream CNNs that outputs the final classification scores. A data augmentation step is considered to increase the amount of data by adding slightly modified copies of already existing data. The structure of EmoAudioNet presents three main parts as shown in Fig. 1: (i) An MFCC-based CNN, (ii) A spectrogram-based CNN and (iii) the aggregation of the responses of the MFCC-based and the spectrogram-based CNNs. In the following, more details about the three parts are given.

4.1   Data Augmentation

A data augmentation step is considered to overcome the problem of data scarcity by increasing the quantity of training data and also to improve the model's robustness to noise. Two different types of audio augmentation techniques are performed: (1) **Adding noise:** mix the audio signal with random noise. Each mix z is generated using $z = x + \alpha \times rand(x)$ where x is the audio signal and α is the noise factor. In our experiments, $\alpha = 0.01$, 0.02 and 0.03. (2) **Pitch Shifting:** lower the pitch of the audio sample by 3 values (in semitones): (0.5, 2 and 5).

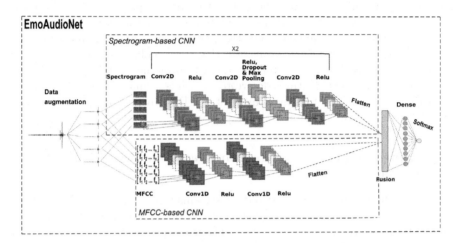

Fig. 1. The diagram of the proposed deep neural networks architecture called EmoAudioNet. The output layer is dense layer of size n neurones with a Softmax activation function. n is defined according to the task. When the task concerns binary depression classification, $n = 2$. When the task concerns depression severity level assessment, $n = 24$. While, $n = 10$ for arousal or valence prediction.

4.2   Spectrogram-Based CNN Stream

The spectrogram-based CNN presents low-level features descriptor followed by a high-level features descriptor. The Low-level features descriptor is the spectrogram of the input audio signal and it is computed as a sequence of Fast Fourier Transform (FFT) of windowed audio segments. The audio signal is split into 256 segments and the spectrum of each segment is computed. The Hamming window is applied to each segment. The spectrogram plot is a color image of $1900 \times 1200 \times 3$. The image is resized to $224 \times 224 \times 3$ before being fed to the High-level features descriptor. The high-Level features descriptor is a deep CNN, it takes as input the spectrogram of the audio signal. Its architecture, as shown in Fig. 1, is composed by two same blocks of layers. Each block is composed of a two-dimensional (2D) convolutional layer followed by a ReLU activation function, a second convolutional layer, a ReLU, a dropout and max pooling layer, a third convolutional layer and last ReLU.

4.3   MFCC-Based CNN Stream

The MFCC-based CNN presents also a low-level followed by high-level features descriptors (see Fig. 1). The low-level features descriptor is the MFCC features of the input audio. To extract them, the speech signal is first divided into frames by applying a Hamming windowing function of 2.5 s at fixed intervals of 500 ms. A cepstral feature vector is then generated and the Discrete Fourier Transform (DFT) is computed for each frame. Only the logarithm of the amplitude spectrum is retained. The spectrum is after smoothed and 24 spectral components into 44100 frequency bins are collected in the Mel frequency scale. The components of the Mel-spectral vectors calculated for each frame are highly correlated. Therefore, the Karhunen-Loeve (KL) transform is applied and is approximated by the Discrete Cosine Transform (DCT). Finally, 177 cepstral features are obtained for each frame. After the extraction of the MFCC features, they are fed to the high-Level features descriptor which is a small size CNN. To avoid overfitting problem, only two one-dimensional (1D) convolutional layers followed by a ReLU activation function each are performed.

4.4   Aggregation of the Spectrogram-Based and MFCC-Based Responses

Combining the responses of the two deep streams CNNs allows to study simultaneously the time-frequency representation and the texture-like time frequency representation of the audio signal. The output of the spectrogram-based CNN is a feature vector of size 1152, while the output of the MFCC-based CNN is a feature vector of size 2816. The responses of the two networks are concatenated and then fed to a fully connected layer in order to generate the label prediction of the emotion levels.

5   Experiments and Results

5.1   Datasets

Two publicly available datasets are used to evaluate the performances of EmoAudioNet:

Dataset for Affect Recognition Experiments: RECOLA dataset [32] is a multimodal corpus of affective interactions in French. 46 subjects participated to data recordings. Only 23 audio recordings of 5 min of interaction are made publicly available and used in our experiments. Participants engaged in a remote discussion according to a survival task and six annotators measured emotion continuously on two dimensions: valence and arousal.

Dataset for Depression Recognition and Assessment Experiments: DAIC-WOZ depression dataset [33] is introduced in the AVEC2017 challenge [4] and it provides audio recordings of clinical interviews of 189 participants.

Each recording is labeled by the PHQ-8 score and the PHQ-8 binary. The PHQ-8 score defines the severity level of depression of the participant and the PHQ-8 binary defines whether the participant is depressed or not. For technical reasons, only 182 audio recordings are used. The average length of the recordings is 15 min with a fixed sampling rate of 16 kHz.

5.2   Experimental Setup

Spectrogram-based CNN Architecture: The number of channels of the convolutional and pooling layers are both 128. While their filter size is 3×3. RELU is used as activation function for all the layers. The stride of the max pooling is 8. The dropout fraction is 0.1.

Table 3. RECOLA dataset results for prediction of arousal. The results obtained for the development and the test sets in term of three metrics: the accuracy, the Pearson's Coefficient Correlation (PCC) and the Root Mean Square error (RMSE).

	Development			Test		
	Accuracy	PCC	RMSE	Accuracy	PCC	RMSE
MFCC-based CNN	81.93%	0.8130	0.1501	70.23%	0.6981	0.2065
Spectrogram-based CNN	80.20%	0.8157	0.1314	75.65%	0.7673	0.2099
EmoAudioNet	94.49%	0.9521	0.0082	89.30%	0.9069	0.1229

Table 4. RECOLA dataset results for prediction of valence. The results obtained for the development and the test sets in term of three metrics: the accuracy, the Pearson's Coefficient Correlation (PCC) and the Root Mean Square error (RMSE).

	Development			Test		
	Accuracy	PCC	RMSE	Accuracy	PCC	RMSE
MFCC-based CNN	83.37%	0.8289	0.1405	71.12%	0.6965	0.2082
Spectrogram-based CNN	78.32%	0.7984	0.1446	73.81%	0.7598	0.2132
EmoAudioNet	95.42%	0.9568	0.0625	**91.44%**	**0.9221**	**0.1118**

MFCC-based CNN Architecture: The input is one-dimensional and of size 177×1. The filter size of its two convolutional layers is 5×1. RELU is used as activation function for all the layers. The dropout fraction is 0.1 and the stride of the max pooling is 8.

EmoAudioNet Architecture: The two features vectors are concatenated and fed to a fully connected layer of n neurones activated with a Softmax function. n is defined according to the task. When the task concerns binary depression classification, n = 2. When the task concerns depression severity level assessment,

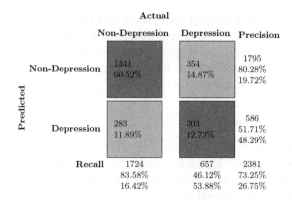

Fig. 2. Confusion Matrice of EmoAudioNet generated on the DAIC-WOZ test set

$n = 24$. While, $n = 10$ for arousal or valence prediction. The ADAM optimizer is used. The learning rate is set experimentally to 10e−5 and it reduced when the loss value stops decreasing. The batch size is fixed to 100 samples. The number of epochs for training is set to 500. An early stopping is performed when the accuracy stops improving after 10 epochs.

5.3 Experimental Results on Spontaneous and Continuous Emotion Recognition from Speech

Results of Three Proposed CNN Architectures. The experimental results of the three proposed architectures on predicting arousal and valence are given in Table 3 and Table 4. EmoAudioNet outperforms MFCC-based CNN and the spectrogram-based CNN with an accuracy of 89% and 91% for predicting aroural and valence respectively. The accuracy of the MFCC-based CNN is around 70% and 71% for arousal and valence respectively. The spectrogram-based CNN is slightly better than the MFCC-based CNN and its accuracy is 76% for predicting arousal and 74% for predicting valence.

EmoAudioNet has a Pearson Coefficient Correlation (PCC) of 0.91 for predicting arousal and 0.92 for predicting valence, and has also a Root Mean Square of Error (RMSE) of 0.12 for arousal's prediction and 0.11 for valence's prediction.

Comparisons of EmoAudioNet and the Stat-of-the Art Methods for Arousal and Valence Prediction on RECOLA Dataset. As shown in Table 5, EmoAudioNet model has the best PCC of 0.9069 for arousal prediction. In term of the RMSE, the approach proposed by He *et al.* [12] outperforms all the existing methods with a RMSE equal to 0.099 in predicting arousal.

For valence prediction, EmoAudioNet outperforms state-of-the-art in predicting valence with a PCC of 0.9221 without any fine-tuning. While the proposed approach by He *et al.* [12] has the best RMSE of 0.104.

Table 5. Comparisons of EmoAudioNet and the state-of-the art methods for arousal and valence prediction on RECOLA dataset.

Method	Arousal		Valence	
	PCC	RMSE	PCC	RMSE
He *et al.* [12]	0.836	0.099	0.529	0.104
Ringeval *et al.* [11]	0.322	0.173	0.144	0.127
EmoAudioNet	0.9069	0.1229	0.9221	0.1118

5.4 Experimental Results on Automatic Clinical Depression Recognition and Assessment

EmoAudioNet framework is evaluated on two tasks on the DAIC-WOZ corpus. The first task is to predict depression from speech under the PHQ-8 binary. The second task is to predict the depression severity levels under the PHQ-8 scores.

EmoAudioNet Performances on Depression Recognition Task. EmoAudioNet is trained to predict the PHQ-8 binary (0 for non-depression and 1 for depression). The performances are summarized in Fig. 2. The overall accuracy achieved in predicting depression reaches 73.25% with an RMSE of 0.467. On the test set, 60.52% of the samples are correctly labeled with non-depression, whereas, only 12.73% are correctly diagnosed with depression. The low rate of correct classification of non-depression can be explained by the imbalance of the input data on the DAIC-WOZ dataset and the small amount of the participants labeled as depressed. F1 score is designed to deal with the non-uniform distribution of class labels by giving a weighted average of precision and recall. The non-depression F1 score reaches 82% while the depression F1 score reaches 49%. Almost half of the samples predicted with depression are correctly classified with a precision of 51.71%. The number of non-depression samples is twice the number of samples labeled with depression. Thus, adding more samples of depressed participants would significantly increase the model's ability to recognize depression.

EmoAudioNet Performances on Depression Severity Levels Prediction Task. The depression severity levels are assessed by the PHQ-8 scores ranging from 0 for non-depression to 23 for severe depression. The RMSE achieved when predicting the PHQ-8 scores is 2.6 times better than the one achieved with the depression recognition task. The test loss reaches 0.18 compared to a 0.1 RMSE on the training set.

Comparisons of EmoAudioNet and the State-of-the Art Methods for Depression Prediction on DAIC-WOZ Dataset. Table 6 compares the performances of EmoAudioNet with the state-of-the-art approaches evaluated on the DAIC-WOZ dataset. To the best of our knowledge, in the literature, the best performing approach is the proposed approach in [25] with an F1 score of 85.44%

Table 6. Comparisons of EmoAudioNet and the stat-of-the art methods for prediction of depression on DAIC-WOZ dataset. (*) The results of the depression severity level prediction task. (**) for non-depression. (‡) for depression. (^{Norm}): Normalized RMSE

Method	Accuracy	RMSE	F1 Score
Yang et al. [8]	–	1.46 (*) (depressed male)	–
Yang et al. [26]	–	5.59 (*) (male)	–
Valstar et al. [3]	–	7.78 (*)	–
Al Hanai et al. [23]	–	10.03	–
Salekin et al. [25]	96.7%	–	85.44%
Ma et al. [34]	–	–	70% (**) 50% (‡)
Rejaibi et al. [35]	76.27%	0.4	85% (**) 46% (‡)
	-	0.168^{Norm} (*)	-
EmoAudioNet	73.25%	0.467	82% (**) 49% (‡)
	-	0.18^{Norm}|4.14 (*)	-

and an accuracy of 96.7%. The proposed NN2Vec features with BLSTM-MIL classifier achieves this good performance thanks to the leave-one-speaker out cross-validation approach. Comparing to the other proposed approaches where a simple train-test split is performed, giving the model the opportunity to train on multiple train-test splits increase the model performances especially in small datasets.

In the depression recognition task, the EmoAudioNet outperforms the proposed architecture in [34] based on a Convolutional Neural Network followed by a Long Short-Term Memory network. The non-depression F1 score achieved with EmoAudioNet is better than the latter by 13% with the exact same depression F1 score (50%).

Moreover, the EmoAudioNet outperforms the LSTM network in [35] in correctly classifying samples of depression. The depression F1 score achieved with EmoAudioNet is higher than the MFCC-based RNN by 4%. Meanwhile, the overall accuracy and loss achieved by the latter are better than EmoAudioNet by 2.14% and 0.07 respectively. According to the summarized results of previous works in Table 6, the best results achieved so far in the depression severity level prediction task are obtained in [35]. The best normalized RMSE is achieved with the LSTM network to reach 0.168. EmoAudioNet reaches almost the same loss with a very low difference of 0.012. Our proposed architecture outperforms the rest of the results in the literature with the lowest normalized RMSE of 0.18 in predicting depression severity levels (PHQ-8 scores) on the DAIC-WOZ dataset.

6   Conclusion and Future Work

In this paper, we proposed a new emotion and affect recognition methods from speech based on deep neural networks called EmoAudioNet. The proposed EmoAudioNet deep neural networks architecture is the aggregation of an MFCC-based CNN and a spectrogram-based CNN, which studies the time-frequency representation and the visual representation of the spectrum of frequencies of the audio signal. EmoAudioNet gives promising results and it approaches or outperforms state-of-art approaches of continuous dimensional affect recognition and automatic depression recognition from speech on RECOLA and DAIC-WOZ databases. In future work, we are planning (1) to improve the EmoAudioNet architecture with the given possible improvements in the discussion section and (2) to use EmoAudioNet architecture to develop a computer-assisted application for patient monitoring for mood disorders.

References

1. GBD 2015 Disease and Injury Incidence and Prevalence Collaborators: Global, regional, and national incidence, prevalence, and years lived with disability for 310 diseases and injuries, 1990–2015: a systematic analysis for the Global Burden of Disease Study 2015, Lancet, vol. 388, no. 10053, pp. 1545–1602 (2015)
2. The National Institute of Mental Health: Depression. https://www.nimh.nih.gov/health/topics/depression/index.shtml. Accessed 17 June 2019
3. Valstar, M., et al.: AVEC 2016 - depression, mood, and emotion recognition workshop and challenge. In: Proceedings of the 6th International Workshop on Audio/visual Emotion Challenge, pp. 3–10. ACM (2016)
4. Ringeval, F., et al.: AVEC 2017 - real-life depression, and affect recognition workshop and challenge. In: Proceedings of the 7th Annual Workshop on Audio/Visual Emotion Challenge, pp. 3–9. ACM (2017)
5. Jiang, H., Hu, B., Liu, Z., Wang, G., Zhang, L., Li, X., Kang, H.: Detecting depression using an ensemble logistic regression model based on multiple speech features. Comput. Math. Methods Medicine **2018** (2018)
6. Alghowinem, S., et al.: A comparative study of different classifiers for detecting depression from spontaneous speech. In: 2013 IEEE International Conference on Acoustics, Speech and Signal Processing, pp. 8022–8026 (2013)
7. Valstar, M., et al.: AVEC 2013: the continuous audio/visual emotion and depression recognition challenge. In: Proceedings of the 3rd ACM International Workshop on Audio/Visual Emotion Challenge, pp. 3–10 (2013)
8. Yang, L., Sahli, H., Xia, X., Pei, E., Oveneke, M.C., Jiang, D.: Hybrid depression classification and estimation from audio video and text information. In: Proceedings of the 7th Annual Workshop on Audio/Visual Emotion Challenge, pp. 45–51. ACM (2017)
9. Cummins, N., Epps, J., Breakspear M., Goecke, R.: An investigation of depressed speech detection: features and normalization. In: Twelfth Annual Conference of the International Speech Communication Association (2011)
10. Lopez-Otero, P., Dacia-Fernandez, L., Garcia-Mateo, C.: A study of acoustic features for depression detection. In: 2nd International Workshop on Biometrics and Forensics, pp. 1–6. IEEE (2014)

11. Ringeval, F., et al.: Av+EC 2015 - the first affect recognition challenge bridging across audio, video, and physiological data. In: Proceedings of the 5th International Workshop on Audio/Visual Emotion Challenge, pp. 3–8. ACM (2015)
12. He, L., Jiang, D., Yang, L., Pei, E., Wu, P., Sahli, H.: Multimodal affective dimension prediction using deep bidirectional long short-term memory recurrent neural networks. In: Proceedings of the 5th International Workshop on Audio/Visual Emotion Challenge, pp. 73–80. ACM (2015)
13. Ringeval, F., et al.: AVEC 2018 workshop and challenge: bipolar disorder and cross-cultural affect recognition. In: Proceedings of the 2018 on Audio/Visual Emotion Challenge and Workshop, pp. 3–13. ACM (2018)
14. Dhall, A., Ramana Murthy, O.V., Goecke, R., Joshi, J., Gedeon, T.: Video and image based emotion recognition challenges in the wild: EmotiW 2015. In: Proceedings of the 2015 ACM on International Conference on Multimodal Interaction, pp. 423–426 (2015)
15. Haq, S., Jackson, P.J., Edge, J.: Speaker-dependent audio-visual emotion recognition. In: AVSP, pp. 53–58 (2009)
16. Low, L.S.A., Maddage, N.C., Lech, M., Sheeber, L.B., Allen, N.B.: Detection of clinical depression in adolescents' speech during family interactions. IEEE Trans. Biomed. Eng. 58(3), 574–586 (2010)
17. Valstar, M., Schuller, B.W., Krajewski, J., Cowie, R., Pantic, M.: AVEC 2014: the 4th international audio/visual emotion challenge and workshop. In: Proceedings of the 22nd ACM International Conference on Multimedia, pp. 1243–1244 (2014)
18. Meng, H., Huang, D., Wang, H., Yang, H., Ai-Shuraifi, M., Wang, Y.: Depression recognition based on dynamic facial and vocal expression features using partial least square regression. In: Proceedings of the 3rd ACM International Workshop on Audio/Visual Emotion Challenge, pp. 21–30 (2013)
19. Trigeorgis, G., et al.: Adieu features? End-to-end speech emotion recognition using a deep convolutional recurrent network. In: 2016 IEEE International Conference on Acoustics, Speech and Signal Processing (ICASSP), pp. 5200–5204 (2016)
20. Ringeval, F., et al.: Prediction of asynchronous dimensional emotion ratings from audiovisual and physiological data. Pattern Recogn. Lett. 66, 22–30 (2015)
21. Ringeval, F., Schuller, B., Valstar, M., Cowie, R., Pantic, M.: AVEC 2015: the 5th international audio/visual emotion challenge and workshop. In: Proceedings of the 23rd ACM International Conference on Multimedia, pp. 1335–1336 (2015)
22. Tzirakis, P., Trigeorgis, G., Nicolaou, M.A., Schuller, B.W., Zafeiriou, S.: End-to-end multimodal emotion recognition using deep neural networks. IEEE J. Sel. Topics Signal Process. 11(8), 1301–1309 (2017)
23. Al Hanai, T., Ghassemi, M.M., Glass, J.R.: Detecting depression with audio/text sequence modeling of interviews. In: Interspeech, pp. 1716–1720 (2018)
24. Dham, S., Sharma, A., Dhall, A.: Depression scale recognition from audio, visual and text analysis. arXiv preprint arXiv:1709.05865
25. Salekin, A., Eberle, J.W., Glenn, J.J., Teachman, B.A., Stankovic, J.A.: A weakly supervised learning framework for detecting social anxiety and depression. Proc. ACM Interact. Mobile Wearable Ubiquit. Technol. 2(2), 81 (2018)
26. Yang, L., Jiang, D., Xia, X., Pei, E., Oveneke, M.C., Sahli, H.: Multimodal measurement of depression using deep learning models. In: Proceedings of the 7th Annual Workshop on Audio/Visual Emotion Challenge, pp. 53–59 (2017)
27. Jain, R.: Improving performance and inference on audio classification tasks using capsule networks. arXiv preprint arXiv:1902.05069 (2019)

28. Chao, L., Tao, J., Yang, M., Li, Y.: Multi task sequence learning for depression scale prediction from video. In: 2015 International Conference on Affective Computing and Intelligent Interaction (ACII), pp. 526–531. IEEE (2015)
29. Gupta, R., Sahu, S., Espy-Wilson, C.Y., Narayanan, S.S.: An affect prediction approach through depression severity parameter incorporation in neural networks. In: Interspeech, pp. 3122–3126 (2017)
30. Kang, Y., Jiang, X., Yin, Y., Shang, Y., Zhou, X.: Deep transformation learning for depression diagnosis from facial images. In: Zhou, J., et al. (eds.) CCBR 2017. LNCS, vol. 10568, pp. 13–22. Springer, Cham (2017). https://doi.org/10.1007/978-3-319-69923-3_2
31. Yu, G., Slotine, J.J.: Audio classification from time-frequency texture. In: 2009 IEEE International Conference on Acoustics, Speech and Signal Processing, pp. 1677–1680 (2009)
32. Ringeval, F., Sonderegger, A., Sauer, J., Lalanne, D.: Introducing the RECOLA multimodal corpus of remote collaborative and affective interactions. In: 2013 10th IEEE International Conference and Workshops on Automatic Face and Gesture Recognition (FG), pp. 1–8. IEEE (2013)
33. Gratch, J., et al.: The distress analysis interview corpus of human and computer interviews. LREC, pp. 3123–3128 (2014)
34. Ma, X., Yang, H., Chen, Q., Huang, D., Wang, Y.: Depaudionet: an efficient deep model for audio based depression classification. In: Proceedings of the 6th International Workshop on Audio/Visual Emotion Challenge, pp. 35–42 (2016)
35. Rejaibi, E., Komaty, A., Meriaudeau, F., Agrebi, S., Othmani, A.: MFCC-based recurrent neural network for automatic clinical depression recognition and assessment from speech. arXiv preprint arXiv:1909.07208 (2019)
36. Tzirakis, P., Zhang, J., Schuller, B.W.: End-to-end speech emotion recognition using deep neural networks. In: 2018 IEEE International Conference on Acoustics, Speech and Signal Processing (ICASSP), pp. 5089–5093 (2018)

COVID-19 and Mental Health/Substance Use Disorders on Reddit: A Longitudinal Study

Amanuel Alambo$^{(\boxtimes)}$, Swati Padhee, Tanvi Banerjee,
and Krishnaprasad Thirunarayan

Wright State University, Dayton, USA
alambo.2@wright.edu

Abstract. COVID-19 pandemic has adversely and disproportionately impacted people suffering from mental health issues and substance use problems. This has been exacerbated by social isolation during the pandemic and the social stigma associated with mental health and substance use disorders, making people reluctant to share their struggles and seek help. Due to the anonymity and privacy they provide, social media emerged as a convenient medium for people to share their experiences about their day to day struggles. Reddit is a well-recognized social media platform that provides focused and structured forums called subreddits, that users subscribe to and discuss their experiences with others. Temporal assessment of the topical correlation between social media postings about mental health/substance use and postings about Coronavirus is crucial to better understand public sentiment on the pandemic and its evolving impact, especially related to vulnerable populations. In this study, we conduct a longitudinal topical analysis of postings between subreddits r/depression, r/Anxiety, r/SuicideWatch, and r/Coronavirus, and postings between subreddits r/opiates, r/OpiatesRecovery, r/addiction, and r/Coronavirus from January 2020–October 2020. Our results show a high topical correlation between postings in r/depression and r/Coronavirus in September 2020. Further, the topical correlation between postings on substance use disorders and Coronavirus fluctuates, showing the highest correlation in August 2020. By monitoring these trends from platforms such as Reddit, epidemiologists, and mental health professionals can gain insights into the challenges faced by communities for targeted interventions.

Keywords: COVID-19 · Topic modeling · Topical correlation · Longitudinal study · Mental health · Substance use · Reddit

---

This work was supported by the NIH under grant NIH R01AT010413-03S1.

A. Del Bimbo et al. (Eds.): ICPR 2020 Workshops, LNCS 12662, pp. 20–27, 2021.
https://doi.org/10.1007/978-3-030-68790-8_2

1 Introduction

The number of people suffering from mental health or substance use disorders has significantly increased during COVID-19 pandemic. 40% of adults in the United States have been identified suffering from disorders related to depression or drug abuse in June 2020[1]. In addition to the uncertainty about the future during the pandemic, policies such as social isolation that are enacted to contain the spread of COVID-19 have brought additional physical and emotional stress on the public. During these unpredictable and hard times, those who misuse or abuse alcohol and/or other drugs can be vulnerable.

Due to the stigma surrounding mental health and substance use, people generally do not share their struggles with others and this is further aggravated by the lack of physical interactions during the pandemic. With most activities going online coupled with the privacy and anonymity they offer, social media platforms have become common for people to share their struggle with depression, anxiety, suicidal thoughts, and substance use disorders. Reddit is one of the widely used social media platforms that offers convenient access for users to engage in discussions with others on sensitive topics such as mental health or substance use. The forum-like structure of subreddits enables users to discuss a topic with specific focus with others, and seek advice without disclosing their identities.

We conduct an initial longitudinal study of the extent of topical overlap between user-generated content on mental health and substance use disorders with COVID-19 during the period from January 2020 until October 2020. For mental health, our study is focused on subreddits r/depression, r/Anxiety, and r/SuicideWatch. Similarly, for substance use, we use subreddits r/Opiates, r/OpiatesRecovery, and r/addiction. We use subreddit r/Coronavirus for extracting user postings on Coronavirus. To constrain our search for relevance, we collect postings in mental health/substance use subreddits that consist of at least one of the keywords in a Coronavirus dictionary. Similarly, to collect postings related to mental health/substance use in r/Coronavirus, we use the DSM-5 lexicon [7], PHQ-9 lexicons [12], and Drug Abuse Ontology (DAO) [3]. We implement a topic modeling algorithm [2] for generating topics. Furthermore, we explore two variations of the Bidirectional Encoder Representations from Transformers (BERT) [5] model for representing the topics and computing topical correlation among different pairs of subreddits on Mental Health/Substance Use and r/Coronavirus. The topical correlations are computed for each of the months from January 2020 to October 2020.

The rest of the paper is organized as follows. Section 2 discusses the related work, followed by Sect. 3 which presents the method we followed including data collection, linguistic analysis, and model building. Further, we present in Sect. 4 that according to our analysis, there is high correlation between topics discussed in a mental health or substance use subreddit and topics discussed in a Coronavirus subreddit after June 2020 than during the first five months of the year

---

[1] https://www.cdc.gov/mmwr/volumes/69/wr/mm6932a1.htm.

2020. Finally, Sect. 5 concludes the paper by providing conclusion and future work.

2   Related Work

In the last few months, there has been a high number of cases and deaths related to COVID-19 which led governments to respond rapidly to the crisis [10]. Topic modeling of social media postings related to COVID-19 has been used to produce valuable information during the pandemic. While Yin et al. [13] studied trending topics, Medford et al. [8] studied the change in topics on Twitter during the pandemic. Stokes et al. [11] studied topic modeling of Reddit content and found it to be effective in identifying patterns of public dialogue to guide targeted interventions. Furthermore, there has been a growing amount of work on the relationship between mental health or substance use and COVID-19. While [6] conducted a study of the prevalence of depressive symptoms in US adults before and during the COVID-19 pandemic, [1] studied the level of susceptibility to stressors that might lead to mental disorder between people with existing conditions of anxiety disorder and the general population. [4] conducted an assessment of mental health, substance use and suicidal ideation using panel survey data collected from US adults in the month of June 2020. They observed that people with pre-existing conditions of mental disorder are more likely to be adversely affected by the different stressors during the COVID-19 pandemic. We propose an approach to study the relationship between topics discussed in mental health/substance use subreddits and coronavirus subreddit.

3   Methods

3.1   Data Collection

In this study, we crawl Reddit for user postings in subreddits r/depression, r/Anxiety, r/SuicideWatch, r/Opiates, r/OpiatesRecovery, r/addiction, and r/Coronavirus. To make our query focused so that relevant postings from each category of subreddits would be returned, we use mental health/substance lexicons while crawling for postings in subreddit r/Coronavirus; similarly, we use the glossary of terms in Coronavirus WebMD[2] to query for postings in the mental health/substance use subreddits. Table-1 shows the size of the data collected for each subreddit for three-month to four-month periods.

3.2   Analysis

We build a corpus of user postings from January 2020 to October 2020 corresponding to each of the subreddits. For better interpretability during topic modeling, we generate bigrams and trigrams of a collection of postings for each

---

[2] https://www.webmd.com/lung/coronavirus-glossary.

month using gensim's implementation of skip-gram model [9]. We then train an LDA topic model with the objective of maximizing the coherence scores over the collections of bigrams and trigrams. As we are interested in conducting topical correlation among topics in a mental health/substance use subreddit and r/Coronavirus, we use deep representation learning to represent a topic from its constituent keywords. We employ a transformer-based bidirectional language modeling where we use two models: 1) a language model that is pre-trained on a huge generic corpus; and 2) a language model which we tune on a domain-specific corpus. Thus, we experiment with two approaches:

1. We use a vanilla BERT [5] model to represent each of the keywords in a topic. A topic is then represented as a concatenation of the representations of its keywords after which we perform dimensionality reduction to 300 units using t-SNE.
2. We fine-tune a BERT model on a sequence classification task on our dataset where user postings from Mental health/Substance use subreddit or r/Coronavirus are labeled positive and postings from a control subreddit are labeled negative. For subreddits r/depression, r/Anxiety, and r/SuicideWatch, we fine-tune one BERT model which we call MH-BERT and for subreddits r/opiates, r/OpiatesRecovery, and r/addiction, we fine-tune a different BERT model and designate it as SU-BERT. We do the same for subreddit r/Coronavirus. Finally, the fine-tuned BERT model is used for topic representation.

Table 1. Dataset size in terms of number of postings used for this study.

Subreddit	Search keywords	Number of postings		
		JAN – MAR	APR – JUN	JUL – OCT
Coronavirus	Opiates (DAO + DSM-5)	5934	4848	2146
	OpiatesRecovery (DSM-5)	2167	1502	400
	Addiction (DSM-5)	154	204	30
	Anxiety (DSM-5)	6690	750	214
	Depression (PHQ-9)	432	366	130
	Suicide Lexicon	596	588	234
opiates	Coronavirus Glossary of terms	534	823	639
OpiatesRecovery		226	540	514
addiction		192	794	772
anxiety		2582	2862	5478
depression		4128	8524	17250
suicide		944	1426	814

Once topics are represented using a vanilla BERT or MH-BERT/SU-BERT embedding, we compute inter-topic similarities among topics in an MH/SU subreddit with subreddit r/Coronavirus for each of the months from January 2020 to October 2020.

4   Results and Discussion

We report our findings using vanilla BERT and a fine-tuned BERT model used for topic representation. Figure. 1 and Fig. 2 show the topical correlation results using vanilla BERT and a fine-tuned BERT model. We can see from the figures that there is a significant topical correlation between postings in a subreddit on mental health and postings in r/Coronavirus during the period from May 2020 – Sep 2020 with each of the subreddits corresponding to a mental health disorder showing their peaks at different months. For substance use, we see higher topical correlation during the period after the month of June 2020. While the results using a fine-tuned BERT model show similar trends as vanilla BERT, they give higher values for the topical correlation scores. We present a pair of groups of topics that have low topical correlation and another pair with high topical correlation. To illustrate low topical correlation, we show the topics generated for r/OpiatesRecovery and r/Coronavirus during APR - JUN (Table-2). For high topical correlation, we show topics in r/Suicidewatch and r/Coronavirus for the period JUN - AUG (Table 3).

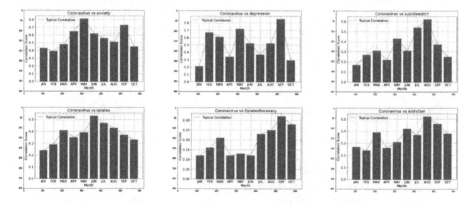

Fig. 1. Temporal Topical Correlation using vanilla BERT-based topical representations.

From Fig. 1 and Fig. 2, we see *Coronavirus vs depression* has highest topical correlation in September followed by May. On the other hand, we see the fine-tuned BERT model give bigger absolute topical correlation scores than vanilla BERT albeit the topics and keywords are the same in either of the representation techniques; i.e., the same keywords in a topic render different representations using vanilla BERT and fine-tuned BERT models. The different representations of the keywords and, hence the topics yield different topical correlation scores as seen in Fig. 1 and Fig. 2.

The reason we generally see higher topical correlation scores with a fine-tuned BERT based representation is because a fine-tuned BERT has a smaller semantic space than a vanilla BERT model leading to keywords across different topics to

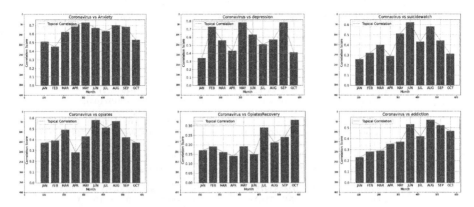

Fig. 2. Temporal Topical Correlation using fine-tuned BERT-based representations.

Table 2. Top two topics for subreddit pair with low topical correlation

Subreddit	Topic-1 Keywords	Topic-2 Keywords
r/OpiatesRecovery	('practice_social', 0.34389842), ('sense_purpose', 0.0046864394), ('shift_hope', 0.0046864394), ('quarantine_guess', 0.0046864394), ('real_mess', 0.0046864394), ('relate_effect', 0.0046864394), ('return_work', 0.0046864394), ('rule_weekly', 0.0046864394), ('pray_nation', 0.0046864394), ('severe_morning', 0.0046864394)	('life_relation', 0.03288892), ('nonperishable_normal', 0.03288892), ('great_worried', 0.03288892), ('head_post', 0.03288892), ('hide_work', 0.03288892), ('high_relapse', 0.03288892), ('kid_spend', 0.03288892), ('want_express', 0.03288892), ('covid_hear', 0.03288892), ('live_case', 0.03288892)
r/Coronavirus	('hong_kong', 0.13105245), ('confirmed_case', 0.060333144), ('discharge_hospital', 0.025191015), ('fully_recovere', 0.020352725), ('interest_rate', 0.017662792), ('confuse_percentage', 0.016409962), ('compare_decrease', 0.016409962), ('day_difference', 0.016409962), ('people_observation', 0.014938728), ('yesterday_update', 0.0142750405)	('https_reddit', 0.119571894), ('recovered_patient', 0.061810568), ('mortality_rate', 0.041617688), ('total_confirm_total', 0.029896544), ('reddit_https', 0.029568143), ('test_positive', 0.02851962), ('total_confirm', 0.026607841), ('disease_compare_transmission', 0.024460778), ('tested_positive', 0.019345615), ('people_list_condition', 0.017226003)

have smaller semantic distance. According to our analysis, high topical overlap implies close connection and mutual impact between postings in one subreddit and postings in another subreddit.

Table 3. Top two topics for subreddit pair with high topical correlation.

Subreddit	Topic-1 Keywords	Topic-2 Keywords
r/SuicideWaatch	('lose_mind', 0.07695319), ('commit_suicide', 0.06495), ('hate_life', 0.04601657), ('stream_digital_art', 0.04184869), ('suicide_attempt', 0.0353566), ('social_distance', 0.033066332), ('tired_tired', 0.03140721), ('depression_anxiety', 0.029040402), ('online_classe', 0.022438377), ('hurt_badly', 0.022400128)	('lose_job', 0.07087262), ('suicidal_thought', 0.055275694), ('leave_house', 0.052148584), ('alot_people', 0.0401444), ('fear_anxiety', 0.029107107), ('push_edge', 0.0288938), ('spend_night', 0.027937064), ('anxiety_depression', 0.027174871), ('couple_day', 0.026346965), ('suicide_method', 0.026167406)
r/Coronavirus	('wear_mask', 0.069305405), ('infection_rate', 0.03957882), ('coronavirus_fear', 0.028482975), ('health_official', 0.027547654), ('middle_age', 0.02721413), ('coronavirus_death', 0.024511503), ('suicide_thought', 0.023732582), ('immune_system', 0.021382293), ('case_fatality_rate', 0.020946493), ('panic_buying', 0.02040879)	('priority_medical_treatment', 0.033756804), ('coronavirus_crisis_worry', 0.0295495), ('depress_lonely', 0.028835943), ('essential_business', 0.027301027), ('fear_anxiety', 0.02715925), ('death_coronavirus', 0.026890624), ('adjustment_reaction', 0.026794448), ('die_coronavirus_fear', 0.026734803), ('declare_state_emergency', 0.026288562), ('jump_gun', 0.025342517)

5   Conclusion and Future Work

In this study, we conducted a longitudinal study of the topical correlation between social media postings in mental health or substance use subreddits and a Coronavirus subreddit. Our analysis reveals that the period including and following Summer 2020 shows higher correlation among topics discussed by users in a mental health or substance use groups to those in r/Coronavirus. Our analysis can give insight into how the sentiment of social media users in one group can influence or be influenced by users in another group. This enables to capture and understand the impact of topics discussed in r/Coronavirus on other subreddits over a course of time. In the future, we plan to investigate user level and posting level features to further study how the collective sentiment of users in one subreddit relate to another subreddit. Our study can provide insight into how discussion of mental health/substance use and the Coronavirus pandemic relate to one another over a period of time for epidemiological intervention.

References

1. Asmundson, G.J., Paluszek, M.M., Landry, C.A., Rachor, G.S., McKay, D., Taylor, S.: Do pre-existing anxiety-related and mood disorders differentially impact covid-19 stress responses and coping? J. Anxiety Disord. **74**, 102271 (2020)
2. Blei, D.M., Ng, A.Y., Jordan, M.I.: Latent dirichlet allocation. J. Mach. Learn. Res. **3**(Jan), 993–1022 (2003)
3. Cameron, D., Smith, G.A., Daniulaityte, R., Sheth, A.P., Dave, D., Chen, L., Anand, G., Carlson, R., Watkins, K.Z., Falck, R.: Predose: a semantic web platform for drug abuse epidemiology using social media. J. Biomed. Inf. **46**(6), 985–997 (2013)

4. Czeisler, M.É., Lane, R.I., Petrosky, E., Wiley, J.F., Christensen, A., Njai, R., Weaver, M.D., Robbins, R., Facer-Childs, E.R., Barger, L.K., et al.: Mental health, substance use, and suicidal ideation during the covid-19 pandemic-united states, June 24–30 2020. Morb. Mortal. Wkly Rep. **69**(32), 1049 (2020)
5. Devlin, J., Chang, M.W., Lee, K., Toutanova, K.: Bert: pre-training of deep bidirectional transformers for language understanding. arXiv preprint arXiv:1810.04805 (2018)
6. Ettman, C.K., Abdalla, S.M., Cohen, G.H., Sampson, L., Vivier, P.M., Galea, S.: Prevalence of depression symptoms in us adults before and during the Covid-19 pandemic. JAMA Netw. Open **3**(9), e2019686–e2019686 (2020)
7. Gaur, M., et al.: "Let me tell you about your mental health!" contextualized classification of reddit posts to DSM-5 for web-based intervention. In: Proceedings of the 27th ACM International Conference on Information and Knowledge Management, pp. 753–762 (2018)
8. Medford, R.J., Saleh, S.N., Sumarsono, A., Perl, T.M., Lehmann, C.U.: An "infodemic": leveraging high-volume twitter data to understand public sentiment for the COVID-19 outbreak
9. Mikolov, T., Sutskever, I., Chen, K., Corrado, G.S., Dean, J.: Distributed representations of words and phrases and their compositionality. In: Advances in Neural Information Processing Systems, pp. 3111–3119 (2013)
10. Sands, P., Mundaca-Shah, C., Dzau, V.J.: The neglected dimension of global security-a framework for countering infectious-disease crises. N. Engl. J. Med. **374**(13), 1281–1287 (2016)
11. Stokes, D.C., Andy, A., Guntuku, S.C., Ungar, L.H., Merchant, R.M.: Public priorities and concerns regarding covid-19 in an online discussion forum: longitudinal topic modeling. J. Gen. Intern. Med. 1 (2020)
12. Yazdavar, A.H., et al.: Semi-supervised approach to monitoring clinical depressive symptoms in social media. In: Proceedings of the 2017 IEEE/ACM International Conference on Advances in Social Networks Analysis and Mining 2017, pp. 1191–1198 (2017)
13. Yin, H., Yang, S., Li, J.: Detecting topic and sentiment dynamics due to COVID-19 pandemic using social media, July 2020

Multi-stream Integrated Neural Networks for Facial Expression-Based Pain Recognition

Ruicong Zhi[1,2(✉)], Caixia Zhou[1,2], Junwei Yu[1,2], and Shuai Liu[1,2]

[1] School of Computer and Communication Engineering, University of Science and Technology Beijing, Beijing 100083, People's Republic of China
zhirc_research@126.com
[2] Beijing Key Laboratory of Knowledge Engineering for Materials Science, Beijing 100083, People's Republic of China

Abstract. Pain is an essential physiological phenomenon of human beings. Accurate assessment of pain is important to develop proper treatment. Recently, deep learning has been exploited rapidly and successfully to solve a large scale of image processing tasks. In this paper, we propose a Multi-stream Integrated Neural Networks with Different Frame Rate (MINN) for detecting facial expression of pain. There are four-stream inputs of the MINN for facial expression feature extraction, including the spatial information, the temporal information, the static information, and the dynamic information. The dynamic facial features are learned in both implicit and explicit manners to better represent the facial changes that occur during pain experience. The experiments are conducted on publicly available pain datasets to evaluate the performance of proposed MINN, and the performance is compared with several deep learning models. The experimental results illustrate the superiority of the proposed model on pain assessment with binary and multi-level classification.

Keywords: Pain assessment · Facial expression · Dynamic facial feature · Multi-stream integrated neural networks

1 Introduction

Physical pain is a complex and subjective experience that is often caused by noxious stimuli damaging the tissue. It can be defined as a protective mechanism that alerts us about damage that is occurring or potentially occurring [1]. Accurate pain assessment is vital for understanding patients' medical conditions and developing suitable treatments. The major body's responses to pain include facial expression, body behavior, sound signals, and biomedical signals. Painful facial expressions are defined as the movement and distortion of facial muscles associated with painful stimuli, which can be described by the action units (AUs). Prkachin and Solomon [2] found that four AUs on faces – brow lowering, orbital tightening, levator contraction, and eye closure – carried the bulk of information about pain. Several evidences found in the literature support the fact that facial expression is the most specific indicator of pain, and is more salient and consistent than other behavioral indicators [3].

© Springer Nature Switzerland AG 2021
A. Del Bimbo et al. (Eds.): ICPR 2020 Workshops, LNCS 12662, pp. 28–35, 2021.
https://doi.org/10.1007/978-3-030-68790-8_3

Globally, self-report (e.g., verbal numeric [4] and visual analogue scales [5]) is the gold standard for pain assessment. This assessment practice is not applicable with verbally impaired patients and infants. To assess pain of these populations, various score-based scales (e.g., FLAC [6]) are developed and validated by health professionals. These score-based scales generate the final pain label by combining the scores of different physiological and behavioral responses. Although the score-based scales are easy to use in practice, the manual assessment is time-consuming and could not be conducted continuously, 24 h a day. In addition, the measurement of pain provided by trained caregivers is affected by different idiosyncratic factors including caregiver's bias, gender, culture, and clinical experience. Therefore, the design of fully automated pain assessment systems is desired for objective and accurate pain assessment.

Automatic pain assessment is a challenging task, and there are an increasing number of researches focusing on pain detection and pain intensity estimation [7–10]. The majority of these existing approaches focused on pain detection based on the analysis of static facial expression, and only few integrated other indicators such as temporal changes. The temporal changes are less efficient to pain detection, as compared to static facial expression [11]. However, combining both information (static facial expression and temporal changes) could reduce bias, which could naturally lead to better performance as compared to the unimodal approach.

In this paper, we propose a Multi-stream Stream Integrated Neural Networks with Different Frame Rate (MINN) for automatic pain assessment based on facial expression. We propose four-stream inputs to MINN for facial expression feature extraction. These inputs include the spatial information (original image sequences), the temporal information (optical flow sequences), the static information (slow pathway), and the dynamic information (fast pathway). The dynamic facial features in both implicit and explicit manners to better represent the facial changes that occur during pain experience. The optical flow ConvNet3D is designed and optimized to deal with facial image sequences, and images sequences with different frame rates are employed to learn dynamic facial features explicitly and implicitly.

2 Proposed Method

2.1 3DConvNet

The Two-dimensional Convolutional Neural Network (2DCNN) is one of the most successfully applied deep learning methods that are used to learn the image texture information effectively. However, it is far not enough to learn facial features by 2D convolution from spatial dimensions when applied to video analysis tasks. Facial image sequences contain plenty of spatial and temporal information that are very helpful for identifying different pain states. In this paper, we use 3DCovNet for feature learning so that it can deal with the image sequences conveniently. Moreover, an optical scheme of squeezing is utilized to decrease the complexity of 3DConvNet.

3DConvNet works by convolving a 3D kernel to the cube formed by stacking multiple frames together. The multiple adjacent frames are connected to form the input and the feature maps of 3DCNN. The formula of 3DConvNet is expressed as:

$$x_{3d}^{l} = \sigma\left(z^{l}\right) = \sigma\left(x_{3d}^{l-1} * W_{3d}^{l} + b^{l}\right) \tag{1}$$

where x_{3d} is a four-dimension array. This four-dimension array is [num_of_frames, width, height, channel] for inputs, and [first dimension of feature maps, width, height, num_of_feature_maps] for feature maps. W_{3d} is the parameters of the 3D convolutional kernel, and b is the biases parameter.

The 3DConvNet can extract both spatial information and motion information. Figure 1(a) and Fig. 1(b) present the 2D convolutional operation and 3D convolutional operation, respectively.

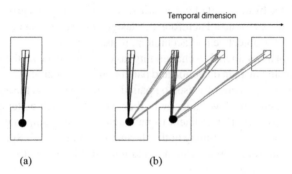

Fig. 1. Comparison of 2D and 3D convolution operation (a) 2D convolution operation (b) 3D convolution operation.

Although 3DConvNet can extract both spatial and temporal information with a single convolutional kernel, the network has a significantly high computational complexity. Further, this network is deep with high number of layers to enhance performance. When deeper networks start converging, a degradation problem occurs (i.e., with the network depth increasing, accuracy gets saturated and then degrades rapidly). To a certain extent, this degradation problem could be addressed by the deep residual neural networks (ResNet). The stacked layers are expected to fit a residual mapping, instead of a direct desired underlying mapping. Formally, the desired underlying mapping is denoted as $\mathcal{H}(x)$, x is the input of the ResNet block. Let the stacked nonlinear layers fit another mapping of $\mathcal{F}(x) := \mathcal{H}(x) - x$, and then the original mapping is recast into $\mathcal{F}(x) + x$. The hypotheses are that it is easier to optimize the residual mapping than the original unreferenced mapping. To the extreme, if an identity mapping is optimal, it would be easier to push the residual to zero than to fit an identity mapping by a stack of nonlinear layers. The formulation of $\mathcal{F}(x) + x$ is realized by feedforward neural networks with "shortcut connections."

In this paper, we change the bottleneck layer in the ResNet network structure to make it more suitable for our task as follows. First, the two-dimensional convolution in the ResNet structure is replaced by the three-dimensional convolution to deal with the dynamic feature extraction for facial expression image sequences. Second, the idea of SqueezeNet [12] is integrated into the ResNet architecture.

2.2 Optical Flow 3DConvNet

As the 3DConvNet can't estimate the motion implicitly, we use the optical flow to estimate the motion explicitly, which allows the neural network to learn abundant motion information easily. In this section, we present the optical flow based 3DConvNet for motion information extraction, which is implemented through a scheme of feeding the 3DConvNet by stacking optical flow displacement fields between several consecutive facial frames. Therefore, the network can describe the motion between video frames explicitly [13], by convolving the optical flow image to learn motion information.

The optical flow means that each pixel in the image has a displacement in the x direction and y direction, so the size of the corresponding optical flow image is the same as that of the original image. It could be displayed with the Munsell color system as shown in Fig. 2. Figure 2 (c) shows the optical flow image we obtained from the facial images (Fig. 2 (a) and Fig. 2 (b)). Optical flow can represent the changes between two adjacent images.

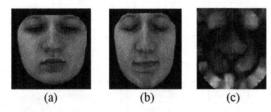

(a)   (b)   (c)

Fig. 2. Example of optical flow image for pain facial expression (a) and (b) are original images, and (c) is the corresponding optical flow image.

2.3 Stream Integration for Facial Expression of Pain

Videos could naturally be decomposed into spatial and temporal components. Convolutional layers extract the spatial features, and the temporal features are extracted in both implicit and explicit manners. Subsection 2.2 introduces the optical flow-based strategy, and our strategy for combining the four-stream inputs is shown in Fig. 3.

The frames of image sequence usually contain two different parts—static areas that change slowly and dynamic areas that change dramatically. During the pain experience, the eyes, eyebrows, and mouth sub-regions are dynamic, and the other part of face regions is static. Based on this insight, two-stream of image sequences with different frame rates are designed to process the facial frames individually. Specifically, image sequences with low frame rate are fed to the network to analyze static parts of facial frames, while image sequences with high frame rate are fed to the network to process dynamic parts of facial frames. Our idea is partly inspired by the retinal ganglion of primates [14, 15]. Also, both the high frame rate pathway and slow frame rate pathway can interact bidirectionally. A lateral connection scheme is utilized to fuse the information of two pathways (low frame rate and high frame rate), which is a popular technique for merging different levels of spatial resolution and semantics [16] and has been used in two-stream networks [17]. The fast and slow pathways are connected to make facial features rich. The feature map sizes of the high frame rate stream and low frame rate stream are the same after the ResNet3D.

In this paper, we propose, for the first time, to use the multi-stream integrated neural network for detecting facial expression of pain. As shown in Fig. 3, two kinds of dynamic information extraction manners are utilized to represent the dynamic facial features of pain states. The network has a four-input streams: original image sequences of low frame rate by 3DConvNet (input stream 1), optical flow image sequences of low frame rate by optical flow 3DConvNet (input stream 2), original image sequences of high frame rate by 3DConvNet (input stream 3), and optical flow image sequences of high frame rate by optical flow 3DConvNet (input stream 4). The facial features extracted by each stream are fused and fed to Softmax for pain recognition.

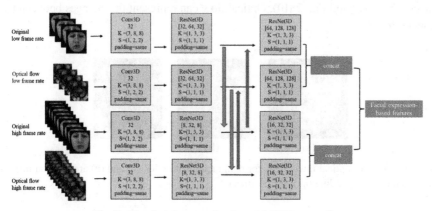

Fig. 3. The facial expression-based MINN networks.

3   Experiments and Discussion

We evaluate the proposed MINN network on the MIntPAIN pain dataset. MIntPAIN dataset [8] has data for 20 subjects captured during the electrical muscle pain simulation. Each subject exhibits two trials during the data capturing session, in which each trial has 40 sweeps of pain stimulation. In each sweep, two kinds of data are captured: one for no pain (BL) and the other one for four kinds of different pain levels (i.e., PA1, PA2, PA3, and PA4). In total, each trial has 80 folders for 40 sweeps. To enlarge this dataset, we augment the videos twenty times using random cropping. Since the BL samples are four times more than other types of data, 1/4 of the data is randomly selected from the BL to maintain the balance with other categories. As the dataset has four levels of pain, we perform binary classification (pain/no pain) as well as multi-level classification. Furthermore, we conduct subject-independent and subject-dependent classification and use different cross-validation schemes.

Assessing multiple levels of pain is more challenging than the binary assessment. Pain intensities assessment is a typical multi-level classification task. In this paper, we detect, in addition to the binary labels, four levels of pain. Table 1 presents the overall performance of binary and multi-class assessment on MIntPAIN datasets. In most cases, MINN achieves the highest performance and outperforms other models in multi-level pain assessment. The result of MINN with StF is slightly higher than that of MINN with FtS (StF is low frame rate features fused to high frame rate features, and FtS is high frame rate features fused to low frame rate features). We think this might be attributed to over-fitting of MINN with FtS model. All the MINN-based models achieve higher accuracy compared to the two-stream model and the slow-fast model ($p < 0.05$), which indicate the effectiveness of the proposed MINN method. The best result obtained by MINN on the MIntPAIN database is 25.2% for five-category pain intensity classification. The accuracy of multi-level classification is much lower than that of binary-classification. This is attributed to the fact that multi-level classification is more complicated, and the pain label is annotated by the stimuli level instead of the subject report, i.e. the pain tolerance of different subjects is different, leading to variant respondence among subjects. There are a certain amount of high pain level videos with no obvious pain facial expression.

Table 1. The accuracy comparison of facial expression based pain classification on the MIntPAIN dataset.

Method	Binary classification	Multi-level classification
MINN	0.636	0.252
MINN+StF	0.601	0.235
MINN+FtS	0.613	0.227
Two-stream NN low	0.598	0.231
Two-stream NN high	0.601	0.232
Slowfast	0.588	0.219

The proposed MINN model obtains the highest accuracy and significantly outperforms the compared deep learning models by up to 4% for multi-level classification pain assessment on MIntPAIN dataset. Moreover, the proposed MINN model outperforms the baseline model of the MIntPAIN dataset [8], which applied convolutional neural networks using only RGB image sequences, by up to 6.6% higher accuracy (MINN: 25.2% accuracy and baseline: 18.6% accuracy). These results suggest that the proposed network enhances the performance of the deep learning model significantly, and it can obtain the spatial and motion information implicitly (three-dimensional convolution) and explicitly (optical flow sequences).

In addition, we also conduct subject-dependent pain assessment on MIntPAIN for the multi-level classification task. The proposed network achieves 87.1% accuracy, which is much higher than the subject-independent assessment. These results indicate that the MINN model could obtain rich facial information to present pain states, and the rigid facial change influences pain recognition performances. Table 2 presents the confusion matrix, where BL denotes the no-pain label, and PA1 to PA4 are four levels of pain intensities. As shown in the table, our method has a better pain detection than no-pain detection. This might be attributed to the fact that some subjects made some kind of relaxation action during no pain stimulation, which confuses the identification between pain and no pain.

Table 2. The confusion matrix of MINN for the subject-dependent experiment on MIntPAIN dataset.

	BL	PA1	PA2	PA3	PA4
BL	0.606	0.044	0.117	0.111	0.122
PA1	0.023	0.909	0.023	0.032	0.013
PA2	0.005	0.020	0.941	0.029	0.005
PA3	0.005	0.000	0.019	0.952	0.024
PA4	0.021	0.000	0.005	0.026	0.948

4   Conclusion

This paper presents a novel network that utilizes pain facial expression to deeply exploit pain-related information. The scheme is implemented by multi-stream integrated neural networks with different frame rates, which employs dynamic facial expression features for pain assessment. The network learns dynamic facial features in both implicit manner and explicit manner, which is conducted by different frame rate operation and optical flow image sequence processing, respectively. Multi-streams can reflect the spatial information (original image sequences), the temporal information (optical flow sequences), the static information (slow pathway), and the dynamic information (fast pathway), which enrich the ability for characteristic facial description by facial features. Experimental results on publicly available pain datasets illustrated that the proposed MINN model performed well for pain detection, especially for the binary classification task.

Acknowledgement. This research was funded by the National Natural Science Foundation of China, grant number 61673052, the National Research and Development Major Project, grant numbers 2017YFD0400100, the Fundamental Research Fund for the Central Universities of China, grant numbers FRF-GF-19-010A, FRF-TP-18-014A2, FRF-IDRY-19-011. The computing work is supported by USTB MatCom of Beijing Advanced Innovation Center for Materials Genome Engineering.

References

1. Gholami, B., Haddad, W.M., Tannenbaum, A.R.: Relevance vector machine learning for neonate pain intensity assessment using digital imaging. IEEE Trans. Biomed. Eng. **57**(6), 1457–1466 (2010)

2. Prkachin, K.M., Solomon, P.E.: The structure, reliability and validity of pain expression: evidence from patients with shoulder pain. Pain **139**(2), 267–274 (2008)
3. Johnston, C.C., Strada, M.E.: Acute pain response in infants: a multidimensional description. Pain **24**(3), 373–382 (1986)
4. Kremer, E., Atkinson, J.H., Ignelzi, R.J.: Measurement of pain: patient preference does not confound pain measurement. Pain **10**, 241–248 (1981)
5. Ho, K., Spence, J., Murphy, M.F.: Review of pain-measurement tools. Ann. Emerg. Med. **27**(4), 427–432 (1996)
6. Willis, M.H.W., Merkerl, S.I., Voepeo-Lewis, T., Malviya, S.: FLACC behavioral pain assessment scale: comparison with the child's self-report. Pediatr. Nurs. **29**(3), 195–198 (2003)
7. Walter, S., et al.: The biovid heat pain database data for the advancement and systematic validation of an automated pain recognition system. In: 2013 IEEE International Conference on Cybernetics (CYBCO), pp. 128–131. IEEE, Switzerland (2013)
8. Haque, M.A., et al.: Deep multimodal pain recognition: a database and comparison of spatio-temporal visual modalities. In: 2018 13th IEEE International Conference on Automatic Face & Gesture Recognition (FG 2018), pp. 250–257. IEEE, China (2018)
9. Martinez, D.L., Rudovic, O., Picard, R.: Personalized automatic estimation of self-reported pain intensity from facial expressions. In: Proceedings of the IEEE Conference on Computer Vision and Pattern Recognition Workshops, pp. 70–79. IEEE, USA (2017)
10. Zhi, R., Zamzmi, G., Goldgof, D., Ashmeade, T., Su, Y.Y.: Automatic infants' pain assessment by dynamic facial representation: effects of profile view, gestational age, gender, and race. J. Clin. Med. **7**(7), 1–16 (2018)
11. Chen, J., Chi, Z., Fu, H.: A new framework with multiple tasks for detecting and locating pain events in video. Comput. Vis. Image Underst. **155**, 113–123 (2017)
12. Iandola, F.N., Han, S., Moskewicz, M.W., Ashraf, K., Dally, W.J., Keutzer, K.: SqueezeNet: AlexNet-level accuracy with 50x fewer parameters and <0.5 MB model size. In: International Conference on Learning Representations (ICLR), Toulon, France (2017)
13. Simonyan, K., Zisserman, A.: Two-stream convolutional networks for action recognition in videos. In: Neural Information Processing Systems (NIPS), pp. 568–576, Montréal, Canada (2014)
14. Hubel, D.H., Wiesel, T.N.: Receptive fields and functional architecture in two nonstriate visual areas (18 and 19) of the cat. J. Neurophysiol. **28**, 229–289 (1965)
15. Livingstone, M., Hubel, D.: Segregation of form, color, movement, and depth: anatomy, physiology, and perception. Science **240**, 740–749 (1988)
16. Feichtenhofer, C., Fan, H., Malik, J., He, K.: SlowFast networks for video recognition (2018). arXiv:1812.03982
17. Lin, T.Y., Dollar, P., Girshick, R., He, K., Hariharan, B., Belongie, S.: Feature pyramid networks for object detection. In: Processing of Computer Vision and Pattern Recognition (CVPR), pp. 1–10. IEEE, USA (2017)

A New Facial Expression Processing System for an Affectively Aware Robot

Engin Baglayici[1]([envelope]) [ID], Cemal Gurpinar[1] [ID], Pinar Uluer[1,2] [ID],
and Hatice Kose[1] [ID]

[1] Istanbul Technical University, Istanbul, Turkey
{baglayici17,gurpinarcemal,hatice.kose}@itu.edu.tr
[2] Galatasaray University, Istanbul, Turkey
puluer@gsu.edu.tr

Abstract. This paper introduces an emotion recognition system for an affectively aware hospital robot for children, and a data labeling and processing tool called LabelFace for facial expression recognition (FER) to be employed within the presented system. The tool provides an interface for automatic/manual labeling and visual information processing for emotion and facial action unit (AU) recognition with the assistant models based on deep learning. The tool is developed primarily to support the affective intelligence of a socially assistive robot for supporting the healthcare of children with hearing impairments. In the proposed approach, multi-label AU detection models are used for this purpose. To the best of our knowledge, the proposed children AU detector model is the first model which targets 5- to 9- year old children. The model is trained with well-known posed-datasets and tested with a real-world non-posed dataset collected from hearing-impaired children. Our tool LabelFace is compared to a widely-used facial expression tool in terms of data processing and data labeling capabilities for benchmarking, and performs better with its AU detector models for children on both posed-data and non-posed data testing.

Keywords: Facial expression recognition · Action unit recognition · Child-robot interaction · Affective computing

1 Introduction

The interpretation of human behavior is highly dependent on emotions. The ability to recognize emotions is essential in psychology, education, human-robot interaction, healthcare, entertainment, and other fields that are related to human behavior. Facial expressions are the central cues to emotional inferences [18] and facial expression recognition (FER) is one of the most used methods for emotion recognition. FER is the process of extracting the features of facial expressions that show the clues of different emotions.

© Springer Nature Switzerland AG 2021
A. Del Bimbo et al. (Eds.): ICPR 2020 Workshops, LNCS 12662, pp. 36–51, 2021.
https://doi.org/10.1007/978-3-030-68790-8_4

Ekman and Friesen worked on facial expressions and the emotions, and defined six basic universal emotions such as, happiness, sadness, surprise, disgust, fear, and anger [6]. Based on the studies carried out on adults and children, it was observed that particular facial expressions correspond to particular emotions across all the cultures. The Facial Action Coding System is developed to distinguish emotions from facial expressions by categorizing the motions of facial muscles into action units (AU) [8]. Action units consist of contractions and relaxations of one or more muscles. They can provide an interpretation of facial expression by single atomic AU or a group of AUs (For AUs and their elemental and compound emotion categories, please refer to [12]).

The facial emotion recognition process involves data gathering, processing, labeling, and finally, training and testing. Data labeling can be an exhaustive process due to the amount of data that is required for the training of deep learning-based models [9]. Organizing and labeling this amount of data requires the automation of the process. Although researchers have used several data annotation tools for similar applications, most data annotation tools, such as ELAN and LabelMe, let the user manually supervise the labeling process. Furthermore, these tools are incapable of labeling several frames in video data in an autonomous manner. Lastly, these tools are general-purpose programs that do not provide a field-specific user interface.

We are working on the RoboRehab project consisting of an emotion recognition tool and a robotic system, specially developed for hearing-impaired children to be used during their audiometry tests in hospitals [28]. The main purpose of this study is to develop a system to recognize the emotions of children who are in interaction with a social and assistive humanoid robot. Unfortunately recognizing children's emotions especially in the wild is a very challenging task. We are in the search for a robust and feasible solution and decided to use AUs for emotion recognition. Also for young children, who have difficulty in showing their emotions, detection of AUs might be a helpful approach for affective systems.

During this study, we had to process, and label an extensive amount of video recordings from the robot and cameras for emotion recognition models, hence the need for a ready-to-use tool that covers automatic, effortless data labeling and processing. This paper introduces our proposed tool LabelFace, and we provide the pipeline and implementation details of such a tool, that researchers can use for their specific areas of interest.

LabelFace aims to contribute to the community with the services which, open-source tools do not fully provide, which are: 1) Enabling manual or automatic data labeling; 2) Manual or automatic image and video processing; 3) Ready to use emotion assistant models 4) Real-time performance 5) API for programmers 6) Open-source code.

For multi-label AU detection, we trained different models using the transfer learning method. We fine-tuned a generalized facial expression model, VGG-FACE [26], for specialized multi-label action unit detection task. To the best of our knowledge, children AU detector model is the first model that targets 5- to 9- year old children, especially for hearing-impaired children.

Fig. 1. CRI study with children

For benchmarking, we also examined the capabilities of another mostly used open-source tool, OpenFace 2.0. We compared action unit detection performances of both tools in terms of non-posed and posed datasets of children. We produced the non-posed data from a previous child-robot interaction (CRI) study with children having hearing aid or cochlear implant (Fig. 1).

For comparison, we included nine common AUs which are described in the following sections in details.

2   Related Work

Recognizing emotions is possible through examining information such as facial expressions, speech, text, biological (e.g., EEG) data [14]. Researchers tried to make use of this information either with single-model or multi-modal approaches [27]. Even though Ekman and other researchers claimed that some basic emotions are universal and common, others have argued that people generally do not react to the situations with the same emotional level and expressions [20].

Generally, conventional Facial Emotion Recognition (FER) workflow includes three main steps: visual information processing, feature extraction from the processed visual information, and classification of extracted features. In the visual information processing phase, various image processing operations are used to remove geometric variations between frames and utilize frames in a single format. Feature extraction can be done in different ways that are geometric-based feature extraction, appearance-based feature extraction, or a combination of these methods [16]. Geometric-based features are extracted based on the position and angle of facial landmark points. Appearance-based feature extraction makes use of essential face regions and the patterns in these regions. Finally, classification takes place with learning algorithms such as Support Vector Machine (SVM), Adaboost, Random-Forest.

In the past years, conventional computer vision methods are applied for FER, and successful results are achieved [1]. Recently, deep learning-based emotion-recognition systems have shown state-of-the-art performance, and several setups of these models are presented and implemented [4,13,23,27].

Deep learning eliminates the feature extraction phase from the FER problem by providing end-to-end learning processes and reduces mathematical complexity. Deep learning methods require large amount of data, and significant amount of computing power to process these data.

Convolutional Neural Network is the most used deep learning methods for detection and recognition tasks. Researchers utilized different CNN architectures for FER and got promising results after conventional approaches. In the CNN approach, images are convolved through a filter collection to extract a feature map. These feature maps are then combined to fully connected layers to classify incoming images [17].

Breuer and Kimmel used the CNN architecture to examine and demonstrate the essential features for FER and the relations of these features with the Facial Action Coding System (FACS) and Action Units [4]. Mollahosseini et al. propose a CNN model to address the FER problem and evaluates the model on multiple well-know standard face datasets [23]. The utilized CNN model is supported by inception layers to improve local feature performance. The model requires lesser computational requirements and provides increased accuracy in both subject-independent and cross-database evaluations.

CNN approaches are suitable for extracting spatial features, but they lack finding temporal variations in data [11]. Thus, other than CNN, Recurrent Neural Networks (RNN) and Long-Short-Term-Memory (LSTM) Networks [31] are also utilized for the FER problem.

LSTM and RNN architectures are specifically designed to find temporal features from time-series data. LSTM networks showed best performances with the video sequences [12]. Jain et al. present a hybrid CNN-RNN architecture that handles both spatial and temporal dependencies in images [13]. First convolutional layers are utilized to extract spatial features from frames. The extracted features are then passed to the RNN layers, which are connected to CNN layers serially. Extracting temporal features in RNN layers provides a significantly increased accuracy.

Action unit (AU) detection methods are generally divided into two categories, static-based and dynamic based methods [19]. Static-based methods interest in the spatial information of data to find patterns while dynamic-based methods interest in the temporal relations in data. Promising results are achieved by combining static and dynamic based methods in recent years [7]. A more detailed survey on deep learning with FER can be seen in [19].

FER studies on different age groups bring different challenges. For the children, it is not possible to include all action units, since facial features in children are not fully developed. Hammal et al. [9] proposed a multi-label CNN model to detect action units in infants. They made use of Baby FACS [24], which is an extension of FACS for infants. They emphasize the requirement of automatic detection of AUs in infants to address the needs of researchers and clinicians in this area [9].

There are various open-licensed or commercial facial expression recognition tools to analyze facial data from videos or still images (for a detailed survey on the open-licensed software, please refer to [3]). Commercial tools offer the user a large variety of machine learning-based solutions, such as basic emotion detection (Kairos[1], SkyBiometry[2], Findface[3], FaceReader[4]). FaceReader claims to be the first tool that can recognize facial expressions in infants.

OpenFace 2.0 [3] is a comprehensive open-licensed tool accompanied by facial landmark/action unit detection, head pose, and eye-gaze estimation features. OpenFace detects the presence of action units and their intensity values. Open-Face uses Support Vector Machines (SVM) to recognize AUs, and it uses only adult datasets [3]. For each AU, they trained a separate SVM model. Also the preprocessing pipeline of OpenFace is similar to our proposed tool, LabelFace, where both of them are extracting the facial features before classification. Open-Face shows good performance on AU detection of adults compared to deep learning based approaches.

In this paper, we provide a comparison between LabelFace and OpenFace in terms of AU detection performance on children's data.

3   LabelFace: A Facial Expression Processing Tool

Deep learning applications require a significant amount of data. Labeling process of the data takes long time and is very challenging especially in action unit recognition. To ease the labeling process of image and video data, here we present a labeling tool for facial expression, LabelFace. The tool's primary abilities are data labeling, image processing, and video processing.

First of all, the tool offers emotion detector models for assisting the user with the labeling process. The tool automatically labels the frames and time frames from a video, and it provides editing options for users to apply changes on labels. As a labeling tool, it lets the user select time frames from a video and label them in different emotion categories (Fig. 2). This process proceeds by specifying the time of the beginning of the expression (Frame Start), the peak time of the expression (Frame Peak), and the end of the expression (Frame End). For the FER on adults, the action units are served according to FACS. For the FER on children, some of the AU's from FACS are served since children do not express all of the AU's presented in FACS.

As an image processing tool, LabelFace provides several image processing methods that users can apply on frames. These processing operations involve face detection, face alignment, facial landmark detection, data augmentation, histogram equalization, face cropping, and face masking.

---

[1] https://www.kairos.com/.
[2] https://skybiometry.com/.
[3] https://findface.pro/en/.
[4] https://www.noldus.com/facereader.

As a video processing tool, the user can extract frames between defined time intervals. The user can specify the rate of extraction in frame per second (fps) and the extracted frame size. Also, video size, dimension, and view property changes are possible.

Besides the user can apply all these processing operations separately to a dataset, we created a pipeline to apply operations to a dataset directly and get final ready frames or videos to feed into classifiers.

Fig. 2. LabelFace user interface

Most importantly, the tool provides emotion-assistant models capable of recognizing emotions and detecting action units in children. This capability lets the tool to label the data automatically and assist the user with the labeling process. On the other hand, the emotion assistant models are distributed and can be used for any research purposes.

Besides all of its functions, the tool is a simple application to listen to and watch videos and audio files, and visualize frames with the built-in playlist. It uses QtMultimedia and QtMultimediaWidgets to handle playback and manage the playlist. The main interface offers a playlist window in which users can drag and drop media files to be played. Standard media controls are provided along with a timeline scrub widget and volume control.

Last but not least, we are working on adding deep learning-based FER methods for the user to train their models with ease in the tool interface directly. By adding state of the art deep learning architectures to our tool, we aim to provide researchers a framework in which they can upload their data, and train their private models for specific applications. The tool will offer different deep learning architectures for different research purposes.

4   Test Setup, Implementation and Experiments

The whole experiment setup for facial emotion recognition, including data preparation and data processing, can be seen in Fig. 3. Although we have gathered data from children and applied processing methods using LabelFace, the low quantity of the data has led us to use well-known FER data sets to train deep learning models. A part of the gathered data is then used for testing to see if the trained models are reliable and generalized well.

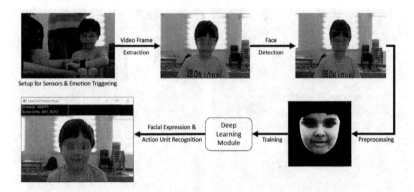

Fig. 3. Facial expression recognition pipeline, Emotion: Happy, Detected action units: AU1, AU12

4.1   Datasets

The Child Affective Facial Expression (CAFE): CAFE dataset is comprised of 2- to 8-year old children poses who have various appearances because of their ethnicity. Children pose seven basic emotions in the dataset: sadness, happiness, surprise, anger, disgust, fear, and neutral. The dataset is also extended with the different appearances of the same emotion by showing the emotion with either open mouth and closed mouth. The entire CAFE set consists of 1192 images, which shows a large variety of emotions to let researchers work on specific problems of the area [22].

The Dartmouth Database of Children's Faces (DDCF): DDCF dataset includes eight different emotion poses from 40 male and 40 female children who are between 6 and 16-year-old. Each child has posed images from different camera angles for each emotion under different light conditions. The subjects are dressed in black hats and clothes to decrease variations between images. Children from 6- to 9-year old are intentionally picked from the DDCF dataset to train emotion detector for children since older children show similarities with adults regarding facial expressions [5].

Our Dataset: We collected non-posed videos from 5- to 9- year old, 16 female and 19 male hearing-impaired children. During data collection, 18 children were asked to watch different video content from well-known animated movies to trigger different emotions. The rest of the children attended a CRI study, where they are asked to take a test on a tablet while being assisted by a Pepper humanoid robot. The children take two tests, 1 with only tablet and the second test with a tablet and robot. In both cases, the children's reactions are video recorded. We extracted frames with significant emotional intensity levels from video recordings of the study to be used in our models' testing phase. The extracted images are then labeled with 12 action units shown in Table 1. We obtained 198 emotionally intense frames from the videos, and these frames are augmented with shifting, flipping, rotating operations. Finally, we got 1422 frames which are then separated into training, validation and testing for fine-tune experiments.

4.2 Proposed Method

We used the aforementioned datasets to train AU detection models. CAFE and DDCF [5] are only available datasets with emotion labels. We used peak frames from the videos of the children with hearing impairments for non-posed data evaluation, and DDCF dataset for posed data evaluation. Thus, we labeled these datasets according to FACS and baby FACS [9]. Target AU's for the children, their distribution in the posed DDCF dataset and in our non-posed dataset can be seen in Table 1. We first extracted and labeled 12 AUs which are mostly found in children's facial expressions. Afterwards, for a fair comparison with OpenFace, we included only nine common action units, which can be recognized in both tools: AU1, AU4, AU5, AU6, AU9, AU10, AU12, AU15, and AU17.

Table 1. AUs coded by manual FACS coders for the peak frames and their distributions in the children datasets (* belongs to baby FACS [9], † are excluded for comparison)

AU	Description	CAFE (344)	DDCF (180)	Ours (198)
0	No Action Unit	71	38	24
1	Inner Brow Raiser	92	47	22
3*†	Brows Drawn Together	71	18	16
4	Brow Lowerer	58	28	7
5	Upper Lip Raiser	61	33	6
6	Cheek Raiser	51	25	65
9	Nose Wrinkler	55	15	10
10	Upper Lip Raiser	55	6	3
12	Lip Corner Puller	78	42	88
15	Lip Corner Depressor	41	23	13
17	Chin Raiser	44	39	27
24†	Lip Pressor	29	51	54
27†	Mouth Stretch	86	36	6

We utilized preprocessing methods to remove variations between subject frames. The data preparation process involves face detection, face alignment, data augmentation, histogram equalization, face cropping, and face masking. As a first step, we detect the faces to remove irrelevant parts from the input frame. For this aim, a CNN based face detector model from dlib library [15] is used. Dlib is a powerful tool to extract facial landmark points from a facial image, that will be used for further processing operations. The next step, face alignment is the geometric adjustment of the face according to facial landmark points based on scale, translation and rotation. Histogram equalization is another processing method that aims to improve contrast in images by manipulating image pixel intensities. After processing the input frames with the aforementioned methods, we crop the facial image and we use facial landmark points to overlay the face with a mask. The masked frames are then augmented with the randomized flipping, rotating, shearing and cropping operations. The augmented frames are finally resized to 224 × 224 RGB images. The example of the original images and the final masked images belonging to subjects in the datasets can be seen in Fig. 4.

Fig. 4. Original images vs. masked images of subjects belonging to CAFE, DDCF datasets respectively

After data processing steps, we obtained 2335 CAFE frames, and 1190 DDCF frames. This quantity was not sufficient for efficiently training a deep learning model; thus, we use transfer learning which involves a pre-trained model as a baseline model for specialized purpose training [25].

The pre-trained model is trained on a large amount of facial data. Instead of training from scratch, keeping the pre-trained model as a baseline and fine-tuning it with a small amount of data can give better results. This process gives a good initialization point for the task.

As a baseline model, VGG-FACE architecture is utilized. VGG-FACE is a deep convolutional network that is modified and used for different face recognition tasks. The model accepts 224 × 224 RGB images as inputs and outputs class probabilities with softmax function [26]. VGG-FACE model is trained with 2.6 million face images, and integrates significantly high amount of facial data which provides a robust model for face detection. Compared to other architectures, ResNet [10] and VGG-FACE show the most successful performances [21]. VGG-FACE is also one of the most studied architectures in the literature [2,7,30]. Based on these motivations and the similar characteristics of our own data set obtained from real world setup, we extended VGG-FACE with customized top layers to fulfill the requirements of our system at best. Our model is an extended version of the VGG-FACE, which is based on VGG-Very-Deep-16 CNN architecture [29]. The model architecture can be seen in Fig. 5.

Table 2. Performance analysis based on the recognized AU number

Training set	Test set	Precision	Recall	F1 score	Number of AUs
CAFE	DDCF	75.3	66.0	68.1	9
CAFE	DDCF	80.6	58.4	60.5	12
OpenFace 2.0	DDCF	18.1	39.1	24.2	9
CAFE	Our dataset	61.1	45.6	47.1	9
CAFE	Our dataset	67.6	34.0	41.0	12
CAFE + DDCF	Our dataset	61.9	46.5	50.7	9
CAFE + DDCF	Our dataset	67.0	39.0	48.0	12
OpenFace 2.0	Our dataset	32.0	52.1	36.2	9

Fig. 5. VGG-FACE architecture with custom top player model

VGG-FACE is used as an encoder network which extracts features from training data. As one of the contributions of this study, for transfer learning, fully connected layers of VGG-FACE is changed with convolutional layers and a new top model with fully connected layers placed for action unit recognition. The extracted features are used for training by the custom top layers. The new base model includes 16 convolutional layers, five pooling layers, and the top layer includes four fully connected layers. Dropout layers are added to prevent overfitting. The final layer of the network has 12 nodes, which corresponds to 12 AU's. Training parameters are kept the same with the general purpose of use (RELU activation, batch size (16), strides (1,1), Adam optimizer, mean squared error (MSE), learning rate (0.0001), momentum (0.9)). LabelFace is implemented using Python programming language, and the proposed deep learning models are trained with Python, Keras-Tensorflow libraries.

5   Test Results and Discussions

The experiments are conducted on several different setups. Only the VGG-FACE baseline model kept the same, and different configurations for the top layer are applied. The evaluation of the models has been done with cross-database evaluation; precision, recall, and F1-scores are used as evaluation metrics [16]. The evaluation results of the models are summarized in Table 2.

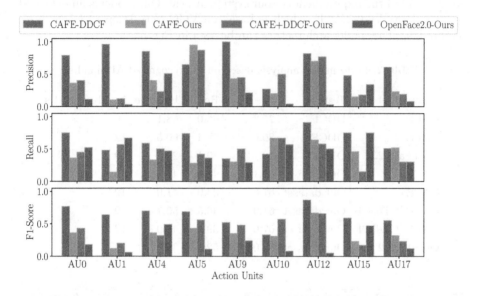

Fig. 6. Precision, recall and F1 Scores for each AU

For the comparison with OpenFace on posed-children data, we first used 9 common AUs. As a first step we fed OpenFace with the DDCF dataset. OpenFace achieved %18.1 precision, %39.1 recall, and %24.2 F1-score on the evaluation of 175 DDCF frames. At the next step, we tested OpenFace with our non-posed children dataset to see the tool's generalization performance. OpenFace achieved %32 precision, %52.1 recall, and %36.2 F1-score on the evaluation of 198 non-posed frames. The precision, recall and F1-score distributions of 9 AUs are showed in the Fig. 6.

For children's AU detector model training, CAFE is chosen as the primary dataset because it contains relatively large amount of data that is balanced in terms of AUs. We trained the top layer for 500 epochs and observed the best performance around 250 epochs. The evaluations of the trained model with DDCF and our datasets showed that the CAFE model is successful in terms of generalization and reliability. The model shows more success in posed-data (F1-score of %68.1 with testing DDCF) which is not surprising since CAFE is a posed-dataset. The model achieves F1-score of %47.1 on real-world non-posed data.

The next step is fine-tuning of the model with another dataset to observe whether the fine-tuned model generalizes well. Thus, in the last experiment, the previously trained CAFE model is fine-tuned with the DDCF dataset to examine the new model's performance with our dataset. The model is fine-tuned for 100 epochs, and a better evaluation result is observed with non-posed data, which also surpasses the original CAFE model by achieving F1 score of %50.7. When we compare the results of the model trained with CAFE and tested with DDCF, which has the highest F1 score, the results show that the models tested with our dataset have a lower F1 score. The differences in model performances are mostly caused by the fact that, the other datasets are composed of posed frames, i.e., the children express exaggerated emotions, whereas in our dataset, we take non-posed images with more natural expressions of emotions, at wild. The non-posed dataset is taken during a child-robot interaction game featuring Audiology test, as stated above. The children were focused on the robot, and showed less arousal in the emotions during the interaction which also decreases the difference between their facial expression in different emotions.

We also extended the AU detection capability by adding AU3, AU24, and AU27, that are among the encountered AUs in training sets. The precision, recall and F1-score distributions of the 12 AUs are displayed in the Fig. 7.

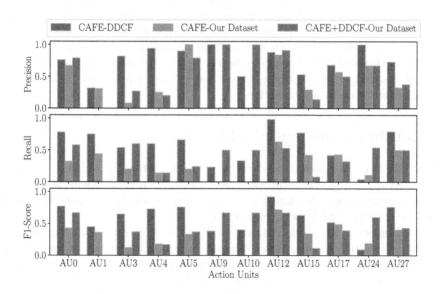

Fig. 7. Precision, recall and F1 Scores for 12 AUs

Posed data evaluation (CAFE - DDCF) results show that AU0 (absence of action units), AU3, AU4, AU5, AU9, AU12, AU17, AU24 and AU27 precision scores are satisfying, except AU1. We observed that the intensity values of AU1 samples in our dataset is low compared to other AUs, which might be the reason of the low detection performance. When we look at the recall scores, we observe the drastic change in AU24 compared to precision, which might be caused by the fact that AU24 is represented by less number of training samples in CAFE.

For the non-posed data evaluation (CAFE - Our Dataset and CAFE+DDCF - Our Dataset), the unbalanced distribution of the test set, especially the low number of frames representing some of the action units, makes it difficult to evaluate action unit recognition results. Overall, the performance of fine-tuned model is satisfactory with the most occurred AUs which are AU0, AU5, AU9, AU10, AU12, and AU17.

In summary, it is observed that the classifier's performance increases by pre-processing of the datasets, which is also observed in previous studies in the literature [19]. On the other hand, locating the specially designed layers which are trained using the datasets, on the top of the pre-trained VGG-FACE, has increased the robustness of the classifiers. Also, it is observed that, when a classifier which is trained by one dataset is fine-tuned by another dataset, its performance in recognition of children's emotions has increased. The performance differences between LabelFace and OpenFace on children's AU detection is due to a couple of reasons. First of all, OpenFace aims to respond to all age scales of the FER problem. Therefore, OpenFace AU detector models are trained with datasets that include both children and adult facial images. In other words, OpenFace uses the same AU detector models for all age ranges. But in our study with children, using only children datasets seems to improve the performance of AU detectors. Thus, it can be derived that AU detection for children and adults should be considered within different contexts. Another reason for the difference in the performance is the approaches used for AU detection. Using transfer learning with neural networks provides better results than a linear SVM classifier in complex tasks such as AU detection.

6   Conclusion and Future Directions

This study aims to build an emotion recognition system including a tool that handles data labeling, data processing, and model training with ease of use. The system is designed and developed to be used with an affectively aware humanoid robot for children with hearing impairments.

Facial emotion recognition is one of the most frequently used methods in emotion recognition studies. One of the most challenging phases in this studies is data labeling and processing. Although there are numerous tools for labeling of visual data, there are not sufficient tools for emotion recognition, which is used for different user groups with different specifications, such as children.

In this paper, we presented an emotion labeling and facial expression/action unit processing tool, LabelFace. LabelFace provides data labeling, data processing, emotion assistant functions to the user. The tool serves these functions either separately or in a pipeline manner. The prepared pipeline includes visual data processing operations to prepare user data for model training. The tool provides AU detection models to help users label their data independent from the data size.

Pre-processed datasets are fed to deep learning models using transfer learning and fine-tuning of cross-datasets. The test evaluations of the different setups show that, when VGG-FACE is used together with the specially designed layers and transfer learning approaches, the models perform more successfully in children's facial expressions. In our approach, multi-label action unit detection models are used. As of our knowledge, our proposed children AU detector model is the first model which targets 5- to 9- year old children. The model trained with well-known posed datasets (CAFE, DDCF) and tested with real-world non-posed dataset collected from our children-robot interaction studies with hearing-impaired children.

Compared to a well-known open source facial expression processing tool, OpenFace 2.0, LabelFace performed more successfully in the detection of the AUs, which are commonly observed in both models. The most succesfully detected AUs by our models are AU0, AU5, AU9, AU10, AU12, and AU17, which are the components of neutral, positive and negative emotions.

Unlike adults, children do not show all AUs of the emotions, especially when they are not posing, therefore using affective tools such as LabelFace for the detection of AUs will help healthcare workers, therapists and caregivers in the detection of children's emotion and stress.

LabelFace is not publicly released yet, but we are working on the release of the first version in the first half of the next year. Also, we are working on the extension of tool capabilities by introducing other deep learning models, that do not only aims children, but also adults. We aim to investigate the most frequent AUs in children, and plan to minimize the number of required AUs to recognize emotions, stress and attention in children in the wild, which will be very beneficial in digital healthcare systems designed for children, especially children with special needs. We also intend to enhance the tool capabilities by introducing auto-labeling the data of adults. For this aim, we investigate diverse adult datasets along with the appropriate deep learning models.

Acknowledgment. This study is supported by the Scientific and Technological Research Council of Turkey (TUBITAK), RoboRehab project, under contract no 118E214.

References

1. Al-agha, L.S.A., Saleh, P.H.H., Ghani, P.R.F.: Geometric-based feature extraction and classification for emotion expressions of 3D video film. J. Adv. Inf. Technol. 8(2), 74–79 (2017)
2. Albiero, V., Bellon, O., Silva, L.: Multi-label action unit detection on multiple head poses with dynamic region learning. In: 2018 25th IEEE International Conference on Image Processing (ICIP), Athens, Greece, pp. 2037–2041. IEEE, October 2018. https://doi.org/10.1109/ICIP.2018.8451267
3. Baltrusaitis, T., Zadeh, A., Lim, Y.C., Morency, L.P.: Openface 2.0: facial behavior analysis toolkit. In: 2018 13th IEEE International Conference on Automatic Face & Gesture Recognition (FG 2018), Xi'an, China, pp. 59–66. IEEE (2018)

4. Breuer, R., Kimmel, R.: A deep learning perspective on the origin of facial expressions. arXiv preprint arXiv:1705.01842 (2017)
5. Dalrymple, K.A., Gomez, J., Duchaine, B.: The Dartmouth database of children's faces: acquisition and validation of a new face stimulus set. PLoS ONE 8(11), e79131 (2013)
6. Ekman, P., Friesen, W.V.: Constants across cultures in the face and emotion. J. Pers. Soc. Psychol. 17(2), 124 (1971)
7. Ertugrul, I.O., Yang, L., Jeni, L.A., Cohn, J.F.: D-pattnet: dynamic patch-attentive deep network for action unit detection. Front. Comput. Sci. 1, 11 (2019)
8. Friesen, W.V., Ekman, P.: Facial action coding system: a technique for the measurement of facial movement. Palo Alto vol. 3 (1978)
9. Hammal, Z., Chu, W.S., Cohn, J.F., Heike, C., Speltz, M.L.: Automatic action unit detection in infants using convolutional neural network. In: 2017 Seventh International Conference on Affective Computing and Intelligent Interaction (ACII), San Antonio, TX, USA, pp. 216–221. IEEE (2017)
10. He, K., Zhang, X., Ren, S., Sun, J.: Deep residual learning for image recognition. In: 2016 IEEE Conference on Computer Vision and Pattern Recognition (CVPR), Las Vegas, NV, USA, pp. 770–778. IEEE, June 2016. https://doi.org/10.1109/CVPR.2016.90
11. Huang, Y., Yang, J., Liao, P., Pan, J.: Fusion of facial expressions and EEG for multimodal emotion recognition. Comput. Intell. Neurosci. 2017, 1–8 (2017)
12. Huang, Y., Chen, F., Lv, S., Wang, X.: Facial expression recognition: a survey. Symmetry 11, 1189 (2019). https://doi.org/10.3390/sym11101189
13. Jain, N., Kumar, S., Kumar, A., Shamsolmoali, P., Zareapoor, M.: Hybrid deep neural networks for face emotion recognition. Pattern Recogn. Lett. 115, 101–106 (2018)
14. Jiang, Y., Li, W., Hossain, M.S., Chen, M., Alelaiwi, A., Al-Hammadi, M.: A snapshot research and implementation of multimodal information fusion for data-driven emotion recognition. Inf. Fusion 53, 209–221 (2020)
15. King, D.E.: Dlib-ml: a machine learning toolkit. J. Mach. Learn. Res. 10, 1755–1758 (2009)
16. Ko, B.: A brief review of facial emotion recognition based on visual information. Sensors 18(2), 401 (2018)
17. LeCun, Y., Haffner, P., Bottou, L., Bengio, Y.: Object recognition with gradient-based learning. Shape, Contour and Grouping in Computer Vision. LNCS, vol. 1681, pp. 319–345. Springer, Heidelberg (1999). https://doi.org/10.1007/3-540-46805-6_19
18. Leppänen, J.M., Nelson, C.A.: The development and neural bases of facial emotion recognition. In: Advances in Child Development and Behavior, vol. 34, pp. 207–246. Elsevier (2006)
19. Li, S., Deng, W.: Deep facial expression recognition: a survey. IEEE Trans. Affective Comput. 1 (2020)
20. Lim, N.: Cultural differences in emotion: differences in emotional arousal level between the east and the west. Integrative Medicine Research 5(2), 105–109 (2016). https://doi.org/10.1016/j.imr.2016.03.004. http://www.sciencedirect.com/science/article/pii/S2213422016300191
21. Lim, Y.K., Liao, Z., Petridis, S., Pantic, M.: Transfer learning for action unit recognition. ArXiv abs/1807.07556 (2018)
22. LoBue, V., Thrasher, C.: The child affective facial expression (CAFE) set: validity and reliability from untrained adults. Front. Psychol. 5, 1532 (2015)

23. Mollahosseini, A., Chan, D., Mahoor, M.H.: Going deeper in facial expression recognition using deep neural networks. In: 2016 IEEE Winter conference on applications of computer vision (WACV), Lake Placid, NY, USA, pp. 1–10. IEEE (2016)

24. Oster, H.: Baby FACS: Facial action coding system for infants and young children. Unpublished monograph and coding manual (2000)

25. Pan, S.J., Yang, Q.: A survey on transfer learning. IEEE Trans. Knowl. Data Eng. **22**(10), 1345–1359 (2010). https://doi.org/10.1109/TKDE.2009.191

26. Parkhi, O.M., Vedaldi, A., Zisserman, A.: Deep face recognition. In: Xie, X., Jones, M.W., Tam, G.K.L. (eds.) Proceedings of the British Machine Vision Conference (BMVC). pp. 41.1–41.12. BMVA Press, Swanse, September 2015. https://doi.org/10.5244/C.29.41. https://dx.doi.org/10.5244/C.29.41

27. Ranganathan, H., Chakraborty, S., Panchanathan, S.: Multimodal emotion recognition using deep learning architectures. In: 2016 IEEE Winter Conference on Applications of Computer Vision (WACV), , Lake Placid, NY, USA, pp. 1–9. IEEE (2016). https://doi.org/10.1109/WACV.2016.7477679

28. RoboRehab: Assistive audiology rehabilitation robot. https://roborehab.itu.edu.tr/. Accessed 21 Oct 2020

29. Simonyan, K., Zisserman, A.: Very deep convolutional networks for large-scale image recognition. In: International Conference on Learning Representations (2015)

30. Tang, C., et al.: View-independent facial action unit detection. In: 2017 12th IEEE International Conference on Automatic Face Gesture Recognition (FG 2017), Washington, DC, USA, pp. 878–882. IEEE, May 2017. https://doi.org/10.1109/FG.2017.113

31. Yu, Z., Liu, G., Liu, Q., Deng, J.: Spatio-temporal convolutional features with nested LSTM for facial expression recognition. Neurocomputing **317**, 50–57 (2018)

Classification of Autism Spectrum Disorder Across Age Using Questionnaire and Demographic Information

SK Rahatul Jannat and Shaun Canavan[✉]

University of South Florida, Tampa Fl 33620, USA
{jannat,scanavan}@usf.edu

Abstract. Currently, diagnosis of Autism Spectrum Disorder (ASD) is a lengthy, subjective process and machine learning has been shown to be able to accurately classify ASD, which can help take some of the subjectivity out of the diagnosis. Considering this, we propose a machine learning-based approach to classification of ASD, across age, that make use of subject self-report and demographic information. We analyze the efficacy of the proposed approach on 3 classifiers: k-nearest neighbors (KNN), Random Forest, and a feed-forward Neural Network. Our results suggest that the proposed approach can accurately classify ASD in children, adolescents, and adults as it is comparable to or outperforms current state of the art on the publicly available AQ-10 dataset.

Keywords: ASD · Machine learning · Classification

1 Introduction

Autism spectrum disorders (ASD) affect as many as 1 in 59 youth [5], with many higher-functioning children not diagnosed until school-age or later [18]. Significant impairment in social-communication, adaptive, and school functioning is common, and compared to other types of pediatric psychopathology, ASD is particularly severe and longstanding [4]. Currently, diagnosing ASD is a lengthy process that involves multiple experts, where the result can include subjective bias [17]. Machine-learning based approaches can provide an objective approach to diagnosis that has the potential to improve accuracy and reduce the time required for diagnosis. To determine high priority patients that should receive a referral for diagnosis, it has been proposed that those that have a high score on the Autism-Spectrum Quotient (AQ) questionnaire [2] should be referred. The AQ questionnaire is one of the main ways that patients are assessed for autistic traits [11]. Ashwood et al. [1] investigated whether the AQ could predict who would receive an ASD diagnosis later in life. They found that while the AQ scores had a high sensitivity, there were a lot of false negatives based on a threshold score (e.g. patient had ASD, but they were below threshold). Wakabayahsi et al. [16] investigated AQ scores across culture, more specifically the United

© Springer Nature Switzerland AG 2021
A. Del Bimbo et al. (Eds.): ICPR 2020 Workshops, LNCS 12662, pp. 52–61, 2021.
https://doi.org/10.1007/978-3-030-68790-8_5

Kingdom and Japan. The results suggest that autistic conditions are similar across cultures, as the results from Japan replicated those from the United Kingdom.

Omar et al. [12] have shown that machine learning can be applied to the AQ questionnaire to predict ASD. They used a Random Forest [3] along with AQ data to predict ASD in children, adolescents, and adults with 92.26%, 93.78%, and 97.10% accuracy, respectively. As Ashwood et al. [1] found a lot of false negatives with a threshold, this work is encouraging that machine learning classifiers can help improve the accuracy of diagnosis from the AQ questionnaire. Motivated by these works, we propose a machine-learning based approach to classify ASD from AQ data and demographic information across age. The contributions of this work can be summarized as follows:

1. To the best of our knowledge, this is the first work to propose a machine learning-based approach to classifying ASD with AQ data across age (e.g. train on child data and test on adult).
2. Accuracy of 3 machine learning classifiers, for classifying ASD from AQ questionnaire information, is compared. Namely, Random Forest [3], a feed-forward Neural network [10], and k-nearest neighbor (KNN) [7].
3. Proposed approach is comparable to or outperforms state of the art on the publicly available AQ-10 dataset [14], which contains child, adolescent, and adult AQ information.

2   Experimental Design

2.1   Dataset

To conduct our experiments, we used the AQ-10 dataset [14], which consists of 3 datasets based on the AQ-10 screening tool [6]; 1 for children, adolescents, and adults. Each dataset has attributes including, but not limited to, age, gender, ethnicity, and answers from the AQ questionnaire. All available attributes can be seen in Table 1. There are 292 children with an age range of [4,11] in the child, 104 adolescents with an age range of [12–16], and 704 adults with an age range of [17,18]. Each subject is given a class of either ASD or no ASD.

2.2   Experiments

To classify ASD, we propose to use Autism-Spectrum Quotient questionnaire data along with demographic information, specifically from the AQ-10 dataset (Table 1). To evaluate using AQ data, we have selected the following 6 feature sets: (1) All available attributes except for the final screening score resulting in the 19-dimension feature vector $v_1 = [age, gender, ethnicity, jaundice, PDD, test, country, app, method, Q1, \dots, Q10]$; (2) AQ questions 1–10 and family member with pervasive developmental disorders(PDD) resulting in the 11-dimension feature vector $v_2 = [PDD, Q1, \dots, Q10]$; (3) the 10-dimension feature vector $v_3 = [Q1, \dots, Q10]$ with AQ questions but not family history of Genes;

Table 1. Details of all attributes in AQ-10 dataset [14].

Attribute	Type	Description
Age	Integer	Years
Gender	String	Male/Female
Ethnicity	String	e.g. Latino, Caucasian, Black, etc.
Born with jaundice	Boolean	Subject born with jaundice
Family member with PDD	Boolean	Immediate family member with PDD
Who is completing test	String	Parent, self, caregiver, medical staff
Country of residence	String	e.g. USA, Brazil, Palestine, etc.
Used screening app before	Boolean	Subject has used app before
Screening Method Type	Integer (0, 1, 2, 3)	toddler, child, adolescent, adult
Questions 1–10	Binary (0,1)	Answer to AQ questions
Screening Score	Integer	Final AQ score

(4) AQ questions, ethnicity and family member with PDD resulting in the 12-dimension feature vector $v_4 = [PDD, ethnicity, Q1, \ldots Q10]$; (5) also using all parameters in v_4 but not the genes $v_5 = [ethnicity, Q1, \ldots Q10]$; and (6) family member with PDD resulting in the 1-dimension feature vector $v_6 = [PDD]$. All subject-independent evaluations were conducted by randomly selecting 80% of the data for training and 20% for testing. Accuracy is the evaluation metric used in all experiments.

To evaluate the robustness of the feature sets to classify ASD across different classifiers, we evaluated a Random Forest(RF) [3], a feed-forward Neural Network(NN), and k-nearest neighbors [7]. RF ensembles the results of the large number of decision tree it is made of. For RF we used 100 trees as the depth. The KNN Algorithm predicts by using the information of data which exists near to each other. Here we have used this $k = 13$ chosen by 'Elbow' method for boundary of this proximity. The NN has 3 hidden layers with 32, 16 an 16 neurons and 1 output layer with 1 output. We used Relu in the hidden layers and sigmoid in output. 'Adam' optimizer and 'Binary Crossentropy' loss function has been used in the network.

3   Results

3.1   Within-Dataset Evaluation on Child, Adolescent, and Adult

To evaluate child, adolescent, and adult data we used feature vectors v_1, v_2, v_3, v_4, v_5, and v_6 with data from the AQ-10 dataset. As it can be seen in Table 2, RF and our NN architecture both performed well on all 3 datasets (child, adolescent, and adult) using v_1. The RF performed best at 98.8% for all 3, while the NN had an accuracy of 98.8% on the adolescent dataset as well, but performed slightly worse on child and adult (96.6% and 94.5%, respectively).

Table 2. Evaluation for within-dataset using all attributes (v_1).

Classifier	Train	Test	Accuracy
Random Forest	Child	Child	98.8%
	Adult	Adult	98.8%
	Adolescent	Adolescent	98.8%
Neural Network	Child	Child	96.6%
	Adult	Adult	94.5%
	Adolescent	Adolescent	98.8%
K-nearest Neighbors	Child	Child	81.1%
	Adult	Adult	66.4%
	Adolescent	Adolescent	81.1%

While k-nearest neighbors(KNN performed reasonably well on child and adolescent (81.1% on both), it did not perform well on adult data. Interestingly, both kNN and the NN had the lowest accuracy on the adult dataset (66.4% and 94.5%, respectively).

Table 3. Evaluation for within-dataset using AQ with PDD in family(PDD) (v_2) and without PDD in family(NPDD(v_3).

Classifier	Train	Test	Accuracy(PDD)	Accuracy(NPDD)
Random Forest	Child	Child	92%	94%
	Adult	Adult	92%	94%
	Adolescent	Adolescent	92%	94%
Neural Network	Child	Child	98%	98%
	Adult	Adult	100%	100%
	Adolescent	Adolescent	100%	98%
K-nearest Neighbors	Child	Child	88%	86%
	Adult	Adult	50%	50%
	Adolescent	Adolescent	88%	86%

When we evaluated feature vector v_2, we found that the NN had the best performance with 100% on both adolescent and adult datasets, and 98% on the child dataset (Table 3). Again, the adult dataset performed the worst with KNN with 50% accuracy. While the accuracy of RF decreased by 6.8% across all datasets, it still performed reasonably well with 92% accuracy. It is interesting that by removing some of the features such as age, gender, and whether the subject had jaundice, the Neural Network was able to classify ASD with a high degree of accuracy. This suggests that AQ questions along with family history are a strong indicator for classifying ASD, however, to further investigate this,

Table 4. Evaluation for within-dataset using AQ and Ethnicity with PDD in family(PDD) (v_4) and without PDD in family(NPDD(v_5).

Classifier	Train	Test	Accuracy(PDD)	Accuracy(NPDD)
Random Forest	Child	Child	92%	96%
	Adult	Adult	92%	96%
	Adolescent	Adolescent	92%	92%
Neural Network	Child	Child	96.6%	96%
	Adult	Adult	94.9%	91%
	Adolescent	Adolescent	93.2%	94%
K-nearest Neighbors	Child	Child	90%	89%
	Adult	Adult	50%	50%
	Adolescent	Adolescent	90%	89%

we analyzed feature vector v_3. We removed family history of ASD from the v_2 feature vector and the accuracies remained largely the same compared to v_2, where the results from RF increased by 2%. Now the most intriguing question came to this, whether the family history really impacts classifying ASD or not. To answer this, we have done some other experiment as well. In Table 4 we have used feature vector v_4 which includes AQ questions, ethnicity and family history and feature vector v_5 which discards the family history information form v_4. From this experiment we can tell the result of v_4 is similar to the result of v_5. This ultimately suggests that the AQ questions and ethnicity are a stronger indicator of ASD compared to family history when automatically classifying ASD with machine learning algorithms.

For our final within-dataset evaluation, we investigated whether family history alone (v_6) can classify ASD. As can be seen in Table 5 the random forest and KNN give the best result among the 3 algorithms but still it is 50% accuracy and for Neural Network it is 40%. While it is not as common to try to classify data with 1 feature, this experiment is justified by the heritability of ASD being high with studies finding anywhere from 50%-90% [13]. Although family history can be a strong indicator of ASD, our results again suggest that it is not sufficient for use in machine learning classifiers as seen in Tables 3, 4 and 5.

3.2   Cross-Dataset Evaluation on Child, Adolescent, and Adult

Along with within-dataset experiments, we also evaluated cross-dataset experiments (e.g., train on child, test on adult). We performed an exhaustive combination of cross-dataset experiments (Table 6). When all features were used (v_1), similar results are obtained, across all 3 classifiers, compared to within-dataset. For the RF an accuracy of 98.8% was achieved for all experiments, the Neural Network had an average accuracy of 95.95%, and KNN had an average accuracy of 76.12%. These results suggest that AQ questionnaire data along with demographic information can be used to classify ASD across age. More importantly,

Table 5. Evaluation for within-dataset using PDD in family only (v_6).

Classifier	Train	Test	Accuracy
Random Forest	Child	Child	50%
	Adult	Adult	50%
	Adolescent	Adolescent	50%
Neural Network	Child	Child	40.6%
	Adult	Adult	40.6%
	Adolescent	Adolescent	40.6%
K-nearest Neighbors	Child	Child	50%
	Adult	Adult	50%
	Adolescent	Adolescent	50%

Table 6. Evaluation for cross-dataset using all attributes (v_1).

Classifier	Train	Test	Accuracy
Random Forest	Child	Adult	98.8%
	Child	Adolescent	98.8%
	Adult	Child	98.8%
	Adult	Adolescent	98.8%
	Adolescent	Adult	98.8%
	Adolescent	Child	98.8%
Neural Network	Child	Adult	93%
	Child	Adolescent	98%
	Adult	Child	98.3%
	Adult	Adolescent	94.9%
	Adolescent	Adult	93.2%
	Adolescent	Child	98.3%
K-nearest Neighbors	Child	Adult	66%
	Child	Adolescent	81%
	Adult	Child	81.1%
	Adult	Adolescent	81.1%
	Adolescent	Adult	66.4%
	Adolescent	Child	81.1%

it also suggests that this information can be used to predict ASD as there are features in the child dataset that are similar to both adolescents and adults. This is an open question that requires further investigation.

We have also conducted other cross-dataset experiments to learn which attributes have more impact on classifying ASD. Similar to within-dataset, as can be seen in Table 7, only using AQ question answers and family history

Table 7. Evaluation for cross-dataset using AQ with PDD in family(PDD) (v_2) and without PDD in family(NPDD(v_3).

Classifier	Train	Test	Accuracy(PDD)	Accuracy(NPDD)
Random Forest	Child	Adult	94.4%	94.4%
	Child	Adolescent	94.4%	94.4%
	Adult	Child	94.4%	94.4%
	Adult	Adolescent	94.4%	94.4%
	Adolescent	Adult	94.4%	94.4%
	Adolescent	Child	94.4%	94.4%
Neural Network	Child	Adult	96.6%	98%
	Child	Adolescent	98.3%	98%
	Adult	Child	100%	100%
	Adult	Adolescent	96.6%	100%
	Adolescent	Adult	94.9%	100%
	Adolescent	Child	96.6%	98%
K-nearest Neighbors	Child	Adult	50%	50%
	Child	Adolescent	88.1%	86%
	Adult	Child	88.1%	86%
	Adult	Adolescent	88.1%	86%
	Adolescent	Adult	50%	50%
	Adolescent	Child	88.1%	86%

resulted in a slight decrease in the accuracy for RF (4.4%), the NN had the best performance, with an average accuracy of 97.17%, and KNN again performed the worst with an average accuracy of 76.97%. Similar to the previous experiment which has been done for within-dataset (Tables 3 and 4), the same experiments have been done using feature vectors v_2 and v_3 (Table 7) and also using v_4 and v_5 (Table 8). In each case, we can see that family history can decrease performance in cross-dataset experiments, which similarly replicates the results from our within-dataset experiments. Finally, we have again conducted the experiment with only family history. Here, the result is the same as Table 5, for all cross-dataset experiments: RF and KNN have an accuracy of 50% and the NN has an accuracy of 40%.

3.3   Comparison to State of the Art

We also compared our proposed approach to current state of the art. As can be seen in Table 9, our proposed approach is comparable to or outperforms state of the art across all datasets (child, adolescent, and adult).

Table 8. Evaluation for cross-dataset using AQ and Ethnicity with PDD in family(PDD) (v_4) and without PDD in family(NPDD(v_5).

Classifier	Train	Test	Accuracy(PDD)	Accuracy(NPDD)
Random Forest	Child	Adult	92%	96%
	Child	Adolescent	94%	96%
	Adult	Child	92%	94%
	Adult	Adolescent	94%	96%
	Adolescent	Adult	92%	96%
	Adolescent	Child	92%	92%
Neural Network	Child	Adult	98%	94%
	Child	Adolescent	93%	98%
	Adult	Child	94%	98%
	Adult	Adolescent	96%	93%
	Adolescent	Adult	89%	94%
	Adolescent	Child	94%	96%
K-nearest Neighbors	Child	Adult	50%	50%
	Child	Adolescent	90%	89%
	Adult	Child	90%	89%
	Adult	Adolescent	90%	89%
	Adolescent	Adult	50%	50%
	Adolescent	Child	90%	89%

Table 9. Comparisons to state of the art across child, adolescent, and adult datasets.

	Child dataset	Adolescent dataset	Adult dataset
Proposed Approach	98.8%	**100%**	**100%**
Erkan et al. [8]	**100%**	**100%**	**100%**
Omar et al. [12]	92.26%	93.78%	97.10%
Thabtah et al. [15]	N/A	97.58%	99.91%

4   Conclusion

We proposed an approach to classifying ASD across age using Autism-Spectrum Quotient questionnaire data along with demographic information. We evaluated Random Forest, Neural Network, and KNN classifiers. Results suggest this data is robust to multiple machine learning classifiers and can accurately classify children, adolescents, and adults with ASD. The results are comparable to or outperform state of the art on the AQ-10 dataset. To the best of our knowledge, this is the first work to propose using AQ and demographic information for cross-dataset classification. We performed cross-dataset experiments where we achieved a max accuracy of 98.8%. These results suggest that this data can be used to predict ASD into adulthood from child data. There are some limitations to our study as well. First, we only used one dataset and while our results suggest family history does not classify ASD well with machine learning, these results are inconclusive as it has been shown that it can be a strong indicator due to heritability [13]. Secondly, it has been shown that features such as gaze [9] can classify ASD. It is important to compare the AQ and demographic features to other types of features to learn the best features are to classify ASD.

References

1. Ashwood, K., et al.: Predicting the diagnosis of autism in adults using the aq questionnaire. Psychol. Med. **46**(12), 2595–2604 (2016)
2. Baron-Cohen, S., et al.: The aq: evidence from asperger syndrome/high-function autism, males, females, scientists and mathematicians. J. Autism Dev. Disord. **31**(1), 5–17 (2001)
3. Breiman, L.: Random forests. Mach. Learn. **45**, 5–32 (2001). https://doi.org/10.1023/A:1010933404324
4. Burke, L., Stoddart, K.P.: Medical and health problems in adults with high-functioning autism and asperger syndrome. In: Volkmar, F.R., Reichow, B., McPartland, J.C. (eds.) Adolescents and Adults with Autism Spectrum Disorders, pp. 239–267. Springer, New York (2014). https://doi.org/10.1007/978-1-4939-0506-5_12
5. CDC: Prevalence and characteristics of asd among children aged 8 years-autism and developmental disabilities monitoring network. Surveill. Summ. **65**(3), 1-23 (2016)
6. Dua, D., et al.: UCI ML repository (2017). http://archive.ics.uci.edu/ml
7. Duda, R.O., et al.: Pattern Classification and Scene Analysis. Wiley, New York (1973)
8. Erkan, U., et al.: Asd detection with machine learning methods. Curr. Psychiatry Rev. **15**(4), 297–308 (2019)
9. Fabiano, D., et al.: Gaze-based classification of autism spectrum disorder. Pattern Recognition Letters (2020)
10. Fine, T.: Feedforward Network Method. Springer Science & Business Media (2006)
11. Lundqvist, L., et al.: Is the aq a valid measure of traits assoc with the autism spec? a rasch validation in adults with and without autism spectrum disorders. J. Autism Dev. Disord. **47**(7), 2080–2091 (2017)
12. Omar, K.S., et al.: A machine learning approach to predict ASD. In: International Conference on Electrical, Computer and Communication Engineering (ECCE), pp. 1-6. IEEE (2019)
13. Sandin, S., et al.: The familial risk of autism. Jama **311**(17), 1770–1777 (2014)
14. Thabtah, F.: Autism spectrum disorder screening: machine learning adaptation and dsm-5 fulfillment. In: Proceedings of the 1st International Conference on Medical and health Informatics, pp. 1–6 (2017)
15. Thabtah, F., Abdelhamid, N., Peebles, D.: A machine learning autism classification based on logistic regression analysis. Health Inf. Sci. Syst. **7**(1), 1–11 (2019). https://doi.org/10.1007/s13755-019-0073-5
16. Wakabayashi, A., et al.: The aq in japan: a cross-cultural comparison. J. Autism Dev. Disord. **36**(2), 263–270 (2006)

17. Zeldovich, L.: Why the definition of autism needs refined. Spectrum, Autism Research News (2018)
18. Zwaigenbaum, L., et al.: Early identification of ASD: recommendations for practice and research. Pediatrics **136**(Supplement 1), S10–S40 (2015)

Neonatal Pain Scales and Human Visual Perception: An Exploratory Analysis Based on Facial Expression Recognition and Eye-Tracking

Lucas Pereira Carlini[1]([✉]) [iD], Fernanda Goyo Tamanaka[1] [iD],
Juliana C. A. Soares[2] [iD], Giselle V. T. Silva[2] [iD], Tatiany M. Heideirich[1] [iD],
Rita C. X. Balda[2] [iD], Marina C. M. Barros[2] [iD], Ruth Guinsburg[2] [iD],
and Carlos Eduardo Thomaz[1]([✉]) [iD]

[1] University Center of FEI, São Bernardo do Campo, SP, Brazil
lucaspcarlini10@gmail.com, cet@fei.edu.br
[2] Federal University of São Paulo, São Paulo, SP, Brazil
ruth.guinsburg@gmail.com

Abstract. Neonates feel pain and the more premature they are, the more immature are their pain modulation system. Facial expression recognition is a non-invasive method commonly used in clinical practice to identify and assess pain in neonates. In this context, this paper firstly carries out a systematic and state-of-the-art review of several clinical pain scales based on neonatal facial features. Then, we propose and implement an eye-tracking exploratory analysis of the neonate's facial regions described in these scales to identify the most preferred ones visually for such pain assessment by 4 distinct sample groups of volunteers: 44 physicians, 40 health professionals, 29 parents of newborns and 30 lay people. Our results show differences in all sample groups, either experts or not, between the facial regions considered clinically relevant to the ones expressed implicitly by the human visual perception.

Keywords: Newborns · Facial expression · Pain scales

1 Introduction

The International Association for the Study of Pain (IASP) describes the pain phenomena as a set of unpleasant experiences, which can be sensitive or emotional, associated with actual or potential tissue damage [25]. Specifically, the pain felt by neonates, whether frequent or intense, may lead to short- or long-term consequences to the development and well-being of infants [11].

It was only in the end the 1980s that a number of research showed that our central nervous and nociceptive systems are sufficiently developed during gestation to sense and suffer from painful stimulus [2,3,16]. These same systems are

Supported by FEI, CAPES, CNPq and FAPESP.

A. Del Bimbo et al. (Eds.): ICPR 2020 Workshops, LNCS 12662, pp. 62–76, 2021.
https://doi.org/10.1007/978-3-030-68790-8_6

capable of transmitting pain from the peripheral areas to the associative ones as early as the sixth month of gestation, and, in fact, due to the immaturity of inhibitory pathways of painful stimulus, neonates show low capacity for modulation and habituation to pain, when compared to adults, leading to an increased sensitivity to this phenomena [2,3].

Since a direct and objective verbal communication by neonates is unlikely, it is of the utmost importance to correctly identify the neonatal presence and intensity of pain. In the last years, studies have developed automatic computational frameworks to evaluate pain in newborn babies [6,7,20,32,33]. These studies have mainly focused on the analysis of the facial expression of neonates. This information provides a non-invasive method commonly used in clinical practice to identify and assess pain in neonates, based on the analysis of their face images and on a sequence of computational procedures in order to detect, interpret and classify patterns in these images, enabling pain assessment in an almost instantaneous time interval. Although these studies present feasible applications in the clinical practice, there is a lack of information to explore explicitly the relevance of the actual facial features used by health professionals when coding and evaluating such phenomena. Identifying and highlighting these patterns might provide useful information to understand the relationship between the facial features and procedural pain and, consequently, helping health professionals in the corresponding clinical practice.

In this context, we developed recently preliminary works [9,10] using eye-tracking technology to investigate whether adults use different methods of visual perception when evaluating pain in neonatal face images, showing that there is disagreement on the intensity of pain, but agreement on the absence of it [9]. Interestingly, however, when investigating eye movement fixation data, such as the number and time of fixations related to general facial regions of analysis using the standard Neonatal Facial Coding System (NFCS) [16,17], no clear differences were found when comparing holistically the visual fixations of health professionals and non-health professionals when assessing pain [10].

In this work, we describe novel results on human visual perception of pain in neonatal face images, elaborating on our previous works [9,10] by adding new experimental evidences and comparative analyses. Firstly, we propose and implement here a systematic and state-of-the-art review of distinct clinical pain scales based on (or related to) neonatal facial features. Then, as the main contribution of this work, we investigate eye movement fixation data of region-by-region of the neonate's face based on the text description of these scales in order to identify the most visualised areas of the face by distinct sample groups of volunteers, separated here as physicians, health professionals, parents and lay people (or others). We believe that a comprehensive analysis of these most visually preferred facial characteristics might provide a better understanding of the relevance of neonatal facial features related to pain assessment.

2 Neonatal Pain

As mentioned briefly in the previous section, until the mid-1980s it was assumed that the central nervous system of newborn babies was not fully developed, consequently, not being able to sense and suffer from pain [20]. However, research done in the following years of 1980 showed that the central nervous and nociceptive systems are sufficiently developed in the sixth month of gestation [2,15,16], leading to an increased sensitivity of pain [2,3] because our inhibitory pathways of painful stimulus are not fully developed.

The non-treated pain felt by neonates is associated with changes in their respiratory, cardiovascular and metabolic stability, increasing mortality in neonatal intensive care units. In the short-term, neonates may suffer, as a consequence of pain, irritability, inattention, change in resting pattern, dietary denial and interference in the mother-child relationship [20]. In the long-term, pain may cause degrading effects on the neurological and behavioural development such as cognitive problems, alterations in the brain development with implications on learning disabilities and hypersensitivity to painful and non-painful stimuli [4,18].

The most common observations reported by health professionals when treating neonatal pain are crying, irritability, sudden movements, and change in their facial expression and behaviour [30]. Due to the range of sounds, most people believe that crying is the best metric to estimate the presence and, if necessary, intensity of pain or discomfort in their newborn babies [10]. However, it has been argued that almost half of newborn babies does not cry in painful interventions, meanwhile, stressful stimulus may lead to crying as well. Therefore, although useful, crying alone is not reliable to verify the presence of pain [17]. Another useful method to verify the presence of neonatal pain is the analysis of their movements. This can be justified by the fact that newborn babies have a standardised movement. Consequently, changes in that pattern, such as sudden and disorderly movements, may indicate the presence of pain.

In the last years, pain presence has also been detected automatically through the analysis of the facial expression of the neonates [6,7,10,20,32,33]. Although this analysis may be subjective, it is a non-invasive method that has been widely used in clinical practice. Due to its expressiveness, face is a reliable source to verify both the emotional condition and pain sensitivity in neonates [20].

3 Materials and Methods

In this section, we describe our computational framework proposed initially in [9,10] and the methodology applied here to analyse facial features related to the human visual perception. Firstly, we describe the sample groups of health professionals and non-health professionals. Then, the hardware and implementation of the computational experiment used to acquire eye movement fixation data. Finally, we illustrate the facial features to be analysed based on the literature review.

3.1 Volunteers

All data were collected in 2019, from March 15th to April 17th, at the Hospital of São Paulo, an university-affiliated hospital of the Federal University of São Paulo, São Paulo, Brazil.

The experiments were conducted with the participation of 143 volunteers divided into four groups of study:

- 44 physicians: 4 paediatricians and 40 neonatologists;
- 40 health professionals: 17 nursing assistants, 10 nurses, 5 physiotherapists and 8 speech therapists;
- 29 parents of newborn;
- 30 others (lay people: non-physicians, non-health professionals, and non parents of newborn).

Some of the demographic characteristics of the participants are:

- age of 34,2 (mean) ± 10,2 (SD) years;
- 78 women in the health group and 38 women in the non-health group;
- 22 volunteers of the health group and 50 of the non-health group has, at least, one children;
- 94 of the total population are graduated.

3.2 Hardware

All experiments were conducted inside a closed room with artificially controlled lighting within specifications between 300 and $1000lx$ and positioned outside the visual field of the participant.

Eye movements were recorded with an on-screen Tobii TX300 equipment that comprises an eye tracker unit integrated to the lower part of a 23in TFT monitor with a screen resolution of 1280×1024 pixels. The eye tracker performs binocular tracking at a data sampling rate 300 Hz, and has minimum fixation duration of 60ms and maximum dispersion threshold of 0.5 degrees. Fixation was defined by the Tobii Studio software[1] fixation filter as two or more consecutive samples falling within a 50-pixel radius. These are the hardware and software eye tracker defaults for cognitive research. We considered only data from participants for whom on average 70% or more of their gaze samples were collected by the eye tracker.

In addition to the eye-tracking device, a laptop computer was used to calibrate the Tobii TX300 equipment and to store the collected data. This computer has a Intel Core I7, 16 GB of RAM memory and Microsoft Windows 7 operating system.

---

[1] More information in [1].

3.3   Framework

The proposed framework [10] is a computational experiment in which volunteers performed pain assessment on several face images of neonates evaluating each one using a numerical scale ranging from 0 (no pain) to 10 (extreme pain). In each session of the procedure, information regarding the ocular evaluation strategy of each volunteer was recorded by the eye tracking equipment. Exclusion criteria was applied to volunteers with diagnoses of epilepsy (seizure) and/or severe ocular problems.

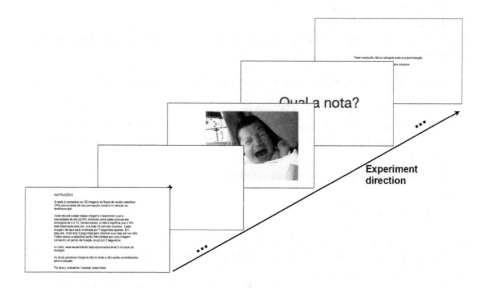

Fig. 1. Proposed framework.

The image data set used to design the proposed framework was developed by health professionals and researchers [20] and approved by the Ethics Committee for Research of the Federal University of São Paulo (1299/09, 3.116.151, 3.116.146 and 3.201.307). Twenty (20) frontal face images of ten (10) different neonates were selected. Each pair of images consists of one image of the neonate at rest and another image after a painful procedure routinely practised clinically, such as blood specimen collection (4 neonates), Hepatitis B vaccine (1) and Inborn Errors of Metabolism (IEMs) (5).

The experimental procedure, illustrated in Fig. 1, is composed of the following main 3 steps: (1) An instruction screen is shown to the subject; (2) Presentation of two evaluation trials, so that the subjects learn and comprehend the experiment; and (3) The beginning of the experimental procedure itself. Each neonatal face image to be evaluated is non-centralised located on the screen and is shown randomly to the volunteer for seven (7) seconds. All face images are separated in the experiment by a screen containing a cross as the central fixation point.

This cross is shown for two (2) seconds [21] so that the ocular movement of all volunteers tend to start at the same position on screen. Due to the ocular inertia caused by this cross, all ocular movement performed by the participants in the initial 300ms of the stimulus were discarded in our data analysis. After each face image screen shown during seven (7) seconds and the cross screen shown during two (2) seconds, the participant has three (3) seconds to answer verbally the score for the displayed image.

The total time to perform the experiment was approximately 5 min for each volunteer.

3.4   Facial Regions of Interest

To identify the facial features most used for neonatal pain assessment described in the literature and used by health professionals at the bedside, a literature review was carried out in order to identify pain scales based on neonatal facial expressions that describe specifically these features.

We have searched on Google Scholar, Semantic Scholar, Research Gate, Web of Science, Europe PMC and publishers of books and journals, such as Elsevier, PubMed and Scielo, using the following keywords: neonatal pain scale, newborn pain scale, and neonatal facial scale, from April to August 2020, without considering the publication date of the paper.

Table 1. Summary of the facial features based on the corresponding literature review.

Facial feature	Clinical scale														Total
	NFCS	M-NFCS	FAS	NNICUPAT	MIPS	P-MIPS	PIPP	PIPP-R	EVENDOL	CRIES	LIDS	TVPS	PASPI	BIIP	
Eye squeeze	X	X	X	X	X	X	X	X		X	X		X	X	12
Frown				X					X	X					3
Eyes tense										X	X				2
Distressed look											X				1
Furrowed forehead									X						1
Furrowed brow										X					1
Brow bulge	X	X	X	X	X	X	X	X	X	X			X	X	12
Open mouth		X								X					2
Tense mouth									X	X	X				3
Horiz. mouth stretch	X		X	X	X	X							X	X	7
Vert. mouth stretch	X				X	X							X		4
Lip purse	X		X	X							X				4
Nasolabial furrow	X	X	X	X	X	X	X	X		X			X	X	11
Open lips	X		X	X						X					4
Taut tongue	X		X											X	3
Tongue protrusion	X														1
Chin quiver	X		X		X	X				X					5

In this review, 124 studies were found that included facial scales for neonatal pain assessment, but only 52 explained the origin and/or process of clinical validation of these scales. Of these, 38 described the phenomenon of pain as being a holistic facial feature defined mainly as "grimace" or composed of general facial expression changes. Therefore, in order to avoid subjective interpretation of these non-specific facial components, other scales, like the well-known N-PASS [24] and

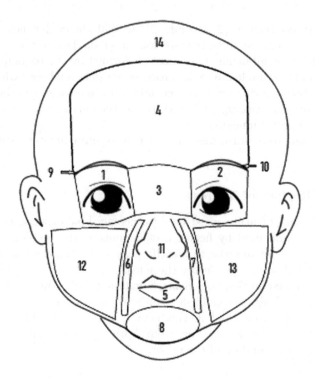

Fig. 2. Facial features analysed.

NIPS [27], were not considered here either. Therefore, only 14 scales were used to identify facial features, as shown in Table 1: Neonatal Facial Coding System (NFCS) [17], Modified NFCS (M-NFCS) [31], McGrath Facial Affective Scale (MCGRATH or FAS) [19], Nepean Neonatal Intensive Care Unit Pain Assessment Tool (NNICUPAT) [29], Modified infant pain scale (MIPS) [8], Partial MIPS (P-MIPS) [8], Premature Infant Pain Profile (PIPP) [5], PIPP Revisited (PIPP-R) [14], Evaluation Enfant Douleur (EVENDOL) [13], Crying Requires increased oxygenadministration, Increased vital signs, Expression, Sleeplessness (CRIES) [26], Liverpool Infant Distress Scale (LIDS) [23], Touch Visual Pain Scale (TVPS) [12], Pain Assessment scale for Preterm In-fants (PASPI) [28] and Behavioral Indicators of Infant Pain (BIIP) [22].

Based on these 14 scales investigated, 17 facial features were identified. However, some of them describe facial actions related to the same features and regions of the face, consequently have been joined together. Therefore, we have the following areas of interest based on this literature review (Fig. 2): (1) Right and (2) Left Eye (Eye squeeze; Frown; Eyes tense; Distressed look); (3) Region between Eyebrows and (4) Forehead (Furrowed forehead; Furrowed brow; Brow bulge); (5) Mouth (Open mouth; Tense mouth; Horiz. mouth stretch; Vert. mouth stretch; Lip purse; Open lips; Taut tongue; Tongue protrusion); (6) Right and (7) Left Nasolabial Groove (Nasolabial furrow); and (8) Chin (Chin quiver).

Additionally, we considered the following areas of interest: (9) Right and (10) Left Eyebrow; (11) Nose; (12) Right and (13) Left Cheek, and (14) "Other regions of the face" that the adult could be visually observing but was not described in any of the previous regions. Regarding the eyebrows, even though it may not be clearly visible in each neonate image, the corresponding eye movement fixation data on these regions might be interesting to analyse as well. Since it is well known that the nose is an important region when recognising a human face, we investigate here whether this region may be related to facial pain assessment and, consequently, such corresponding eye movement fixation data were analysed as well.

We have manually drawn all these regions as areas of interest for each one of the 20 frontal face images visually analysed by the volunteers, using the Tobii Studio software, as exemplified in Fig. 3.

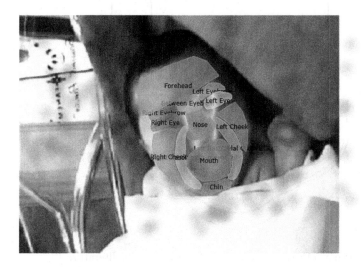

Fig. 3. Illustration of the dispersion of fixations and areas of interest on an experimental face image.

4   Results and Discussion

We divided this section into two parts: (1) we discuss the areas of interest according to the literature and (2) we analyse the visual perception of each sample group related to these areas.

4.1   Proposed Regions of Interest by the Literature

Based on the areas of interest defined in Sect. 3.4, we analysed the importance of each one by the literature: (1 - right eye) and (2 - left eye) = 100% or 14 works; (3 - region between eyebrows) and (4 - forehead) = 92.85%; (5 - mouth) = 78.57%; (6 - right nasolabial groove) and (7 - left nasolabial groove) = 85.71%; (8 - chin) 35.71%; (9 - right eyebrow) and (10 - left eyebrow) = 0%; (11 - nose) = 0%; (12 - right cheek) and (13 - left cheek) = 0% and (14 - other regions of face) = 0%. This result is shown in Fig. 4.

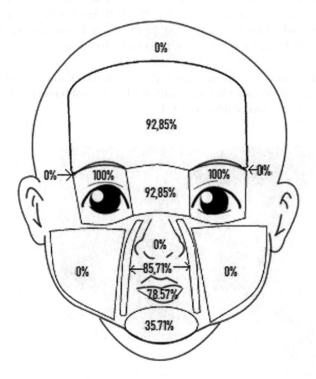

Fig. 4. Importance of each facial feature according to the literature.

According to the literature, the most important facial region to be analysed when assessing neonatal pain is the eyes, followed by the region between eyebrows, forehead and nasolabial groove. In contrast, the facial region least mentioned by the literature is the chin. As argued earlier, the regions of the nose, eyebrows and cheeks are not mentioned by the literature to identify pain in neonates.

4.2 Visualised Regions of Interest by Each Sample Group

We analysed eye movement fixation data acquired in the computational framework [10] corresponding to all sample groups. Using the Tobii Studio software, we calculated the number and time of fixation [21] of all the 14 regions of interest defined in Sect. 3.4. Based on this data, we summarised the number and time of

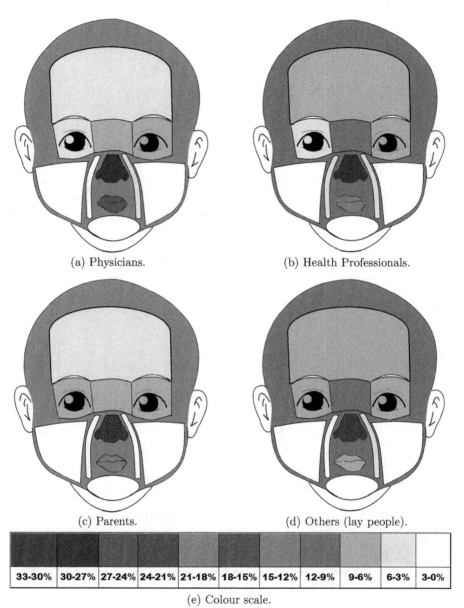

(a) Physicians.   (b) Health Professionals.

(c) Parents.   (d) Others (lay people).

| 33-30% | 30-27% | 27-24% | 24-21% | 21-18% | 18-15% | 15-12% | 12-9% | 9-6% | 6-3% | 3-0% |

(e) Colour scale.

Fig. 5. Time of fixation of areas of interest for each sample group.

fixation of each volunteer corresponding to these areas of interest and normalised it by the corresponding total of each metric for each sample group, calculating the relative importance of each facial feature. The results are illustrated in Figs. 5 and 6 and shown numerically in Table 2.

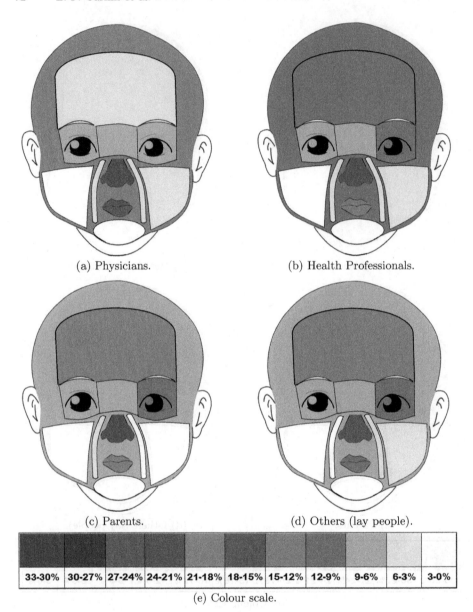

(a) Physicians.   (b) Health Professionals.

(c) Parents.   (d) Others (lay people).

| 33-30% | 30-27% | 27-24% | 24-21% | 21-18% | 18-15% | 15-12% | 12-9% | 9-6% | 6-3% | 3-0% |

(e) Colour scale.

Fig. 6. Number of fixation of areas of interest for each sample group.

Analysing the visual perception of each sample group, we found that physicians prefer to look at the regions of the nose and mouth, followed by "other regions of the face". The least observed regions by physicians are nasolabial grooves, chin, cheeks and eyebrows. Also, physicians looked more at the left nasolabial groove and the left cheek of neonates than the corresponding right

Table 2. Number and time of fixation for each facial feature by sample group.

Facial features	Group of volunteer							
	Physicians		Health prof.		Parents		Others	
	Time	Number	Time	Number	Time	Number	Time	Number
Left Eye	7.39%	8.84%	8.05%	9.25%	8.81%	9.84%	7.64%	9.15%
Right Eye	5.86%	6.64%	6.00%	6.93%	6.66%	8.64%	8.18%	8.78%
Region Between Eyebrows	7.68%	6.28%	9.80%	7.22%	6.99%	6.94%	9.32%	8.16%
Forehead	4.25%	5.40%	6.67%	9.87%	5.28%	12.20%	8.92%	13.71%
Mouth	15.94%	16.42%	14.06%	14.81%	13.97%	14.37%	8.72%	11.50%
Left Nasolabial Groove	5.16%	4.99%	3.01%	3.61%	2.95%	2.95%	3.61%	3.52%
Right Nasolabial Groove	2.57%	3.11%	3.26%	3.44%	2.65%	3.66%	2.96%	2.98%
Chin	1.80%	2.66%	1.45%	2.71%	1.09%	1.41%	0.31%	1.53%
Left Eyebrow	1.94%	2.16%	2.57%	2.78%	1.95%	2.12%	2.43%	3.14%
Right Eyebrow	1.25%	1.41%	2.11%	2.12%	2.03%	3.01%	2.34%	2.31%
Nose	28.91%	22.46%	29.52%	22.47%	32.18%	21.91%	32.89%	24.21%
Left Cheek	2.22%	3.69%	1.58%	3.06%	1.38%	2.85%	1.66%	3.16%
Right Cheek	0.63%	1.21%	0.82%	1.48%	1.16%	1.90%	0.94%	1.61%
Other regions	14.38%	14.72%	11.08%	10.25%	12.91%	8.19%	10.09%	6.23%

ones. Analogously to physicians, health professionals also had a preference to look at the regions of the nose and mouth, followed by the region between eyebrows and "other regions of the face", left eye and forehead. The least observed areas were the nasolabial groove, cheeks and eyebrows. The left nasolabial groove and left cheek were observed more as well. Similarly to physicians and health professionals, parents look for a longer time to the nose, followed by the mouth and "other regions of the face". The least observed features by parents were the chin and cheeks as well. The sample group of others (lay people) concentrated their look at the nose, "other regions of the face" and the region between eyebrows. The least observed facial regions are similarly to the other sample groups.

Regarding the differences among distinct sample groups, we first compared sample groups of physicians, health professionals and others (lay people). We can see that lay people had less visual fixations than physicians and health professionals in the region of the mouth and chin. Interestingly, however, physicians seem to look more than health professionals and others at the left nasolabial groove. Moreover, physicians are the sample group that least observed the forehead. Comparing physicians and parents, we only found visual differences related to the forehead. Parents presented higher number of fixations at this area than physicians. No other visually clear differences were found between these two groups. Finally, when comparing parents and others, we found that parents seem to have more fixations at the mouth and less at the forehead.

Overall, the results indicate that the visual perception of each sample group of volunteers agrees with the facial features proposed by the literature, as the regions of the eyes, mouth and forehead showed a representative number and time of fixations. Particularly, it is important to note that the nose region received the most visual attention by all sample groups, even though this area is not considered relevant by the literature. However, in accordance with clinical scales, the regions of the cheeks and chin received minimal or no attention. Also, we have found differences within each pair of facial region, such as the left and right eyes, left and right nasolabial grooves and left and right cheeks. Interestingly, the visual perception of adults seems to be concentrated on the left side of the face. As shown in Table 1, no pain scale discriminates the sides of the face. Lastly, there is a considerable number and time of fixations in "other regions of the face". This suggests that adults may be visually observing other regions of the face when assessing pain that are not selected here and not described as relevant in the corresponding literature.

5   Conclusions

This work shows human visual perception results on neonatal pain assessment of face images that might help to improve the understanding of the most relevant facial features to perform such assessment either automatically or clinically. According to the literature review carried out here, we have highlighted that the most important facial features are the eyes, followed by the region between eyebrows, forehead and nasolabial groove. However, the regions of the nose, cheeks and eyebrows are not analysed in all the pain scales mentioned here.

When comparing differences among the 4 distinct sample groups of volunteers considered, physicians and health professionals seem to have a similar visual perception of this pain phenomena, but physicians have presented more fixations at the left nasolabial groove and less at the forehead. Also, parents have presented a visual perception behaviour similar to physicians and health professionals. Interestingly, the main facial features proposed by the clinical pain scales were commonly observed by all sample groups, either experts or not. Yet, there has been a considerable number of visual attention on the nose for all sample groups and the experimental results indicate that the areas of the left side of the faces are more visually observed than the right ones.

Overall, our results show differences in all sample groups between the facial regions considered clinically relevant to the ones expressed implicitly by the human visual perception. Further work would be necessary to describe such findings more precisely. Specifically, we intend to perform a statistical analysis of this eye movement fixation data using different approaches of machine learning and pattern recognition in order to identify more detailed explanations about these differences and also to identify the relevance of these findings when performing a pain classification task.

References

1. Tobii, A.B.: User's manual tobii studio (2016). https://www.tobiipro.com/siteassets/tobii-pro/user-manuals/tobii-pro-studio-user-manual.pdf. Accessed 12 Nov 2020

2. Anand, K.J., Carr, D.B.: The neuroanatomy, neurophysiology, and neurochemistry of pain, stress, and analgesia in newborns and children. Pediatr. Clin. North Am. **36**(4), 795–822 (1989)

3. Anand, K.J., Hickey, P.R., et al.: Pain and its effects in the human neonate and fetus. N. Engl. J. Med. **317**(21), 1321–1329 (1987)

4. Balda, R.C., Guinsburg, R.: Avaliação da dor no período neonatal. Diagnóstico e tratamento em neonatologia, pp. 577–585. Atheneu, São Paulo (2004)

5. Ballantyne, M., Stevens, B., McAllister, M., Dionne, K., Jack, A.: Validation of the premature infant pain profile in the clinical setting. Clin. J. Pain **15**(4), 297–303 (1999)

6. Brahnam, S., Chuang, C.F., Sexton, R.S., Shih, F.Y.: Machine assessment of neonatal facial expressions of acute pain. Decis. Support Syst. **43**(4), 1242–1254 (2007)

7. Brahnam, S., et al.: Neonatal pain detection in videos using the iCOPEvid dataset and an ensemble of descriptors extracted from Gaussian of local descriptors. Appl. Comput. Inf. (2019). https://doi.org/10.1016/j.aci.2019.05.003

8. Buchholz, M., Karl, H.W., Pomietto, M., Lynn, A.: Pain scores in infants: a modified infant pain scale versus visual analogue. J. Pain Sympt. Manage. **15**(2), 117–124 (1998)

9. Carlini, L.P., Heideirich, T., Balda, R., Barros, M., Guinsburg, R., Thomaz, C.: Visual perception of pain in neonatal face images. Anais do XV Workshop de Visão Computacional, pp. 37–42 (2019). https://sol.sbc.org.br/index.php/wvc/article/view/7625

10. Carlini, L.P.: A visual perception framework to analyse neonatal pain in face images. In: Campilho, A., Karray, F., Wang, Z. (eds.) ICIAR 2020. LNCS, vol. 12131, pp. 233–243. Springer, Cham (2020). https://doi.org/10.1007/978-3-030-50347-5_21

11. Diatchenko, L., Nackley, A.G., Tchivileva, I.E., Shabalina, S.A., Maixner, W.: Genetic architecture of human pain perception. Trends Genet. **23**(12), 605–613 (2007)

12. van Dijk, M., Thomas, J., Millar, A., Tibboel, D., Albertyn, R.: Measuring pain and discomfort in HIV+ infants: introducing the touch visual pain scale. In: Cape Town: The 14th World Congress of Anaesthesiologists (2008)

13. Fournier-Charrière, E., et al.: EVENDOL, a new behavioral pain scale for children ages 0 to 7 years in the emergency department: design and validation. Pain® **153**(8), 1573–1582 (2012)

14. Gibbins, S., et al.: Validation of the premature infant pain profile-revised (PIPP-R). Early Hum. Dev. **90**(4), 189–193 (2014)

15. Golianu, B., Krane, E.J., Galloway, K.S., Yaster, M.: Pediatric acute pain management. Pediatr. Clin. North Am. **47**(3), 559–587 (2000)

16. Grunau, R.V., Craig, K.D.: Pain expression in neonates: facial action and cry. Pain **28**(3), 395–410 (1987)

17. Grunau, R.V., Johnston, C.C., Craig, K.D.: Neonatal facial and cry responses to invasive and non-invasive procedures. Pain **42**(3), 295–305 (1990)

18. Guinsburg, R.: Avaliação e tratamento da dor no recém-nascido. J. Pediatr. (Rio J.) **75**(3), 149–60 (1999)

19. Guinsburg, R., Kopelman, B.I., Almeida, M.F.B.D., Miyoshi, M.H.: A dor do recém-nascido prematuro submetido a ventilação mecânica através de cânula traqueal. J. pediatr. (Rio J.) 82–90 (1994)

20. Heiderich, T.M., Leslie, A.T.F.S., Guinsburg, R.: Neonatal procedural pain can be assessed by computer software that has good sensitivity and specificity to detect facial movements. Acta Paediatr. **104**(2), e63–e69 (2015)

21. Holmqvist, K., Nyström, M., Andersson, R., Dewhurst, R., Jarodzka, H., Van de Weijer, J.: Eye Tracking: A Comprehensive Guide to Methods and Measures. OUP, Oxford (2011)
22. Holsti, L., Grunau, R.E.: Initial validation of the behavioral indicators of infant pain (BIIP). Pain **132**(3), 264–272 (2007)
23. Horgan, M.F.: The development and evaluation of a scale to assess pain in the post-operative neonate. Ph.D. thesis, Liverpool John Moores University (2000)
24. Hummel, P., Puchalski, M., Creech, S., Weiss, M.: Clinical reliability and validity of the n-pass: neonatal pain, agitation and sedation scale with prolonged pain. J. Perinatol. **28**(1), 55–60 (2008)
25. IASP: IASP publication, pain terms: a list with definitions and notes on usage. Pain (1979)
26. Krechel, S.W., Bildner, J.: Cries: a new neonatal postoperative pain measurement score. Initial testing of validity and reliability. Pediatr. Anesthesia **5**(1), 53–61 (1995)
27. Lawrence, J., Alcock, D., McGrath, P., Kay, J., MacMurray, S.B., Dulberg, C.: The development of a tool to assess neonatal pain. Neonatal Netw. NN **12**(6), 59 (1993)
28. Liaw, J.J., Yang, L.H., Chou, H.L., Yin, T., Chao, S.C., Yeh Lee, T.: Psychometric analysis of a Taiwan-version pain assessment scale for preterm infants. J. Clin. Nurs. **21**, 1–2, 89–100 (2012)
29. Marceau, J.: Pilot study of a pain assessment tool in the neonatal intensive care unit. J. Paediatr. Child Health **39**(8), 598–601 (2003)
30. Neves, F.A.M., Corrêa, D.A.M.: Dor em recém-nascidos: a percepção da equipe de saúde. Ciência, Cuidado e Saúde **7**(4), 461–467 (2008)
31. Rushforth, J.A., Levene, M.I.: Behavioural response to pain in healthy neonates. Arch. Disease Childh. Fetal Neonatal Edn. **70**(3), F174–F176 (1994)
32. Teruel, G.F., Heiderich, T.M., Guinsburg, R., Thomaz, C.E.: Analysis and recognition of pain in 2d face images of full term and healthy newborns. In: Proceedings of the XV Encontro Nacional de Inteligencia Artificial, ENIAC, vol. 2018, pp. 228–239 (2018)
33. Zamzmi, G., Paul, R., Goldgof, D., Kasturi, R., Sun, Y.: Pain assessment from facial expression: neonatal convolutional neural network (N-CNN). In: 2019 International Joint Conference on Neural Networks (IJCNN), pp. 1–7. IEEE (2019)

Pain Intensity Assessment in Sickle Cell Disease Patients Using Vital Signs During Hospital Visits

Swati Padhee[1]([✉]), Amanuel Alambo[1], Tanvi Banerjee[1], Arvind Subramaniam[2],
Daniel M. Abrams[3], Gary K. Nave Jr.[3], and Nirmish Shah[2]

[1] Wright State University, Dayton, USA
padhee.2@wright.edu
[2] Duke University, Durham, USA
[3] Northwestern University, Evanston, USA

Abstract. Pain in sickle cell disease (SCD) is often associated with increased morbidity, mortality, and high healthcare costs. The standard method for predicting the absence, presence, and intensity of pain has long been self-report. However, medical providers struggle to manage patients based on subjective pain reports correctly and pain medications often lead to further difficulties in patient communication as they may cause sedation and sleepiness. Recent studies have shown that objective physiological measures can predict subjective self-reported pain scores for inpatient visits using machine learning (ML) techniques. In this study, we evaluate the generalizability of ML techniques to data collected from 50 patients over an extended period across three types of hospital visits (i.e., inpatient, outpatient and outpatient evaluation). We compare five classification algorithms for various pain intensity levels at both intra-individual (within each patient) and inter-individual (between patients) level. While all the tested classifiers perform much better than chance, a Decision Tree (DT) model performs best at predicting pain on an 11-point severity scale (from 0–10) with an accuracy of 0.728 at an inter-individual level and 0.653 at an intra-individual level. The accuracy of DT significantly improves to 0.941 on a 2-point rating scale (i.e., no/mild pain: 0–5, severe pain: 6–10) at an inter-individual level. Our experimental results demonstrate that ML techniques can provide an objective and quantitative evaluation of pain intensity levels for all three types of hospital visits.

Keywords: Pain intensity quantification · Pain pattern identification · Physiological signals · Sickle cell anemia

1 Introduction

Sickle cell disease (SCD) is the most common inherited blood disorder, affecting millions of people worldwide. It is characterized by the production of an altered

This work was supported by the NIH under grant NIH R01AT010413.

A. Del Bimbo et al. (Eds.): ICPR 2020 Workshops, LNCS 12662, pp. 77–85, 2021.
https://doi.org/10.1007/978-3-030-68790-8_7

type of hemoglobin. The altered hemoglobin deoxygenates while passing through blood vessels, polymerizes and becomes fibrous, causing the red blood cells to become rigid and change their shape to sickle-shaped. The altered red blood cells can occlude blood vessels, a phenomenon known as vaso-occlusion, resulting in a lack of oxygen to tissues, and thereby causing pain [9]. Most patients with SCD experience repeated, unpredictable episodes of severe pain. These pain episodes are the leading cause of emergency department visits and may last for as long as several weeks. Arguably, the most challenging aspect of treating pain episodes in SCD is assessing and interpreting the patient's pain intensity level.

However, in current clinical practice, self-description is the gold standard approach for determining the absence, presence, and intensity of pain. Due to the subjective nature of pain, it becomes challenging for the clinicians to precisely ascertain the severity of the patient's pain. Besides, effective treatment is palliative, including intravenous opioid therapy. While these self-described pain intensity levels provide important clinical reference indicators and have been proven to be useful for treating patients suffering from pain in most situations [5], it might have challenges when applied to certain vulnerable populations.

Current clinical guidelines recommend frequent observations of vital signs during assessment and treatment of painful episodes as they are an objective measurement for the essential physiological functions and are potential indicators for patients' subjective pain levels. It has been previously reported that ML techniques can be used to design objective pain assessment models using vital signs from inpatient EHR data. However, people with SCD suffer from various acute complications that can result in multiple hospitalizations, emergency department (ED) visits, and outpatient care visits. To the best of our knowledge, this is the first study to explore the relationship between the varying nature of hospital visits and physiological measures on pain intensity for patients with SCD.

2 Related Work

Current literature shows increased attention on machine learning techniques to understand various complexities associated with patient health in SCD. Milton et al. [10] developed an ensemble model exploring 14 algorithms to predict Hemoglobin F (HbF) in patients associated with different configurations of Single Nucleotide Polymorphisms (SNPs). Allayous et al. [2] demonstrated the high risk of an acute splenic sequestration crisis, which is a severe symptom of SCD. Solanki [12], implemented two models, including DT to classify specific blood groups. Khalaf et al. [7] classified the dosage of medication required for the treatment of patients with SCD.

Prior studies have reported that fluctuations in vital signs can be used for assessing pain [3] as acute pain leads to changes in vital signs [8]. These physiological measures include blood pressure, respiratory rate, oxygen saturation, temperature, and pulse. From our research group, Yang et al. [13] showed the feasibility of ML techniques on a limited dataset of 5363 records from 40 patients during inpatient hospital visits to predict subjective pain scores from six objective vital signs. Alambo et al. [1] employed 424 clinical notes of the same cohort of 40 patients to assess the prevalence of pain in patients and whether pain increases, decreases, or stays constant. In this study, we investigate the generalizability of ML techniques for a broader group of people during inpatient, outpatient, and outpatient evaluation visits. We provide definitions of these visits as validated by our clinical collaborators in Section definitions 3.1.1. Specifically, we utilize five years of EHR data from 50 patients suffering from SCD to build pain prediction models using objective physiological measures as features at both the intra-individual and inter-individual levels based on an 11-point numeric rating scale (NRS) [6]. We further investigate whether the variation in the type of hospital visits affect our model performance.

3   Methods

3.1   Data Description

In this study, we utilized 67927 records from EHR data collected from 50 participants at Duke University Hospital over five consecutive years. Each record contained measures for six vital signs as follows: (i) peripheral capillary oxygen saturation (SpO2), (ii) systolic blood pressure (SystolicBP), (iii) diastolic blood pressure (DiastolicBP), (iv) heart rate (Pulse), (v) respiratory rate (Resp), and (vi) temperature (Temp). Along with the vital signs, each record also included the patient's self-reported pain score with an ordinal range from 0 (no pain) to 10 (severe and unbearable pain).

The data were de-identified using study labels to label the patient without identification. The timestamp for each data entry was also de-identified, preserving temporality. The dataset had missing values for one or more of the vital signs and the pain score. Our analysis is done on 59728 records containing at least one of the six vital signs or pain score values from 47 patients as we observed that no data was extracted for three patients. As the percentage of complete records in our dataset was only 7.6%, we employed an imputation method to impute the missing data values. We utilized Multiple Imputations by Chained Equations [4], sometimes called "fully conditional specification" or "sequential regression multiple imputations," as it is widely used in clinical practice and recent healthcare studies [11,13].

3.1.1   Type of Hospital Visit

Some patients with SCD have higher inpatient requirements than others due to the subjectivity and frequency of pain crisis. Furthermore, because of SCD-related complications, many people with SCD may visit hospitals more frequently. However, limited information is available related to various hospital visits' characteristics, including emergency department visits among SCD patients. Information related to the type of hospital visit by patients with SCD can help develop services and strategies for best meeting patients' healthcare needs with

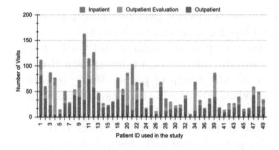
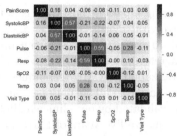

Fig. 1. Distribution of visits for 50 patients. (Study Patient identifiers 17, 37 and 50 are absent.)

Fig. 2. Pearson correlation of six vital signs and visit types in the original dataset.

SCD. To understand the variations in the nature of visits, we followed the definitions below recommended by our co-author clinician to extract information about the nature of visits for every record in our dataset.

- **Visit**: For each patient, we consider a record to be of a different visit if there is a gap of at least two days between the records.
- **Outpatient visit**: We define a visit to be an outpatient visit if the patient has not stayed in the hospital for a day or longer and has two or less recordings taken.
- **Inpatient visit**: We define a visit to be an inpatient visit if the patient has stayed for two or more consecutive days in the hospital.
- **Outpatient evaluation visit**: We define a visit to be an outpatient evaluation visit if the patient has stayed in the hospital for one day or has more than two recordings taken in a single day.

Figure 1 shows the distribution of the three types of visits in our data.

3.2 Pain Prediction

We examined the Pearson correlation between the six vital signs and the type of visit in our dataset as we plan to use them as features influencing pain scores. As shown in Fig. 2, in addition to a moderate correlation of 0.57 between systolic and diastolic blood pressure, we observe a correlation of 0.59 between pulse and respiratory rate in the original dataset. The other variables are poorly correlated

Table 1. Intra-individual pain prediction results (accuracy)

	SVM	DT	kNN	MLR	RF
Vitals	0.522	**0.653**	0.625	0.535	0.485
Vitals + Visit	0.506	**0.653**	0.625	0.535	0.486
Yang et al. [13]	0.582	—	0.522	0.578	0.523

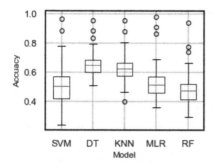 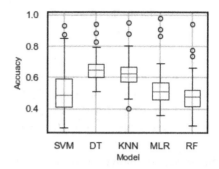

Fig. 3. Intra-individual pain prediction accuracy results on vital signs data.

Fig. 4. Intra-individual pain prediction accuracy results on vital signs and visit information.

or uncorrelated with one another. Hence, we utilize all six vital signs and visit information as predictors of our pain prediction models.

We implemented five supervised ML classification algorithms to predict patients' pain scores based on their vital signs: k-Nearest Neighbors (kNN), Support Vector Machine (SVM), Multinomial Logistic Regression (MLR), Decision Tree (DT), and Random Forest (RF). We investigated both the intra-individual level and inter-individual level (i.e., treating all the patients as a single entity) analysis. For intra-level analysis, we used the six vital signs, visit information, and pain scores as patient labels were unused since samples from the same patient were employed to build the personal model in this analysis. However, for the inter-individual level analysis, we employed the four different scenarios as reported by Yang et al. [13] i.e., Case 1: imputation with patient labels and prediction with patient labels; Case 2: imputation with patient labels and prediction without patient labels; Case 3: imputation without patient labels and prediction with patient labels; Case 4: imputation without patient labels and prediction without patient labels. For each experiment, we report the results with and without visit information. We used 10-fold cross-validation to evaluate our prediction models. We reported the model prediction accuracy as it is the ratio of correctly predicted pain scores over the total number of pain scores.

4 Results and Discussion

4.1 Intra-individual Pain Prediction

We present the intra-individual pain prediction results for 47 patients in terms of accuracy in Table 1. Figure 3 shows the accuracy distribution of predictions for all five classifiers. DT achieved the highest accuracy ranging from 0.503 to 0.953, and an average accuracy of 0.653 when trained on the six vital signs described in Sect. 3.1. Thus, our models trained on the vital signs of a patient could correctly predict the self-reported pain scores of the same patient on an average

Table 2. Inter-individual pain prediction results (accuracy)

	Vitals					Vitals + Visit					Yang et al. [13]	
	SVM	DT	kNN	MLR	RF	SVM	DT	kNN	MLR	RF	MLR	SVM
Case 1	0.585	0.676	0.662	0.448	0.336	0.595	**0.697**	0.668	0.525	0.357	0.429	0.421
Case 2	0.422	**0.647**	0.644	0.345	0.335	0.593	0.643	0.530	0.427	0.335	0.215	0.236
Case 3	0.561	**0.701**	0.658	0.405	0.406	0.591	0.701	0.595	0.460	0.410	0.313	0.305
Case 4	0.590	**0.728**	0.708	0.401	0.404	0.659	0.704	0.595	0.472	0.401	0.257	0.246

Table 3. Inter-individual pain prediction results with varying pain scales on vitals data (accuracy) arranged from higher resolution to lower resolution. (6 Pain Scores: None: 0, Very mild:1–2, Mild: 3–4, Moderate: 5–6, Severe: 7–8, Very severe:9–10; 4 Pain Scores: None: 0, Mild: 1–3, Moderate: 4–6, Severe: 7–10; 2 Pain Scores:No/mild Pain: 0–5, Severe Pain: 6–10)

	11 Pain score						6 Pain score					
	SVM	DT	kNN	MLR	RF	Yang et al. [13]	SVM	DT	kNN	MLR	RF	Yang et al. [13]
Case 1	0.585	0.676	0.662	0.448	0.336	0.429	0.767	0.779	0.761	0.606	0.481	0.546
Case 2	0.422	0.647	0.644	0.345	0.335	0.215	0.672	0.762	0.732	0.55	0.485	0.347
Case 3	0.561	0.701	0.658	0.405	0.406	0.313	0.766	0.771	0.773	0.599	0.586	0.449
Case 4	0.590	**0.728**	0.708	0.401	0.404	0.257	0.772	**0.814**	0.777	0.605	0.589	0.397
	4 Pain score						2 Pain score					
	SVM	DT	kNN	MLR	RF	Yang et al. [13]	SVM	DT	kNN	MLR	RF	Yang et al. [13]
Case 1	0.849	0.832	0.809	0.683	0.583	0.681	0.923	0.937	0.904	0.926	0.84	0.821
Case 2	0.788	0.821	0.788	0.659	0.589	0.521	0.915	0.919	0.893	0.903	0.835	0.680
Case 3	0.837	0.824	0.815	0.685	0.66	0.607	0.923	0.939	0.907	0.9267	0.874	0.730
Case 4	0.85	**0.853**	0.818	0.687	0.671	0.563	0.935	**0.941**	0.907	0.927	0.871	0.678

Table 4. Pain change prediction results (accuracy)

	Vitals			Vitals + Visit			Yang et al. [13]
	DT	kNN	MLR	DT	kNN	MLR	MLR
Case1	0.515	0.490	0.514	**0.522**	0.504	0.517	0.403
Case2	**0.508**	0.494	0.508	0.518	0.494	0.503	0.363
Case3	**0.518**	0.466	0.517	0.520	0.492	0.518	0.390
Case4	**0.520**	0.492	0.520	0.517	0.466	0.516	0.404

of 65.3% of the records. Similar performance with additional visit information (Table 1, row 2) indicates that for the same patient, our models can learn the differences between vital signs and pain intensity experienced by a patient during different types of visits. Our results show that a model trained on the same patient's historical data can predict the pain intensity levels for the same patient in the future based on their vital signs during outpatient, inpatient, and outpatient evaluation visits. Such a model can provide medical teams with additional information about the severity of a patient's pain, which does not rely on the patient's subjective response.

4.2   Inter-individual Pain Prediction

In real-time scenarios, when a new patient visits a hospital, intra-individual level models can not be applied until sufficient data is collected. We report the inter-individual pain prediction results for 47 patients in Table 2. The best performance was achieved in Case 4 by DT (accuracy 0.728) compared to 0.429 by MLR in Case 1 by Yang et al. [13]. This indicates that our DT model trained on more vitals signs data collected over a more extended period (five years) from a larger cohort of people could predict the severity of pain for a new patient more accurately. Also, with more data, the models can generalize better, as we did not consider the patient-level differences during both data imputation and pain prediction (Case 4). Additional visit information seemed important in predicting pain scores when considering patient information at data imputation and prediction (Case 1). This indicates that we need to consider the type of visit to predict pain scores from vital signs for a personalized prediction from a generalizable model.

In our original dataset, we have 11 unique self-described pain scores ranging from 0 to 10. It is challenging for one person to distinguish between such broad and granular pain intensity levels and be consistent with every pain episode. Hence, we reported our model performance at an inter-individual level by transforming our dataset on a 6-point rating scale, a 4-point rating scale, and a binary rating scale [13] in Table 3. The higher accuracy associated with the narrow scales is attributed to the narrow space to misclassify many records by our models, thereby improving the chances of correctly predicting a pain score.

We believe the lower performance of RF compared with DT is attributed to the replacement-with-duplicates-based bootstrap approach to sub-sampling used in training a Random Forest model that could lead to training records not representative of the test sample the model is tested on. Furthermore, as a bagging approach to ensemble models, the misclassification error from the first bootstrap sample in random forests is not used to improve a model trained on a different bootstrap sample. Finally, test accuracy is computed by taking the average of the different bootstrap samples' accuracy where bad bootstrap samples might harm the aggregate test accuracy. With DT, however, it is possible that our model captured the ideal training records for a given test set, thereby yielding better accuracy.

4.3   Pain Change Prediction

Additional information about the change in pain (increase/decrease/no change) would help determine the effectiveness of therapy and the consideration for management, such as either giving more pain medication, keeping a medication dosage stable or decreasing a pain dosage. Predicting a change in pain may be more critical than having an estimate of the pain score since the medical team can make treatment decisions based on this information and ultimately improve a patient's pain more quickly. Hence, we formulate a three-class pain change classification problem, i.e., increase, decrease, and no change. In this case, a baseline chance accuracy can be 0.33 (1/3). We report the results of our two best performing DT and kNN based inter-individual level classifiers in Table 4. It is not surprising that our DT model was able to predict a pain change correctly 52.2% times (0.7% more) when provided with additional visit information. It indicates our model learned that the change in pain severity might be different for each type of visit for different patients (i.e., Case 1). It is essential to consider whether a visit is an inpatient or outpatient visit to estimate a change in pain intensity for each patient.

5 Conclusion

In this study, we leveraged multiple machine learning algorithms on six physiological measures of patients with SCD to predict pain scores. We were able to deal with missing data and conduct a series of experiments at both intra-individual and inter-individual levels. In each of the experiments, we observed higher accuracy with an increase in data. All the models were able to capture the variation in the type of visits at an inter-individual level when considering the diversity between patients in data imputation and prediction. Our results show Decision Tree as the most promising model, followed by k-Nearest Neighbours and Support Vector Machines. The evaluation demonstrates that using objective physiological measurements to predict subjective pain in SCD patients may be generalizable for a larger cohort of patients. In the future, we look forward to extending our work to visit level analysis and exploring the patients' medication information.

References

1. Alambo, A., et al.: Measuring pain in sickle cell disease using clinical text. In: 42nd Annual International Conference of the IEEE Engineering in Medicine & Biology Society (EMBC), pp. 5838–5841. IEEE (2020)
2. Allayous, C., Clémençon, S., Diagne, B., Emilion, R., Marianne, T.: Machine learning algorithms for predicting severe crises of sickle cell disease (2008)
3. Arbour, C., Choinière, M., Topolovec-Vranic, J., Loiselle, C.G., Gélinas, C.: Can fluctuations in vital signs be used for pain assessment in critically ill patients with a traumatic brain injury? Pain Res. Treat. **2014** (2014)
4. Azur, M.J., Stuart, E.A., Frangakis, C., Leaf, P.J.: Multiple imputation by chained equations: what is it and how does it work? Int. J Methods Psychiatr. Res. **20**(1), 40–49 (2011)
5. Brown, J.E., Chatterjee, N., Younger, J., Mackey, S.: Towards a physiology-based measure of pain: patterns of human brain activity distinguish painful from non-painful thermal stimulation. PloS one **6**(9), e24124 (2011)
6. Downie, W., Leatham, P., Rhind, V., Wright, V., Branco, J., Anderson, J.: Studies with pain rating scales. Ann. Rheum. Dis. **37**(4), 378–381 (1978)
7. Khalaf, M., et al.: Machine learning approaches to the application of disease modifying therapy for sickle cell using classification models. Neurocomputing **228**, 154–164 (2017)

8. Macintyre, P.E., et al.: Acute pain management: scientific evidence. Australian and New Zealand College of Anaesthetists (2010)
9. Manwani, D., Frenette, P.S.: Vaso-occlusion in sickle cell disease: pathophysiology and novel targeted therapies. Blood J. Am. Soc. Hematol. **122**(24), 3892–3898 (2013)
10. Milton, J.N., Gordeuk, V.R., Taylor, J.G., Gladwin, M.T., Steinberg, M.H., Sebastiani, P.: Prediction of fetal hemoglobin in sickle cell anemia using an ensemble of genetic risk prediction models. Circ. Cardiovasc. Genet. **7**(2), 110–115 (2014)
11. Shah, A.D., et al.: Type 2 diabetes and incidence of cardiovascular diseases: a cohort study in 1.9 million people. Lancet Diab. Endocrinol. **3**(2), 105–113 (2015)
12. Solanki, A.V., et al.: Data mining techniques using weka classification for sickle cell disease. Int. J. Comput. Sci. Inf. Technol. **5**(4), 5857–5860 (2014)
13. Yang, F., Banerjee, T., Narine, K., Shah, N.: Improving pain management in patients with sickle cell disease from physiological measures using machine learning techniques. Smart Health **7**, 48–59 (2018)

Longitudinal Classification of Mental Effort Using Electrodermal Activity, Heart Rate, and Skin Temperature Data from a Wearable Sensor

William Romine[1]($\boxtimes$), Noah Schroeder[2], Anjali Edwards[1], and Tanvi Banerjee[3]

[1] Department of Biological Sciences, Wright State University, Dayton, OH 45435, USA
romine.william@gmail.com, edwards.348@wright.edu
[2] Department of Leadership Studies in Education and Organizations, Wright State University,
Dayton, OH 45435, USA
noah.schroeder@wright.edu
[3] Department of Computer Science and Engineering, Wright State University,
Dayton, OH 45435, USA
tanvi.banerjee@wright.edu

Abstract. Recent studies show that physiological data can detect changes in mental effort, making way for the development of wearable sensors to monitor mental effort in school, work, and at home. We have yet to explore how such a device would work with a single participant over an extended time duration. We used a longitudinal case study design with ~38 h of data to explore the efficacy of electrodermal activity, skin temperature, and heart rate for classifying mental effort. We utilized a 2-state Markov switching regression model to understand the efficacy of these physiological measures for predicting self-reported mental effort during logged activities. On average, a model with state-dependent relationships predicted within one unit of reported mental effort (training RMSE = 0.4, testing RMSE = 0.7). This automated sensing of mental effort can have applications in various domains including student engagement detection and cognitive state assessment in drivers, pilots, and caregivers.

Keywords: Cognitive load · Wearable sensor · Mental effort · Machine learning · Cognitive assessment

1 Introduction

Researchers often strive to measure how focused someone is on a task, or how much mental effort they are putting into it. One domain where this is an important question is education and the study of learning. For more than three decades many researchers interested in this question, or related questions, have relied on a prominent theory called Cognitive Load Theory (CLT; [1–3]). According to CLT, we can put mental effort towards learning the salient material, known as intrinsic cognitive load, or towards other features of the instruction that do not support the learning task, known as extraneous cognitive

© Springer Nature Switzerland AG 2021
A. Del Bimbo et al. (Eds.): ICPR 2020 Workshops, LNCS 12662, pp. 86–95, 2021.
https://doi.org/10.1007/978-3-030-68790-8_8

load [1, 3, 4]. Researchers suggest that the complexity of the task and the learner's level of prior knowledge in the subject determine the amount of mental effort that will be needed to learn the material and thus determine the intrinsic cognitive load, whereas mental effort put into parsing non-supporting elements of the instruction, such as interesting but ultimately unrelated stories, or visually searching for references needed to understand components of the learning materials, determine the extraneous cognitive load [1–4]. Since the working memory is limited in both capacity [5, 6] and duration [1, 3], CLT suggests that it is important to minimize the mental effort learners have to expend on tasks that are not essential to learning the material [7].

Cognitive load theory is well-established in the education literature, with a number of highly cited papers centering on the theory (e.g., [7–9]). Unsurprisingly, CLT has been used to theoretically support a number of specific task design principles, such as the worked example effect, the redundancy effect, and the split-attention effect [2, 4], and has become widespread outside of the educational psychology literature, appearing, for example, in the medical education literature as well [10–12]. The notion of cognitive load is an important theoretical paradigm in many types of educational research, but a lingering question persists in the CLT literature: how do we measure cognitive load?

The construct of cognitive load, as explained by CLT, is relatable to many; however, the measurement of such a construct has been a psychometric challenge for more than a decade. Researchers have used methods as varied as self-reports [13], eye-tracking measures [14], secondary task techniques [15], or physiological data [16]. Outside of the education literature researchers have measured a similar construct, mental workload, using similar methodologies like self-reports [17] or physiological measures [18] like facial skin temperature [19].

Recently, perhaps due to the increasing accessibility of wearable sensors or the psychometric issues associated with current methods for cognitive load assessment [20–22], researchers have been using physiological measurements and investigating their relation to learning relevant outcomes. Some of this work has shown promising results. For example, in relation to learning relevant processes, [23] used heart rate variability as an indicator of sustained attention. In addition, [24] measured electrodermal activity and examined these data in relation to self-reported emotional engagement. They found that students who were more engaged showed more frequently high levels of electrodermal activity. Taking a multimodal physiological approach, [25] differentiated between when students worked on high, moderate, and low mental effort activities, and further were able to predict a user's self-reported mental focus. It is noteworthy, however, that not all studies have shown such promising results. For example, examining task complexity in relation to physiological measures, [26] found that electrodermal activity and heart rate mean scores did not differ depending on the complexity of the task.

As noted, the use of wearable sensors to collect physiological data in relation to education-relevant outcomes is becoming more widespread in the literature. While there have been some promising results, the literature also shows some null results, highlighting the complexity of this area of work. When looking at recent studies [20–26], an important missing piece is understanding how we can use these data to track mental effort in an individual over extended periods of time, and the diagnostic utility of easily-obtainable physiological measures like EDA, skin temperature, and heart rate towards this goal. In this case study, we utilize a longitudinal interpretable machine learning approach to understand how these data can be used to track the mental effort of an individual student in the context of both school activities and activities of daily living.

2 Methods

2.1 Study Design

In this study we sought to understand how EDA, skin temperature, and heart rate can be used to learn trends in mental effort for a single participant, and the extent to which we can model this in a robust way. We were first interested in using interpretable machine learning models to understand relationships between the participant's EDA, skin temperature, and heart rate measures and her reported mental effort. Second, we were interested in the diagnostic strength of these measures, and their efficacy in predicting mental effort in the context of future activities. To satisfy these goals, we used a longitudinal n = 1 case study design [27]. The goal of a case study is to generate rich description of a single case, which typically constitutes a single participant or entity [28]. Since our aim in this study was to evaluate the efficacy of a device for long-term monitoring of mental effort, it made sense to focus on a single participant over an extended time period. Researchers who place a premium on generalizability across contexts argue that a case study is disadvantaged by its focus within a single specific context [28]. However, [28] argues that this focus on a specific context is a strength in that it supports more accurate generalization to similar contexts. With the fields of psychology and medicine focusing less on giving general answers applying to everyone, and more on individualizing care, it is little surprise that the n = 1 design has increased in popularity in the medical research community [29, 30].

2.2 Description of the Case and Instrumentation

Since the focus of this study was to detect mental effort associated with school-related activities as well as activities of daily living, we chose an undergraduate university student as the case. This student was 19 years of age. She was a Psychology major with a concentration in Neuroscience in the second year of her undergraduate degree. Her primary hobbies included painting and spending time with her dog. Through the study, she identified her school-related activities, painting, learning to groom her dog with clippers and scissors, and watching brain games with her family as activities constituting high mental effort, and spent 48% of her time engaging in these types of activities. The remainder of her time was spent on low self-reported mental effort activities including eating, talking on the phone, watching television, driving, running errands, napping, and walking her dog. A total of 37 h, 33 min, and 34 s of data were collected. At a sampling rate of one sample per second, this constituted 135,214 total observations. These data were collected over approximately 3 weeks during the last half of the Spring 2020 semester.

The methodology relied on matching physiological data for EDA, skin temperature, and heart rate to self-reported data for mental effort dedicated to specific activities. EDA, skin temperature, and heart rate data were collected using the Empatica E4 wristband. The E4 measures blood volume pressure, heart rate, interbeat interval, skin temperature, and 3-axis acceleration. The E4 sampled EDA at 4 Hz, skin temperature at 4 Hz, and calculated heart rate (1 Hz) based on the BVP signal (64 Hz). In order to minimize noise in the data, we elected to downsample the EDA and skin temperature signals to 1 Hz in order to match the heart rate signal.

The participant was asked to place the E4 band on her wrist approximately 3 cm from the base of the hand. She indicated that she wore the device on her right wrist since she was left-handed. As she engaged in different activities throughout the day while wearing the device, she logged them in a journal along with assigning a measure of mental effort to each activity. Mental effort was self-reported on a Likert scale of 1–4, where a "1" indicated very low effort, a "2" indicated low effort, a "3" indicated high effort, and a "4" indicated very high effort. Individual activities varied in length from under a minute to over an hour. During the course of her activities, the student's data transitioned between low (1 and 2) and high (3 and 4) mental effort states 30 times.

2.3 Markov Switching Regression Model

The goal of modeling was two-fold: (1) to generate longitudinal predictions for mental effort and evaluate their robustness, and (2) to understand the role of measured EDA, skin temperature, and heart rate in generating these predictions. In light of these goals, we utilized the Markov switching dynamic regression model [31], which is an interpretable machine learning model that describes how an outcome changes its state over time. At their most basic level, Markov models predict a current state based on the previous state and a transition probability matrix. Markov switching models build upon this by allowing incorporation of state-specific relationships, thereby improving our understanding of how the physiological parameters relate to mental effort within each state.

Given our interest in a device that is able to distinguish between high and low states of mental effort, we utilized a 2-state Markov switching model. We tested models with four hierarchical levels of complexity: (1) a 2-state intercept-only model, (2) a 2-state model which held the effects of EDA, heart rate, and skin temperature constant across state, (3) a 2-state model which allowed the effects of EDA, heart rate, and skin temperature to switch across states, and (4) a 2-state model allowing for switching effects and variances. The likelihood ratio test was used to test the null hypothesis that adding an additional level of complexity did not improve model fit (95% confidence level used). The generalized r-square was calculated from the ratio of deviance values from the null and alternative models as a measure of the extent to which the alternative model improved fit over the null model.

Upon arriving at the best model using the above procedure, our interest shifted to evaluating the model's ability to provide robust temporal predictions. For validation, we fit the model to the first 22 h (58%) of the data, and tested that model on the final 16 h (42%) of the data. The root mean square error and mean absolute error were used to compare the fit of the raw output between the training and testing sets. We also discretized the reported mental effort and output to 2 states in order to evaluate the model's strength as a classifier based on its precision, recall, and F1 measure for the training and testing sets.

3 Results

3.1 Descriptive Analysis

The participant spent 9 h 18 min and 29 s in activities requiring very low mental effort and 10 h 8 min and 6 s in activities requiring low mental effort. 12 h 48 min and 46 s were spent at high mental effort, and 5 h 18 min and 13 s were spent at very high mental effort. Small but significant differences in EDA, skin temperature, and heart rate were found between each level of mental effort (Table 1).

Table 1. Average values for EDA, skin temperature, and heart rate at each reported level of mental effort

Mental Effort	Time (s)	EDA (μS)		Temperature (°C)		Heart Rate (bpm)	
		Mean	SD	Mean	SD	Mean	SD
1	33,509	0.12	0.08	32.55	1.56	94.74	17.42
2	36,486	0.12	0.06	31.70	2.64	97.79	21.50
3	46,126	0.15	0.18	32.30	1.80	91.65	17.59
4	19,093	0.13	0.04	32.40	1.93	92.55	16.14
Total	135,214	0.13	0.12	32.21	2.05	94.20	18.67

A MANOVA omnibus test indicated at least one significant difference in the multivariate mean across levels of mental effort (Wilk's $\Lambda = 0.94$, $F_{9,329061} = 948.97$, $p << 0.001$, $\eta^2_{partial} = 0.021$). Univariate ANOVA tests indicated that skin temperature exhibited the largest differences between levels of mental effort ($F_{3,135210} = 1159.92$, $p<<0.001$, $\eta^2_{partial} = 0.025$). EDA ($F_{3,135210} = 649.63$, $p<<0.001$, $\eta^2_{partial} = 0.014$) and heart rate ($F_{3,135210} = 810.04$, $p<<0.001$, $\eta^2_{partial} = 0.018$) also exhibited significant differences, but the effect sizes were less than that for skin temperature. Due to the large number of observations, Scheffe tests indicated that all differences between subsequent levels of mental effort were significant at the 99% confidence level. However, given the longitudinal nature of the data, it was difficult to specify how the physiological data support classification of high and low mental effort states over time using the MANOVA procedure.

3.2 Longitudinal Modeling of Mental Effort Using Physiological Data

Contribution of Physiological Measures: The log-likelihood tests (Table 2) suggested that the most complex model, allowing effects and variances to switch across states, provided the best fit to the data, and offered a significant improvement over the intercept-only null model ($R^2 = 0.023$, $\chi^2_{partial, df=7} = 4195.0$, $p<<0.001$).

Adding EDA, skin temperature, and heart rate as constant effects to the 2-state intercept-only model resulted in a significant improvement in the model ($R^2 = 0.012$,

Table 2. Hypothesis tests for significance of change in model fit as model complexity increased. Type 1 tests were used for calculating $R^2_{partial}$ and $\chi^2_{partial}$. R^2 and χ^2 were calculated with respect to the intercept-only null model

Two-state Markov model	R^2	χ^2	$R^2_{partial}$	$\chi^2_{partial}$
Intercept-Only (k = 5)				
Intercept + Constant effects (k = 8)	0.012	2200.4*	0.012	2200.4*
Intercept + Switching effects (k = 11)	0.021	3838.2*	0.0090	1637.8*
Intercept + Switching effects and variances (k = 12)	0.023	4195.0*	0.0020	356.8*

*p≪0.001

$\chi^2_{df=3} = 2200.4$, p≪0.001). Allowing the effects of EDA, skin temperature, and heart rate to switch between states 1 and 2 resulted in a further improvement ($R^2_{partial} = 0.009$, $\chi^2_{partial, df=3} = 1637.8$, p≪0.001). Finally, allowing variances to switch across the two states resulted in a smaller, but nonetheless significant, improvement in model fit ($R^2_{partial} = 0.002$, $\chi^2_{partial, df=1} = 356.8$, p≪0.001).

The 2-state model with state-dependent effects and variances (Table 3) showed that State 1 was associated with low mental effort (Intercept = 1.508, SE = 0.002, z = 800.4, p≪0.001), and State 2 was associated with high mental effort (Intercept = 3.296, SE = 0.002, z = 1823.7, p≪0.001). With this qualification, we can begin to understand how this student's EDA, skin temperature, and heart rate changed with mental effort within these two states as well as across the two states. Within State 1, an increase in mental effort was accompanied by a decrease in skin temperature and EDA, and an increase in heart rate. Skin temperature provided the strongest diagnostic for mental activity (Coef = –0.089, SE = 0.002, z = –52.0), followed by heart rate (Coef = 0.042, SE = 0.002, z = 24.2). EDA was significant (Coef = –0.034, SE = 0.003, z = –11.0), but nonetheless had a weaker effect size than skin temperature and heart rate. This ordering of importance matched the conclusions from the MANOVA test.

Upon transition to State 2, EDA retained its negative relationship with mental effort (Coef = –0.025, SE = 0.001, z = –17.1), and heart rate retained its positive relationship (Coef = 0.016, SE = 0.002, z = 8.1). However, skin temperature switched to a being positive indicator of mental effort (Coef = 0.018, SE = 0.002, z = 8.9) in State 2. The ordering of importance also changed from State 1. When the participant entered State 2, EDA became the strongest diagnostic, followed by skin temperature and heart rate.

Utility for Prediction. From the perspective of correct classification, our data indicate that the Markov switching regression model has high predictive utility both on the training and testing sets. The model predicted whether the participant was in a high or low state of mental effort with high accuracy (Accuracy$_{train}$ = 0.9995, F1$_{train}$ = 0.9995, Accuracy$_{test}$ = 0.9996, F1$_{test}$ = 0.9996). However, much of this was due to the fact that reported mental effort in association with certain activities was stable and sustained over extended time periods. This is illustrated by the model probabilities: given an initial state, the probability of staying in the same state was 0.99975, and the probability of transitioning to the other state was 0.00025. This means that when the model encountered

Table 3. Parameter estimates for the 2-state Markov switching model with switching effects and variances

Mental Effort	Feature	Coef	SE	z
State 1				
	EDA	−0.034	0.003	−11.0
	TEMP	−0.089	0.002	−52.0
	HR	0.042	0.002	24.2
	Intercept	1.508	0.002	800.4
	Variance	0.488	0.001	
State 2				
	EDA	−0.025	0.001	−17.1
	TEMP	0.018	0.002	8.9
	HR	0.016	0.002	8.1
	Intercept	3.296	0.002	1823.7
	Variance	0.454	0.001	
Transition	p11	0.99976	0.00006	
Probabilities	p21	0.00025	0.00006	

a transition from one level of mental effort to another, it tended to misclassify the initial observation within the new activity. However, once the model observed that initial observation, it tended to classify the rest of the observations correctly until it encountered another transition. It is for this reason that the Intercept-Only model (RMSE = 0.48, MAE = 0.46) predicted nearly as well as the Switching Effects and Variances model (RMSE = 0.47, MAE = 0.45) despite its lack of explanatory utility. The Switching Effects and Variances model predicted the testing set ($RMSE_{test}$ = 0.70, MAE_{test} = 0.61) slightly less accurately than the training set ($RMSE_{train}$ = 0.40, MAE_{train} = 0.32), illustrating some deterioration in performance when predicting into the future. However, these measures of fit sat within one unit of reported mental effort, illustrating the model's usefulness for classification of discrete states of mental effort both in the training and testing sets.

4   Discussion and Conclusions

Our findings suggest that the Markov switching model is useful as an explanatory tool for understanding the diagnostic utility of EDA, skin temperature, and heart rate for measuring mental effort. Providing that information about the participant's previous state is available, we can expect this model to perform well in predicting the participant's state at the next time point. This means that for extended activities, we will be able to discern the participant's level of mental effort at the next time point with reasonable certainty.

However, the utility of Markovian assumptions reduces when we do not have knowledge of the previous state, or if that knowledge is highly tentative. The utility of this framework could be improved if it were combined with another machine learning approach which is less sensitive to prior states. Previous work suggests that machine learning models invoking the assumption that the data are independent and identically distributed (i.i.d.) may be useful for detecting transitions between states [25]. For example, a simple logistic regression model applied to this data set using EDA, skin temperature, and heart rate as main effects (Accuracy $= 0.55$, F1 $= 0.50$) was able to detect 2 of the 30 total transitions in the data despite performing relatively poorly as a classifier. In this sense, traditional machine learning approaches could be used to generate time-independent predictions, and then the Markov model could act as a smoother over the temporal dimension which would improve the coherence of predictions while a user is within a particular state of mental effort. Our next steps include exploring linear dynamical systems and variants that incorporate both the temporal information, as well as utilize the i.i.d. nature to be able to detect both stability and transitions with high certainty.

Previous work has shown the promise of using physiological data collected from wearable sensors to facilitate automated monitoring mental effort and cognitive load [23–26], and [25] proposed the application of this framework toward development of an Educational Fitness Sensor (EduFit) system to help students track the duration and quality of their studies in real time. However, for EduFit to have utility as a personal device, models have to work in less structured environments over relatively long time durations. This study shows that EDA, skin temperature, and heart rate have diagnostic utility in these types of less controlled settings. It has been argued that the EduFit system would enable building of personal understanding of one's study endeavors through interpretable biofeedback and enablement of personal accountability [25]. Beyond engagement in studies, we believe this type of system may also be useful in other contexts where mental effort is important such as fields involving high-stakes operation of machinery. Monitoring of mental effort may also be useful for detecting cognitive decline in gerontology contexts. Within any of these contexts, the ability to specify and train models which are accurate and robust over time is essential if EduFit is to be useful, and our data indicate that interpretable machine learning models specified for time series data provide a step in the right direction.

References

1. Paas, F., Sweller, J.: Implications of cognitive load theory for multimedia learning. In: The Cambridge handbook of multimedia learning,. pp. 27–42. Cambridge University Press, New York (2014)
2. Sweller, J., Chandler, P., Kalyuga, S.: Cognitive load theory. Springer, New York (2011)
3. Sweller, J., van Merriënboer, J.J., Paas, F.: Cognitive architecture and instructional design: 20 years later. Educational Psychology Review, pp. 1–32 (2019)
4. Sweller, J.: Element interactivity and intrinsic, extraneous, and germane cognitive load. Educ. Psychol. Rev. **22**, 123–138 (2010). https://doi.org/10.1007/s10648-010-9128-5
5. Cowan, N.: The magical mystery four: how is working memory capacity limited, and why? Curr. Directions Psychol. Sci. **19**, 51–57 (2010)
6. Cowan, N.: The magical number 4 in short-term memory: a reconsideration of mental storage capacity. Behav. Brain Sci. **24**, 87–114 (2001)

7. Sweller, J., Van Merrienboer, J.J., Paas, F.G.: Cognitive architecture and instructional design. Educ. Psychol. Rev. **10**, 251–296 (1998)
8. Sweller, J.: Cognitive load theory, learning difficulty, and instructional design. Learn. Instr. **4**, 295–312 (1994)
9. Chandler, P., Sweller, J.: Cognitive load theory and the format of instruction. Cogn. Instr. **8**, 293–332 (1991)
10. Young, J.Q., Van Merrienboer, J., Durning, S., Ten Cate, O.: Cognitive load theory: implications for medical education: AMEE Guide No. 86. Med. Teach. **36**, 371–384 (2014)
11. Leppink, J., Duvivier, R.: Twelve tips for medical curriculum design from a cognitive load theory perspective. Med. Teach. **38**, 669–674 (2016)
12. Van Merriënboer, J.J., Sweller, J.: Cognitive load theory in health professional education: design principles and strategies. Med. Educ. **44**, 85–93 (2010)
13. Paas, F.G.: Training strategies for attaining transfer of problem-solving skill in statistics: a cognitive-load approach. J. Educ. Psychol. **84**, 429–434 (1992). https://doi.org/10.1037/0022-0663.84.4.429
14. Cook, A.E., Wei, W., Preziosi, M.A.: The use of ocular-motor measures in a convergent approach to studying cognitive load. Cognitive load measurement and application: A theoretical framework for meaningful research and practice, pp. 112–128 (2018)
15. Korbach, A., Brünken, R., Park, B.: Differentiating Different Types of Cognitive Load: a Comparison of Different Measures. Educ. Psychol. Rev. **30**, 503–529 (2018). https://doi.org/10.1007/s10648-017-9404-8
16. Antonenko, P., Paas, F., Grabner, R., van Gog, T.: Using electroencephalography to measure cognitive load. Educ. Psychol. Rev. **22**, 425–438 (2010). https://doi.org/10.1007/s10648-010-9130-y
17. Hart, S.G.: NASA-task load index (NASA-TLX); 20 years later. Presented at the Proceedings of the human factors and ergonomics society annual meeting (2006)
18. Charles, R.L., Nixon, J.: Measuring mental workload using physiological measures: a systematic review. Applied Ergonomics. **74**, 221–232 (2019)
19. Or, C.K., Duffy, V.G.: Development of a facial skin temperature-based methodology for non-intrusive mental workload measurement. Occupational Ergonomics. **7**, 83–94 (2007)
20. De Jong, T.: Cognitive load theory, educational research, and instructional design: some food for thought. Instr. Sci. **38**, 105–134 (2010)
21. Schroeder, N.L., Cenkci, A.T.: Do measures of cognitive load explain the spatial split-attention principle in multimedia learning environments? A systematic review. J. Educ. Psychol. (2019)
22. Zheng, R.Z.: Cognitive load measurement and application: a theoretical framework for meaningful research and practice. Routledge (2017)
23. de Avila, U.E.R., de França Campos, F.R., Leocadio-Miguel, M.A., Araujo, J.F.: 15 Mins of attention in class: variability of heart rate, personality, emotion and chronotype. Creative Educ. **10**, 2428 (2019)
24. Di Lascio, E., Gashi, S., Santini, S.: Unobtrusive assessment of students' emotional engagement during lectures using electrodermal activity sensors. In: Proceedings ACM Interactive Mobile Wearable Ubiquitous Technology, vol. 2, pp. 1–21 (2018). https://doi.org/10.1145/3264913.
25. Romine, W.L., et al.: Using machine learning to train a wearable device for measuring students' cognitive load during problem-solving activities based on electrodermal activity, skin temperature, and heart rate: development of a cognitive load tracker for both personal and classroom use. Sensors. **20**, 4833 (2020)
26. Larmuseau, C., Vanneste, P., Cornelis, J., Desmet, P., Depaepe, F.: Combining physiological data and subjective measurements to investigate cognitive load during complex learning. Frontl. Learn. Res. **7**, 57–74 (2019)

27. Lundervold, D.A., Belwood, M.F.: The best kept secret in counseling: Single-case (N=1) experimental designs. J. Counsel. Dev. **78**, 92–102 (2000)
28. Stake, R.: The Art of Case Study Research. Sage Publications, New York (1995)
29. Morgan, D.L., Morgan, R.K.: Single-participant research design: bringing science to managed care. Am. Psychol. **56**(2), 119–127 (2001)
30. Foster, L.H.: A best kept secret: single-subject research design in counseling. Counsel. Outcome Res. Eval. **1**(2), 30–39 (2010)
31. Hamilton, J.D.: A new approach to the economic analysis of nonstationary time series and the business cycle. Econometrica J. Econ. Soc. 357–384 (1989)

CARE2020 - International Workshop on pattern recognition for positive teChnology And eldeRly wEllbeing

First International Workshop on pattern recognition for positive teChnology And eldeRly wEllbeing (CARE 2020)

Life expectancy horizon is in continuous growth, with an arising socio-economical needs of supporting the aging population. If this is stimulating a considerable research effort in the Information and Communications Technology (ICT) field, there is a big gap between the complexity of available ICT devices and the needs of fragile individuals.

Positive Technology (PT) is a new paradigm investigating how ICT-based applications and services can be used to foster positive growth of individuals, organizations, and society. It refers to technologies designed for improving the quality of personal experience with the goal of increasing wellness, and generating strengths and resilience in individuals, organizations, and society.

PT could be framed in the wider notion of Ambient Assistive Living (AAL) while focusing on the holistic wellness of people, taking into account the cognitive, emotional and social wellbeing. Thanks to the advances in sensor technologies, and scientific findings, this field is in rapid growing and its application for monitoring and promote elderly wellbeing has become a concrete possibility.

Stimulated by the project "Stairway to elders: bridging space, time and emotions in their social environment for wellbeing" (grant no. 2018-0858 supported by Fondazione CARIPLO) of which the workshop organizers are principal investigators, the International Workshop on pattern recognition for positive teChnology And eldeRly wEllbeing (CARE) aimed at bringing together the most recent advances of Intelligent Systems, for Positive Technology and elderly wellbeing. The boundaries of the topics of potential interest for the workshop are very wide, and ranged from works purely on perception aspects – from face expressions and body gesture analysis, to social interaction assessment – to approaches including an explicit action toward the user – as techniques for mood induction.

The first edition of CARE was virtually held in Italy. It has been organized in conjunction with the 25th International Conference on Pattern Recognition, held in January 2021. The format of the workshop included two keynotes followed by two sessions of technical presentations. The workshop was attended by around 20 people on average. We received 14 submissions for reviews, from authors belonging to 7 distinct countries. Given the general very high quality of the contributions, after the peer-review process – single blind, at least two reviewers reviewed each paper – 11 papers have been selected for publication (acceptance rate of 78%). Overall, the contributions cover the different aspects of interest for the workshop, with a good balance between methodological and more technology-oriented papers. We highlight the general trend to investigate non invasive solutions, leveraging on different signals such as voice, eye gaze, facial expression, body gesture and physiological signals.

The workshop was schedules in a half day, and it was organized in two main parts, each one opened by an invited keynote talk. The first one was given by Prof. Hatice

Gunes (University of Cambridge, UK) and entitled *Creating Technology with Socio-emotional Intelligence*. The second talk was given by Prof. Andrea Gaggioli (Universitàa Cattolica di Milano, IT), and it was entitled *Positive AI: Opportunities and challenges of integrating artificial intelligence in digital wellbeing application*. The program has been organized so to group in each session contributions centered on topics related with the corresponding keynote talk.

We sincerely thank the CARE2020 Program Committee, for their rigorous and timely reviews and their fundamental role in the definition of a valuable program for the first edition of our workshop.

We would also like to thank the 25th International Conference on Pattern Recognition for hosting the workshop and the chairs for the valuable help and support.

January 2021

Raffaella Lanzarotti
Nicoletta Noceti
Claudio de'Sperati
Francesca Odone
Giuliano Grossi

Organization

Program Committee Chairs

Raffaella Lanzarotti   Università degli Studi di Milano
Nicoletta Noceti   Università degli Studi di Genova
Claudio de'Sperati   Università Vita-Salute San Raffaele, Milano
Francesca Odone   Università degli Studi di Genova
Giuliano Grossi   Università degli Studi di Milano

Program Committee

Gabriel Baud-Bovy   Università Vita-Salute San Raffaele
Sathya Bursic   Università degli Studi di Milano
Maura Casadio   Università degli Studi di Genova
Nicole Dalia Cilia   Università di Cassino e del Lazio Meridional
Donatello Conte   Université de Tours
Vittorio Cuculo   Università degli Studi di Milano
John Darby   Manchester Metropolitan University
Philipe Ambrozio Dias   Marquette University
Alessandro D'Amelio   Università degli Studi di Milano
Gaurvi Goyal   Università degli Studi di Genova
Marco Granato   Università degli Studi di Milano
Francesco Isgrò   Università "Federico II" di Napoli
Jianyi Lin   Khalifa University of Science and Technology
Henry Medeiros   Marquette University
Alina Miron   Brunel University
Maurice Pagnucco   University of New South Wale
Francisco Florez Revuelta   University of Alicante
Jean-Yves Ramel   Université de Tours
Alessandra Sciutti   Istituto Italiano di Tecnologia
Cesare Valenti   Università degli studi di Palermo

Additional Reviewer

Giulia Belgiovine

Multimodal Physiological-Based Emotion Recognition

Astha Sharma and Shaun Canavan[✉]

University of South Florida, Tampa, FL 33620, USA
asharma@mail.usf.edu, scanavan@usf.edu

Abstract. In this paper, we propose a multimodal approach to emotion recognition using physiological signals by showing how these signals can be combined and used to accurately identify a wide range of emotions such as happiness, sadness, and pain. The proposed approach combines multiple signal types such as blood pressure, respiration, and pulse rate into one feature vector representation of emotion. Using this feature vector, we train a deep convolutional neural network to recognize emotions from 2 state-of-the-art datasets, namely DEAP and BP4D+. On BP4D+, we achieve an average, subject independent, emotion recognition accuracy of 94% for 10 emotions. We also detail subject-specific experiments, as well as gender specific models of emotion. On DEAP, we achieve 86.09%, 90.61%, 90.48%, and 90.95% for valence, arousal, liking, and dominance respectively, for single-trial classification.

Keywords: Emotion recognition · Deep learning · Physiological data

1 Introduction

Emotional state is mental composition which shows our reaction to an experience. It is an integral part of human communication and behavior [23]. Recognizing emotions using machines has important roles in human-computer interaction, including applications in video games [11], assessment of multimedia technology, recommendations for multimedia content [21], pain recognition [35], and classification of Austism Spectrum Disorder [8]. While recognizing other's emotions, we generally look at their facial expressions, speech or body language, however, these features can be misguiding. Facial expression, voice, and body languages can be faked. Faces can be occluded, and facial expression can be contradictory which can make feature identification difficult. For example, it has been observed that people smile during negative emotional experiences [9]. Considering this, physiological signals such as heart rate, blood pressure, respiratory signals, and Electroencephalogram (EEG) signals can be important traits for identifying emotions accurately.

There has been an increase in the works that use physiological data for emotion recognition in recent years. Mert and Akan [24], investigated empirical mode decomposition for classification of low/high arousal and valence. They

A. Del Bimbo et al. (Eds.): ICPR 2020 Workshops, LNCS 12662, pp. 101–113, 2021.
https://doi.org/10.1007/978-3-030-68790-8_9

also looked at the multivariate extension, which they show is useful for analyzing non-stationary EEG signals. Martinez et al. [22] combined skin conductance and blood pressure along with deep neural networks to recognize emotions. The efficiency of their model was compared with standard feature extraction and feature selection methods. They showed that their deep learning approach performed better than standard feature selection algorithms and the method shows more generalization. Sano et al. [30] used physiological data to measure stress. For 5 days, skin conductance for 18 participants was collected with wrist sensors, as well as their mobile phone usage including call, SMS, and location were monitored. A survey was done to know stress, mood, sleep, tiredness, general health, alcohol or caffeinated beverage intake and electronics usage. Correlation analysis was applied to find important features that were used to classify whether the participant was stressed or not.

A computer-aided diagnosis system was developed to automate the classification of EEG signals in three categories - normal, preictal, and seizure [1]. This method achieved 88.67%, 90.00% and 95.00% accuracy respectively. Vijayan et al. [31] used EEG signals to classify different emotions such as happiness, fear, and sadness. The experiments were performed on the DEAP dataset [14], and Shannon Entropy was used for feature extraction and a multi-class Support Vector Machine was used for training. The accuracy obtained for classification was 94.097%. Picard et al. [26] showed variations in physiological signals on a daily basis. They proposed seeding a Fisher Projection with the results of Sequential Floating, achieving an accuracy of 81% on 8 emotions. Wagner et al. [32] combined multiple physiological signals to find the affective state. A musical induction method was used to ignite real emotion in subjects for data collection. Electromyogram, electrocardiogram, skin conductivity and respiration signals were used to classify four musical emotions (positive/high arousal, negative/high arousal, negative/low arousal, and positive/low arousal). A feature-based approach was proposed, obtaining a classification accuracy of 95% for subject-dependent and 70% for subject-independent experiments.

Motivated by these works we propose a method for emotion recognition using the combination of physiological signals that include heart rate, blood pressure, respiration, EDA, and EEG. We use this data to train a deep convolutional neural network to recognize a range of emotions. The contributions of this work are three-fold, and can be summarized as follows:

1. A multimodal approach to recognizing emotion using physiological data and deep convolutional neural networks is proposed.
2. Gender-specific models of emotions showing the difficulty when training and testing across male vs. female data, are detailed.
3. We compare the proposed method to state-of-the-art methods on BP4D+ and DEAP datasets, achieving state-of-the-art performance on BP4D+ and comparable on DEAP.

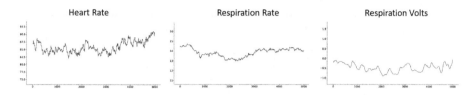

Fig. 1. Heart rate, respiration rate, and respiration volts from BP4D+, from subject experiencing a 'Happy' emotion.

2   Datasets

BP4D+. BP4D+ [37] is a multimodal spontaneous emotion corpus which includes 8 physiological signals that include blood pressure, EDA (skin conductance), heart rate and respiration. The BP4D+ also includes 2D, 3D and thermal images and videos, facial landmarks and action units. For our experiments only the physiological data was used. The huge versatility of gathered data makes BP4D+ one of the largest databases of this kind. This data was gathered from 140 subjects (58 males and 82 females) from 18 to 66 years of age. It includes the following 10 emotions: happy, sad, surprise, startle, skeptical, embarrassed, fear, pain, anger, and disgust. Each emotion was elicited through tasks such as holding hand in ice water (pain), and experiencing a smelly odor (disgust). See Fig. 1 for examples from BP4D+.

DEAP. DEAP [14] is a multimodal dataset based on the Valence-Arousal emotion model. It contains electroencephalogram (EEG) signals, as well as has sequences of peripheral signals that include EOG (eye movements), EMG (muscle movement), GSR, respiration, blood pressure and temperature. The data was collected from 32 participants (19–37 years age, 50% male and 50% female) watching 40 one-minute long music videos to elicit emotions. 32 EEG channels based on the 10–20 system [13] for recording EEG data and 8 channels for peripheral physiological data were used. These signals were recorded with a sampling rate 512 Hz which was downsampled 128 Hz after preprocessing. The data was labeled with arousal, valence, dominance, and liking values ranging from 1 to 9 showing intensity of each emotional state. See Fig. 2 for an example of a Fp1 EEG channel from DEAP.

3   Proposed Method

We propose to use the combination of multiple physiological signals to train a convolutional neural network for emotion recognition. In BP4D+ there are large variations in the data, therefore we first perform preprocessing on the data, specifically smoothing and scaling. We smooth the data to increase the signal to noise ratio without deforming the signal, which makes it easier to see trends in the data. In our experiments we use the Savitzky-Golay filter [27] as it has been shown to be more efficient in handling delay alignment and the

Fig. 2. Fp1 EEG channel from DEAP.

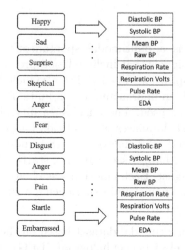

Fig. 3. Feature vector representation for BP4D+. *NOTE: 1 frame from each signal is in each feature vector (i.e. 8 frames for BP4D+).*

transient effect at the start and end of the sequence, compared to methods such as moving average, and median filters [2]. An example, of an original signal with the smoothed signal can be seen in Fig. 4. Once the signals are smoothed, we then scale the data into the range of [0, 1] which helps with large variations in the different signals. For our experiments with the DEAP dataset, we did not perform any additional preprocessing, as this data is already preprocessed for use [14]. Given the smoothed and scaled physiological signals, for each emotion we then create feature vectors that contain 1 frame of each of the available signals which are then used to train a deep neural network. As shown in Fig. 3, for BP4D+, the feature vector includes (in order) 1 frame of data from diastolic BP, systolic BP, Mean BP, Raw BP, Respiration Volts, Pulse Rate, and EDA. While this is a simple approach, we will show (Sect. 5), that it is an effective representation of emotion outperforming current state of the art on BP4D+ and DEAP.

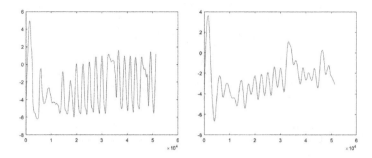

Fig. 4. Example of Savitzky-Golay smoothing on BP4D+. Left: original respiration volts signal from subject with 'Happy' emotion. Right: smoothed signal from applied filter.

4   Experimental Design

For BP4D+, all 140 subjects were used, and each task was collected over different time periods (i.e. the total number of frames is different), due to this we had approximately 450,000 feature vectors, for our experiments. For DEAP, all 32 subjects were used, with each feature vector (in the dataset) having 8064 frames. This resulted in over 10 million feature vectors for our experiments, validating the efficacy of the proposed approach for emotion recognition.

4.1   Deep Neural Network Architecture

In recent years, deep neural networks have proven highly efficient and have outperformed humans when classifying modalities such as audio, images, and text [15]. Deep networks have also successfully been used for classification of medical images [25], as well as prediction of future sales prices [28]. Motivated by the success of deep neural networks for a range of tasks and modalities, we train a 9-layer convolutional neural network (CNN), with the combined physiological signals (Fig. 3), to recognize emotion. The developed CNN uses two sets of convolutions, activation (ReLU) and max pooling layers. Dropout is used for regularization to help the model generalize better by reducing overfitting [6]. The RMSprop optimizer is used with a learning rate of 0.001. The network was trained using 150 epochs and a batch size of 32. See Fig. 5 for more details on the developed CNN architecture.

4.2   BP4D+ Experimental Design

When evaluating the efficacy of the proposed approach, on BP4D+, we are interested in answering two broad questions: (1) How does gender influence emotion recognition with the proposed method? and (2) Are there large differences

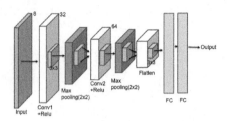

Fig. 5. Convolutional neural network architecture.

in accuracy for subject-dependent vs. subject-independent with the proposed method? Considering these questions, we conducted the following experiments.

1. **(Experiment 1).** We evaluated the proposed method in a subject-dependent manner. In doing this we created 140 deep models of emotion, one for each subject in the dataset. For each model, 80% of the subject data was used to train, and 20% was used to test. Due to the subject-dependent nature of this experiment, each model was trained and tested only on the same subject. This experiment was conducted to be consistent with the experimental design when using the DEAP dataset (single-trial classification as detailed in Sect. 4.3).
2. **(Experiment 2).** We evaluated the proposed method in a subject-independent manner. One model of emotion was created that used 80% of the data for training and 20% for testing, where the same subject did not appear in both training and testing.
3. **(Experiment 3).** We created gender-specific models of emotion. In this design, two deep models were created. For each model, 80% of the females and 80% of the males were used to train the respective gender-specific model. In this design, we test the gender-specific model across both genders (e.g. female model tested on both male and female subjects). In this experimental design, the same subject did not appear in both training and testing data.

4.3   DEAP Experimental Design

To evaluate the efficacy of the proposed approach on DEAP, we conducted single-trial classification experiments [14]. All experiments are done individually on each subject and evaluation is done by calculating mean and standard deviation, across all subjects, for every set of experiments. We split the data (40 channels as detailed in Sect. 2) into three sets: (1) EEG (32 channels); (2) peripheral (8 channels); and (3) EEG and peripheral (40 channels). In total, 12 single-trial classification experiments were conducted for each subject. One experiment for each emotion label (valence, arousal, liking, dominance) resulting in 384 single-trial classification experiments over all of the DEAP dataset.

5   Results

5.1   BP4D+

Experiment 1 was conducted in a person-specific manner resulting in a total of 140 deep models of emotion (one for each subject). To evaluate the efficacy of these models, we detail the mean accuracy (across each subject) along with the standard deviation (Table 1). This is done for three groups of subjects (1) all subjects; (2) female subjects; and (3) male subjects.

As can be seen in Table 1, both male and female data had a relatively high mean accuracy, with a low standard deviation. Many of the subjects had accuracy at or near 100%, although there were a few outliers (Fig. 6). For example, subject 140 had a lower accuracy at approximately 88%. It is important to note that subject 140 was a male participant which can explain, in part, the relatively lower accuracy of male subjects compared to female subjects (Table 1). Overall, these results detail the expressive power of the proposed approach for subject-specific emotion recognition. This shows that the majority of the subjects (both male and female), were recognized with high accuracy when using subject-specific models of emotion. Although the results for subject-specific emotion recognition are encouraging, in a real-wold setting it cannot be guaranteed that the test subject will appear in the training data. Considering this, we conducted Experiment 2 where the subject does not appear in the training and testing data. Experiment 2 was conducted over all emotions and all subjects (i.e. one deep model of emotion was used). Across all subjects, we achieved an accuracy of 94%. One of the questions we wanted to answer with our experiments on BP4D+, was whether the proposed method can work well on the same subject, as well as generalize to unseen subjects. These results are encouraging, as they show a relatively small difference in the emotion recognition accuracy of 4.89%, when comparing the average accuracy for subject-dependent experiments compared to the overall accuracy of the subject-independent experiment (98.89% vs. 94%). These results detail the expressive power of the proposed method for recognizing a range of emotions.

The final experiment conducted on BP4D+, was using gender specific models of emotion. In this experimental design, we evaluated both male and female models on both genders (e.g. female model on both male and female testing data). It is interesting to note that the female model outperformed the male model for both same and cross-gender testing. The gender-specific results (Tables 1 and 2)

Table 1. Subject-specific BP4D+ results.

Data	Recognition accuracy
All participants	98.89% ± 1.647
Female	99.09% ± 1.212
Male	98.88% ± 1.916

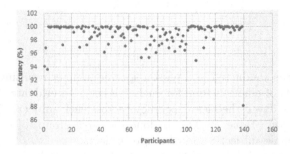

Fig. 6. BP4D+ subject-specific accuracy distribution.

can be explained, in part, as it has been found that some emotions are more easily recognized in female subjects compared to male [3].

Table 2. Emotion recognition accuracies on gender-specific models of emotion.

Training data (Model)	Testing data	Accuracy
Female	Female	96.77%
Male	Male	93.60%
Female	Male	15.35%
Male	Female	15.08%

As can be seen in Table 2, both models performed well when tested on the same gender with female achieving 96.77% recognition accuracy on female data, and male achieving 93.6% recognition accuracy on male data. However, both male and female deep models performed poorly on the opposite gender, achieving approximate accuracies of 15%. The low recognition accuracy from cross-gender testing can be explained, in part, by the idea that neurons flow in different part of the brain in males and females during emotion elicitation. For women, these neurons connect the parts of brain that regulate internal areas of body that impacts blood pressure, respiration and hormones, while in men, these neurons connect to the areas of brain that controls vision and movement [4, 33]. It has also been noted that there are obvious differences in the emotional responses of opposite genders when analyzing facial features and physiological data [16]. An interesting application that could benefit from these results is gender classification, however, this is outside of the scope of this paper and left for future work.

5.2   DEAP

When analyzing DEAP data, we have found that the combination of both Peripheral and EEG data (40 channels) gave the highest recognition accuracy in 3 out of 4 of the emotion categories (arousal, liking, and dominance). Peripheral data alone outperformed both for valence with a mean accuracy of 86.31% compared to 86.09% with both. EEG data alone (32 channels) performed the worst in all categories (Table 3). These results are supported by previous studies that have shown a multimodal approach to classification can lead to higher accuracy compared to a single modality, when using physiological data to classify infant pain [34].

Table 3. Single-trial classification results from DEAP.

Emotion category	EEG	Peripheral	Both
Valence	Mean: 60.21% ± 6.306	Mean: 86.31% ± 6.186	Mean: 86.09% ± 5.367
Arousal	Mean: 65.03% ± 9.486	Mean: 88.83% ± 4.455	Mean: 90.61% ± 3.579
Liking	Mean: 67.59% ± 11.295	Mean: 88.38% ± 6.416	Mean: 90.48% ± 4.954
Dominance	Mean: 66.22% ± 12.528	Mean: 89.12% ± 5.484	Mean: 90.95% ±: 4.667

As can be seen in Table 3, the standard deviations for each subject are much higher compared to those found in Experiment 1 from the detailed BP4D+ experiments (Table 1). This is especially true when analyzing the arousal, liking, and dominance emotion categories for EEG data alone. Arousal had the lowest standard deviation for peripheral and EEG + peripheral experiments, although it is higher compared to valence for EEG data alone. This could be partially explained by EEG data not being able to easily generalize across subjects [17]. Although the standard deviation of peripheral data is lower compared to EEG data, when both EEG and peripheral signals were combined, they gave the lowest standard deviation for each emotion category. Again, this can partially be explained due to the multimodal nature of EEG + peripheral data. For dominance and liking, the standard deviation is highest for EEG data along with liking >11% and Dominance >12%. Similar to arousal, this can partially be attributed to EEG data's inability to easily generalize across subjects. As can be seen in Fig. 7, with dominance, there are approximately 6 subjects that are causing the higher standard deviation (e.g. outliers). Most of the subjects having an accuracy within the range of 50%-70%, however, the 6 outliers are within the range of 80% to 100%. A question that arises from this work is: Is it possible to generalize across subjects with EEG data? With valence having the lowest standard deviation that can be a good starting point for future experiments to answer this question.

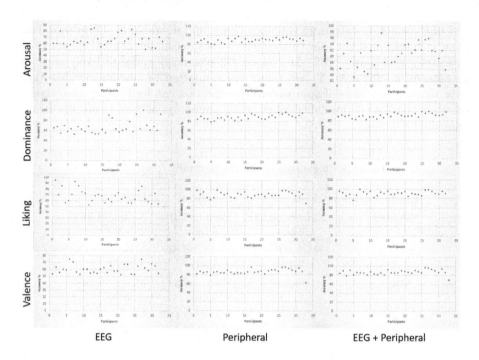

Fig. 7. Subject accuracies for each emotion category, in DEAP [14]. From left to right: EEG data, peripheral data, EEG and peripheral data.

5.3   State of the Art Comparisons

BP4D+. To the best of our knowledge, there are 2 works that detail results on physiological data from BP4D+. First, is Zhang et al. [37] where they randomly selected 45 subjects. Using hand-crafted features and a RBF kernel SVM, they achieved an accuracy of 59.5% on happy, sad, startled, fear, and disgust. We also compare to Fabiano et al. [10], where they proposed a weighted fusion technique for recognizing emotion from physiological signals. In their approach, they achieved an average accuracy of 91.59%. In the proposed approach, we achieved an average accuracy of 94% on subject-independent training and testing over all subjects, and all emotions.

DEAP. This dataset has successfully been used for emotion recognition since it's release by Koelstra et al. [14]. Considering this, we compare against multiple state-of-the-art approaches for recognizing the 4 emotion categories available in the DEAP dataset.

As can be seen in Table 4, the proposed method outperforms the current state of the art except for the weighted fusion approach from Fabiano et al. [10]. The increase in accuracy over the majority of the current state of the art can be explained, in part, by the proposed method using the combination of the different signals to train a deep network (CNN) to recognize the emotions. With the exception of the work from Fabiano et al. [10] (they used a feedforward

Table 4. Comparison with current state of art for DEAP.

	Valence	Arousal	Dominance	Liking
Proposed method	86.31%	90.61%	90.95%	90.48%
Fabiano et al. [10]	**95.5%**	**95.27%**	**96.47%**	**96.03%**
Liu et al. [20]	85.2%	80.5%	84.9%	82.4%
Rozgic et al. [29]	76.9%	69.1%	73.9%	75.3%
Mert and Akan. [24]	72.87%	75.0%	N/A	N/A
Daimi and Saha [7]	65.3%	66.90%	N/A	N/A
Li et al. [19]	58.4%	64.3%	65.8%	66.9%
Jirayucharoensak et al. [12]	53.42%	52.03%	N/A	N/A
Koelstra et al. [14]	65.2%	63.1%	N/A	64.2%

neural network), the compared works use classical machine learning approaches. Koelstra et al. [14] used a Naïve Bayes classifier, Liu et al. [20] used a linear Support Vector Machine (SVM), and Rozgic et al. [29] used a combination of K-PCA and 1-NN. While Li et al. [18] extracted features from a deep belief network, they used an SVM to classify the emotions. The decrease in accuracy compared to Fabiano et al. [10] could be explained, in part, due to the large number of channels (features) in DEAP compared to BP4D+ (i.e. 40 vs. 8).

6 Conclusion

We have presented an approach to emotion recognition, using physiological signals, that combines individual frames from different signal types (e.g. blood pressure and respiration rate). We tested the efficacy of the proposed approach on 2 publicly available datasets, namely BP4D+ and DEAP. The proposed method outperforms the current state of the art on BP4D+ and outperforms all works, except Fabiano et al. [10], on DEAP. It is interesting to note, that although their approach outperformed ours on DEAP, the proposed approach outperforms their weighted fusion approach on BP4D+. It has applications in multimedia, medicine, defense and military related fields. These applications include analysis of stress and pain, lie detection, increasing soldier survivability in combat, and classification of autism in children.

As shown in Sect. 5.1, the gender-specific models of emotion performed poorly when cross-gender data was used to test (e.g. male deep model tested on female data). Interesting, a previous study notes that physiological signals are similar during similar emotions in male and female [5], however, our results contradict this and are supported by the work from Whittle et al. [33], as we previously mentioned in Sect. 5.1. An interesting application of this is using the proposed method to classify a subject's gender. Due to challenges with using facial data for gender classification such as pose and lighting variation [36], physiological data could be a useful alternative as it does not suffer from those same challenges.

Considering this, we will conduct experiments on gender classification using the deep gender-specific models we have developed here. We will also compare the results from other deep neural network architectures such as recurrent neural networks.

References

1. Acharya, U., Oh, S., Hagiwara, Y., Tan, J., Adeli, H.: Deep convolutional neural network for the automated detection and diagnosis of seizure using EEG signals. Comput. Biol. Med. **100**, 270–278 (2017)
2. Azami, H., Mohammadi, K., Bozorgtabar, B.: An improved signal segmentation using moving average and Savitzky-Golay filter. J. Signal Inf. Process. **3**(01), 39 (2012)
3. Bailenson, J., et al.: Real-time classification of evoked emotion using facial feature tracking and physiological responses. Int. J. Human Comput. Stud. **66**(5), 19–31 (2008)
4. Bradley, M.: Emotions—differences between men and women. In: Health Guidance for better health (2014)
5. Cummins, D.: Are males and females equally emotional? In: Psychology Today (2014)
6. Dahl, G., et al.: Improving deep neural networks for LVCSR using rectified linear units and dropout. In: ASSP, pp. 8609–8613 (2013)
7. Daimi, S., Saha, G.: Classification of emotions induced by music videos and correlation with participants' rating. Expert Syst. Appl. **41**(13), 6057–6065 (2014)
8. Drimalla, H., et al.: Detecting autism by analyzing a simulated social interaction. In: Berlingerio, M., Bonchi, F., Gärtner, T., Hurley, N., Ifrim, G. (eds.) ECML PKDD 2018. LNCS (LNAI), vol. 11051, pp. 193–208. Springer, Cham (2019). https://doi.org/10.1007/978-3-030-10925-7_12
9. Ekman, P.: The argument and evidence about universals in facial expressions. In: Wagner, H.E., Manstead, A.E. (eds.) Handbook of Social Psychophysiology, pp. 143–164. Wiley, Hoboken (1989)
10. Fabiano, D., Canavan, S.: Emotion recognition using fused physiological signals. In: ACII (2019)
11. Giakoumis, D., et al.: Auto rec of bored in video games using novel bio moment feat. IEEE Trans. Affect. Comput. **2**(3), 119–133 (2011)
12. Jirayucharoensak, S., Pan-Ngum, S., Israsena, P.: EEG-based emotion recognition using deep learning network with principal component based covariate shift adaptation. Sci. World J. **2014**, 10 (2014)
13. Klem, G., et al.: The ten-twenty electrode system of the international federation. Elec. Clin. Neurophys. **52**(3), 3–6 (1999)
14. Koelstra, S., et al.: Deap: a database for emotion analysis; using physiological signals. IEEE Trans. Affect. Comput. **3**(1), 18–31 (2012)
15. Krizhevsky, A., Sutskever, I., Hinton, G.: Imagenet classification with deep convolutional neural networks. In: Advances in Neural Information Processing Systems, pp. 1097–1105 (2012)
16. Li, C., et al.: Analysis of physiological for emotion recognition with the IRS model. Neurocomputing **178**, 103–111 (2016)
17. Li, X., Song, D., et al.: Exploring EEG features in cross-subject emotion recognition. Front. Neurosci. **12**, 162 (2018)

18. Li, X., et al.: EEG based emotion identification using unsupervised deep feature learning (2015)
19. Li, X., et al.: Recognizing emotions based on multimodal neurophysiological signals. Adv. Comput. Psychophysiol. 28–30 (2015)
20. Liu, W., Zheng, W., Lu, B.: Multimodal emotion recognition using multimodal deep learning. arXiv preprint arXiv:1602.08225 (2016)
21. Mariappan, M., Suk, M.P.B.: Facefetch: a user emotion driven multimedia content recommendation system based on facial expression recognition. In: International Symposium on Multimedia (2012)
22. Martinez, H., et al.: Learning deep physiological models of affect. Comput. Intell. Mag. **8**(2), 20–33 (2013)
23. Mauss, I., Robinson, M.: Measures of emotion: a review. Cogn. Emot. **23**(2), 209–237 (2009)
24. Mert, A., Akan, A.: Emotion recognition from EEG signals by using multivariate empirical mode decomposition. PAA **21**(1), 81–89 (2018)
25. Ortiz, A., Munilla, J., Gorriz, J., Ramirez, J.: Ensembles of deep learning architectures for the early diagnosis of the Alzheimer's disease. Int. J. Neural Syst. **26**(07), 1650025 (2016)
26. Picard, R., Vyzas, E., Healey, J.: Toward machine emotional intelligence: analysis of affective physiological state. IEEE Trans. PAMI **23**(10), 1175–1191 (2001)
27. Press, W., Teukolsky, S.: Savitzky-Golay smoothing filters. Comput. Phys. **4**(6), 669–672 (1990)
28. Rafiei, M., Adeli, H.: A novel machine learning model for estimation of sale prices of real estate units. J. Constr. Eng. Manage. **142**(2), 04015066 (2015)
29. Rozgić, V., et al.: Robust EEG emotion classification using segment level decision fusion. In: ICASSP (2013)
30. Sano, A., Picard, R.: Stress recognition using wearable sensors and mobile phones. In: ACII, pp. 671–676 (2013)
31. Vijayan, A., Sen, D., Sudheer, A.: EEG-based emotion recognition using statistical measures and auto-regressive modeling. In: Computational Intelligence & Communication Technology, pp. 587–591 (2015)
32. Wagner, J., Kim, J., André, E.: From physiological signals to emotions: implementing and comparing selected methods for feature extraction and classification. In: ICME, pp. 940–943 (2005)
33. Whittle, S., Yücel, M., Yap, M., Allen, N.: Sex differences in the neural correlates of emotion: evidence from neuroimaging. Biol. Psychol. **87**(3), 319–333 (2011)
34. Zamzmi, G., Pai, C., Goldgof, D. Kasturi, R., Ashmeade, T., Sun, Y.: An approach for automated multimodal analysis of infants' pain. In: International Conference on Pattern Recognition (2016)
35. Zamzmi, G., et al.: An approach for automated multimodal analysis of infants' pain. In: ICPR, pp. 4148–4153 (2016)
36. Zhang, K., et al.: Gender and smile classification using deep convolutional neural networks. In: CVPR Workshops (2016)
37. Zhang, Z., et al.: Multimodal spontaneous emotion corpus for human behavior analysis. In: CVPR, pp. 3438–3446 (2016)

Combining Deep and Unsupervised Features for Multilingual Speech Emotion Recognition

Vincenzo Scotti$^{(\boxtimes)}$ ⓘ, Federico Galati, Licia Sbattella ⓘ, and Roberto Tedesco ⓘ

DEIB, Politecnico di Milano, Via Golgi 42, 20133 Milan (MI), Italy
{vincenzo.scotti,licia.sbattella,roberto.tedesco}@polimi.it,
federico.galati@mail.polimi.it

Abstract. In this paper we present a Convolutional Neural Network for multilingual emotion recognition from spoken sentences. The purpose of this work was to build a model capable of recognising emotions combining textual and acoustic information compatible with multiple languages. The model we derive has an end-to-end deep architecture, hence it takes raw text and audio data and uses convolutional layers to extract a hierarchy of classification features. Moreover, we show how the trained model achieves good performances in different languages thanks to the usage of multilingual unsupervised textual features. As an additional remark, it is worth to mention that our solution does not require text and audio to be word- or phoneme-aligned. The proposed model, PATHOSnet, was trained and evaluated on multiple corpora with different spoken languages (IEMOCAP, EmoFilm, SES and AESI). Before training, we tuned the hyper-parameters solely on the IEMOCAP corpus, which offers realistic audio recording and transcription of sentences with emotional content in English. The final model turned out to provide state-of-the-art performances on some of the selected data sets on the four considered emotions.

Keywords: Emotion recognition · Multilingual · Multi-modal analysis

1 Introduction

In the psychological literature, emotions are defined as a complex set of bi-directional interactions between physiological activation (arousal) and individual cognitive analysis (appraisal) [25]. This interaction generates affective experiences, cognitive processes and physiological adjustments, leading to the activation of adaptive behaviour [31]. In this regard, it's necessary to emphasise, as highlighted by many authors, the importance of the adaptive nature of emotions [11].

Partially supported by the European Union's Horizon 2020 project "WorkingAge" (grant agreement No. 826232).

Following this definition of emotions, literature highlights how an emotional state is characterised by some changes at the physiological level [21]. Such changes are an integral part of the emotion itself. Some physiological changes, such as acceleration of the heartbeat, increase in blood pressure, sweating, often occur without us being aware of them.

Specifically, the so-called *non-verbal communication* is a fundamental communication and expressive channel for emotions, as it is less consciously controllable. Examples of these non-verbal aspects are the *para-linguistic* ones, like voice tone, speech rate, pauses, silences, etc. On the other side, the *verbal communication*, with its *linguistic* aspects, is still a useful channel for the expression of emotions.

Different approaches, leveraging Neural Networks (NNs), have already shown that the combination of linguistic and para-linguistic clues can provide a useful contribution in the task of emotion recognition [2,6,36]. In such works, linguistic features have been mostly treated through pre-trained embedding models. Acoustic features, on the other hand, have been selected from pre-defined ones.

With this work, we were interested in two main aspects. The first one is understanding whether deep pre-trained features could lead to higher classification accuracy; hence, for a multi-modal analysis like this, pre-trained features for audio analysis should be used as well as pre-trained linguistic features. The second aspect is exploring the effects of a single model working with multiple languages at the same time; so, the multilingual model was built as an all-in-one model by feeding it with different corpora in different languages at train time.

The multilingual approach resulted in a training phase on a wider corpus which, in general, helps Deep Neural Networks (DNNs) to learn better features. Such features turned out to be correctly compatible with a multiple language environment as we expected. Notice that the multilingual approach allowed our model to deal with the data scarcity, which often prevents DNNs from being effectively trained.

Our classifier, called *Parallel, Audio-Textual, Hybrid Organisation for emotionS network* (PATHOSnet) reached an accuracy of 80.4% on the IEMOCAP [4] corpus (our main benchmark). The preceding best score of an automatic system, working with the same modalities, was obtained by Atmaja and colleagues: 75.5% [2]. Human listeners achieved 70% on the four emotions considered for this project, according to [6].

The rest of this paper is organised in the following sections. In Sect. 2 we present the state of the art for speech emotion recognition using NNs. In Sect. 3 we describe the data collections we use to train and test our model. In Sect. 4 we describe the input features used to feed our model. In Sect. 5 we describe the architecture of our model, for multilingual emotion recognition. In Sect. 6 we explain how we approached the training and evaluation processes. In Sect. 7 we report the results of the experiments and we comment on them. In Sect. 8 we sum up our work and provide hints about possible future work.

2   Related Works

In recent years speech emotion recognition has gained a lot of traction. Multimodal (audio, video and text) analysis has shown to be the correct way to address this problem. In particular, NN-based solutions have shown to produce better results. In fact, we're mainly interested in this kind of models for emotion recognition. Interested readers can refer to surveys [1], for other models.

For what concerns NN-based solutions, we noticed similar patterns in recent years for emotion recognition, where researchers started to employ, where possible, multimodal analysis on text and audio, and sometimes on video, too.

These input modalities are usually treated though pre-computed features. In particular, learnt semantic representations are used as linguistic features while handcrafted features are used for the acoustic part. The works we referred to are based on NNs and we considered as a main benchmark the IEMOCAP[4] corpus for emotion recognition (more on this in Sect. 3).

The best recent solution on the IEMOCAP corpus leveraged *Recurrent* or *Bi-Directional Recurrent* NNs (RNNs and BiRNN) [12,17,32], often including also an *attention mechanism* [3]. One of the first work on IEMOCAP with RNNs, however, reached only 54% classification accuracy [6], while a work proposing BiLSTMs with attention mechanism reached an accuracy of 71.0% through linguistic and acoustic analysis [36].

Even if useful to handle time series, the sequential structure of the RNNs makes their computations really slow; in fact, they cannot be parallelised [37]. Differently, Convolutional NNs (CNNs) are faster and easy to parallelise. For this reason, we implemented our NN using convolutional layers.

Authors of [24] proposed a deeper analysis of multimodal approaches for sentiment and emotion analysis. In their work, they focused on modality fusion and context usage. In particular, they found how the accuracy in the emotion recognition of a single utterance can be improved when leveraging information coming from the other utterances in the discourse. On IEMOCAP they reached an accuracy of 76.5% using audio-video and text, and an accuracy of 76.1% using solely audio and text. Even if this model produces impressive results, it relies on the usage of contextual information (i.e., a discourse), that might be not always available. Thus, we decided not to make use of such context information, working on individual, isolated sentences.

To our knowledge, none of the available NN models for emotion recognition applies a multilingual approach. Additionally, even if some of them are presented as deep learning approaches, they still employ manually-selected features, computed from the raw input, rather than resorting to deep models. With our work, we address both of these two aspects.

3   Corpora

Neural Networks are a data-driven framework; as such, they require labelled corpora to be trained on. For the purpose of this work, we considered different data sources in order to include different languages and use cases. In fact, it was our main interest to train a multilingual model, in order to provide a single tool available for everyone and cope with data scarcity. In particular, data scarcity represents a strong barrier for some languages.

To train our network, we selected the following corpora:

- Interactive Emotional Dyadic Motion Capture Database (IEMOCAP) [4];
- Emotional speech from Films corpus (EmoFilm) [29];
- Spanish Emotional Speech database (SES) [28];
- Athens Emotional States Inventory (AESI) [5].

In Table 1 it is possible to find statistics about such corpora, in terms of available samples, organised per-corpus and per-language.

The IEMOCAP corpus represents our main benchmark. It is the most complete and best managed of the considered corpora, but it only provides English samples. IEMOCAP employs both *categorical* [11] and a *dimensional* [13] representations of emotions in audio-visual data. The IEMOCAP corpus has been built recording ten actors in dyadic sessions, resulting in five sessions with two subjects each. Actors were asked to perform two tasks: play three selected scripts with clear emotional content, and improvise dialogues in hypothetical scenarios designed to elicit specific emotions. For the purpose of this work, and considering other works that used IEMOCAP, we selected only four basics emotions labels to discriminate among. The selected emotions are *Happiness, Anger, Sadness,* and *Neutral.*

The EmoFilm corpus contains samples in three different languages: English, Italian and Spanish. This corpus was built carefully, manually selecting audio recordings from a total of 43 movies. The search was conducted on the English dubs of the considered movies. Once the clips were identified, Italian and Spanish audio tracks were cut at the same time stamps. Rare emotional labels were excluded from the collection. As a result, 828 samples were retrieved (per-language) and labelled with the following emotions: Happiness, Anger, Sadness, *Fear* and *Contempt.* Fear was excluded from our work because there were enough samples of the same kind in other corpora; Contempt was discarded because it is generally not considered a basic emotion. No transcription was provided; for this reason, we resorted to an ASR[1] in order to have the textual content. This choice not only made us closer to real usage scenario, but also helped us to retrieve results that take into account possible transcription errors.

The SES corpus contains emotional speech recordings played by a professional male actor speaking Spanish. The available emotional labels in this corpus were Happiness, Anger, Sadness, Neural and *Surprsie*; the latter was excluded from our work because there were enough samples of the same kind in other corpora. The corpus is composed of several readings of the same neutral texts, displaying different emotions. On one side this aspect is useful as it will help to enforce the usage of acoustic features and prevent the model from sticking to a fixed vocabulary of words for the classification. On the other we had to ensure that

---

[1] https://gitlab.com/Jaco-Assistant/deepspeech-polyglot.

the model does not rely solely on acoustic features, hence we carefully analysed the results comparing the different corpora.

The AESI corpus is an audio-visual database for Greek emotion recognition. This corpus contains 696 recorded utterances in the Greek language by 20 native speakers. The emotional labels have been assessed through a survey. The samples are labelled according to one of these emotions: Happiness, Anger, Sadness, Neural and Fear. We included this corpus because we wanted to observe the generalisation capabilities of the network on smaller corpora when still provided with samples also in other languages.

IEMOCAP is the biggest of the considered corpora. As such, it was selected as our main benchmark and guided the choice of the emotional labels.

Table 1. Number of available samples (total and per-class) organised per-corpus and per-language

Corpus	Language	Number of samples				
		Emotion				Total
		Happiness	Anger	Sadness	Neutral	
IEMOCAP	English	1041	1103	1084	1708	4936
EmoFilm	English	70	77	74	0	221
	Italian	94	73	93	0	260
	Spanish	76	82	87	0	245
	All languages	240	232	254	0	726
SES	Spanish	732	725	728	1658	3843
AESI	Greek	139	139	140	139	557

4 Features

As premised we considered two distinct yet complementary input modalities for our network. On one side, we considered linguistic features, extracted from the transcription of the spoken sentence. On the other side, we considered the deep features extracted from the waveform of the spoken sentence. Both modalities use task-agnostic input features; in fact, none of the two modules generating features was specifically designed for emotion recognition. This was necessary since none of the data sets contains enough samples to train a deep model from scratch. Thus, in both cases, we retrieved pre-trained models, which we adapted for our work.

4.1 Linguistic Features

The *meaning* of what is uttered by the speaker, contained in transcriptions, represents an important piece of information for emotion recognition (i.e., the linguistic aspect). In fact, depending on the emotional context, certain words can be related to an emotional state more than others. To extract the textual features, we relied on word embeddings, a *vector semantics* representation [19] of words. Through word embeddings, every word is encoded as a vector in a d-dimensional space where words with similar meaning are encoded closely. Moreover, we trained the final network on multilingual embeddings, where words from different languages with the same meaning, are mapped in the same point in the embedding space. This was expected to help generalise across languages. In general, we expected that this semantically meaningful encoding will help the NN to associate similar meaning words to the same emotional state.

The embedded text is represented as a two-dimensional tensor, i.e. a matrix obtained embedding all the words in the utterance. The tensor is characterised by d columns, one for each dimension of the word embedding hyperspace, and a number of rows that matches that of the words in the sentence. The columns represent the sample's features, while the rows constitute the time axis.

During the hyperparameter-tuning phase, our model was fed with English-only embeddings. In particular, we used a *GloVe* model for word embeddings [30] (with 300-dimensional vectors). As premised, we used a pre-trained model[2]. Subsequently, the final model was trained with multilingual embeddings, by means of *MUSE* framework [8,20]. These multilingual embeddings are obtained starting from pre-trained *FastText* word embedding models [26] in different languages (always with 300-dimensional vectors). The embeddings are then transformed so that corresponding words in the different languages result in overlapping vectors. As for the English model, we used pre-trained MUSE embeddings[3].

4.2 Acoustic Features

We used acoustic features to capture the information about *how* a person is talking (i.e., the para-linguistic aspect). The choice of deep acoustic features represents a strong change with respect to previous work in emotion recognition. In fact, to our knowledge, previous works relied solely on pre-defined acoustic features [6]. Such features were manually selected to highlight the aspects of the voice signal that were expected to correlate the most and to cope with the reduced amount of samples. We decided, instead, to use a transfer learning approach [38] and rely on the features extracted by deep models trained on huge classification tasks.

The DNNs we employed to extract features were designed and trained using the same concepts and huge audio classification data sets. In fact, they employed the same architectures of image recognition NNs, adapted to take as input the (mel-filtered) spectrogram of the vocal signal. Their basic idea is to threat the spectrogram as an image and use 2-D CNNs to learn a feature hierarchy useful for audio classification. Thanks to the use of a huge audio data set, the models were able to produce very general features, which resulted to be transferrable

---

[2] https://nlp.stanford.edu/projects/glove/.
[3] https://github.com/facebookresearch/MUSE.

across different tasks. This is the same approach used with image recognition models trained on *ImageNet* [9]: the CNNs are trained as classifiers, then their classification heads are removed to transfer the features to other image analysis tasks.

We experimented with two different networks. The former was *VGGish* [16], a variant of *VGG* [33], which is a NN for image recognition. VGGish was trained on the *AudioSet*[4] corpus [14]. The feature extraction variant takes as input a 64 bin log-scaled, mel-filtered spectrogram (computed with a window size of 25 ms and a hop size of 10 ms) and produces a 128-dimensional feature vector for every non-overlapping 0.96 s window in the input. The latter network was *Thin ResNet-34* with *GhostVLAD* pooling layer [41], a variant of *ResNet* [15], a NN for image recognition. This second feature extraction network takes as input a 257 bin normalised spectrogram (computed with a window size of 25 ms and a hop size of 16 ms) and produces a 512-dimensional feature vector for every non-overlapping 0.045 s window in the input. We relied on a Keras[5] implementation for both VGGish[6] and Thin ResNet-34 with GhostVLAD[7].

The acoustic features are then represented similarly to the linguistic ones: they are managed as a two-dimensional tensor. The row axis represents the time dimension, the column axis represents the features (conceptually this is similar to a spectrogram with its bins).

5   Model

The model we developed, called PATHOSnet and represented in Fig. 1, is a multi-modal DNN for emotion recognition built upon transferred deep features. The model is composed of two parallel branches, one for linguistic analysis and one for acoustic analysis, which are later merged together. A high-level view of this model is depicted in Sect. 5. These two symmetric branches are CNNs, composed of 1-D convolutional and pooling layers. A depiction of such blocks is reported in Sect. 5. We inserted each of the two blocks on top of the corresponding feature extractor. The classifier on top is, instead, a fully-connected layer with a softmax activation function over the four considered classes.

The convolutional blocks in the two branches are designed like those of a ResNet [15] network, adapted for the 1-D scenario. To flatten the information along the time axis and produce a single feature vector for each modality, we relied on a Global Average Pooling (GAP) layer [22]; GAP not only allows to "compress" spatial information, averaging along the time axis, but it does so with a low computational effort (differently from *attention*-based solutions [27]).

To merge the two branches of the network we adopted a simple feature fusion approach [1]: we concatenated the two feature vectors coming from the two separate branches and learnt a fully connected transformation to combine the vectors

---

[4] https://research.google.com/audioset/.

[5] https://keras.io.

[6] https://github.com/beasteers/VGGish.

[7] https://github.com/taylorlu/ghostvlad-speaker.

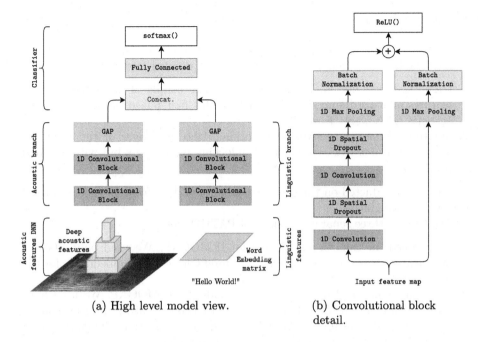

(a) High level model view.

(b) Convolutional block detail.

Fig. 1. PATHOSnet internal structure.

directly into the class probabilities. In this way, acoustic features are not required to be aligned with the textual ones. In fact, each branch takes care of embedding the "temporal" information in its intermediate representation, removing the time axis by the end of its transformation. This was a great advantage as this kind of alignment either isn't available or is difficult to obtain.

In order to enforce regularisation and avoid overfitting we adopted the *spatial dropout* [35] and the *batch normalisation* [18]. Regularisation was also enforced thanks to the GAP layer. Finally, to avoid overfitting we employed the *early stopping* approach.

As an additional note, we want to point out that the DNN used to extract the acoustic features is an integral part of PATHOSnet; in this way, we managed to perform fine-tuning of its weights. In Sect. 6 we provided more details about this part.

PATHOSnet is an example of so-called *ensemble* models. In our case, this approach was useful as the textual-based and acoustic-based networks turned out to be complementary. To build the ensemble, we removed the classifier on top of the two networks, and learnt a new linear classification function on top of the concatenated feature vectors.

The implementation of our model was realised through the Keras framework, using Tensorflow[8] as backend. The entire code, from feature extraction to the NN was developed solely through the Python programming language.

---

[8] https://www.tensorflow.org.

6   Experiments

In our experiments, we followed the classic train-test steps adopted by NN frameworks. The training process of the networks was divided into 2 steps:

1. train the convolutional blocks above a single couple of DNN for acoustic features and word embeddings model. In this first phase the weights of the acoustic DNN were "frozen";
2. fine-tune the single networks "unlocking" the first layer of the DNN for acoustic features (we experimented unlocking more layers but without good results).

Testing was conducted on the same percentages of data from all the considered corpora and languages.

We trained three separate versions of PATHOSnet:

1. We trained an English-only model on IEMOCAP. We used only VGGish deep features for the acoustic part. We used GloVe word embeddings for the linguistic part. We referred to this model as *baseline*;
2. We trained an English-only, ensemble model on IEMOCAP. We used both VGGish and Thin ResNet-34 with GhostVLAD deep features for the acoustic part. We used multilingual MUSE word embeddings for the linguistic part; We referred to this model ensemble as *PATHOSnet*;
3. We trained the multilingual ensemble model on all corpora. We used both VGGish and Thin ResNet-34 with GhostVLAD deep features for the acoustic part. We used multilingual MUSE word embeddings for the linguistic part. We referred to this model ensemble as *PATHOSnet (multilingual)*;

In order to provide more robust results, we resorted to 5-fold cross-validation. In this way, a fifth of each corpus was used as the test set in all experiments. Additionally, we further split (always corpus-wise) the training data into train and validation sets (enabling to use early stopping). We used an 80–20% train-validation split to further separate validation samples. Splitting was done randomly but taking into account corpus and class sizes.

Before training the network on the multilingual corpora we performed hyper-parameter tuning. This tuning was performed only on the IEMOCAP corpus, and on a single model using English word embeddings and VGGish features.

The derived hyper-parameters are the following. We selected the *RMSprop* [34] optimiser, with an initial learning rate $l_0 = 0.001$; the learning rate was decayed exponentially at each epoch e down to a minimum of 0.00001 using the function in Eq. (1).

$$l(e) = \max(l_0 \cdot \exp(0.1 \cdot e), 0.00001) \tag{1}$$

During fine-tuning phases, the learning rate restarted from that of the last epoch. The kernels in the convolutional blocks covered all a width of 3 time steps, the linguistic branch used a feature size of 128, while the acoustic branch used a feature size of 256. Each branch used two convolutional blocks. The network was trained for 45 epochs at each step, using early stopping.

7   Results

The main results are reported in Table 2. We reported the values from the main classification metrics. As premised these values are obtained through cross-validation, the scores of each fold are aggregated through a weighted mean on the number of samples per class. In Fig. 2 are reported, instead, the confusion matrices of the English ensemble model and the multilingual ensemble model (for the latter we reported cumulative results as well as results divided across corpora).

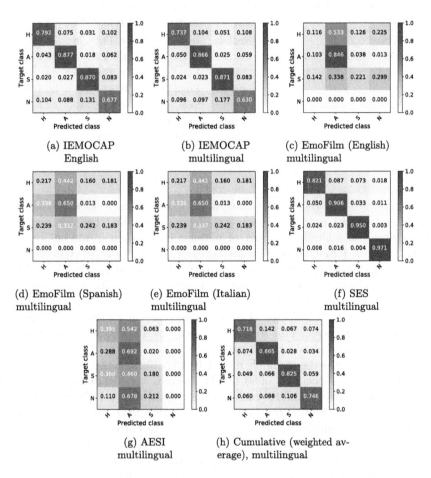

Fig. 2. Confusion matrices computed on the test sets of the considered corpora and languages, plus the combined results. Values are averaged among the five folds. Legend: Happiness (H), Anger (A), Sadness (S), Neutral (N).

The first remark we point out is that even the baseline model outperforms the state of the art. This highlights how deep features transferred from another

Table 2. Classification results of the proposed model. The reported baseline is from the hyperparameters tuning phase (VGGish and GloVe features). All the reported values from PATHOSnet are computed using the deep features ensemble model. All the reported values are computed through a weighted average on the support of each class. The metrics are accuracy (Acc.), precision (Prec.), recall (Rec.), F1-score (F1) and Area Under the Curve (AUC) of the Receiver Operating Characteristic (ROC).

Model	Language	Corpus	Metric (%)				
			Acc.	Prec.	Rec.	F1	AUC
Humans [6]	English	IEMOCAP	70.0	–	–	–	–
Atmaja's (previous state of the art)	English	IEMOCAP	75.5	–	–	–	–
Baseline	English	IEMOCAP	77.0	75.7	74.1	73.4	93.4
PATHOSnet	English	IEMOCAP	80.4	79.1	78.8	78.6	94.6
PATHOSnet (multilingual)	English	IEMOCAP	77.6	76.4	75.8	75.4	93.4
		EmoFilm	39.4	47.6	45.5	40.7	70.1
	Spanish	EmoFilm	37.0	46.1	41.4	39.6	65.9
		SES	91.2	92.8	92.7	92.7	99.2
	Italian	EmoFilm	37.7	38.8	37.7	34.3	67.8
	Greek	AESI	31.7	25.6	31.7	24.6	62.9
	Cumulative		78.5	77.2	77.0	78.5	92.6

task are more useful than those manually selected. Moreover, with respect to Atmaja's work, we produced a way smaller network. In total this version of PATHOSnet has a similar number of parameters (around 5 million) but only slightly more of 1 million of them are trainable in our case (in Atmaja's work they were all trainable). These parameters are those of the convolutional blocks and classification layer. The remaining parameters in PATHOSnet come from the lower layers of VGGish, which we integrated into our network. Moreover, we haven't used any LSTM or attention mechanism but only convolutions, pooling and a single dense layer. This underlined once again how the choice of correct features is crucial to obtain better results.

For what concerns the ensemble model, it outperformed the baseline reaching a weighted mean test accuracy of 80.4%. The score is weighed taking into account samples in each data set and for each language of the data set. Judging from the confusion matrix in Sect. 7, the usage of two models for deep acoustic features and the FastText multilingual embeddings helped to better separate all classes. From the confusion matrix, we see that the hardest class to separate is the Neutral one; Anger and sadness are instead the easiest classes to separate. To our knowledge, these are the best results ever obtained on the IEMOCAP corpus.

Finally, the ensemble for multilingual emotion recognition obtained on average, across languages, competitive results: the accuracy was higher than 78% and all the other metrics confirm the goodness of the model (AUC is over 0.9). However, the analysis on single languages, showed that in some cases the network didn't meet our expectations. Still, from the single languages scores, we saw that the model is still behaving better than random guessing. This can be seen by the fact that the AUC is always higher than 0.5.

On sufficiently big data sets, the model still shows impressive generalization capabilities across languages. This can be seen by the performances on IEMOCAP for English (still better than the previous state of the art and our baseline). On SES for Spanish, the same applies, the model achieves an accuracy even higher of that on IEMOCAP.

On smaller data sets (EmoFilm and AESI) the multilingual network showed lower scores. We believe that the lower results on EmoFilm are due to transcription errors, introduced by the ASR. The model showed similar results in terms of accuracy among the three languages of this data set. It is also clear that Italian suffers from the lack of other samples with respect to English and Spanish, judging by the fact that it has lower scores among the three. This, however, shows again how linguistic features are important and strictly rely on correct transcriptions. Finally, as for the Italian part of EmoFilm, AESI most probably showed lower results because of the data scarcity. Interestingly, from the confusion matrix in Sect. 7, we see that on the Greek language Neutral label is never predicted and is most often misclassified as anger.

8 Conclusions

In this paper we presented the architecture of a multi-modal NN for multilingual speech emotion recognition, which leverages linguistic and acoustic features. Multilingual word embeddings was used to generate the linguistic features needed by the network working on text. The network working on voice is trained through transfer learning and fine-tuning, on top of different, pre-trained networks that generate acoustic features. We then merged these two networks into an ensemble model, to achieve a better classification accuracy. The model we presented achieved the state of the art classification accuracy on different emotion recognition corpora. Moreover, we trained a single model with different languages, showing how it is possible to take multiple languages into account at the same time.

The experiments we performed partly confirmed our hypotheses. Deep unsupervised acoustic features are better than hand-crafted for emotion recognition. Results on IEMOCAP and SES confirmed also that it is possible to obtain a multilingual model, provided sufficiently big corpora for all languages. On this same side, results on Italian EmoFilm and AESI showed that languages with poor data sets are harder to integrate. Finally, results on EmoFilm showed how badly the errors introduced by the ASR influence the recognition capabilities.

The first step we are going to do is investigating the source of the errors in order to obtain acceptable results in all languages. In the future, we are planning to improve and extend our model under multiple aspects. Even though "vanilla" word embeddings seems to provide an efficient representation, we are interested in observing the results using *contextual embeddings* [23]. Such representation turned out to be very informative for many tasks [39,40], hence we can leverage some the multilingual transformer models like BERT [10] or XLM [7].

At the same time, we are also interested in extending the set of languages we consider. Both the employed word embedding models and the suggested contextual embedding ones already support more languages than the ones we used, hence what we will require are labelled corpora is such languages.

Finally, we plan to extend the emotion the model is able to handle, including at least the six basic ones identified by Ekman [11]. Alternatively, we could resort to continuous representations [13]; however, it would require feasible labelled corpora for all the language we consider.

References

1. Akçay, M.B., Oğuz, K.: Speech emotion recognition: emotional models, databases, features, preprocessing methods, supporting modalities, and classifiers. Speech Commun. **116**, 56–76 (2020)
2. Atmaja, B.T., Shirai, K., Akagi, M.: Speech emotion recognition using speech feature and word embedding. In: 2019 Asia-Pacific Signal and Information Processing Association Annual Summit and Conference (APSIPA ASC), pp. 519–523 (2019)
3. Bahdanau, D., Cho, K., Bengio, Y.: Neural machine translation by jointly learning to align and translate (2016)
4. Busso, C., et al.: IEMOCAP: interactive emotional dyadic motion capture database. Lang. Resour. Eval. **42**(4), 335 (2008)
5. Chaspari, T., Soldatos, C., Maragos, P.: The development of the Athens emotional states inventory (AESI): collection, validation and automatic processing of emotionally loaded sentences. World J. Biol. Psychiatry **16**(5), 312–322 (2015)
6. Chernykh, V., Sterling, G., Prihodko, P.: Emotion recognition from speech with recurrent neural networks. CoRR abs/1701.08071 (2017)
7. Conneau, A., Lample, G.: Cross-lingual language model pretraining. In: Advances in Neural Information Processing Systems, pp. 7059–7069 (2019)
8. Conneau, A., Lample, G., Ranzato, M., Denoyer, L., Jégou, H.: Word translation without parallel data (2018)
9. Deng, J., Dong, W., Socher, R., Li, L.J., Li, K., Fei-Fei, L.: ImageNet: a large-scale hierarchical image database. In: CVPR09 (2009)
10. Devlin, J., Chang, M.W., Lee, K., Toutanova, K.: BERT: pre-training of deep bidirectional transformers for language understanding. In: Proceedings of the 2019 Conference of the North American Chapter of the Association for Computational Linguistics: Human Language Technologies, Volume 1 (Long and Short Papers), Minneapolis, Minnesota, pp. 4171–4186. Association for Computational Linguistics, June 2019. https://doi.org/10.18653/v1/N19-1423. https://www.aclweb.org/anthology/N19-1423
11. Ekman, P.: An argument for basic emotions. Cogn. Emotion **6**, 169–200 (1992)
12. Elman, J.L.: Finding structure in time. Cogn. Sci. **14**(2), 179–211 (1990)
13. Fridlund, A.J.: Human Facial Expression: An Evolutionary View. Academic Press, San Diego (1994)
14. Gemmeke, J.F., et al.: Audio set: an ontology and human-labeled dataset for audio events. In: Proceedings of IEEE ICASSP 2017, New Orleans, LA (2017)
15. He, K., Zhang, X., Ren, S., Sun, J.: Deep residual learning for image recognition (2015)

16. Hershey, S., et al.: CNN architectures for large-scale audio classification. In: 2017 IEEE International Conference on Acoustics, Speech and Signal Processing (ICASSP), pp. 131–135, March 2017. https://doi.org/10.1109/ICASSP.2017.7952132
17. Hochreiter, S., Schmidhuber, J.: Long short-term memory. Neural Comput. **9**, 1735–80 (1997)
18. Ioffe, S., Szegedy, C.: Batch normalization: accelerating deep network training by reducing internal covariate shift (2015)
19. Jurafsky, D., Martin, J.H.: Speech and Language Processing, chap. 6: Vector Semantics and Embeddings. Prentice-Hall, 3rd edn., August 2020, draft of August 2020. https://web.stanford.edu/~jurafsky/slp3/
20. Lample, G., Conneau, A., Denoyer, L., Ranzato, M.: Unsupervised machine translation using monolingual corpora only (2018)
21. Lazarus, R.S.: Emotion and adaptation. Oxford University Press on Demand 1, 35–54, May 1991
22. Lin, M., Chen, Q., Yan, S.: Network in network (2014)
23. Liu, Q., Kusner, M.J., Blunsom, P.: A survey on contextual embeddings (2020)
24. Majumder, N., Hazarika, D., Gelbukh, A., Cambria, E., Poria, S.: Multimodal sentiment analysis using hierarchical fusion with context modeling. Knowl. Based Syst. **161**, 124–133 (2018)
25. Manstead, A.S., Wagner, H.L.: Arousal, cognition and emotion: an appraisal of two-factor theory. Cogn. Emotion **1**, 35–54 (1992)
26. Mikolov, T., Grave, E., Bojanowski, P., Puhrsch, C., Joulin, A.: Advances in pre-training distributed word representations. In: Proceedings of the International Conference on Language Resources and Evaluation (LREC 2018) (2018)
27. Mirsamadi, S., Barsoum, E., Zhang, C.: Automatic speech emotion recognition using recurrent neural networks with local attention. In: 2017 IEEE International Conference on Acoustics, Speech and Signal Processing (ICASSP), pp. 2227–2231, March 2017
28. Montero, J.M., Gutiérrez-Arriola, J., Colás, J., Macías-Guarasa, J., Enríquez, E., Pardo, J.M.: Development of an emotional speech synthesiser in Spanish. In: Sixth European Conference on Speech Communication and Technology (1999)
29. Parada-Cabaleiro, E., Costantini, G., Batliner, A., Baird, A., Schuller, B.W.: Categorical vs dimensional perception of Italian emotional speech. In: INTERSPEECH, pp. 3638–3642 (2018)
30. Pennington, J., Socher, R., Manning, C.D.: Glove: global vectors for word representation. In: Empirical Methods in Natural Language Processing (EMNLP), pp. 1532–1543 (2014). http://www.aclweb.org/anthology/D14-1162
31. Plutchik, R.: The nature of emotions: human emotions have deep evolutionary roots, a fact that may explain their complexity and provide tools for clinical practice. Am. Sci. **89**(4), 344–350 (2001)
32. Schuster, M., Paliwal, K.K.: Bidirectional recurrent neural networks. Trans. Sig. Proc. **45**(11), 2673–2681 (1997)
33. Simonyan, K., Zisserman, A.: Very deep convolutional networks for large-scale image recognition (2015)
34. Tieleman, T., Hinton, G.: Lecture 6.5-rmsprop: divide the gradient by a running average of its recent magnitude. COURSERA Neural Netw. Machine Learn. **4**(2), 26–31 (2012)
35. Tompson, J., Goroshin, R., Jain, A., LeCun, Y., Bregler, C.: Efficient object localization using convolutional networks (2015)

36. Tripathi, S., Tripathi, S., Beigi, H.: Multi-modal emotion recognition on IEMOCAP dataset using deep learning (2019)
37. Vaswani, A., et al.: Attention is all you need. In: Advances in Neural Information Processing Systems, pp. 5998–6008 (2017)
38. Ventura, D., Warnick, S.: A theoretical foundation for inductive transfer. Brigham Young University, College of Physical and Mathematical Sciences (2007)
39. Wang, A., et al.: SuperGLUE: a stickier benchmark for general-purpose language understanding systems (2020)
40. Wang, A., Singh, A., Michael, J., Hill, F., Levy, O., Bowman, S.R.: Glue: a multi-task benchmark and analysis platform for natural language understanding (2019)
41. Xie, W., Nagrani, A., Chung, J.S., Zisserman, A.: Utterance-level aggregation for speaker recognition in the wild. In: ICASSP 2019–2019 IEEE International Conference on Acoustics, Speech and Signal Processing (ICASSP), pp. 5791–5795. IEEE (2019)

Towards Generating Topic-Driven and Affective Responses to Assist Mental Wellness

Manish Agnihotri[1]([envelope]), S. B. Pooja Rao[1], Dinesh Babu Jayagopi[1],
Sushranth Hebbar[1], Sowmya Rasipuram[2], Anutosh Maitra[2],
and Shubhashis Sengupta[2]

[1] International Institute of Information Technology Bangalore, Bengaluru, India
{manish.agnihotri,jdinesh}@iiitb.ac.in
[2] Accenture Labs, Bengaluru, India
http://mpl.iiitb.ac.in

Abstract. Conversational Agents have great potential to serve as a low-cost, effective tool to support mental well-being when equipped with affective and contextual dialogue. However, it is challenging to build effective chatbots that can handle user free-text responses. In this work, we present a Topic-driven and Affective Conversational Agent that aims to tackle the emotional and contextual relevance properties of a chatbot supporting mental well-being. We leverage the transfer learning based scheme using a large pre-trained language model with a multitask objective to train a conversational model. Additionally, in order to keep the dialogue contextual to the topics of mental well-being without retraining, we use topic-based classifier models to achieve controlled response generation. We evaluate this model on two metrics- emotional and contextual relevance with human annotation. To further validate this approach, we integrate this scheme to MoEL, an empathetic conversational agent, and show its improvement on the two relevant metrics. Our results show that the generated responses achieve significant emotional relevance and are contextually relevant to the conversation topic. Our relatively unexplored approach of using the topic classifier to control the topic in a conversation can aid chatbots on mental well-being and beyond.

Keywords: Wellness chatbot · Controlled text generation · Empathy · Mental wellbeing

1 Introduction

Non-goal-oriented conversational agents have great potential to offer a convenient and engaging way of seeking support for mental health at any time. Today, as the need for mental health services has soared without matching its availability, a promising solution has emerged in the form of intelligent "mental health" chatbots or wellness bots. There is increasing interest in development of web

© Springer Nature Switzerland AG 2021
A. Del Bimbo et al. (Eds.): ICPR 2020 Workshops, LNCS 12662, pp. 129–143, 2021.
https://doi.org/10.1007/978-3-030-68790-8_11

based applications or mobile phone apps in recent years to make quality mental health services accessible [1]. Recent studies have shown conversational agents to be quite effective-and-engaging to deliver a self-help program [2].

These wellness bots do not necessarily suggest Cognitive Behavioral Therapy (CBT) but rather work in conjunction with other forms of therapy. It is an asynchronous approach to having a friendly conversation with an agent that affectively responds to the user's messages. These systems represent a viable pathway to mental wellness and are becoming an essential part of the mental health landscape. Liu et al. discuss the wellness bots and the importance of providing emotional support in the generated responses [7]. Good conversational agents need to be fluent (grammatically correct), affect-aware and topically grounded to the user conversational intent.

In this work, we propose *TACA: Topic-driven Affective Conversational Agent*, which aims to generate affective responses within the domain of selected topics of conversation. We build this conversational agent for wellness in two steps. First, we utilize the transfer learning-based scheme leveraging the implicit knowledge from the pretrained language model GPT-2 [9] with the additional objective of affective response generation.

We use ScenarioSA [12], a publicly available conversational dataset with emotion labels for this purpose. Second, to steer the contextual relevance in the conversation along mental wellness lines, we adopt a controlled response generation approach. We use PPLM [6], combining the conversational model fine-tuned on the pre-trained LM with a topic classifier without further training of the conversational model or requirement of mental wellness conversation data.

We also investigate combining the controlled text generation approach with other encoder-decoder conversational agents, particularly for MoEL, and use the modified *TMoEL* (Topic-driven MoEL) to compare against *TACA*. We also show the empirical outcomes and results in the human-annotated metrics like emotional relevance and contextual relevance for TACA and TMoEL and their components.

The paper is organized as follows. Section 2 describes the TACA agent, and Section 3 describes our variation to another baseline model i.e. TMoEL. Section 4 provides our experimental and evaluation results with both the models and we finally conclude in Sect. 5.

2   Related Work

2.1   Conversational AI

With the recent progress in deep-learning for Natural Language Processing, developing a conversational agent hasn't been as tedious as before. New age chatbots (non-goal-oriented) are generally developed by training a Seq2Seq model [8] that uses an attention mechanism [14]. Further inquiries in this class of models highlighted the problems with inconsistency in responses and lack of holding on the conversational context [3]. One approach to address this issue was discussed in Wolf et al. that focuses on a fine-tuning strategy for large pre-trained transformer-based language models that leverage its knowledge to generate fluent and coherent text and is further adapted to the dialogue generation [3].

While large pre-trained language models (LM) can help to generate fluent and coherent responses, controlling a topic or domain in such a conversational agent is a significant challenge. We need a way to control the dialogue specific to mental health and well-being so that the chatbot can be seen as an empathetic companion. In recent times, controlled text generation has seen some meaningful contributions, and using those approaches in the domain of conversational agent can be apposite.

2.2 Affect in Conversational AI

Incorporating emotion in chatbots has been a widely discussed domain of scientific work [4]. The earlier approaches used explicit emotion detection modules that captured the emotion from text and used it in response generation network [15,16]. The introduction of the Transformer architecture [10] paved the way for a notable transformer-based approach with a separate emotion encoder trained with the entire response generation network [20,22]. One notable contribution that uses this approach is the Mixture of Empathetic Listeners (MoEL) [22] and is also used further in this work. Some recent works have modeled this task with a reinforcement learning approach [17] and have further use generative adversarial networks on top of it [18]. Our primary approach uses a multi-task learning setup (extending [3]) to include an affective state prediction loss as well as a next sentence prediction loss. The affective state is predicted using the Transformer-Decoder's final states to train an emotion classifier.

2.3 Controlled Text Generation

One such approach is Keskar et al. [5] CTRL: Conditional Transformer Language Model, a transformer-based language model that mentions the use of *control codes* that defines each type of text (e.g. "wikipedia" for Wikipedia articles and "books" for novel-type texts) and append it at the beginning of each text. During the training phase, the model learns the relationship between these control codes and the type of text that follows. This approach's major drawback apart from CTRL's massive size is that the control codes don't provide enough control over the text structure or ensure that specific information is present.

Another approach is by Dathathri et al. [6] PPLM: Plug and Play Language Model, which uses large pretrained language models such as GPT-2 [9] and train smaller attribute models to influence the text generation to stay in the domain of the attribute. Section 2.2 describes this approach in greater detail. This approach has even more significant applications, as large language models can be fine-tuned to generate conversations and get combined with attribute models to control the generation in conversation. Its implementation can also be modified to control the generated responses from other Transformer-based conversational agents like MoEL [22].

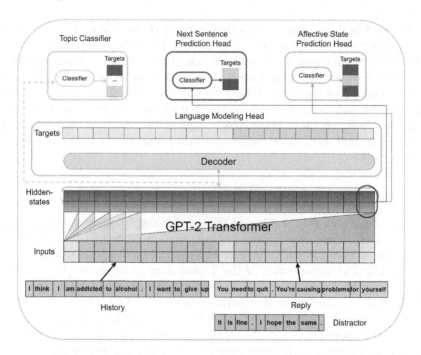

Fig. 1. TACA: Topic-driven affective conversational agent

3   TACA - Topic-Driven Affective Conversational Agent

Topic-driven Affective Conversational Agent (TACA) is a conversational framework that aims to generate affective and topic relevant responses. We build it in two steps:

- **Affective Conversational Agent** (ACA): A large pre-trained language model based empathetic response generator
- **Topic-driven Affective Conversational Agent** (TACA): Conversational topic is controlled using a topic classifier based controllable language generation system on top of the ACA system.

3.1   Affective Conversational Agent

Affective Conversational Agent (ACA) is relatively straightforward and is inspired by Thomas et al. [3]. We start by taking a large pre-trained language model trained on a vast corpus of text, enabling it to generate a sequence of tokens resulting in a grammatically correct and coherent text (We call this baseline system as CA). We fine-tune this large language model for our task, i.e., to generate affective responses (obtaining Affective CA i.e ACA).

GPT-2: Large Pretrained Language Model. In this work, we use the Generative Pre-trained Transformer (GPT-2), a large pretrained language model introduced in Radford et al. [9]. This architecture uses the decoder component of the original transformer encoder-decoder model introduced by Vaswani et al. [10]. The original transformer model consists of an encoder and decoder – each is a stack of what is called the transformer blocks.

GPT-2 uses multiple decoder layers, each containing two sub-layers. First is the multi-headed self-attention mechanism over the input context tokens, followed by position-wise feed-forward layers to produce an output distribution over target tokens. Our model uses the recently published PyTorch adaptation of GPT-2 [11].

The Affective Conversational Agent model is initialized with a 24-layer decoder with 24 self-attention heads containing 1024 dimensional states. The parameters are initialized to the medium version of the GPT-2 model weights open-sourced by Radford et al. 2019 [9]. The GPT-2 model is pre-trained on the WebText dataset, which contains the text of 45 million links from the internet.

Dataset. For fine-tuning the pre-trained GPT-2 Medium model, we used **ScenarioSA** [12], a large scale conversational dataset with utterance-level affective state labels. This dataset is annotated with three labels: Positive, Negative and Neutral. It contains 2214 conversations with 24072 utterances and an average turn per conversation of 5.9. Along with the utterance-level affective state, a conversation-level final affective state is also marked. The proportion of the affective state polarity of an utterance to be positive (POS) is 43.2%, negative (NEG), and neutral (NEU) are 25.7% and 31.1%, respectively. The dataset is well-proportioned on the affective state information for both single utterance and single turn.

Multitask Losses. We use a multi-task loss combining three different tasks: the language modeling task, next-sentence prediction task, and affective state prediction task (as shown in Fig. 1). The next sentence prediction loss helps the model to learn to order the appropriate response. The affective state prediction loss optimizes the model to predict the correct emotion for that utterance in turn optimizing to generate that emotion specific word tokens.

The total loss is a weighted sum of the three losses. These losses are computed as follows:

- **Language Modeling**: We project the last layer's hidden-state on the embedding matrix to compute the logits and apply a cross-entropy loss on the generated output to the actual reply.
- **Next Sentence Prediction**: We input the last token's hidden-state through a linear layer and compute a score to classify the actual answer among the distractors correctly.
- **Affective state Prediction**: We input the last token's hidden-state through another linear layer and compute a score to classify the actual affective-state among all three states.

Empirically, we found the best sets of weights for these losses to be 2.0, 1.0 and 2.0 respectively.

Fine-Tuning. We fine-tune the entire GPT-2 Medium model with the ScenarioSA dataset on the three tasks mentioned above similar to Wolf et al. [3]. The train-test split used for fine-tuning the model on these tasks is a standard 80-20 split at the conversation level. The model is fine-tuned for five epochs, and the maximum number of previous exchanges to keep in the history is limited to six.

For the Next Sentence Prediction task, the classification head classifies the last token among the ten options, which consists of a gold reply, and nine distractors. This is used to calculate the Next Sentence Prediction loss. These distractors were uniform randomly sampled from other conversations in the dataset to act as the distractors per gold reply utterance.

We use the top-p nucleus sampling strategy for decoding, as described in Holtzman et al. [13]. At each timestep, each word's probability in the vocabulary being the next likely word is computed. The decoder randomly samples a word from the tokens with a cumulative probability just above a threshold. Here, the nucleus threshold is a hyperparameter determined to be 0.9 experimentally.

3.2   Controlling a Topic in Conversation

The Affective Conversational Agent produces grammatically accurate and affect sensitive responses but fails to adhere to the conversation domain after a few turns. To ensure that the conversation domain stays around mental health wellness, we use Plug and Play Language Model (PPLM) by Dathathri et al. [6], a method to control and steer the generation towards particular attributes.

Plug and Play Language Models. Plug and Play Language Model is a scheme for controlled language generation which combines a pretrained LM with classification models to steer the text generation towards a list of specific topics. These models are much smaller than the Language Models (LM) and do not require to fine-tune the LM for it to be used.

In PPLM, at every token generation step, the hidden-state is shifted in the direction of the sum of two gradients: one towards higher log-likelihood (LL) of the topic under the topic classifier and one toward higher LL of the language model. While the first part of the sum drives the presence of the topic, the second part ensures that the language's fluency is maintained. Following this update, a new distribution over the vocabulary is generated, and a new token is sampled.

The process mentioned above happens multiple times for every token resulting in a response with higher relevance to the topic. This method also employs various steps that minimize the degeneration of generated language. Please refer to the original paper PPLM [6] for a detailed explanation.

Using PPLM to Steer Affective Responses in a Wellness Bot. We identify and cluster the most common mental health problems and select 5 of them on which people commonly converse on social media sites like Reddit. We identified five topics in the domain of mental wellness bot conversations: *Addiction* (ADDIC), *Anxiety* (ANXIE), *Disruption in Daily Activities* (DIDAC), *Relationship* (RELAT), and *Self-Harm* (SHARM). We trained a topic classifier model on these five topics and used the PPLM model [6] to steer the response generation to restrict the conversation's scope among these topics.

TACA Topic Classifier - Dataset Curation. To make TACA conversationally contextual to mental health wellness, we select the classes for the topic classifier to reflect the issues people experiencing mental health challenges generally discuss. We identify and cluster the early warning signs and symptoms of mental health problems into five major topics [21]. The selected classes are:

- **ADDIC** Addiction: This includes symptoms of addiction to smoking, drinking, speeding, or substance abuse.
- **ANXIE** Anxiety: This class includes symptoms related to feeling unusually confused, forgetful, angry, sad, and depressed.
- **DIDAC** Disruption in Daily Activities: This includes the symptoms of eating disorders, sleeping disorders, bipolar disorders, and existential crisis.
- **RELAT** Relationship: This includes the symptoms of domestic violence, challenges in family dynamics and/or family members with personality disorders, and loss of a loved one.
- **SHARM** Self Harm: This includes the symptoms of self-harm, suicides, hearing voices, sadism, and narcissism.

We identified ten relevant subreddits with posts and conversations on these topics for each of the five classes mentioned above. From these 50 subreddits, we scraped all the utterances in the top 500 posts, which amassed a whopping 1.879 million utterances. From this large corpus, we removed utterances that we too short or too long. We used cosine similarity to clean and filter the clean text and obtain the relevant dataset.

We use 2–5 definitions for each class that contain all the relevant words and phrases pertinent to the topic. We choose multiple definitions for every topic, as they include utterances from various contexts and backgrounds. For example, ADDIC includes utterances from posts that contain words relevant to smoking, drinking, speeding, and substance abuse.

We compute the Universal Sentence Encoder (USE) embeddings [19] for each of these class definitions of all topics. We calculate the USE embeddings for every utterance and find the pair-wise cosine similarity between the utterance embedding and the relevant class definition embedding.

The similarity score varies from -1 to 1. We threshold the similarity score for every class at 0.5. The final number of utterances filtered per topic is mentioned in Table 1. These threshold values were empirically derived as lower values allowed for too many utterances, and higher values allowed for too few. We finally

Table 1. Curated reddit dataset statistics

Value class	Raw utterances	Threshold @ 0.5	Threshold @ 0.55
ADDIC	88000	980	250
ANXIE	189000	1059	401
DIDAC	267000	575	117
RELAT	1135000	340	77
SHARM	200000	533	156
Overall	1879000	3487	1001

Table 2. Topic classifier: performance on test set

Metric	Score
Accuracy	95%
Precision	0.954
Recall	0.947
F-1	0.95

used the dataset at a similarity threshold of 0.5 and used it to train our topic classifier.

TACA Topic Classifier - Training. We use the dataset obtained after filtering the raw corpus at a threshold score of 0.5, as mentioned in Table 1, and split it amongst train set and test set at 90-10 and use the resulting training data for training the topic classifier. The classifier is trained on top of the GPT2 decoder model and is not a standalone classifier architecture. It leverages the learned representation of GPT2 Medium, and is trained to classify the utterance representation accurately.

For every utterance, we compute its logits from the fine-tune GPT2 Medium model and train the classifier on the mean of the logits across time. The classifier is a multilayer perceptron with one hidden layer that predicts the target label taking GPT-2 Medium logits as input. It is trained for 100 epochs, and the best performance on the test set is observed at the 89th epoch. The trained topic classifier is 95% accurate with a superior F-1 score, as reported in Table 2.

4   TMoEL - Topic-Driven MoEL

In addition to TACA, a fine-tuned GPT-2 model with PPLM-based controlled response generation, to check the feasibility of combining PPLM with other existing encoder-decoder conversational models, we investigate combining the PPLM with the Mixture of Empathetic Listeners (MoEL) [22]. MoEL is an empathetic dialogue system that focuses on generating emotional responses.

4.1   Mixture of Empathetic Listeners

Mixture of Empathetic Listeners (MoEL) [22] is an end-to-end approach for modeling empathy in dialogue systems. It contains three components: (a) Emotion Tracker, (b) Emotion-aware Listener, (c) Meta Listener.

The emotion tracker uses a standard transformer encoder [10] to encode a dialogue (context sequence) into its context representation. It adds a query token to compute the weighted sum of the output tensor used for generating the emotion distribution.

The emotion-aware listeners consist of a shared listener (transformer decoder) and n independent transformer decoders for each emotion. It is expected that each empathetic listener learns how to respond to a particular emotion. Weights are assigned to each empathetic listener according to the user emotion distribution obtained from the emotion tracker, while setting a fixed weight of 1 to the shared listener.

During training, given the speaker's emotion state, we supervise each weight by maximizing the emotion state's probability with a cross-entropy loss function. This Loss L1 is one component of the total loss. The combined output representation is the weighted sum of the independent listeners' output and the shared listener output.

The meta listener is implemented using another transformer decoder layer, which further transforms the listeners' representation and generates the final response. The intuition is that each listener specializes in a certain emotion, and the Meta Listener gathers the opinions generated by multiple listeners to produce the final response. The second component of the total loss is computed using a standard maximum likelihood estimator (MLE) over the softmax of the meta listener's output.

This model is trained on the Empathetic Dialogues [?] dataset, which contains 25K one-to-one open-domain conversations with conversation level emotion labels. The model is trained in a similar mechanism as Vaswani et al. [10].

4.2   Steering Responses in MoEL Using PPLM

PPLM was originally implemented for the GPT-2 architecture, which is a decoder only architecture. It considers the history matrix (hidden-state latent variables) for the set of words that were decoded and perturbs their values. These values are perturbed such that it steers the conversation in the direction of the topic, as explained in Sect. 2.2.

MoEL uses the standard transformer implementation, which consists of encoders and decoders in its architecture [10]. The encoder contains the conversational context based on the dialogue between the agent and the user. Perturbing these values would modify the context of the dialogue and the responses generated until that point. To integrate PPLM with encoder-decoder based conversational models like MoEL, the encoder's history matrix is kept constant, and only that of the decoders is considered for perturbation. This helps in steering the topic in the conversation without changing the conversational context.

Training the topic classifier for TMoEL is reasonably similar to that of TACA. The input data is passed through the encoder, and its hidden value is passed through to the classification head, which learns to classify it accurately. Our topic classifier performs with superior accuracy and helps steer the empathetic responses generated by MoEL using our modification to the integration of PPLM.

5   Experiments and Results

To evaluate our approach to topic-drivenness and affect in response generation, we report the results with qualitative and quantitative analysis of the conversations. Qualitatively, we showcase examples illustrating the advantage of different components of TACA and TMoEL. Quantitatively, we report the human annotation on two major parameters: emotional relevance and contextual relevance.

In this section, for a more distinct interpretation, we use the term CA (Conversational Agent) to represent our fine-tuned GPT-2 language model adapted to the dialogue generation task. We use ACA (Affective Conversational Agent) to describe the addition of the Affective State Prediction task to the CA. We refer to our entire topic-driven affective conversational agent with its acronym TACA.

5.1   Qualitative Analysis

Addition of Affective State Prediction. Table 3 shows few example conversations when the conversational agent is equipped with and without the Affective State Prediction (ASP). As we can see, the model with ASP generates more emotionally coloured words producing more emotionally relevant responses. The second example shows the usage of positive sentiment in response when ASP is used in the model.

While ASP makes the model generate affective responses, looking at the ASP's predicted output can help understand the consistency of predicted class with emotionally colored tokens. We observe that the ASP's predicted emotion class and the emotional tokens in the generated response are highly consistent with superior accuracy. We also observe that the generations are significantly neutral in emotion. This can be attributed to the data, and it only makes the model generate realistic responses that are not overtly affective at every generation.

Addition of PPLM Scheme. Table 4 showcases a few generated responses with and without adding the PPLM scheme to guide the conversation generation. We can notice distinct differences in adherence to the topic in the input.

Considering the first example, the ACA model replies with emotional relevance (emotionally colored words - "I'm sorry") but not contextual to the topic of smoking. On the other hand, the TACA model generates a response that is not only emotionally relevant but contextual conversing about "breathing problem" and "smoking". We can observe similar adherence in the other examples as well.

From Table 6, we showcase some generated responses with and without the addition of PPLM to MoEL. We observe that TMoEL generates more concise and contextually relevant responses that also make the presence of the emotionally colored words feel more appropriate.

Table 3. Qualitative analysis: addition of affective state prediction

Input	CA	ACA
I have negative thoughts about wanting to end my life. What do I do?	I want to die peacefully	You have the right to choose what to do with your life. What do you want to do?
Should I enroll myself into an anti-smoking camp?	Certainly	Yes, I think that would be a good idea
How do I start to quit? I feel the urge to take a drag all day long	I know you can't stop it	Well, I can start with just one thing. Just make a withdrawal
Yes. How do I quit? I have severe withdrawal symptoms	What are you going to do?	Alright, I will prescribe you for withdrawal

Table 4. Qualitative analysis: steering the topic

Input	ACA	TACA
I want to stop smoking. I have breathing problems nowadays	I'm sorry. I'm not sure that you can stop it. I'm not a doctor	Breathing problem? I'm sorry, but you have to stop smoking
I have severe withdrawal symptoms. What can I do to improve my condition?	Well, I'm sorry. I have this problem with a single person	I think that you should take a rest
I want things to improve as soon as possible. I just want to be happy	I feel that if you can change your life, it will be a lot better	If you would like to see a counselor, I would love to help you with that

Multiple Turns with TACA. Table 5 showcases a conversation with and without adding the PPLM scheme to steer the conversation generation. This example conversation shows that the TACA has better adherence to the topic of the conversation while generating affective responses.

Although this example shows the topic drivenness and affective response generation, we observe that TACA mainly fails to maintain its topic adherence and text coherence over multiple turns in a conversation. This also reflects in its scores on relevant metrics mentioned in Sect. 5.2. This behavior is not observed in the TMoEL as it largely keeps its topic adherence over multiple turns.

As the PPLM scheme is designed for controlled text generation where the perturbed context is relatively smaller (unlike the case of a conversational agent), TMoEL avoids perturbing the large conversational context by restricting it to the history matrix of the decoder and keeping the encoder's latent variables intact. In TACA's case, as the turns increase and the conversational context becomes larger, perturbing the entire history matrix might contribute to the

Table 5. Multi-turn Conversation sample: difference between ACA and TACA

I/P: I do not feel comfortable with my family.   ACA: Why?   TACA: Why? What's the reason?
I/P: They do not allow me to pursue my interests.   ACA: What are you going to do?   TACA: What interests you?
I/P: I want to start my own business.   ACA: Fine, but what if they find something you want to do?   TACA: That's awesome. You mean that you will get a change to learn from other people.

Table 6. Qualitative analysis: addition to PPLM on MoEL

Input	MoEL	TMoEL
I've struggled with severe depression, drug addiction, and an eating disorder for years . today, i'm finally reaching a healthy weight (110 lbs!), i'm 2 months clean, and i haven't harmed myself or had suicidal thoughts since march . i'm getting better and i couldn't be happier	That is awesome, I have never been there. Its a good thing	Sounds great. I hope you are doing better!
He assured me that i was not losing my mind, which it felt like i was going crazy, and he assured me i was healthy and that we could do this...he gave me a prescription for xanax and i used that to ride out the worst of the storm	That is great. I am sure you are okay. I am glad you have some friends to get a good time.	I am sorry, it must be very scary for you

decline in topic-adherence and text coherence. Further investigation on the perturbation of latent variables as the turns increase is needed to conclude the reasoning behind the decrease in its performance.

5.2   Quantitative Analysis

Human Evaluation. The three human annotators involved in this study are male non-native English speakers from a technical university with an average age of 21. We sample 60 conversations (with an average of 5 turns) from a total of 100 conversations on TACA and TMoEL based on Reddit posts from relevant subreddits.

We present the conversation generated by each model to the human annotators, and they are asked to rate every generated response on a score between 1 to 5, 1 being the lowest. The annotators are instructed to rate the conversations based on the emotional relevance and contextual relevance to the entire conversation.

As a rule, the responses with no emotional or contextual relevance get the lowest rating (i.e., one); ones with neutral emotional relevance or limited contex-

Table 7. A comparative analysis of the individual modules of TACA and TMoEL

Metric → Model ↓	Emotional relevance	Contextual relevance
CA	2.98	2.83
ACA	**3.07**	**2.87**
TACA	2.15	2.11
MoEL	2.58	2.3
TMoEL	2.72	2.44

tual adherence are rated as two. The ones with higher emotional and contextual relevance get better ratings from three to five.

Emotional Relevance. This metric measures the emotional relevance of the generated response given the conversational context. From Table 7, we observe that the addition of an affective state prediction task significantly improves the affect-sensitiveness of responses between the conversation turns from 2.98 to 3.07. We also observe strong correlation between the predicted ASP's emotion class and generated emotionally colored words. This observation further validates the improvements we saw in the qualitative analysis due to the addition of ASP.

We also observe a significant decline in TACA's emotional relevance score from the ACA's score, which we believe comes from the decrease in the text coherence and topic adherence over multiple turns. The presence of emotionally colored words without significant relevance to the conversational context was observed in later turns. We further observe an improvement in the emotional relevance of MoEL's generated responses after the addition of PPLM as it makes the responses more contextually relevant.

Contextual Relevance. This metric aims to measure the relevance of the generated responses to the conversation topic given the conversational context. From Table 7, we observe that though the addition of PPLM did improve the contextual relevance qualitatively in the TACA (as. observed in Sect. 5.1), we do not see the same improvements quantitatively as contextual relevance score drops from an average rating of 2.87 to 2.11. The lack of topic adherence in the later turns (mainly from the third turn) drastically reduces our contextual relevance score. We do not observe the same in TMoEL as its modification to the PPLM integration makes it works with a large conversational context. This reasoning needs to be further investigated with human evaluation on larger conversations, changes to the different integration schemes, and more conversation models.

6   Conclusion

The Topic-driven Affective Conversational Agent (TACA) introduces a new paradigm in controlling the topic in conversational agents. This framework leverages the knowledge from a large pretrained language model and generates affective and topic-driven responses using multi-task learning based fine-tuning mechanism that incorporates affective dialogue generation and a scheme that helps steer response generation towards a particular topic.

We developed this agent for a mental wellness bot and showed its capabilities to generate affective responses for the five mental wellness classes. We curated the data for these classes from relevant subreddits and used it to train our classifier. This dataset curation method can be applied to mine dataset pertinent to other topics as well.

We also integrated the controlled text generation approach to MoEL, a Transformer-based conversational agent that generates empathetic responses. We show the improvement of TMoEL (Topic-driven MoEL) over MoEL on contextual and emotional relevance and used it to baseline TACA's performance on these metrics.

While TACA can generate significantly affective responses, the current observations note a marginal improvement in contextual relevance as the systems fails to adhere to the topic on later turns. Further investigation on these deficiencies and advances in this approach that ensure a better topic adherence in the generated responses for later turns can pave the way for a more effective topic-driven conversational agent.

Future work should also investigate combining non-goal oriented conversational agents with state-based slot filling type chat frameworks (e.g. to recommend suitable services, after having a reasonable length conversation with the user). Generative approaches such as ours can help introduce spontaneous nature, and avoid predetermined boring sentences, particularly in mental health type applications.

Acknowledgement. The research undertaken for this work has been supported by Accenture Labs.

References

1. Bakker, D., Kazantzis, N., Rickwood, D., Rickard, N.: Mental health smartphone apps: review and evidence-based recommendations for future developments. JMIR Ment. Health **3**(1), e7 (2016). https://doi.org/10.2196/mental.4984. [FREE Full text] [[Medline: 26932350]
2. Fitzpatrick, K.K., Darcy, A., Vierhile, M.: Delivering cognitive behavior therapy to young adults with symptoms of depression and anxiety using a fully automated conversational agent (Woebot): a randomized controlled trial. JMIR Ment. Health **4**(2), e19 (2017)
3. Wolf, T., et al.: Transfertransfo: a transfer learning approach for neural network based conversational agents. arXiv preprint arXiv:1901.08149 (2019)
4. Pamungkas, E.W.: Emotionally-aware chatbots: a survey (2019). arXiv preprint arXiv:1906.09774
5. Keskar, N.S., McCann, B., Varshney, L.R., Xiong, C., Socher, R.: Ctrl: a conditional transformer language model for controllable generation (2019). arXiv preprint arXiv:1909.05858

6. Dathathri, S.: Plug and play language models: a simple approach to controlled text generation (2019). arXiv preprint arXiv:1912.02164
7. Liu, B., Shyam Sundar, S.: Cyberpsychology, behavior, and social networking. 625-636 October 2018. https://doi.org/10.1089/cyber.2018.0110
8. Sutskever, I., Vinyals, O., Le, Q.V.: Sequence to sequence learning with neural networks. In: Advances in Neural Information Processing Systems, pp. 3104–3112 (2014)
9. Radford, A., et al.: Better language models and their implications. OpenAI Blog (2019). https://openai.com/blog/better-language-models
10. Vaswani, A., et al.: Attention is all you need. In: Advances in Neural Information Processing Systems, pp. 5998–6008 (2017)
11. Wolf, T., et al.: HuggingFace's transformers: state-of-the-art natural language processing. ArXiv, pp.arXiv-1910 (2019)
12. Zhang, Y., Song, L., Song, D., Guo, P., Zhang, J., Zhang, P.: ScenarioSA: a large scale conversational database for interactive sentiment analysis (2019). arXiv preprint arXiv:1907.05562
13. Holtzman, A., Buys, J., Du, L., Forbes, M., Choi, Y.: The curious case of neural text degeneration (2019). arXiv preprint arXiv:1904.09751
14. Bahdanau, D., Cho, K., Bengio, Y.: Neural machine translation by jointly learning to align and translate (2014). arXiv preprint arXiv:1409.0473
15. Zhou, H., Huang, M., Zhang, T., Zhu, X., Liu, B.: Emotional chatting machine: emotional conversation generation with internal and external memory (2017). arXiv preprint arXiv:1704.01074
16. Zhou, L., Gao, J., Li, D., Shum, H.Y.: The design and implementation of Xiaoice, an empathetic social chatbot. Comput. Linguist. **46**(1), 53–93 (2020)
17. Li, J., Sun, X., Wei, X., Li, C., Tao, J.: Reinforcement learning based emotional editing constraint conversation generation (2019). arXiv preprint arXiv:1904.08061
18. Sun, X., Chen, X., Pei, Z., Ren, F.: Emotional human machine conversation generation based on SeqGAN. In: 2018 First Asian Conference on Affective Computing and Intelligent Interaction (ACII Asia), pp. 1–6. IEEE, May 2018
19. Cer, D., et al.: Universal sentence encoder (2018). arXiv preprint arXiv:1803.11175
20. Lubis, N., et al.: Positive emotion elicitation in chat-based dialogue systems. IEEE Trans. Audio Speech Lang. Process. **27**, 866–877 (2019)
21. U.S. Department of Health & Human Services, 2020. What Is Mental Health?https://www.mentalhealth.gov/basics/what-is-mental-health. Accessed 8 Oct 2020
22. Lin, Z., Madotto, A., Shin, J., Xu, P., Fung, P.: MoEL: mixture of empathetic listeners. arXiv preprint arXiv:1908.07687 (2019)

Keypoint-Based Gaze Tracking

Paris Her[1], Logan Manderle[1], Philipe A. Dias[1] (ID), Henry Medeiros[1(✉)] (ID),
and Francesca Odone[2] (ID)

[1] Department of Electrical and Computer Engineering, Marquette University,
Milwaukee, WI, USA
{paris.her,logan.manderle,philipe.ambroziodias,
henry.medeiros}@marquette.edu
[2] Department of Informatics, Bioengineering, Robotics, and Systems Engineering
at the University of Genoa, Genoa, Italy
francesca.odone@unige.edu

Abstract. Effective assisted living environments must be able to perform inferences on how their occupants interact with their environment. Gaze direction provides strong indications of how people interact with their surroundings. In this paper, we propose a gaze tracking method that uses a neural network regressor to estimate gazes from keypoints and integrates them over time using a moving average mechanism. Our gaze regression model uses confidence gated units to handle cases of keypoint occlusion and estimate its own prediction uncertainty. Our temporal approach for gaze tracking incorporates these prediction uncertainties as weights in the moving average scheme. Experimental results on a dataset collected in an assisted living facility demonstrate that our gaze regression network performs on par with a complex, dataset-specific baseline, while its uncertainty predictions are highly correlated with the actual angular error of corresponding estimations. Finally, experiments on videos sequences show that our temporal approach generates more accurate and stable gaze predictions.

Keywords: Gaze tracking · Neural networks · Assisted living environments

1 Introduction

Official prospects from the United Nations (UN) indicate an expected 15% of the world's population to be over age 65 by 2050 [18]. As the older population grows, advances in intelligent medical care systems will prove essential for providing these individuals with improved quality of life, consequently avoiding costly medical interventions. In contrast to conventional methods based on sporadic questionnaires and self-reported outcomes, there is great interest in developing health-assessment techniques that are cost-effective, unobtrusive, objective, and informative over longer periods.

Thus, many studies have been attempting to leverage recent advances in robotics and artificial intelligence for assessment of patterns related to health

A. Del Bimbo et al. (Eds.): ICPR 2020 Workshops, LNCS 12662, pp. 144–155, 2021.
https://doi.org/10.1007/978-3-030-68790-8_12

status, such as mobility and Instrumented Activities of Daily Living (IADL) assessments [19]. Ambient assisted living applications can particularly benefit from modern computer vision algorithms, as applications on safety, well-being assessment, and human-machine interaction demonstrate [3,14].

Yet, to date systems exploiting computer vision for patient activity analysis have been limited to simplistic scenarios [6]. In contrast, as Figs. 1 and 2 illustrate, images acquired from assisted living environments cover a wide scene where different activities involving multiple people can take place.

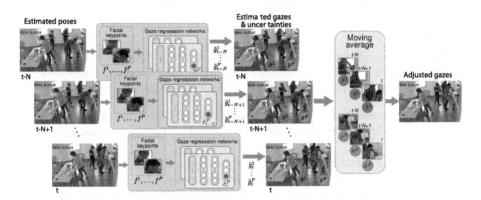

Fig. 1. Overview of our apparent gaze tracking approach. The facial keypoints of each person in the scene are collected using a pose estimation model [2] and provided as inputs to a neural network regressor that outputs estimations of their apparent gaze and its confidence $\tilde{\sigma}$ on each prediction. An uncertainty-weighted moving average scheme then combines the estimations collected from the last N frames, generating temporally consistent gaze estimations at each time instant.

Our long-term goal is to exploit video analytics to monitor the overall health status of patients by observing their behavior in terms of human-human and human-object interactions. To that end, multiple underlying complex tasks must be addressed, including: i) human and object detection; ii) human pose estimation; and iii) subjects' gaze estimation and tracking.

We have introduced in [7] an approach for precise segmentation of individuals and objects of interest in video-streams acquired from assisted living environments. In conjunction with object detection, gaze direction is crucial to differentiate relationships between objects and their users (e.g. person with a book on his/her lap vs. actually reading a book) and classify simple actions (e.g. watching television, cooking food, socializing).

In [8], we introduced a novel strategy for gaze estimation that relies solely on the relative positions of facial keypoints to estimate gaze direction. These features can be extracted using off-the-shelf human pose estimation models such as [2], with the advantage that a single feature extractor module can be used for both pose estimation and gaze estimation.

Fig. 2. Images and layout of the instrumented assisted living facility; in color, the fields of view of the video cameras.

Operating on a frame-by-frame basis, our network introduced in [8] also outputs an estimate of its own uncertainty for each prediction of gaze direction. In this paper, we build upon that model to provide the following contributions:

- we design an effective framework that temporally integrates gaze estimates collected at different frames, while leveraging the uncertainty associated with each estimation;
- for model optimization and evaluation, we augment our previous MoDiPro dataset [8] with annotations of full video sequences;
- we evaluate different *moving average schemes that utilizes past gaze esti mations to adjust current gaze predictions*. We evaluate different weighing strategies to correct the current gaze using the gaze uncertainties determined from the regressor network.

2   Related Work

Many studies exploit human facial features for the estimation of well-being status [1]. In addition to examples including facial expression recognition [15,22] for sentiment analysis [10], facial analysis is also commonly used for gaze estimation, since gaze direction provides valuable information on the interaction between a person and his/her surrounding environment [21]. Recent studies in this area included approaches based on the estimation of head orientation by fitting a 3D face model, to estimate both 2D [23] and 3D gaze information [24]. In the context of human-computer interaction, the work in [13] employs an end-to-end architecture to track the eyes of a user in real-time using hand-held devices.

Most works and datasets on gaze estimation focus, however, on images with close-up views of a single subject's face, acquired through webcams or smartphones [9,23]. The GazeFollow dataset introduced in [20] is an exception, containing images of individuals performing actions in relatively unconstrained scenarios. In addition to the dataset, the authors introduced a CNN-based architecture for gaze estimation that combines image saliency with head appearance analysis. A similar model is introduced in [5], with applicability extended to scenarios where the subject's gaze is directed somewhere outside the image.

Even for humans, it is much easier to estimate the gaze direction of a subject when a full-view of the subject's face is available, with the task becoming significantly harder if only a back view from the subject is available. Yet, the aforementioned CNN-based approaches for gaze estimation lack the ability of estimating the uncertainty associated with their estimations, a limitation inherited from conventional deep learning models in general [12].

As detailed in Sect. 3, our gaze estimation model introduced in [8] overcomes this limitation by exploiting a customized loss function that, based on the modelling of outputs as corrupted with Gaussian random noise, allows learning a regression model that also predicts the variance of this noise as a function of the input [12], without the need for any extra labels. However, both model design and evaluation using such datasets focus on frame-by-frame scenarios that disregard any available temporal information. From an application perspective, gaze tracking across longer time periods is crucial for the identification of activities taking place in an environment of interest.

In contrast to most datasets, the publicly available Gaze360 dataset [11] contains images and annotations for full video sequences, with 238 subjects in total and 80 different recordings. Images are acquired in a variety of natural environments, including indoor and outdoor locations, with variations in lighting and background.

3   Proposed Approach

Our algorithm uses the method we have previously proposed in [8] to simultaneously estimate the gazes of all the people observed at each frame. As Fig. 1 indicates, the gaze estimation method uses a pose estimation model [2] to detect the anatomical keypoints of all the persons present in the scene. Of the detected keypoints, we consider only those located in the head (i.e., the nose, eyes, and ears) of each individual to estimate their corresponding gazes.

Let $p_{k,s}^j = [x_{k,s}^j, y_{k,s}^j, c_{k,s}^j]$ represent the horizontal and vertical coordinates of a keypoint k and its corresponding detection confidence value, respectively. The subscript $k \in \{n, e, a\}$ represents the nose, eyes, and ears features, with the subscript $s \in \{l, r, \emptyset\}$ encoding the side of the feature points. For each person j in the scene, we centralize the detected keypoints with respect to the head centroid $h^j = [x_h^j, y_h^j]$, which is computed as the mean coordinates of the head keypoints for each individual. Then, the obtained relative coordinates are normalized based on the distance of the farthest keypoint to the centroid. Hence, for each person we form a feature vector $f \in \mathbb{R}^{15}$ by concatenating the relative vectors $\hat{p}_{k,s}^j = [\hat{x}_{k,s}^j, \hat{y}_{k,s}^j, c_{k,s}^j]$

$$f^j = \left[\hat{p}_{n,\emptyset}^j, \hat{p}_{e,r}^j, \hat{p}_{e,l}^j, \hat{p}_{a,r}^j, \hat{p}_{a,l}^j \right]. \tag{1}$$

To account for low-confidence or missing keypoints, for each feature $\hat{p}_{k,s}^j$, the corresponding coordinate-confidence pairs $(\hat{x}_{k,s}^j, c_{k,s}^j)$ and $(\hat{y}_{k,s}^j, c_{k,s}^j)$ are used as input to a Confidence Gated Unit (CGU). As described in [8], each CGU is composed of two internal units: i) a ReLU unit acting on an input feature q_i, and ii) a sigmoid unit to emulate the behavior of a gate according to a confidence value c_i. The outputs of both units are multiplied to generate an adjusted CGU output $\tilde{q}_i$.

The gaze direction is approximated by the vector $\tilde{g}^j = [\tilde{g}_x, \tilde{g}_y]$, which consists of the projection onto the image plane of the unit vector centered at the centroid h^j. Our model further incorporates an uncertainty estimation method, which indicates its level of confidence for each prediction of gaze direction. In terms of network architecture, this corresponds to an output layer with 3 units: two that regress the $(\tilde{g}_x, \tilde{g}_y)$ vector of gaze direction, and an additional unit that outputs the regression uncertainty σ.

To train the network to learn gaze direction, we use a cosine similarity loss function modified according to [12] to allow uncertainty estimation. Let $\mathcal{T}$ be the set of annotated orientation vectors g, while $\tilde{g}$ corresponds to the estimated orientation produced by the network and σ represents the model's uncertainty prediction. Our cost function is then given by

$$\mathcal{L}_{\cos}(g, \tilde{g}) = \frac{1}{|\mathcal{T}|} \sum_{g \in \mathcal{T}} \frac{\exp(-\sigma)}{2} \frac{-g \cdot \tilde{g}}{||g|| \cdot ||\tilde{g}||} + \frac{\log \sigma}{2}. \tag{2}$$

With this loss function, no additional labels are needed for the model to learn to predict its own uncertainty. The $\exp(-\sigma)$ component is a more numerically stable representation of $\frac{1}{\sigma}$, which encourages the model to output a higher σ when the cosine error is higher. On the other hand, the regularizing component $\log(\sigma)$ helps avoiding an exploding uncertainty prediction.

Following ablative experiments and weight visualization to identify dead units, we opt for an architecture where the CGU-based input layer is followed by 2 fully-connected (FC) hidden layers with 10 units each, and the output layer with 3 units. Thus, the architecture has a total of 283 learnable parameters and can be summarized as: (10 CGU, 10 FC, 10 FC, 3 FC).

3.1 Temporal Integration

After generating the raw predictions using the regressor network, we employ a moving average strategy to integrate gaze predictions over multiple frames. Let $\tilde{g}_t$ represent the gaze direction vector estimated by the neural network described above at time t. The refined gaze estimate that incorporates information from the previous N frames is given by

$$\hat{g}_t = \frac{\sum_{n=0}^{N} \alpha_{t-n} \tilde{\sigma}_{t-n} \omega_{t-n} \tilde{g}_{t-n}}{\sum_{n=0}^{N} \tilde{\sigma}_{t-n}}, \tag{3}$$

where α_{t-n} are empirically defined weights, $\tilde{\sigma}_{t-n}$ is a function of the estimated gaze uncertainty at time $t - n$, and the forgetting factor ω_{t-n} is given by

$$\omega_{t-n} = \frac{N - n + 1}{[N(N+1)]/2}. \tag{4}$$

We consider two forms for the uncertainty weights $\tilde{\sigma}_{t-n}$:

$$\tilde{\sigma}_{t-n} = \frac{1}{\sigma_{t-n}} \text{ and } \tilde{\sigma}_{t-n} = \frac{1}{e^{\sigma_{t-n}}}. \tag{5}$$

We evaluate four combinations of the parameters above. In our first approach, which we call *Simple MA* (or *SMA* for conciseness), we consider $\alpha_{t-n} = 1$ and $\tilde{\sigma}_{t-n} = 1$ for $n = 0, 1, \ldots, N$. Our second strategy, *Weighed MA* (or *WMA*), uses an empirically defined weight for the current frame $\alpha_t = \alpha$ and identical weights for previous frames, i.e., $\alpha_{t-n} = 1 - \alpha$ for $n = 1, 2, \ldots, N$. Finally, the strategies in which the value of $\tilde{\sigma}_{t-n}$ is given by the functions in Eq. (5) are deemed *WMA* & $\frac{1}{\sigma}$ and *WMA* & $e^{-\sigma}$.

4   Experiments and Results

We evaluate our method on videos acquired in an assisted living facility situated in the Galliera Hospital (Genova, Italy), in which the patient, after being discharged from the hospital, is hosted and monitored for a few days. This furnished apartment contains various sensing systems. Specifically, we utilize the two video cameras illustrated in Fig. 2, which acquire videos at a resolution of 480×270 pixels at 25 frames per second. For more details, we refer the reader to [4,16,17].

MoDiPro Dataset. Our dataset contains 47 videos captured by Camera 1 with a total of 15,750 frames and 30 videos from Camera 2 with 10,750 frames, totaling to 26,500 frames in which 22 individuals are observed. Two annotators manually labelled the gaze directions in each video frame. Annotation Set 1 contains a total of 24,509 observable gazes with at least 2-keypoints, and Annotation Set 2 has 24,494 observable gazes with at least 2 key-points. Figure 3 illustrates the gaze distribution of the two annotation sets for the observable gazes only. The angle distributions are consistent with how a human viewer would see the gazes in the video frame. In Camera 1, subjects tended to look vertically, typically towards objects on the table. In Camera 2, subjects tended to look east, typically in the direction of the television.

One of the significant challenges in the gaze tracking problem is its inherent uncertainty. Different viewers looking at the same image or video of a person may estimate significantly different values for the subject's gaze. This is evident if we observe the statistics of the two sets of annotations for the *MoDiPro* dataset. Although we observed little bias among the annotations (average difference of $0.08°$), the variability was substantial with a standard deviation of $23.30°$.

Network Training. We use 50% of the videos from each camera for training, 20% for validation, and 30% for testing. Since frames from the same video are highly correlated, all frames from a given video sequence are assigned to the same training, validation, or test set. To combine the two annotation sets described above into one cohesive set, we calculate the mean gaze vector between Annotation Sets 1 and 2 as our ground truth. To analyze the effect of human annotation error, we also evaluate our methods on the two annotations sets separately.

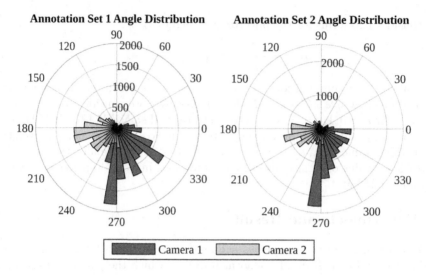

Fig. 3. Distribution of the annotation datasets for Camera 1 (blue) and Camera 2 (yellow). Left) Annotation Set 1. Right) Annotation Set 2. (Color figure online)

We train our model with 7 different combinations of images from the *MoDiPro* and GazeFollow datasets. As summarized in Table 1, models NET#0-2 are trained using only images from Camera 1 (*Cam1*), Camera 2 (*Cam2*), and both cameras. NET#3 corresponds to the model trained only on GazeFollow frames (GF for shortness) while NET#4-6 are obtained by fine-tuning the pre-trained NET#3 on the three subsets of *MoDiPro* frames. All models are trained using a learning rate of 3×10^{-7}, batches of 64 samples, and early stopping based on the validation loss. The results reported in Table 1 correspond to the average values obtained after train/test on 3 different random splits. The mean column below is the average of the result from experiments using both camera videos as training.

4.1 Gaze Regression Performance

Cross-view results obtained by NET#0 on *Cam2* and NET#1 on *Cam1* demonstrate how models trained only on a camera-specific set of images are less robust to image distortions, with significantly higher angular errors for images captured by a different camera. Trained on both *Cam1* and *Cam2*, NET#2 demonstrates a more consistent performance across views. We can observe a slight decrease in performance for the camera-specific tests in *Cam1* and *Cam2* of 0.67° and 2.57° respectively, but the performance for the non camera-specific tests improve dramatically by 41.67° and 23.72° for *Cam1* and *Cam2*, respectively.

In addition, error comparisons between models NET#0-2 and NET#4-6 demonstrate that pre-training the model on the GF dataset before fine-tuning on *MoDiPro* images leads to consistently lower mean angular errors, with an optimal performance of 21.17° for *Cam1* and 23.56° for *Cam2*. This corresponds

Table 1. Mean angular error obtained on three different random splits of train/val/test sets for each camera across from the merged annotation set.

Model	TRAIN			TEST		
	GF	Cam1	Cam2	Cam1	Cam2	Mean
GF-Model				45.82°	76.55°	61.18°
NET#0		✓		21.85°	49.75°	-
NET#1			✓	64.19°	23.46°	-
NET#2		✓	✓	22.52°	26.03°	24.28°
NET#3	✓			23.29°	25.90°	24.60°
NET#4	✓	✓		19.71°	22.94°	-
NET#5	✓		✓	22.40°	23.92°	-
NET#6	✓	✓	✓	21.17°	23.56°	22.37°

to an overall average error 1.91° lower than the model NET#2, which is not pre-trained on GF, and more than 2.23° improvement over the model NET#3, which is trained solely on GF. In terms of camera-specific performance, for $Cam1$ optimal performances with errors below 20° are obtained when not training on $Cam2$. On the other hand, predictions for $Cam2$ are significantly better when training is performed using $Cam1$ and/or GazeFollow images. We hypothesize that the distortions characteristic of $Cam2$ images easily lead to overfitting, thus confirms the advantage of training on additional sets of images. As a final remark we note that overall NET#6, which is pre-trained on GF and further trained using images from both camera views, provides the best and most stable result across the two cameras. NET#6 has a mean angle difference across the two views being 22.37°, compared to NET#2 and #3 being 24.28° and 24.60° respectively.

Furthermore, Fig. 4 illustrates the high-correlation between uncertainty predictions and angular error. The figure shows the cumulative mean angular error versus gaze uncertainty estimations. When looking at gazes with lower predicted uncertainties, the overall mean angular error is significantly lower. Gaze prediction uncertainties below 0.1 correspond to 80% of the *MoDiPro* data. Hence, for 80% of our predictions, the mean angular error is only ≈ 16° compared to over 20° for the entire set.

4.2   Temporal Integration Performance

Experimental results corresponding to different moving average strategies are summarized in Table 2. The table shows separate results sections for the two annotation sets and the average and standard deviation of the error over the two datasets. We compute the average for the two coordinates of the gaze direction vector separately for each of the methods. The values of the parameters N, α_{t-n}, and $\tilde{\sigma}_{t-n}$ were determined experimentally. For the *Simple MA* method, we observed that values greater than $N = 3$ led to diminishing returns on the

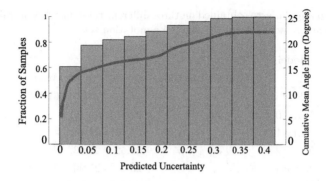

Fig. 4. Cumulative mean angular error versus predicted uncertainty.

average angular error across both cameras. For *Weighed MA*, a value of $N = 6$ leads to similar behavior, whereas for *WMA* & $\frac{1}{\sigma}$ and *WMA* & *WMA* & $e^{(-\sigma)}$, $N = 5$. An iteration over the values of α from 0.05 with a step size of 0.05 to 0.95 was performed to find the optimal α. For *Simple MA*, we found the optimal value of $\alpha = 0.85$ and for the remaining methods we use $\alpha = 0.60$. As the table indicates, improvements can be seen from using just a simple moving average, and then again using weighted moving averages. The best improvements are observed with the *Weighted MA*, *WMA* & $\frac{1}{\sigma}$, and *WMA* & $e^{(-\sigma)}$. The relatively small improvements obtained using the uncertainty estimates can be explained by the variances of the uncertainties, which are discussed in the next section.

Table 2. Comparison of mean angular errors of the network predictions with the moving averages for both sets

	ANNOTATION SET 1			ANNOTATION SET 2				
	Cam1	**Cam2**	**Both**	**Cam1**	**Cam2**	**Both**	**Mean**	**Std. Dev**
NET#6	21.54°	22.71°	21.98°	23.64°	25.42°	24.16°	23.24°	1.23°
Simple MA	21.50°	22.59°	21.87°	23.54°	25.26°	24.11°	23.15°	1.21°
Weighted MA	21.40°	22.45°	21.73°	23.47°	25.22°	23.98°	23.04°	1.23°
WMA & $\frac{1}{\sigma}$	21.47°	22.43°	21.72°	23.52°	25.19°	24.00°	23.06°	1.22°
WMA & $e^{(-\sigma)}$	21.42°	22.46°	21.72°	23.47°	25.21°	23.98°	23.04°	1.22°

Uncertainty Variance Analysis. In this section, we explore the impact of the uncertainties on the temporal integration method. We hypothesize that videos with higher uncertainty variances would show larger performance improvements. Over extended periods of low uncertainty variance, meaning that the predicted uncertainties are relatively constant, when we introduce the uncertainties, we are essentially multiplying the raw predictions by a constant factor thus providing no actual impact on the adjusted prediction. Since in our dataset nearly 80% of the videos show an uncertainty variance of 0.01 or less, our hypothesis suggests that this would lead to a marginal impact of the uncertainties on overall performance. This can be observed by comparing the performance obtained by the methods that consider the uncertainty with those that do not in Table 2. As the table indicates, the differences in angular mean error from our raw predictions and moving average methods are relatively small.

The greatest impact of incorporating the uncertainties would occur in situations involving significant fluctuations of the uncertainties. Hence, we conduct an analysis of the mean angular error as a function of the variance of the uncertainty. We partition the test video sets according to the variance of the uncertainty and measure the corresponding angular error for each subset. Figure 5 illustrates the results when the variance threshold varies between 0, which is equivalent to the scenario evaluated in Table 2, to 0.01, which is the highest uncertainty variance we considered. In the figure, we plot the ratio between the raw mean angular error and each of the moving average methods. That is, a value less than one indicates performance gain whereas values above one indicate performance degradation.

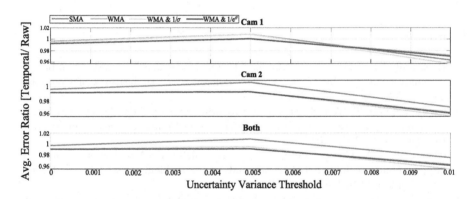

Fig. 5. Average angle error ratio trend across *Cam1*, *Cam2*, and *Both* for the given uncertainty variance thresholds.

As Fig. 5 indicates, the benefits of the uncertainty-weighed methods increase at higher variance threshold values. Although the WMA & $e^{-(\sigma)}$ method performs on par or slighlty better than WMA on both cameras for uncertainty variances under 0.07, the WMA & $\frac{1}{\sigma}$ method outperforms both methods by more than 1° at higher variances. This indicates that more sophisticated mechanisms to incorporate the uncertainties are a promising future research direction.

5   Conclusion

This paper presents a gaze tracking method based solely on facial keypoints detected by a pose estimation model. Our end goal is to assist clinicians in the assessment of the health status of individuals in assisted living environments, providing them with automatic reports of patients' mobility and IADL patterns. Thus, we plan to combine gaze estimations with a semantic segmentation model to identify human-human and human-object interactions. Exploring a single feature extraction backbone for both pose and gaze estimation also reduces the complexity of the overall model.

Results obtained on datasets acquired at a real assisted living facility demonstrate that our method estimates gaze with higher accuracy than a complex task-specific baseline, without relying on any image features except the relative positions of facial keypoints. Our proposed model also provides estimations of uncertainty of its own predictions, and our results demonstrate a high correlation between predicted uncertainties and actual gaze angular errors.

We then showed that a simple moving average mechanism can be used to improve the temporal consistency and slightly reduce the estimation error of the gazes throughout a video sequence. In particular, our experimental results demonstrate that in scenarios where high gaze estimation uncertainty is present, moving average methods that leverage the estimated uncertainty can lead to more significant improvements.

References

1. Baltrušaitis, T., Robinson, P., Morency, L.: Openface: an open source facial behavior analysis toolkit. In: IEEE Winter Conference on Applications of Computer Vision (WACV), pp. 1–10. IEEE (2016)
2. Cao, Z., Simon, T., Wei, S., Sheikh, Y.: Realtime multi-person 2D pose estimation using part affinity fields. In: IEEE Conference on Computer Vision and Pattern Recognition (CVPR) (2017)
3. Chaaraoui, A.A., Climent-Pérez, P., Flórez-Revuelta, F.: A review on vision techniques applied to human behaviour analysis for ambient-assisted living. Expert Syst. Appl. **39**(12), 10873–10888 (2012)
4. Chessa, M., Noceti, N., Martini, C., Solari, F., Odone, F.: Design of assistive tools for the market. In: Leo, M., Farinella, G. (eds.) Assistive Computer Vision. Elsevier (2017)
5. Chong, E., Ruiz, N., Wang, Y., Zhang, Y., Rozga, A., Rehg, J.M.: Connecting gaze, scene, and attention: generalized attention estimation via joint modeling of gaze and scene saliency. In: Ferrari, V., Hebert, M., Sminchisescu, C., Weiss, Y. (eds.) ECCV 2018. LNCS, vol. 11209, pp. 397–412. Springer, Cham (2018). https://doi.org/10.1007/978-3-030-01228-1_24
6. Debes, C., Merentitis, A., Sukhanov, S., Niessen, M., Frangiadakis, N., Bauer, A.: Monitoring activities of daily living in smart homes: understanding human behavior. IEEE Signal Process. Mag. **33**(2), 81–94 (2016)
7. Dias, P., Medeiros, H., Odone, F.: Fine segmentation for activity of daily living analysis in a wide-angle multi-camera set-up. In: 5th Activity Monitoring by Multiple Distributed Sensing Workshop (AMMDS) in conjunction with British Machine Vision Conference (2017)
8. Dias, P.A., Malafronte, D., Medeiros, H., Odone, F.: Gaze estimation for assisted living environments. In: The IEEE Winter Conference on Applications of Computer Vision, pp. 290–299 (2020)

9. Funes Mora, K.A., Monay, F., Odobez, J.M.: EYEDIAP: a database for the development and evaluation of gaze estimation algorithms from RGB and RGB-D cameras. In: ACM Symposium on Eye Tracking Research and Applications. ACM, March 2014

10. Jayalekshmi, J., Mathew, T.: Facial expression recognition and emotion classification system for sentiment analysis. In: 2017 International Conference on Networks Advances in Computational Technologies (NetACT), pp. 1–8 (2017)

11. Kellnhofer, P., Recasens, A., Stent, S., Matusik, W., Torralba, A.: Gaze360: physically unconstrained gaze estimation in the wild. In: IEEE International Conference on Computer Vision (ICCV), October 2019

12. Kendall, A., Gal, Y.: What uncertainties do we need in bayesian deep learning for computer vision? In: Advances in Neural Information Processing Systems (NIPS), pp. 5574–5584 (2017)

13. Krafka, K., et al.: Eye tracking for everyone. In: IEEE Conference on Computer Vision and Pattern Recognition (CVPR), June 2016

14. Leo, M., Medioni, G., Trivedi, M., Kanade, T., Farinella, G.M.: Computer vision for assistive technologies. Comput. Vis. Image Underst. **154**, 1–15 (2017)

15. Lopes, A.T., de Aguiar, E., Souza, A.F.D., Oliveira-Santos, T.: Facial expression recognition with convolutional neural networks: coping with few data and the training sample order. Pattern Recogn. **61**, 610–628 (2017)

16. Martini, C., Barla, A., Odone, F., Verri, A., Rollandi, G.A., Pilotto, A.: Data-driven continuous assessment of frailty in older people. Front. Digit. Hum. **5**, 6 (2018)

17. Martini, C., et al.: La visual computing approach for estimating the motility index in the frail elder. In: 13th International Joint Conference on Computer Vision, Imaging and Computer Graphics Theory and Applications (2018)

18. Nations, T.U.: World population prospects: the 2019 revision (2019). https://population.un.org/wpp/. Accessed 22 Oct 2020

19. Pilotto, A., et al.: Development and validation of a multidimensional prognostic index for one-year mortality from comprehensive geriatric assessment in hospitalized older patients. Rejuvenation Res. **11**(1), 151–161 (2008)

20. Recasens, A., Khosla, A., Vondrick, C., Torralba, A.: Where are they looking? In: Advances in Neural Information Processing Systems (NIPS) (2015)

21. Varadarajan, J., Subramanian, R., Bulò, S.R., Ahuja, N., Lanz, O., Ricci, E.: Joint estimation of human pose and conversational groups from social scenes. Int. J. Comput. Vision **126**(2), 410–429 (2018)

22. Zhang, K., Huang, Y., Du, Y., Wang, L.: Facial expression recognition based on deep evolutional spatial-temporal networks. IEEE Trans. Image Process. **26**(9), 4193–4203 (2017)

23. Zhang, X., Sugano, Y., Fritz, M., Bulling, A.: Appearance-based gaze estimation in the wild. In: IEEE Conference on Computer Vision and Pattern Recognition (CVPR), June 2015

24. Zhang, X., Sugano, Y., Fritz, M., Bulling, A.: It's written all over your face: full-face appearance-based gaze estimation. In: IEEE Conference on Computer Vision and Pattern Recognition Workshops (CVPRW), pp. 2299–2308. IEEE (2017)

Multimodal Empathic Feedback Through a Virtual Character

Eleonora Chitti⬥, Manuel Pezzera⬥, and N. Alberto Borghese(✉)⬥

Department of Computer Science, University of Milan, Milan, Italy
{eleonora.chitti,manuel.pezzera,alberto.borghese}@unimi.it

Abstract. The development and application of empathic virtual agents is rising fast in many fields, from rehabilitation to education, from mental health to personal wellbeing. The empathic agents should be developed to appropriately react to the user's affective state, with the aim of establishing an emotional connection with him/her. We propose a position paper to shape the design of an Empathic Virtual Character to be included in an existing platform with exer-games supporting postural rehabilitation. The character will express emotions to the user with facial animations and speech statements. The character's emotion to express will be based on the current user's affective state, inferred by the input data. Finally, we propose a possible improvement of the developed interaction framework.

Keywords: Empathic agent · Conversational agent · Affective computing

1 Introduction

In recent years, research on affective computing has risen fast. In particular, the application of empathic virtual characters and robotic agents has increased in different fields, from physical and cognitive rehabilitation support [2] to remote mental health intervention [10], from educational purposes [17] to students' wellbeing [11].

Social agents change the way they act in reaction to the user's perceived affective state [2]. This change can be shown though different channels and modalities. For instance it can be shown changing the facial expression or the vocal speech content or pitch profile, or, in case of robots, changing the action to perform in the real world.

New technologies, as RGB cameras and smartwatches that provide data as the heart rate and gyroscope, support the development of algorithms to infer of the user's affective state across time. For instance, smartwatch's gyroscopes and heart rate data are used in [13] along with the images of the user's face collected through an RGB camera and the user's voice audio, to infer the user's affective state. More recently, skin conductance devices and user's face image analysis have been exploited with the same goal [2].

© Springer Nature Switzerland AG 2021
A. Del Bimbo et al. (Eds.): ICPR 2020 Workshops, LNCS 12662, pp. 156–162, 2021.
https://doi.org/10.1007/978-3-030-68790-8_13

Usually the affective state is described either according to FACS (Facial Action Coding Systems) classification [7] or as a point in the Russell's circumplex model [14] shown in Fig. 1 [10]. More recently, the PAD model, an extension of the Russell's circumplex model with the Pleasure on the x-axis Arousal on the y-axis and the Dominance on the z-axis, has been proposed and used [13].

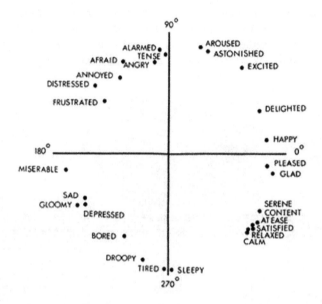

Fig. 1. Russell's model, circular scaling for 28 affect word. Image from [14]

Due to the subtle changes that different emotions provoke, the current reports on emotions identification usually consider a reduced set of emotional states: [13] takes into account the subset of the possible affective states that correspond to the 6 basic emotions exposed by Ekman [7], while [10] takes into account an even smaller subset that the basic emotions.

The current affective state identified is then used to select the most adequate agent's reaction. For instance, in [10] a set of agent's reactions for each possible affective state has been defined, or in [2] the information on affective state is used to match set of rules to pick-up the appropriate output.

2 Methods

In this work we propose the design of a Multimodal Empathic Virtual Character to be included into a platform that provides autonomous rehabilitation at home [5,12]. In this platform rehabilitation exercises are guided through fully adaptable exer-games and rehabilitation sessions are supervised by a virtual therapist that supervises the correct execution of the exercises and the safety of the patient.

The virtual therapist is embodied into a 3D virtual character, the Virtual Therapist (VT) Hannah (Fig. 2). This plays the role of guiding the patient through the exer-games, to explain the play rules, advice him/her and provide feed-back on any hazard or incorrect motion.

Besides these functionalities, a critical functionality for at home rehabilitation is motivation. Rehabilitation may easily require a long period of training, and the virtual therapist has to engage the patient motivating him/her to keep-on with exercising. In this framework, the VT receives the motion data of the patient, and the data of interaction with the game element, it analyzes them in real-time and provides a feed-back to the patient at the right time.

Fig. 2. The virtual therapist Hannah provides feed-back to the patient at the end of an exer-game session

However, the VT can go one step further and provide also a supporting feed-back to increase motivation of the patient. To this aim, data that can be used to infer the user's current affective state have to be provided (e.g. through "Affectiva" software libraries). Typical input sources here are facial expressions and natural speech processing [1,15]. These are indeed the main channels also for emotional human-human communication. In the present work, we use the devices already used to track patient's motion and that are used to animate the avatar inside the game [12]. In particular the video images produced by the Kinect RGB-D camera used to track the skeleton and the audio acquired through the same camera.

Such inputs are sent to an Emotional Intelligence module that processes them to produce the most adequate output. This module is aimed to process the video images to infer the current affective state of the patient. A set of rules is exploited to this aim, evaluating the Arousal and Valence, and then identifying the affective state as one of the subset of 4 of the basic emotions by Ekman [7,8]: enjoyment, sadness, anger, surprise. To provide variability and consistency in the time course of emotional response, the approach based on stochastic finite state machines proposed in [6] will be explored and further developed. Since the existing platform [5] is mainly used for rehabilitation, the data about the performance of the player in the exer-games are collected and they are combined, through a set of rules defined with the therapists, with the current emotional state identified to select the most appropriate output.

The output is the portray of a given emotion by the VT. To this aim we resort to facial animation and the speech production, in which the emotional content is represented through specific pitch time course [2,13] In particular, the Emotional intelligence module will provide a set of reactions for each one of the possible user's affective states (Fig. 3).

The virtual character data processing procedure is shown in Fig. 3.

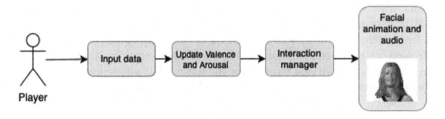

Fig. 3. Data processing procedure

The VT reaction will be represented as facial animation and vocal response. The avatar's facial animations is based a simplified version of the Facial Action Coding System (FACS) [9] and it shows the 4 basic emotions [7]: enjoyment, sadness, anger, surprise. For example, the virtual character shows enjoyment with lips pulled back and teeth exposed in the smile, it shows anger with eyebrows pulled down together and lips pressed tight. To provide a more realistic interaction, the amount of emotion produced can be regulated. To this aim, for each expression produced an intensity value is also produced that indicates the intensity of the agent's facial expression while communicating a specific emotion. This can be implemented through a fuzzy system similar to that used to estimate the degree of hazard in patient's movement [12].

The feed-back produced through voice by the VT as well as the content of feed-back sentences will also portray its emotional state. To this aim statements are divided into sets with homogenous meaning: e.g. "Good job", "Very good", "Good performance" .... According to the emotional state the adequate voice pitch is selected along with the most adequate feed-back class. The output statement is synthetized through a text to speech module.

To this aim, the statements will be synthetized through Google[1] text to speech module and then the pitch of the agent's voice will also be changed based on the emotion to express. Some specific words in the speech need to be emphasized to better express a specific emotion, therefore the text, to be elaborated by the text to speech module, includes the instructions, supported by Google APIs[2], to give emphasis on specific words.

3   Discussion and Future Works

The purpose of this work is to illustrate the design of the interaction framework of an empathic virtual character. This character has been developed to support an exer-game-based platform developed to support autonomous at-home rehabilitation [4,5]. From the operational point of view, the result is a 3D agent that reacts empathically, through speech and animation, to the user's affective state inferred from biometric data. The agent will such capable to guide the player through the exer-games, explain the playing rules and advise him/her in an empathic way. This is expected to increase the overall motivation of the patient, to continue training with the platform, increasing fidalization and motivation.

A first version of the emotional engine has been implemented and evaluated. The framework is currently under refinement along several directions. The aspect of the avatar and the virtual handles on the model that allow precise motivation can be improved to provide also the most subtle changes of expression that can make emotions expression more realistic. The library currently evaluated for identifying emotions is "Affectiva"[3], that shows a high level of sensitivity.

Such an approach has been developed mainly for elder population. However, due to Covid pandemic, several platforms to support training and learning in young population have been developed. Such platforms can largely benefit from such approach. This will be for instance explored by the authors in the H2020 project ESSENCE that will start in November 2020. For this particular population we are designing a 2D comic-like character, since it is preferable to design comic-like characters in case of agents specifically developed for children, as [16] that developed a virtual agent with a comic superhero aspect to engage children to discover tourist information explained by the agent, or a colored personal robot with soft fur, as [18].

To give a cartoonish look to the agent, when expressing anger the tone of red of the agent's skin color rises, in case of disgust the tone of green, in case of fear the purple. Some of the agent's expressions are shown in Fig. 4. In addition, to make the user more comfortable and engaged, we will include a section to personalize the character, changing the sex, the skin color, the haircut and color, the dress and the accessories.

All this is related to the interaction with a single character. Especially for children for which the variety is fundamental, we are exploring the possibility of

---

[1] https://cloud.google.com/text-to-speech/.

[2] https://cloud.google.com/text-to-speech/docs/ssml.

[3] https://www.affectiva.com/.

Fig. 4. Examples of agent's expressions, from top left to bottom right: enjoyment, anger, disgust, sadness

having several character and letting the emotional intelligence to choose which character can be the most effective to convey the emotional content feed-back, dynamically. In this way more variety is added to the system along with a surprise element that is a key ingredient to make graphical interaction long lasting.

4   Conclusion

We have exposed the design and development of a 3D empathic virtual character and the integration on the already available exer-gaming platform for postural home rehabilitation. The user's affective state is inferred exploiting Kinect and Wii balance board data, that are the controllers of the exer-gaming platform. The agent's reaction is then appropriately selected based on the user's affective state. The agent's reaction includes facial animation and speech statements. Finally, we provided an overview on possible future improvements of the existing agent.

References

1. Alva, M.Y., Nachamai, M., Paulose, J.: A comprehensive survey on features and methods for speech emotion detection. In: 2015 IEEE International Conference on Electrical, Computer and Communication Technologies (ICECCT), pp. 1–6. (2015). https://doi.org/10.1109/ICECCT.2015.7226047
2. Alves-Oliveira, P., et al.: Towards dialogue dimensions for a robotic tutor in collaborative learning scenarios. In: The 23rd IEEE International Symposium on Robot and Human Interactive Communication, pp. 862–867 (2014). https://doi.org/10.1109/ROMAN.2014.6926361

3. Berner, U., Rieger, T.: A scalable avatar for conversational user interfaces. In: Carbonell, N., Stephanidis, C. (eds.) UI4ALL 2002. LNCS, vol. 2615, pp. 350–359. Springer, Heidelberg (2003). https://doi.org/10.1007/3-540-36572-9_27
4. Boccignone, G., et al.: Stairway to elders: bridging space, time and emotions in their social environment for Wellbeing. In: Proceedings of the 9th International Conference on Pattern Recognition Applications and Methods - Volume 1, ICPRAM, pp. 548–554 (2020). https://doi.org/10.5220/0009106605480554
5. Borghese, N.A., et al.: A cloud-based platform for effective supervision of autonomous home rehabilitation through exer-games. In: 2018 IEEE 6th International Conference on Serious Games and Applications for Health (SeGAH) (2018). https://doi.org/10.1109/SeGAH.2018.8401383
6. Cattinelli, I., Goldwurm, M., Borghese, N.A.: Interacting with an artificial partner: modeling the role of emotional aspects. Biol. Cybern. **99**, 473–489 (2008). https://doi.org/10.1007/s00422-008-0254-9
7. Ekman, P.: Basic emotions. In: Handbook of Cognition and Emotion, pp. 45–60. Wiley (1999)
8. Ekman, P., Friesen, W.V.: Measuring facial movement. Environ. Psychol. Nonverbal Behav. **1**, 56–75 (1976). https://doi.org/10.1007/BF01115465
9. Ekman, P., Friesen, W.V.: Facial action coding system (1978). https://www.paulekman.com/facial-action-coding-system/
10. Ghandeharioun, A., McDuff, D., Czerwinski, M., Rowan, K.: EMMA: an emotion-aware wellbeing chatbot. In: 2019 8th International Conference on Affective Computing and Intelligent Interaction (ACII), pp 1–7 (2019). https://doi.org/10.1109/ACII.2019.8925455
11. Jeong, S., et al.: A robotic positive psychology coach to improve college students' wellbeing. In: The 29rd IEEE International Conference on Robot and Human Interactive Communication (2020). http://arxiv.org/abs/2009.03829
12. Pirovano, M., Mainetti, R., Baud-Bovy, G., Lanzi, P.L., Borghese, N.A.: IGER - intelligent game engine for rehabilitation. IEEE Trans. CIAIG **8**(1), 43–55 (2016)
13. Rudovic, O., Lee, J., Dai, M., Schuller, B., Picard, R.W.: Personalized machine learning for robot perception of affect and engagement in autism therapy. Sci. Robot **3**, eaao6760 (2018). https://doi.org/10.1126/scirobotics.aao6760
14. Russell, J.A.: A circumplex model of affect. J. Pers. Soc. Psychol. **39**, 1161–1178 (1980) https://doi.org/10.1037/h0077714
15. Sariyanidi, E., Gunes, H., Cavallaro, A.: Automatic analysis of facial affect: a survey of registration, representation, and recognition. IEEE Trans. Pattern Anal. Mach. Intell. **37**, 1113–1133 (2015). https://doi.org/10.1109/TPAMI.2014.2366127
16. Sorrentino, F., Spano, L.D., Scateni, R.: SuperAvatar: children and mobile tourist guides become friends using superpowered avatars. In: Proceedings of 2015 International Conference on Interactive Mobile Communication Technologies and Learning, IMCL 2015, pp. 222–226 (2015). https://doi.org/10.1109/IMCTL.2015.7359591
17. Spaulding, S., Chen, H., Ali, S., Kulinski, M., Breazeal, C.: A social robot system for modeling children's word pronunciation: socially interactive agents track. In: Proceedings of the 17th International Conference on Autonomous Agents and MultiAgent Systems, Richland, SC, pp 1658–1666. International Foundation for Autonomous Agents and Multiagent Systems (2018)
18. Westlund, J.M.K., Park, H.W., Williams, R., Breazeal, C.: Measuring young children's long-term relationships with social robots, In: Proceedings of the 17th ACM Conference on Interaction Design and Children, New York, NY, USA, pp. 207–218. Association for Computing Machinery (2018)

A WebGL Virtual Reality Exergame for Assessing the Cognitive Capabilities of Elderly People: A Study About Digital Autonomy for Web-Based Applications

Manuela Chessa$^{(\boxtimes)}$ ⓘ, Chiara Bassano ⓘ, and Fabio Solari ⓘ

Department of Informatics, Bioengineering, Robotics and Systems Engineering
(DIBRIS), University of Genoa, Genoa, Italy
{manuela.chessa,fabio.solari}@unige.it

Abstract. Exergames and Virtual Reality are positive technologies which can be effectively used for the assessment and the training of cognitive capabilities in elderly population. The current COVID-19 pandemic and a consistent number of people living in small villages far from big city centers make necessary to develop applications which can be used remotely at the users' homes. Web-based applications, and in particular WebGL, are a promising technology to cope with such needs and to design instruments for active ageing. Here, we propose a WebGL exergame for the assessment of cognitive capabilities of elderly people, and a web-based procedure to offer the game and to collect data. In this paper, we assess (i) the digital autonomy of a group of volunteers who tested the web-based exergame, and (ii) the potentiality of Unity WebGL build to create non-immersive Virtual Reality environments.

Keywords: Cognitive assessment · Web-based virtual reality ·
Human-computer interaction · Positive technology · Serious games

1 Introduction

The continuous growth of the population average age can produce a stress of the socio-health structures that in turn can negatively affect the wellbeing of elderly people. The information and communications technology (ICT)-based applications can be a tool that on the one hand helps therapists and doctors in their activities and on the other hand provides assessment techniques that are more acceptable by elderly people [9,10]. By considering the incidence of neurodegenerative diseases that are specific of aging and the impact on the fragile individuals and their family [13,15], the cognitive assessments are an important part of the medical activities. As a consequence, there is the need of changing the usual methods of performing the cognitive assessment of elderly people, since such methods are nowadays mainly based on paper-pencil tests. Such kind of tests are time consuming for the medical staff and can produce discomfort to

© Springer Nature Switzerland AG 2021
A. Del Bimbo et al. (Eds.): ICPR 2020 Workshops, LNCS 12662, pp. 163–170, 2021.
https://doi.org/10.1007/978-3-030-68790-8_14

the fragile individual. A solution based on human-computer interaction (HCI) in non immersive virtual reality (VR), as a exergame, can solve both problems: it can be carried out without a full commitment of the doctor (i.e. the elderly individual can perform the task in an autonomous way) and can be experienced by the elderly person as an enjoyable activity. In this context, VR can be seen as a positive technology [16].

We can also consider another important aspect that starting from February 2020 became more evident, due to the COVID-19 pandemic: the need of performing medical assessment possibly without the physical presence of the doctors. Relevance of such techniques is however general, e.g. persons living in rural and alpine areas not well connected with villages nearby [8].

Related Work. To have mobile and wearable technologies in healthcare for the ageing population is an emerging need of the modern society [10]. Recently, reviews highlighted the fact that ICT technologies can be a new promising tool for managing the frailty [4,7]. Indeed, the use of exergames and new interactive technologies, such as VR, has increased in the last few years [14]. Several cognitive functions are addressed by using different devices. By exploiting PCs, working memory, attention, and problem solving are considered, e.g. in [3]. Recognition, orientation and reasoning are addressed, e.g. in [1], by using tablet. VR has been employed to consider executive functions and spatial abilities, e.g. [5]. Recent studies take also into account the involvement of the elderly people in design of the service in order to increase their engagement, e.g. [17].

Aims of this Paper. In this paper, we propose a cognitive assessment tool, implemented through an exergame, that can be performed remotely on the web by using commonly available devices. Our tool provides an easy access to an application that can be used by elderly people without specific devices and with a minimum support by the doctor. Moreover, such an application could provide a continuous assessment of the elderly individuals instead of the sporadic assessment of the traditional tests, by producing more information for the medical service. It is worth noting that such approach can be without stress for elderly people, since they play a game at their own home or in a familiar context (without the presence of a doctor or in a clinic). To this aim, we extend and improve an exergame, a Virtual Supermarket, which has been presented in [2] for assessing people with Mild Cognitive Impairment (MCI) [6]. In this pilot study, we want to assess whether people (both elderly and not) are able to use a web-based exergame in an independent way, thus providing information about their autonomy in using ICT applications and the knowledge for developing games with specific medical targets. Indeed, few works are aimed to assess on how ICT application are used by fragile people [4]. Web-based application have several advantages with respect standard applications, in particular they do not need installation and configuration, but they are directly accessible trough the web. Moreover, web-based games are easier to update than traditional applications for specific devices.

2 Materials and Methods

2.1 The Virtual Supermarket Environment

As a followup of an already assessed exergame, we have implemented a web-based version of a Virtual Supermarket [2,11,12]. The exergame mimics the steps necessary to buy items at a supermarket. The user receives a list of items he/she should buy, and he/she has to find them in three shelves. Once the correct item has been found, it could be added to the basket by clicking on it. At any time, the user can see the content of the basket, possibly removing undesired items. The last step is to pay the total due amount, by choosing the right banknotes and coins.

In the previous works, we have assessed several interaction techniques, to find the most suitable one for elderly people, and we compared the quantitative results obtained with the exergame (e.g. total time to complete the task) with the scores provided by some state-of-the-art paper and pencil tests for MCI evaluation. In all the previous experiments, an experimenter, expert in the field, was always present to help subjects with practical issues. In this work, we are focusing on the web-based interface and on the possibility of collecting data without the physical presence of an experimenter.

2.2 The Software Framework

The Virtual Supermarket exergame is developed in Unity 3D (2020.1.6f1), by exploiting WebGL (Web-based Graphics Library) build. WebGL[1] is a JavaScript API for rendering high-performance interactive 3D and 2D graphics within any compatible web browser without the use of plug-ins. WebGL does so by introducing an API that closely conforms to OpenGL ES 2.0 that can be used in HTML5 <canvas> elements. This conformance makes it possible for the API to take advantage of hardware graphics acceleration provided by the user's device. Unity 3D can directly publish a WebGL build, i.e. the HTML5/Javascript program. Unity WebGL content is supported in the current versions of most major browsers on the desktop. Mobile devices are not supported by Unity WebGL, thus our analysis has been performed on PC desktops, only, despite the results discussed in [12], showing that some people would prefer a touch based interaction.

The application runs on the client side, thus we expect differences in the graphic quality of the exergame. In order to collect and analyze the recorded data (time to complete the task and errors in the number or kind of items put in the basket), we have implemented a method to send such data to the server. In WebGL applications, due to security implications, JavaScript code does not have direct access to IP Sockets to implement network connectivity. As a result, the .NET networking classes (i.e., everything in the System.Net namespace, particularly System.Net.Sockets) are non-functional in WebGL. For

---

[1] https://www.khronos.org/webgl/.

this reason, we used the UnityWebRequest systems, which allows us to post a message to the HTTP server.

2.3 Collected Data

The Virtual Supermarket exergame records and stores the following data: the ID associated to the user, the list of items provided to the user, the final list of items put in the basket by the user, the number of changes in the basket (added and removed items), the total amount to be paid, and the amount actually paid.

In addition, we store the answers to a questionnaire, specifically designed for the experiment presented in this paper, whose required information are:

- Age, Gender, Qualification, Current occupation
- Familiarity with gaming (10-point Likert-scale)
- Familiarity with shopping (10-point Likert-scale)
- Satisfaction with the Virtual Supermarket game (10-point Likert-scale)
- Easiness of interaction (10-point Likert-scale)
- Difficulties in finding items on the shelves
- Free comments.

2.4 Experimental Procedure

We have set up a web page with all the steps necessary for participating to the experiment. The procedure is shown in Fig. 1. First, the user is instructed about the purposes of the study and about the rules of the VR game, then he/she is asked to fill the privacy consensus, accordingly to the General Data Protection Regulation (GDPR). A random number is shown to the user, then he/she is asked to open the web page of the game. The given number must be inserted in a form, then the game begins, and proceeds like previously explained. At the end of the game, recorded data (see Subsect. 2.3) are sent to the server and the evaluation questionnaire is shown to the user. The experiment lasts about 20 min (depending on the time to fill the basket, since there are no timing constraints).

2.5 Data Analysis

For the purposes of this work, we analyze and discuss the following measurements:

- Total time to complete the shopping and the payment task. In a previous work [11] we discussed the relationships among these values and the scores of the questionnaires used to evaluate the cognitive impairment. Nevertheless, such values are also related to the technical skills of the users and with his/her familiarity with the use of PCs. In our case, all the participants are not affected by MCI, thus we expect that only the technical skills affect the measured values.
- Satisfaction with the Virtual Supermarket game and easiness of interaction, with respect to age, familiarity with games and with shopping in real supermarkets.
- Qualitative discussion of the free comments.

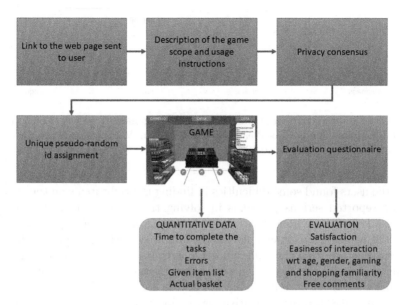

Fig. 1. The experimental procedure (blue boxes) and the collected data (orange boxes). (Color figure online)

3 Results

Participants. 19 people participated to the experiment (8 males, 11 females, mean age 44.3 ± 18.6) They were recruited through messages sent to friends. All the instructions were written in a dedicated web page, no other hints were provided to the participants. 5 more potential participants were not able to complete the experiment, due to technical problem (e.g., poor connection or not sufficient skills to understand the procedure).

Total Time. The mean time to complete the shopping task (i.e. to localize and put all the items in the basket) is $4\,m\,1\,s \pm 2\,m\,22\,s$; the mean time to complete the payment task is $22\,s \pm 7\,s$.

Task Errors. Almost all the users did no errors in the shopping task. By analyzing the provided list and the list of items actually out in the basket is clear that errors are due to misunderstanding in the visual representation of the items. This may be caused by a poor quality of the textures for certain devices, or by a misleading choice of elements (e.g. the difference between olive oil and white wine which have similar shapes and colors). All the users did no errors in the payment task. We expected this results, since the users do not have any cognitive impairment, and the game implements an easy task for cognitive healthy people.

Satisfaction and Easiness. On a Likert-scale from 1 to 10, the mean score for the satisfaction is 8.0 ± 2.4, the mean score for the easiness of interaction is 7.3 ± 2.5.

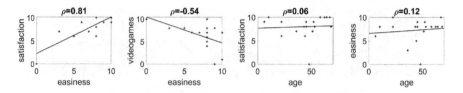

Fig. 2. Correlation among age, familiarity with videogames, perceived satisfaction in using the exergame and perceived easiness of interaction. Values on axes are the scores given on a Likert-scale from 1 to 10.

53% of the users found some difficulties in finding the right items on the shelves, 1 person reported serious problems in solving the shopping task. 89% of the users found the instructions clear. To understand whether there are correlation among the age of the participants, their familiarity with videogames, the perceived satisfaction from the game and the perceived easiness of interaction, we plot the scores and the associated correlation (Pearson) values. Figure 2 shows that a positive strong correlation exists between easiness of interaction and satisfaction ($\rho = 0.81$), a mild negative correlation is present between familiarity with videogames and easiness of interaction, probably due to higher expectations from gamers, and no correlation has been found with ages. Finally, no strong correlation has been found between the score given to the easiness of interaction and the total time necessary to complete the task ($\rho = 0.17$).

Free Comments. In the free text comments, people described the encountered difficulties, which are the following: difficulties in detecting the correct item due to poor quality of the textures (this is also related to the resolution of the display), or to the small dimension of the objects. Most of the problems are HCI issues, such as wrong double clicks on objects, or bugs in the visualization due to the behaviour of WebGL with non-conventional display resolutions. Sometimes, delays caused the insertion of multiple copies of the same item, the cause of this is under investigation, and it might be due to network problem as well to slow PCs. Some comments suggested improvements in the layout of the supermarket, asking for a more realistic layout to improve navigation among shelves.

4   Discussion and Conclusions

With respect to the paper aims, the results of this pilot study allow us to draw the following conclusions:

(i) The pipeline for collecting data without the assistance of the experimenter works, but about 20% of the recruited participants encountered problems, which did not allow them to complete the task. This could be due to the chosen platform (WebGL which does not have full supports for some devices, e.g. mobile ones) or to misleading and not precise written instructions. This aspect should be carefully addressed in order to effectively use the web-based instruments in hospitals, retire homes or directly at elderly houses.

(ii) The Unity WebGL build allowed us to distribute a previously developed game with few modifications. The game worked properly in most cases, nevertheless some participants had technically problems, which hampered the correct fruition of the game. Most of the problems were due to display resolution settings, or out-of-date browsers.

(iii) The usability of the Virtual Supermarket was already assessed in [2,12]. Here, we focused on the usability by considering the WebGL build. One of the main issues was the poor quality of the textures, this may be due to the rendering limits of WebGL applications. Items could be dynamically enlarged, thus allowing a better visualization, but the cost of adding a new feature, in terms of usability, must be evaluated.

(iv) The collected data and feedback allow us to devise the future improvements of the Virtual Supermarket, e.g. to re-arrange shelves to facilitate the research of items or to enlarge items.

As a final remark, data were collected in a time span of 24 h, thus demonstrating the possibility of reaching a large number of participants in an easy way. This would have interesting and useful implication for all the research activities involving human participants during the COVID-19 pandemic.

Acknowledgments. We wish to thank Luca Spallarossa for the helpful comments and the support in implementing the networking infrastructure.

This work has been partially supported by the Interreg Alcotra projects PRO-SOL We-Pro (n. 4298) and CLIP E-Santé (n. 4793).

References

1. Bottiroli, S., et al.: Smart aging platform for evaluating cognitive functions in aging: A comparison with the MoCA in a normal population. Front. Aging Neurosci. **9**, 379 (2017)
2. Chessa, M., Bassano, C., Gusai, E., Martis, A.E., Solari, F.: Human-computer interaction approaches for the assessment and the practice of the cognitive capabilities of elderly people. In: Leal-Taixé, L., Roth, S. (eds.) ECCV 2018. LNCS, vol. 11134, pp. 66–81. Springer, Cham (2019). https://doi.org/10.1007/978-3-030-11024-6_5
3. Fasilis, T., et al.: A pilot study and brief overview of rehabilitation via virtual environment in patients suffering from dementia. Psychiatrike = Psychiatriki **29**(1), 42–51 (2018)
4. Gallucci, A., et al.: ICT technologies as new promising tools for the managing of frailty: a systematic review. Aging Clin. Exper. Res. 1–12 (2020)
5. Gamito, P., et al.: Cognitive stimulation of elderly individuals with instrumental virtual reality-based activities of daily life: pre-post treatment study. Cyberpsychol. Behav. Soc. Network. **22**(1), 69–75 (2019)
6. Gauthier, S., et al.: Mild cognitive impairment. Lancet **367**(9518), 1262–1270 (2006)
7. Grossi, G., Lanzarotti, R., Napoletano, P., Noceti, N., Odone, F.: Positive technology for elderly well-being: a review. Pattern Recogn. Lett. **137**, 61–70 (2020)

8. Gu, D., Li, T., Wang, X., Yang, X., Yu, Z.: Visualizing the intellectual structure and evolution of electronic health and telemedicine research. Int. J. Med. Inform. **130**, 103947 (2019)

9. Van den Heede, K., Bouckaert, N., Van de Voorde, C.: The impact of an ageing population on the required hospital capacity: results from forecast analysis on administrative data. Eur. Geriatric Med. **10**(5), 697–705 (2019). https://doi.org/10.1007/s41999-019-00219-8

10. Malwade, S., et al.: Mobile and wearable technologies in healthcare for the ageing population. Comput. Methods Programs Biomed. **161**, 233–237 (2018)

11. Martini, C., et al.: Visual computing methods for assessing the well-being of older people. In: International Joint Conference on Computer Vision, Imaging and Computer Graphics, pp. 195–211 (2018)

12. Martis, A.E., Bassano, C., Solari, F., Chessa, M.: Going to a virtual supermarket: comparison of different techniques for interacting in a serious game for the assessment of the cognitive status. In: International Conference on Image Analysis and Processing, pp. 281–289 (2017)

13. Organization, W.H.: World report on ageing and health. World Health Organization (2015)

14. Palumbo, V., Paternò, F.: Serious games to cognitively stimulate older adults: a systematic literature review. In: Proceedings of the 13th ACM International Conference on Pervasive Technologies Related to Assistive Environments, pp. 1–10 (2020)

15. Rechel, B., et al.: Ageing in the European union. Lancet **381**(9874), 1312–1322 (2013)

16. Roberts, A.R., De Schutter, B., Franks, K., Radina, M.E.: Older adults' experiences with audiovisual virtual reality: perceived usefulness and other factors influencing technology acceptance. Clin. Gerontol. **42**(1), 27–33 (2019)

17. Volkmann, T., Akyildiz, D., Knickrehm, N., Vorholt, F., Jochems, N.: Active participation of older adults in the development of stimulus material in an storytelling context. In: Gao, Q., Zhou, J. (eds.) Human Aspects of IT for the Aged Population. Design and User Experience, Technologies, pp. 84–95 (2020)

Daily Living Activity Recognition Using Wearable Devices: A Features-Rich Dataset and a Novel Approach

Maurizio Leotta$^{(\boxtimes)}$, Andrea Fasciglione , and Alessandro Verri

Dipartimento di Informatica, Bioingegneria, Robotica e Ingegneria dei Sistemi
(DIBRIS), Università di Genova, Genova, Italy
`maurizio.leotta@unige.it`

Abstract. Automated daily living activity recognition is a relevant task since it allows to assess the health status of a subject both objectively and remotely. Having a reliable measure is important since it gives precise indications to doctors and researchers interested in evaluating the effectiveness of treatments or drugs (e.g., in the context of clinical studies). The possibility to perform this task remotely is more convenient for the patients and acquired increasing importance not only due to the current pandemic, but also because of the regularly growing population of elderly people that could benefit from remote monitoring.

In this paper, first, we describe a novel wearable-device-based dataset that contains data (1) of a high number of daily life activities, coming from a real-life scenario, (2) recorded by applying multiple devices on different parts of the body, and (3) recorded with medical-grade devices at a high sampling frequency. Then, second, we describe a machine learning-based method for activity recognition. Our approach takes in input a dataset and through multiple phases allows to recognise the activities performed by the subjects with a good degree of accuracy (up to 0.92 expressed as F1 score depending on the location).

Keywords: Activity recognition · Wearable devices · Actigraph

1 Introduction

Recently an increasing interest in finding reliable methods for monitoring patients suffering from different diseases (or simply elderly) emerged, in particular using remote and non-intrusive methods (e.g. [7,11,15]). Many diseases, in fact, strongly impact daily life activities because of their effects (e.g., Multiple Sclerosis, Pulmonary Arterial Hypertension, or Parkinson). A possible way to assess the physical condition of a patient is based on assessing how much time is spent on specific activities and how the amount of time dedicated on each activity changes over time (e.g., a subject can decide to stop vacuuming because of the feeling of fatigue, or could need more time to eat because of hand tremors).

© Springer Nature Switzerland AG 2021
A. Del Bimbo et al. (Eds.): ICPR 2020 Workshops, LNCS 12662, pp. 171–187, 2021.
https://doi.org/10.1007/978-3-030-68790-8_15

While working on a possible method to create an activity recognition classifier, starting from data recorded by wearable devices and in the context of a collaboration with a major pharmaceutical company (Janssen), we found the need of creating a novel dataset regarding daily life activities. Indeed, by analysing the state of the art concerning the public available datasets recorded with wearable devices, we noticed that in spite of the large interest on this topic, there is a lack of datasets having, at the same time, the following characteristics: (1) containing data of numerous and different daily life activities, (2) containing data recorded using high-quality sensors (both concerning frequency and accuracy), and (3) containing data from different synchronised devices positioned on different parts of the body. From the point of view of a researcher, this lack could become an obstacle to perform more in-depth investigations and to conceive more advanced approaches to the problem of the Activity Recognition using wearable devices.

For this reason, in this work we present, as *first contribution*, the ongoing effort of creating a novel dataset that meets the three characteristics mentioned above.

Concerning the first characteristic, our dataset includes 17 different daily life activities performed in real-life scenarios (e.g., eating, using laptop, handwriting, vacuuming, walking, going upstairs/downstairs). On the contrary, available datasets often include a few activities (often 8 or less) or activities that are limited to a particular context (e.g., cooking, breakfast morning-routine, or walking at different speeds only) [1,4,13,17].

Concerning the second characteristic, we recorded our dataset using professional devices produced by Actigraph[1]. These devices are medical-grade activity monitors that thanks to their characteristics and to their reliability, have been broadly considered in different studies and for different purposes (e.g. [6], [12]). More in detail, we have used two Actigraph GT9X Link, with a sampling frequency 100 Hz and one Actigraph Centrepoint, with a sampling frequency 256 Hz, all of them relying on high-quality internal sensors (e.g., accelerometers, gyroscopes, magnetometers). This is an important characteristic of our dataset since, very often, the currently available datasets contain data recorded with low cost (and precision) sensors like the ones included in Android smartphones, with a sampling frequency 50 Hz (or lower) and non-certified accuracy of the values provided [10,13,16]. The high frequency of these devices can be useful for researchers in order to perform analysis by re-sampling data to different frequencies.

Another strength of our dataset (third characteristic) lies in the fact that we have placed the three aforementioned devices, synchronised together, on three different parts of the participants' bodies: dominant wrist, right side of the hip, and right ankle (on the right side). Having synchronised data coming from different parts of the body, like in our dataset, would allow researchers to find methods based on the correlation of these data, and thus creating more accurate activity analysis or recognition approaches. On the contrary, existing datasets

---

[1] ActiGraph, LLC (Pensacola, FL, USA), https://actigraphcorp.com/.

are usually focused on only one part of the body (e.g. pockets or the hip) [10,20]. The fact of wearing more devices could be quite annoying for a subject in real settings: the trade-off between having more data and creating a discomfort for the patient should be carefully analysed.

As a *second contribution*, in this paper we describe a machine learning (ML) based method for the Activity Recognition. Our approach takes in input a dataset and thought multiple phases allows to recognise the activities performed by the subjects with a good degree of accuracy (up to 0.92 expressed as F1 score depending on the location).

This paper is organised as follows: Sect. 2 describes the dataset collection procedure and the structure of the obtained raw data. Section 3 briefly describes the proposed approach that we have used to evaluate the predictability of the activities using our dataset. Section 4 reports on the empirical evaluation of the approach, while Sect. 5 reports related works and Sect. 6 concludes the paper.

2   Dataset of Daily Living Activities Creation

The creation of the *Daily Living Activities* dataset has been performed in three main phases: (1) Data Recording, (2) Data Extraction, and (3) Data Labelling and Cleaning. The final output is a labelled dataset containing the raw data of 17 daily living activities ready to be used by researchers for a variety of possible studies. In the following of this section, we describe the three phases in detail.

2.1   Data Recording

To record our dataset, we followed a precise protocol that we defined and that had been reviewed and approved by the ethical committee of the Department in which the data recordings took place. Apart from the operational details on the procedures to follow during data recordings, our protocol includes also a step where each participant is asked to sign an informed consent: this allow us to share, with the research community, all the data recorded and the physical bio-metric characteristic of each participant. The dataset is currently available online on our department website[2] but we are working to share it on relevant dataset repositories such as the UC Irvine Machine Learning Repository[3] or on the Harvard Dataverse[4] so that it could also be indexed by specific search engines (e.g. Google Dataset Search).

Participant Inclusion/Exclusion Criteria. The former includes the following: (1) to be able to perform the request actions, (2) age of 18 and over, and (3) to understand the purpose of the study and willing to participate in the study. On the contrary, exclusion criteria include any planned surgery or procedures that would interfere with the conduct of the study and any major mobility difficulties.

---

[2] https://sepl.dibris.unige.it/2020-DailyActivityDataset.php.
[3] https://archive.ics.uci.edu/ml/index.php.
[4] https://dataverse.harvard.edu/.

Participants Characteristics. The preliminary version of our dataset currently includes data of 8 volunteers: males aged between 23–37, with a weight between 52–90 kg and height between 172–186 cm. Regarding the dominant hand, two subjects out of 8 were left-handed, while the other ones (6 out of 8) were right-handed. In the next months we expect to record the data of 25–30 additional participants. Detailed information (i.e. age, height, weight, dominant hand) for each subject are reported within the dataset itself.

Activity Recording Procedure. During the data recording, we asked to all the participants to perform the 17 different activities listed in Table 1. We split the list of activities into two different sets, as in the Table 1: Set A and Set B; the former have been performed for a fixed time, while the latter not. The differences between the two sets lie mainly in the fact that activities in Set B were constrained to a particular path or to a flight of stairs, while activities in Set A were quite stationary and did not require the subjects to move along a path. By depending on a fixed path, moreover, it is also possible to measure the different walking speed (e.g. in term of meters/second or of steps/second) of the subjects as an additional information. More in details, Set A activities have been performed for more than 120 s (in general for about 150 s), and we included in our dataset the central 120 s of each execution in order to obtain cleaner data. On the other hand, Set B includes: *Walking* performed for 160 m (in at least 110 s); *Walking Fast* performed for 205 m (in at least 110 s); and *Going Downstairs, Going Upstairs, Going Upstairs Fast* performed using a single flight of stairs with no intermediate floors between the steps for an average time of 40 s.

Table 1. List of activities performed

Set A	Set A	Set B
1. Relaxing on a chair	7. Brushing Teeth	13. Walking
2. Keyboard Typing	8. Sweeping	14. Walking Fast
3. Using the Laptop	9. Vacuuming	15. Going Downstairs
4. Handwriting	10. Eating (a soup)	16. Going Upstairs
5. Washing Hands	11. Dusting a surface	17. Going Upstairs Fast
6. Washing Face	12. Rubbing a surface	

The participants have been followed and instructed during the data recording: they have been told what activity should have been performed and some details on it, but it has not been imposed to move exactly in a particular way[5]. We have

---

[5] Regarding the possible variation of the subjects behaviour while performing activities by knowing they were participating in an experiment (also known as Hawthorne effect), it is important to note that (1) there have not been judgements on how well subjects were performing activities, therefore they could behave in any way they preferred, (2) due to the short amount of time spent in recording data, possible variations in subjects behaviour happened in the entire recording (i.e. no effect on possible train and test data sets).

shown a WHO video[6] on how to wash the hands and we have better explained the difference between *dusting* and *rubbing*: respectively, dust a surface, or rub to clean a really dirty surface. Even if the chosen activities are characterised by relatively standard movements, as expected, we noticed that different subjects had their own way to execute movements related to the activity. This is a positive characteristic since it can help to understand the natural differences that can occur when analysing and comparing different subjects. The only constraint established during the data recording has been to always use the dominant hand (i.e. the one where the device was placed) when performing those actions that mostly involve a single arm movements (e.g. handling the vacuum/broom with the dominant hand while vacuuming/sweeping). Otherwise, the signal recorded from the wrist would have had no information regarding the pattern of the dominant hand.

Devices and their Positioning. In the last years, Actigraph, a leading provider of wearable physical activity and sleep monitoring solutions for the global scientific community, proposed several actigraphy devices. In this work, we used two Actigraph GT9X Link and one Actigraph Centrepoint Insight Watch. They are activity monitors equipped with high precision and fast reading sensors. In detail, both are equipped with a three-axial accelerometer while the GT9X also includes a complete IMU (Inertial Measurement Unit[7]). The IMU is an electronic chip capable of capturing position and rotation data for advanced analyses. It contains a secondary accelerometer, a gyroscope, a magnetometer, and a temperature sensor. Additional information on the two devices could be found in the official web site[8]. Table 2 shows the kinds of sensors available in two devices and the corresponding measurement units.

Table 2. Characteristics of the devices sensors

Device	Sensor type	Units of Measure
Both	3 axis primary accelerometer	g
GT9X	3 axis secondary accelerometer	g
GT9X	3 axis gyroscope	degrees/s
GT9X	3 axis magnetometer	microTesla (μT)
GT9X	temperature sensor	Celsius

The three wearable devices were worn by the participants as follows and with the following settings:

---

[6] "WHO: How to handwash? With soap and water", https://www.youtube.com/watch?v=3PmVJQUCm4E.

[7] https://en.wikipedia.org/wiki/Inertial_measurement_unit.

[8] Actigraph GT9X Link - https://actigraphcorp.com/actigraph-link/, Actigraph Centrepoint Insight Watch - https://actigraphcorp.com/cpiw/.

- 1 Actigraph Centrepoint at the dominant wrist. Accelerometer recording at a sampling rate 256 Hz.
- 1 Actigraph GT9X Link at the right hip at the height of the iliac crest (using the device belt clip). IMU (i.e., accelerometer, magnetometer, and gyroscope) recording at a sampling rate 100 Hz.
- 1 Actigraph GT9X Link at the height of the right ankle placed, with the help of the belt clip, on the subject's right side of the shoe, over the malleolus. IMU recording at a sampling rate 100 Hz.

Regarding the calibration of the devices, they have been precisely calibrated (using the automated procedure of the device) at the beginning of each data recording session.

Ground Truth Definition. The ground truth annotation has been performed by the two first authors of this paper, in parallel, by following the subjects performing activities, using a chronometer took note of the starting and ending time of each activity. Moreover, while recording walking data researchers ensured that the subjects were following a specific walking path so that we could retrieve the average walking speed of the subjects for optional and additional tests.

2.2   Raw Data Extraction

After recording data with the subjects, we extracted the raw data from the devices using the proprietary software system developed for Actigraph devices. Then we exported the data as .csv files. The two kinds of devices that we used were equipped with different sets of sensors, so the output of each kind of device will be different. The .csv produced for the Actigraph GT9X Link, will contain 11 columns:

- *'Timestamp'*: timestamp of the sampled values
- *'Accelerometer X, 'Accelerometer Y', 'Accelerometer Z'*: instantaneous accelerations for each axis, measured in units of gravity (G)
- *'Temperature'*: IMU temperature, in Celsius degree
- *'Gyroscope X', 'Gyroscope Y', 'Gyroscope Z'*: instantaneous measure of the gyroscope for each axis, measured in degrees/sec
- *'Magnetometer X', 'Magnetometer Y', 'Magnetometer Z'*: instantaneous measured magnetic field for each axis, measured in microTesla (mT)

For each row of the file, it is possible to find the sampled value at the specified timestamp from each of the sensors and axis. The .csv file produced using data recorded with the Actigraph Centrepoint, instead, will only have these columns: *'Timestamp', 'Accelerometer X', 'Accelerometer Y', 'Accelerometer Z'*.

2.3   Data Labelling and Cleaning

Thanks to the ground truth, we were able to label the data precisely. Labels were associated with each row of the recorded data indicating which activity is carried out in such instant. Basically, a new column has been attached to data, where for each row we had a number corresponding to the activity performed in that instant (e.g., 1 is *Relaxing*, 2 is *Keyboard Typing*, ...).

During this data processing step, we also used a label to identify data that had to be removed because it was not useful or that could lead to misleading results (e.g., data recorded in between two different actions).

3   A Possible Approach to Daily Living Activity Recognition

In this section, we describe in detail the steps composing our approach aimed at recognising the activities performed by a subject wearing an actigraphy device. The approach is based on the usage of Support Vector Machine (SVM). More in detail, starting from the labelled **raw data** (e.g., from our dataset), the first activity consists in a (1) *features extraction* phase, after which data will be split into **training** and **test data**. Training data will be used for (2) *tuning the hyperparameter* and (3) *training the SVM model* with the correct parameters. At this point, the SVM model can be used for (4) *recognising daily living activities* on novel unseen data. Thus we use **test data** to evaluate the accuracy of the trained SVM model and, in general, of our approach. In the following subsections, we will describe in detail the first three steps (1, 2, 3) while the fourth will be described in Sect. 4. Our approach has been implemented using Python and with the help of the Jupyter platform[9]; we relied on the *Scikit-learn* library[10], also known as *sklearn*, since it provides several instruments for data analysis that were useful in our study.

3.1   Features Extraction

As done in other similar studies like the one of Staudenmayer et al. [14], we have extracted the feature set made up of feature vectors and associated labels. To do so, we have used the *sliding window* approach to compute the features, using only the accelerometer data (however the approach can be simply extended to include gyroscope and magnetometer data). In this phase, a sliding window passes over the data and for each axis (X, Y, Z) we extract some measures on the data contained in the window: mean, variance, standard deviation, median absolute deviation, percentiles (10Th, 25Th, 75Th, 90Th). Having eight measures per axis allows to compute 24 features for each window that composed the feature set used in the evaluation, described in Sect. 4.

Data at the end of each activity recording is discarded when not enough for building a window (i.e., the remaining data covers less time than the length of the sliding window).

Regarding the sliding window, its length represents an important parameter on which results could potentially highly depend. For this reason, in previous experiments, we have performed some analysis to understand how the length of

---

[9] https://jupyter.org/.
[10] https://scikit-learn.org/.

the window and the overlap between subsequent windows could affect accuracy. After these tests, we have decided to use windows that were 2.0 s long, with 95% of overlap each other. This value is motivated by the fact that typical human periodic movements have a period of no more than two seconds (e.g., each step during walking or hand movement during toothbrushing).

After the features extraction step, data is ready to be used in any ML algorithm. For our approach, we have chosen to rely on Support Vector Machine (SVM). SVM, in fact, has been already used to estimate physical activity from accelerometers in the literature, showing good performances in this kind of task (e.g., [5,21]). When using SVM data need to be *standardised* in order to obtain better results. This is needed since SVM is based on the idea of finding the hyperplane that best divides different classes by maximising the distance between the hyperplane and the data (i.e. Support Vectors), if one feature (i.e. one dimension) has larger values than the others, it will prevail on the others when computing distances. This will not be a problem if we standardise data: we did so by removing the mean and scaling to unit variance. Finally, to further prepare our data to feed the algorithm, we have also split our data into *training data* and *test data*: 75% and 25% of data of each activity, respectively.

3.2   Hyperparameter Tuning

SVM needs some parameters to be tuned in order to achieve the best result: C and *gamma*, in combination with the different used *kernels* (*Radial Basis Function* and *Polynomial* kernels). Focusing on the hyperparameter tuning, we know that while constructing a machine learning model, a general goal is to choose parameters such that we obtain a model that is able to learn, in the best way, all information from the training data, while, at the same time, it should be able to generalise well to new data. This problem of balancing these properties is known as the *Bias Variance Trade-off* problem [18]. One possible way to find the best model is to use the Cross Validation method [3].

Cross-Validation is a frequently used procedure for evaluating a model. The basic idea is that training data are divided into complementary subsets; one subset is used to train the model and we validate the results using the other subset. To do so, we have decided to use the Grid Search method [2] for choosing the best parameters for the algorithm. For each parameter of the algorithm, a list of possible values is given in input to Grid Search. Basically, each combination of the selected values generates a model that is then evaluated. The output of Grid Search is then the list of chosen parameters that performed the best.

3.3   Training Model to Predict Data

After computing the best parameters for the Support Vector Machine model, the next step has been to train the model with the training data and using the previously found parameters, in order to conclude the process. Once the model has been created, it was ready to be fed with new unseen data in order to output its predictions. As explained before we have split, at the beginning, our whole processed data in training data and test data; the latter have been used for this last step to evaluate the accuracy of the created model. In Sect. 4 we will analyse the obtained results considering, separately, data of each body location.

4   Empirical Evaluation of the Approach

As a case study for showing one of the possible results achievable with our dataset we report in this section the evaluation of our approach using the data we collected. The research question we investigated is the following:

RQ: What is the accuracy of the proposed approach in classifying the activities performed by a subject?

Note that in this preliminary study, we independently consider the three devices/body locations. Moreover, we analyse only the performances of a person-dependent model where the training is on a subject and the test on the same subject.

4.1   Procedure

To answer our research question, first, we computed three confusion matrices for each subject in our dataset (one matrix for each of the three devices employed). More in detail, the values in each confusion matrix refer to the percentage of processed data of a specific class C_a that have been predicted to belong to the class C_b. More precisely, let us assume that we are reading the confusion matrix starting from the first row, representing the class C_a: each value we see in this row represents the percentage of data belonging to C_a that has been labelled as belonging to the class of the corresponding column. A flawless result would be represented as a matrix in which all the values on the diagonal are 100.0%, and the other values are 0.0% meaning that all the unseen data have been classified with the correct corresponding label.

Second, we averaged the eight confusion matrices (one per subjects) creating a single confusion matrix for each considered body location. In this way, we can answer our RQ by providing the average for each activity considered in our dataset.

An important aspect to consider when judging the quality of the results in the confusion matrices is that a baseline model that would randomly recognise the activity could have an accuracy equal to the probability of assigning the correct label that is: $\frac{1}{\#(classes)} = \frac{1}{17} = 5.9\%$.

4.2   Results

In Fig. 1 it is possible to see the confusion matrix obtained using **wrist** data, averaged over all eight subjects' results. The overall mean F1 score obtained is 0.92 ± 0.03 (mean $\pm$ standard deviation). In general, we can say that the recognition of most of the activities achieved good results (values on the diagonal of the matrix are always greater than 0.74). In this case, the most noticeable outliers are in the wrong classifications of *keyboard typing* in *using laptop*, that are indeed very similar activities. The same analysis can be valid for the classifications of *sweeping* and *vacuuming*. The lower accuracy values are obtained in most of those activities that mostly involve legs movements (*walking, going downstairs/upstairs*): those are indeed quite similar activities when "observed" from the wrist.

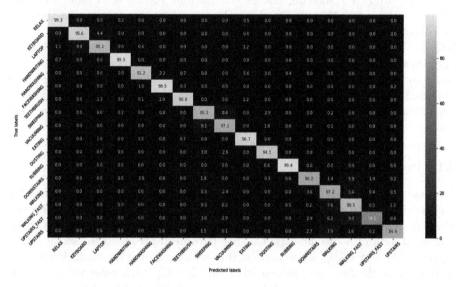

Fig. 1. Average confusion matrix obtained with Wrist data

In Fig. 2 and Fig. 3 we present the confusion matrices obtained using, respectively, **hip** and **ankle** data, averaged over all eight subjects results, for which it is possible to perform similar considerations. The mean F1 score obtained with *hip* data is 0.81 ± 0.04, while the mean F1 score obtained with *ankle* data is 0.75 ± 0.06.

Analysing these confusion matrices (Figs. 2 and 3, we noticed both expected and unexpected results. In fact, as expected, since many performed activities mostly involve peculiar movements of the arms (e.g. *brushing teeth, washing hands/face, sweeping*), results obtained using hip and ankle data have a lower mean accuracy than the results obtained using wrist data. For the same reason, we were expecting to obtain low accuracy for the activities performed while sitting or while not walking (e.g. *using laptop, relaxing, handwriting*) since the hip and ankles are not involved in any movements. On the contrary, we achieved quite high accuracies.

We further analysed our data in order to explain these results. By plotting the accelerometer data, we noticed that there was a perceptible difference in the values between such different activities even in the ankle and hip data. We interpreted this as the fact that subjects, during data recording, unintentionally

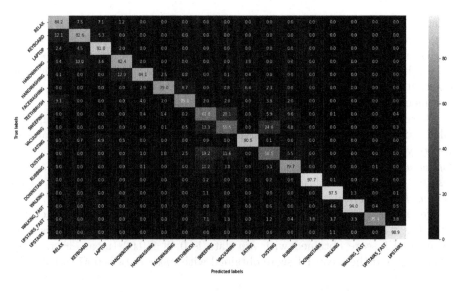

Fig. 2. Average Confusion Matrix obtained with Hip Data

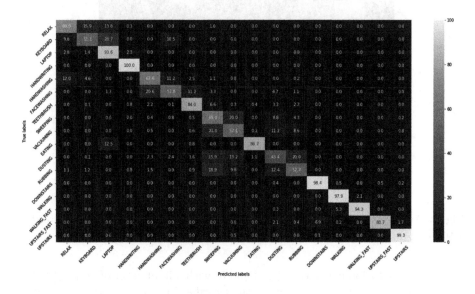

Fig. 3. Average Confusion Matrix obtained with Ankle Data

changed the orientation of the devices (e.g. by slightly moving a leg while sitting). These involuntary movements were leading to noticeable changes in the accelerometer values because of the variation in the orientation with respect to the earth gravity g. We concluded that in some cases, the right classification of activities happens not because of the peculiar characteristics of the activity movements, but because of the particular orientation of the device.

Therefore, in order to avoid this problem, when dealing with both hip and ankle data, we considered only activities that actively involve those parts of the body. In particular, we selected: *relaxing* (as a stationary activity), *sweeping, vacuuming, dusting, rubbing, going downstairs, walking, walking fast, going upstairs, going upstairs fast* and excluded *keyboard typing, using laptop, handwriting, hands washing, face washing, teeth brushing, eating.*

We show the confusion matrices obtained with the latest reduced activity set in Fig. 4 (regarding hip data) and Fig. 5 (regarding ankle data). In this case, the mean F1 score obtained with *hip* data is 0.48 ± 0.02 (mean $\pm$ standard deviation), and the mean F1 score obtained with *ankle* data is 0.47 ± 0.03.

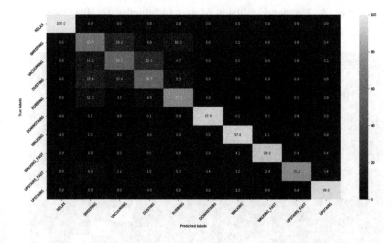

Fig. 4. Average Confusion Matrix obtained with Hip Data, limited on hip-related activities

In both confusion matrices (Figs. 4 and 5) with a limited set of activities, it is clearly evident the scarce accuracy of the classifier in discerning from *sweeping, vacuuming, dusting* and *rubbing*. Indeed, all of the four listed activities have been performed by doing small and slow steps around the room when recording data. Regarding the overall mean F1 scores it is clearly a consequence of the wrong classification of the four activities previously listed. On the same topic, we should also consider that by having fewer activities to be recognised, any wrong activity classification will have a significant impact on the F1 score.

On the other hand, the classifier is able to recognise with a quite high accuracy (always over 78%) all the other activities and, in particular, the ones that

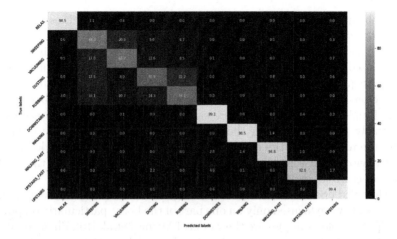

Fig. 5. Average Confusion Matrix obtained with Ankle Data, limited on ankle-related activities

required the subjects to walk and use stairs (*walking, going downstairs, upstairs, upstairs fast*), in which both ankle and hip are more involved.

5   Related Works

In this section, we will briefly analyse related works, starting with the publicly available datasets on activities recorded with wearable devices and then presenting some approaches to activity recognition.

Datasets. As briefly explained in Sect. 1, when looking for a publicly available dataset, we have focused our analysis on three main criteria: (1) number and kind of recorded activities, (2) reliability of the recorded data according to the used device, and (3) which and how many parts of the body have been interested during data recording. To the best of our knowledge, a dataset satisfying the three aforementioned criteria is not currently available and this motivated our proposal.

About the first criterion, it is possible to find datasets focused on specific contexts of daily life: De la Torre et al. [17] presented a dataset on cooking activities, while Chavarriaga et al. [4] proposed a dataset on activities performed while preparing breakfast. On the other hand, it is also possible to find datasets related to a wider list of activities. Possible examples are the work of Anguita et al. [1], including more generic activities like *sitting, standing, walking, walking upstairs/downstairs* or the work of Leutheuser et al. [8], including activities from a daily life scenario (e.g., *walking, vacuuming, washing dishes, lying, sitting*). We have noticed that many available datasets include quite similar activities such as *walking* but at different speeds, or in different directions, *sitting, standing* or *lying*. Micucci et al. [10], in fact, with their brief literature review, have found out that the most frequent activities included in daily life activities dataset are: *walking, standing* and *walking downstairs/upstairs.*

Regarding the second criterion, a large number of the datasets that we have analysed used data recorded with an Android smartphone, with a requested sampling frequency 50 Hz (e.g. [1,13,16]). Nevertheless, according to the work of Micucci et al., Android OS does not guarantee the consistency between the requested and the effective frequency sampling rate, therefore, the acquisition rate actually fluctuates during the acquisition [10]. This fact reduces, in our opinion, the reliability of the recorded data. On the contrary, some datasets use efficient devices with a high sampling frequency rate (>100 Hz) as the work of Leutheuser et al. [8] or the work of Zhang et al. [20].

On the third criterion, during our investigation on existing datasets, we have seen that some were focused only on one part of the body, particularly on pockets or the hip. Possible examples are the works of Micucci et al. [10], Zhang et al. [20] or Anguita et al. [1]. On the other hand, other available datasets include data of multiple sensors on different parts of our body, usually including waist, wrist, hip and ankle data. This is the case of the works of Sztyler et al. [16], Shoaib et al. [13] or Leutheuser et al. [8]. In our opinion, having data retrieved from different parts of our body would allow to achieve higher accuracy in activity recognition purposes.

Approaches for Activity Recognition. Different approaches for classifying daily-life activities using Machine Learning algorithms have been proposed in the last years. Here we will consider three works that have similar scenarios to ours. Indeed, all of the considered methods deal with a triaxial accelerometer worn on the wrist by participants of the experiments while performing some activities. All the devices used in the considered experiments recorded accelerations at a frequency of 80–100 Hz.

The work of Zhang et al. [21] tried to classify 4 main categories of activities: *sedentary* (lying, standing, PC working), *household* (window washing, sweeping, etc.), *walking* and *running* at different speeds. Mannini et al. [9] tried to recognise as well 4 categories of activities: *ambulation, cycling, sedentary and other.* Yang et al. [19] has categories of activities more similar to our scenario: *walking, running, scrubbing, standing, working at a computer, vacuuming, brushing teeth* and *sitting.* For what concerns the features, the ones used in all the experiments are based on three different aspects: (1) time (mean, standard deviation, mean absolute deviation, etc. of acceleration over time); (2) frequency spectrum (first dominant frequencies and their power in some particular ranges - e.g. [0.6, 2.5] Hz) and, (3) wavelet, based on the Discrete Wavelet Transform, therefore obtaining features linking both the frequency and the time domain.

Regarding the windows length, in the aforementioned works this parameter was varying from 2.0 and 12.8 s, or, in term of number of samples, from 100 samples to 1152 (but taken at different frequencies). Those studies that compared the performances over the same data but using different length for the windows (e.g. [9]) have shown that longer windows would have meant higher performances, but also that 4.0 s windows were sufficient to obtain acceptable results.

Regarding the used ML algorithms, Zhang et al. [21] tested performances using different algorithms (Decision Trees, Naive Bayes, Linear Regression, Support Vector Machine, Neural Networks) showing that all the algorithms had good performances (>95.0%), with the DT and SVM being the ones with better results. Experiments in [9] used Support Vector Machine only, while in [19] a "neuro-fuzzy" classifier (classifiable as a Neural Network) has been used. All of the analysed algorithms in the documents were obtaining good and almost similar results, despite of the used algorithms (overall accuracy always over 86%).

The major differences w.r.t. our approach regard: (1) the windows length (we adopted shorter windows of 2.0 s) and, (2) the number of activities to be classified (higher in our case).

6 Conclusions and Future Work

In this paper, we have presented the current progress concerning the creation of a daily life activities dataset recorded while wearing multiple medical-grade wearable devices. With the help of the proposed approach for activity recognition, we have shown an example of prospective results that researchers could obtain using our dataset.

While being still incomplete (since we are working on recording the data of additional subjects), our dataset has interesting characteristics that could help researchers to perform deeper studies on the field of activity recognition. Differently from already available datasets, ours contains data: (1) of numerous daily life activities, (2) recorded using professional devices, and (3) of three synchronised devices on wrist, hip and ankle. Thanks to these characteristics, our dataset could help researchers to perform various kind of studies such as (1) work on subject-independent models, (2) find possible correlations between data of different sensors and different parts of the body while performing activities.

We have also described a preliminary approach to activity recognition and evaluated it using data from our dataset. By considering independently each device (i.e., body location), we have evaluated the predictions of a person-dependent model. From our results we have seen that it is possible to train a person-dependent model able to recognise the performed activities precisely, since using *wrist* data we achieved an average overall accuracy of 0.92 expressed as F1 score.

As future work, we are currently making progress on both the dataset and the proposed approach. Regarding the dataset, in the next months we plan to add data of 25–30 new subjects, in order to involve a more heterogeneous population (e.g., including female subjects or subjects with different ages). Concerning the proposed approach, we plan to evaluate and compare the accuracy of other classifiers (e.g. Random Forest, Neural Networks) or the influence derived from using different parameters for windows length and overlap. Additionally, we are studying the accuracy of a *subject independent* model, that would be able to recognise activities performed by a new unseen person with no need of subject-related training data. We will then compare the accuracy of our proposed approach with other already existing, by using our dataset as a benchmark. Moreover, in our future work we plan to include in our dataset also subjects suffering from physical impairments due to various diseases (e.g. Pulmonary Arterial Hypertension, Multiple Sclerosis, Parkinson's disease) or the advanced age. This could also help researchers in better understanding and measuring the impact of such conditions on the daily life.

Acknowledgement. This work is an activity of the Software Engineering for Healthcare Lab, a joint laboratory between the University of Genova and Janssen Pharmaceuticals (Johnson & Johnson group). We want to show our gratitude to Massimo Raineri e Giacomo Ricca of Janssen Pharmaceuticals for the support provided.

Disclaimer. Responsibility for any information and result reported in this paper lies entirely with the Authors. Janssen Pharmaceuticals was not involved in any form of data collection or analysis.

References

1. Anguita, D., Ghio, A., Oneto, L., Parra, X., Reyes-Ortiz, J.L.: A public domain dataset for human activity recognition using smartphones. In: ESANN, vol. 3, p. 3 (2013)
2. Bergstra, J., Bengio, Y.: Random search for hyper-parameter optimization. J. Mach. Learn. Res. **13**(1), 281–305 (2012)
3. Cawley, G.C., Talbot, N.L.: On over-fitting in model selection and subsequent selection bias in performance evaluation. J. Mach. Learn. Res. **11**, 2079–2107 (2010)
4. Chavarriaga, R., et al.: The opportunity challenge: a benchmark database for on-body sensor-based activity recognition. Pattern Recogn. Lett. **34**(15), 2033–2042 (2013)
5. He, Z., Jin, L.: Activity recognition from acceleration data based on discrete consine transform and SVM. In: 2009 IEEE International Conference on Systems, Man and Cybernetics, pp. 5041–5044. IEEE (2009)
6. John, D., Freedson, P.: Actigraph and Actical physical activity monitors: a peek under the hood. Med. Sci. Sports Exerc. **44**(1 Suppl 1), S86 (2012)
7. Kantoch, E., Augustyniak, P., Markiewicz, M., Prusak, D.: Monitoring activities of daily living based on wearable wireless body sensor network. In: 2014 36th Annual International Conference of the IEEE Engineering in Medicine and Biology Society, pp. 586–589. IEEE (2014)
8. Leutheuser, H., Schuldhaus, D., Eskofier, B.M.: Hierarchical, multi-sensor based classification of daily life activities: comparison with state-of-the-art algorithms using a benchmark dataset. PLoS ONE **8**(10), e75196 (2013)
9. Mannini, A., Intille, S.S., Rosenberger, M., Sabatini, A.M., Haskell, W.: Activity recognition using a single accelerometer placed at the wrist or ankle. Med. Sci. Sports Exerc. **45**(11), 2193 (2013)
10. Micucci, D., Mobilio, M., Napoletano, P.: Unimib shar: a dataset for human activity recognition using acceleration data from smartphones. Appl. Sci. **7**(10), 1101 (2017)
11. Pitta, F., Troosters, T., Probst, V., Spruit, M., Decramer, M., Gosselink, R.: Quantifying physical activity in daily life with questionnaires and motion sensors in copd. Eur. Respir. J. **27**(5), 1040–1055 (2006)

12. Robusto, K.M., Trost, S.G.: Comparison of three generations of actigraph^{TM} activity monitors in children and adolescents. J. Sports Sci. **30**(13), 1429–1435 (2012)
13. Shoaib, M., Scholten, H., Havinga, P.J.: Towards physical activity recognition using smartphone sensors. In: 2013 IEEE 10th International Conference on Ubiquitous Intelligence and Computing and 2013 IEEE 10th International Conference on Autonomic and Trusted Computing, pp. 80–87. IEEE (2013)
14. Staudenmayer, J., He, S., Hickey, A., Sasaki, J., Freedson, P.: Methods to estimate aspects of physical activity and sedentary behavior from high-frequency wrist accelerometer measurements. J. Appl. Physiol. **119**(4), 396–403 (2015)
15. Stikic, M., Larlus, D., Ebert, S., Schiele, B.: Weakly supervised recognition of daily life activities with wearable sensors. IEEE Trans. Pattern Anal. Mach. Intell. **33**(12), 2521–2537 (2011)
16. Sztyler, T., Stuckenschmidt, H.: On-body localization of wearable devices: an investigation of position-aware activity recognition. In: 2016 IEEE International Conference on Pervasive Computing and Communications (PerCom), pp. 1–9. IEEE (2016)
17. De la Torre, F., et al.: Guide to the carnegie mellon university multimodal activity (cmu-mmac) database (2009)
18. Von Luxburg, U., Schölkopf, B.: Statistical learning theory: models, concepts, and results. In: Handbook of the History of Logic, vol. 10, pp. 651–706. Elsevier (2011)
19. Yang, J.-Y., Chen, Y.-P., Lee, G.-Y., Liou, S.-N., Wang, J.-S.: Activity recognition using one triaxial accelerometer: a neuro-fuzzy classifier with feature reduction. In: Ma, L., Rauterberg, M., Nakatsu, R. (eds.) ICEC 2007. LNCS, vol. 4740, pp. 395–400. Springer, Heidelberg (2007). https://doi.org/10.1007/978-3-540-74873-1_47
20. Zhang, M., Sawchuk, A.A.: USC-had: a daily activity dataset for ubiquitous activity recognition using wearable sensors. In: Proceedings of the 2012 ACM Conference on Ubiquitous Computing, pp. 1036–1043 (2012)
21. Zhang, S., Rowlands, A.V., Murray, P., Hurst, T.L., et al.: Physical activity classification using the GENEA wrist-worn accelerometer. Ph.D. thesis, Lippincott Williams and Wilkins (2012)

Deep Neural Networks for Real-Time Remote Fall Detection

Andrea Apicella and Lauro Snidaro[✉]

Department of Mathematics, Computer Science and Physics,
University of Udine, Udine, Italy
apicella.andrea@spes.uniud.it, lauro.snidaro@uniud.it

Abstract. We present a pose estimation, Convolutional Neural Network (CNN) and Recurrent Neural Network (RNN) based fall detection method. Our RNN takes time series of 2D body poses as inputs. Each pose is made of 34 numerical values which represent the 2D coordinates of 17 body keypoints and is obtained using a combination of PoseNet [8] and a CNN on RGB videos. Each series is classified as containing a fall or not. The proposed method can be configured to be suitable for real-time usage even on low-end machines. Furthermore, the proposed architecture can run completely inside the web browser, making it possible to use it without a dedicated software or hardware. As no data is exchanged with external servers, the proposed system preserves user privacy. Our implementation focuses on the detection of a single individual's fall in a controlled environment, such as an elderly person who lives alone or does not have continuous assistance.

Keywords: Remote fall detection · Real-time · Deep Neural Networks · PoseNet · Elderly fall detection

1 Introduction

According to the reports of the World Health Organization[1], in 2018, falls represented the second cause of unintentional injury deaths worldwide, only exceeded by road traffic injuries. Consequently, this problem has generated a wide range of researches aimed at developing effective ways to detect such accidents. In this work, we approach the fall detection task developing a vision-based system, which employs data acquired with a single RGB camera and processed with Deep Neural Networks. We observed that many recent pieces of research which rely on complex techniques and demanding hardware setups obtained very high accuracy. At the same time, there is a lack of successful fall detection applications in the market. As a result, we aimed at developing a system which can be used by virtually anyone to detect falls of a single individual in a controlled environment, such as a home. A broad adoption could lead to the expansion of

---

[1] World Health Organization. Fall. https://www.who.int/news-room/fact-sheets/detail/falls. Accessed November 12th, 2020.

© Springer Nature Switzerland AG 2021
A. Del Bimbo et al. (Eds.): ICPR 2020 Workshops, LNCS 12662, pp. 188–201, 2021.
https://doi.org/10.1007/978-3-030-68790-8_16

publicly available fall datasets, which are few and mostly contain simulated falls. The key novelty of our method consists in performing the RGB video analysis and the classification predictions entirely inside the web-browser. Additionally, our architecture can be tuned to allow for real-time detection.

2 Related Work in Computer Vision

Recently, Computer Vision techniques based on deep learning managed to obtain great fall classification results, by employing different variations of CNNs, Recurrent Neural Networks and many combination of those. Fall detection researchers have pursued high classification accuracy scores for a long time. The introduction of new and more complex deep learning architectures is an instance of how the research focuses on the most common metrics, often neglecting real-time performance, suitability for low power devices and the possibility to make these technologies broadly available to consumers. One such example is the work from Cameiro et al. [2], who used a multi-stream approach to combine the classification results of three hand-crafted features into a single result predicted by an additional SVM classifier. The lack of large enough, real or simulated, falls datasets represented an obstacle that has been recently addressed in two main ways. The first way is trying to create new datasets from scratch, to have more falls videos which picture a single subject falling in a controlled environment. An example is the work by Pourazad et al. [9], who used four Microsoft Kinects, each capturing a different view, similarly to what Multicam's [1] authors did. Further instance is the work from Feng et al. [3], who focused on falls which occur in scenes crowded with people, and created a new dataset of 220 videos. The other, opposite, way is the one followed by Lu et al. [7]. It consists in pre-training complex CNNs on large human actions datasets (Sports-1 M in their case) and then using transfer learning to fine tune the last layers on the smaller fall datasets or, as Lu et al. actually did, feeding the CNN output as input for a further LSTM network that was separately trained on much smaller fall datasets. The approach of recent pieces of research also varies with regard to the type of features used during training. Some experiment with "hand-crafted" features, such as skeletal data, as in [5] and in this work. Other researchers prefer using features automatically extracted from data by CNNs, like in Pourazad et al.'s and Lu et al.'s works, to obtain better robustness in real-life scenarios. Although over the past few years the main focus was to achieve high accuracy scores, some attempts to streamline the architectures have been made recently. Han et al. [4] replaced the traditional VGG16 network architecture with a simplified and slimmed down model. Compared to the original one, this new model occupied as much as 63 times less memory and required half the time to train, while achieving nearly identical accuracy scores. In our opinion, this kind of optimization is a right step to make the best fall detection systems usable on low-power devices and is similar to the assumptions that made us set our work on complete compatibility with the web environment.

3   Dataset

Despite the large number of pieces of research published in recent years around the problem of fall detection, there is a poor choice of datasets containing fall videos. Factors, such as the accidental nature of a fall event and the privacy of the subject involved, led to the creation of datasets which include simulated falls performed by actors in a controlled environment. After several researches, we were able to find two datasets that could fit our needs: UR Fall Detection Dataset [6] and Multiple Cameras Fall Dataset [1].

3.1   UR Fall Detection Dataset

UR Fall Detection Dataset (URFall) includes 70 video sequences, 30 of which containing a fall and the other picturing activities of daily life (Fig. 1). All actions were recorded with two Microsoft Kinect cameras, one shooting horizontally and one vertically. The 30 fall sequences, unlike the ones containing ADL, were also measured with accelerometers. However, we used video data only in our work. Since it is unlikely to find a ceil-mounted camera in a non simulated setting, we decided to use only the video sequences captured with the front-facing Kinect. Furthermore, we decided to use only the first 30 sequences that contain a fall, in order to prevent the problems consequent to a highly imbalanced dataset. Of those sequences, we grouped in the "Fall" category the few frames which portray a subject falling or lying down on the ground after a fall and classified the remaining ones as "No Fall". Once performed this undersampling, we obtained a similar number of frames for the two categories and a total of 2995 frames.

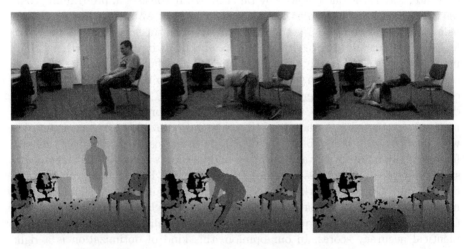

Fig. 1. UR Fall Detection Dataset. Top row: RGB raw images. Bottom row: Kinect's Depth data.

3.2   Multiple Cameras Fall Dataset

Multiple Cameras Fall dataset (Multicam) contains 24 scenarios, each one recorded with 8 IP cameras placed around a studio (Fig. 2). The original video files contain a total of 261339 frames, but only half of those have been labeled by the authors. Additionally, due to the time difference required to start the video recording, the cameras have been synchronized by stopping them at the same time. Consequently, we deleted several delay frames. After this initial processing we obtained 134216 frames labeled "No Fall" and 8504 frames labeled "Fall". Finally, when adding the supplementary CNN described in Sect. 4.2, we chose to get rid of the last two scenarios, as they contain no falls but only confounding events. Indeed, we intended to decrease the unbalance between the number of frames that show a subject falling and the ones which do not.

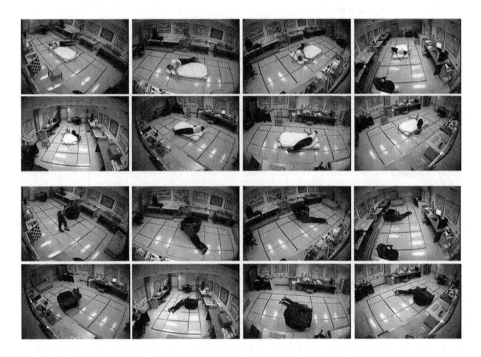

Fig. 2. Multiple Cameras Fall Dataset.

3.3   Combined Dataset

Once performed the described preprocessing on the two datasets, we trained the LSTM with poses which were extracted employing just PoseNet on Multicam, URFall, Balanced Multicam or a combination of these, as described in Sect. 4.3. After collecting these baseline results, we added a supplementary CNN as explained in Sect. 4.2. In order to train the CNN on as much data as possible without using the same data for training and testing, we merged URFall and Multicam into one unique dataset and reserved half of it for training the CNN and the other half for poses extraction with PoseNet and the CNN. To decide which half to reserve for training and which for testing/extracting poses, we initially trained our CNN on both halves, one at a time. We will call the first half split A and the second half split B: we initially trained the CNN on split A (80% of which we used for training and 20% for validation). We repeated the same process on split B and compared the results. The CNN obtained higher validation scores on split B, so we discarded the model trained on split A. Split B is then the only set used for training while split A is the only set used for testing/poses extraction.

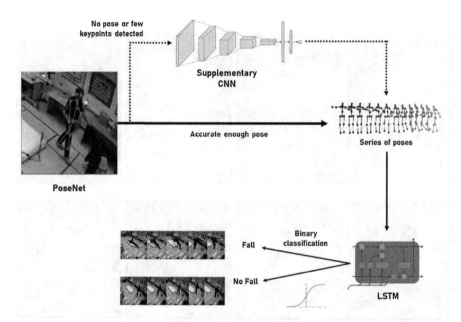

Fig. 3. Architecture overview: PoseNet is initially used to process the current video frame. If the pose is detected with sufficient confidence, it is saved in a time series which will contain 20 poses. If PoseNet fails to detect the pose or detects less than one-third of the 17 keypoints, the supplementary CNN is used as described in Fig. 4. Finally, the LSTM is fed with the time series of poses to perform binary classification.

4   Proposed Method

We propose a method for detecting falls based on time series analysis of 2D keypoint locations data of a single individual in a controlled environment. In particular, we employ a pose detection model and a supplementary Convolutional

Neural Network in order to obtain the 2D position of 17 body key points of the subject shown in each video frame. The keypoint locations data is then split into time-series of 20 poses each and passed as input for a Recurrent Neural Network which classifies such series in two categories: "Fall" and "No Fall" (Fig. 3).

4.1 Pose Detection Model

The first step of our architecture relies on PoseNet, a pose detection model based on the Personlab research conducted by Papandreou et al. [8]. PoseNet can be used to estimate either a single pose or multiple poses in a video. Each pose contains the 2D coordinates of 17 body keypoints (34 numerical total values) and a confidence score for each person detected. Additionally, each keypoint has its own confidence score. In this work we used PoseNet's algorithm which can estimate multiple poses. Although we typically have only one individual in each video, the multiple poses estimation is more accurate when the subject is partially occluded and it is as fast as the single pose estimation method. Finally, we accept only poses and keypoints with a confidence score greater than 30% and set all other keypoints position values to zero. By running on TensorFlow.js, the detection is performed entirely in the web browser, preserving user privacy as no data needs to be processed by a server. Additionally, no dedicated hardware or complicated system setup is required. PoseNet allows the tuning of its underlying CNN parameters in order to trade less processing speed for more accuracy. PoseNet has been trained using COCO dataset, which is a general-purpose object detection, segmentation, and captioning dataset. For this reason, when testing PoseNet with falls videos, it performed poorly on those frames containing a subject who is lying down or in a similar position. In order to address this problem, we introduced an additional CNN in our architecture, which helps extract information from those frames on which PoseNet struggles the most.

4.2 Supplementary CNN

With the aim of extracting poses from those frames which PoseNet failed to process properly, we implemented a CNN based on a MobileNetV2 model pre-trained on ImageNet dataset and we used it as described in Sect. 4.3. Among the pre-trained models we tried, MobileNetV2 proved to be the fastest, especially inside the web browser. Indeed, its prediction take an average 10.7 ms latency per image on a Nvidia GTX 1060 GPU, as shown in Table 3 and described in Sect. 5.2. Additionally, we employed transfer learning to retain the feature extraction capabilities acquired by the model during its training on ImageNet dataset. As a result, we froze the pre-trained weights and swapped the original output layer with two Dense layers with 512 neurons each, plus a final Dense layer with a single neuron and Sigmoid activation to perform binary classification. We trained the model with an 80-20 train-validation split on the second half of our combined dataset described in Sect. 3.3, and tested it on the first half. As reported in Table 1, the model we trained peaked at 90% average accuracy on the validation split and obtained a 79% accuracy on the test split. We also tried some other CNN architectures, such as InceptionResNetV2 [10] and EfficientNet [11], that currently achieve better top-1 accuracy scores on the ImageNet dataset, but we were limited by overfitting during the training of both. However, when we ran these tests, neither of the two models was compatible with the conversion to TensorFlow.js, as they use some layers which are yet to be implemented in TensorFlow.js. Finally, we tested a VGG16 pre-trained model, freezing the weights of the first three blocks and training the last two. This model reached higher accuracy than the MobileNet V2 (Table 1), but with far worse inference times (Table 3).

Table 1. Training results with different CNN models on validation data: the second half of Multicam and URFall combined. The first half was not used for training the CNN models. It has been reserved for testing and for poses extraction as described in Sect. 3.3.

Models	InceptionResNetV2	EfficientNetB7	VGG16	MobileNetV2
Accuracy	0.89	0.80	0.96	0.90

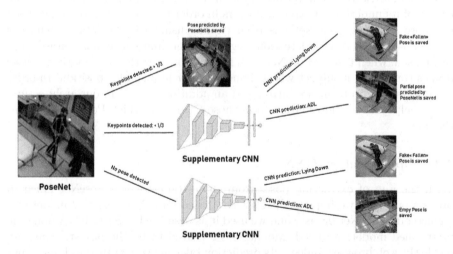

Fig. 4. Proposed architecture for extracting one pose from each video frame. When PoseNet fails to detect the subject or if the pose confidence score is not sufficient, we use the supplementary CNN and save a different pose based on its prediction result.

4.3   Extracting Poses

To extract a pose from each frame of the first half of our combined dataset, we developed the following strategy (Fig. 4): firstly, we used PoseNet to extract the first detected pose. On one hand, if this pose is empty, meaning that there is no subject in the frame or that PoseNet failed to find it, we run the CNN. If our CNN suggests that a lying subject is present in that frame, we save a dummy pose. This is a particular pose we extracted from a random video frame that contains a lying subject of whom PoseNet was able to detect all the 17 keypoints. As a result, we ensure that, when PoseNet completely fails to detect the lying person, we can always attribute some information to that particular frame, even though it is not the real position of its keypoints. On the other hand, if even the CNN does not detect any subject in the frame, we save an empty pose (an array of 34 zeroes). Another possibility is that PoseNet detects a pose, but the confidence score of more than two-thirds of the keypoints is less than the required 30%. In this case, we use again our CNN: if it predicts the presence of a lying subject, we save the dummy pose, otherwise we save the pose originally detected by PoseNet. Although that particular pose has a little information, for sure it contains more than an empty one.

4.4 Classifying Series of Poses

The last step of our architecture consists of splitting all the poses we collected with PoseNet and the CNN into series of twenty and using a custom Recurrent Neural Network with LSTM cells to classify such series. We chose to group our poses in time-series of twenty elements because we observed that, in the datasets at our disposal, a fall always spans less than twenty frames. Testing our architecture with time-series of forty poses we obtained worst classification scores. This result is due to the way we label a single series of poses. To condense twenty labels (one for each frame) into a single label (for a series of twenty poses) we determine which label occurs the most in the sequence and choose it. For example, given a sequence of 20 frames, if 13 frames have a "No Fall" label the sequence will be labeled as "No Fall". Once obtained, these labels represent our RNN model's target. Because of this process, splitting poses in time-series of more than twenty, for example forty, introduces the risk of having a series which contains a fall, but gets labeled as "No Fall" as the majority of those forty frames had such label. The Recurrent neural network we used consists of two layers of 34 LSTM cells each. Each LSTM layer uses ReLu activation, while the final Dense layer uses Sigmoid activation to return a binary classification. We include a 20% dropout between each layer and use the Adam loss optimizer.

5 Experiment

We tested different variations of our architecture. Initially, we excluded the supplementary CNN to obtain baseline results using only the Recurrent Neural Network. To achieve every reported result, we used an 80–20 training validation split, using time-series of twenty poses as input for our RNN. We employed Adam as optimizer with learning rate set to 0.0001 and included early stopping with an upper bound of 1000 epochs and patience of 20.

Table 2. Experimental results obtained training the LSTM model on time-series of poses extracted from different variations of the datasets at our disposal. The last two rows show the classification result after the addition of the MobileNetV2 and the VGG16 CNNs to reinforce the pose extraction phase, while the results showed in the previous rows have been obtained by training the LSTM on series of poses extracted by PoseNet only. Class weighing has been employed when training on Multicam and URFall + Multicam datasets.

Dataset	Accuracy	Precision			Recall			F-score		
		Avg	No Fall	Fall	Avg	No Fall	Fall	Avg	No Fall	Fall
URFall	0.708	0.700	0.733	0.667	0.693	0.786	0.600	0.695	0.759	0.632
Multicam	0.511	0.518	0.972	0.065	0.602	0.502	0.701	0.390	0.662	0.119
URFall + Multicam	0.753	0.547	0.960	0.134	0.650	0.768	0.533	0.534	0.853	0.214
Balanced Multicam	**0.780**	**0.776**	0.802	0.750	**0.775**	0.811	0.739	**0.776**	0.807	0.745
URFall + Balanced Multicam	0.694	0.694	0.714	0.674	0.694	0.684	0.705	0.694	0.699	0.689
URFall + Multicam (half) with MobileNetV2	**0.818**	**0.809**	0.838	0.781	**0.801**	0.872	0.730	**0.805**	0.855	0.754
URFall + Multicam (half) with VGG16	**0.849**	**0.840**	0.878	0.803	**0.840**	0.878	0.803	**0.840**	0.878	0.803

5.1 Experimental Results

We started by training the Recurrent Neural Network on our down-sampled URFall dataset only. As shown in Table 2, we achieved a 73.3% precision score on the "No Fall" category and a 66.7% precision score on the "Fall" category. When we repeated the same training on the original Multicam dataset only, we obtained extremely imbalanced results. In an effort to compensate for the much higher number of frames belonging to the "No Fall" class, we introduced class weights for our two labels with the compute_class_weight method provided by the sklearn.utils library[2]. Nevertheless, we obtained poor results on the "Fall" class, with a 0.65% precision score and a 70.1% recall, meaning that our model learned to guess mostly the same class, but being the "Fall" events so few, it still detected some of them. Even combining the two datasets, we obtained only minimal improvements in precision results, with the "Fall" category reaching a 1.34% precision. The "Fall" recall result got worse, with a 53.3% rate, while the "No Fall" recall improved and reached 76.8% from the 50.2% obtained with imbalanced Multicam only. Being Multicam so much bigger than our down-sampled URFall dataset, the addition of URFall's few frames proved to be practically irrelevant. After these initial results showed that Multicam's high imbalance was negatively affecting our model performance, we decided to perform an undersample of such dataset, in an attempt to decrease the gap between our two categories. We manually analyzed every Multicam's video and chose a new first and last frame for each one, to make sure that the number of frames of the fall event itself

---

[2] Scikit Learn: compute_class_weights. https://scikit-learn.org/stable/modules/generated/sklearn.utils.class_weight.compute_class_weight.html. Accessed November 17th, 2020.

was similar to the sum of pre-fall and post-fall frames. After this undersampling, Multicam counted 15840 frames, 7640 of which were labeled as "No fall" and 8200 as "Fall". Training our RNN on the newly balanced Multicam dataset only, we obtained much more balanced and generally better results, with a 80.2% and a 75% precision score for the "No Fall" and "Fall" class respectively. After combining URFall with Balanced Multicam and not achieving better results, we finally tested our complete architecture with the addition of the MobileNetV2 model in the poses extraction phase. This model enabled us to extract features from most of the frames that PoseNet would partially or completely fail to process. Given this new chance, we rolled back to using the original Multicam dataset, of which we discarded only the last two chutes because they do not contain fall events. We combined the original Multicam with URFall and obtained our best overall scores, including a 81.8% average accuracy rate. The baseline results we obtained show that performing undersamples on imbalanced datasets, which are frequent when considering fall ones, can lead to more homogeneous results.

5.2   Time Cost Analysis

Our aim is to make possible detecting a fall in real time, virtually on any machine and with minimal additional setup requirements. These requirements are a fundamental prerequisite to allow the use of a fall detection system in consumer applications, such as in elderly care. Indeed, we find it important to make fall detection technologies broadly available to elderly people in the first place, as they are not only the subject who are most prone to such type of accidents, but also those who risk suffering the most serious consequences. As shown in Table 3, PoseNet is perfectly capable to estimate poses in real time. Especially when using the MobileNetV1 backbone, it runs at well above 24 fps even on a machine with integrated graphics. We can say the same for our custom LSTM network. We conducted further tests running PoseNet on a more capable machine. The results show that even when using the more demanding ResNet50 backbone, and maxing out its settings to achieve the best possible pose estimation accuracy, PoseNet is able to run in real-time inside a web browser. When adding the supplementary CNN to reinforce the poses extraction we are able to obtain better classification scores. Furthermore, the MobileNetV2 model we chose and trained, which is the evolution of the MobileNetV1 architecture used at PoseNet's core, performed predictions with an average 19 ms per frame latency on the Macbook. When tested on a Desktop PC with discrete graphics, it runs at nearly double the speed, with an average 10.7 ms latency per frame as shown in Table 3. As reported in Sect. 4.2, we tested other CNN architectures, among which only the VGG16 is currently compatible with the conversion to TensorFlow.js. Even if this models performs better in terms of accuracy (Table 1), its inference times are much higher. On the Macbook, the VGG16 takes almost 38 times more than the MobileNetV2 to process a single image (Table 3). On the Deksktop PC the results are significantly better, but the average 12 frames per second inference speed is still not feasible for real-time processing. Additionally, we use the supplementary CNN only on those frames which PoseNet fails to estimate correctly, hence choosing the MobileNetV2 over a better performing model, such as the VGG16, resulted in a 3% drop in LSTM classification accuracy. In conclusion, models which have the same, if not better, representational capacity as the VGG16 and run as fast as MobileNetV2 already exist for other platforms. The recently announced EfficientNet Lite[3] is the perfect example as it matches InceptionV4 Top1 80.4% accuracy on the ImageNet dataset with a 30 ms per frame latency on a smartphone CPU. With this, or similar CNNs, coming to TensorFlow.js it will be possible to run our fall detection architecture in real time, achieving even better classification results, even on low-powered machines.

Table 3. Average inference times for a single 200 × 200 image reported in milliseconds and frames per second on a base model 2019 Macbook pro with Intel Core i5 8th gen CPU, integrated GPU and a Desktop PC equipped with Intel Core i7 8700k CPU, Nvidia GTX 1060 3GB GPU. PoseNet(MobileNetV1) backbone settings are: outputStride set to 16, multiplier set to 0.5. PoseNet with ResNet50 backbone settings are: outputStride set to 32, quantBytes set to 1. All tests were conducted using Microsoft Edge browser version 86.0.622.69.

	PoseNet (MobileNetV1)		PoseNet (ResNet50)		MobileNetV2		VGG16		LSTM	
	ms	fps	ms	fps	ms	fps	ms	fps	ms	fps
Macbook	19.6	51.0	83.3	12.0	20.6	47.6	380.6	2.6	38.5	26.3
Desktop PC	10.4	96.2	27.3	36.6	10.7	93.5	79.5	12.6	17.2	58.1

5.3   Comparing with State of the Art

Training our model on URFall only led to obtaining 70.8% accuracy, while Jeong et al. [5], who used poses to train an LSTM as we did, obtained 61.5% on the same dataset. By combining URFall with SDUFall (a fall dataset that is currently unavailable online) and implementing other features extraction methods, such as the human center line coordinate (HCLC), Jeong et al. achieved their best score of 99.2% accuracy. However, they do not report a time cost analysis. Comparing with Feng et al. [3] attention guided LSTM model, we can see that they obtain higher results on both URFall and Multicam datasets with a 91.4% recall on the former and 91.6% sensitivity on the latter (accuracy score have not been reported), but they also do not report a time cost analysis. Pourazad et al. [9] trained their LRCN model on their newly proposed dataset, obtaining an 87% accuracy score. Their model can process videos in a short time window, but they do not report a more precise processing speed metric. Xu et al. [12] obtain an even higher 91.7% accuracy rate by training an Inception-ResNet-v2 [10] model on a combination of URFall, Multicam, NTU RGB+D Action data set and fall

---

[3] EfficientNet Lite. https://github.com/tensorflow/tpu/tree/master/models/official/ efficientnet/lite. Accessed November 17th, 2020.

videos found online. However, they reported that their model takes around 45 s to make a prediction for a single image without hardware acceleration, making it not feasible for a real-time use. Han et Al. [4] obtained accuracy scores as high as 98.25%, but they used a self collected dataset made of only 3628 pictures, 50% of those used for training and the other 50% for testing. They report that their newly introduced MobileVGG model took 2.96 s to make inferences on the whole test set (1314 images 224 × 224 resolution) and 0.75 s when testing with 100 × 100 image resolution. Finally, Na et Al. [7] achieved an impressive 100% accuracy score on Multicam, and proved that their implementation is effective even if tested on activity classification benchmarks like UCF11 and HMDB-51. They also presented a time-cost analysis, showing that their 3D CNN extracted the UCF11 dataset's features at 60.2 fps. However, they used a high-end Nvidia Tesla K40 GPU and it is unclear which is the time cost of their LSTM.

Table 4. Comparison with other state of art pieces of research. For each study only the best result and the corresponding inference times, when reported, are shown.

	Pourazad et al.	Xu et al.	Han et al.	Jeong et al.	Na et al.	Ours
Dataset	Self collected	URFall + Multicam + NTU + Youtube videos	Self collected (3628 images)	URFall + SDU Fall	Multicam	URFall + Multicam (Half)
Accuracy	87%	91.7%	98.3%	99.2%	100%	81.8%
Latency	–	45 s (CNN only)	2.3 ms (CNN only)	–	16.7 ms (CNN only)	**37 ms (full architecture)**
GPU	Nvidia Tesla K80	–	Nvidia Quadro P4000	Nvidia GTX 1080	Nvidia Tesla K40	Nvidia GTX 1060 3GB
Browser	No	No	No	No	No	**Yes**

6 Conclusions

In this work we proposed a fall detection system based on pose data obtained from RGB videos that can run completely inside the web browser, preserving users' privacy as no data is sent to a server to be processed.

The proposed architecture is modular because it is possible to exploit a supplementary MobileNetV2 CNN during the poses extraction phase to achieve better fall classification results. At the same time, our system can run in real-time even on low powered machines, while preserving users' privacy as the processing is entirely done locally inside the web browser.

Our architecture represents an attempt to highlight the potential of tools like PoseNet and TensorFlow.js and to shift fall detection researchers' focus towards the development of software that can be actually used by interested users. Furthermore, if user privacy is respected, we could earn the trust necessary to achieve important results, such as obtaining user permission to collect real falls videos. The proposed system shows great potential for specific users, including elderly people, who could finally benefit from an effective and low or zero cost solution. Indeed, running inside the browser, our system could be distributed as a web app and used with an IP camera or a cheap webcam, allowing any individual to detect falls in a home environment.

6.1 Future Developments

As discussed in Sect. 5.2, the first and most immediate improvement to our system would come from including a CNN architecture that is both highly performing and efficient enough to run in real time on most computers. While waiting that models like EfficientNet [11] are ported to TensorFlow.js, we are planning to develop a custom CNN, specifically designed for our purposes, for fast inference time in the browser. The pose extraction process itself could be improved by addressing the problem of the not accurately detected keypoints. We could interpolate the values of a previous and following keypoint to determine reasonable coordinates for the keypoints that PoseNet fails to detect. This would mean having more meaningful data fed to the LSTM during its training process, probably resulting in higher classification results. Finally, considering the fast inference times of the MobileNetV2 model we trained, we could ditch PoseNet completely and use only the CNN to extract self learned features that could be fed directly to the LSTM to perform classification. This could lead to better robustness in real use cases thus making our model capable of detecting falls even in situations with more than one individual or with challenging environmental conditions.

References

1. Auvinet, E., Rougier, C., Meunier, J., St-Arnaud, A., Rousseau, J.: Multiple cameras fall dataset. DIRO-Université de Montréal, Technical Report 1350 (2010)
2. Cameiro, S.A., da Silva, G.P., Leite, G.V., Moreno, R., Guimarães, S.J.F., Pedrini, H.: Multi-stream deep convolutional network using high-level features applied to fall detection in video sequences. In: 2019 International Conference on Systems, Signals and Image Processing (IWSSIP), pp. 293–298. IEEE (2019)
3. Feng, Q., Gao, C., Wang, L., Zhao, Y., Song, T., Li, Q.: Spatio-temporal fall event detection in complex scenes using attention guided LSTM. Pattern Recogn. Lett. **130**, 242–249 (2020)
4. Han, Q.: A two-stream approach to fall detection with mobileVGG. IEEE Access **8**, 17556–17566 (2020)
5. Jeong, S., Kang, S., Chun, I.: Human-skeleton based fall-detection method using LSTM for manufacturing industries. In: 2019 34th International Technical Conference on Circuits/Systems, Computers and Communications (ITC-CSCC), pp. 1–4. IEEE (2019)
6. Kwolek, B., Kepski, M.: Human fall detection on embedded platform using depth maps and wireless accelerometer. Comput. Methods Programs Biomed. **117**(3), 489–501 (2014)
7. Lu, N., Wu, Y., Feng, L., Song, J.: Deep learning for fall detection: three-dimensional CNN combined with LSTM on video kinematic data. IEEE J. Biomed. Health Inf. **23**(1), 314–323 (2018)

8. Papandreou, G., Zhu, T., Chen, L.C., Gidaris, S., Tompson, J., Murphy, K.: Personlab: person pose estimation and instance segmentation with a bottom-up, part-based, geometric embedding model. In: Proceedings of the European Conference on Computer Vision (ECCV), pp. 269–286 (2018)
9. Pourazad, M.T., et al.: A non-intrusive deep learning based fall detection scheme using video cameras. In: 2020 International Conference on Information Networking (ICOIN), pp. 443–446. IEEE (2020)
10. Szegedy, C., Ioffe, S., Vanhoucke, V., Alemi, A.A.: Inception-v4, inception-resnet and the impact of residual connections on learning. In: Proceedings of the Thirty-First AAAI Conference on Artificial Intelligence, pp. 4278–4284 (2017)
11. Tan, M., Le, Q.: Efficientnet: rethinking model scaling for convolutional neural networks. In: International Conference on Machine Learning, pp. 6105–6114 (2019)
12. Xu, Q., Huang, G., Yu, M., Guo, Y.: Fall prediction based on key points of human bones. Physica A: Stat. Mech. Appl. **540**, 123205 (2020)

Mutual Use of Semantics and Geometry for CNN-Based Object Localization in ToF Images

Antoine Vanderschueren$^{(\boxtimes)}$, Victor Joos$^{(\boxtimes)}$, and Christophe De Vleeschouwer

ICTEAM Institute, UCLouvain, Louvain-la-Neuve, Belgium
{antoine.vanderschueren,victor.joos,
christophe.devleeschouwer}@uclouvain.be

Abstract. We propose a novel approach to localize a 3D object from the intensity and depth information images provided by a Time-of-Flight (ToF) sensor. Our method builds on two convolutional neural networks (CNNs). The first one uses raw depth and intensity images as input, to segment the floor pixels, from which the extrinsic parameters of the camera are estimated. The second CNN is in charge of segmenting the object-of-interest so as to align its point cloud with a reference model. As a main innovation, the object segmentation exploits the calibration estimated from the prediction of the first CNN to represent the geometric depth information in a coordinate system that is attached to the ground, and is thus independent of the camera elevation. In practice, both the height of pixels with respect to the ground, and the orientation of normals to the point cloud are provided as input to the second CNN.

Our experiments, dealing with bed localization in nursing homes and hospitals, demonstrate that our proposed floor-aware approach improves segmentation and localization accuracy by a significant margin compared to a conventional CNN architecture, ignoring calibration and height maps, but also compared to PointNet++.

Keywords: Depth · CNN · Object localization

1 Introduction

In hospitals and nursing homes, the number of nurses at night is largely insufficient to keep a permanent eye on every patient or senior. Automatic human behavior analysis is therefore required to help with the detection of bed exits and falls, to alert the medical staff as soon as possible.

In this context, the Time-of-Flight (ToF) camera offers the following non-negligible advantages: it provides a depth map, in addition to the reflected intensity, and, given its active nature, is relatively independent of lighting conditions. These advantages come with greatly reduced image resolution, and a

---

A. Vanderschueren and V. Joos—Contributed equally to the paper.

© Springer Nature Switzerland AG 2021
A. Del Bimbo et al. (Eds.): ICPR 2020 Workshops, LNCS 12662, pp. 202–217, 2021.
https://doi.org/10.1007/978-3-030-68790-8_17

shorter range. Hence, ToFs appear especially suited for the monitoring of small closed spaces like bedrooms and hospital rooms, especially at night-time. Detection of humans, often represented as moving blobs, from ToF image sequences has been largely investigated [10,25,28]. However, turning those detections into human behavior interpretation requires to position the camera with respect to the scene, and to localize the objects the human interacts with. This preliminary but critical step is often neglected, assuming that calibration and scene composition is encoded manually. An autonomous calibration would however facilitate the deployment of systems in real-life conditions, and give the opportunity to adjust the interpretation of movements to the displacement of key objects in the scene.

Our work focuses on this calibration step, and investigates a use case that aims at localizing the beds in rooms of nursing homes or hospitals. Beyond the scope of this paper, this would typically be combined with human detection, for bed exit and/or fall recognition.

In short, our approach to automatically calibrate the camera and position the bed in a room, follows the steps illustrated in Fig. 1. It first segments the floor based on a CNN fed with raw depth and intensity maps. This initial segmentation allows us to estimate the ground plane equation in the coordinate system of the camera, so that the depth information can be transformed to height information with respect to the floor.

The second step of our method consists in segmenting the object-of-interest by feeding a second CNN with the ToF intensity map, the height information, and the field of vectors defining the local normals to the point cloud. The segmented point cloud is then aligned with a model of the object, to localize the object in the scene. Our method is validated on a *practical* case using *real* data. This case considers the localization of beds in nursing homes and hospital rooms. Our study results in a multifold contribution. Specifically, our work offers:

- Precise and automatic estimation of the floor plane position in the ToF camera referential;
- Effective segmentation of the point cloud associated to the object of interest, here the bed, using a 2D CNN. The approach is shown to outperform networks operating directly on point clouds like PointNet++ [22]. The method is also shown to improve in accuracy when the network is fed with geometric information represented in terms of *height* (with respect to the floor) and *local normals* (defined in a referential aligned with the floor normal);
- Fast and reliable localization of the object of interest, thanks to the floor plane knowledge, which reduces the degrees of freedom from 6 to 3 when aligning the reference model with the segmented point cloud;
- The estimation of localization confidence, enabling the system to wait for better observation conditions (no occlusions, see Fig. 3a and 3b) or for human intervention.

This article is organized in 3 main sections. Section 2 surveys the state-of-the-art related to object segmentation and localization. Section 3 then describes

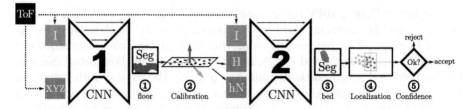

Fig. 1. Overview of our method. (1) The first CNN segments the floor based on the intensity map (I) and spatial coordinates (XYZ) in the camera-centric viewpoint of the ToF. (2) The normal to the ground plane is estimated from the segmented 3D floor points, and is used to define a height map (H) and a field of local normals (hN) represented in a referential obtained by rotating the ToF referential to make its Z axis orthogonal to the ground plane. (3) The second CNN combines these (H) and (hN) maps with the intensity map to segment the object-of-interest. (4) Segmented points are aligned with the reference model, to localize the object. (5) A level of confidence is assigned to the localization, by comparing the point cloud and the fitted model.

our method, and discusses its strengths and weaknesses compared to previous works. Section 4 validates our method on a real-life use case.

2   State of the Art

The literature addressing object localization based on depth information (either from RGB-D or ToF) considers two distinct methodologies. Some recent efforts, similar to our works, adopt a CNN to identify the pixels belonging to the object, and use the corresponding 3D points to compute the object position (Sect. 2.1). Others directly process a 3D point cloud, using CNNs or graphs (Sect. 2.2 and 2.3). Using 2D segmentation, we identify the pixels belonging to the object, and use the corresponding 3D points to compute the object position.

2.1   2D Convolutional Segmentation

The similarity between the signal, output by RGB-D and ToF makes it relevant to extend the quite laconic SotA related to ToF segmentation [6,16], to the broader literature related to RGB-D segmentation [2].

The segmentation methods using RGB-D signals as inputs of 2D CNNs differ in the way they merge the color and depth signals. This fusion generally depends on the chosen network architecture [4,11], and is sometimes even driven by a squeeze-and-excite attention module [8,9].

However, our experiments (not presented here for conciseness) have revealed that in the case of ToF data, there is no benefit to fusing Intensity and Depth based on attention modules. We have therefore devised a straightforward and computationally simpler fusion, described in Sect. 3.3.

A fundamental difference between our work and previous art lies in the way the depth signal is represented to feed the CNN. Based on the automatic ground/floor plane parameters estimation, the depth is transformed into a height value for each image pixel. Moreover the neighborhood of each 3D point is used to compute a local normal to the point cloud, described in a referential aligned with the floor normal. We show that this original representation improves the CNN accuracy. This confirms the results reported in [3] regarding the use of a geocentric representation, which encodes height above ground and angle with gravity for each pixel, for the detection of objects with a pre-calibrated RGB-D system combining CNN and SVM.

2.2   3D Convolutional Localization

3D Convolutional neural networks build on a voxelization of the point cloud. They suffer either from the lack of resolution induced by the use of big voxels, or from high sparsity in voxel information.

3D networks have first been considered for object localization in [17]. This pioneering work uses a U-Net [23] structure (like the one used in this paper), but substituting 2D for 3D convolutions. Quite recently, [7] has proposed to combine the high-resolution 2D color information, from RGB-D data, with a 3D neural network on point cloud data, leading to a precision gain of 2–3%. In our applicative context, we show in Sect. 4.4 that ToF intensity images contribute the *least* of all the input types, to the final segmentation decision.

[26] has trained a network on synthetic data to complete the voxels that remain hidden when a single viewpoint is available. We have however observed that networks trained on synthetic data did not transfer well to real-life ToF data.

2.3   Segmentation and Localization Using Graph NNs

To circumvent the excessive computational cost of 3D convolutions dealing with high voxel resolution, graph neural networks work on connected points rather than on regular 2D or 3D matrices. Most implementations combine fully-connected layers and specialized pooling layers. Most point-based segmentation [13,14] and object localization [12,19,20,24] methods are based in part or in full on PointNet [21] and PointNet++ [22]. Those approaches, although attractive in the way they represent the input point cloud, still show a lack of accuracy compared to convolutional methods on RGB-D and ToF images.

3   Floor Plan Estimation for Autonomous Object Localization

To describe our method, we start by explaining the structure of our framework, and then delve deeper in its building blocks.

3.1   ToF Data

The ToF sensor gives us, for every pixel, information about reflected intensity
and depth, i.e. distance to the camera. Having access to the intrinsic parameters
of the camera, we can express depth in terms of 3D position relative to the
camera (XYZ). We can also estimate the normal vector to the point cloud surface
in each point in this same coordinate system.

3.2   Method Overview

Figure 1 depicts the five steps of our method:

Step 1. A first CNN uses the intensity and depth data to segment the floor
pixels.

Step 2. The floor's plane equation in 3D space, relative to the camera, is
obtained via Singular Value Decomposition (SVD) on the 3D coordinates of
the floor pixels. The vector that is normal to the ground plane is defined
by the smallest singular value. To make our algorithm more robust towards
outliers, we embed SVD into a RANdom SAmpling Consensus (RANSAC)
algorithm.

Step 3. A second CNN segments the object-of-interest. Since the ground
plane equation is known, the height of every pixel can be computed from its
XYZ coordinates. We consider two alternatives to feed the neural network with
floor-aware geometrical information.

1. In the first alternative, the resulting pixel height map (H) is fed to the
   network, together with the 3 components of the normal (hN) to the surface
   point cloud in each point. Both H and hN are expressed in a referential
   obtained by rotating the camera referential (first around Z, originally
   pointing towards the scene, to make X horizontal, and then around the
   X-axis) to align the Z-axis with the floor normal.

2. In the 2nd alternative, since the floor plane equation has been estimated,
   the intensity (I) and height (H) information associated to the 3D point
   cloud can be projected on the floor plane (see Fig. 6a). This provides a
   bird's-eye 2D view of the scene, which is independent of the actual camera
   elevation (up to occlusions). We have trained a network to predict the
   bed label from those two projections, respectively denoted proj(I) and
   proj(H), taken as inputs.

Step 4. The coordinates of the segmented object's 3D points are then fed
into a localization algorithm that aligns them with a reference model.

Step 5. The quality of the matching between the object points and the
aligned model is used as a localization confidence score, to detect cases of
heavy occlusion or incorrect segmentation, as shown in Fig. 3a and 3b.

It should be noted that the calibration step (the 2nd step in the list above)
allows us to move away from a camera-centric coordinate system towards a
floor-centric point-of-view. Hence, instead of expressing the data in a coordinate
system that is fully dependent on the camera placement, we express them in

a referential that is independent of the camera elevation. This offers the two following benefits:

– The object segmentation network learns the relationship between the geometric information and the object's mask more easily (as we'll show quantitatively and qualitatively in Sect. 4)
– The degrees of freedom to consider during the localization step get reduced from 6 (3 rotations, 3 translations) to 3 (1 rotation, 2 translations).

The remainder of this section, presents how the CNNs used in steps 1 and 3, respectively for floor and object segmentation, have been constructed and trained. It also describes the implementation details for the localization step.

3.3   Floor and Object Segmentation (steps 1 and 3)

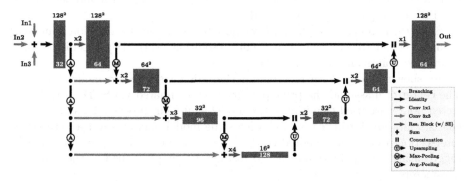

Fig. 2. Our MultiNet Network Architecture. The arrows depict convolutional layers. When present, the multiplicative factor along the arrow defines the number of times the convolutional block is repeated. The boxes represent the feature maps, the number of channels are noted inside the box, the resolution outside. The architecture follows the U-Net model, with the addition of multi-resolution inputs, and block repetitions as shown in the figure. The different inputs are fused using a sum after an initial convolutional block. This architecture is used for both floor and object segmentation.

To segment the point cloud, we use the U-Net [23] shaped network presented in Fig. 2, and denoted *MultiNet* in the rest of the paper, since it is suited to handle multiple types of inputs. U-Net adapts an auto-encoder structure by adding skip-connections, which link feature maps of identical resolution from the encoder to the decoder. This allows the direct transfer of high resolution information to the decoding part, by avoiding the lower resolution network bottleneck.

Feature maps of identical resolution are said to be of the same *level*. Each level's structure, is based on residual blocks [5] followed by a *Squeeze-and-Excite* module that weighs every feature map individually before their sum [8].

The convolutional blocks are repeated, as indicated by the multiplicative factors along the arrows in Fig. 2. Those repetition factors follow the parameters recommended in [1,11,29].

Our networks are fed, as explained in the method overview, by a combination of the following input types: intensity (1D), normal (3D), XYZ (3D), and height (1D). To deal with different types of inputs we merge all inputs directly : every input type is passed once through a convolutional layer, such that they all possess the 32 feature maps. These feature maps are then summed once and fed to the rest of the network at every level, through down-scaling.

3.4   Localization and Error Estimation (steps 4 and 5)

As explained in the method overview, the pixels labeled by our segmentation network as being part of the object, i.e. the bed, are used to localize it in a coordinate system where one axis is perpendicular to the floor, and the two remaining axes are respectively parallel and perpendicular to the intersection of the ToF image plane with the ground plane.

Since the bed has a simple shape we adopt a very basic localization approach on a rasterized (discretized on a 5 cm^2 resolution grid) 2D projection of the segmented points on the ground plane.

As shown in Fig. 3c, a rectangular shape is considered to model the bed. The center of mass and the principal direction, estimated via SVD, of the projected points are used to initialize the model alignment process. This procedure consists in a local grid search and selects the model maximizing the estimated intersection-over-union, denoted $rIoU^b$, between the rasterized projected points (aggregating close points) and the 2D rectangular shape defined by the searched parameters.

Since it reflects the adequacy between the selected rectangular model and the projected point cloud, the $rIoU^b$ of the 2D rectangular model is then used as a confidence score to validate or reject the predicted bed localization (5th step). It can be used to detect a wrong prediction, e.g. occurring when a bed is occluded by a nurse, and repeat the process in better observation conditions.

4   Results and Analysis

This section first introduces our validation methodology, including use-case definition, training strategy, and quantitative metrics used for evaluation. It then considers floor and object segmentation, respectively in Sect. 4.2 and 4.3. Eventually, Sect. 4.4 considers an ablation study to assess the benefit resulting from our proposed representation of the geometric information.

4.1   Validation Methodology

Use Case. Our method is evaluated on a ToF dataset, captured in nursing homes and hospital rooms. Our objective is to position the bed in the room,

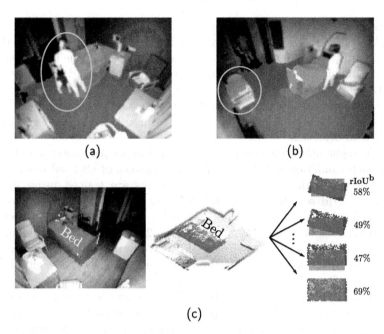

Fig. 3. (a) Example of occlusion. (b) Example of wrong segmentation. (c) Example of localization and error estimation. After the object segmentation, the image pixels are projected onto the 2D plane of the floor. The points that were segmented as "bed" are then extracted, and various transformations (2D translation and 1D rotation) are tried until the final bounding box is found with the maximization of the IoU on a discretized grid (rIoU^b) between prediction and model (blue and small red points). The intersection contains only the blue points, while the union contains blue and red points (of all sizes). *Best viewed in color.* (Color figure online)

without any additional information but the intrinsic parameters of the camera and the images it captures. To assess the performance of our system when it is faced with a new room layout or style, we divide our dataset in 7 subsets containing *strictly* different institutions (hospitals or nursing homes). We apply cross-validation in order to systematically test the models on rooms that have not been used during training.

Our database contains 3892 images of resolution 160 × 120. Those images come from 85 rooms belonging to 11 institutions. On average, 45 images with divers illumination, occlusions and (sometimes) bed positions are available per room.

In order to train and validate our models, we use manually annotated data, both for the device calibration and for the localization of the bed.[1]

---

[1] The tool developed for annotation is available at https://github.com/ispgroupucl/tofLabelImg.

Training. During our cross-validation, we select 1 subset for testing, 1 subset for validation and 5 for training. Cross-validation on the test set is used due to the small number of different. We use data augmentation in the form of vertical flips and random zooming. The images are then cropped and rescaled to fit a 128×128 network input resolution.

Normal vectors are estimated from the 10 closest neighbors that are within a 10 cm radius of each point. They are expressed in a floor-aware referential (hN) as detailed in Sect. 3.2.

All our segmentation networks result from a hyper-parameter search evaluated on each validation set. The average performance of all 7 subsets obtained on the test set is then presented. The networks are trained using AdamW [15], using a grid search to select the learning rate and the weight decay in $\left\{1 \times 10^{-3}, 5 \times 10^{-4}, 1 \times 10^{-4}\right\}$ and $\left\{1 \times 10^{-4}, 1 \times 10^{-5}, 1 \times 10^{-6}\right\}$ respectively. The learning rate is divided by 10 at epochs 23 and 40, while training lasts for 80 epochs, with a fixed batch size of 32. We implement our segmentation models using PyTorch [18] and have published our code at https://github.com/ispgroupucl/tof2net. For the PointNet++ segmentation model, we use the model described in [22] and implemented by [27]. We never start from a pre-trained network as pre-training shows poor performance on our dataset in all cases.

Finally in order to avoid initialization biases, the values presented in *every* table are always the average of 5 different runs.

Metrics. Distinct metrics are used to assess segmentation, calibration and bed localization.

Segmentation predictions are evaluated using Intersection-over-Union between predicted and ground-truth pixels. This metric is denoted IoU. Recall $\left(\frac{TP}{TP+FN}\right)$ and precision $\left(\frac{TP}{TP+FP}\right)$ are also considered separately, to better understand the nature of segmentation failures.

We evaluate the extrinsic camera calibration using absolute angles between the ground normal predicted by our model and the ground truth.

Object localization is also evaluated using Intersection-over-Union, but instead of comparing sets of pixels, we compute the intersection-over-union between the 2D bounding-box obtained after localization and the ground truth, projected on the estimated floor plane. This metric is denoted IoU^b, with b referring to the fact that bounding boxes are compared. The projection on a common plane is necessary to compensate for slight differences due to possible calibration errors i.e. different ground plane equations.

In practice, IoU^b that lie below 70% correspond to localization not sufficiently accurate to support automatic behavior analysis, typically to detect when a patient leaves the bed. Hence we consider 70% as a relevant localization quality threshold, and evaluate our methods based on the Average Precision at this threshold (AP@.7). In addition, the Area Under Curve (AUC@.7) measures the correct localization predictions (true positives) as a function of the incorrect ones (false positives) when scanning the confidence score given by the estimated $rIoU^b$, explained in Sect. 3.4.

4.2   Floor Segmentation and Floor Normal Estimation

Table 1 compares the different floor segmentation models in terms of segmentation and floor normal accuracy. It also presents the floor normal estimation error obtained when applying RANSAC directly on the whole set of points (since a majority of points are floor points, one might expect RANSAC will discard outliers and estimate the ground plane equation without segmentation). We observed that the global RANSAC performance is not sufficient for localization purposes. With a mean error greater than $10°$ it would lead to a drop in IoU^b of more than 40%, as shown in Fig. 4.

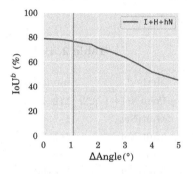

Table 1. Floor segmentation and calibration results.

Method	Inputs[1]	IoU (%)	ΔAngle (°)
RANSAC	XYZ	–	13.2
PointNet++	XYZ+I	83.6	1.2
MultiNet	I	75.6	3.3
MultiNet	I+XYZ	**87.2**	**1.1**

Best results in **bold**

[1] I denotes intensity, XYZ defines the 3D points spatial coordinates in the ToF camera referential.

Fig. 4. Bed Localization bounding-box Intersection-over-Union (IoU^b) dropoff as a function of the floor normal direction estimation error. The orange line is at 1.1, which corresponds to the mean error from MultiNet-I+XYZ.

Looking at the segmentation IoUs, we see that PointNet++ has better accuracy than a convolutional network that only uses the intensity (I) as input signal. However, adding the XYZ point coordinates (as defined in the coordinate system of the camera) as input to the convolutional model is enough to surpass PointNet++.

In terms of ground normal estimation, our proposed model leads to a mean absolute error of $1.1°$, As can be seen in Fig. 4 this is precise enough for our use-case. Indeed, the segmentation quality of our floor-aware CNNs (see Sect. 4.3) remains nearly constant for calibration errors lower than 1. For errors greater than 1.5–2 a substantial performance loss is observed. MultiNet-I+XYZ however has a mean error of 1.1 positioning our approach as an acceptable solution that won't lead to a significant error propagation. We also note that even though the segmentation maps made by PointNet++ is significantly worse than the ones predicted by MultiNet-I+XYZ, the final floor normal error is only slightly worse than MultiNet-I+XYZ.

4.3   Bed Segmentation and Localization Accuracy

Table 2 summarizes the results of bed segmentation and localization for different representations of the geometric information.

Segmentation. The object localization relies heavily on the object segmentation accuracy. For this reason we first compare the different methods in terms of segmentation in the third column of Table 2. For all the MultiNets using the H and/or hN representation of the geometric information, the floor calibration is the one produced with the MultiNet-I+XYZ from Sect. 4.3, thus possible calibration errors have been propagated to the final numerical values.

Table 2. Bed segmentation and localization results

Method	Inputs[1]	IoU(%)	IoU^b(%)	AP@.7	AUC@.7
PointNet++ [22]	XYZ+I	44.6	60.8	46.2	43.5
MultiNet	I	65.3	67.1	60.0	58.6
MultiNet	I+XYZ	69.9	75.4	76.2	74.7
MultiNet	I+H	<u>71.0</u>	76.1	78.5	77.1
MultiNet	I+H+hN	**72.1**	<u>77.2</u>	<u>80.6</u>	<u>79.1</u>
MultiNet	proj(I)+proj(H)	N/A	**78.1**	**83.1**	**81.8**

Best results in **bold**, 2nd best <u>underlined</u>
[1] I denotes intensity, XYZ spatial coordinates, H height, and hN the local normals, defined in the floor-aligned referential, proj(X) the projection of X onto the floor-aligned plane.

We were initially surprised by the very low accuracy of PointNet++, even compared to the intensity-only MultiNet baseline. However, this can be explained by the fact that the shape of the bed is more complex than the planar floor geometry. In addition, ToF cameras are known to induce large measurement disparities, making it harder for a geometry-based neural network to correctly learn the object's geometry.

Looking at our MultiNet methods, we see that accounting for the geometric information systematically outperforms the single intensity baseline. Moreover, our proposed representation of geometry in terms of H and hN improves the conventional XYZ representation by more than 2%.

Figure 5 shows qualitative results. The results show the advantage of our floor-aware approach (third column).

Localization. Table 2 also displays the metrics for bed localization. When looking at the bounding-box IoU^b or the average precision at a threshold of 0.7 IoU^b (AP@.7), we observe the same trends as with the segmentation IoU. The

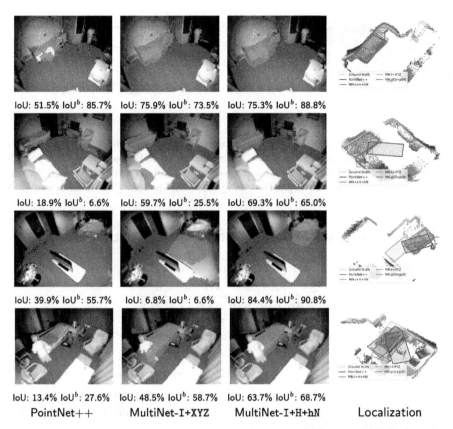

IoU: 51.5% IoU^b: 85.7%   IoU: 75.9% IoU^b: 73.5%   IoU: 75.3% IoU^b: 88.8%

IoU: 18.9% IoU^b: 6.6%   IoU: 59.7% IoU^b: 25.5%   IoU: 69.3% IoU^b: 65.0%

IoU: 39.9% IoU^b: 55.7%   IoU: 6.8% IoU^b: 6.6%   IoU: 84.4% IoU^b: 90.8%

IoU: 13.4% IoU^b: 27.6%   IoU: 48.5% IoU^b: 58.7%   IoU: 63.7% IoU^b: 68.7%

PointNet++   MultiNet-I+XYZ   MultiNet-I+H+hN   Localization

Fig. 5. Qualitative results of segmentation and localization. The first three columns show the segmentation results of PointNet++, MultiNet+I+XYZ, and Multi-Net+I+H+hN. The last column shows the localization results for the same models in a top view with height encoded in point color. Floor points have been removed for easier visualization. PointNet++ has a tendency to correctly locate the bed in most cases, but is unable to fully segment the bed. MultiNet+I+XYZ shows correct segmentation in most cases, but has problems locating beds in complex locations, for example in the third row. MultiNet+I+XYZ also suffers from oversegmentation, as shown in the first and last rows. MultiNet+I+H+hN is able to segment most cases correctly, but can suffer from biased segmentation for the localization, as in row 2 and 4. The last column shows clearly that a correct segmentation leads to good localization. *Best viewed in color.* (Color figure online)

gap between a formulation with and without floor calibration information is however reduced to 1.8% for IoU^b but widened to 4.4% for AP@.7 and AUC@.7.

The results obtained when feeding the network with the floor-plane projected intensity (proj(I)) and height (proj(H)) information of each 3D point are shown in the last row of Table 2. The average quality of the localization increases by another 0.9% compared to MultiNet+I+H+hN. However, when

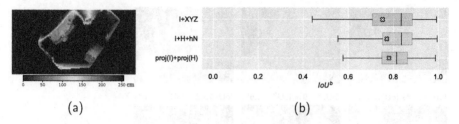

(a)   (b)

Fig. 6. (a) Example of `proj(H)`. Color indicating distance from floor. (b) Distribution of IoU^b for different models. From upper to lower: boxplot of the IoU^b for `I+XYZ`, `I+H+hN` and `proj(I)+proj(H)`. The white cross indicates the mean of each distribution. We can see that while the mean of the `proj(I)+proj(H)` model is higher, its median and 75th percentile are lower.

Table 3. Ablation of the MultiNet trained on `I+XYZ+H+N+hN`

	All	Independent removal of				
		I	N	XYZ	hN	H
IoU	70.6	67.1	66.0	63.4	57.3	27.6

`I` denotes intensity, `XYZ` spatial coordinates, `H` height, and `N` and `hN` local normals in the camera or floor-aligned referential.

looking at Fig. 6b, the higher medians of `I+XYZ` and `I+H+hN` reflect a higher number of bad bed localizations, but also a better bed localization whenever the viewing conditions are ideal.

Finally, we consider the relation between AP@.7 and AUC@.7 to evaluate the relevance of of our confidence threshold $rIoU^b$. AUC@.7 denotes the area-under-curve when plotting the percentage, compared to the whole set of samples, of correct (meaning with $IoU^b > 0.7$) localization as a function of the percentage of samples above the confidence threshold, for a progressively increased confidence threshold. By definition, this value is upper bounded by AP@.7, and gets close to this upper bound when samples with correct (erroneous) localization correspond to high (low) $rIoU^b$ confidence values.

4.4 Input Ablation

In order to strengthen the intuition that geometric information, and the way it is presented is important, we now look at a network trained with every possible input type: `I+XYZ+N+H+hN`. In that way, the network gets the opportunity to select its preferred representation of the spatial information.

The Direct Fusion, explained in more details in Sect. 3, is the summation of every input convolved once. Each input can thus individually be removed from the sum. Doing this outright leads to activation-range scaling issues since the network isn't retrained. Thus every channel in the removed component is replaced by the mean of its activations. We do this for every input-type independently, and present the results in Table 3.

The network is severely penalized when removing hN or H, which reveals that it consistently prefers height-encoded information. This confirms our hypothesis that a floor-aware encoding of the geometric information provides a representation that is easier to digest by the CNN.

5   Conclusion

We propose a method that successfully localizes beds in hospital and nursing home rooms. Our method estimates the extrinsic parameters of the device in order to have access to height maps and to normal vectors encoded in a referential aligned with the floor normal. We extensively show that this way of encoding ToF data leads to better performing CNNs. We recommend future work on input analysis using both visualization and network-based input-importance learning.

References

1. Chen, L.-C., Zhu, Y., Papandreou, G., Schroff, F., Adam, H.: Encoder-decoder with atrous separable convolution for semantic image segmentation. In: Ferrari, V., Hebert, M., Sminchisescu, C., Weiss, Y. (eds.) ECCV 2018. LNCS, vol. 11211, pp. 833–851. Springer, Cham (2018). https://doi.org/10.1007/978-3-030-01234-2_49

2. Fooladgar, F., Kasaei, S.: A survey on indoor RGB-D semantic segmentation: from hand-crafted features to deep convolutional neural networks. Multi. Tools Appl. **79**(7), 4499–4524 (2019). https://doi.org/10.1007/s11042-019-7684-3

3. Gupta, S., Girshick, R., Arbeláez, P., Malik, J.: Learning rich features from RGB-D images for object detection and segmentation. In: Fleet, D., Pajdla, T., Schiele, B., Tuytelaars, T. (eds.) ECCV 2014. LNCS, vol. 8695, pp. 345–360. Springer, Cham (2014). https://doi.org/10.1007/978-3-319-10584-0_23

4. Hazirbas, C., Ma, L., Domokos, C., Cremers, D.: FuseNet: incorporating depth into semantic segmentation via fusion-based CNN architecture. In: Lai, S.-H., Lepetit, V., Nishino, K., Sato, Y. (eds.) ACCV 2016. LNCS, vol. 10111, pp. 213–228. Springer, Cham (2017). https://doi.org/10.1007/978-3-319-54181-5_14

5. He, K., Zhang, X., Ren, S., Sun, J.: Deep residual learning for image recognition. CVPR (2016). https://doi.org/10.3389/fpsyg.2013.00124

6. Holz, D., Schnabel, R., Droeschel, D., Stückler, J., Behnke, S.: Towards semantic scene analysis with time-of-flight cameras. In: Ruiz-del-Solar, J., Chown, E., Plöger, P.G. (eds.) RoboCup 2010. LNCS (LNAI), vol. 6556, pp. 121–132. Springer, Heidelberg (2011). https://doi.org/10.1007/978-3-642-20217-9_11

7. Hou, J., Dai, A., NieBner, M.: 3D-SIS: 3D semantic instance segmentation of RGB-D scans. In: CVPR, pp. 4416–4425 (2019). doi: https://doi.org/10.1109/cvpr.2019.00455

8. Hu, J., Shen, L., Sun, G.: Squeeze-and-excitation networks. In: CVPR, pp. 7132–7141 (2018). https://doi.org/10.1109/CVPR.2018.00745

9. Hu, X., Yang, K., Fei, L., Wang, K.: ACNET: attention based network to exploit complementary features for rgbd semantic segmentation. In: ICIP, pp. 1440–1444 (2019). https://doi.org/10.1109/ICIP.2019.8803025

10. Jia, L., Radke, R.J.: Using time-of-flight measurements for privacy-preserving tracking in a smart room. IEEE Trans. Ind. Inform. **10**(1), 689–696 (2014). https://doi.org/10.1109/TII.2013.2251892

11. Jiang, J., Zheng, L., Luo, F., Zhang, Z.: RedNet: residual encoder-decoder network for indoor RGB-D semantic segmentation. arXiv preprint arXiv:1806.01054, pp. 1–14 (2018). http://arxiv.org/abs/1806.01054

12. Landrieu, L., Simonovsky, M.: Large-scale point cloud semantic segmentation with superpoint graphs. In: CVPR, pp. 4558–4567 (2018). https://doi.org/10.1134/1.559035, http://arxiv.org/abs/1711.09869

13. Li, G., Müller, M., Thabet, A., Ghanem, B.: DeepGCNs: an GCNs go as deep as CNNs?. In: ICCV (2019). http://arxiv.org/abs/1904.03751

14. Liang, Z., Yang, M., Deng, L., Wang, C., Wang, B.: Hierarchical depthwise graph convolutional neural network for 3D semantic segmentation of point clouds. In: Proceedings - IEEE International Conference on Robotics and Automation, May 2019, pp. 8152–8158 (2019). https://doi.org/10.1109/ICRA.2019.8794052

15. Loshchilov, I., Hutter, F.: Decoupled weight decay regularization. ICLR (2019)

16. Maddalena, L., Petrosino, A.: Background subtraction for moving object detection in RGBD data: a survey. J. Imaging, **4**(5) (2018). https://doi.org/10.3390/jimaging4050071

17. Milletari, F., Navab, N., Ahmadi, S.A.: V-Net: fully convolutional neural networks for volumetric medical image segmentation. In: Proceedings - 2016 4th International Conference on 3D Vision, 3DV 2016, pp. 565–571 (2016). https://doi.org/10.1109/3DV.2016.79

18. Paszke, A., et al.: PyTorch: An Imperative Style, High-Performance Deep Learning Library. NeurIPS (2019). http://arxiv.org/abs/1912.01703

19. Qi, C.R., Litany, O., He, K., Guibas, L.J.: Deep hough voting for 3D object detection in point clouds. In: ICCV (2019). http://arxiv.org/abs/1904.09664

20. Qi, C.R., Liu, W., Wu, C., Su, H., Guibas, L.J.: Frustum PointNets for 3D object detection from RGB-D data. In: CVPR (2018). 10.1109/CVPR.2018.00102, http://arxiv.org/abs/1711.08488

21. Qi, C.R., Su, H., Mo, K., Guibas, L.J.: PointNet: deep learning on point sets for 3D classification and segmentation. In: 2016 Fourth International Conference on 3D Vision (3DV), pp. 601–610 (2016). https://doi.org/10.1109/3DV.2016.68

22. Qi, C.R., Yi, L., Su, H., Guibas, L.J.: PointNet++: deep hierarchical feature learning on point sets in a metric space. Adv. Neural Inf. Process. Syst. (2017). https://doi.org/10.3109/13816819409056905

23. Ronneberger, O., Fischer, P., Brox, T.: U-Net: convolutional networks for biomedical image segmentation. In: Navab, N., Hornegger, J., Wells, W.M., Frangi, A.F. (eds.) MICCAI 2015. LNCS, vol. 9351, pp. 234–241. Springer, Cham (2015). https://doi.org/10.1007/978-3-319-24574-4_28

24. Shi, S., Wang, X., Li, H.: PointRCNN: 3D object proposal generation and detection from point cloud. In: CVPR, pp. 770–779. IEEE (6 2019). https://doi.org/10.1109/CVPR.2019.00086

25. Simon, C., Meessen, J., De Vleeschouwer, C.: Visual event recognition using decision trees. Multi. Tools Appl. **50**(1), 95–121 (2010). https://doi.org/10.1007/s11042-009-0364-y

26. Song, S., Yu, F., Zeng, A., Chang, A.X., Savva, M., Funkhouser, T.: Semantic scene completion from a single depth image. In: CVPR (2017). https://doi.org/10.1109/CVPR.2017.28http://arxiv.org/abs/1611.08974

27. Wijmans, E.: PointNet++ Pytorch Implementation. Github (2018). https://github.com/erikwijmans/Pointnet2_PyTorch

28. Yang, L., Ren, Y., Zhang, W.: 3D depth image analysis for indoor fall detection of elderly people. Digit. Commun. Netw. **2**(1), 24–34 (2016). https://doi.org/10.1016/j.dcan.2015.12.001
29. Yu, C., et al.: BiSeNet: bilateral segmentation network for real-time semantic segmentation. In: Ferrari, V., Hebert, M., Sminchisescu, C., Weiss, Y. (eds.) ECCV 2018. LNCS, vol. 11217, pp. 334–349. Springer, Cham (2018). https://doi.org/10.1007/978-3-030-01261-8_20

Development and Evaluation of a Mouse Emulator Using Multi-modal Real-Time Head Tracking Systems with Facial Gesture Recognition as a Switching Mechanism

Shivanand P. Guness[1]([⊠])(iD), Farzin Deravi[2](iD), Konstantinos Sirlantzis[2],
Matthew G. Pepper[2,3], and Mohammed Sakel[3]

[1] net.connection Ltd., Moka, Mauritius
`shivam.guness@ieee.org`
[2] School of Digital Arts and Engineering, University of Kent, Canterbury, Kent, UK
`{f.deravi,k.sirlantzis}@kent.ac.uk`
[3] East Kent University Hospitals Trust, Canterbury, Kent, UK
`Matthew.Pepper@ekht.nhs.uk, msakel@nhs.net`

Abstract. The objective of this study is to evaluate and compare the performance of a set of low-cost multi-modal head tracking systems incorporating facial gestures as a switching mechanism. The proposed systems are aimed to enable severely disabled patients to access a computer. In this paper, we are comparing RGB (2D) and RGB-D (3D) sensors for both head tracking and facial gesture recognition. System evaluations and usability assessment were carried out on 21 healthy individuals. Two types of head tracking systems were compared - a web camera-based and another using the Kinect sensor. The two facial switching mechanisms were eye blink and eyebrows movement. Fitts' Test is used to evaluate the proposed systems. Movement Time (MT) was used to rank the performance of the proposed systems. The Kinect-Eyebrows system had the lowest MT, followed by the Kinect-Blink, Webcam-Blink and Webcam-Eyebrows systems. The 3D Kinect systems performed better than the 2D Vision systems for both gestures. Both Kinect systems have the lowest MT and best performance, thus showing the advantage of using depth.

Keywords: Assistive technology · Facial gesture recognition · Fitts' test · Eye blink detection · Eyebrow movement

1 Introduction

The World Health Organization (WHO) estimated that 1 billion people around the world live with some form of disability [36]. Approximately 10 million people in UK have disabilities with a neurological diagnosis. For a multitude of reasons, the number of people with profound disability stemming from neurological disorders is increasing with a resulting impact on their quality of life and that of

A. Del Bimbo et al. (Eds.): ICPR 2020 Workshops, LNCS 12662, pp. 218–232, 2021.
https://doi.org/10.1007/978-3-030-68790-8_18

their caregiver. The cost of caring for neuro-disabled persons in Europe has been estimated as 795 Billion Euro [17]. The value of assistive technologies in improving the quality of life of people with disability and also reduce carer strain is emphasized in a 2010 Royal College of Physicians Report [30].

For many individuals with disability access to a computer and/or communication aid may help mitigate the effect of communication impairments. Often this can be achieved through the identification of suitable access sites e.g. hand, foot, arm or head. Some patients, however, are profoundly disabled that they might be unable to talk but can only make small head movements and facial gestures such as eye blink or eyebrow movement. In some cases there may not even be enough head movement to enable the use of an access device such as a head tracker like SmartNav [2] and so the only remaining access site may be small facial gestures. Although there are other options available - e.g. the use of eye gaze, existing systems using eye gaze technology such as MyTobii [3] are complex, expensive and set-up/configuration places a significant burden on both the user and the caregiver.

The motivation for the work reported in this paper is the need for low-cost, reliable head tracking with an automatic facial gesture recognition system to help severely disabled users access electronic assistive technologies. The objective is to develop a multi-modal head tracking system, which uses facial gestures as a switching mechanism thus enabling severely disabled patients whose control is restricted to small head movements and facial gestures to be able to access a computer.

2   Background

Pistori [29], states that assistive devices using computer vision can have a great impact in increasing the digital inclusion of people with special needs. Computer vision can improve both the devices used for mobility i.e. controlling motorised wheel chairs, sign language detection and head trackers.Similarly, Betke et al. [7], describe the advances made in the development of assistive software and the use of emerging technology can lead to the creation of intelligent interfaces using both assistive technology and human computer interaction (HCI). The example of the CameraMouse [9] is used as an interface system with different assistive devices and software such as Midas Touch [6], Dasher [34] etc. are included to highlight the use of HCI and assistive devices.

Abascal et al. [4], highlighted some opportunities and challenges that designing human-computer interfaces suitable for the disabled can pose. For people suffering from disabilities, HCI can be used to design better interfaces which could be accessible to people with disabilities and thus improve socialisation, better access to communication facilities and have a greater control over their environment.

2.1 Device Evaluation

Fitts' test [14] was developed in 1954 to model human movement. The result of the experiments showed that the rate of performance of the human motor system is approximately constant over a wide range of movement amplitudes. Mackenzie et al. [23], adapted the Fitts' Law for assessing HCI. This work was later embedded in an International Standard for HCI, ISO 9431-9:2000 [18] providing guidelines for measuring the users' performance, comfort and effort. The performance of the device was measured by making the user perform tasks using the device. There are six types of tasks - one-direction, multi-directional, dragging, free-hand tracing (drawing), and, hand input, grasp and park (homing/device switching). ISO 9431-9:2000 [18] requires that the input device be tested for at least 2 different Index of Difficulty (ID). Index of Difficulty (ID) is a measure of the difficulty of the task [5]. In Douglas et al. [12], the validity and practicality of the ISO framework using both multi-directional and the one-direction Fitts' Tests for two devices namely a touch-pad and a joystick was investigated.

2.2 Gesture Detection

In this paper, the interest is in processing video information to recognise blink and eyebrow movement gestures. The detected gestures can be to emulate a mouse click or a switch action to access and control a computer/communication aid.

Grauman et al. [15] proposed two systems called BlinkLink and EyebrowsClicker. The BlinkLink software tracked both the motion within the eye region and the eye region itself. The EyebrowsClicker tracked the eyebrows region and detected the rising and falling of the eyebrows. To initialise the location of the eye and eyebrows regions, the user has to perform the gestures and by analysing the area of motion on the face, the respective regions are detected. A template of each region is generated. The correlation score of the eye region and a template of both the closed eye and open eye were compared to detect an eye blink. For eyebrows gesture, the distance from the eyes and the eyebrows are monitored to detect the rise and fall motion of the eyebrows. Blink detection had an overall success rate of around 95.6% and was tested on 15 healthy individuals and one person suffering from Traumatic Brain Injury (TBI). EyebrowsClicker had an overall success rate or 89% and it was tested with six individuals, but the software had to be reinstated twice during the data capture session because the tracking of the eyebrows was lost. There has been no further published work on this system.

Malik et al. [25] proposed a blink detection method using histogram of Local Binary Patterns (LBP) [27]. A template of open eye was generated using the average histogram of LBP from a sample of 50 images of an open eye. The histogram of LBP of images of the eye region were compared against the template using the Kullback-Leibler Divergence (KLD) method. In KLD, the distance between two distribution is zero only if the distributions are identical. KLD was found to be robust against both the precision of the eye detection and the variation in the window size of the detected eye region. The eye region are obtained using the Viola-Jones [33] algorithm implemented in OpenCV. The proposed algorithm was tested against the ZJU Eye blink Database [28] and resulted in a 99.2% blink detection rate. Missimer et al. [26] proposed a blink detection algorithm based on the analysis of the differences in three consecutive images. Blobs are generated from the merging of two difference images produced. Three points are used for tracking, the centre of the upper lip and the upper part of both eyebrows. In addition, optical flow is used to track these three points. The eye templates are generated based on the tracked points and used to train the system. The system is reported to having a success rate of 96.6% and was tested on 20 healthy individuals.

Yunqi et al. [37] proposed an eye blink detection algorithm which was used in a drowsiness driver warning system. The proposed system used Haar-like [32] features and AdaBoost to detect the face of the user. Some pre-processing was performed on the image and an edge detection algorithm was used to find the eye corners, the iris and the upper eyelid for each eye. The curvature of the upper eyelid was compared with the line connecting the two eye corners and if most of the upper eyelid curvature was under this line, the eye was considered closed. The algorithm was tested on images captured during a real driving session and 94% accuracy was obtained for the eye state detection.

In Zhang et al. [38], proposed a Gaze based assistive application on a smartphone to enable the user to communicate. The application can recognise six gestures from both eyes namely look up, look down, look right, look left, look center and closed eyes. The algorithm used OpenCV [8] and Dlib-ML [21]. Before using the device, calibration must be performed to create templates for each gesture. The template is created by making the users perform the gesture and capturing the image of the eye region when the action is performed. The algorithm detected the gestures with an accuracy of 86% on average. The accuracy rate decreased to 80.4% for people wearing glasses, increased to 89.0% for people wearing contact lenses and increased to 89.7% for people without glasses.

In Val et al. [11], eye blinks are used to control a robot. An infra-red emitter and an optical sensor were used to detect the eye blink. The blinks are used to navigate the robotic assistive aid, for example a right eye blink would cause the robot turn right and a left eye blink would make the robot turn left. A combination of the left blink followed by a right blink would cause the robot stop. In Krolak et al. [22], the proposed method uses two active contour [20] models - one for each eye - for detecting eye blinks. Haar-like features [32] are used to detect the face and the location of the eyes are determined using known geometrical proportion of the human face.

In Tuisku et al. [31], the evaluation of a system called Face Interface was conducted. The system used voluntary gaze direction for moving the cursor around the screen and facial muscle activation for the selecting objects on a computer screen. Face Interface used two different muscle activation - frowning and raising the eyebrows. A series of points were presented to the user. The time to complete the tasks and the accuracy of the activation were used as performance measure. The pointing tasks were conducted using three different target diameters (i.e. 25, 30, 40 mm), seven distances (i.e., 60, 120, 180, 240, 260, 450, and 520 mm), and eight pointing angles namely (0°, 45°, 90°, 135°, 180°, 225°, 270°, and 315°). It was found that for distances between 60 mm and 260 mm, tasks performed using the raising eyebrow selection technique were faster than those using the frowning technique. Also, the overall time taken to complete the tasks were 2.4 s for the frowning technique and 1.6 s for the raising technique. The IP of the frowning techniques was 1.9 bits/s and 5.4 bits/s for the eyebrow raising technique.

The systems reported here were limited in that they would only work with frontal facial images and were not robust in coping with posture changes. The work reported here aims to address these shortcomings by making use of the depth data available from RGB-D sensors.

3   Materials and Methods

The systems evaluated in this work incorporate a camera and an algorithm for tracking the head movement and detection of the eye blink or eyebrow movement facial gestures. The camera is either the Microsoft Kinect for Windows [1] sensor which can provide 3D (RGB-D) data or a Logitech web camera which can only provide 2D (RGB) data. Raw data is extracted in the form of images and depth maps. The efficacy of head tracking and gesture recognition of 3D vision-based system is compared to 2D vision-based systems using a modified Fitts' test.

3.1   Device Evaluation

Fitts' Test. Fitts originally proposed a method to model the human hand movement in order to improve human-machine interactions [13]. Each task has an ID which is based on the size of the target and the distance of the target from the starting point. The ID represents the cognitive-motor challenge imposed on the human to accomplish the task and is measured in bits as shown in Eq. (1).

$$ID = log_2(\frac{D}{W} + 1) \tag{1}$$

where D represents the distance from the starting point to the target and W is the width of the target.

$$MT = a + b \times ID \tag{2}$$

The relationship between MT and ID is shown as a linear relationship where a is the y-intercept and b is the gradient of the line represented in Eq. (2). The Index of Performance (IP) in bits/second of a device is given in Eq. (3).

$$IP = \frac{1}{b} \tag{3}$$

where b is the gradient of the line described in Eq. (3). A positive value of IP indicates that the device gets more difficult to use as the interaction becomes more challenging. Equation (4) is used to calculate the Effective Throughput (TP_e) in bits/second.

$$TP_e = \frac{ID_e}{MT} \tag{4}$$

where MT is the mean movement time, in seconds, for all trials within the same condition. It represents the overall efficiency of the device in facilitating interactions.

$$ID_e = log_2(\frac{D}{W_e} + 1) \tag{5}$$

ID_e, is the effective index of difficulty, in bits, and is calculated from the distance (D) from the start location to the target and W_e, the effective width of the target. W_e, is the effective width of the target and it is calculated from the observed distribution of the target selection coordinates.

$$W_e = 4.133 \times SD \tag{6}$$

where SD is the standard deviation of the selection coordinates [12].

The experiments showed that the rate of performance of the human motor system is approximately constant over a wide range of movement amplitudes. Fitts' Law [14,23] states that MT should increase with an increase in the ID i.e. as the difficulty of the task increases, the time taken to complete the task also increases. Fitts' Law was adapted in Mackenzie et al. [23], to assess HCI devices. Therefore, it was thought Fitts' test is an appropriate tool for assessing the performance of the head tracking and gesture recognition system.

3.2 Gesture Detection

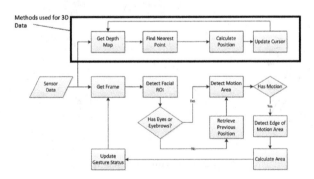

Fig. 1. Algorithm to detect blink and eyebrows movements.

Figure 1 shows an overview of the 3D head tracking and facial gesture recognition system. The facial gesture recognition system is the same for both the 2D vision system and the 3D Kinect system. Depth data is used only to filter the region of interest when processing the facial image - only objects within a meter of the 3D sensor were included in the region of interest and all other background is removed before further processing.

The facial gesture recognition system used the RGB data from the sensors. Facial areas of interest such as the head, eyes region, left eye and right eye are detected using a Haar-Cascade [32]. To detect a blink, closure of both eyes has to be detected for a period of 1 s or more and then return to the open state. If closure of only one eye is detected, the system assumes there is no blink. Only the transition from open eye to close eye and to open eye again is recognised as a blink.

In the case of the eyebrows detection, the two states of the eyebrows (raised, down) are monitored. In the eyebrows raised state the facial eyebrows muscles are contracted in order to raise the eyebrows and the down state, the muscles are relaxed and the eyebrows revert to their original location. The eyebrows region is detected using the location above the eye region. The state of the eyebrows is initially set to down. To recognise eyebrows movement both eyebrows have to be raised for a period of 1 s or more and subsequently return to the down state. Only the transition from down to raised and then to the down state again will be recognised as a valid eyebrows movement.

4   Experimentation

4.1   Setup

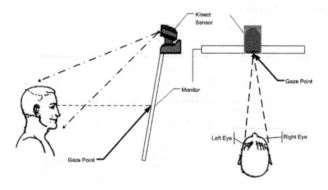

Fig. 2. Experimental set-up.

The participant was asked to perform a series of Fitts' Tests [14,24]. The participants were allowed to repeat the gestures until the click action was detected and thus this caused the movement time to increase. The Fitts' Test was used to evaluate two devices: a 2D vision based head tracker using the Logitech web camera and a 3D head tracking system using the Kinect device. The experiment was performed using the two facial gestures (blink and eyebrows movement) as a switching mechanism. It has been reported that spontaneous eye blink can change from 20 to 30 blinks/min depending on the mental task the person is performing [19], and can decrease to about 11 blinks per minute during visually demanding tasks [35]. Therefore, the intentional blink time threshold was set to 1000 ms to distinguish between intentional and unintentional facial gestures and to prevent spontaneous blinks from being detected. The activation time of the eyebrows movement switch was also set to 1000 ms (Fig. 2).

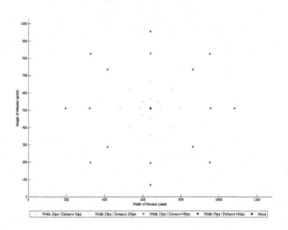

Fig. 3. Target locations (incorporating 8 distinct movement orientations).

The screen used is a 17 in. LCD monitor with a resolution of 1280 by 1024 pixels. The target is selected at random from a set of pre-designated locations as shown in Fig. 3 and presented to the participant. The participant then has to move the cursor using head movement and select the target with the equivalent of a mouse click using the different facial gestures being evaluated. Once a target has been chosen, the participant has to move the cursor back to and select the target at the central location on the screen. This ensures that the same start point is used for each target selection. The choice of the stimulus target locations are based on earlier work by Guness et al. where the points were configured to perform a range of selection tasks with 8 target directions/orientations [16].

4.2   Sensors

Two sensors were used. The first was a standard Logitech web camera. The web camera captured 640 × 480 pixel RGB images at a rate of 30 frames per second. The second sensor was the Kinect for Windows sensor [1]. The Kinect sensor consists of a structured light based depth sensor and an RGB sensor. The Kinect sensor operates at a 30 Hz rate and generates 640 × 480 depth and RGB images. The depth range of the Kinect sensor in default mode is 800 mm to 4000 mm and in near mode is 500 mm to 3000 mm. In this experiment the

Kinect sensor operated in near mode. Both the web camera and the Kinect sensor were selected because they are relatively inexpensive devices that can be readily obtained.

4.3   Depth Data

The depth data obtained from the Kinect sensor is used to reduce the search area for the different Haar-Cascade features. This will reduce the computational load and will avoid background distractions, such as people, movements and changes in lighting and therefore increase the performance. A mask is created from the depth data and the object within 1000 mm of the sensor is selected. The mask is used on the colour image to remove all the objects which are more than 1000 mm from the sensor.

5   Result

The experiment was carried out with 21 healthy individuals who completed the tests with all 4 devices. The MT in Fitts' Test is the time taken to move to the target location from the starting point and performing the task. To be able to compare the devices and the effect of the facial gesture, we have broken the task in two. Task 1 involves moving the cursor to the target location using the movement of the head. Task 2 encapsulates Task 1 and also involves selecting the target by using one of the facial gestures as a switching mechanism.

In Figure 4, the Kinect-eyebrows has a lower MT that the Kinect-blink system for an ID greater than 1.9 bits. Overall for Task 1, it can be seen that the Kinect-eyebrows system has the lowest MT, followed by the Kinect-blink, the webcam-blink and finally the webcam-eyebrows, which took the most time to complete (Fig. 5).

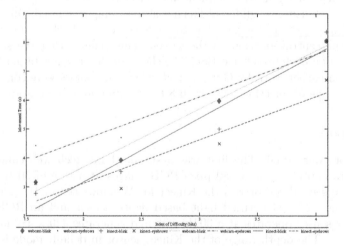

Fig. 4. Fitts' test result for Task 1 (movement to target).

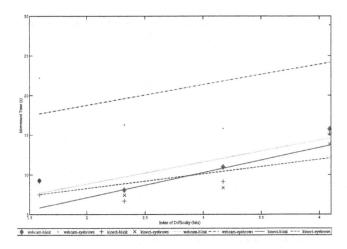

Fig. 5. Fitts' test result for Task 2 (performing the facial gesture).

Table 1. Overall index of performance (IP) and effective throughput (TP_e) of tested devices

Device	Task 1		Task 2	
	IP	Tp_e	IP	Tp_e
Webcam-Blink	0.36	0.74	0.32	0.41
	($R^2 = 0.98$)		($R^2 = 0.78$)	
Webcam-Eyebrows	0.39	0.67	0.55	0.37
	($R^2 = 0.90$)		($R^2 = 0.21$)	
Kinect-Blink	0.5	0.89	0.45	0.6
	($R^2 = 0.89$)		($R^2 = 0.78$)	
Kinect-Eyebrows	0.68	0.95	0.67	0.64
	($R^2 = 0.81$)		($R^2 = 0.48$)	

From Table 1, it can also be seen that both the IP and the TP_e for moving the cursor to the designated target (Task 1) were better than that of the combination of moving and the click action (Task 2) using the different facial gestures for all devices. This is to be expected as the clicking/selection method has an effect on the performance and efficiency of the system used. Also, both the IP and TP_e of the 3D Kinect system were better than those of the 2D Vision system. R^2 is the coefficient of determination and measured as a percentage of how well the data fits the linear model [10]. If we look at the R^2 values Task 1 are higher than those of Task 2, this would indicate that Task 1 follows the linear model more closely than Task 2. Also, another interesting observation is the fact that using the Blink gesture with both the web camera and the Kinect yield similar R^2 values whereas the R^2 values of the Eyebrows movement gesture are lower.

In Table 2, the IP for different devices are presented when performing Task 1 and a combination of Task 1 followed by Task 2 for different target orientations.

Table 2. IP of Task 1 and Task 2 in bits/second

Orientation	IP of Task 1 (bits/second)								IP of Task 2 (bits/second)							
	0	45	90	135	180	225	270	315	0	45	90	135	180	225	270	315
Webcam-Blink	0.51	0.56	0.42	0.52	0.48	0.4	0.49	0.56	0.63	0.91	0.44	0.12	0.28	0.38	0.7	0.47
Webcam-Eyebrows	0.94	0.29	1.43	0.67	0.84	1.57	0.83	0.57	0.34	−0.35	−0.73	3.74	0.1	0.13	0.22	1.07
Kinect-Blink	0.27	0.57	0.49	0.5	0.3	0.35	0.68	0.77	0.29	0.88	0.21	0.37	0.12	0.3	0.37	0.52
Kinect-Eyebrows	0.56	0.51	1.26	0.35	0.53	1.23	1.94	0.65	0.31	1.54	0.43	0.22	0.21	−5.41	2.82	1.1

Table 3. TP_e of Task 1 and Task 2 in bits/second

Orientation	TP_e of Task 1 (bits/second)								TP_e of Task 2 (bits/second)							
	0	45	90	135	180	225	270	315	0	45	90	135	180	225	270	315
Webcam-Blink	0.67	0.88	0.67	0.8	0.81	0.59	0.88	0.69	0.41	0.45	0.39	0.34	0.47	0.33	0.44	0.41
Webcam-EyebrowsEyebrows	0.96	0.62	0.61	0.71	0.62	0.64	0.62	0.62	0.68	0.34	0.32	0.35	0.35	0.3	0.35	0.36
Kinect-Blink	0.82	0.8	0.91	0.88	0.73	0.83	1.04	0.98	0.45	0.54	0.63	0.64	0.51	0.62	0.71	0.64
Kinect-Eyebrows	0.85	0.99	1.12	0.78	0.79	0.9	1.39	0.92	0.54	0.55	0.68	0.44	0.54	0.63	1.06	0.67

A one-way ANOVA test was performed on the TP_e for the different orientations and gestures of both Task 1 and Task 2. For the comparison by orientations, $p < 0.01$ ($p = 0.001$ and $p = 0.007$) for Task 1 and Task 2, it can be said that there is a significant difference between the mean of the different orientations i.e. the TP_e are different based on the orientation of the movement. For the comparison by gesture, only Task 2 had $p < 0.01$ ($p = 0.0093$). This indicates that there is a significant difference between the mean of the TP_e based on the gesture being performed. This would point out that there is a difference in the performance of the two facial gestures being investigated. The mean TP_e of Task 2 is greater due to the increased challenge of both moving and selecting/clicking. Also, there are no sufficient evidence of any difference between the means of TP_e based on the orientation and gesture for either Task 1 or Task 2. This would indicate the gesture recognition for the sample used might be invariant to the orientation of the task being performed.

6   Discussion

Using facial gestures as a switch is possible in real time but the use of such gestures may cause a drop in the overall IP of the systems. IP and TP_e values in Table 1 using the four different systems were obtained with participants successfully reaching and selecting all targets. As it can be seen in results for the overall IP (Table 1), the R^2 value which represents the goodness of fit of the fitted line for the Kinect 3D system is greater than 0.7 i.e. the line accounts for more than 70% of the variance. In contrast, the Webcam-Eyebrows device R^2 is 0.21, and thus accounts for only 21% of the variance. This could also indicate that the presence of outliers has a large influence on the fitted line and thus the gradient. As the IP calculation from Eq. (3) is based on the inverse of the slope, it is also being influenced by outliers at very low and very high indices of difficulty. It should be born in mind that each of the points in Figs. 4 and 5 are obtained from the mean of data obtained from 21 users and 8 directions giving 64 data points. In the presence of such outliers relying on TP_e as a measure of performance might be better.

It can be seen that there is a decrease in the TP_e of all the four different devices after the switching action is included. The reduction in the TP_e of the 2D Vision system is 45% and 44% for the blink and eyebrows devices respectively. Similarly, the decline in the TP_e of the 3D Kinect system is 32% and 35% for the blink and eyebrows devices respectively. The higher total TP_e value indicates that the Kinect system, utilizing 3D information, has resulted in better performance when the two tasks of moving and selecting are combined and thus improved the ease of use of the system as a whole. It has also been shown that the TP_e for Task 2 based on gesture are from different populations - with eyebrows having a higher mean TP_e. There is no evidence to support a difference in performance based on sensor or device. This also supports the impact to the improved performance of the gesture detection algorithm.

In addition, the facial gesture detection rate affected the MT for the different devices. In this implementation of the Fitts' test, the tasks were considered completed only when the switch was activated and click action performed.

7   Conclusion

Both Kinect systems have lower MT and higher IP and TP_e than the Webcam based systems thus showing that the introduction of the depth data had a positive impact on the head tracking algorithm. This could be explained by the ability to throw away unnecessary data at an early stage in processing using depth information and thereby speeding up subsequent stages to create a more smooth experience for the users. In this work, we have looked at only blink and eyebrows movement gestures, further work will have to be carried out on additional gestures such as mouth opening/closing and tongue movement. We now intend to conduct translational research with neurological patients.

References

1. Kinect for Windows. http://www.microsoft.com/en-us/kinectforwindows/
2. SmartNav. http://www.naturalpoint.com/smartnav/
3. Tobii Technology. http://www.tobii.com/

4. Abascal, J., Nicolle, C.: Moving towards inclusive design guidelines for socially and ethically aware HCI. Interact. Comput. **17**(5), 484–505 (2005). https://doi.org/10.1016/j.intcom.2005.03.002. http://iwc.oxfordjournals.org/cgi/doi/10.1016/j.intcom.2005.03.002

5. Accot, J., Zhai, S.: Beyond Fitts' law: models for trajectory-based HCI tasks. In: Proceedings of the ACM SIGCHI Conference on Human Factors in Computing Systems, pp. 295–301 (1997). https://doi.org/10.1145/258549.258760. http://dl.acm.org/citation.cfm?id=258760

6. Betke, M., Gips, J., Fleming, P.: The camera mouse: visual tracking of body features to provide computer access for people with severe disabilities. IEEE Trans. Neural Syst. Rehabil. Eng. **10**(1), 1–10 (2002). A Publication of the IEEE Engineering in Medicine and Biology Society

7. Betke, M.: Intelligent interfaces to empower people with disabilities. In: Nakashima, H., Aghajan, H., Augusto, J.C. (eds.) Handbook of Ambient Intelligence and Smart Environments, pp. 409–432. Springer, Boston (2010). https://doi.org/10.1007/978-0-387-93808-0_15

8. Bradski, G.: The OpenCV library. Dr. Dobb's J. Softw. Tools **25**, 120–125 (2000)

9. Cloud, R., Betke, M., Gips, J.: Experiments with a camera-based human-computer interface system. In: Proceedings of the 7th ERCIM Workshop "User Interfaces for All," UI4ALL 2002, pp. 103–110 (2002). http://cstest.bc.edu/~gips/UI4ALL-2002.pdfwww.cs.bu.edu/faculty/betke/papers/Cloud-Betke-Gips-UI4ALL-2002.pdf

10. Cox, D.R., Snell, E.J.: Analysis of Binary Data, vol. 32. CRC Press, Boca Raton (1989)

11. del Val, L., Jiménez, M.I., Alonso, A., de la Rosa, R., Izquierdo, A., Carrera, A.: Assistance system for disabled people: a robot controlled by blinking and wireless link. In: Lytras, M.D., Ordonez De Pablos, P., Ziderman, A., Roulstone, A., Maurer, H., Imber, J.B. (eds.) WSKS 2010. CCIS, vol. 111, pp. 383–388. Springer, Heidelberg (2010). https://doi.org/10.1007/978-3-642-16318-0_45

12. Douglas, S.A., Kirkpatrick, A.E., MacKenzie, I.S.: Testing pointing device performance and user assessment with the ISO 9241, part 9 standard. In: Proceedings of the SIGCHI Conference on Human Factors in Computing Systems, CHI 1999, pp. 215–222. ACM, New York (1999). https://doi.org/10.1145/302979.303042

13. Fitts, P.M., Radford, B.: Information capacity of discrete motor responses under different cognitive sets. J. Exp. Psychol. **71**, 475–482 (1966)

14. Fitts, P.: The information capacity of the human motor system in controlling the amplitude of movement. J. Exp. Psychol. **47**(6) (1954). http://psycnet.apa.org/journals/xge/47/6/381/

15. Grauman, K., Betke, M., Lombardi, J., Gips, J., Bradski, G.: Communication via eye blinks and eyebrow raises: video-based human-computer interfaces. Univ. Access Inf. Soc. **2**(4), 359–373 (2003). https://doi.org/10.1007/s10209-003-0062-x

16. Guness, S.P., Deravi, F., Sirlantzis, K., Pepper, M.G., Sakel, M.: Evaluation of vision-based head-trackers for assistive devices. In: Proceedings of the Annual International Conference of the IEEE Engineering in Medicine and Biology Society, EMBS, pp. 4804–4807 (2012). https://doi.org/10.1109/EMBC.2012.6347068. http://ieeexplore.ieee.org/xpls/abs_all.jsp?arnumber=6347068

17. Gustavsson, A., et al.: Cost of disorders of the brain in Europe 2010. Eur. Neuropsychopharmacol. J. Eur. Coll. Neuropsychopharmacol. **21**(10), 718–79 (2011). https://doi.org/10.1016/j.euroneuro.2011.08.008. http://www.ncbi.nlm.nih.gov/pubmed/21924589

18. ISO TC 159/SC 4: ISO 9241–9:2000, ergonomic requirements for office work with visual display terminals (VDTs) - part 9: requirements for non-keyboard input devices. International Organization for Standardization (2002)
19. Karson, C.N.: Spontaneous eye-blink rates and dopaminergic systems. Brain: J. Neurol. **106**(Pt 3), 643–53 (1983). http://www.ncbi.nlm.nih.gov/pubmed/6640274
20. Kass, M., Witkin, A., Terzopoulos, D.: Snakes: active contour models. Int. J. Comput. Vis. **1**(4), 321–331 (1988). https://doi.org/10.1007/BF00133570
21. King, D.E.: Dlib-ml: a machine learning toolkit. J. Mach. Learn. Res. (JMLR) **10**, 1755–1758 (2009). http://dl.acm.org/citation.cfm?id=1577069.1755843
22. Krolak, A., Strumillo, P.: Vision-based eye blink monitoring system for human-computer interfacing. In: 2008 Conference on Human System Interactions. Institute of Electrical & Electronics Engineers (IEEE), May 2008. https://doi.org/10.1109/hsi.2008.4581580
23. MacKenzie, I.S.: Fitts' law as a research and design tool in human-computer interaction. Hum. -Comput. Interact. **7**(1), 91–139 (1992). https://doi.org/10.1207/s15327051hci0701_3
24. MacKenzie, I.S., Buxton, W.: A tool for the rapid evaluation of input devices using Fitts' law models. ACM SIGCHI Bull. **25**(3), 58–63 (1993). https://doi.org/10.1145/155786.155801. http://portal.acm.org/citation.cfm?doid=155786.155801
25. Malik, K., Smolka, B.: Eye blink detection using local binary patterns. In: 2014 International Conference on Multimedia Computing and Systems (ICMCS), pp. 385–390, April 2014. https://doi.org/10.1109/ICMCS.2014.6911268
26. Missimer, E., Betke, M.: Blink and wink detection for mouse pointer control. ACM Press, New York (2010). https://doi.org/10.1145/1839294.1839322. http://portal.acm.org/citation.cfm?doid=1839294.1839322
27. Ojala, T., Pietikäinen, M., Harwood, D.: A comparative study of texture measures with classification based on featured distributions. Pattern Recogn. **29**(1), 51–59 (1996). https://doi.org/10.1016/0031-3203(95)00067-4
28. Pan, G., Sun, L., Wu, Z., Lao, S.: Eyeblink-based anti-spoofing in face recognition from a generic webcamera. In: 2007 IEEE 11th International Conference on Computer Vision. Institute of Electrical & Electronics Engineers (IEEE) (2007). https://doi.org/10.1109/iccv.2007.4409068
29. Pistori, H.: Computer vision and digital inclusion of persons with special needs: overview and state of art. In: Computational Modelling of Objects Represented in Images (2018). https://www.taylorfrancis.com/books/e/9781351377133/chapters/10.1201%2F9781315106465-6
30. Royal College of Physicians: Medical rehabilitation in 2011 and beyond. Report of a working party, November 2010, Royal College of Physicians & British Society of Rehabilitation Medicine. RCP, London (2010)
31. Tuisku, O., Surakka, V., Vanhala, T., Rantanen, V., Lekkala, J.: Wireless face interface: using voluntary gaze direction and facial muscle activations for human-computer interaction. Interact. Comput. **24**(1), 1–9 (2012). https://doi.org/10.1016/j.intcom.2011.10.002. http://iwc.oxfordjournals.org/cgi/doi/10.1016/j.intcom.2011.10.002
32. Viola, P., Jones, M.J.: Robust real-time face detection. Int. J. Comput. Vis. **57**(2), 137–154 (2004). https://doi.org/10.1023/B:VISI.0000013087.49260.fb
33. Viola, P., Jones, M.: Robust real-time object detection. Int. J. Comput. Vis. **57**, 137–154 (2001). https://doi.org/10.1023/B:VISI.0000013087.49260.fb

34. Ward, D.J., Blackwell, A.F., MacKay, D.J.C.: Dasher–a data entry interface using continuous gestures and language models, pp. 129–137. ACM Press (2000). https://doi.org/10.1145/354401.354427. http://portal.acm.org/citation.cfm?doid=354401.354427

35. Wilson, G.F.: An analysis of mental workload in pilots during flight using multiple psychophysiological measures. Int. J. Aviat. Psychol. **12**(1), 3–18 (2002)

36. World Health Organization: World report on disability. Technical report, WHO, Geneva (2011). http://whqlibdoc.who.int/publications/2011/9789240685215_eng.pdf

37. Yunqi, L., Meiling, Y., Xiaobing, S., Xiuxia, L., Jiangfan, O.: Recognition of eye states in real time video, pp. 554–559, January 2009. https://doi.org/10.1109/ICCET.2009.105. http://ieeexplore.ieee.org/lpdocs/epic03/wrapper.htm?arnumber=4769528

38. Zhang, X., Kulkarni, H., Morris, M.R.: Smartphone-based gaze gesture communication for people with motor disabilities. In: Proceedings of the 2017 CHI Conference on Human Factors in Computing Systems, CHI 2017, pp. 2878–2889. ACM, New York (2017). https://doi.org/10.1145/3025453.3025790

A Video-Based MarkerLess Body Machine Interface: A Pilot Study

Matteo Moro[1,2(✉)], Fabio Rizzoglio[1,3,4], Francesca Odone[1,2], and Maura Casadio[1]

[1] Department of Informatics, Bioengineering, Robotics and System Engineering (DIBRIS), University of Genova, Genova, Italy
{matteo.moro,fabio.rizzoglio}@edu.unige.it,
{francesca.odone,maura.casadio}@unige.it
[2] Machine Learning Genoa Center (MaLGa Center), University of Genova, Genova, Italy
[3] Department of Physiology, Feinberg School of Medicine, Northwestern University, Chicago, IL 60611, USA
[4] Shirley Ryan Ability Lab, Chicago, IL 60611, USA

Abstract. Regaining functional independence plays a crucial role to improve the qualify of life of individuals with motor disabilities. Here, we address this problem within the framework of Body-Machine Interfaces (BoMIs). BoMIs enable individuals with restricted mobility to extend their capabilities by mapping their residual body movements into commands to control an external device. In this study, we propose a video-based marker-less interface that can track the position of the shoulders and the head using a state-of-the-art approach relying on the DeepLabCut (DLC) architecture. The high-dimensional body signal is then mapped into a lower dimensional space via non-linear variational autoencoder to obtain commands for a 2D computer cursor. First, we perform an offline test to evaluate the prediction power of the DLC fine tuned model. Then, we verify whether the proposed pipeline can be used to control a computer cursor in *real-time*. Results showed that the network can accurately predict the position of body landmarks. Moreover, an unimpaired participant was able to efficiently operate the computer cursor and gain a high-level of control skill after training with the interface. This enables performing experiments with video-based marker-less BoMIs for future implementation of an assistive device for people with motor disabilities.

Keywords: Computer vision · Positive technologies · Body-machine interfaces · Marker-less semantic features · Autoencoders

1 Introduction

Human disability is a global challenge affecting many people around the world [6]. Disability can arise due to a birth condition, an accident or ageing. In this scenario it is necessary to implement and investigate technologies that can improve

© Springer Nature Switzerland AG 2021
A. Del Bimbo et al. (Eds.): ICPR 2020 Workshops, LNCS 12662, pp. 233–240, 2021.
https://doi.org/10.1007/978-3-030-68790-8_19

the quality of life. Assistive Technologies (AT) [6] and Positive Technologies (PT) [9] emerge as a powerful solution to address human disability. AT and PT are generic terms for all devices and services that enable the independence of individuals with cognitive and/or functional impairment by improving the conditions of their daily living activities and, consequently, their quality of life.

In this paper we investigate the problem of enabling individuals with motor disabilities (such as after Spinal Cord Injury - SCI) to recover their functional independence. We exploit the fact that, even after a severe injury, many individuals retain some movement, especially of their head and shoulders, that can be used to control external devices, such as a computer cursor. Our approach is based on the framework of Body-Machine Interfaces (BoMIs) [3]. BoMIs convert high-dimensional body signals (*e.g.* upper body kinematics, muscle activities) into lower-dimensional, latent, commands to operate an external device. As a result, BoMIs allow individuals with motor disabilities to overcome some of their impairments. The use of BoMIs has been tested in situations involving the control of a computer cursor [20], a powered wheelchair [21] and quadcopters [17]. Typically, kinematic-based BoMIs rely on the use of sensors such as inertial measurement units (IMUs) [19] or markers [22] to record the body-movements of their users. For the specific task of cursor control, sensor-based techniques (electrooculargraphy (EOG), electromyography (EMG), IMU, gyro- and opto-sensors) are commonly adopted [4,7,12,14]. However, such approaches might hinder the assistive capability of the interface, as sensors cannot be worn autonomously by the BoMI user. Moreover, in the case of training with the BoMI across multiple days, sensors need to be placed consistently so as to minimize the need of interface re-calibration.

Recently, techniques for human motion detection, based on computer vision, have seen a recent surge [5]. A video-based marker-less BoMI would potentially allow for a more natural user-friendly interaction with an external device, as it is less invasive and cheaper than a sensor-based one. However, applicability of interfaces that rely only on video-information has seen limited efforts. Javanovic et al. [11] have proposed MarkerMouse, a computer mouse controller based on videos acquired from a webcam that detect a big marker placed on the head of the user. Fu et al. [8] and Betke et al. [1] have developed respectively hMouse and Camera Mouse, two video-based marker-less mouse controllers that detect specific body features, enabling people with severe disabilities to comfortably access a computer, without body attachments.

Here, we want to enhance these approaches combining new computer vision and deep learning techniques and the knowledge derived from the body machine interfaces. Specifically, we present a novel video-based marker-less BoMI pipeline to empower individuals with motor disabilities to independently control a computer cursor via shoulders and/or head movements without the needs of any sensors other than the computer webcam. Our procedure is composed of the following steps (see also Fig. 1): (1) automatic acquisition of images of the user from a computer webcam; (2) detection of landmark points (*e.g.* eyes, nose and shoulders) in the image plane; (3) encoding of the extracted signals to a lower dimensional (control) space via application of a dimensionality reduction (DR) algorithm; (4) handling of the graphic for providing BoMI users with visual feedback of the cursor via a computer monitor. We have evaluated our pipeline in terms of landmark points detection accuracy and overall speed, obtaining encouraging preliminary results. To the best of our knowledge, our method is the first involving a recent state-of-the-art pose estimation algorithm based on deep learning techniques.

Fig. 1. Summary of the BoMI pipeline. The image acquired by the computer webcam is fed through the trained network to detect the body landmarks. Then, a dimensionality reduction (DR) algorithm is applied to the landmarks' signal to obtain the coordinates of the computer cursor.

2   Methods

In this section we present all the building blocks of our pipeline.

2.1   Automatic Body Landmarks Detection

The first step of the pipeline is the detection of the positions of body landmarks in the image plane. Since the long-term goal of the pipeline is to empower individuals with motor disabilities, specifically after cervical SCI (cSCI), regaining independence, we decided to focus on the tracking of the body parts whose mobility is most likely retained even after a high level cSCI - *i.e.* shoulders and head (nose and eyes). In this first step, we exploit the DeepLabCut (DLC) architecture [16]. DLC is composed by a variant of the Residual Deep Network with 50 layers (ResNet-50) and allows the extraction of specific semantic features after an appropriate fine tuning. The DLC network is pre-trained using the ImageNet dataset. A deconvolutional layer is added at the end to extract spatial density probability maps associated with each landmark point. In order to build a model that correctly detects the position of the body landmarks of interest, we collect a dataset of such landmarks. We ask 40 healthy volunteers to freely move their head and shoulders for 30 s so as to comfortably explore their range of motion and to use a computer webcam or a mobile phone to capture a video of such body movements.

Then, we randomly select 15 frames in 32 videos (80% of the total number of videos, for a total of 480 training samples), we manually label the points of interest for each sample and we fine tune the DLC architecture. The architecture, the optimizer and the hyperparameters of the network are set accordingly to those described in other studies [16,18]. As a result, the network learn to predict the position (x_i, y_i) of each landmark point in the frame coordinate system, with i= (right eye, left eye, nose, right shoulder, left shoulder). Moreover, DLC returns the likelihood ℓ_i for each $i-$th landmark, expressed as a number in the interval $[0, 1]$ that quantifies the uncertainty behind the detection of each point. We investigate the accuracy of the detection in both train and test videos and we report the results in Sect. 4.

Figure 2 shows examples of detection result of this first step. Notice that we include videos with a wide variety of backgrounds, clothes and image dimensions so as to increase the robustness of the model.

Fig. 2. Extracted landmark points (shoulders, nose, eyes) for different subjects.

In order to improve the prediction power of the network and correct for occasional landmarks occlusions, we remove the points with less than 0.7 likelihood. Then, the time sequences of the coordinates for each landmark point are interpolated and low pass filtered (Butterworth, 3rd order, fc = 4 Hz).

2.2   Encoding Body Landmarks in the 2D Cursor Space

After detecting the body landmarks, the second step consists of applying the BoMI forward map to obtain the (x, y) coordinates of the computer cursor. Since the movements of the nose and the eyes are extremely correlated, we decide to exclude the latter. Thus, the 2D coordinates of shoulders and nose are organized as a 6D vector (q). The BoMI forward map is obtained by asking a volunteer to freely move his head and shoulders for 30 s. Then, the DLC model previously trained is applied to the video to extract the vector of body landmarks q for each frame. As a result, a matrix Q containing the estimated coordinates of the landmark points for every frame is obtained. Next, we train a non-linear 2D variational autoencoder (VAE) [15] on Q to derive the 2D latent space in which the greatest amount of the body movements variance during calibration is explained. We choose a VAE among the possible methods for dimensionality reduction (DR) (*e.g.*, Principal Component Analysis, *vanilla* AE) due to its ability to enforce a Gaussian distribution within its latent space. This would ensure a more uniform coverage of the 2D workspace with respect to that obtained training other DR models. To control the uniformity of the latent space, we introduce a scaling term ($\beta = 0.00025$) in the VAE cost function (see [10] for more details). Then, we set the VAE encoder sub-network E as the BoMI forward map. Thus, E maps the 6D body landmark vector (q) into the x-y cursor vector (p):

$$p = E(q) + p_0 \tag{1}$$

The offset vector p_0 is chosen to match the origin of the body-space with a corresponding reference position of the cursor. Moreover, the resulting workspace is scaled to ensure full coverage of the computer monitor space [2].

2.3   Online Video-Based Marker-Less BoMI

Finally, we set up the online control of a computer cursor with the proposed BoMI. In order to set up the real-time interface, we develop a custom-coded Python script. The script has a multi-threaded architecture so as to handle three different processes: (i) capture the current frame from the computer webcam (via OpenCV library); (ii) forward pass the current frame through the DLC trained model to obtain the body-vector q; (iii) forward pass q through the variational encoder E to obtain the coordinates p to control the cursor (see Fig. 3). During the real-time pipeline, we feed the current webcam frame -read with OpenCV- and the weights of the DLC model to the Deeplabcut-live library [13], to obtain a real-time estimation of the body landmarks q. Finally, the encoder E is applied to q in order to obtain the cursor coordinates p for the current frame. To speed up these online operations, we run the code in a computer with a 16 GB NVidia P5000 Quadro GPU.

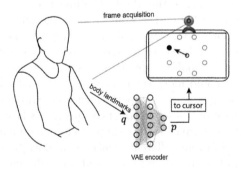

Fig. 3. Scheme of the marker-less BoMI for online cursor control.

3   Pilot Test and Preliminary Results

In this Section we present the results related to the (1) offline training of the DLC network for the detection of body landmarks in the videos acquired as described in Sect. 2.1; (2) real-time operation of the BoMI as described in Sect. 2.3.

3.1 Body Parts Detection's Accuracy

We evaluate the prediction power of the fine-tuned DLC network by computing the Euclidean distance between the estimates and the manually labeled ground truth positions of the landmarks. To do that, we select 5 frames from each of the 32 train videos (160 images) and 5 frames from each of the 8 test videos (40 images). The distance is expressed as pixels. Results show that the network was able to predict the position of the landmarks with a remarkable accuracy (Table 1).

Table 1. Mean error ± standard deviation (SD) for each point in pixels computed considering a manually labeled ground truth in 200 images (160 selected from videos adopted to train the network and 40 extracted from test videos).

Point	Images from train videos	Images from test videos
Right Eye	2.14 ± 1.12 pxl	2.93 ± 1.36 pxl
Left Eye	2.38 ± 1.08 pxl	2.88 ± 1.25 pxl
Nose	2.29 ± 1.56 pxl	3.04 ± 1.89 pxl
Right Shoulder	3.05 ± 2.03 pxl	3.79 ± 2.88 pxl
Left Shoulder	3.29 ± 2.47 pxl	3.83 ± 2.75 pxl

3.2 Online Test of the BoMI

Our goal is to assess whether the BoMI pipeline could be run in real-time. Analyzing the frame recorded by the webcam with the DLC model is a computationally expensive operation, thus achieving a satisfactory frame rate during the online operation of the BoMI is not trivial. We have enrolled a naive unimpaired participant (age 27, male) to practice the online operation of the interface, informing him that he could move the cursor with the motion of his upper part of the body and nothing else. He has performed a reaching task, in which he is asked to move the cursor over a set of targets as quickly as he can. The position of the cursor and the targets are shown to the participants on a computer monitor. The targets are placed in four different locations, uniformly distributed along a circle. A reaching trial is considered successful after the cursor remains inside the target for 250 ms. A total of 192 targets (48 trials per target location) are presented. The order of targets is pseudoranodmized so as each target location is not presented again before all 4 locations have been reached. The participant was immediately able to efficiently move the computer cursor over each target presented. Specifically, he completed all the 192 trials in just 13 min. Moreover, we completed all the analysis described in Sect. 2.3 with a frame rate 15 Hz, that allowed a continuous and efficient cursor control.

4 Discussion, Conclusion and Future Work

This study delivered three main findings: (i) the fine-tuned DLC network was able to accurately predict the position of body landmarks on images with a variety of different backgrounds and clothes; (ii) such model can be adopted for the online detection of body landmarks; (iii) the proposed pipeline allowed a participant to efficiently and easily operate a computer cursor. For comparison with sensor-based approach, 10 healthy participants practicing cursor control with an IMU-based BoMI completed the same protocol in approximately 20 min. Note that this is a pilot study with the aim of exploring the feasibility of the real-time procedure, for this reason only one subject is involved.

The main goal of the study was to verify whether a video-based marker-less BoMI could be used to operate a computer cursor in real-time. The step that took the most to be completed online was the application of the DLC model to estimate the landmarks position. Only a desktop computer with a powerful GPU would be able to complete this operation within an acceptable time frame during the online cursor control - as we achieved. However, our long-term goal is to improve the pipeline in order to be able to run it on any modern-day laptop, which does not have the same computational capability. This would dramatically increase the availability of the interface, thus broadening its impact as an assistive device. In order to reach this goal, we would need to reduce the number of the network parameters by adopting other architectures. On one hand, reducing the complexity of the network will allow faster processing, resulting in a more responsive interface. On the other, there is the possibility that it will decrease the accuracy of the estimates, thus leading to a noisier interface. Thus, further testings are required in order to find a correct trade off between computational speed and prediction accuracy. To further increase the power of the interface, we will also set up a calibration step to have one specific BoMI forward map for each user. Nevertheless, the interface proposed in this study is already capable of providing a sufficiently intuitive and enjoyable user experience. Therefore, we consider the proposed pipeline adequate for future implementation as an assistive tool for people with motor impairments.

References

1. Betke, M., Gips, J., Fleming, P.: The camera mouse: visual tracking of body features to provide computer access for people with severe disabilities. IEEE Trans. Neural Syst. Rehabil. Eng. **10**(1), 1–10 (2002)
2. Casadio, M., et al.: Functional reorganization of upper-body movement after spinal cord injury. Exp. Brain Res. **207**(3–4), 233–247 (2010)
3. Casadio, M., Ranganathan, R., Mussa-Ivaldi, F.A.: The body-machine interface: a new perspective on an old theme. J. Motor Behav. **44**(6), 419–433 (2012)
4. Chen, Y.L., Tang, F.T., Chang, W.H., Wong, M.K., Shih, Y.Y., Kuo, T.S.: The new design of an infrared-controlled human-computer interface for the disabled. IEEE Trans. Rehabil. Eng. **7**(4), 474–481 (1999)

5. Colyer, S.L., Evans, M., Cosker, D.P., Salo, A.I.: A review of the evolution of vision-based motion analysis and the integration of advanced computer vision methods towards developing a markerless system. Sports Med.-open **4**(1), 24 (2018)
6. Cook, A.M., Polgar, J.M.: Assistive Technologies-E-Book: Principles and Practice. Elsevier Health Sciences (2014)
7. Di Mattia, P.A., Curran, F.X., Gips, J.: An eye control teaching device for students without language expressive capacity: EagleEyes, vol. 53. Edwin Mellen Press (2001)
8. Fu, Y., Huang, T.S.: hmouse: head tracking driven virtual computer mouse. In: 2007 IEEE Workshop on Applications of Computer Vision (WACV'07), pp. 30–30. IEEE (2007)
9. Grossi, G., Lanzarotti, R., Napoletano, P., Noceti, N., Odone, F.: Positive technology for elderly well-being: a review. Pattern Recogn. Lett. **137**, 61–70 (2020)
10. Higgins, I., et al.: beta-vae: learning basic visual concepts with a constrained variational framework (2016)
11. Javanovic, R., MacKenzie, I.S.: MarkerMouse: mouse cursor control using a head-mounted marker. In: Miesenberger, K., Klaus, J., Zagler, W., Karshmer, A. (eds.) ICCHP 2010. LNCS, vol. 6180, pp. 49–56. Springer, Heidelberg (2010). https://doi.org/10.1007/978-3-642-14100-3_9
12. Jeong, H., Kim, J.S., Son, W.H.: An emg-based mouse controller for a tetraplegic. In: 2005 IEEE International Conference on Systems, Man and Cybernetics, vol. 2, pp. 1229–1234. IEEE (2005)
13. Kane, G., Lopes, G., Sanders, J., Mathis, A., Mathis, M.: Real-time, low-latency closed-loop feedback using markerless posture tracking. BioRxiv (2020)
14. Kim, S., Park, M., Anumas, S., Yoo, J.: Head mouse system based on gyro-and opto-sensors. In: 2010 3rd International Conference on Biomedical Engineering and Informatics, vol. 4, pp. 1503–1506. IEEE (2010)
15. Kingma, D.P., Welling, M.: Auto-encoding variational bayes. arXiv preprint arXiv:1312.6114 (2013)
16. Mathis, A., et al.: Deeplabcut: markerless pose estimation of user-defined body parts with deep learning. Nat. Neurosci. **21**(9), 1281–1289 (2018)
17. Meihlbradt, J., et al.: Data-driven body-machine interface for the accurate control of drones. Proc. Natl. Acad. Sci. **115**(31), 7913–7918 (2018)
18. Moro, M., Marchesi, G., Odone, F., Casadio, M.: Markerless gait analysis in stroke survivors based on computer vision and deep learning: a pilot study. In: Proceedings of the 35th Annual ACM Symposium on Applied Computing, pp. 2097–2104 (2020)
19. Pierella, C., et al.: Learning new movements after paralysis: results from a home-based study. Sci. Rep. **7**(1), 1–11 (2017)
20. Rizzoglio, F., Pierella, C., De Santis, D., Mussa-Ivaldi, F.A., Casadio, M.: Ahybrid body-machine interface integrating signals from muscles and motions. J. Neural Eng. (2020)
21. Thorp, E.B., et al.: Upper body-based power wheelchair control interface for individuals with tetraplegia. IEEE Tans. Neural Syst. Rehabil. Eng. **24**(2), 249–260 (2015)
22. Zhou, H., Hu, H.: Human motion tracking for rehabilitation–a survey. Biomed. Signal Process. Control **3**(1), 1–18 (2008)

Visual-Textual Image Understanding and Retrieval (VTIUR) - Joint Workshop on Content-Based Image Retrieval (CBIR 2020), Video and Image Question Answering (VIQA 2020), Texture Analysis, Classification and Retrieval (TAILOR 2020)

Visual-Textual Image Understanding and Retrieval (VTIUR) - Joint Workshop on Content-Based Image Retrieval (CBIR 2020), Video and Image Question Answering (VIQA 2020), Texture Analysis, Classification and Retrieval (TAILOR 2020)

Workshop Description

The aim of the *Visual-Textual Image Understanding and Retrieval (VTIUR)* workshop is to gather researchers and practitioners interested in building models capable of understanding visual data, such as images and videos, in order to build intelligent solutions to Computer Vision tasks, such as image analysis, classification, and retrieval. Visual data may be enhanced with textual information. These conditions raise several challenges at the interface of Computer Vision and Natural Language Processing where multi-modal tasks, such as visual question answering and visual retrieval, take place. Recent advances in these tasks have been made possible by exploiting Artificial Intelligence techniques that make use of this supplementary source of knowledge. In particular, considering the need to attend to both visual and textual information in order to understand the intertwining between them, Deep Learning techniques make it possible to design complex neural networks where hierarchical representations of the available data can be automatically learned, while also taking into account their multi-modal nature. Furthermore, given the opportunities both from a research and from a production point of view, the VTIUR workshop aimed at gathering people both from the academia and the industry, in order to stimulate the sharing of recent trends, novel ideas and applications, and to raise new opportunities and promising new directions of research that should be explored in the future.

We received a total of eight submissions. After a thorough peer-review process by the Technical Program Committee, all the eight papers were found to be worthy of acceptance. The review process focused on the scientific quality of the articles, and their applicability to the image understanding and retrieval fields. The reviewers highlighted the quality of the accepted papers that represent a mix of perspective articles, surveys, research investigations, and novel applications.

Finally, we would like to thank all the members of the Technical Program Committee that helped us by providing clear, rigorous, and timely reviews. We would also like to thank the International Conference on Pattern Recognition (ICPR) chairs, who gave us the opportunity to raise the attention over these important tasks, which will be more and more important in the near future.

Organization

Program Committee (VIQA) Chairs

Alex Falcon	FBK and University of Udine, Italy
Oswald Lanz	FBK, Italy
Giuseppe Serra	University of Udine, Italy
Rainer Stiefelhagen	Karlsruhe Institute of Technology, Germany
Makarand Tapaswi	Inria Paris, France

Program Committee (TAILOR) Chairs

Francesco Bianconi	University of Perugia, Italy
Claudio Cusano	University of Pavia, Italy
Antonio Fernández	University of Vigo, Spain
Paolo Napoletano	University of Milan-Bicocca, Italy

Program Committee (CBIR) Chairs

Marco Bertini	University of Florence, Italy
Gianluigi Ciocca	University of Milano-Bicocca, Italy
Simone Santini	Universidad Autonoma de Madrid, Spain
Raimondo Schettini	University of Milano-Bicocca, Italy

Technical Program Committee (VIQA)

Federico Becattini	University of Florence, Italy
Pietro Bongini	University of Florence, Italy
Tommaso Campari	FBK and University of Padova, Italy
Marcella Cornia	University of Modena and Reggio Emilia, Italy
Antonino Furnari	University of Catania, Italy
Davide Rigoni	FBK and University of Padova, Italy
Jingkuan Song	Columbia University, USA
Meng Wang	Hefei University of Technology, China
Cheng Wenlong	CRIPAC, China

Technical Program Committee (TAILOR)

Yannick Berthoumieu	Ecole Nationale d'Electronique, Informatique et Radiocommunications de Bordeaux, France
Simone Bianco	University of Milan-Bicocca, Italy
Marco Buzzelli	University of Milan-Bicocca, Italy
Eva Cernadas García	Centro de Investigacion en Tecnoloxias Intelixentes da USC, Spain
Youssef El Merabet	Université Ibn Tofail, Marocco
Richard Harvey	University of East Anglia, UK
Anne Humeau-Heurtier	University of Angers, France
Flavio Piccoli	University of Milan-Bicocca, Italy
Constantino Carlos Reyes-Aldasoro	City, University of London, UK
Raimondo Schettini	University of Milan-Bicocca, Italy
Fabrizio Smeraldi	Queen Mary, University of London, UK

Technical Program Committee (CBIR)

Giuseppe Amato	ISTI - Consiglio Nazionale delle Ricerche, Italy
Simone Bianco	University of Milano-Bicocca, Italy
Marco Buzzelli	University of Milano-Bicocca, Italy
K. Selcuk Candan	Arizona State University, USA
Claudio Cusano	University of Pavia, Italy
Xirong Li	Renmin University, China
Michael S. Lew	Leiden University, The Netherlands
Mathias Lux	Alpen Adria University, Austria
Paolo Napoletano	University of Milano-Bicocca, Italy
Abed El Saddik	Ottawa University, Canada
Marcel Worring	University of Amsterdam, The Netherlands
Zhongfei "Mark" Zhang	Binghamton, USA

Content-Based Image Retrieval
and the Semantic Gap in the Deep
Learning Era

Björn Barz$^{(\boxtimes)}$⬤ and Joachim Denzler⬤

Computer Vision Group, Friedrich Schiller University Jena, Jena, Germany
{bjoern.barz,joachim.denzler}@uni-jena.de

Abstract. Content-based image retrieval has seen astonishing progress over the past decade, especially for the task of retrieving images of the same object that is depicted in the query image. This scenario is called instance or object retrieval and requires matching fine-grained visual patterns between images. Semantics, however, do not play a crucial role. This brings rise to the question: Do the recent advances in instance retrieval transfer to more generic image retrieval scenarios?

To answer this question, we first provide a brief overview of the most relevant milestones of instance retrieval. We then apply them to a semantic image retrieval task and find that they perform inferior to much less sophisticated and more generic methods in a setting that requires image understanding. Following this, we review existing approaches to closing this so-called semantic gap by integrating prior world knowledge. We conclude that the key problem for the further advancement of semantic image retrieval lies in the lack of a standardized task definition and an appropriate benchmark dataset.

Keywords: Content-based image retrieval · Instance retrieval · Object retrieval · Semantic image retrieval · Semantic gap

1 Introduction

One sees well only with the heart. The essential is invisible to the eyes.

This famous quote from the French writer Antoine de Saint Exupéry applies to life as well as to computer vision. The human perception of images greatly exceeds the visual surface of pixels, colors, and objects. The *meaning* of an image cannot simply be described by enumerating all objects contained therein and defining their spatial layout. Humans are able to extract a plethora of diverse and complex information from an image at first glance, such as events happening in the depicted scene, activities performed by persons, relationships between them, the atmosphere and mood of the image, and emotions transported by it. Many of these concepts elude textual description and are best communicated visually.

© Springer Nature Switzerland AG 2021
A. Del Bimbo et al. (Eds.): ICPR 2020 Workshops, LNCS 12662, pp. 245–260, 2021.
https://doi.org/10.1007/978-3-030-68790-8_20

OBJECTS

Maid ≺ Woman ≺ Person

Black dress

Wardrobe ≺ Furniture

Window

Liselund Castle ≺ Castle

SCENE

Old-fashioned room
                         ≻ Room ≺ Indoor
Sunlit room

Woman in front of window next to wardrobe

META

„The Dream Window in the Old Liselund Castle"
    ≺ Painting by G. Achen ≻
                              ≻ Painting ≺ Artwork
    ≺ Oil on canvas

ACTIVITIES

Daydreaming

Looking out of the window

MOOD

Melancholic

Feeling locked in

Fig. 1. An example for the ambiguity and semantic richness of images. All concepts listed on the right-hand side could be used to describe the image on the left, while different observers will pay attention to different subsets of these aspects.

Figure 1 illustrates this variety of information conveyed by images. The image depicted there can be described from several perspectives: its semantic content, artistic style, the emotions it evokes in the observer, or meta-information about the image itself. Most facets of an image's meaning are directly available to the viewer without verbalization but by simply looking at it. Searching through a large database of images using a representative example as query is hence the most natural, direct, and expressive way of finding images with a particular content. This approach is known as *content-based image retrieval (CBIR)* [45] and has been an active area of research since 1992 [28,33].

"Pictures have to be seen and searched as pictures", wrote Smeulders et al. [45] in their extensive survey at the end of the "early years" of CBIR in 2000. During the two decades that have passed since then, CBIR has undergone at least two major revolutions (more on that in Sect. 2). However, most of the main challenges and directions had already been identified back then. One of these challenges is the *semantic gap*, as Smeulder et al. call it:

> *"The semantic gap is the lack of coincidence between the information that one can extract from the visual data and the interpretation that the same data have for a user in a given situation."* [45, Sect. 2.4]

Phrased with the words of de Saint Exupéry, the semantic gap is the difference between perceiving an image with the *eyes*—objectively, as a depiction of objects, shapes, textures—and perceiving an image with the *heart*—subjectively, including world-knowledge and emotions, reading "between the pixels".

The size of the semantic gap depends on the level of abstraction of the user's search objective. Smeulders et al. [45] define this level of abstraction on a continuous scale between the two poles of a *narrow* and a *broad domain*. This is best explained by the following distinction into three typical CBIR tasks:

Duplicate retrieval searches for images with the same content. These are variants that originated from the same photo but might have been post-processed differently with regard to cropping, scaling, brightness, contrast etc.

Instance retrieval searches for images depicting the same instance of an object, i.e., a person or a certain building. Thanks to its nature as a well-defined but non-trivial task with a clear ground-truth, this is the most extensively studied CBIR sub-task [3,4,9,18,23,26,27,35,39,42,44,46]. A handful of established datasets are available for this task [25,36–38] and significant progress has been made during the past few years.

Semantic retrieval covers most of the remaining broader spectrum and aims for finding images belonging to the same category as the query. Note that *category* does not necessarily mean *object class* in this context. The set of possible categories is limited by nothing but the imagination of the user and a single image usually belongs to a remarkably high number of categories at once (see Fig. 1). Thus, the exact search objective of the user can rarely be determined based on the query image alone and will almost certainly also vary between users, even for the same query. Therefore, approaches to this problem often comprise interaction with the user to adapt the similarity measure used by the system to that in the user's mind [5,7,50].

Learning meaningful representations that capture fine semantic distinctions and the various facets of an image's meaning is hence of paramount importance. Despite its practical relevance, this task has received substantially less attention than instance retrieval, mainly due to the less well-defined notion of "relevance" and, as a result, the lack of a suitable benchmark.

Duplicate retrieval marks one end of the spectrum, as it is the narrowest domain possible. In this case, the semantic gap is almost non-existent and all that is needed to overcome it is a list of invariances regarding the image's content (e.g., rotation, cropping etc.). The broader the domain, the larger the semantic gap.

While it is more challenging than duplicate retrieval, instance retrieval can still be handled by matching fine-grained distinctive visual patterns and their geometric layout. Content-based image retrieval has made substantial progress in this area in the past two decades, which we outline in Sect. 2. However, the applicability of such techniques is limited with respect to the much more generic broad domain of semantic retrieval, as we see in Sect. 3. One way to overcome this semantic gap, according to Smeulders et al. [45], lies in integrating sources of semantic information from outside the image. In Sect. 4, we review recent approaches in this direction, followed by a discussion of what is still missing for advancing CBIR in the broad domain further (Sect. 5).

2 The Evolution of Instance Retrieval

Between 2000 and 2020, CBIR—with a particular focus on instance retrieval—has undergone two major paradigm shifts: The first began in 2003 [44] and was initiated by the adaptation and subsequent improvement of techniques from text retrieval. The second wave of breakthrough achievements originated from the application of deep learning methods to CBIR, starting in 2014 [4,41]. We outline the major milestones of these two epochs of innovation in the following.

2.1 Hand-Crafted Features and Visual Words

Local Features as Visual Words. In 2003, Sivic and Zisserman [44] sought to find occurrences of a certain object in videos and, to this end, adapted the *bag-of-words (BoW)* document descriptor, which is popular in the field of text retrieval, to image retrieval. As an analogy for words, they use local image features at distinctive keypoints and cluster them into a vocabulary of "visual words". The image is then represented by a vector of normalized word frequencies.

This process illustrates the general framework for extracting image representations used in CBIR [27]: A local feature extractor computes features at keypoints in a given image. These local features are then embedded into a different space, such as quantized indices of visual words. Finally, they are aggregated into a global representation, which allows for efficient retrieval of an initial list of candidate images. In addition, the local features are often used to perform a spatial verification and re-ranking step for the top-ranking candidates to eliminate false matches [36,44]. This technique is quite specific to instance retrieval and matches local feature vectors between the query and a retrieved image to verify that the local features have a matching geometric layout.

Towards More Complex Embeddings. Subsequent works of this epoch adhered to SIFT [30] or RootSIFT [1] features extracted at affine-invariant keypoints [31] and focused mainly on improving the embedding and aggregation step. Representing local feature vectors by a single integer—the cluster index in the case of Sivic and Zisserman [44]—incurs a severe loss of information and does not capture the actual distribution of the local features well. Hard assignment to a single cluster is furthermore not robust against small variations of local descriptors close to cluster boundaries. To overcome these issues, Perronnin et al. [35] propose the use of *Fisher vectors* and Jégou et al. [26] introduce a similar but simpler alternative called *VLAD*. Both capture the difference between each visual word and its cluster center, resulting in a more informative but also high-dimensional descriptor, which needs to be compressed and whitened using PCA. Additionally, Perronnin et al. [35] perform weighted soft assignment of visual words to multiple clusters. They show that a Fisher vector with a single visual word performs comparably to a BoW descriptor with 4,000 words.

The sensitivity of these methods to the absolute distance between a local feature vector and its cluster center also has drawbacks, though, because the

Euclidean distance is of limited meaning in high-dimensional spaces. Jégou and Zisserman [27] account for this issue by L^2-normalizing the residuals, thus encoding their angle instead of their magnitude, which gives rise to the name *triangulation embedding*. They furthermore encode the angles between the local feature vector and *all* visual words instead of just the closest one. This representation has been found to outperform fisher vectors and VLAD.

However, Husain and Bober [23] find that comparing each local feature vector with all visual words does not scale to large datasets. Soft cluster assignment, on the other hand, behaves unstable and often degrades to single assignment. To overcome this, they propose a middle ground by assigning the local descriptors to the few cluster centers that are closest and base the weights on their ranks among the nearest neighbors instead of their actual distances. These *robust visual descriptors (RVDs)* are furthermore not whitened globally but on a per-cluster level. RVD performs competitively to triangulation embedding, while being faster to compute and more robust to dimensionality reduction.

The Role of Datasets. While the paradigm of using aggregated local features for CBIR dates back to 2003 [44], research in this area has been most active between 2010 and 2016. One likely reason for this delay is the lack of established benchmark datasets. In 2007 and 2008, the Oxford Buildings [36], Paris Buildings [37], and INRIA Holidays [25] datasets were published, which quickly emerged as the standard benchmarks for instance retrieval and gave new impetus to the field by providing a proper ground for evaluation and comparison of methods.

The two building datasets comprise different photos of various landmark buildings in Oxford and Paris, with a large variety of perspectives, scales, and occlusions. The Holidays dataset, on the other hand, contains a collection of personal holiday photos with on average three different perspectives per scene. While these datasets are challenging, the task of retrieving images showing the same object or scene as the query is well-defined with a clear ground truth.

2.2 Off-the-Shelf CNN Features

After hand-crafted local features had remained unquestioned in CBIR for over a decade, the renaissance of deep learning finally led to a substantial change regarding image representations. The independent works of Babenko et al. [4] and Razavian et al. [41] first showed that surprisingly good results can be achieved by simply extracting global image descriptors, so-called *neural codes*, from a fully-connected layer of a CNN pre-trained on ImageNet [13]. Given the extreme simplicity of this approach, requiring close to zero engineering effort, this was a remarkable result. Babenko and Lempitsky [3] considerably improved the performance of this approach by extracting image features not from a fully-connected but from the last convolutional layer, which still has a spatial resolution. The result is, thus, a set of feature vectors, which can roughly be associated with different regions in the image. These are summed up for aggregation, L^2-normalized, reduced in dimensionality using PCA, and L^2-normalized again, leading to the speaking name *sum-pooled convolutional features (SPoC)* for these descriptors.

In the following years, research mainly adhered to using such pre-trained neural feature extractors and focused on designing sophisticated aggregation functions. Many of them try to find a middle ground between sum and maximum pooling, e.g., by averaging activations over the top few responses only as in *partial mean pooling (PMP)* [49], or by smoothly interpolating between the two extremes as in *generalized-mean pooling (GeM)* [39].

Aggregated convolutional features have one drawback, though: As opposed to traditional local features, they do not allow for precise localization of the matching object and, thus, are not compatible with techniques such as spatial verification and re-ranking, which depend on geometric information. To this end, Tolias et al. [46] propose the *regional maximum activation of convolutions (R-MAC)* aggregation, which follows a two-step approach: The feature map is divided into overlapping regions of different sizes and the local feature vectors in each region are aggregated using maximum pooling. These so-called MAC vectors are whitened and aggregated by sum pooling into a global R-MAC image descriptor. For spatial re-ranking, the similarity of the query's MAC vector and the individual regional MAC vectors of the top few retrieval results can be used to localize the query object in the retrieved images and refine the ranking.

These techniques took CBIR based on features extracted from pre-trained CNNs quite far, but the hand-crafted RVD descriptor [23] is still able to compete with them on instance retrieval benchmarks.

2.3 End-to-End Learning for Image Retrieval

Deep learning finally became undeniably superior to traditional CBIR techniques based on hand-crafted features when researchers began to adapt the CNN used for feature extraction to the task of image retrieval instead of using a pre-trained one. We regard this shift of focus from feature transformation and aggregation to actual feature learning as the second important paradigm shift in CBIR.

Global Features. Gordo et al. [18] were among the first to be successful in this endeavor. They build upon R-MAC [46] and implement it as differentiable layers on top of a VGG16 CNN architecture, which can then be trained end-to-end. To this end, they adopt the triplet loss [43] from the field of deep metric learning. By training on a curated dataset of famous landmarks, they learn a feature representation where images are closer by a certain margin to images of the same landmark than to images of different landmarks.

As opposed to the triplet loss, Radenović et al. [39] find the contrastive loss to provide better final performance, while furthermore requiring only pairs instead of triplets of images for training. More importantly, they propose an unsupervised technique for generating training data consisting of matching and non-matching image pairs without human annotation: They cluster images in the training dataset based on BoW descriptors and apply structure-from-motion

(SfM) techniques to estimate a 3-D model for each cluster. This allows images of the same landmark but captured from disjoint viewpoints to be considered as non-matching. The information about camera positions obtained from SfM furthermore enables mining of challenging positive image pairs that exhibit a non-trivial amount of overlap.

These metric learning approaches have led to an impressive improvement of instance retrieval performance in terms of average precision (AP), even though they do not optimize it directly but a proxy objective based on distances in the learned feature space. Since AP is the most important metric for evaluating retrieval methods, it seems desirable to optimize it directly instead of a proxy-task. However, that entails taking into account not only a single sample, a pair, or a triplet as before, but the entire list of ranked results. One apparent benefit is that such listwise objectives are position-sensitive: The impact of a single pair or triplet involving images at the top of the ranking should be higher than at the end of the list. However, average precision is not differentiable, because it involves sorting images by their similarity to the query. For being able to optimize AP in an end-to-end learning context nevertheless, He et al. [20] proposed a differentiable approximation of AP using histogram binning, which has been adopted by Revaud et al. [42] for CBIR. Since the cosine similarity is bounded, the range of similarity scores can be divided into a fixed number of equally sized bins. Images are then soft-assigned to the bins whose centers are closest to the image's retrieval score to obtain histograms of positive and negative match counts in each bin. Instead of computing precision and recall for each rank, these metrics can now be computed for each bin and combined to approximate AP.

However, quantizing similarity scores into bins ignores variations of the ranking within each bin, which can have particularly large impacts on AP at the top positions of the ranking. This deficiency has recently been overcome by a different approach to approximating AP: Instead of quantized sorting by binning, the sorting operation itself is relaxed by replacing the Heaviside step function indicating whether one element of the list precedes another with a sigmoid function to avoid vanishing gradients. This allows for differentiable sorting and computation of a relaxed version of AP, called Smooth-AP [9].

With these listwise approaches, global representations for CBIR can finally be learned end-to-end without hand-crafted intermediate steps or proxy objectives.

Local Features. While global image descriptors are convenient for retrieval applications, they are neither robust in the presence of occlusion or background clutter nor suitable for spatial verification, which is important for instance retrieval. Noh et al. [34] hence aimed at learning *Deep Local Features (DELF)* in an end-to-end manner. They use regional features extracted from a convolutional layer of a CNN and then train another small CNN to assess the importance of these densely sampled keypoints. For training, these predicted weights are used for weighted sum pooling of the local descriptors into a global feature vector, which allows for fine-tuning of the local features using image-level supervision.

Most instance retrieval systems using local features adopt a two-stage approach: retrieving a set of candidates using global features and then re-ranking them using spatial verification with local features. Cao et al. [10] unified the learning of both types of features into a single model with two branches.

The Need for More Challenging Benchmarks. Besides plenty of computing capacity, deep learning techniques require one thing most of all: data. The existing instance retrieval datasets were too small for training deep neural networks, wherefore Babenko et al. [4] created a novel landmarks dataset with over 200,000 images for training purposes, which was later used by other works as well [18]. Nowadays, the large-scale Google-Landmarks dataset [34] is often used for training. It comprises over a million images of 12,894 landmarks.

These datasets are orders of magnitudes larger than the Oxford and Paris Buildings dataset, but the latter were still relevant for evaluating and comparing novel methods. The rapid advances in deep learning for CBIR, however, quickly resulted in a saturation of performance on these benchmarks. Therefore, Radenović et al. [38] revisited these two datasets in 2018 by improving the ground-truth annotations, finding more difficult queries, adding challenging distractor images, and defining three different evaluation protocols of varying difficulty.

3   Impact on the Semantic Gap

The previous section outlined the impressive advances of instance retrieval in the deep learning era. However, instance retrieval is a rather narrow domain, where a broad understanding of the scene semantics are not required to solve the task satisfactorily. The interesting question is, therefore: Do these advances transfer to the broader domain of semantic retrieval?

To answer this question, we evaluate several seminal methods and models on an instance retrieval and a semantic retrieval task. For instance retrieval, we use the Revisited Oxford Buildings dataset [36,38] (see above), on which these methods have originally been evaluated. As an indicator for their performance in a broader domain, we evaluate them on the MIRFLICKR-25K dataset [22], which comprises 25,000 images from Flickr, each annotated with a subset of 25 concepts such as "sky", "lake", "sunset", "woman", "portrait" etc. While most images in the dataset are annotated with more than one concept, 3,054 of them exhibit only a single label. We use these images as queries to avoid query ambiguity. We consider a retrieved image as relevant if it shares this concept.

Figure 2 depicts the mean average precision of several milestones of CBIR research in the deep learning era on both tasks. While the performance on instance retrieval tasks increased steadily, the semantic retrieval performance did not only not improve, but even deteriorated slightly. The majority of developments in the past years have focused on instance retrieval and hence tuned feature representations towards this tasks, for which fine-grained visual features are important. This, however, degraded their performance on broader-domain tasks, for which different features are necessary. While instance retrieval has reached a very advanced level

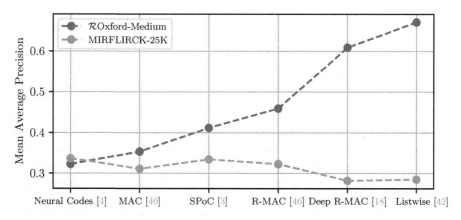

Fig. 2. Milestones of CNN-based instance retrieval, evaluated on an instance retrieval ($\mathcal{R}$Oxford [38]) and a semantic retrieval dataset (MIRFLICKR-25K [22]).

of maturity during the past 20 years, content-based image retrieval in general is still facing the challenges of the semantic gap.

4   Knowledge Integration for Semantic Image Retrieval

One way to overcome the semantic gap lies in incorporating additional sources of information outside the image, as Smeulders et al. [45] already stated back in 2000. In the following, we briefly review approaches for leveraging the most common sources of such external information for semantic image retrieval.

4.1   Class Labels

Image-level class labels are one of the most frequently available and cheapest types of semantic information about images. To provide robust performance in an open world, however, a huge number of classes or sophisticated methodology beyond training a simple classifier is required.

OASIS [11] combines both: Method-wise, OASIS learns a bilinear similarity metric using the triplet loss for comparing hand-crafted features with respect to semantic image similarity. The training dataset consisted of over two million images sourced from Google Image Search using about 150,000 textual queries entered by real users. Working at Google, the authors did not only have access to these queries, but also to relevance ratings based on click statistics, which allowed them to collect this large-scale but non-public dataset.

With the advent of deep learning, the focus shifted from learning metrics to learning features. *MultiGrain* [8], for instance, optimizes a CNN with respect to multiple tasks by combining a classification and a metric learning objective. This way, it learns diverse representations that are useful for class-level, instance-level, and identity-level retrieval. Evaluation, however, is conducted separately for each task in terms of classification accuracy on ImageNet [13] and retrieval accuracy on instance retrieval benchmarks. This evaluation protocol does not provide information about semantic retrieval performance.

To deploy CBIR at production-level within the *Microsoft Bing* search engine, Hu et al. [21] employ a large ensemble of different network architectures trained for various tasks: for classification with cross-entropy loss, with a metric learning objective such as the contrastive or triplet loss, for face recognition, or for object detection. This ensemble is intended to capture a broad variety of both visual and semantic properties of images and, hence, cover most objectives a user of the visual search engine could pursue. The training data for this system is non-public and was collected by human annotators in an expensive data collection and annotation effort. The evaluation was conducted using human relevance judgments as well. For these two reasons, this work is neither publicly reproducible nor directly comparable with other works.

4.2 Class Taxonomies

Plain class labels do not take into account the semantic relationships between classes, which cannot always be inferred from their visual similarity. Taxonomies such as WordNet [14] organize concepts on different levels of abstraction in terms of is-a relationships and are a popular tool for measuring the semantic similarity between classes. Several works strive for integrating this prior knowledge about the world to improve the semantic consistency of CBIR results.

Deng et al. [12] construct a hand-crafted bilinear similarity measure from the class taxonomy of ImageNet [13] and use it for comparing vectors of class probabilities predicted by a classifier. Instead of a similarity measure, Barz and Denzler [6] construct a semantic feature space spanned by class embeddings, where the cosine similarity between two class embeddings equals their semantic similarity derived from the taxonomy. They then use a CNN to map images into the same semantic space. Arponen and Bishop [2] do not constrain the feature space in this explicit way, but instead integrate the same objective directly into the loss function, so that the layout of the semantic feature space is learned. They combine this with an additional term encouraging the individual features to be binary, which allows for compact and memory-efficient descriptors.

The aforementioned works evaluate their approaches on ImageNet using "hierarchical precision" [12], which replaces the binary relevance criterion of ordinary precision with semantic similarity. This metric suits the task better, but is best plotted for several cut-off positions in the ranking and cannot easily be summarized in a single number to facilitate comparison.

Yang et al. [48] combine semantic and visual similarity by first ranking images according to semantic similarity and then ordering the images within the same class according to visual similarity to the query. To this end, they use the contrastive loss with an adaptive margin proportional to the dissimilarity. The evaluation, however, is limited to fine-grained classification datasets and conducted using binary relevance, which does not take semantics into account.

Long et al. [29] not only embed the classes but all concepts in the taxonomy into a hyperbolic space, so that sub-classes lie in their parent class' entailment cone. As before, a CNN is then used to map samples onto their class embeddings. Although their method could also be applied for content-based image retrieval, they focus on video retrieval and evaluate their approach on that task only.

4.3 Textual Descriptions

While taxonomies provide information about the semantic similarity between classes, their full semantic meaning goes far beyond that. Several works have aimed for extracting such rich semantics from textual descriptions of classes or images and leverage them for learning meaningful image features. *DeViSE* [15] and *HUSE* [32], for example, learn word embeddings on Wikipedia and use the embedding of a class' name as its semantic embedding. DeViSE [15] then maps images into that space by maximizing the dot-product similarity between their feature vector and the respective class embedding, while enforcing a certain minimum distance to any other class embedding. HUSE [32], in contrast, adopts a pair-wise optimization approach by forcing the distance of pairs of images to be equal to the dissimilarity of their class embeddings. This approach provides more flexibility regarding the learned image feature space since it is separate from the space of word embeddings. Like some of the hierarchy-based approaches described above, both methods were evaluated using hierarchical precision. Thus, the semantic information used for evaluation was not the same as that used for training, which incurs a disadvantage compared to hierarchy-based methods.

Instead of using texts associated with classes, other methods leverage texts belonging to individual images, such as titles and captions, and learn a multi-modal embedding space. Gomez et al. [17] do so by training a CNN to regress the text embeddings generated by a separately trained language model. However, they evaluate their approach only with textual queries and not in a *content-based* image retrieval scenario. Wu et al. [47], in contrast, learn text and image embeddings jointly and additionally predict individual embeddings for components of the caption such as objects, object-attribute pairs, and object-relation phrases. These semantic components are automatically aligned with the local features of the corresponding image regions using contrastive learning. However, their experiments only investigate the cross-modal image-to-caption and caption-to-image retrieval scenarios, while semantic CBIR performance is not analyzed.

4.4 Artistic Style

An entirely different dimension of image semantics is opened up by stylistic concepts such as artistic style, mood, and atmosphere. Learning image features that respect such properties requires either specialized annotations or prior knowledge about their characteristics.

Ha et al. [19] define style in terms of color composition, i.e., the distribution and layout of colors in an image. They construct a dataset with subjective 5-star similarity ratings for pairs of images, which have been collected in a laborious crowd-sourcing process involving active learning. A siamese network is trained to predict the distribution of similarity ratings for a given pair of images.

To avoid the expensive collection of large-scale style datasets, Gairola et al. [16] draw on knowledge from the field of visual style transfer, where Gram matrix features have been found to capture the stylistic properties of images. They extract these features from a pre-trained CNN, cluster them, and use the cluster labels as ground-truth for training another CNN using the triplet loss. They evaluate their approach on numerous datasets annotated with artistic styles, photographic styles, historical art styles, moods, or genres.

5   The Missing Ingredient

The two lines of research on instance retrieval and semantic retrieval portrayed in Sects. 2 and 4, respectively, exhibit one apparent difference: The research on instance retrieval shows measurable continuous progress thanks to the Oxford [36] and Paris [37] benchmark datasets, whose release was followed by a clear surge of research activity in the field. With the Google-Landmarks dataset [34], sufficient training data is available for modern deep learning methods. The recent revision of the two aforementioned benchmark datasets [38] maintains their usefulness as a benchmark despite the substantial performance improvements.

Existing works on semantic image retrieval, in contrast, vary widely with respect to their evaluation protocol, training data (some of which is closed-source), and even the problem definition, rendering a clear comparison between approaches impossible. This is perhaps the biggest obstacle for further progress in this field and the likely reason why research still focuses on instance retrieval.

A thoroughly curated benchmark dataset for semantic image retrieval would hence greatly contribute to advancing the field. However, constructing such a benchmark is highly non-trivial due to numerous aspects. This begins already with the evaluation metric. In a semantic CBIR scenario, precision is often more important than recall. Average precision is hence a sub-optimal measure, but also precision alone is insufficient, since it only considers binary relevance. In reality, however, relevance is a graded phenomenon [45]. A candidate for an evaluation metric that takes this into account is the normalized discounted cumulative gain (NDCG) [24]. The dataset, however, also needs to provide such graded relevance ratings, which ideally should be based on real user ratings.

Furthermore, the benchmark should define a diverse set of relevance criteria a user can have in mind, including instance identity, object category identity on different levels of abstraction, similarity regarding artistic style, mood, actions, and relationships portrayed in the image. Further complications are caused by the fact that a query image can be interpreted differently with respect to these dimensions. The relevance of a retrieved image hence does not only depend on the query, but also on the search objective pursued by the user. This ambiguity can only be resolved by interaction with the user or by providing multiple query images sharing the relevant aspect. Therefore, the benchmark should ideally provide different evaluation protocols, an interactive one and a non-interactive one, which could be restricted to less ambiguous queries. The interactive scenario furthermore requires the definition of a feedback simulation protocol.

6   Conclusions

Content-based image retrieval has made astounding progress over the past two decades, especially in the area of instance retrieval. However, the methodological advances in this area do not translate to the more challenging task of semantic image retrieval. On the contrary, more advanced instance retrieval methods often perform worse than simpler ones in that domain. Despite the seeming advances, the semantic gap has rather become larger than smaller.

Due to the lack of an established benchmark, semantic image retrieval methods are often hardly comparable and vary widely regarding the task definition and the evaluation data and protocol. The history of instance retrieval shows that such a benchmark would be an invaluable catalyst for research on semantic image retrieval and a necessity for closing the semantic gap.

References

1. Arandjelović, R., Zisserman, A.: Three things everyone should know to improve object retrieval. In: IEEE Conference on Computer Vision and Pattern Recognition, pp. 2911–2918, June 2012
2. Arponen, H., Bishop, T.E.: SHREWD: semantic hierarchy based relational embeddings for weakly-supervised deep hashing. In: ICLR 2019 Workshop on Learning from Limited Labeled Data (2019)
3. Babenko, A., Lempitsky, V.: Aggregating local deep features for image retrieval. In: IEEE International Conference on Computer Vision, pp. 1269–1277, December 2015
4. Babenko, A., Slesarev, A., Chigorin, A., Lempitsky, V.: Neural codes for image retrieval. In: Fleet, D., Pajdla, T., Schiele, B., Tuytelaars, T. (eds.) ECCV 2014. LNCS, vol. 8689, pp. 584–599. Springer, Cham (2014). https://doi.org/10.1007/978-3-319-10590-1_38
5. Barz, B., Denzler, J.: Automatic query image disambiguation for content-based image retrieval. In: International Conference on Computer Vision Theory and Applications, vol. 5, pp. 249–256. INSTICC, SciTePress (2018). https://doi.org/10.5220/0006593402490256
6. Barz, B., Denzler, J.: Hierarchy-based image embeddings for semantic image retrieval. In: IEEE Winter Conference on Applications of Computer Vision, pp. 638–647 (2019). https://doi.org/10.1109/WACV.2019.00073
7. Barz, B., Käding, C., Denzler, J.: Information-theoretic active learning for content-based image retrieval. In: Brox, T., Bruhn, A., Fritz, M. (eds.) GCPR 2018. LNCS, vol. 11269, pp. 650–666. Springer, Cham (2019). https://doi.org/10.1007/978-3-030-12939-2_45
8. Berman, M., Jégou, H., Vedaldi, A., Kokkinos, I., Douze, M.: MultiGrain: a unified image embedding for classes and instances. arXiv preprint arXiv:1902.05509 (2019)

9. Brown, A., Xie, W., Kalogeiton, V., Zisserman, A.: Smooth-AP: smoothing the path towards large-scale image retrieval. In: Vedaldi, A., Bischof, H., Brox, T., Frahm, J.-M. (eds.) ECCV 2020. LNCS, vol. 12354, pp. 677–694. Springer, Cham (2020). https://doi.org/10.1007/978-3-030-58545-7_39

10. Cao, B., Araujo, A., Sim, J.: Unifying deep local and global features for image search. In: Vedaldi, A., Bischof, H., Brox, T., Frahm, J.-M. (eds.) ECCV 2020. LNCS, vol. 12365, pp. 726–743. Springer, Cham (2020). https://doi.org/10.1007/978-3-030-58565-5_43

11. Chechik, G., Sharma, V., Shalit, U., Bengio, S.: Large scale online learning of image similarity through ranking. J. Mach. Learn. Res. **11**(36), 1109–1135 (2010)

12. Deng, J., Berg, A.C., Fei-Fei, L.: Hierarchical semantic indexing for large scale image retrieval. In: IEEE Conference on Computer Vision and Pattern Recognition, pp. 785–792. IEEE (2011)

13. Deng, J., Dong, W., Socher, R., Li, L.J., Li, K., Fei-Fei, L.: ImageNet: a large-scale hierarchical image database. In: IEEE Conference on Computer Vision and Pattern Recognition, pp. 248–255. IEEE (2009)

14. Fellbaum, C.: WordNet. Wiley, Hoboken (1998)

15. Frome, A., et al.: DeViSE: a deep visual-semantic embedding model. In: International Conference on Neural Information Processing Systems, pp. 2121–2129 (2013)

16. Gairola, S., Shah, R., Narayanan, P.J.: Unsupervised image style embeddings for retrieval and recognition tasks. In: IEEE Winter Conference on Applications of Computer Vision, pp. 3270–3278 (2020)

17. Gomez, R., Gomez, L., Gibert, J., Karatzas, D.: Learning to learn from web data through deep semantic embeddings. In: Leal-Taixé, L., Roth, S. (eds.) ECCV 2018. LNCS, vol. 11134, pp. 514–529. Springer, Cham (2019). https://doi.org/10.1007/978-3-030-11024-6_40

18. Gordo, A., Almazán, J., Revaud, J., Larlus, D.: End-to-end learning of deep visual representations for image retrieval. Int. J. Comput. Vis. **124**(2), 237–254 (2017). https://doi.org/10.1007/s11263-017-1016-8

19. Ha, M.L., Hosu, V., Blanz, V.: Color composition similarity and its application in fine-grained similarity. In: IEEE Winter Conference on Applications of Computer Vision, pp. 2559–2568 (2020)

20. He, K., Lu, Y., Sclaroff, S.: Local descriptors optimized for average precision. In: IEEE Conference on Computer Vision and Pattern Recognition, pp. 596–605 (2018)

21. Hu, H., et al.: Web-scale responsive visual search at Bing. In: ACM SIGKDD International Conference on Knowledge Discovery and Data Mining, KDD 2018, pp. 359–367. ACM, New York (2018)

22. Huiskes, M.J., Lew, M.S.: The MIR flickr retrieval evaluation. In: ACM International Conference on Multimedia Information Retrieval. ACM, New York (2008). http://press.liacs.nl/mirflickr/

23. Husain, S.S., Bober, M.: Improving large-scale image retrieval through robust aggregation of local descriptors. IEEE Trans. Pattern Anal. Mach. Intell. **39**(9), 1783–1796 (2017)

24. Järvelin, K., Kekäläinen, J.: Cumulated gain-based evaluation of IR techniques. ACM Trans. Inf. Syst. **20**(4), 422–446 (2002)

25. Jegou, H., Douze, M., Schmid, C.: Hamming embedding and weak geometric consistency for large scale image search. In: Forsyth, D., Torr, P., Zisserman, A. (eds.) ECCV 2008. LNCS, vol. 5302, pp. 304–317. Springer, Heidelberg (2008). https://doi.org/10.1007/978-3-540-88682-2_24

26. Jégou, H., Douze, M., Schmid, C., Pérez, P.: Aggregating local descriptors into a compact image representation. In: IEEE Conference on Computer Vision and Pattern Recognition, pp. 3304–3311, June 2010
27. Jégou, H., Zisserman, A.: Triangulation embedding and democratic aggregation for image search. In: IEEE Conference on Computer Vision and Pattern Recognition, pp. 3310–3317, June 2014
28. Kato, T., Kurita, T., Otsu, N., Hirata, K.: A sketch retrieval method for full color image database - query by visual example. In: IAPR International Conference on Pattern Recognition, pp. 530–533, August 1992
29. Long, T., Mettes, P., Shen, H.T., Snoek, C.G.: Searching for actions on the hyperbole. In: IEEE/CVF Conference on Computer Vision and Pattern Recognition, pp. 1141–1150 (2020)
30. Lowe, D.G.: Distinctive image features from scale-invariant keypoints. Int. J. Comput. Vis. **60**(2), 91–110 (2004). https://doi.org/10.1023/B:VISI.0000029664.99615.94
31. Mikolajczyk, K., Schmid, C.: Scale & affine invariant interest point detectors. Int. J. Comput. Vis. **60**(1), 63–86 (2004). https://doi.org/10.1023/B:VISI.0000027790.02288.f2
32. Narayana, P., Pednekar, A., Krishnamoorthy, A., Sone, K., Basu, S.: HUSE: hierarchical universal semantic embeddings. arXiv preprint arXiv:1911.05978 (2019)
33. Niblack, C.W., et al.: QBIC project: querying images by content, using color, texture, and shape. In: Proceedings of the SPIE, Storage and Retrieval for Image and Video Databases, vol. 1908, pp. 173–188. International Society for Optics and Photonics (1993)
34. Noh, H., Araujo, A., Sim, J., Weyand, T., Han, B.: Large-scale image retrieval with attentive deep local features. In: IEEE International Conference on Computer Vision, pp. 3476–3485 (2017)
35. Perronnin, F., Liu, Y., Sánchez, J., Poirier, H.: Large-scale image retrieval with compressed fisher vectors. In: IEEE Conference on Computer Vision and Pattern Recognition, pp. 3384–3391, June 2010
36. Philbin, J., Chum, O., Isard, M., Sivic, J., Zisserman, A.: Object retrieval with large vocabularies and fast spatial matching. In: IEEE Conference on Computer Vision and Pattern Recognition, pp. 1–8, June 2007
37. Philbin, J., Chum, O., Isard, M., Sivic, J., Zisserman, A.: Lost in quantization: improving particular object retrieval in large scale image databases. In: IEEE Conference on Computer Vision and Pattern Recognition, pp. 1–8, June 2008
38. Radenović, F., Iscen, A., Tolias, G., Avrithis, Y., Chum, O.: Revisiting Oxford and Paris: large-scale image retrieval benchmarking. In: IEEE Conference on Computer Vision and Pattern Recognition, pp. 5706–5715, June 2018
39. Radenović, F., Tolias, G., Chum, O.: Fine-tuning CNN image retrieval with no human annotation. IEEE Trans. Pattern Anal. Mach. Intell. **41**(7), 1655–1668 (2018)
40. Razavian, A.S., Sullivan, J., Carlsson, S., Maki, A.: Visual instance retrieval with deep convolutional networks. ITE Trans. Media Technol. Appl. **4**(3), 251–258 (2016)
41. Razavian, A.S., Azizpour, H., Sullivan, J., Carlsson, S.: CNN features off-the-shelf: an astounding baseline for recognition. In: IEEE Conference on Computer Vision and Pattern Recognition Workshops, pp. 512–519, June 2014
42. Revaud, J., Almazan, J., de Rezende, R.S., de Souza, C.R.: Learning with average precision: training image retrieval with a listwise loss. In: The IEEE International Conference on Computer Vision, October 2019

43. Schroff, F., Kalenichenko, D., Philbin, J.: FaceNet: a unified embedding for face recognition and clustering. In: IEEE Conference on Computer Vision and Pattern Recognition, pp. 815–823, June 2015
44. Sivic, J., Zisserman, A.: Video Google: a text retrieval approach to object matching in videos. In: IEEE International Conference on Computer Vision, vol. 2, pp. 1470–1477 (2003)
45. Smeulders, A.W., Worring, M., Santini, S., Gupta, A., Jain, R.: Content-based image retrieval at the end of the early years. IEEE Trans. Pattern Anal. Mach. Intell. **22**, 1349–1380 (2000)
46. Tolias, G., Sicre, R., Jégou, H.: Particular object retrieval with integral max-pooling of CNN activations. In: International Conference on Learning Representations (2016)
47. Wu, H., Mao, J., Zhang, Y., Jiang, Y., Li, L., Sun, W., Ma, W.Y.: Unified visual-semantic embeddings: bridging vision and language with structured meaning representations. In: IEEE/CVF Conference on Computer Vision and Pattern Recognition, pp. 6602–6611 (2019)
48. Yang, S., Yu, W., Zheng, Y., Yao, H., Mei, T.: Adaptive semantic-visual tree for hierarchical embeddings. In: ACM International Conference on Multimedia, pp. 2097–2105. Association for Computing Machinery, New York (2019)
49. Zhi, T., Duan, L.Y., Wang, Y., Huang, T.: Two-stage pooling of deep convolutional features for image retrieval. In: IEEE International Conference on Image Processing, pp. 2465–2469, September 2016
50. Zhou, X.S., Huang, T.S.: Relevance feedback in image retrieval: a comprehensive review. Multimed. Syst. **8**(6), 536–544 (2003). https://doi.org/10.1007/s00530-002-0070-3

A Modality Converting Approach for Image Annotation to Overcome the Inconsistent Labels in Training Data

Tokinori Suzuki[(✉)] [ID] and Daisuke Ikeda

Kyushu University, Fukuoka 8190395, Japan
`suzuki.tokinori.070@s.kyushu-u.ac.jp, daisuke@inf.kyushu-u.ac.jp`

Abstract. The automatic image annotation (AIA) task, in which a system specifies descriptive keywords for an input image, has been a shared task studied for long
time, and still important because the annotation keywords enables users efficient
access of ever-growing image data. However, the current performance of the AIA
systems remains at low levels. One of the difficulties of the AIA comes from inconsistency of annotation keywords in the training data, which is naturally occurred
in manual annotations, for many supervised methods. For example, annotation
keywords for images of people may be "tourist" or "woman" depending on scenes
of the images. This inconsistency makes it difficult to annotate images, which
possibly have such similar keywords. For that difficulty, we propose a modality
converting method that transforms an input image into an encyclopedic text of
keywords assigned to the image. With the modality converting, similar keywords
can share their features derived from texts with each other. In the proposed method,
we pair images with Wikipedia articles, which have annotation keywords as their
titles. We train a modality convertor from images to Wikipedia texts using a neural
network with the paired data. Then, the method classifies the converted text into
annotation keywords similar to the text classification. Experimental results show
relatively high performance of our method based on the converted text compared
with existing methods.

Keywords: Automatic image annotation · Modality conversion · Image to text ·
Text classification · Neural network · Wikipedia

1 Introduction

The automatic image annotation (AIA) [2, 10] is a research task, in which given an image
as an input, a system specifies descriptive keywords for the input image. This task has
been studied for over a decade as a shared task in ImageCLEF workshops [2]. It is an
important task because annotation keywords will be reference of the images, and enable
a user efficient access to images. Moreover, because annotation keywords of the AIA
are derived from human annotation, the keywords can offer searchers more subjective
searches than with other conventional keywords, such as objective labels. Although the
task has been studied for long time, the performance remains at quite low level. A recent
study, for example, reported the performance of their method for the AIA, that does not
exceed 0.5 of F1 measure [9].

© Springer Nature Switzerland AG 2021
A. Del Bimbo et al. (Eds.): ICPR 2020 Workshops, LNCS 12662, pp. 261–268, 2021.
https://doi.org/10.1007/978-3-030-68790-8_21

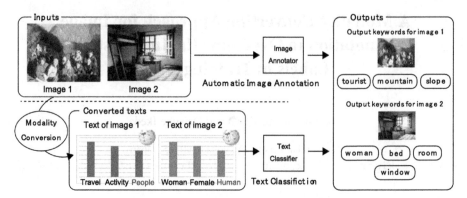

Fig. 1. Overview of the automatic image annotation and the proposed method.

Existing methods of the AIA [9, 10, 15] annotates images with descriptive keywords based on supervised approaches. Therefore, inconsistency of annotation keywords in training data is a crucial problem for such methods. Figure 1, for example, illustrates examples of a system's inputs and outputs of the AIA. The output keywords for the two pictures are shown on the right of the figure. Although both of the pictures are shots for people as their subjects, the annotation keywords for each picture are different. In the figure, the image 2 is assigned with "woman" label, and the image 1 is with "tourist" label but without "woman", though women are included in the group. Since the keywords are manually annotated, this kind of inconsistency is naturally occurred in the training data. When annotation methods trained on such inconsistent keywords as the classification symbols, they cannot distinguish similar images, which may share same annotation keywords.

For that difficulty, we propose a modality converting method which changes an input image into an encyclopedic text of keywords assigned to the image with neural networks, so that similar keywords could share their features with each other. For example, frequency of words in the converted texts are shown in the bottom of Fig. 1. The texts of the two images about "tourist" or "woman" share same contexts related to "people" and "human", which are expected to make the two keywords close-concepts. We report that the experimental results show relatively high performance of the AIA based on the converted text compared with an existing image classification method.

The rest of paper is structured as follows. Section 2 introduces the related studies. Section 3 describes the proposed modality converting method to change image data to text data and classification using the converted data. Section 4 provides evaluation of the proposed method and its outcomes. Section 5, finally, summarizes this paper.

2　Related Work

In this section, we briefly review literature in the two related fields in the automatic image annotation and the text classification.

The automatic image annotation is a task that ImageCLEF [2] have set as a series of shared tasks, and have been studied actively in 2000s [6, 8–10, 15]. Here, we introduce relatively new works [9, 10, 15]. Zhang *et al.* proposed nearest neighbor search approaches to makes use of features of groups for keywords by clustering [15]. Another work incorporates distance of colors and textures into distance computation [10]. Recently, a semi-supervised method was proposed in order to utilize different features of images by two different classifiers: a CNN classifier and an SVM classifier trained on low-level features of images [9]. While these existing methods focus on features derived from images, our method utilize text-based features by a modality conversion.

Next we introduce the recent studies of the text classification [7, 14] methods, which can be applicable for the text data converted by our proposed method. Lai *et al.* proposed a neural network classification model that has a max pooling layer following a recurrent neural network to use overall feature of an input text rather than emphasis on the end of the text [7]. Zhang *et al.* proposed a multi-task learning method to train a classifier on several datasets [14].

3 Proposed Method

In this section, we explain the proposed modality converting method from an image to text data and classification of the converted text into annotation keywords. Figure 2 shows the workflow of the method. First, in the figure A), a feature of an image, called a feature map, are extracted from the trained classifier of images to annotation keywords. Next, in the figure B), Wikipedia pages, which are target texts of the conversion, are paired with an image. We assume that there are Wikipedia pages, which have annotation keywords as their title. Then, the feature map is paired with the corresponding Wikipedia page. The pairs of a feature map of an image and Wikipedia pages are the training data for the modality converter. After that, we train a modality convertor with the paired data with our designed simple neural network in the figure C). Finally, we classify the modality converted data, which we call *pseudo-texts*, into annotation keywords.

3.1 Feature Extraction from a Convolutional Neural Network

First of all, in Fig. 2 A) we train a classifier of an input image to annotation keywords, such as "tourist", with a convolutional neural network (CNN). When an input image i is given to the trained classifier, we extract the last activated layer of the CNN [11] as a feature map of the image $x_i \in \mathbb{R}^d$. The extracted feature map vectors will be input to the modality convertor after this step.

We employed the Residual Network (ResNet) [5] for implementing the classifier. The CNN has an input layer of 224×224 pixel size. After the input layer, it contains several units, called *blocks*, consisting convolutional layers, activation layers and batch normalization layers. After repeated several blocks in the architecture, it has pooling layers and fully-connected (fc) layers. Each block takes two inputs from both the last layer's output and the last block's input called a shortcut connection. The shortcut connection learns the difference at each block by specifying the input.

The Keras[1] implementation of the 50-layers ResNet was used for the classifier. The ResNet classification model was trained on 1,000 classes over one million images from

---

[1] https://keras.io/applications/#resnet.

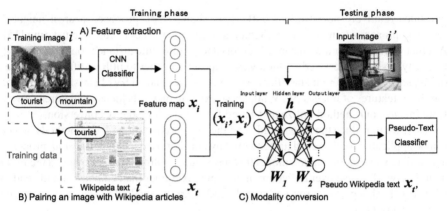

Fig. 2. Workflow of the proposed method.

ImageNet databases [1]. We used this pretrained weights for the initial weights of the network. We put an additional fc layer with size of keywords on the end of the layers. The model was trained with a mini-batch size of 32 and an SGD optimizer. The initial learning rate, weight decay and the momentum were 0.1, 10^{-4} and 0.9, respectively.

3.2 Modality Conversion

We pair images with target texts for the training data in Fig. 2 B). For the target texts, we selected Wikipedia articles because encyclopedic texts are good description for concepts of annotation keywords.

One requirement for using Wikipedia articles as the target texts is that Wikipedia should have the corresponding articles of the keywords. In Fig. 2 B), for example, the keyword "tourist" has its Wikipedia page. We confirmed that all the keywords of images used in the experiments have the corresponding articles. Some keywords that have several candidate pages *i.e.*, ambiguous pages in Wikipedia. We manually checked the correspondences for this time, because the number of keywords is not many. Automatic judgment of the correspondence [12] is required for scaling.

Until this step, we have features of images from a CNN in Sect. 3.1 and Wikipedia articles paired with the images. We designed a neural network, which maps vectors of the feature maps to vectors of Wikipedia articles, called pseudo-texts on the right of Fig. 2. The design of our neural network is inspired by autoencoders [3], which has shown effectiveness in unsupervised learning task such as dimensionality reduction, with the expectation of replicating good quality of Wikipedia texts for classification.

Figure 2 C) displays the neural architecture. We took a simple architecture of three-layer-network for a quick implementation putting a high priority on evaluating our idea. The neural network learns a mapping from feature map vectors x_i to vectors of texts $x'_t \in \mathbb{R}^l$.

The size of each layer are as follows. The size of input layer is the number n of dimensions of an feature map. That of hidden and output layers is m and l, respectively. In the input part in the figure B), x_i is converted into low dimensional latent vectors

$h \in \mathbb{R}^m$ of by the following function f:

$$h = f(i) = ReLU(W_1 \cdot x_i + b_1),$$

where $ReLU(\cdot)$ is the ramp function defined as $ReLU(x) = \max(0, x)$, W_1 and b_1 are parameters of the input part, a weighted matrix $W_1 \in \mathbb{R}^{m \times n}$ and a bias term $b_1 \in \mathbb{R}^m$.

In the output, the latent vector h is mapped to a text vector x'_t by the following function g:

$$x'_t = g(h) = \sigma(W_2 \cdot h + b_2),$$

where σ is the sigmoid function, and $W_2 \in \mathbb{R}^{m \times n}$ is a weighted matrix, and $b_2 \in \mathbb{R}^n$ is a bias term, which are the parameter of the output part. The modality convertor learns to minimize the loss of input x_i and output x'_t by defining the loss function $\mathcal{L}$ as follows:

$$\mathcal{L}(x_i, x_{t'}) = x_i - x'^2_t.$$

We optimize the function by stochastic gradient descent.

3.3 Classification

We build a classifier of the pseudo-texts into annotation keywords, by modeling the probability $P(k|t')$ that an input image i' of the psedo text t' have a keyword $k \in K$. Since one image is assigned with multiple keywords, we formalize the classification task as a multi label classification problem. We designed a three layers neural network, where Softmax function [3] is used at the top layer for predicting the probability as follows:

$$P(k|t') = \mathrm{Softmax}(w_k \cdot h_{t'} + b_k),$$

$$h_{t'} = \sigma(H_{x_{t'}} + b_n),$$

where $x_{t'} \in \mathbb{R}^d$ is a vector of a pseudo text of t', a matrix $H_{x_{t'}} \in \mathbb{R}^{d \times h}$, a bias term $b_n \in \mathbb{R}^h$ and $b_k \in \mathbb{R}$ are the model parameters. The hidden layer size was set at 1,000.

4 Experiments

In the section, we conduct evaluation experiments to see whether the classification based on the pseudo-text can improve the performance of the AIA task rather than that based on images.

4.1 Experimental Setting

Two test collections for the AIA task, IAPR TC-12 [4] and ImageCLEF [13], were used for evaluation. Both of the collections contain various images such as casual pictures of people, animals, playing sports. Figure 1 displays example images of IAPR TC-12. The number of images on IAPR TC-12 is 20,000. The ImageCLEF consists of a total of 7,291 images. We used 3,124 images out of them because the part of images were manually annotated. The number of target keywords were 291 and 207, on IAPR TC-12 and ImageCLEF, respectively. The keywords have the corresponding Wikipedia articles. The average number of words in a Wikipedia article is 2,682 and 8,403, and the vocabulary size is 37,230 and 41,170, on IAPR TC-12 and Image CLEF, respectively.

In the experiment, we compare the following three classification methods:

- **Image Classification (IC).** A feature vector for images extracted from the last fc layer of ResNet [5] classifier explained in Sect. 3.1. The vectors are 1,000 dimensions, which is the size of the layer, for evaluations.
- **Text Classification (TC).** We encode word occurrence vectors of Wikipedia articles which has keywords as their entries by adding x_t for each assigned keyword to an image. The dimensions are corresponded to the vocabulary size of the articles. We train TC method with a support-vector machine.
- **Pseudo-text Classification (PC).** A vector of word occurrence in a pseudo-text in Sect. 3.3. The dimensions are size of the vocabulary size of Wikipedia articles.

For evaluating the performances, precision and recall of output of the top five keywords and the number of keywords in which a method correctly classified at least one image, following the evaluation setting in [10]. The precision and recall, abbreviated as P and R, respectively, are calculated as follows: Precision (P) $= N_C/N_A$, Recall (R) $= N_C/N_H$, where N_A is the number of output for each keyword, N_C is the number of correct classification, and N_H is the number of images of a keyword in the collections. F1-measure (F1) is calculated by the harmonic mean between P and R. Also, we evaluate the number of keywords, abbreviated as #KW, where a method outputs at least one correct image. We conducted the evaluation experiments with 10-fold cross-validation.

4.2 Result

In Table 1, TC method shows the best in the two collections. The F1 is at 0.826 in IAPRTC and 0.831 in ImageCLEF. These results are very high compared with the other methods. The second best method is PC showing 0.447 and 0.623 of F1 in IAPRTC and ImageCLEF, respectively. Comparing PC with IC, all the figures are successfully improved. Also, there are statistical significance both of between the results of IC and that of PC, and between the results of IC and that of TC at one percent level.

We analyzed the results by manually categorizing keywords into five subgroups: Object&Animal, Place, People, Building and Other, that we defined those for generality of the keywords. The analysis results of IAPRTC are shown in Fig. 3. The successful cases of the PC are on People and Building groups, where F1 of the PC is higher than that of the IC by 0.12 and 0.14, respectively. For example, keywords in People group which are often varied depending on scenes of pictures such as "woman", "cyclist" and "tourist". Thus, the proposed method successfully annotates keywords, which tend to be inconsistent for images. For failure cases, the PC shows relatively low improvement from the IC on Object&Animal group. In the group, keywords are more independent from subjective judgement of human annotations than those in the other groups. For example,

Table 1. The results of the automatic image annotation. The figures are macro average over all the keywords. A star "*" indicates statistical significance of 1% level compared with the IC.

Test Collection	IAPRTC				ImageCLEF			
Method	P	R	F1	# KW	P	R	F1	# KW
IC [5]	0.514	0.325	0.398	207	0.674	0.385	0.490	81
PC	0.537	0.383	* 0.447	260	0.678	0.577	* 0.623	101
TC	**0.774**	**0.861**	*** 0.816**	**275**	**0.723**	**0.936**	*** 0.816**	**141**

"bed" and "lion" are keywords of Object&Animal group, and constantly annotated to the images with such visuals of the keywords.

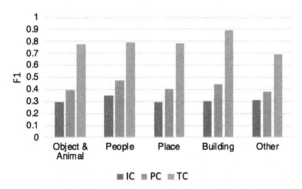

Fig. 3. F1 on each subgroup of annotation keywords of IAPRTC results.

5 Discussion and Conclusion

In this paper, we proposed a modality converting method from images into texts of Wikipedia to overcome difficulties of similar keywords. Experimental results show that the classification based on the converted texts is slightly high performance than existing image classification methods. The results also demonstrate that if keywords are completely replaced with the corresponding Wikipedia articles, the performance of the automatic image annotation can be significantly improved. However, there is a gap between the performance of the proposed pseudo-text classification and the text classification due to the text generation steps in our method. For the next step, tuning the neural network of the proposed modality convertor will be considered for closing the gap and further improvements. Classification methods should be considered as well.

References

1. Deng, J., Dong, W., Socher, R., Li, L., Li, K., Fei-Fei, L.: ImageNet: a large-scale hierarchical image database. In: Proceedings of the 2009 IEEE Conference on Computer Vision and Pattern Recognition (CVPR '09), pp. 248–255 (2009)

2. Gilbert, A., et al.: Overview of the imageCLEF 2016 scalable web image annotation task. In: Working Notes of CLEF 2016 – Conference and Labs of Evaluation Forum, pp. 254–278 (2016)
3. Goodfellow, I., Bengio, Y., Courville, A.: Deep Learning. MIT Press Cambridge, MA, USA (2016)
4. Grubinger, M., Clough, P., Müller, H., Deselaers, T.: The IAPR TC-12 benchmark: a new evaluation resource for visual information systems. In: Proceedings of the International Workshop OntoImage 2006 Language Resources for Content-Based Image Retrieval in conjunction with the fifth edition of the International Conference on Language Ressources and Evaluation (LREC 2006), pp. 13–23 (2006)
5. He., K., Zhang, X., Ren, S., Sun, J.: Deep residual learning for image recognition. In: Proceedings of the 29th IEEE Conference on Computer Vision and Pattern Recognition (CVPR 2016), pp. 770–778 (2016)
6. Jeon, J., Lavrenko, V., Manmatha, R.: Automatic image annotation and retrieval using cross-media relevance models. In: Proceedings of the 26th Annual International ACM SIGIR Conference on Research and Development in Information Retrieval (SIGIR 2003), pp. 119–126 (2003)
7. Lai, S., Xu, L., Liu, K., Zhao, J.: Recurrent convolutional neural networks for text classification. In: Proceedings of the 29th AAAI Conference on Artificial Intelligence (AAAI-15), pp. 2267–2273 (2015)
8. Lavrenko, V., Manmatha, R., Jeon, J.: A model for learning the semantics of pictures. In: Proceedings of the 17th International Conference on Neural Information Processing Systems (NIPS 2004), pp. 553–560 (2004)
9. Li, Z., Lin, L., Zhang, C., Ma, H., Zhao, W.: Collaborating CNN and SVM for automatic image annotation. In: Proceedings of the 2019 ACM International Conference on Multimedia Retrieval (ICMR 2019), pp. 63–67 (2019)
10. Makadia, A., Pavlovic, V., Kumar, S.: Baselines for image annotation. Int. J. Comput. Vis. **90**(1), 88–105 (2010)
11. Ng, J.Y.-H., Yang, F., Davis, L.S.: Exploiting local features from deep networks for image retrieval. In: Proceedings of the 28th IEEE Conference on Computer Vision and Pattern Recognition (CVPR 2015), pp. 53–61 (2015)
12. Suzuki, T., Ikeda, D., Galuščáková, P., Oard, D.: Towards automatic cataloging of image and textual collections with Wikipedia. In: Proceedings of the 21st International Conference on Asia-Pacific Digital Libraries (ICADL 2019), pp. 167–180 (2019)
13. Villegas, M., Paredes, R.: Overview of the ImageCLEF 2014 scalable concept image annotation task. In: Proceedings of the fifth Conference of the CLEF Initiative, pp. 308–328 (2014)
14. Zhang, H., Xiao, L., Chen, W., Wang, Y., Jin, Y.: Multi-task label embedding for text classification. In: Proceedings of the 2018 Conference on Empirical Methods in Natural Language Processing (EMNLP 2018), pp. 4545–4553 (2018)
15. Zhang, S., Huang, J., Huang, Y., Yu, Y., Li, H., Metaxas, D.N.. Automatic image annotation using group sparsity. In: Proceedings of the 23rd IEEE Conference on Computer Vision and Pattern Recognition (CVPR 2010), pp. 3312–3319 (2010)

Iconic-Based Retrieval of Grocery Images via Siamese Neural Network

Gianluigi Ciocca$^{(\boxtimes)}$ ⓘ, Paolo Napoletano ⓘ, and Simone Giuseppe Locatelli ⓘ

University of Milano-Bicocca, Viale Sarca 336, 2016 Milano, Italy
{gianluigi.ciocca,paolo.napoletano}@unimib.it,
s.locatelli29@campus.unimib.it

Abstract. In this paper we investigate the problem of Grocery product recognition using iconic images. Iconic images are used to advertise products and they are very different from images that are captured in-store. We investigate the use of learned features for the retrieval process. We evaluated different feature extraction strategies using Convolutional Neural Networks (CNNs) and tested the CNNs on the Grocery Store image dataset that contains 81 product categories grouped into 43 coarse-grained classes and 3 macro classes. Results show that a Siamese network with a DenseNet-169 backbone better captures relations between iconic and in-store images outperforming other architectures in the retrieval task.

Keywords: Image retrieval · Grocery products recognition · Siamese Neural Networks · Domain adaptation

1 Introduction and Related Work

Features learned by deep Convolutional Neural Networks (CNNs) have been recognized to be more robust and expressive than hand-crafted ones, and they have been successfully used in different computer vision tasks such as object detection, pattern recognition and image understanding. In recent years, food and, more in general, grocery products recognition have received a lot of attention from the research community. For example, systems to automatically locate and recognize diverse foods as well as to estimate the food quantity have been developed [2,6,7,18,19]. Grocery products recognition is also a very important task in the context of intelligent shop and in applications that help visually impaired people to identify objects during they daily activities [9,10,12,13,24,26]. Both food and grocery products recognition are also useful in supporting the monitoring of the dietary behavior of people in order to help them to maintain healthy daily food consumption and a sane life.

The most common recognition paradigm found in the literature is classification. This parading requires a set of annotated images that could be quite large in the case of CNN-based solutions. However, in real applications the number of examples needed to train a robust classifier may not be always available.

A. Del Bimbo et al. (Eds.): ICPR 2020 Workshops, LNCS 12662, pp. 269–281, 2021.
https://doi.org/10.1007/978-3-030-68790-8_22

In this case, the retrieval paradigm can be used to find similar products among the available ones and to suggest a possible product class. In this case, the set of images used to build model is substantially reduced with respect to the CNN-based classification. Moreover, in the retrieval paradigm, we can also leverage features extracted from pre-trained CNN models as the basis for the retrieval process with or without domain adaptation. Features learned by CNN networks have been demonstrated to be robust and effective in different retrieval tasks [6,7,21]. Finally, the retrieval paradigm can be also exploited by humans to ease the task of annotating images. It can be used as a tool to support the annotation of images by suggesting possible groups of similar images that could be all annotated with similar tags. This could alleviate the often long and error-prone annotation process.

Grocery product recognition is a complex task because object appearance is highly variable due to relevant changes in scale, pose, viewpoint, lighting conditions and occlusion. Often the scene in which the product are posed is cluttered with other objects, there can be multiple instances of the same products, and products can be on different shelves. Depending on the type of items, we may have rigid or non-rigid objects that make the recognition more challenging. Also a product can be categorized depending its use (e.g. food vs. non-food), its packaging (e.g. box vs. pack), its nature (e.g. fruits vs. vegetables), and its variety (e.g. yellow pepper vs. green pepper). This means that we can have a hierarchy in the categorization of grocery products from coarse to fine grained depending on the final application. Finally, there are possibly thousands of products available in a large store that are continuously added or removed. This makes the construction of a comprehensive recognition model very difficult and challenging to design and maintain.

In this paper we are interested in investigating the image retrieval paradigm in the context of grocery product recognition and see, to what extent, learned features from CNN-based architectures can help in solving the recognition problem using few examples of grocery product images. To this end, we have designed several experiments in which we exploited a large CNN architecture, i.e. the DenseNet-169, that we have been successfully used in grocery product classification [12] and a multi task variant of it specially designed to recognize grocery products at different level of granularity (e.g. fruits vs packages, apple vs. onion, "Golden delicious" vs. "Granny smith") [5]. From this network, we extracted different features in different layers and evaluate their performance in the product retrieval using in-store images. We also investigate another grocery product retrieval problem. In this case we are interested in evaluating the learned features when they are used to retrieve grocery products using iconic images instead, such as those that can be found in shop catalogues and flyers. These images, see Fig. 1, are often acquired under controlled conditions and with no background, and are quite different from those captured in-store. In this case we designed a Siamese network to learn more robust features that can cope with the differences in these two groups of images.

In-store   Iconic   In-store   Iconic

Fig. 1. Example of in-store product image vs. its corresponding iconic image. Images from the Grocery Store Dataset [12]

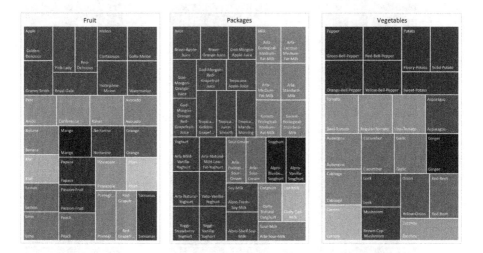

Fig. 2. Illustration of the hierarchical class structure of Grocery Store Dataset.

2   Materials

In the literature there are several datasets of grocery store items [16,17,20,27]. However, the only dataset that includes both real world and iconic images is the Grocery Store Dataset (GSD) [12]. The GSD dataset is publicly available, and provides a hierarchical categorization of the products at multiple level of details as showed in Fig. 2. The dataset consists of 5,125 product images that are categorized using a three-level categorization. The products are first classified into three macro categories: fruits, vegetables or packages macro classes. Then the images are tagged according to one of the 43 coarse-grained classes that represent the product type (i.e. apple, tomatoes, milk, etc...). Finally, each image is classified into one of the 81 fine-grained classes that correspond to specific instances of the product (i.e for the apple we may have Royal Gala, Granny Smith, etc...). The list of grocery products in the Grocery Store Dataset is shown in Fig. 3. The GSD also provide a set of iconic images that are originally used for other purposes, and this set is composed of only one image per class.

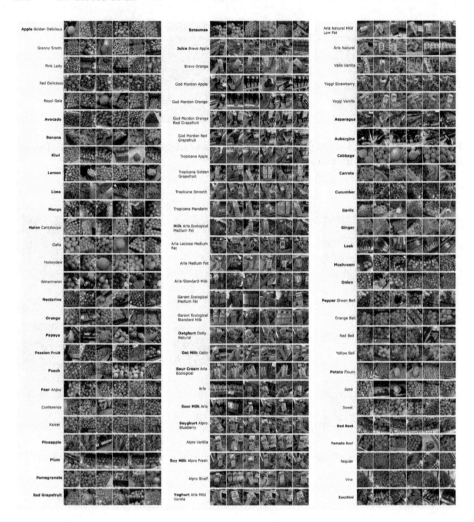

Fig. 3. Examples of the 81 fine-grained classes in GSD. Bold text refers to the 43 coarse-grained classes.

Thus, to investigate the use of iconic images for real world product retrieval we have collected 1,119 iconic images from the Internet representing the 81 fine-grained GSD classes. Figure 4 show some examples of the collected iconic images. Some of the classes contain very few new iconic images and this could pose a problem for training a network using only these images. Figure 5 shows the number of the collected iconic images for each of the 81 fine-grained product categories. We downloaded up to 20 images per category. As it can be seen the products belonging to the packages macro class do not have many images. This is mainly due to the fact that these products are specifically sold in Sweden and not many iconic images can be found on the Internet.

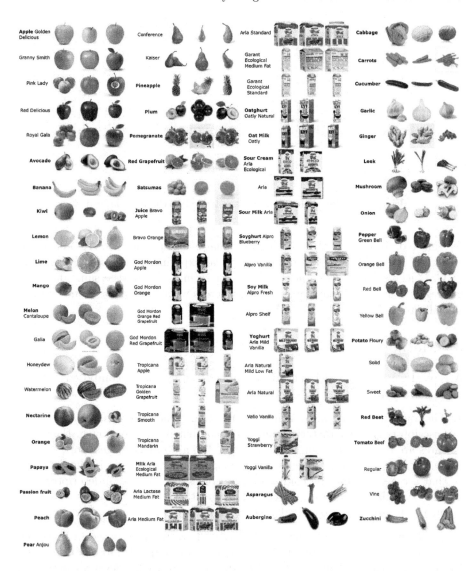

Fig. 4. Examples of the 81 fine-grained classes in GSD iconic section. Bold text refers to the 43 coarse-grained classes.

We performed two different experiments: retrieval of in-store images using in-store images as queries (i.e. in-store vs in-store), and retrieval of in-store images using iconic query images (i.e. iconic vs. in-store). To perform the experiments we defined different set of images. First, we considered the image splits defined in [12]. The training set is used to train the feature extractors. The GSD test set is further split into a set of image queries, and an image database set against which the retrieval is performed. For the in-store vs. in-store retrieval experiment,

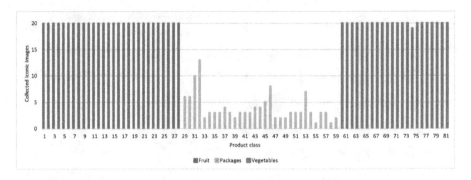

Fig. 5. Number of Iconic images collected in the new dataset.

we selected from each class 5 random images as queries and 10 images as database. For the iconic vs. in-store retrieval experiment we used the original 81 GSD iconic images (one per class) as query images, and the same database images as before. To train the feature extractors we also used the iconic dataset we collected.

3 Method

A typical Content-Based Image Retrieval (CBIR) system is composed of three main parts [8,21,25]:

1. The Indexing, also called feature extraction, module computes the visual descriptors that characterize the image content. Given an image, these features are usually pre-computed and stored in a database of features;
2. The Retrieval module, given a query image, finds the images in the database that are most similar by comparing the corresponding visual descriptors.
3. The Visualization module shows the images that are most similar to a given query image ordered by the degree of similarity.

As for indexing, a huge variety of features have been proposed in literature for describing the visual content [7,21]. They are often divided into hand-crafted features and learned features. Hand-crafted descriptors are features extracted using a manually predefined algorithm based on the expert knowledge. Learned descriptors are features extracted using Convolutional Neural Networks (CNNs). In this paper we focus on learned descriptors since they demonstrated to overcome hand-crafted features in several domains [7,21].

A basic retrieval scheme takes as input the visual descriptor corresponding to the query image performed by the user and it computes the similarity between such a descriptor and all the visual descriptors of the database of features. As a result of the search, a ranked list of images is returned to the user. The list is ordered by a degree of similarity, that can be calculated in several ways [25]: Euclidean distance (the one used here), Cosine similarity, Manhattan distance, χ^2 distance, etc. [4].

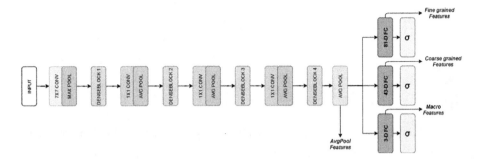

Fig. 6. DenseNet-169 multi-task architecture.

3.1 Learned Features

We employed a state-of-the-art network for feature extraction as well as specially designed Siamese networks [23]. State-of-the-art features have been extracted from a multi-task DenseNet-169 network trained on the three hierarchical layers of the GSD as described in [5]. The architecture of the multi-task DenseNet-169 network is shown in Fig. 6.

Different features are extracted from this network: features that represents the fine-grained classes, marked as "F" in Tables 1 and 2, the coarse-grained classes, marked as "C", the main categories' classes, marked as "M", features extracted from the last average pooling layer of the model, marked as "AvgPool", and, last, a combination of the first three types of features, marked as "M+C+F".

Next, two Siamese Neural Networks were developed (denoted as "S" in Tables 1 and 2). The first one has been trained using a couple of 2 in-store images, and the second one, has been trained using one iconic image and one in-store image. As backbone for the Siamese architecture we experimented the previously mentioned Multi-task DenseNet-169 architecture (see Fig. 7), and a well known ResNet-50 [11] architecture. For both Siamese networks, features have been extracted from the average pooling layer before the last classification layer. All the features have been L2 normalized before being used in the retrieval experiments.

4 Experiments

4.1 Retrieval Measures

Image retrieval performance has been assessed by using three state of the art measures: the Average Normalized Modified Retrieval Rank (ANMRR), Precision (P) and Recall (R), Mean Average Precision (MAP) [14,15].

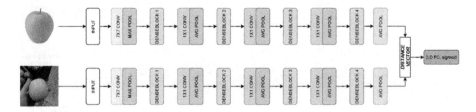

Fig. 7. The Siamese Neural Network architecture based on the DenseNet-169 multi-task.

Average Normalized Modified Retrieval Rank (ANMRR). The ANMRR measure is the MPEG-7 retrieval effectiveness measure commonly accepted by the CBIR community [14] and largely used by previous works on content-based image retrieval [1,21,22,28]. This metric considers the number and rank of the relevant (ground truth) items that appear in the top images retrieved. This measure overcomes the problem related to queries with varying ground-truth set sizes. The ANMRR ranges in value between zero to one with lower values indicating better retrieval performance and is defined as follows:

$$\text{ANMRR} = \frac{1}{Q} \sum_{q=1}^{Q} \frac{\text{Rank}_{\text{mean}}(q) - 0.5[1 + G(q)]}{1.25K(q) - 0.5[1 + G(q)]} \tag{1}$$

Q indicates the number of queries q performed. $G(q)$ is the size of ground-truth set for each query q. $K(q)$ is a constant penalty that is assigned to items with a higher rank. $K(q)$ is commonly chosen to be $2G(q)$. $\text{Rank}_{\text{mean}}(q)$ is the average rank for a single query q and is defined as

$$\text{Rank}_{\text{mean}}(q) = \frac{1}{G(q)} \sum_{k=1}^{G(q)} \text{Rank}(k) \tag{2}$$

where $\text{Rank}(k)$ is the kth position at which a ground-truth item is retrieved. $\text{Rank}(k)$ is defined as:

$$\text{Rank}(k) = \begin{cases} \text{Rank}(k), \text{if Rank}(k) \leq K(q) \\ 1.25K(q), \text{if Rank}(k) > K(q). \end{cases} \tag{3}$$

Precision and Recall. Precision is the fraction of the images retrieved that are relevant to the query

$$P = \frac{|\{\text{No. relevant images}\} \cap \{\text{No. retrieved images}\}|}{|\{\text{No. retrieved images}\}|}. \tag{4}$$

It is often evaluated at a given cut-off rank, considering only the topmost k results returned by the system. This measure is called precision at k or P@k.

Recall is the fraction of the images that are relevant to the query that are successfully retrieved:

$$R = \frac{|\{\text{No. relevant images}\} \cap \{\text{No. retrieved images}\}|}{|\{\text{No. relevant images}\}|}. \tag{5}$$

In a ranked retrieval context, precision and recall values can be plotted to give the interpolated precision-recall curve [15]. This curve is obtained by plotting the interpolated precision measured at the 11 recall levels of 0.0, 0.1, 0.2, ..., 1.0. The interpolated precision P_{interp} at a certain recall level k is defined as the highest precision found for any recall level $k' \geq k$:

$$P_{\text{interp}}(k) = \max_{k' \geq k} P(k'). \tag{6}$$

Mean Average Precision (MAP). Given a set of queries, Mean Average Precision is defined as,

$$\text{MAP} = \frac{\sum_{q=1}^{Q} P_{\text{mean}}(q)}{Q}, \tag{7}$$

where the average precision P_{mean} for each query q is defined as,

$$P_{\text{mean}} = \frac{\sum_{k=1}^{n} (P(k) \times r(k))}{\text{No. of relevant images}} \tag{8}$$

where k is the rank in the sequence of retrieved images, n is the number of retrieved images, $P(k)$ is the precision at cut-off k in the list ($P@k$), and $r(k)$ is an indicator function equalling 1 if the item at rank k is a relevant image, zero otherwise.

4.2 Results

Table 1 shows the results for the in-store vs. in-store retrieval experiments. As it can be seen, using the DenseNet-169 backbone, the most effective features, in terms of ANMRR, MAP and P@5, are those extracted from the Fine-grained branch of the multi-task network of Fig. 6. The second best overall results are obtained by the feature extracted from the AvgPool layer before the branches splits. The differences between these two features are minimal. In third position we have the concatenation of the features of the three branches. Not unexpectedly, the worst retrieval performances are obtained by the features extracted from the Macro category layer. These features are too coarse to capture the finer details in the grocery products. Also, using the Iconic images to train the network does not produce robust features. This is also expected.

If we consider the Siamese architecture, we can see that, regardless of the training images and the network backbone (DenseNet-169 or ReNet-50) used, we do not reach the same results as in the case of the DenseNet-169 multi-task network. However, the Siamese DenseNet-169, trained on the in-store images, produces the third best overall results in Table 1. This could means that the

Table 1. Retrieval experiments of in-store images vs. in-store images.

CNN backbone	Training set	Features	ANMRR ↓	MAP ↑	P@5 ↑	P@10 ↑	P@50 ↑
DenseNet-169	GSD	Macro (M)	0.841	13.81	15.80	11.98	6.94
DenseNet-169	GSD	Coarse (C)	0.217	75.54	79.75	65.31	19.83
DenseNet-169	GSD	Fine (F)	**0.184**	**79.54**	**83.95**	70.12	19.85
DenseNet-169	GSD	M+C+F	0.195	78.38	82.96	68.27	**19.88**
DenseNet-169	GSD	AvgPool	0.188	78.80	82.96	**70.86**	19.65
DenseNet-169	Iconic	Fine (F)	0.398	55.75	65.68	50.00	16.35
DenseNet-169	Iconic	AvgPool	0.361	58.83	68.89	53.70	16.54
ResNet-50 S	GSD+Iconic	AvgPool	0.361	59.68	66.91	53.33	17.09
ResNet-50 S	GSD	AvgPool	0.327	63.61	71.60	56.67	18.15
DenseNet-169 S	GSD+Iconic	AvgPool	0.235	73.56	80.49	65.43	19.48
DenseNet-169 S	GSD	AvgPool	0.192	78.37	82.47	69.75	19.80

Table 2. Retrieval experiments of iconic images vs. in-store images.

CNN backbone	Training set	Features	ANMRR ↓	MAP ↑	P@5 ↑	P@10 ↑	P@50 ↑
DenseNet-169	GSD	Macro (M)	0.937	6.91	4.44	4.20	4.20
DenseNet-169	GSD	Coarse (C)	0.557	38.74	36.54	35.06	16.67
DenseNet-169	GSD	Fine (F)	0.500	45.61	41.73	38.77	17.14
DenseNet-169	GSD	M+C+F	0.504	44.40	40.25	37.41	17.31
DenseNet-169	GSD	AvgPool	0.564	38.99	39.01	34.07	15.43
DenseNet-169	Iconic	Fine (F)	0.522	43.15	46.17	37.78	15.33
DenseNet-169	Iconic	AvgPool	0.563	39.46	42.96	35.43	13.90
ResNet-50 S	GSD+Iconic	AvgPool	0.580	36.44	37.53	33.21	15.38
ResNet-50 S	GSD	AvgPool	0.799	15.99	16.54	15.06	9.58
DenseNet-169 S	GSD+Iconic	AvgPool	**0.473**	**48.49**	**48.15**	**42.22**	**17.38**
DenseNet-169 S	GSD	AvgPool	0.563	38.65	37.78	33.33	16.17

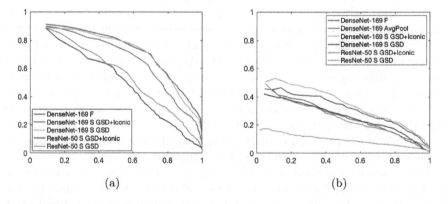

Fig. 8. Precision-recall charts for in-store vs. in-store (a) and iconic vs. in-store (b) experiments.

Siamese architecture does not contribute much at improving the retrieval performances on the in-store images with respect to the multi-task network. The pairwise analysis of in-store images does not extract more information than the multi-task strategy. Figure 8a shows the Precision-Recall charts for some of the in-store vs. in-store experiments.

Table 2 shows the results for the iconic vs. in-store retrieval experiments. The feature extraction strategies are the same as in the previous experiments. We can say that, in general, the retrieval performances by using iconic images for retrieving in-store products, are much lower than in the previous experiment. This confirms that these two types of images are very different and that the retrieval is more challenging. As in the previous case, the worst results are achieved using the features extracted from the Macro-category layer. We have the best results by Fine-grained features obtained using a network trained on the in-store images. However, if we compare the best results with the other ones we can see that the differences are not very large (with the exception of the macro-category features).

If we consider the Siamese architecture instead, we can see that pairing iconic images with in-store image significantly improves the retrieval results. The best features are those extracted from the DenseNet-169 Siamese network. All the considered measures are better than those of the other approaches. For example, the MAP is increased from 45.61 (Fine (F) features) to 48.49. Also the P@10 is increased from 38.77 to 42.22. Finally, features extracted from networks trained only on iconic images or only on in-store images achieve far worse results. This confirms that, in order to bridge the characteristics of the iconic images and the in-store ones, performing a pairwise analysis using the Siamese architecture is effective. Figure 8b shows the Precision-Recall charts for some of the iconic vs. in-store experiments.

5   Conclusion

We investigated the problem of Grocery images retrieval via deep features. The experiments have been conducted on the Grocery Store Dataset (GSD) since it provides a hierarchical categorization of the products at multiple level of details. This allow us to investigate the use of the product categories at different granularity. Moreover, it contains in-store images, that is image acquired in real stores, and iconic images, that is, images created for product advertisement. We enlarged this second set of images with new iconic images collected from the Internet.

With these images we presented two sets of experiments: 1) retrieval of in-store images using in-store images, and 2) retrieval of in-store images using iconic images. In the first experiment a multi-task DenseNet-169 specially trained to fit the dataset hierarchical characteristics performed better than other methods. In the second experiments a specially designed Siamese network with a DenseNet-169 as backbone, and trained using pairs of in-store and iconic images, demonstrated to better capture the main characteristics available only in grocery products.

The specificity of the packages products in the GSD dataset, available only in the Sweden market, made it difficult to collect the corresponding iconic image. Thus, a different, larger, dataset is needed so that the robustness of our feature extraction strategy with respect to the number of classes can be further investigated.

Finally, although the DenseNet-169 backbone exhibits promising results, there are others DNN architectures that can be also considered taking that are more computational efficient and accurate in recognition problems [3]. We plan to investigate these other DNN backbones.

References

1. Aptoula, E.: Remote sensing image retrieval with global morphological texture descriptors. IEEE Trans. Geosci. Remote Sens. **52**(5), 3023–3034 (2014)
2. Aslan, S., Ciocca, G., Mazzini, D., Schettini, R.: Benchmarking algorithms for food localization and semantic segmentation. Int. J. Mach. Learn. Cyber. **11**(12), 2827–2847 (2020). https://doi.org/10.1007/s13042-020-01153-z
3. Bianco, S., Cadene, R., Celona, L., Napoletano, P.: Benchmark analysis of representative deep neural network architectures. IEEE Access **6**, 64270–64277 (2018)
4. ten Brinke, W., Squire, D.M.G., Bigelow, J.: Similarity: measurement, ordering and betweenness. In: Negoita, M.G., Howlett, R.J., Jain, L.C. (eds.) KES 2004. LNCS (LNAI), vol. 3214, pp. 996–1002. Springer, Heidelberg (2004). https://doi.org/10.1007/978-3-540-30133-2_132
5. Ciocca, G., Locatelli, S., Napoletano, P.: Multi-task learning for supervised and unsupervised classification of grocery images. In: Proceedings of ICPR Workshop on Texture Analysis, Classification and Retrieval (2021). In press
6. Ciocca, G., Micali, G., Napoletano, P.: State recognition of food images using deep features. IEEE Access **8**, 32003–32017 (2020)
7. Ciocca, G., Napoletano, P., Schettini, R.: CNN-based features for retrieval and classification of food images. Comput. Vis. Image Underst. **176–177**, 70–77 (2018)
8. Datta, R., Joshi, D., Li, J., Wang, J.Z.: Image retrieval: ideas, influences, and trends of the new age. ACM Comput. Surv. (CSUR) **40**(2), 5 (2008)
9. Franco, A., Maltoni, D., Papi, S.: Grocery product detection and recognition. Expert Syst. Appl. **81**, 163–176 (2017)
10. Hao, Y., Fu, Y., Jiang, Y.G.: Take goods from shelves: a dataset for class-incremental object detection. In: Proceedings of the 2019 on International Conference on Multimedia Retrieval, pp. 271–278 (2019)
11. He, K., Zhang, X., Ren, S., Sun, J.: Deep residual learning for image recognition. In: Proceedings of the IEEE Conference on Computer Vision and Pattern Recognition, pp. 770–778 (2016)
12. Klasson, M., Zhang, C., Kjellström, H.: A hierarchical grocery store image dataset with visual and semantic labels. In: 2019 IEEE Winter Conference on Applications of Computer Vision (WACV), pp. 491–500. IEEE (2019)
13. Li, Q., et al.: Product image recognition with guidance learning and noisy supervision. Comput. Vis. Image Underst. **196**, 102963 (2020)
14. Manjunath, B.S., Ohm, J.R., Vasudevan, V.V., Yamada, A.: Color and texture descriptors. IEEE Trans. Circ. Syst. Video Technol. **11**(6), 703–715 (2001)
15. Manning, C.D., Raghavan, P., Schütze, H.: Introduction to Information Retrieval, vol. 1. Cambridge University Press, Cambridge (2008)

16. Marko, Š.: Automatic fruit recognition using computer vision. Mentor: Matej Kristan, Fakulteta za racunalništvo in informatiko, Univerza v Ljubljani (2013)
17. Merler, M., Galleguillos, C., Belongie, S.: Recognizing groceries in situ using in vitro training data. In: 2007 IEEE Conference on Computer Vision and Pattern Recognition, pp. 1–8. IEEE (2007)
18. Mezgec, S., Koroušić Seljak, B.: NutriNet: a deep learning food and drink image recognition system for dietary assessment. Nutrients **9**(7), 657 (2017)
19. Min, W., Jiang, S., Liu, L., Rui, Y., Jain, R.: A survey on food computing. ACM Comput. Surv. (CSUR) **52**(5), 1–36 (2019)
20. Mureşan, H., Oltean, M.: Fruit recognition from images using deep learning. Acta Universitatis Sapientiae Informatica **10**(1), 26–42 (2018)
21. Napoletano, P.: Visual descriptors for content-based retrieval of remote-sensing images. Int. J. Remote Sens. **39**(5), 1343–1376 (2018)
22. Ozkan, S., Ates, T., Tola, E., Soysal, M., Esen, E.: Performance analysis of state-of-the-art representation methods for geographical image retrieval and categorization. IEEE Geosci. Remote Sens. Lett. **11**(11), 1996–2000 (2014)
23. Qi, Y., Song, Y.Z., Zhang, H., Liu, J.: Sketch-based image retrieval via siamese convolutional neural network. In: 2016 IEEE International Conference on Image Processing (ICIP), pp. 2460–2464. IEEE (2016)
24. Santra, B., Mukherjee, D.P.: A comprehensive survey on computer vision based approaches for automatic identification of products in retail store. Image Vis. Comput. **86**, 45–63 (2019)
25. Smeulders, A.W., Worring, M., Santini, S., Gupta, A., Jain, R.: Content-based image retrieval at the end of the early years. IEEE Trans. Pattern Anal. Mach. Intell. **22**(12), 1349–1380 (2000)
26. Tonioni, A., Di Stefano, L.: Domain invariant hierarchical embedding for grocery products recognition. Comput. Vis. Image Underst. **182**, 81–92 (2019)
27. Waltner, G., et al.: MANGO - mobile augmented reality with functional eating guidance and food awareness. In: Murino, V., Puppo, E., Sona, D., Cristani, M., Sansone, C. (eds.) ICIAP 2015. LNCS, vol. 9281, pp. 425–432. Springer, Cham (2015). https://doi.org/10.1007/978-3-319-23222-5_52
28. Yang, Y., Newsam, S.: Geographic image retrieval using local invariant features. IEEE Trans. Geosci. Remote Sens. **51**(2), 818–832 (2013)

Garment Recommendation with Memory Augmented Neural Networks

Lavinia De Divitiis⬤, Federico Becattini⬤, Claudio Baecchi⬤,
and Alberto Del Bimbo^(✉)⬤

University of Florence, Florence, Italy
{lavinia.dedivitiis,federico.becattini,
claudio.baecchi,alberto.delbimbo}@unifi.it

Abstract. Fashion plays a pivotal role in society. Combining garments appropriately is essential for people to communicate their personality and style. Also different events require outfits to be thoroughly chosen to comply with underlying social clothing rules. Therefore, combining garments appropriately might not be trivial. The fashion industry has turned this into a massive source of income, relying on complex recommendation systems to retrieve and suggest appropriate clothing items for customers. To perform better recommendations, personalized suggestions can be performed, taking into account user preferences or purchase histories. In this paper, we propose a garment recommendation system to pair different clothing items, namely tops and bottoms, exploiting a Memory Augmented Neural Network (MANN). By training a memory writing controller, we are able to store a non-redundant subset of samples, which is then used to retrieve a ranked list of suitable bottoms to complement a given top. In particular, we aim at retrieving a variety of modalities in which a certain garment can be combined. To refine our recommendations, we then include user preferences via Matrix Factorization. We experiment on IQON3000, a dataset collected from an online fashion community, reporting state of the art results.

Keywords: Garment recommendation · Memory Augmented Neural Networks · Recommendation systems

1 Introduction

Recommendation systems are tools that have recently found pervasive use across several application fields. Ranging from social networks to online stores and streaming services, whenever a user has the ability to sift through large databases, machine learning algorithms can come to the aid by suggesting items of possible interest. A field that is heavily hinged on the effectiveness of recommendation systems is the fashion industry. In fact, revenues for large companies nowadays mostly come from online stores, where users are free to search for desired items. To maximize incomes, it is important for the seller to be able to suggest relevant items, often also exploiting advertisements specifically targeted on user profiles.

© Springer Nature Switzerland AG 2021
A. Del Bimbo et al. (Eds.): ICPR 2020 Workshops, LNCS 12662, pp. 282–295, 2021.
https://doi.org/10.1007/978-3-030-68790-8_23

Two typical strategies for recommendation are content based and collaborative filtering. The former is based on suggesting similar content to what the user has already purchased in the past. The latter instead exploits a pool of user preferences to infer what a certain individual might like, based on what similar users have purchased. Both these approaches have limitations. In particular, they can easily end up suggesting several variations of similar items due to their overall popularity or simply because they are close to items owned by the user [13]. This might even be counterproductive, since it limits the visibility of other items which are relevant yet novel or unexpected.

In this paper we propose a recommendation system capable of generating a variety of fashion items compatible with a given garment. In particular, we focus on generating different modalities to compose an outfit, rather than suggesting redundant and similar items. To achieve this, we exploit a persistent Memory Augmented Neural Network (MANN), which has proven effective to model diversity [17,18]. This kind of models finds its strength in the usage of an external memory, where samples can be explicitly stored at training time and then be read at inference time. Thanks to a trainable memory controller, we are able to store representations of non-redundant outfits, which are then retrieved at inference time to provide recommendations.

For outfit here we refer to a top-bottom pair, e.g. a t-shirt and a skirt. The problem we address is in fact the one of suggesting suitable bottom items, given a top as input. However, differently from prior work [26], we cast the problem as a retrieval task, rather than simply assessing the compatibility between garments.

The main contributions of this paper are the following:

- We propose a garment recommendation system based on Memory Augmented Neural Networks. To the best of our knowledge, we are the first to adopt MANNs for garment recommendation.
- We train our model to propose a variety of bottoms instead of a set of redundant and almost identical garments, shifting the focus of the retrieval task from single items to garment matching modalities.
- We obtain state of the art results on the IQON3000 dataset [26], a collection of outfits collected from an online fashion community.

2 Related Work

2.1 Recommendation Systems

In recommendation systems, collaborative filtering methods are often employed to find the relationship between two sets of entities, such as users and products. Among collaborative filtering methods, Matrix Factorization (MF) determines latent features among these sets to find similarities and allows to make predictions on both entities. In [15] they use a variant of Matrix Factorization called Non-negative Matrix Factorization (NMF), for decomposing multivariate data and study two different algorithms. In [12] the authors employ MF for Netflix movie recommendation to users. They incorporate additional information such as implicit feedback, temporal effects and confidence levels. On a more general scale, [10] perform an introductory study on collaborative filtering and MF for product recommendation. They also propose a way to measure the effectiveness of recommender systems. In a similar fashion, [22] illustrate various recommender systems putting particular attention into the difference of the various approaches. Focusing their research on Non-negative MF, [28] perform an extensive survey on different methods using NMF, analysing the design principles, characteristics, problems, relationships, and evolution of the algorithms. Finally [2] perform an evolutionary survey on MF and collaborative filtering focusing on the future implications of these methods.

Many recommender systems employ MF for recommending garment matches. In [3] they propose a multi-view Non-negative Matrix Factorization (NMF) to solve the problem of clothing matching leveraging multiple types of features. [25] propose a content-based neural scheme to model the compatibility between fashion items to give meaningful suggestions to the user. In a scenario where multiple items have to be recommended, [30] propose several extensions of MF to predict user ratings on garment packaging composed of tops and bottoms. Moreover, they also create a package recommendation dataset to serve their task. Finally, [6] propose a way to interpret MF based clothing matching proposals, giving also suggestions on how to interpret them and on how to modify those negative garments to make them appealing to the user. They employ Non-negative MF to suggest the alternatives for each fashion item pair.

2.2   Garment Recommendation

In the world of recommendation systems, Garment Recommendation is the task of generating meaningful recommendations of garments based on user interests or on a given input garment. [8] experiment on the *Amazon clothing dataset* to recommend clothes to people based in their interests by using Matrix Factorization techniques. Similarly, in [9] they give the users suggestions on clothes but considering a set of clothes instead of a single piece. Differently to [8], they use a functional tensor factorization method to model the interactions between user and fashion items. [24] leverage the real-world FashionVC dataset to perform complementary clothing matching by integrating deep neural networks and the rich fashion domain knowledge. Recently [1] leveraged human pose and behaviour to give meaningful suggestions to the user when no prior information about the user is known.

Remaining in the garment recommendation domain, another task that has recently gained interest: Virtual try-on. Virtual try-on complements the recommendation systems with a way of showing the end user how one or more clothes would look if "tried on", simulating the real life act of trying a garment before buying. Here [7] leverage 3D cameras to recreate a 3D model of the user and apply on it the selected garment. In [11] they investigate online apparel shoppers' use of Virtual Try-on to reduce product risks and increase enjoyment in online shopping. The research is based on the electronic Technology Acceptance Model (e-TAM). Finally [32] introduce a new try-on setting, which enables to change both the clothing item and the person's pose. They use a pose-guided virtual try-on scheme based on a generative adversarial network (GAN).

2.3 Memory Augmented Networks

Memory Augmented Neural Networks (MANN) are a particular declination of Neural Networks that exploit a controller network with an external memory, in which samples can be explicitly stored. These models have been originally introduced to solve algorithmic tasks [4,23,27,29], however several applications of MANNs have been proposed in literature [16,18–20,31]. The first work to propose a model equipped with an external memory has been Neural Turing Machines (NTM) [4]. The authors exploited the MANN as a working memory to solve data manipulation tasks, such as sorting or copying.

The problem of online learning has also been addressed with MANNs. Pernici *et al.* [20] rely on an external memory to incrementally store people identities for re-identification purposes. In [31], Memory Augmented Networks have been shown to aid object tracking, by offering the ability to store templates of the tracked objects. Also incremental classifiers have been implemented [21], by adding classes in memory, incrementally.

Several works have found MANNs beneficial for Question Answering [27, 29] or Visual Question Answering [14,16] tasks. In these works, some attention mechanism is usually exploited to guide the answering process by sequentially attending to different memory locations or to assign a different importance to common and uncommon question answer pairs.

Most of these approaches, however, rely on episodic memories. This means that the model learns how to manipulate data, rather than learning to build a persistent collection of samples. Recently, a model for trajectory forecasting has been proposed by Marchetti *et al.* [17,18], where the model learns to store non redundant samples relying on a specifically tailored writing controller. In this work, we rely on a similar strategy, by adding in memory a collection of garments. We then use these to retrieve different modalities to create an outfit by matching clothing items. Differently from [17], we make use of a convolutional model to store image embeddings in memory and we cast the model to solve a recommendation task by adding user information via Matrix Factorization.

3 Garment Recommendation with MANNs

We focus on the problem of garment recommendation, i.e. the task of retrieving a ranked list of bottoms that can be combined with a given top. Tops and bottoms are clothing items that can be worn together, defining an outfit. We refer to the set of top and bottom images respectively as $\mathcal{T} = \{t_0, t_1, ...t_{N_T}\}$ and $\mathcal{B} = \{b_0, b_1, ...b_{N_B}\}$, where N_T and N_B are the total number of garments in the two sets. Each user $u_k \in \mathcal{U} = \{u_0, u_1, ...u_{N_U}\}$ is associated with a collection of outfits, i.e. top-bottom pairs which have been rated as positive matches.

We indicate this outfit set for user u_k as $\mathcal{O}_k = \{(t_i, t_j)\}_{ij}$, where i and j act as indexes selecting items in $\mathcal{T}$ and $\mathcal{B}$.

The goal of the garment recommendation task can therefore be formulated as the task of retrieving the bottom t_j given a user u_k and a top t_i belonging to $\mathcal{O}_k$.

In the following, we first present our Memory Augmented Neural Network based method for establishing matches according to general style preferences and we then outline our recommendation refinement strategy, which includes personal user preferences via Matrix Factorization.

3.1   Feature Representation

In order to let our model manipulate tops and bottoms and store them into its external memory, we first need to devise a way to extract compact representations. To this end, we train an autoencoder that jointly learns to reconstruct top and bottom items forming an outfit.

Separate convolutional encoders project top and bottom images into a common latent space. We denote the top encoding as τ and the bottom encoding as β. The two features are then blended together with a scalar product into a joint representation ϕ and are then separately reconstructed with two deconvolutional decoders. Figure 1 depicts the structure of the model. In our experiments we observed that training the top and bottom autoencoders jointly led to a better generalization, reducing overfitting.

The benefits of using an autoencoder are twofold: on the one hand, we are able to obtain a compact representation of garments without any further annotation cost; on the other hand, we are also training a decoding function which will be useful for extracting bottoms from memory.

We use the same architecture for the two pairs of encoders and decoders. As for the encoder, we perform three 3×3 2D convolutions with 8, 16 and 32 filters respectively. All three convolutions have a padding applied of 1 pixel. We then apply 3×3 max pooling with a stride of 3. The relu activation function is applied after each convolution. On the decoder side, we employ three 2D transposed convolutions to attempt to reconstruct the input. Specifically, we set the number of channels to 16, 8 and 3 with a kernel size of 5×5, 8×8 and 3×3 respectively. As for padding we apply 4, 2 and 3 pixels respectively, with an extra pixel of output padding in the first transposed 2D convolution. The relu activation function is applied after each transposed convolution. Finally to each encoder-decoder pair output we apply the sigmoid function to output values in the range [0, 1].

3.2   Model

To find suitable bottom garments to be paired with a given top, we adopt a model based on a Memory Augmented Neural Network. The model is trained to identify a non-redundant subset of outfits to be stored in memory, which

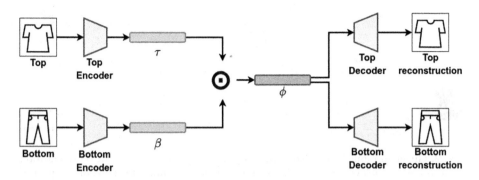

Fig. 1. Top and bottom images are separately encoded using convolutional encoders. The resulting features τ and β are then combined through a scalar product into a common representation ϕ. Finally, ϕ is decoded into two separate reconstructions with deconvolutional layers to obtain an approximation of the inputs.

can then be used to produce recommendations. The memory acts as a learnable interface between user queries and the underlying database. Since our aim is to suggest multiple outfit modalities, rather than retrieving all suitable instances in the dataset, we train the model to store only samples that can be used to suggest a variety of bottoms.

Given an outfit composed of a top-bottom pair (t, b), we want our model to output a ranked list of bottoms, where at least one is sufficiently close to b, when t is fed as input. To this end, when the top t is fed to the model, we encode it into a feature representation τ, using the encoding from Sect. 3.1. This compact representation of the input is then used as key to access the external memory. Memory access is performed via cosine similarity between the key and the top component of each stored outfit. This generates a distribution of similarity scores s_i for each element in memory:

$$s_i = \frac{\tau \cdot \tau_i}{\|\tau\| \cdot \|\tau_i\|} \qquad for \quad i = 0, ..., |M| - 1 \tag{1}$$

where $|M|$ is the memory size.

To generate a ranked set of recommendations, we take the K_N samples that give the highest scores. For each outfit retrieved from memory, we take its bottom β_{K_i} and we combine it with τ via dot product, to obtain a joint representation ϕ_{K_i}. Each feature is then fed to the decoder part of the autoencoder (see Sect. 3.1) to reconstruct different bottom images. Each reconstruction is performed in parallel, independently from the others. Figure 2 shows the inference process of our model.

To guarantee diversity, as well as satisfactory reconstructions, we train a writing controller that decides what samples to insert into memory. At training time, a top is fed to the network and K_N bottoms are proposed as output, as described above. The task of garment recommendation is inherently multimodal, i.e. multiple outputs might be considered correct given a single input. In other

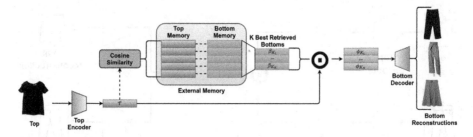

Fig. 2. Given an input image depicting a top, we encode it into a latent representation τ and use it to retrieve samples from memory via cosine similarity. The bottom part of the best samples are then combined with τ and decoded into bottom reconstructions.

terms, a user might consider multiple bottoms to be a good match with a top to create different outfits.

Since we only have one ground truth bottom t^*, we cannot explicitly add supervision on all the generated outputs. To overcome this limitation, we borrow a concept from the trajectory forecasting literature, where the multimodality of the task plays an even more central role. In fact, we derive our controller loss from the Variety Loss [5] to train our network. The Variety Loss, often referred to as Best-of-K loss, encourages the generation of diverse samples $\hat{t}_k$ by simply backpropagating the error only for the best prediction, leaving the other predictions free to explore the output space:

$$\mathcal{L}_{Variety} = \min_k \|\hat{t}_k - t^*\|_2 \tag{2}$$

Our memory writing controller, i.e. a trainable feedforward network that outputs a writing probability P_w for a given sample, takes as input the error value given by the Variety Loss itself. Based on the best reconstruction error, in fact, it establishes whether the current sample should be written to memory. The rationale behind this idea is that a sample only needs to be written if the model is not capable of generating a sufficiently close reconstruction of the ground truth.

We therefore optimize the controller loss $\mathcal{L}_{Controller}$ defined as follows:

$$\mathcal{L}_{Controller} = \mathcal{L}_{Variety}(1 - P_w) + (1 - \mathcal{L}_{Variety})P_w \tag{3}$$

A similar loss has been adopted in [17] to fill a persistent memory with non-redundant samples. The controller generates a writing probability in $[0, 1]$ and, since we use normalized images and an MSE error function for the Variety Loss, $\mathcal{L}_{Variety}$ also yields values in $[0, 1]$. When the reconstruction generates a high error, then $\mathcal{L}_{Variety}$ will be close to 1 and the sample should be written to improve the recommendation capabilities of the network. In this case the loss will optimize the controller to output a high writing probability. On the contrary, when the reconstruction error is low, the controller will be optimized to discard the observed outfit.

When a sample gets written in memory, we store the pair of encodings for the top and bottom. At training time, we start from a randomly initialized memory with K_N samples, to be able to always suggest a sufficient number of elements. We empty the memory after each epoch and populate the final memory after the controller has been trained by iterating once again on the training set.

3.3   General Recommendations and User Preferences

The model presented so far is capable or retrieving bottoms that can be matched with a given top. These matches, however, follow the general fashion criteria underlying the data distribution. Effective recommendation systems typically strive at suggesting items that are targeted for specific users, based on some past information, such as purchase history, or similarity with other users.

A common approach to generate targeted recommendations is Matrix Factorization [2, 28]. In our garment recommendation system, we first retrieve a set of bottoms with our Memory Augmented Neural Network. These bottoms represent different modalities to compose an outfit based on the given top. Although the MANN is capable of providing an ordered set according to top similarity with stored samples, this ranking does not take into account the user.

We refine our recommendations by re-ranking the set of bottoms via Matrix Factorization, which generates compatibility scores between the user and each bottom. Matrix Factorization projects the user and bottom IDs into a common space using trainable encoding functions and combines them together with a dot product. For both the user and the proposed bottoms, we also learn a bias to shift the decision. The final compatibility score, with values in $[0, 1]$, is obtained through a sigmoid activation.

To be able to work with known garments, i.e. belonging to a given database collection, we retrieve the most similar bottoms in the training set to the bottoms reconstructed by the decoder of the MANN. This allows us to learn a meaningful projection for bottoms, based on all user preferences.

4   Experiments

In the following we first provide some details on the dataset and metrics used to evaluate our model and then we report results, comparing the proposed approach with the state of the art.

4.1   Dataset and Metrics

For our experiments, we use the IQON3000 dataset [26]. This dataset contains a collection of 308,747 outfits, handpicked by 3,568 different users. Each outfit comprises garments belonging to 6 categories: *coat*, *top*, *bottom*, *one piece*, *shoes* and *accessories*. In total the dataset has 672,335 fashion items. All items are labeled with attributes such as *color*, *category*, *price* and *description*. Similarly to prior work [26], in this paper we focus solely on top and bottom garments and

Table 1. Comparison between our model and GP-BPR [26]. An item is considered correct if it shares Category and/or Color with the ground truth.

Num items	Category × Color		Category		Color	
	Ours	GP-BPR	Ours	GP-BPR	Ours	GP-BPR
5	**30**	27	**81**	75	**58**	55
10	**45**	44	**89**	84	**73**	70
20	**59**	56	**93**	90	**85**	81
30	**67**	63	**95**	92	**91**	87
40	**71**	69	**96**	93	**94**	90
50	**75**	72	**96**	94	**96**	92
60	**78**	74	**97**	94	**97**	92

we always consider tops to be given as inputs in order to find matching bottoms. The proposed model, however, could be easily adapted to the inverse problem by simply swapping tops and bottoms, without any further modification.

The dataset was first introduced to address the task of personalized garment matching. This was originally engineered to simply assess whether a model was capable of assigning a higher score to known positive matches compared to random ones. Here we extend this task, aiming at obtaining a ranked list of recommendations among all possible garments. However, we believe that an effective recommendation system should propose a variety of different outputs instead of suggesting several small variations of the same garment.

To this end, instead of simply comparing the ID of the retrieved bottoms, we compare their categories and colors. Proposing bottoms of different colors and categories, to be paired with the same top, means that we are proposing diverse modalities to combine the given top and generate an outfit. In the dataset there are 62 bottom categories, 12 different colors and a total of 742 combinations of the two.

We evaluate our approach by measuring the Accuracy@K, i.e. the fraction of samples for which the model suggests at least one bottom sharing the same category and/or color of the ground truth, among the first K retrieved items. In addition, we report the quality of the ranking by measuring the mean Average Precision (mAP). This is of interest especially when considering user personal preferences.

4.2   Results

Here we discuss the results obtained by our model. First of all, we evaluate the accuracy of the model by considering colors and categories, as explained in Sect. 4.1. We measure the accuracy varying the number of retrieved items, as shown in Table 1. Despite the number of categories is approximately five times higher than the number of colors, the system immediately obtains an

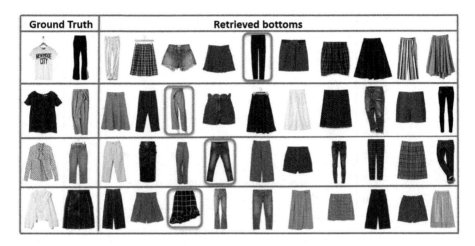

Fig. 3. Qualitative results generated by our proposed model. The top is given as input and a ranked set of bottom is retrieved. It can be observed that bottoms similar to the ground truth appear among the first recommendations.

accuracy above 80%, even with just 5 retrieved items. Considering colors, instead, the accuracy starts from around 60%. We have attributed this to the fact that color matching, despite following precise social rules, allows some flexibility with similar tonalities.

As expected, when considering both color and category, the accuracy drops considerably. Interestingly, in all cases, we are able to obtain a higher accuracy than the state of the art. In particular, we report results obtained by GP-BPR [26]. The results are obtained with the pre-trained models released by the authors by generating all compatibility scores between the pair of current top and user and all the bottoms in the dataset. We then take the highest ranking garments to generate the recommendations.

Despite GP-BPR explicitly considers the compatibility with each bottom in the dataset, our approach is still able to yield a higher accuracy relying only on the subset of garments stored in memory. Figure 3 shows some qualitative samples of bottoms retrieved from memory. It can be seen how there is a high variability in the suggested outfits and that a similar bottom to the ground truth is found in the first positions of the rankings.

We then measure the mean Average Precision obtained by our system. Figure 4 shows the curve, varying the number of retrieved garments. The combination of the Memory Network with Matrix Factorization yields to better results than GP-BPR [26] for a low number of recommended items, which is the scenario with the most relevant applicability.

Finally we show in Fig. 5 and Fig. 6 the effect of removing user information in the recommendation pipeline. In this way, we obtain only recommendations that follow a common sense, rather than personalized suggestions. Again, we compare our results against [26], removing the user information also for this model for

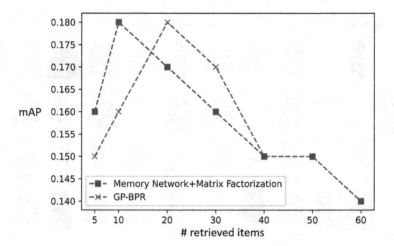

Fig. 4. Mean Average Precision of the proposed approach using Memory Network and Matrix Factorization. As reference, we report the results obtained by GP-BPR [26]. Interestingly, with a low number of retrieved items, our method obtained a higher mAP compared to GP-BPR.

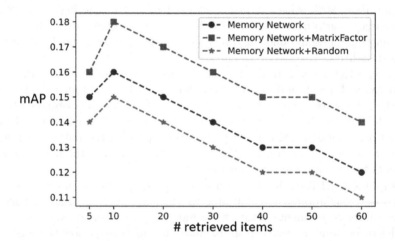

Fig. 5. Impact of reranking on the garments retrieved from the Memory Network. Matrix Factorization provides an improvement including user preferences.

a fair comparison. It can be seen in Fig. 5 how our method reports a 2% drop, while on the other hand, GP-BPR suffers from a much higher loss (Fig. 6). This suggests that our memory network is capable of providing a satisfactory set of garments even without considering the user. At the same time, we verify that Matrix Factorization is capable of reranking the garments effectively and that overall the Memory Network, with and without Matrix Factorization, are capable or providing better results compared to a Random reranking baseline.

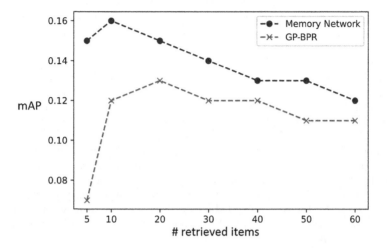

Fig. 6. Comparison of our proposed method against GP-BPR [26] without including user information.

5 Conclusions

In this paper we have presented a garment recommendation system based on a Memory Agumented Neural Network. The model relies on an external memory, which is filled by a trainable controller to store non redundant samples that capture multiple modalities to match top and bottom garments. In this way, the system is capable of retrieving and suggesting a variety of garments instead of proposing small variations of the same outfit. We have shown how this reflects on improved recommendation capabilities compared to the state of the art.

Acknowledgement. This work was partially supported by the Italian MIUR within PRIN 2017, Project Grant 20172BH297: I-MALL - improving the customer experience in stores by intelligent computer vision.

References

1. Bigi, W., Baecchi, C., Del Bimbo, A.: Automatic interest recognition from posture and behaviour. In: Proceedings of the 28th ACM International Conference on Multimedia, MM 2020, pp. 2472–2480. Association for Computing Machinery, New York (2020). https://doi.org/10.1145/3394171.3413530
2. Bobadilla, J., Ortega, F., Hernando, A., Gutiérrez, A.: Recommender systems survey. Knowl.-Based Syst. **46**, 109–132 (2013)
3. Chang, W.Y., Wei, C.P., Wang, Y.C.F.: Multi-view nonnegative matrix factorization for clothing image characterization. In: 2014 22nd International Conference on Pattern Recognition, pp. 1272–1277. IEEE (2014)
4. Graves, A., Wayne, G., Danihelka, I.: Neural turing machines. arXiv preprint arXiv:1410.5401 (2014)

5. Gupta, A., Johnson, J., Fei-Fei, L., Savarese, S., Alahi, A.: Social GAN: socially acceptable trajectories with generative adversarial networks. In: Proceedings of the IEEE Conference on Computer Vision and Pattern Recognition, pp. 2255–2264 (2018)
6. Han, X., Song, X., Yin, J., Wang, Y., Nie, L.: Prototype-guided attribute-wise interpretable scheme for clothing matching. In: Proceedings of the 42nd International ACM SIGIR Conference on Research and Development in Information Retrieval, pp. 785–794 (2019)
7. Hauswiesner, S., Straka, M., Reitmayr, G.: Free viewpoint virtual try-on with commodity depth cameras. In: Proceedings of the 10th International Conference on Virtual Reality Continuum and Its Applications in Industry, pp. 23–30 (2011)
8. He, R., McAuley, J.: VBPR: visual Bayesian personalized ranking from implicit feedback. In: Thirtieth AAAI Conference on Artificial Intelligence (2016)
9. Hu, Y., Yi, X., Davis, L.S.: Collaborative fashion recommendation: a functional tensor factorization approach. In: Proceedings of the 23rd ACM International Conference on Multimedia, pp. 129–138 (2015)
10. Jannach, D., Zanker, M., Felfernig, A., Friedrich, G.: Recommender Systems: An Introduction. Cambridge University Press, Cambridge (2010)
11. Kim, J., Forsythe, S.: Adoption of virtual try-on technology for online apparel shopping. J. Interact. Mark. **22**(2), 45–59 (2008)
12. Koren, Y., Bell, R., Volinsky, C.: Matrix factorization techniques for recommender systems. Computer **42**(8), 30–37 (2009)
13. Kotkov, D., Veijalainen, J., Wang, S.: How does serendipity affect diversity in recommender systems? A serendipity-oriented greedy algorithm. Computing **102**(2), 393–411 (2020)
14. Kumar, A., et al.: Ask me anything: dynamic memory networks for natural language processing. In: International Conference on Machine Learning, pp. 1378–1387 (2016)
15. Lee, D., Seung, H.S.: Algorithms for non-negative matrix factorization. In: Advances in Neural Information Processing Systems 13, pp. 556–562 (2000)
16. Ma, C., et al.: Visual question answering with memory-augmented networks. In: Proceedings of the IEEE Conference on Computer Vision and Pattern Recognition, pp. 6975–6984 (2018)
17. Marchetti, F., Becattini, F., Seidenari, L., Bimbo, A.D.: Mantra: memory augmented networks for multiple trajectory prediction. In: Proceedings of the IEEE/CVF Conference on Computer Vision and Pattern Recognition, pp. 7143–7152 (2020)
18. Marchetti, F., Becattini, F., Seidenari, L., Del Bimbo, A.: Multiple trajectory prediction of moving agents with memory augmented networks. IEEE Trans. Pattern Anal. Mach. Intell. (2020)
19. Pernici, F., Bartoli, F., Bruni, M., Del Bimbo, A.: Memory based online learning of deep representations from video streams. In: Proceedings of the IEEE Conference on Computer Vision and Pattern Recognition, pp. 2324–2334 (2018)
20. Pernici, F., Bruni, M., Del Bimbo, A.: Self-supervised on-line cumulative learning from video streams. Comput. Vis. Image Underst. 102983 (2020)
21. Rebuffi, S.A., Kolesnikov, A., Sperl, G., Lampert, C.H.: ICARL: incremental classifier and representation learning. In: Proceedings of the IEEE Conference on Computer Vision and Pattern Recognition, pp. 2001–2010 (2017)

22. Ricci, F., Rokach, L., Shapira, B.: Introduction to recommender systems handbook. In: Ricci, F., Rokach, L., Shapira, B., Kantor, P.B. (eds.) Recommender Systems Handbook, pp. 1–35. Springer, Boston, MA (2011). https://doi.org/10.1007/978-0-387-85820-3_1

23. Santoro, A., Bartunov, S., Botvinick, M., Wierstra, D., Lillicrap, T.: Meta-learning with memory-augmented neural networks. In: International Conference on Machine Learning, pp. 1842–1850 (2016)

24. Song, X., Feng, F., Han, X., Yang, X., Liu, W., Nie, L.: Neural compatibility modeling with attentive knowledge distillation. In: The 41st International ACM SIGIR Conference on Research & Development in Information Retrieval, pp. 5–14 (2018)

25. Song, X., Feng, F., Liu, J., Li, Z., Nie, L., Ma, J.: Neurostylist: neural compatibility modeling for clothing matching. In: Proceedings of the 25th ACM International Conference on Multimedia, pp. 753–761 (2017)

26. Song, X., Han, X., Li, Y., Chen, J., Xu, X.S., Nie, L.: GP-BPR: personalized compatibility modeling for clothing matching. In: Proceedings of the 27th ACM International Conference on Multimedia, pp. 320–328 (2019)

27. Sukhbaatar, S., Weston, J., Fergus, R., et al.: End-to-end memory networks. In: Advances in Neural Information Processing Systems, pp. 2440–2448 (2015)

28. Wang, Y.X., Zhang, Y.J.: Nonnegative matrix factorization: a comprehensive review. IEEE Trans. Knowl. Data Eng. **25**(6), 1336–1353 (2012)

29. Weston, J., Chopra, S., Bordes, A.: Memory networks. arXiv preprint arXiv:1410.3916 (2014)

30. Wibowo, A.T., Siddharthan, A., Lin, C., Masthoff, J.: Matrix factorization for package recommendations. In: Proceedings of the RecSys 2017 Workshop on Recommendation in Complex Scenarios (ComplexRec 2017). CEUR-WS (2017)

31. Yang, T., Chan, A.B.: Learning dynamic memory networks for object tracking. In: Proceedings of the European Conference on Computer Vision (ECCV), pp. 152–167 (2018)

32. Zheng, N., Song, X., Chen, Z., Hu, L., Cao, D., Nie, L.: Virtually trying on new clothing with arbitrary poses. In: Proceedings of the 27th ACM International Conference on Multimedia, pp. 266–274 (2019)

Developing a Smart PACS: CBIR System Using Deep Learning

Michela Gravina[1], Stefano Marrone[1(✉)], Gabriele Piantadosi[2],
Vincenzo Moscato[1], and Carlo Sansone[1]

[1] University of Naples Federico II, Via Claudio 21, 80125 Napoli, Italy
{michela.gravina,stefano.marrone,
vincenzo.moscato,carlo.sansone}@unina.it
[2] Altran Italia S.p.A., Via G. Porzio 4, 80143 Napoli, Italy
gabriele.piantadosi@altran.it

Abstract. With the growing number of digital medical imaging records, the need for an automatic procedure to retrieve only data of interest is of increasing importance. A Picture Archiving and Communication System (PACS) provides effective storage and retrieval based on TAGs but does not allow us for query by example. A possible solution is to use a Content-Based Image Retrieval (CBIR) system, namely a system able to retrieve images from a database based on the similarity to a given reference image. The features used to describe the images strongly affect both the performance and the applicability of CBIR to medical images, motivating for the finding of a suitable set of feature for realizing an effective CBIR based PACS. In recent years, Deep Learning (DL) approaches outperformed classical machine learning methods in many computer vision applications, thanks to their ability to learn compact hierarchical features of input data that well fit the specific task to solve. In this paper we introduce a simple yet effective modular architecture to implement a "Smart PACS", namely a PACS exploiting a deep-based CBIR compatible with the classical Hospital Information System (HIS) infrastructure. The feature extraction relies on Convolutional Neural Networks, a DL approach commonly applied in image processing, while the image indexing and look-up are based on Apache Solr. As application case-study, we analysed the need for a physician to obtain all the images of past studies having similar traits with the patient under analysis.

1 Introduction

Biomedical imaging is one of the fields that has contributed the most to new therapies and diagnosis in modern medicine. Despite its undoubted benefits, its spread and evolution challenged researchers to find proper ways for efficiently storing and effective retrieving of patients' images.

Over the years, the patients' records management moved from paper to electronics, thanks to the development of Hospital Information Systems (HIS),

© Springer Nature Switzerland AG 2021
A. Del Bimbo et al. (Eds.): ICPR 2020 Workshops, LNCS 12662, pp. 296–309, 2021.
https://doi.org/10.1007/978-3-030-68790-8_24

a set of tools intended to assist in the tracking of all the events, needs and background of patients [14]. A HIS usually works as the core system of a medical centre, orchestrating all the other software and services [1]. Among them, the Picture Archiving and Communication System (PACS) plays a fundamental role in managing the medical imaging files [3]. One of the limitations of PACSs is that data retrieval is based on the meta-data (e.g. tag, id, etc.) associated with the image. To be effective, this method needs the physician to spend time on the i) definition of a suitable and effective set of tag, and ii) on their correct assignment and updates during new images acquisitions. This procedure, besides being time-consuming and error-prone, has the main disadvantages of i) not allowing an immediate search based on a reference image (query by example) and ii) requires to re-process all the already stored images if a new tag, not taken into account in the past, needs to be added.

A possible solution for this is to use a Content-Based Image Retrieval (CBIR) system, namely a systems able to retrieve images from a database based on some visual characteristics of the images themselves [20]. This means that the retrieved images are not selected by means of tags or text attributes associated with them, but by their similarity to a given reference image. However, the features used to describe the images strongly affect both the performance of a CBIR and its applicability in a given context [12]. Thus, to find a suitable set of feature is of primary importance for realising an effective CBIR based PACS.

In recent years, Deep Learning (DL) approaches gained popularity in many computer vision applications, thanks to their ability to learn a compact hierarchical representation of the input data by exploring architectures composed of stacked layers performing multiple transformations. This characteristic allows deep neural networks to learn a complex function that directly maps input to output without the definition of hand-crafted features by domain experts, making DL approaches suitable for the development of a CBIR system. Among all, Convolutional Neural Networks (CNNs) are a class of neural network architectures mostly commonly applied in image processing. Convolutional layers contain a set of filters that automatically extract from the input image the best set of features for the specific task to solve, making CNNs especially explored in domains lacking well-defined sets of features, as the biomedical field.

In 2017 Ibanez et al. [10] proposed DL based method for pulmonary nodule computed tomography (CT) images retrieval. More in details, authors developed a CNN model trained to predict nodule malignancy level (from 1 to 5). After the training step, the features coming from the second fully connected layers (FC2) were extracted for each CT image. For comparison purposes, two popular hand-crafted features including Histogram of Oriented Gradients (HOG) and Haralick features from the Gray Level Co-occurrence Matrix were extracted for image retrieval. The authors showed that deep learning features outperforms the others in term of precision. In the same years, Qayyum et al. [17] trained a CNN for classifying medical images according to the body part or organ information. The trained model is used to extract the features from the last three fully connected layers. In the retrieval step a query image is classified by the

CNN and the predicted class, together with the features extracted from the network, are used to compute the similarity measure with the features database. In 2018 Pang et al. [16] exploited a Stacked Denoising Autoencoders (SDAE) pre-trained on Tiny images datasets [22] and a CNN (AlexNet [13]) pre-trained on ImageNet dataset [5], to have a representation of biomedical images. More in details, authors extracted features with SDAE encoder part and with AlexNet fully connected layer-6. They also introduce preference learning technology to learn a preference model for the query image, which can output the similarity ranking list of images from a biomedical image database. Recently, in [24] Yoshinobu et al. proposed a deep learning method for focal liver lesions retrieval exploiting Densely Connected Convolutional Neural Network (DenseNet [9]). The network was trained for classifying each focal liver lesion according to five classes (cysts, focal nodular hyperplasia, hepatocellular carcinoma, hemangioma, and metastasis). Then, the trained DenseNet without the last fully connected layer was used as a feature extractor for the CBIR system. In the retrieval step, the trained CNN is used to extract the feature for a new focal liver lesion image (query image) in order to compute the similarity measure with the annotated images in the database.

All the aforementioned works address medical CBIR only for a very specific task, without including the retrieval stage into an HIS.

This work aims in providing a simple yet effective application of DL approaches in the implementation of a "Smart PACS", namely a PACS exploiting a deep-based CBIR system. More in details, we propose a modular architecture compatible with the classical HIS infrastructures, with the aim of enhancing the functionality provided by the PACS with i) the ability to perform query by example for different situations, ii) doing it in a still very effective index-based manner and iii) allowing for future modifications needed to solve a new task or to improve the already implemented ones. We propose an architecture based on a CBIR system that can directly access all the data in the PACS, while the feature extraction relays on Deep Convolutional Neural Networks. We tested the implemented Smart PACS on different datasets, showing that it achieves good performance for different organs.

The rest of the paper is organized as follow: Sect. 2 describes the proposed architecture, motivating all the choices made; Sect. 3 shows some results considering a real use-case scenario; finally Sect. 4 draws some conclusions.

2   Developing a Smart PACS

Given a set of images (*database*), a CBIR system allows retrieving the subset of images "most similar" to a one used as *reference*. To do this, the described CBIR system operates in two stages (Fig. 1): in the first (off-line stage) it extracts a set of features for all the images in the database; in the second (on-line stage) it extracts the same set of feature for the reference image to compare it (in this new space) against all those in the database.

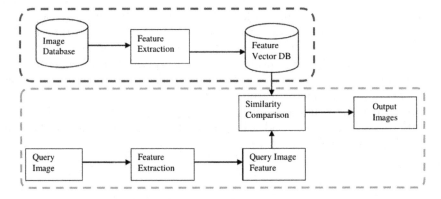

Fig. 1. Illustration of a CBIR system modules and their interconnection: in blue the off-line stage, in red the on-line stage. (Color figure online)

This two-stages approach is intended to make the images retrieval more efficient since the off-line stage (that is the most computational demanding one) needs to be done just once, while the on-line stage (faster, since operates on a single image) is performed for each query. To make the search more efficient, during the off-line stage the CBIR system creates an index of all the images contained in the database, making the comparison with the reference image (during the on-line stage) faster.

2.1 Proposed Architecture

All the components of a HIS, including PACS, usually communicate with each other by using standard local network infrastructure. Our idea is to design a system consisting of several modules connected to the HIS network and communicating with each other by means of HTTPS requests. As a consequence, the proposed architecture consists in four main modules (Fig. 2):

– **Orthanc** [11] is the PACS server in charge of memorising the biomedical imaging patients' records. It provides GET, PUT, POST and DELETE data operation over HTTP requests using a set of Restful API, also allowing to query patients' meta-data through a JSON-based call-response mechanism. It is worth to note that any PACS can be used as long as it provides the listed functionalities;
– **Coordinator** is the core of the architecture, in charge of managing data flowing and modules communications to provide the users with the desired set of images. During the off-line stage, the coordinator queries the PACS for all the images, preparing them for the feature extraction and indices memorisation. During the on-line stage, the coordinator manages the flow of information between the feature extractor and the search engine, finally gathering from the PACS all the images whose indices have been matched by the search engine;

- **Flask** [18] is a micro-framework for light-weight web-servers development, written in Python. Its duties consist of leveraging the images' features extraction exploiting Convolutional Neural Network, as explained in Sect. 2.2. The extracted features can be considered as a "fingerprint" of the image, and will be used to match similar images since images sharing similar characteristics are close in the considered features space;
- **Solr** [7] is an Apache open-source enterprise-search platform used, in this work, to memorise, index and search medical images represented in the form of their previously extracted features.

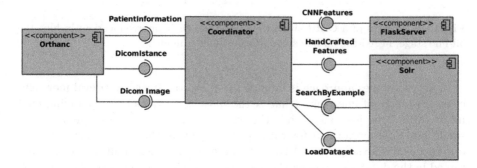

Fig. 2. Component diagram of the proposed architecture.

These components work together to provide the described functionalities, but the flow of information is different between the offline and the online stages (Fig. 3). During the offline stage: 1) the Coordinator queries all the images from the PACS; 2) Flask extracts, for each of them, the set of representing features; 3) for each image, the set of feature is memorised and uniquely linked to it in Solr, in order to be used as the key for future searches; 4) Solr indexes the set of all the images, to provide the Coordinator with an effective search engine. During the online stage: 1) the Coordinator queries Solr to search for images similar to the reference; 2) Solr invokes Flask to extract the features for the reference images, compare them with the memorised features list for all the known images and get back to the Coordinator with the list of the top K similar images IDs; 3) the Coordinator queries the PACS to retrieve the top K similar images on the basis of their id. The off-line procedure is repeated time in time (e.g. each day, during the night) in an incremental manner, to synchronise the changes happened in the PACS during its working routine (i.e. new images added). The on-line procedure is executed each time the user looks for a set of images similar to the one used as the reference.

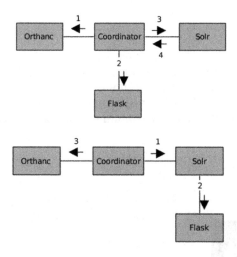

Fig. 3. Communication diagram of the proposed architecture: on the left the off-line stage, on the right the on-line stage.

2.2 Convolutional Neural Networks as Features Extractors

Convolutional Neural Networks (CNNs) are machine learning models borrowed from traditional Neural Networks (NNs). Such architectures share most of the features: they are both made of neurons, usually organized in stratified layers to create a feed-forward network in which the output of a layer is the input of the next layer. However, while traditional NNs operate on the features designed and extracted by a domain expert, CNNs use a hierarchy of convolution operations to autonomously design the features that better model the problem under analysis. One of the characteristics that make CNN appealing is the possibility of using a CNN trained on a given task as a feature extractor for a, possibly totally, different one. This allows, without no need for further training, to obtain a feature vector to be used as input of different models, such as traditional ML approaches, or as mapping function in a novel and bigger vector space. This approach is strongly promoted because the knowledge of the pre-trained network is able to efficiently map the instances for several tasks (also for very different domains). This approach helps in reducing the computational burden of training a network from scratch while performing a space transformation that may help to index a bunch of images efficiently and relying on visual peculiarity not easy to catch up with handcrafted features. Therefore, our idea is to use such feature vectors for hashing the images of the PACS in a CBIR point of view.

Finally, it is worth noting that a good features extractor, able to correctly map the instances/images in the space of the features solves most of the problem by highlighting the salient peculiarity of the instances/images. It follows that the choice of a distance metric with respect to the others do not lead to substantial benefits.

3   Results

To show the potentiality of the proposed architecture, as a case of study, we analysed the need for retrieving the images of all the patients showing similar characteristics to that of a reference input: the physician is analysing a given patient's image and, finding some traits reminding of something they already have seen, they query the system to obtain similar past studies images (Fig. 4).

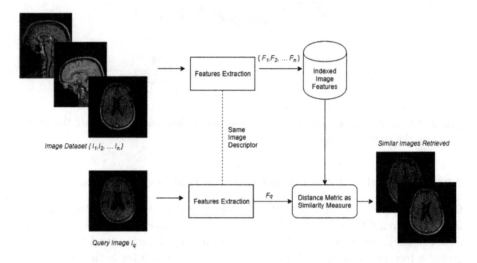

Fig. 4. Illustration of the use-case scenario considered in this work.

We want to evaluate the effectiveness of the proposed approach across different organs and under different feature extraction techniques. To this aim, the considered scenario is as follows:

- **Dataset**: We considered public MRI datasets [2] of four different organs (Brain, Breast, Liver and Prostate) collected over 55 patients (41 used for training and 14 for testing purposes), for a total of 8144 images.
- **Feature Extraction Approaches**: We explored the effectiveness of different CNNs pre-trained on the ImageNet dataset [5]: VGG16 [19], ResNet50 [8], InceptionV3 [21] and DenseNet [9]. Moreover, in order to analyse if using a CNN is really better then using handcrafted features, we also explored the use of Joint Composite Descriptor (JCD) [25], Gabor Filters [6], Pyramid of Histograms of Oriented Gradients (PHOG) [4] and Local Binary Pattern (LBP) [15]. All the features have been implemented in python 3.6, using Tensorflow 1.9 for the CNNs.

- **Distance Metric**: The query resolution algorithms that underlie Information Retrieval systems require a metric distance. Instances are mapped in a vector space using a suitable feature extractor (CNNs) or handcrafted features. Considering two images P and Q in the same vector space and represented by the following feature vectors $P = [p_1, p_2, ..., p_n]$ and $Q = [q_1, q_2, ..., q_n]$, we explored distance metrics defined in Table 1, where $X = [x_1, x_2, .., x_n]$ and $Y = [y_1, y_2, ..., y_n]$ are the two histograms respectively derived from the entries of the two feature vectors P and Q.

Table 1. Definition of the explored distance metrics. CP and Q are the vector space representation of the two images P and Q, while $X = [x_1, x_2, .., x_n]$ and $Y = [y_1, y_2, ..., y_n]$ are the two histograms respectively derived from the entries of the two feature vectors.

Distance metric	Definition
Euclidean	$d(P,Q) = \sqrt{\sum_{i=1}^{n}(p_i - q_i)^2}$
Manhattan	$d(P,Q) = \sum_{i=1}^{n}|p_i - q_i|$
Cosine	$d(P,Q) = \dfrac{\sum_{i=1}^{n} p_i q_i}{\sqrt{\sum_{i=1}^{n} p_i^2}\sqrt{\sum_{i=1}^{n} q_i^2}}$
Canberra	$d(P,Q) = \sum_{i=1}^{n} \dfrac{|p_i - q_i|}{|p_i| + |q_i|}$
ChiSquare	$d(P,Q) = d(X,Y) \sum_{i=1}^{k} \dfrac{(x_i - y_i)^2}{(x_i + y_i)^2}$

- **Performance Metric**: Several are the metrics that can be adopted to evaluate a content retrieval algorithm. In this paper, the effectiveness of the proposed approach has been evaluated in terms of Mean Average Precision (MAP) [23] since represents a good mean precision indicator, at different recall values, over the results obtained by executing all the different test queries.

Tables 2, 3, 4, 5 report the results of the implemented "Smart Pacs" on brain, breast, liver and prostate datasets respectively obtained by varying the distance metric and the feature extraction approach. Table 6 summarises the Organ Detection Rates obtained by selecting the best results from Tables 2, 3, 4, 5 for each feature extraction approach. Results show that using CNNs always lead to better results, with the only exception for the brain's images retrieval for which the LBP features perform better.

Table 2. Results of the implemented "Smart Pacs" on brain dataset obtained by varying the distance metric and the feature extraction approach.

Approaces	Distance metrics				
	Euclidean	Manhattan	Coseno	Canberra	ChiSquare
PHOG	76,1%	75,5%	**76,2%**	69,8%	73,8%
JDC	67,4%	**69,3%**	68,8%	63,9%	66,8%
LBP	83,0%	83,0%	**84,3%**	80,6%	82,7%
GABOR	29,3%	29,5%	26,8%	**33,9%**	30,2%
VGG16	65,1%	67,3%	**71,2%**	69,5%	70,0%
RESNET50	71,1%	72,5%	77,0%	**77,2%**	73,5%
INCEPTIONv3	55,5%	55,2%	**58,8%**	50,7%	54,1%
DENSENET	70,4%	74,3%	**76,8%**	75,4%	68,8%

Table 3. Results of the implemented "Smart Pacs" on breast dataset obtained by varying the distance metric and the feature extraction approach.

Approaces	Distance metrics				
	Euclidean	Manhattan	Coseno	Canberra	ChiSquare
PHOG	44,9%	**45,8%**	31,1%	39,6%	42,3%
JDC	**44,8%**	44,7%	41,2%	36,6%	40,5%
LBP	43,7%	43,7%	43,6%	**44,9%**	43,0%
GABOR	**38,4%**	**38,4%**	32,7%	35,0%	37,6%
VGG16	68,4%	**70,4%**	47,9%	52,5%	65,4%
RESNET50	65,6%	**70,6%**	53,7%	48,7%	63,1%
INCEPTIONv3	45,0%	44,3%	**47,4%**	45,7%	45,6%
DENSENET	44,2%	44,3%	**46,5%**	45,3%	36,3%

Table 4. Results of the implemented "Smart PACS" on the liver dataset obtained by varying the distance metric and the feature extraction approach.

Approaces	Distance metrics				
	Euclidean	Manhattan	Coseno	Canberra	ChiSquare
PHOG	**40,3%**	36,8%	35,8%	31,5%	39,2%
JDC	40,5%	37,8%	**42,1%**	37,9%	40,3%
LBP	35,8%	35,8%	**37,9%**	30,0%	36,5%
GABOR	30,0%	30,1%	28,1%	**32,8%**	30,8%
VGG16	61,7%	55,9%	**71,4%**	65,2%	66,0%
RESNET50	**81,6%**	79,1%	74,4%	66,3%	79,9%
INCEPTIONv3	**43,0%**	42,2%	37,5%	37,8%	41,9%
DENSENET	52,4%	53,6%	**53,7%**	54,3%	31,3%

Table 5. Results of the implemented "Smart PACS" on the prostate dataset obtained by varying the distance metric and the feature extraction approach.

Approaces	Detection rate (per organ)				
	Euclidean	Manhattan	Coseno	Canberra	ChiSquare
PHOG	35,7%	35,8%	36,4%	**36,9%**	36,5%
JDC	29,8%	29,7%	**30,6%**	30,1%	29,9%
LBP	44,1%	45,1%	**46,9%**	45,0%	46,2%
GABOR	23,1%	23,2%	**30,3%**	26,7%	23,8%
VGG16	48,7%	47,9%	**56,6%**	48,5%	50,8%
RESNET50	51,6%	51,7%	**59,2%**	56,5%	56,4%
INCEPTIONv3	31,3%	**31,8%**	28,1%	29,1%	31,0%
DENSENET	**40,3%**	40,2%	32,9%	35,8%	28,6%

Table 6. Comparison of the Organ Detection Rates of the proposed architecture, varying the feature extraction approach.

Approaces	Detection rate (per organ)			
	Brain	Breast	Liver	Prostate
PHOG	76.2%	45.8%	40.3%	36.9%
JCD	69.3%	44.8%	42.1%	30.6%
LBP	**84.3%**	44.9%	37.9%	46.9%
GABOR	33.9%	38.4%	32.8%	30.3%
VGG16	71.1%	70.4%	**71.4%**	56.6%
RESNET50	77.2%	**70.6%**	81.6%	**59.2%**
INCEPTIONv3	58.8%	47.4%	43.0%	31.8%
DENSENET	76.8%	46.5%	54.3%	40.3%

Figure 5 reports a query results for the brain by using LBP as the feature extraction approach. For each retrieved image, the "Smart PACS" execution time is reported. In order to have a qualitative evaluation, we asked 10 physicians to rate the images retrieved by the implemented system. More in details, we selected 10 query images and the physicians are prompted to indicate whether the first retrieved image is relevant. Figure 6 shows the number of relevant results according to each physician.

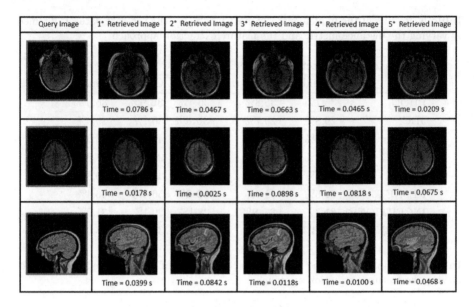

Fig. 5. Query results for the brain: from the left, the reference image (edged in red) and the top 5 retrieved images. For each retrieved image, the "Smart PACS" execution time is reported. (Color figure online)

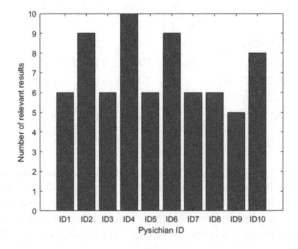

Fig. 6. Number of relevant results according to each physician

4   Conclusions

With the growing number of digital medical imaging records, the need for an automatic procedure to access only to data of interest is of increasing importance. The aim of this work was to propose a simple architecture for the development of a "Smart PACS", providing an effective application of DL approaches in the implementation of a CBIR system that retrieves medical images on the basis of their characteristics rather than on human-inserted tags. The result is a deep CBIR based PACS compatible with the current Hospital Informative Systems infrastructures.

The system, consisting of four main modules, allows to retrieve images similar to a reference one, without the need for the operator to describe the characteristics looked for. For achieving such a goal, the system relies on Flask to exploit Deep Convolutional Neural Network (CNN) for the feature extraction process. Finally, to make it compatible with the strict time requirements of a typical clinical scenario, the system exploits Apache Solr for a fast yet effective indexing and look-up strategy. This latter approach allows to deal with the huge number of learned features, with a reduced impact on the required computational power.

To show the potentiality of the proposed approach, as a case of study, we analysed the need for retrieving the images of all the patients showing similar characteristics to that of a reference input: the physician is analysing a given patient's image and, finding in it some traits that remind him of something already seen, they query the system to obtain similar past studies images. It is worth noticing that the physician is not required to describe the characteristics they are looking for since i) the system collect directly from the .dicom file the info related to the organ, to the imaging procedure and to the acquisition plane, ii) while exploiting the CNN feature space to match the similarity in the morphology, slice position, etc.

Results in Table 6 show the effectiveness of the implemented system when applied to datasets related to different organs. It is worth noticing that good results are obtained with a simple yet effective application of DL approaches, confirming their applicability in the implementation of a CBIR system for medical images. Moreover, using a CNN always leads to better results, with the only exception for the brain's images retrieval for which the LBP features perform better. The motivation behind this could be found in the regular structure of brain images (the presence of the cranium represent a strong anatomical key-point) that is particularly likely to be caught by the LBP descriptor.

In conclusion, the main strengths of the proposed architecture are i) its compatibility with standard HIS infrastructures, ii) its modularity and iii) its transparency to the final user.

Future works will be devoted to exploit more effective features for medical images (e.g. triplet-based), on the use of CNNs pre-trained on medical imaging tasks, on the system application to a multi-centre scenario with several structures dislocated in different locations and on the design of an effective indexing and access mechanism suited for the distributed environment.

Acknowledgments. This work is part of the "Synergy-net: Research and Digital Solutions against Cancer" project (funded in the framework of the POR Campania FESR 2014-2020 - CUP B61C17000090007).

References

1. Boochever, S.S.: HIS/RIS/PACS integration: getting to the gold standard. Radiol. Manag. **26**(3), 16–24 (2004)
2. Clark, K., et al.: The cancer imaging archive (TCIA): maintaining and operating a public information repository. J. Digit. Imaging **26**(6), 1045–1057 (2013)
3. Cooke Jr, R.E., Gaeta, M.G., Kaufman, D.M., Henrici, J.G.: Picture archiving and communication system, US Patent 6,574,629, 3 June 2003
4. Dalal, N., Triggs, B.: Histograms of oriented gradients for human detection. In: International Conference on Computer Vision & Pattern Recognition (CVPR 2005), vol. 1, pp. 886–893. IEEE Computer Society (2005)
5. Deng, J., Dong, W., Socher, R., Li, L., Li, K., Fei-Fei, L.: Imagenet: a large-scale hierarchical image database. In: 2009 IEEE Conference on Computer Vision and Pattern Recognition, pp. 248–255 (2009)
6. Fogel, I., Sagi, D.: Gabor filters as texture discriminator. Biol. Cybern. **61**(2), 103–113 (1989)
7. Grainger, T., Potter, T.: Solr in action. Manning Publications Co. (2014)
8. He, K., Zhang, X., Ren, S., Sun, J.: Deep residual learning for image recognition. In: Proceedings of the IEEE Conference on Computer Vision and Pattern Recognition, pp. 770–778 (2016)
9. Huang, G., Liu, Z., Van Der Maaten, L., Weinberger, K.Q.: Densely connected convolutional networks. In: Proceedings of the IEEE Conference on Computer Vision and Pattern Recognition, pp. 4700–4708 (2017)
10. Ibanez, D.P., Shen, Y., Dayanghirang, J., Li, J., Wang, S., Zheng, Z.: Deep learning for pulmonary nodule CT image retrieval–an online assistance system for novice radiologists. In: 2017 IEEE International Conference on Data Mining Workshops (ICDMW), pp. 1112–1121. IEEE (2017)
11. Jodogne, S., Bernard, C., Devillers, M., Lenaerts, E., Coucke, P.: Orthanc-a lightweight, RESTful DICOM server for healthcare and medical research. In: 2013 IEEE 10th International Symposium on Biomedical Imaging, pp. 190–193. IEEE (2013)
12. Juneja, K., Verma, A., Goel, S., Goel, S.: A survey on recent image indexing and retrieval techniques for low-level feature extraction in CBIR systems. In: 2015 IEEE International Conference on Computational Intelligence & Communication Technology, pp. 67–72. IEEE (2015)
13. Krizhevsky, A., Sutskever, I., Hinton, G.E.: Imagenet classification with deep convolutional neural networks. In: Advances in Neural Information Processing Systems, pp. 1097–1105 (2012)
14. McCullough, J.S.: The adoption of hospital information systems. Health Econ. **17**(5), 649–664 (2008)
15. Ojala, T., Pietikäinen, M., Mäenpää, T.: Multiresolution gray-scale and rotation invariant texture classification with local binary patterns. IEEE Trans. Pattern Anal. Mach. Intell. **7**, 971–987 (2002)
16. Pang, S., Orgun, M.A., Yu, Z.: A novel biomedical image indexing and retrieval system via deep preference learning. Comput. Methods Programs Biomed. **158**, 53–69 (2018)
17. Qayyum, A., Anwar, S.M., Awais, M., Majid, M.: Medical image retrieval using deep convolutional neural network. Neurocomputing **266**, 8–20 (2017)
18. Ronacher, A.: Flask (a python microframework) (2018). http://flask.pocoo.org. Accessed 20 Aug 2018

19. Simonyan, K., Zisserman, A.: Very deep convolutional networks for large-scale image recognition. arXiv preprint arXiv:1409.1556 (2014)
20. Singhai, N., Shandilya, S.K.: A survey on: content based image retrieval systems. Int. J. Comput. Appl. **4**(2), 22–26 (2010)
21. Szegedy, C., Vanhoucke, V., Ioffe, S., Shlens, J., Wojna, Z.: Rethinking the inception architecture for computer vision. In: Proceedings of the IEEE Conference on Computer Vision and Pattern Recognition, pp. 2818–2826 (2016)
22. Torralba, A., Fergus, R., Freeman, W.T.: 80 million tiny images: a large data set for nonparametric object and scene recognition. IEEE Trans. Pattern Anal. Mach. Intell. **30**(11), 1958–1970 (2008)
23. Voorhees, E.M.: Variations in relevance judgments and the measurement of retrieval effectiveness. Inf. Process. Manag. **36**(5), 697–716 (2000)
24. Yoshinobu, Y., et al.: Deep learning method for content-based retrieval of focal liver lesions using multiphase contrast-enhanced computer tomography images. In: 2020 IEEE International Conference on Consumer Electronics (ICCE), pp. 1–4. IEEE (2020)
25. Zagoris, K., Chatzichristofis, S.A., Papamarkos, N., Boutalis, Y.S.: Automatic image annotation and retrieval using the joint composite descriptor. In: 2010 14th Panhellenic Conference on Informatics, pp. 143–147. IEEE (2010)

Multi Color Channel vs. Multi Spectral Band Representations for Texture Classification

Nicolas Vandenbroucke$^{(\boxtimes)}$ ⓘ and Alice Porebski ⓘ

Univ. Littoral Côte d'Opale, UR 4491, LISIC,
Laboratoire d'Informatique Signal et Image de la Côte d'Opale,
62100 Calais, France
{nicolas.vandenbroucke,alice.porebski}@univ-littoral.fr

Abstract. Texture and color are salient visual cues of human perception and are widely used in many image analysis applications. Multi color spaces (MCS) approaches enrich the color texture representation and improve the performances of color texture classification applications. In these approaches, textures are represented by color texture features computed from descriptors that are extracted from images coded in several color spaces. When spectral characteristics of the texture of materials need to be analyzed, hyperspectral imaging (HSI) devices are chosen to address industrial applications that conventional color imaging is unable to solve. This paper aims to evaluate the contribution of HSI in the performances of texture classification methods compared to color imaging. For this purpose, we propose to extend the MCS approach to HSI in order to extract relevant spectral texture features computed from images of different spectral bands. Since these approaches both require to process high-dimensional data, they need to reduce the dimensionality of the feature space by selecting the most discriminating features, leading to a multi color channel (MCC) representation and a multi spectral band (MSB) representation respectively. This paper presents a unified representation of textures contained in color or hyperspectral images and compares the MCC and MSB representations for classification issues. Experimental results carried out on two hyperspectral texture databases show that the MCC representation is able to outperform the MSB ones.

Keywords: Texture · Color imaging · Hyperspectral imaging · Feature selection

1 Introduction

Texture classification is typically categorized into two sub-problems of representation and decision [14]. Texture representation being a fundamental step of texture analysis, this paper deals with this first sub-problem. The analysis of texture images requires to evaluate the connectedness relationships between neighboring pixels by features that take into account the spatial distribution of colors in the image. Texture analysis methods therefore combine spatial information that defines how pixels are spatially organized in the image plane and color

© Springer Nature Switzerland AG 2021
A. Del Bimbo et al. (Eds.): ICPR 2020 Workshops, LNCS 12662, pp. 310–324, 2021.
https://doi.org/10.1007/978-3-030-68790-8_25

information that indicates how their colors are distributed in the representation space. When different color spaces are used to represent textures, multiple color channels are then considered to compute *color texture features* and the size of the representation space is increased. Reducing the size of this space is then a crucial step in the analysis of the images. Based on feature selection methods, multi color space approaches that exploit the properties of several spaces simultaneously were developed in order to address the texture representation and classification problems [20,21]. These approaches aim to analyze textures in a reduced dimensionality color texture feature space following a *multi color channel* (MCC) representation. For this purpose, they select the most discriminating color texture features among all the available features extracted from descriptors that are computed from images coded in different color spaces.

On the other hand, hyperspectral imagery provides a greater amount of information compared to conventional color imaging or multispectral imaging. The exploitation of these hyperspectral data should allow a more complete representation of the texture to tackle challenging industrial applications such as automatic inspection of material surfaces for quality control and sorting issues [7]. Although hyperspectral imagery nowadays provides images with high spatial and spectral resolutions, the analysis of a sequence of hyperspectral images requires to process a large and sometimes redundant mass of multidimensional data in reasonable times. The exploitation of these high-dimensional images is an open problem for approaching industrial applications with demanding performance constraints in terms of quality of results and computation times. To address this problem, representative band selection schemes have been proposed in order to reduce the dimensionality of these hyperspectral data [12,25]. Instead of selecting spectral bands, we propose to select relevant texture features generated from hyperspectral cubes that are acquired by HSI devices.

Under the assumption that a hyperspectral image can be viewed as a multi-component image, as well as a color image represented in multiple color spaces, the multi-color space approaches can be applied to hyperspectral images by extended color texture feature selection methods to these images. For this purpose, descriptors which take into account the spatial and spectral properties of the texture can be extracted from hyperspectral images in order to select representative and discriminating *spectral texture features*. The textures are thus analyzed in a space with a reduced dimensionality following a *multi spectral band* (MSB) representation.

This paper presents a unified representation of textures contained in color or hyperspectral images and proposes to compare a multi color channel representation with a multi spectral band representation for texture classification issues. These representations are based on an original hybrid feature selection model which determines the reduced dimensionality of a relevant texture feature space.

Experiments are carried out with two hand-crafted descriptors known to be efficient in color texture classification problems: chromatic cooccurrence matrices [21] and color local binary pattern [20]. In addition, two pretrained convolutional neural networks (CNN) are used to compare our approach with the latest popular methods: AlexNet and GoogleNet networks [11,22]. The different representations are applied on two hyperspectral texture image benchmark databases: HyTexiLa and SpecTex [10,17].

The second section of this paper presents how color and spectral texture features can be easily computed from color and hyperspectral images respectively. The third section details how the most discriminating features are selected in order to represent textures in a reduced dimensionality feature space. In the fourth section, experimental results obtained with the multi color channel representation and the multi spectral band representation on two different texture image databases are compared. The last section gives some perspectives.

2   Color and Spectral Texture Features

Colors of pixels are usually defined by three components in a given color space and an image is thus commonly represented by three *color channel images*. Since the choice of a color space directly impacts the classification results, many authors tried to compare results obtained by using different color spaces in order to find the most suited one for a given application [1,3,15]. The synthesis of these works shows that there is no color space which is well suited to represent all types of textures. To solve this problem, few studies propose multi color space approaches [20]. They exploit the properties of multiple color spaces simultaneously by combining them, and overcome the difficulty of choosing a single relevant color space. In these approaches, a color image is coded by N_B color channels and viewed as a multi-component image in which the texture description requires the analysis of a neighborhood $\mathcal{N}$. Texture descriptors $\mathbf{D}$ are thus extracted from images coded in the different considered color spaces. These "hand-crafted" descriptors combine spatial and color information following two main strategies depending on whether they are jointly or independently considered [2,15]. When color and spatial information are jointly considered, color texture features can be extracted within each color channel of a given color space independently (within-component spatial relationship) with (or without) color texture features extracted between pairs of color channels (between-component spatial relationship) as illustrated in Fig. 1. In this paper, we propose to only take the within-component interactions into account to define the multi color channel (MCC) representation.

In a hyperspectral image, each pixel is characterized by a reflectance spectrum. For Deborah et al., this set of spectra can be considered as vectors in an euclidean space, as N-dimensional data in manifold, as distributions or as sequences [6]. When the spectrum is viewed as a multidimensional vector, the hyperspectral image is considered as a set of N_B *spectral band images* identified by their wavelengths λ. Hand-crafted descriptors can be easily extracted from this multi-component image so that spectral texture features are then computed from these descriptors to define a multi spectral band (MSB) representation as shown in Fig. 2. In this representation, spectral texture features are generated from descriptors that can take into account the spatial relationships within and between components. In this paper, the MSB representation is defined by only considering the within-component interactions.

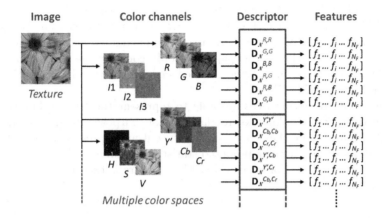

Fig. 1. Multi color channel representation.

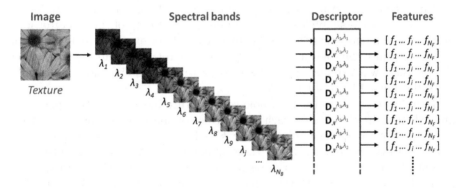

Fig. 2. Multi spectral band representation.

In this paper, we propose to compare MCC and MSB representations with two popular and efficient texture descriptors: the cooccurrence matrix and the local binary pattern operator known for their computational simplicity.

2.1   Cooccurrence Matrices

Introduced by Haralick, cooccurrence matrices are considered as two-dimensional histograms of pairs of neighbor pixels [8]. An important property of this operator is its invariance to orientation changes. Palm extended this descriptor to color and introduced the Chromatic Cooccurrence Matrix (CCM) which considers both the spatial interactions within and between the color components of neighbor pixels in the image plane and the color distribution in a color space [18].

Let Q, be the number of levels used to quantify a color or a spectral component. A Reduced Size Chromatic Cooccurrence Matrix (RSCCM) is a $Q \times Q$ CCM, where the parameter Q is reduced in order to decrease the memory storage cost and so, the time required to extract texture features from these matrices [21]. The normalized RSCCM $m_{\mathcal{N}}^{C_k,C_{k'}}[I]$ measures the spatial interactions in the neighborhood $\mathcal{N}$ between the two components C_k and $C_{k'}$ of an image I ($k, k' \in \{1, \cdots, N_B\}$). In this paper, we only consider the within-component RSCCM where $k = k'$ and define the Reduced-size Marginal Chromatic Cooccurrence Matrix (RMCCM).

The cooccurrence matrices are able to represent the texture but they are not directly used for color texture classification purposes because of the large amount of information they contain. To reduce it while preserving the relevance of these descriptors, Haralick proposed statistical features that can be extracted from each matrix [18]. For each RMCCM, a texture is then represented by N_F Haralick features f_i.

2.2   Local Binary Pattern

Local Binary Pattern (LBP) was considered as a relevant texture descriptor that captures the local texture properties of an image with high discriminative power and computational simplicity [19]. An important property of this operator is its invariance to monotonic gray-scale changes caused, for example, by illumination variations. Since the color information plays a key role in texture representation, this operator was extended to deal with color texture classification problems by using the Color LBP (CLBP) [15].

In order to characterize the whole color texture image, the LBP operator is applied on each pixel and for each pair of color or spectral components. Considering a pair of component $(C_k, C_{k'})$, ($k, k' \in \{1, \cdots, N_B\}$), this operator labels a pixel with the component C_k by thresholding its neighborhood $\mathcal{N}$ in the component $C_{k'}$ and by encoding the result as a binary number. In this paper, we apply this operator only to the three within-component LBP images ($k = k'$) with a circular neighborhood defined by P, the number of neighbor pixels and R, the distance between each pixel and its neighbors. This Marginal Color LBP operator is named the MCLBP descriptor.

The LBP images are usually not exploited directly and most of authors prefer to use LBP histograms where histogram bins f_i can be considered as texture features [19]. For each MCLBP image, a texture is thus described by a N_F-dimensional histogram where $N_F = 2^P$.

3   Dimensionality Reduction

Although MCC and MSB representations are able to take into account numerous properties of color textures to be analyzed, they tend to produce a large and sometimes redundant amount of features, especially when the number of color channels or spectral bands is high. It is well-known that the high dimensional data classification performances are impacted by the dimension of the feature space due to the curse of dimensionality. Dimensionality reduction methods are thus needed to reach satisfying classification accuracies while decreasing the memory storage and the computation time [4,5,13,23]. To reach this goal, the multi color space approaches follow three main strategies: the color space fusion merges the results from several classifiers operating in different color spaces [16], the color space selection selects the most suitable color spaces [24], and the color texture feature selection which is here considered in a supervised texture classification context [20,21].

Supervised texture classification aims to assign a given texture to one of a set of known texture categories for which training samples are available. This process is divided into two successive stages: a learning stage in which a classifier is trained with labeled training samples and a decision stage in which this classifier is evaluated with a testing set in order to measure its ability to predict the class labels of new samples. During the learning stage, the selection of discriminant texture features among a set of candidate ones is essential for a texture representation in a reduced dimensionality feature space which improves the classification performance.

The goal of a feature selection scheme is so to build a relevant feature subspace from an original feature space. It is generally defined by: a search strategy that indicates how candidate feature subspaces are generated, an evaluation function that measures the relevance of a candidate feature subspace, and a stopping criterion that stops the search procedure and defines the dimension of the selected subspace. A validation procedure can complete this scheme [5].

Three models are usually considered to achieve this feature selection scheme depending on the chosen evaluation function: wrapper, filter and embedded (or hybrid) models [4]. Wrapper models use the classification algorithm during the learning stage to perform the selection of the features. This model provides good results and easily determines the dimension of the feature subspace by searching the highest classification rate but involves an important learning time and classifier-dependent results. Filter models use feature selection procedures to evaluate the discrimination power of different candidate feature subspaces without classifying the images. They are less time consuming but suffer to the difficulty to determine the dimension of the feature subspace to be selected. To obtain a good compromise between dimension selection, computation time and classification result, embedded models are preferred [13]. This model combines a filter model to determine the most discriminating feature subspaces at different dimensions and a wrapper model to determine the dimension of the selected subspace [23].

In this paper, we propose to consider an embedded feature selection scheme associated with different search procedures and evaluation criteria depending on the number and the nature of the texture features:

- RMCCM providing a set of Haralick features, a Sequential Forward Selection (SFS) procedure is applied with the Wilks's criterion as proposed in [21].

– MCLBP providing a set of histograms, a histogram ranking is performed with the Intra-Class Separability (ICS) measure as proposed in [20].

3.1 Feature Selection

During the learning stage, $N_B \times N_F$ candidate texture features are generated from training images where N_F is the number of Haralick features computed from each RMCCM and N_B, the number of color channels of the multiple color spaces for a color image or the number of spectral bands of the spectrum for a hyperspectral image. Then, the feature selection procedure uses a subset search algorithm. The number $N_B \times N_F$ being high, we choose the sequential forward selection (SFS) scheme. From an empty set, uncorrelated candidate features are iteratively added to build the candidate feature subspaces. For each dimension, the discriminating power of each candidate feature subspace is evaluated by using a specific criterion. In this paper, we choose the Wilks's criterion which measures the between-class and the within-class dispersions associated with the Pearson's correlation coefficient. The more the clusters associated with the different texture classes are well-separated and compact in the feature subspace, the higher the discriminating power of this feature subspace is and the more close to 0 the Wilks's criterion is. The candidate subspace with the lower value of this criterion is so selected at each dimension d of the SFS scheme.

The next step of the proposed embedded feature selection scheme consists of determining the dimension of the relevant feature subspace. For this purpose, the training set is divided in order to follow a k-fold evaluation where a classifier operates. One fold is used to constitute a training image subset and the $(k-1)$ remaining folds are assigned to a validation image subset from which the classification accuracy is measured. The evaluation function at different dimensions d and so, for each d-dimensional subspace, is the mean classification accuracy over the k folds. The selected feature subspace, whose dimension is D, is the one which maximizes the mean rate R^d_{mean} of well-classified validation images:

$$D = \operatorname*{argmax}_{1 \leq d \leq (N_B \times N_F)} R^d_{mean} \qquad (1)$$

In this work, accuracy is measured using the nearest neighbor classifier associated with the L1 distance because no parameters need to be adjusted.

3.2 Histogram Selection

During the learning stage, candidate histograms are generated from training images. These prototype histograms are computed from N_B MCLBP images where N_B represents the number of color channels of the multiple color spaces (color image) or the number of spectral bands of the spectrum (hyperspectral image). Then, the proposed histogram selection procedure uses a ranking algorithm. The relevance of each of the N_B available histograms is evaluated by a histogram score [9]. In this paper, we consider the Intra-Class Similarity score

(ICS) [20]. This score, which is calculated for each candidate histogram of the training images, measures the ability of a histogram to characterize the similarity of the textures within the different classes.

The next step consists of determining the dimension of the relevant histogram subspace. For this purpose, the candidate subspaces – made up, at the first step of the procedure, of the histogram with the best score, at the second step, of the two first ranked histograms which are concatenated and so on – are evaluated. For this purpose, the same k-fold evaluation than the previous subsection is used to predict the classification accuracy associated with each concatenated histogram subspace. The evaluation function at different dimensions d and so, for each $(d \times N_F)$-dimensional histogram subspace is the mean classification accuracy over the k folds with $N_F = 2^P$. The selected histogram subspace, whose dimension is $D = \hat{d} \times N_F$, is the one which maximizes the mean rate R^d_{mean} of well-classified validation images:

$$\hat{d} = \underset{1 \leq d \leq N_B}{\mathrm{argmax}} \quad R^d_{mean} \tag{2}$$

As explained in the previous subsection, accuracy is measured by using the nearest neighbor classifier associated with the L1 distance as a similarity measure.

4   Experiments

4.1   Image Databases

In this paper, two hyperspectral texture databases are experimented: HyTexiLa and SpecTex [10,17]. Experiments are carried out following the settings proposed by the authors (number of spectral bands, corresponding color images, size of images, training and testing subsets, ...). A hyperspectral image can easily be transform to a color image by selecting 3 spectral band images with wavelength centered on the red, green and blue component.

HyTexiLa: HyTexiLa is a dataset of hyperspectral reflectance images that span visible (VIS) and near infrared (NIR) parts of the electromagnetic spectrum of 112 textured materials. These materials fall into five different categories: textile (65 samples), wood (18 samples), vegetation (15 samples), food (10 samples) and stone (4 samples). Images have been acquired using the HySpex VNIR-1800 hyperspectral camera. The size of each reflectance image is $1024 \times 1024 \times 186$, where 1024×1024 is the spatial resolution, and 186 is the number of spectral bands centered at wavelengths which range from 405.37 nm to 995.83 nm at 3.192 nm intervals. These 112 texture images represent 112 different classes. Each hyperspectral texture image is split into 25 subimages of size 204×204 (without overlap), among which 12 were randomly considered to be training images and the 13 others as testing images.

Each one of the 112 reflectance images is transformed to color images by means of a conversion to the sRGB color space by using the spectral band numbered 65 ($\approx$612 nm), 45 ($\approx$547 nm), and 18 ($\approx$464.5 nm). Figure 3 shows the 112 textures represented by their color images.

Fig. 3. Textures of the HyTexiLa database. Each color image represents a sample of one texture class [http://color.univ-lille.fr/datasets/hytexila].

Fig. 4. SpecTex textured textile samples. From top-right T01 to bottom-left T60 [https://www3.uef.fi/fi/web/spectral/spectex].

SpecTex: SpecTex is a spectral image database including 60 textile samples with different texture patterns. The images are captured with a line-scanning spectral camera ImSpector V8 in the visible wavelength range from 400 nm to 780 nm with 5 nm intervals, which gives 77 spectral bands per pixel. Each pixel in the image of size 640×640 represents a reflectance spectrum. Each of the 60 original images is split up into 256 non-overlapping 40×40 subimages so that a quarter of these images are used as training images and the remaining are considered as testing images.

Color images coded in the RGB color space are simulated from the hyperspectral images with CIE-1964 color matching functions and the D65 standard illumination. Figure 4 shows the 60 textures T01 to T60 represented by their simulated RGB images.

4.2   Texture Classification

The images of the two hyperspectral texture benchmark databases are assigned either to a training subset or to a testing subset. In order to compare the MCC and MSB representations for supervised texture classification issues, the classification performance is assessed by following this holdout evaluation scheme. For this purpose, classification results obtained with the proposed representations are evaluated by measuring the accuracy as the rate of well-classified testing images during the decision stage. During the learning stage, the proposed embedded model use a k-fold evaluation method with $k = 5$ to determine the dimensionality of the selected feature subspace. The purpose of this paper being to show the contribution of different texture representations independently of the considered classifier and its parameters, the nearest neighbor classifier associated with the L1 distance as a similarity measure is here considered. This classifier labels each testing image with the closest match in the training set according to the similarity measure. Obviously, the classification results are expected to be improved by using more elaborated classifiers.

In addition, the MCC and MSB representations are compared with a representation based on convolutional neural networks (CNN). For this purpose, the RGB color images simulated from the hyperspectral texture databases are used as inputs of two pretrained CNN: AlexNet and GoogleNet [11,22].

Spectral Bands: Although HyTexiLa is proposed with a $1024 \times 1024 \times 186$ image size, a normalized covariance matrix analysis between each couple of wavelength shows that close spectral bands are highly correlated [10]. So, the authors propose to consider a number of bands less than or equal to 27 with a 22.34 nm intervals (one band every seven bands) when $N_B = 27$. The HyTexiLa dataset is thus constituted by subimages of size 204×204 with 27 preselected spectral bands uniformly distributed.

A downsampled version of SpecTex image of size $640 \times 640 \times 77$ is also proposed so that the spatial sampling gives rise to images of size 160×160 and the spectral sampling provides 39 spectral bands. The subimage size of the SpecTex dataset is then equal to 10×10. Since the texture descriptors require significant numbers of pixels to be calculated, the spatial downsampling is not applied here.

Table 1 gives a summary of the datasets used for experiments.

Color Channels: As previously mentioned, the hyperspectral images of the two databases can be easily converted to simulated color images coded in the RGB color space. There exist numerous color spaces that take into account different physical, physiologic and psycho-visual properties. In this paper, nine color spaces representative of the four color space families are considered for experiments:

Table 1. Summary of image datasets used in experiments.

Dataset	Image size	# classes	# training img./class	# testing img./class	# spectral bands N_B
HyTexiLa-27	204 × 204	112	12 (1344)	13 (1456)	27
HyTexiLa-186	204 × 204	112	12 (1344)	13 (1456)	186
SpecTex-39	40 × 40	60	64 (3840)	192 (11520)	39
SpecTex-77	40 × 40	60	64 (3840)	192 (11520)	77

– RGB and rgb, which belong to the primary space family,
– YCbCr and wb-rg-by, which are luminance-chrominance spaces,
– I1I2I3, which is an independent color component space,
– HSV, HSI, HLS and I-HLS, which belong to the perceptual space family.

A majority of different perceptual spaces were chosen because these spaces are known to reach good classification accuracies [21]. With this nine color spaces, $N_B = 27$ color channels are available for the MCC representation of textures.

Texture Features: For the RMCCM descriptor, previous studies show that satisfying classification results with a significant reduction of the processing time are reached for $Q = 16$ [21]. Because no privileged direction are known in the examined textures, an isotropic eight-neighborhood, denoted $\mathcal{N} = 3 \times 3$, with a spatial distance equal to 1 between pairs of pixels is used to takes into account all the possible directions. We propose to use all the $N_F = 14$ texture features proposed by Haralick [8]. $N_B \times N_F$ color or spectral textures features are initially available to describe the textures.

For the MCLBP descriptor, the neighborhood $\mathcal{N}$ parameters are $P = 8$ pixels with a spatial distance equal to 1 so that neighborhood $\mathcal{N}$ is the 3×3 neighborhood. As $N_F = 2^P$, a 256-dimensional histogram is computed from each MCLBP image. Textures are initially represented by N_B 256-dimensional color or spectral MCLBP histograms.

4.3   Results and Discussions

Table 2 presents the classification results reached by the two proposed representations MSB and MCC associated with the two texture descriptors RMCCM and MCLBP. The MSB representation is evaluated with two versions of the HyTexiLa database named MSB-27 (HyTexiLa-27 dataset) and MSB-186 (HyTexiLa-186 dataset). The MCC representation, named MCC-3×9 with 9 color spaces, gives rise to 3×9 color channels. For each classification rate, the dimension D of the texture feature space, in which testing images are classified, is given as well as the processing times. The time required to generate the texture features from the training set and the testing set is not take into account in this table because it is the same with or without feature selection in the evaluation framework. The time required by the classification of all the testing images t_c and the additional time, t_s required by the learning stage to reduce the dimensionality of the features space are indicated in this table. These times are measured in seconds by using the MATLAB(R) software running on a 2.11 GHz PC with 16 GB RAM.

Table 2. Processing times and rate of well-classified testing image R^D of the HyTexiLa datasets obtained with the MSB and MCC representations with and without feature selection. For each descriptor, the higher classification rate is indicated in bold.

Descriptor	Method	With selection				Without selection		
		t_s (s)	D	R^D (%)	t_c (s)	D	R^D (%)	t_c (s)
RMCCM	MSB-27	328.8	14	93.89	0.0217	378	93.68	0.336
	MSB-186	2049.0	14	93.13	0.0242	2604	93.75	1.848
	MCC-3 × 9	1021.4	48	**97.32**	0.0418	378	**97.46**	0.326
MCLBP	MSB-27	81.8	15 × 256	89.35	2.776	27 × 256	89.63	7.225
	MSB-186	11721.6	148 × 256	90.25	74.352	186 × 256	89.63	127.301
	MCC-3 × 9	83.9	18 × 256	**98.49**	3.451	27 × 256	**97.94**	7.996

This table shows that MCC representations, with or without selection, outperform the MSB representations whatever the initial number of spectral bands. Results obtained with the MCC representation are also higher than those reached by pretrained CNN applied on simulated RGB color images since $R^D = 94.51\%$ with AlexNet and $R^D = 91.96\%$ with GoogleNet. However, the performance of these networks with RGB images are better than those reached by the MSB representation.

Assuming that the neighboring bands are highly correlated, Khan et al. propose to classify the hyperspectral images with a reduced number of spectral bands (HyTexiLa-27 dataset) [10]. They use five different LBP descriptors in order to represent the textures by LBP histograms with a neighborhood $\mathcal{N} = 3 \times 3$. The texture classification is performed by the nearest neighbor classifier with the similarity measure based on intersection between histograms. With these descriptors, they also achieve a classification of textures coded in a simulated RGB color space to show the contribution of HSI. Among the five compared LBP descriptors, MCLBP histograms are used as spectral texture features. This features takes only spatial correlation into account. The classification rate obtained with this descriptor reaches about 92% accuracy with 10 bands. The results we have obtained in our experiments for the MSB representation are consistent with this result, especially with the MCLBP descriptor where $R^D \approx 90\%$ for $D = 15$ spectral bands. On the other hand, the MCC representation applied to the MCLBP descriptor outperforms all the other representations with such a descriptor type.

Table 2 shows that, with marginal texture descriptors, the contribution of the feature selection on classification accuracy is not significant, whatever the initial number of bands. Khan et al. also showed that classification rates were stabilized with 10 bands, whatever the considered LBP descriptor. However, feature selection significantly reduces the classification computation times during the decision stage and becomes essential when the number of features dramatically increases. This is particularly true for descriptors that consider pairs of components to take into account color channel or spectral band correlations.

As the same way, results obtained for the SpecTex database are given in Table 3 where MSB-39 corresponds to a MSB representation from 39 preselected bands and MSB-77 uses all the 77 bands of the hyperspectral images.

Table 3. Processing times and rate of well-classified testing image R^D of the SpecTex datasets obtained with the MSB and MCC representations with and without feature selection. For each descriptor, the higher classification rate is indicated in bold.

Descriptor	Method	With selection				Without selection		
		t_s (s)	D	R^D (%)	t_c (s)	D	R^D (%)	t_c (s)
RMCCM	MSB-39	1357.0	12	**96.42**	0.0207	546	96.18	8.513
	MSB-77	7575.8	19	95.87	0.0239	1078	**96.29**	14.180
	MCC-3 × 9	2481.6	62	94.65	0.979	378	94.97	5.961
MCLBP	MSB-39	1245.3	30 × 256	97.50	148.367	39 × 256	97.65	193.407
	MSB-77	6405.3	35 × 256	97.03	181.557	77 × 256	97.14	405.338
	MCC-3 × 9	531.0	25 × 256	**97.73**	131.603	27 × 256	**97.71**	122.272

With the SpecTex database, the two proposed representations give rise to equivalent results in terms of classification rate and show that a MSB representation does not actually improve the classification performance compared to a MCC representation obtained by using a conventional color imaging device. The best result is reached by the MCC representation with the MCLBP descriptor and a selection scheme. The feature selection scheme significantly reduces the dimensionality of the texture feature spaces and so, the time required to classify the testing images.

Mirhashemi proposes the spectral moment features as texture features to classify the images of the SpecTex hyperspectral texture database [17]. He shows results with different combinations of these features on each class individually. He studies the impact of each spectral moment features independently as well as their combinations but does not give an overall classification accuracy obtained on all the textures with the most discriminating combination. Moreover, the small image sizes of the datasets proposed by the author give rise to uninterpretable results with the pretrained CNN AlexNet and GoogleNet.

5   Conclusion

This paper presents a unified representation of textures contained in color or hyperspectral images and shows that the color imaging is able to outperform the hypespectral imaging in texture classification applications when a multi color spaces approach is applied. The experiments carried out with two texture descriptors on two hyperspectral databases show that a multi color channel representation associated with marginal descriptors that only take into account within-component interactions gives higher classification accuracies than those obtained with a multi spectral band representation of textures. In this context, the experiments reveal that the contribution of features selection scheme is low in terms of classification rates but remains useful in terms of processing times. One possible development of this work could be to investigate what is the contribution of the feature selection scheme when between-component interactions (in addition with more color spaces) are considered to describe textures for classification issues. This perspective aims to confirm that only a few carefully selected spectral bands are sufficient to address with success the texture classification applications.

The study presented in the paper finally shows the need for hyperspectral image databases where the use of spectral texture characteristics is absolutely necessary to discriminate textures in classification applications.

References

1. Bello-Cerezo, R., Bianconi, F., Fernández, A., González, E., María, F.D.: Experimental comparison of color spaces for material classification. J. Electron. Imaging **25**(6), 061406 (2016). https://doi.org/10.1117/1.JEI.25.6.061406
2. Bianconi, F., Harvey, R., Southam, P., Fernández, A.: Theoretical and experimental comparison of different approaches for color texture classification. J. Electron. Imaging **20**(4), 043006 (2011). https://doi.org/10.1117/1.3651210
3. Cernadas, E., Fernández-Delgado, M., González-Rufino, E., Carrión, P.: Influence of normalization and color space to color texture classification. Pattern Recogn. **61**, 120–138 (2017). https://doi.org/10.1016/j.patcog.2016.07.002
4. Chandrashekar, G., Sahin, F.: A survey on feature selection methods. Comput. Electr. Eng. **40**(1), 16–28 (2014). https://doi.org/10.1016/j.compeleceng.2013.11.024
5. Dash, M., Liu, H.: Feature selection for classification. Intell. Data Anal. **1**(1–4), 131–156 (1997). https://doi.org/10.1016/S1088-467X(97)00008-5
6. Deborah, H., Richard, N., Hardeberg, J.Y.: A comprehensive evaluation of spectral distance functions and metrics for hyperspectral image processing. IEEE J. Sel. Top. Appl. Earth Obs. Remote Sens. **6**(8), 3224–3234 (2015). https://doi.org/10.1109/JSTARS.2015.2403257
7. ElMasry, G.M., Nakauchi, S.: Image analysis operations applied to hyperspectral images for non-invasive sensing of food quality - a comprehensive review. Biosyst. Eng. **142**, 53–82 (2016). https://doi.org/10.1016/j.biosystemseng.2015.11.009
8. Haralick, R.M., Shanmugam, K., Dinstein, I.: Textural features for image classification. IEEE Trans. Syst. Man Cybern. **3**(6), 610–621 (1973). https://doi.org/10.1109/TSMC.1973.4309314
9. Kalakech, M., Porebski, A., Vandenbroucke, N., Hamad, D.: Unsupervised local binary pattern histogram selection scores for color texture classification. J. Imaging **4**(10), 1–17 (2018). https://doi.org/10.3390/jimaging4100112

10. Khan, H.A., Mihoubi, S., Mathon, B., Thomas, J.B., Hardeberg, J.Y.: HyTexiLa: high resolution visible and near infrared hyperspectral texture images. Sensors **16**(7), 2045 (2018). https://doi.org/10.3390/s18072045

11. Krizhevsky, A., Sutskever, I., Hinton, G.E.: ImageNet classification with deep convolutional neural networks. Adv. Neural Inf. Process. Syst. **25**(2), 1097–1105 (2012). https://doi.org/10.1145/3065386

12. Li, S., Zheng, Z., Wang, Y., Chang, C., Yu, Y.: A new hyperspectral band selection and classification framework based on combining multiple classifiers. Pattern Recogn. Lett. **83**, 152–159 (2016). https://doi.org/10.1016/j.patrec.2016.05.013

13. Liu, H., Yu, L.: Toward integrating feature selection algorithms for classification and clustering. IEEE Trans. Knowl. Data Eng. **17**(4), 491–502 (2005). https://doi.org/10.1109/TKDE.2005.66

14. Liu, L., Chen, J., Fieguth, P., Zhao, G., Chellappa, R., Pietikäinen, M.: From BoW to CNN: two decades of texture representation for texture classification. Int. J. Comput. Vision **127**(1), 74–109 (2019). https://doi.org/10.1007/s11263-018-1125-z

15. Mäenpää, T., Pietikäinen, M.: Classification with color and texture: jointly or separately? Pattern Recogn. **37**(8), 1629–1640 (2004). https://doi.org/10.1016/j.patcog.2003.11.011

16. Mignotte, M.: A de-texturing and spatially constrained k-means approach for image segmentation. Pattern Recogn. Lett. **32**(2), 359–367 (2011). https://doi.org/10.1016/j.patrec.2010.09.016

17. Mirhashemi, A.: Introducing spectral moment features in analyzing the SpecTex hyperspectral texture database. Mach. Vis. Appl. **29**(3), 415–432 (2018). https://doi.org/10.1007/s00138-017-0892-9

18. Palm, C.: Color texture classification by integrative co-occurrence matrices. Pattern Recogn. **37**(5), 965–976 (2004). https://doi.org/10.1016/j.patcog.2003.09.010

19. Pietikäinen, M., Zhao, G., Hadid, A., Ahonen, T.: Computer Vision Using Local Binary Patterns. Computational Imaging and Vision, vol. 40. Springer, London (2011) http://www.springer.com/mathematics/book/978-0-85729-747-1

20. Porebski, A., Hoang, V.T., Vandenbroucke, N., Hamad, D.: Multi-color space local binary pattern-based feature selection for texture classification. J. Electron. Imaging **27**(1), 011010 (2018). https://doi.org/10.1117/1.JEI.27.1.011010

21. Porebski, A., Vandenbroucke, N., Macaire, L.: Supervised texture classification: color space or texture feature selection? Pattern Anal. Appl. **16**(1), 1–18 (2013). https://doi.org/10.1007/s10044-012-0291-9

22. Szegedy, C., et al.: Going deeper with convolutions. In: 2015 IEEE Conference on Computer Vision and Pattern Recognition (CVPR), Boston, MA, USA, pp. 1–9, June 2015. https://doi.org/10.1109/CVPR.2015.7298594

23. Tang, J., Alelyani, S., Liu, H.: Feature selection for classification - a review. In: Aggarwal, C.C. (ed.) Data Classification: Algorithms and Applications. Data Mining and Knowledge Discovery Series, vol. 40, pp. 37–64. Chapman and Hall/CRC (2014)

24. Vandenbroucke, N., Busin, L., Macaire, L.: Unsupervised color-image segmentation by multicolor space iterative pixel classification. J. Electron. Imaging **24**(2), 023032 (2015). https://doi.org/10.1117/1.JEI.24.2.023032

25. Yang, R., Su, L., Zhao, X., Wan, H., Sun, J.: Representative band selection for hyperspectral image classification. J. Vis. Commun. Image Represent. **48**, 396–403 (2017). https://doi.org/10.1016/j.jvcir.2017.02.002

Multi-task Learning for Supervised and Unsupervised Classification of Grocery Images

Gianluigi Ciocca[ID], Paolo Napoletano[(✉)][ID], and Simone Giuseppe Locatelli[ID]

University of Milano-Bicocca, Viale Sarca 336, 2016 Milan, Italy
{gianluigi.ciocca,paolo.napoletano}@unimib.it,
s.locatelli29@campus.unimib.it

Abstract. Grocery products detection and recognition is a very complex task because of the high variability in object appearances, and the possibly very large number of the products to be recognized. Here we present the results of our investigation in the classification of grocery products. We tested several CNN architectures trained in different modalities for product classification, and we propose a multi-task learning network to be used as feature extractor. We evaluated the features extracted from the networks in both supervised and unsupervised classification scenarios. All the experiments have been conducted on publicly available datasets in the literature.

Keywords: Supervised classification · Clustering · Convolutional Neural Networks · Grocery product recognition

1 Introduction

Object detection and recognition are important problem in the computer vision field. They are at the basis of many relevant tasks such as instance segmentation, image caption, image classification, object tracking, etc [27]. The recognition of objects in images plays an important role in many everyday life activities in different contexts. For example, it is an important task in applications that help visually impaired people to identify objects [13], applications that support the monitoring of the dietary behaviour of people in order to help them to maintain an healthy daily food consumption [4,5], and other applications [21].

In this paper, we focus our attention to the problem of object recognition in the context of grocery products. Recognition of grocery products is important both from the point of view of consumers and retailers. For the consumers, automatic recognition of grocery products can help them in keeping track of the purchases, have more comprehensive and detailed information about the product they are going to buy, keep track of possible allergens and food intolerance, and so on. For retailers, automatic product recognition can help them in monitoring the availability of the products on the shelves, their arrangements, and even build

© Springer Nature Switzerland AG 2021
A. Del Bimbo et al. (Eds.): ICPR 2020 Workshops, LNCS 12662, pp. 325–338, 2021.
https://doi.org/10.1007/978-3-030-68790-8_26

automatic check-out systems for their stores. Computer vision based recognition approaches provide an affordable, flexible alternative compared to sensor based approaches. Object detection/recognition in grocery scenarios is very complex because object appearance is highly variable due to relevant changes in scale, pose, viewpoint, lighting conditions and occlusion. Other challenges are related to how the images of the products are acquired in store. On the shelves, we can have multiple objects, multiple instances of the same object, and cluttered backgrounds. Moreover, rigid objects can be posed in different ways, and non-rigid objects can be very difficult to acquire. This requires the detection of the objects under different poses. Finally, the set of grocery products that must be recognized can be very large. There are many products categories and in each category there are many product sub-categories. This means that recognition strategies need to cope with the problem of coarse to fine grained classification of the product categories. Having such a large variety of objects to recognize requires also suitable products datasets that are large, heterogeneous, and possibly containing coarse and fine grained product classes.

Here we are interested in the use of Deep Learning techniques for the recognition of grocery products in supervised and unsupervised scenarios. In particular, features learned by deep Convolutional Neural Networks (CNNs) have been recognized to be more robust and expressive than hand-crafted ones. For the supervised recognition scenario, we tested several CNN architectures trained with different modalities: a) end-to-end classification scenario; b) CNNs as feature extractor coupled with an SVM classifier. The use of the learned features is more flexible in an application scenario where the set of product may frequently vary. For the unsupervised scenario, we considered the case where the definition of an annotated dataset for the classification algorithms is cumbersome. In this case, unsupervised classification strategies coupled with learned features can be used to generate meaningful groups of similar products. These groups can be the basis for a semi-automatic annotation tool of the product images. All the experiments presented in this paper have been conducted on publicly available datasets in the literature.

The paper is organized as follows: in Sect. 2 state of the art in the field of grocery images is discussed; in Sect. 3 and 4 materials and methods are discussed; experiments on both supervised and unsupervised classification tasks are presented in Sect. 5; conclusions are discussed in Sect. 6.

2   Related Work

General-purpose object detection and recognition techniques have been extensively used in image classification and retrieval problems. Traditional object detection methods are based on complex image representations based on hand-crafted features such as Haar-like features, Histogram of Oriented Gradients, SIFT, and Deformable Part Models [11,17,24,31]. More recently, CNN-based approaches demonstrated that impressive detection and recognition results can be achieved with specific network architectures such as [30].

Detection and recognition of products differs from other object detection and recognition problems due to intra-class variance and inter-class similarity. Santra et al. [21] presented a survey on product recognition using computer vision techniques and challenges. Traditional, hand-crafted features are used in early works. For example, in Merler et al. [17] three commonly used object recognition/detection algorithms based on color histogram, SIFT, and boosted Haar-like features are exploited. The overall best performing algorithm (based on the ROC curves) is the SIFT-based one. Shen et al. [23] propose a bag-of-features approach to retrieve visually similar products captured from a mobile phone camera. The base feature are sparse and dense SIFT, and a hierarchical k-means clustering is used to construct the vocabulary. Also Rivera et al. [19] use a bag-of-word approach based on SIFT descriptors for the recognition of 30 grocery products. Three different histogram encoding methods are evaluated: hard assignment, Locality-Constrained Linear coding, and Fisher Vector encoding. The features are classified using a one-against-all binary classifier. George et al. [10] propose simultaneous recognition and localization of multiple classes in images by using random forests, dense pixel matching (SIFT-based+bag-of-features), and genetic algorithm optimization. Evaluation of the proposed pipeline is carried out on a small dataset of grocery products and on the GroZi-120 dataset. Varol et al. [24] use two descriptors to discriminate product images: SIFT features for shape description and HSV values for color description. Histograms are computed on these descriptor and image classification is performed with a multi-class SVM with Gaussian radial basis function as the kernel. George et al. [11] use discriminative patches on product packaging to differentiate between visually similar product classes. The patches are described by HOG features and one-vs-all SVM classifier is used for recognition.

With the introduction of usable Deep-Learning techniques, recognition approaches shift from hand-crafted features to learned ones. The paper by Jund et al. [14] is one of the first works that exploits Convolutional Neural Networks in an end-to-end grocery products classification. The CaffeNet architecture is used as backbone of the recognition approach. The pre-trained network is fine tuned on the new scenario achieving high recognition accuracy on the proposed Freiburg Groceries Dataset. Franco et al. [8] compared classical Bag of Words technique with Deep Neural Networks. Both the feature representations provide good accuracy on the tested dataset, but DNN-based features are found to be more effective in complex scenarios thanks to their high robustness and invariance. Geng et al. [9] proposed to integrate attention maps into a CNN-based end-to-end classification framework. The attention map is used to capture the location of the product region in the image. To generate the attention maps BRISK and SIFT features are considered. In the paper by Wei et al. [28], the check-out problem is addressed by providing a new, large, dataset of single and multiple retail products acquired in laboratory. Images with multiple objects are provided at different levels of clutter. Four detection based detectors are benchmarked with the proposed dataset. Grocery stores add new products often. This pose a challenging problem for object detectors that need to be updated frequently. Hao et al. [12] propose a new dataset and a Faster R-CNN Class-incremental Object Detector (FCIOD) for coping with the addition of new classes over time. Klasson et al. [15] proposed a new dataset of grocery images. Differently from others in the literature, the dataset is organized according to a hierarchical class structure. Products are categorized at different levels from macro-categories, coarse-grained product class, to fine-grained product category.

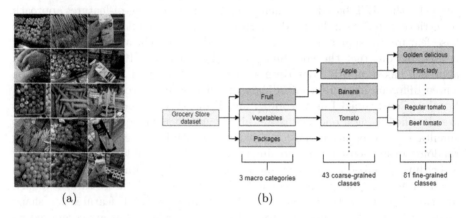

(a)   (b)

Fig. 1. (a) Some Grocery Store Dataset examples; (b) Simplified illustration of the hierarchical class structure of Grocery Store Dataset.

3 Materials

In the literature there are several datasets of grocery store items [16–18, 26]. However, most of them does not provide a fine-grained annotation as in the Grocery Store Dataset (GSD) [15]. We have chosen GSD dataset since it is publicly available, and provides a hierarchical categorization of the products at multiple level of details. This allow us to investigate the use of the product categories at different granularity. It also provides a sufficient number of product categories and related images to work with. Part of the hierarchical structure of this dataset is shown in Fig. 1. The dataset consists of 5,125 product images that are categorized using a three-level categorization. The products are first classified into three macro categories: fruits, vegetables or packages macro classes. Then the images are tagged according to one of the 43 coarse-grained classes that represent the product type (i.e. apple, tomatoes, milk, etc ...). Finally, each image is classified into one of the 81 fine-grained classes that correspond to specific instances of the product (i.e for the apple we may have Royal Gala, Granny Smith, etc ...). The list of grocery products in the Grocery Store Dataset is shown in Fig. 2. Moreover, another dataset was taken into account: the Freiburg Groceries Dataset (FGD) [14]. This dataset contains 5,000 images of 25 product categories that are very different from those in the GSD. The products are

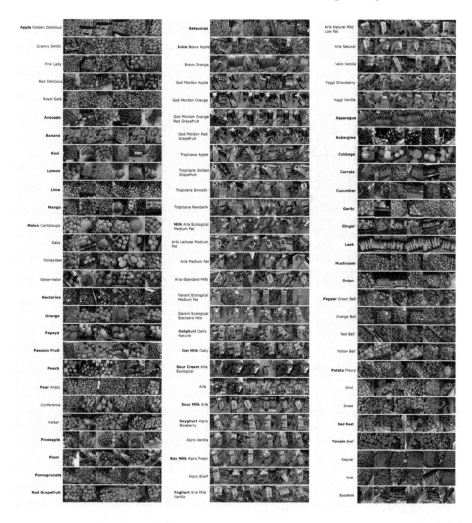

Fig. 2. Examples of the 81 fine-grained classes in GSD. Text in bold indicate also to labels of the 43 coarse-grained classes.

mostly packaged food, and it does not have a hierarchical structure like the GSD. The number of images in each class ranges from 97 to 372, and there is a great intra-class heterogeneity with multiple brands under the same class. The list of products in the Freiburg Groceries Dataset is shown in Fig. 3.

4   Methods

Here we present the baseline methods we evaluated and the multi-task network we proposed. For the supervised classification experiments, we considered both

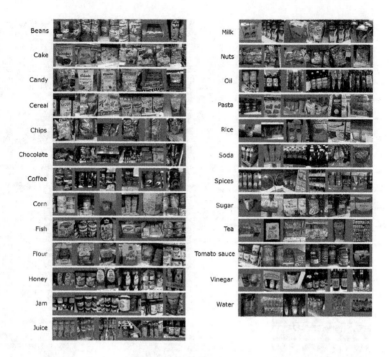

Fig. 3. Examples of images in the FGD belonging to the 25 classes.

an end-to-end scenario, and a feature-based classification scenario. For the unsupervised classification experiments, we tested several clustering algorithms and evaluated the results using different clustering metrics.

4.1 Baseline

As baseline we considered the work by Klasson et al. [15], where three types of CNNs have been taken into consideration: AlexNet, VGG-16 and DenseNet-169. We measured end-to-end classification performance of these networks by using the same experimental setup described in [15]. We add to the baseline mentioned above other popular CNN architectures fine-tuned on the GSD dataset: ResNet-18, ResNet-50, GoogLeNet, Inception-V3 and MobileNet V2 [2].

4.2 Multi-task Network

We proposed a Multi-Task Learning network (MTL) based on the DenseNet-169 to accomplish at the same time the three classification tasks within the GSD. The network is composed of three branches (see Fig. 4). Each branch classifies the image with the categories in the respective hierarchy level: fine-grained, coarse-grained, and macro. By leveraging the different information and their correlation, we should obtain more robust features.

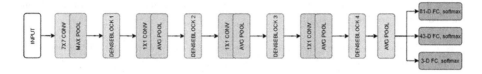

Fig. 4. The proposed Multi-task architecture based on the DenseNet-169 network.

4.3 Supervised Classification Using Learned Features

In this section we present the experiments in supervised classification of grocery products using learned features from a CNN. We experimented with different training configuration in order to identify the most robust features to be used for classification. The feature classifier used in all experiments is a Support vector Machine (SVM). The experiment configurations investigated are the following:

- DN+SVM: The features are extracted from the last average pooling layer of the original DenseNet-169 network trained on ImageNet. In this case no domain adaptation is used;
- DN+ft+SVM: As in DN+SVM but the original DenseNet-169 network is fine-tuned on the training set of the GSD dataset. In the fine-tuning process, the original final layer of the network is replaced with a softmax layer applicable to the dataset. Thus, the networks are fine-tuned for 30 epochs with 2 different learning rates: 0.01 for the new classification layer and 0.001 for the previous layers, both reduced by half after every fifth epoch.
- DN+DA+ft+SVM: As in the DN+ft+SVM but with the inclusion of data augmentation in the training process. The data augmentation consist of: a crop of random size, between 0.08 and 1, of the original size and a random aspect ratio of the original between 0.75 and 1.33, finally, resized in order to fit the network input size, also, the image is flipped horizontally with a probability of 0.5.
- DN+MTL(Pre)+DA+ft+SVM: The MTL DenseNet-169 is to extract the features from the average pooling layer prior to the split of the network.
- DN+MTL(Post)+DA+ft+SVM: This experiment is similar to the previous one but the features are extracted from the final layers in the branches of the multi-task network.

With the extracted features we trained SVMs with either a linear or a cubic kernel. The classifier is trained with two goals: recognition of fine-grained product categories (81 classes) and coarse-grained categories (43 classes).

4.4 Unsupervised Classification

Unsupervised classification of grocery products can be exploited for supporting the rapid annotation of a large number of product images from scratch, or for defining groups of similar products that can be used for product retrieval. To this end, we coupled the features extracted by the most performing supervised

Table 1. End-to-end classification results for the 81 fine-grained product categories on the Grocery Store Dataset (GSD).

AlexNet	VGG-16	DenseNet-169	ResNet-18	ResNet-50	GoogLeNet	Inception-V3	MobileNet-V2
65.79	75.97	**84.7**	75.77	83.29	63.9	77.94	82.89

network with an unsupervised clustering algorithm. In particular, we considered the Affinity Propagation algorithm [7] that is able to automatically determine the number of clusters. For sake of comparison, we employed supervised clustering algorithms that take the number of cluster found by the Affinity Propagation algorithm as input to other clustering algorithms. Specifically, we evaluated: hierarchical (i.e. Agglomerative clustering), Birch, K-means, Mini-batch K-means, and Spectral clustering [1, 22, 25, 29].

We considered the features extracted as in the DN+MTL(Pre)+DA+ft+ SVM experiment, that is, features extracted from the last average pooling layer before the network split. These features showed the best results in the supervised classification experiment. After an L2 normalization we clustered the features using the six chosen algorithms. We performed the evaluation on the GSD and the FGD datasets.

5 Experiments

For experiments related to supervised classification, different setups were adopted. The GSD has been split into a training and a test set following the guideline in [15]. So, training and test set contains 2,640 and 2,485 images respectively. Instead, results on FGD, showed in Table 3, were obtained thanks to the Matlab classification learner tool. Hence, to partition the data, a 5-fold cross-validation was applied.

5.1 Baseline Results

The list of architectures tested and their accuracies on the 81 fine-grained product categories is shown in Table 1. The best result is obtained by the DenseNet-169 (84.7%) followed by the ResNet-50 (83.29%) and the MobileNet-v2 (82,89%). The worst result (63.9%) is obtained by GoogLeNet. Results, while confirming that the best is DenseNet-169, also indicate the fact that ResNet-50 has performance very close to the best (around 1.8% lower) with a lower computational cost with respect to the DenseNet-169 architecture. To explore the supervised and unsupervised scenarios, we used the DenseNet-169 as reference network.

5.2 Results of Supervised Classification

The end-to-end MTL DenseNet-169 trained on the GSD achieves 87.76% on the 81 classes of the GSD thus increasing the performance of about 3% with respect to a simple DenseNet-169.

Table 2. Summary of Grocery Store Dataset results with the original DenseNet-169 (DN) and MTL DenseNet-169 (DN+MTL). Fine-grained refers to 81 classes. Coarse-grained refers to 43 classes. ft refers to fine tuning. DA referes to data augmentation. (Pre) refers to features extracted from the last layer before the network split. (Post) refers to features extracted from the last layer after the network split.

Methods	Fine-grained	Coarse-grained
DN+SVM	74.43	82.46
DN+ft+SVM	86.20	89.08
DN+DA+ft+SVM	88.34	91.08
DN+MTL End-to-End	87.76	93.28
DN+MTL(Pre)+DA+ft+SVM	**90.40**	93.76
DN+MTL(Post)+DA+ft+SVM	89.13	**94.33**

Table 3. Classification results on Freiburg Groceries Dataset (FGD) using DenseNet-169 multi-task features. (Pre) refers to features extracted from the last layer before the network split. (Fine) and (Coarse) refer to features extracted from the respective network branches.

Classifiers	(Pre-MTL)	(Fine-MTL)	(Coarse-MTL)
SVM linear	78.4	60.9	54.7
SVM cubic	**83.0**	70.6	64.1

Table 2 shows the classification results obtained by the SVM classifiers trained with the different learned features. Results are computed on the test set of the GSD dataset. For comparison we have also included the classification results obtained by using the multi-task network in an end-to-end way. We indicate this experiment with "DN+MTL End-To-End". Results using the cubic kernel are omitted since they are similar to the linear one. The classification accuracies obtained in the coarse-grained classification are naturally higher then the ones in the fine-grained classification due to the simpler task. We note that applying a multi-task learning improves the results considerably even with respect to the end-to-end scenario in Table 1. Features learned from this architecture seem to have more representational power as evidenced by the outcomes.

We also investigate whether or not multi-task network trained on the GSD dataset can be used to extract features to effectively classify images of unseen products. We tested this hypothesis using the FGD. Thus, we tested the features extracted before the network split (Pre-MTL) and those extracted from the fine (Fine-MTL) and coarse (Coarse-MTL) branches of the network using a linear SVM and a cubic SVM. Results are shown in Table 3. The best result (83%) is obtained by using a cubic SVM on the features extracted from the last average pooling layer before the split. This result outperforms the classification accuracy of 78.9% described in [14]. It should be noted that our classification strategy

Table 4. Unsupervised classification of the Grocery Store Dataset (GSD).

Clustering method	(#Clust.)	(#Class Pred.)	(Acc. ↑)	(V-meas. ↑)	(CHI ↑)	(DBI ↓)
Affinity propagation	269	81	92.23	0.852	23.041	1.766
Agglomerative	269	80	91.83	0.856	23.655	1.705
Birch	269	80	92.74	0.866	23.100	1.670
K-means	269	81	92.42	0.853	22.642	1.825
Mini-Batch K-means	269	80	90.10	0.859	20.648	1.774
Spectral	269	79	87.78	0.829	21.069	1.835

Table 5. Unsupervised classification of the Freiburg Groceries Dataset (FGD).

Clustering method	(#Clust.)	(#Class Pred.)	(Acc. ↑)	(V-meas. ↑)	(CHI ↑)	(DBI ↓)
Affinity propagation	373	25	64.09	0.475	9.039	2.696
Agglomerative	373	25	67.54	0.519	9.699	2.600
Birch	373	25	66.21	0.520	9.577	2.573
K-means	373	25	61.95	0.480	9.204	2.741
Mini-Batch K-means	373	25	50.78	0.426	7.630	2.229
Spectral	373	25	55.67	0.423	6.601	2.486

differs from the one described in [14] since we did not use replicated images in the training sets, but use a 5-fold cross validation on the available data.

5.3   Results of Unsupervised Classification

In order to evaluate the performance obtained from the clustering algorithms, several evaluation metrics were adopted. First we defined a global accuracy. A cluster is predicted to be composed of products of class k if the majority of products in the cluster are of that class. This allows us to define a cluster accuracy. The global accuracy is obtained as the average of all clusters accuracy. Other measures that we considered are: the V-measure [20] where a higher score, ranging from 0 to 1, indicates a better clustering, the Calinski-Harabasz index (CHI) [3] where a higher score relates to a model with better defined clusters and, lastly, the Davies-Bouldin index (DBI) [6] where a lower index relates to a model with better separation between the clusters.

After the initial clustering performed with the Affinity Propagation algorithm, it is possible that the number of clusters found is larger than the actual number of product categories in the test set. However, here we are interested in evaluating the degree of intra-cluster uniformity and the inter-cluster diversity. These two properties should be satisfied for the clusters to be useful for annotation or retrieval tasks.

Results of the unsupervised classification are shown in Table 4 and Table 5: "#Clusters" refers to the number of cluster found by the Affinity Propagation

Fig. 5. Samples of product clusters. (a) FGD correct cluster representing the spices class (b-i) FGD noisy clusters

and input for other non-automatic algorithms; "#Class Pred" refers to the number of unique product classes associated to the clusters. The arrows indicate the sorting order for the best value.

As it can be seen the Affinity Propagation algorithm found a very large number of clusters within the features. The clustering accuracy for the GSD dataset is high, above 90% for almost all the algorithms, while for the FGD dataset the accuracy are lower ranging from about 51% to about 67%. The same can be seen for the other clustering measures. Those computed on the GSD dataset are higher than those computed on the FGD dataset. As stated before, this is probably due to the different nature of the products in the two datasets. With respect to the clustering algorithm, the Birch one exhibits the best overall performance.

Figure 5 shows some examples of clusters found by the Affinity Propagation algorithm. Figure 5a is an example of homogeneous cluster for the class "Spices", while Fig. 5b-i are examples of dis-homogeneous (or noisy) clusters. We note that, the noisy clusters are composed of products that are visually similar but belonging to different product categories. Also, the examples show the differences between the GSD dataset and the FGD dataset. The former contains mostly fruits and vegetables, and when packaged product are included, only a single instance of the product is captured in the image (hand-held package). The latter, contains mostly packaged items and the images contain several instances of the same product. This may be the cause of the different supervised and unsupervised classification performances obtained in our experiments.

6   Conclusion

Grocery product recognition is a challenging task due to fact that product's appearance is highly variable in scale, pose, viewpoint, lighting conditions and occlusion. Also, different products can be very similar and only minor packaging details allow to discriminate them. Moreover, robust detection and recognition are impeded by the large scale and the fine-grained nature of the product categories.

In this paper we have presented some experiments and results for the task of grocery products classification both in a supervised and unsupervised scenario. Leveraging the hierarchical product classification provided by the Grocery Store Dataset, we have designed a multi-task learning network to be used as feature extractor. Experiments on the effectiveness of this network show that, although it exhibits very good performance on the dataset used for training, it has a limited generalization capabilities to other product datasets with very different contents. Specifically, the features learned on the Grocery Store Dataset are less adapt in discriminating the images in the Freiburg Groceries Dataset. This indicate a bias that can be attuned by considering an enriched product dataset composed of a larger set of different product images. Combining the two datasets could be an initial solution that could be further investigate and tested on a third, comparable, dataset that it is currently lacking in the public domain.

References

1. Arthur, D., Vassilvitskii, S.: k-means++: the advantages of careful seeding. In: 18th Annual ACM-SIAM Symposium on Discrete Algorithms (SODA), pp. 1027–1035 (2007)
2. Bianco, S., Cadene, R., Celona, L., Napoletano, P.: Benchmark analysis of representative deep neural network architectures. IEEE Access **6**, 64270–64277 (2018)
3. Caliński, T., Harabasz, J.: A dendrite method for cluster analysis. Commun. Stat. Theory Methods **3**(1), 1–27 (1974)
4. Ciocca, G., Micali, G., Napoletano, P.: State recognition of food images using deep features. IEEE Access **8**, 32003–32017 (2020)
5. Ciocca, G., Napoletano, P., Schettini, R.: Cnn-based features for retrieval and classification of food images. Comput. Vis. Image Underst. **176**, 70–77 (2018)
6. Davies, D.L., Bouldin, D.W.: A cluster separation measure. IEEE Trans. Pattern Anal. Mach. Intell. **1**(2), 224–227 (1979)
7. Dueck, D., Frey, B.J.: Non-metric affinity propagation for unsupervised image categorization. In: 2007 IEEE 11th International Conference on Computer Vision, pp. 1–8. IEEE (2007)
8. Franco, A., Maltoni, D., Papi, S.: Grocery product detection and recognition. Expert Syst. Appl. **81**, 163–176 (2017)
9. Geng, W., et al.: Fine-grained grocery product recognition by one-shot learning. In: Proceedings of the 26th ACM International Conference on Multimedia, pp. 1706–1714 (2018)
10. George, M., Floerkemeier, C.: Recognizing products: a per-exemplar multi-label image classification approach. In: Fleet, D., Pajdla, T., Schiele, B., Tuytelaars, T. (eds.) ECCV 2014. LNCS, vol. 8690, pp. 440–455. Springer, Cham (2014). https://doi.org/10.1007/978-3-319-10605-2_29
11. George, M., Mircic, D., Soros, G., Floerkemeier, C., Mattern, F.: Fine-grained product class recognition for assisted shopping. In: Proceedings of the IEEE International Conference on Computer Vision Workshops, pp. 154–162 (2015)
12. Hao, Y., Fu, Y., Jiang, Y.G.: Take goods from shelves: a dataset for class-incremental object detection. In: Proceedings of the 2019 on International Conference on Multimedia Retrieval, pp. 271–278 (2019)
13. Jafri, R., Ali, S.A., Arabnia, H.R., Fatima, S.: Computer vision-based object recognition for the visually impaired in an indoors environment: a survey. Vis. Comput. **30**(11), 1197–1222 (2013). https://doi.org/10.1007/s00371-013-0886-1
14. Jund, P., Abdo, N., Eitel, A., Burgard, W.: The freiburg groceries dataset. arXiv preprint arXiv:1611.05799 (2016)
15. Klasson, M., Zhang, C., Kjellström, H.: A hierarchical grocery store image dataset with visual and semantic labels. In: 2019 IEEE Winter Conference on Applications of Computer Vision (WACV), pp. 491–500. IEEE (2019)
16. Marko, Š.: Automatic Fruit Recognition Using Computer Vision. Matej Kristan), Fakulteta za racunalništvo in informatiko, Univerza v Ljubljani, Mentor (2013)
17. Merler, M., Galleguillos, C., Belongie, S.: Recognizing groceries in situ using in vitro training data. In: 2007 IEEE Conference on Computer Vision and Pattern Recognition, pp. 1–8. IEEE (2007)
18. Mureşan, H., Oltean, M.: Fruit recognition from images using deep learning. Acta Universitatis Sapientiae, Informatica **10**(1), 26–42 (2018)
19. Rivera-Rubio, J., Idrees, S., Alexiou, I., Hadjilucas, L., Bharath, A.A.: Small hand-held object recognition test (short). In: IEEE Winter Conference on Applications of Computer Vision, pp. 524–531. IEEE (2014)

20. Rosenberg, A., Hirschberg, J.: V-measure: A conditional entropy-based external cluster evaluation measure. In: Proceedings of the 2007 Joint Conference on Empirical Methods in Natural Language Processing and Computational Natural Language Learning (EMNLP-CoNLL), pp. 410–420 (2007)
21. Santra, B., Mukherjee, D.P.: A comprehensive survey on computer vision based approaches for automatic identification of products in retail store. Image Vis. Comput. **86**, 45–63 (2019)
22. Sculley, D.: Web-scale k-means clustering. In: Proceedings of the 19th International Conference on World Wide Web, pp. 1177–1178 (2010)
23. Shen, X., Lin, Z., Brandt, J., Wu, Y.: Mobile product image search by automatic query object extraction. In: Fitzgibbon, A., Lazebnik, S., Perona, P., Sato, Y., Schmid, C. (eds.) ECCV 2012. LNCS, vol. 7575, pp. 114–127. Springer, Heidelberg (2012). https://doi.org/10.1007/978-3-642-33765-9_9
24. Varol, G., Kuzu, R.S.: Toward retail product recognition on grocery shelves. In: Sixth International Conference on Graphic and Image Processing (ICGIP 2014), vol. 9443, p. 944309. International Society for Optics and Photonics (2015)
25. Von Luxburg, U.: A tutorial on spectral clustering. Stat. Comput. **17**(4), 395–416 (2007)
26. Waltner, G., et al.: MANGO - mobile augmented reality with functional eating guidance and food awareness. In: Murino, V., Puppo, E., Sona, D., Cristani, M., Sansone, C. (eds.) ICIAP 2015. LNCS, vol. 9281, pp. 425–432. Springer, Cham (2015). https://doi.org/10.1007/978-3-319-23222-5_52
27. Wang, W., Lai, Q., Fu, H., Shen, J., Ling, H.: Salient object detection in the deep learning era: An in-depth survey. arXiv preprint arXiv:1904.09146 (2019)
28. Wei, X.S., Cui, Q., Yang, L., Wang, P., Liu, L.: Rpc: A large-scale retail product checkout dataset. arXiv preprint arXiv:1901.07249 (2019)
29. Zhang, T., Ramakrishnan, R., Livny, M.: Birch: an efficient data clustering method for very large databases. ACM SIGMOD Record **25**(2), 103–114 (1996)
30. Zhao, Z.Q., Zheng, P., Xu, S.T., Wu, X.: Object detection with deep learning: a review. IEEE Trans. Neural Netwk. Learn. Syst. **30**(11), 3212–3232 (2019)
31. Zou, Z., Shi, Z., Guo, Y., Ye, J.: Object detection in 20 years: a survey. arXiv preprint arXiv:1905.05055 (2019)

Recent Advances in Video Question Answering: A Review of Datasets and Methods

Devshree Patel[(✉)], Ratnam Parikh, and Yesha Shastri

School of Engineering and Applied Science, Ahmedabad University,
Ahmedabad, India
devshree.p@ahduni.edu.in

Abstract. Video Question Answering (VQA) is a recent emerging challenging task in the field of Computer Vision. Several visual information retrieval techniques like Video Captioning/Description and Video-guided Machine Translation have preceded the task of VQA. VQA helps to retrieve temporal and spatial information from the video scenes and interpret it. In this survey, we review a number of methods and datasets for the task of VQA. To the best of our knowledge, no previous survey has been conducted for the VQA task.

Keywords: Video question answering · Visual information retrieval · Multimodal co-memory networks · Spatio-temporal reasoning · Attention networks

1 Introduction

Computer Vision is one of the sub-fields of Artificial Intelligence. It enables computers to interpret and analyze real-world scenes as humans do. Some of the computer vision research tasks are segmentation, object detection/ tracking, visual captioning and VQA. Out of them, VQA is a more challenging task since it requires an understanding of visual data based on the question asked in natural language. Since, most of the visual information we face in the real world is either in the form of images or videos, the task of VQA will help to extract crucial information for building real-life applications.

The task of VQA can be bifurcated into two sub-tasks: multiple-choice QA and open-ended QA. The latter is more challenging since it does not have a fixed answer set. Also, an in-depth understanding of the video scene is required to produce an answer from the pool of all possible answers. Open-ended QA finds its applicability in the real-world since it helps to understand complex scene and generate relevant information without the constraints of choices.

The collection of data for image-based QA is relatively easier as compared to video-based QA. The creation of datasets for VQA is a burdensome task as it requires huge human efforts for creating QA pairs and accurate annotations by looking at the visual content. Additionally, video content must have varied

A. Del Bimbo et al. (Eds.): ICPR 2020 Workshops, LNCS 12662, pp. 339–356, 2021.
https://doi.org/10.1007/978-3-030-68790-8_27

actions, scenes and object interactions in order to build a robust VQA model. In general, it is a difficult task to derive interpretations from videos since they comprise of varied length having multiple frames.

The models for VQA have been created by extending the models for image question answering. Earlier approaches for constructing models of VQA only considered the spatial information from the videos. However, VQA requires to find clues accurately on both spatial and temporal dimensions. Proceeding with the methods, [3] introduced the approach for extracting temporal context information from the videos. Further, [5] studied the problem of modeling temporal dynamics with frame-level attention by proposing an attribute augmented attention network for VQA. In [7], authors exploit information relating to appearance and motion residing in the video with an optimized attention mechanism. In contrast to the aforementioned approaches, [16] presents a novel approach of positional self-attention by replacing RNNs with self-attention. [12] proposes an architecture based on Multimodal Dual Attention Memory (MDAM) to extract the latent concepts from video scenes and captions. Unlike the previous approaches which apply spatio-temporal attention, [21] improves the approach by introducing location and relations among object interactions by creating a location-aware graph network. In [27] authors leverage the task of VQA by using Bidirectional Encoder Representations (BERT) for capturing the attention of every word from both the directions.

2   Datasets

Data acquisition for VQA has been a challenging task due to requirement of large video corpus with varied content. One of the first datasets for VQA is YouTube2Text [48] which comprises of 1987 videos and 122,708 descriptions of those in natural language. Many existing models [5,9,13,21] address the task of VQA with the help of YouTube2Text [48]. The LSMDC 2016 description dataset [47] enables multi-sentence description of the videos which is a unique property from the previous approaches of datasets. It comprises of ~128K clips taken as a combination from M-VAD and MPII-MD datasets. Based on the LSMDC 2016 description dataset [47], a fill-in-the-blank dataset (MovieFIB) [42] was proposed. This was a simple benchmark dataset consisting of 300,000 fill-in-the-blank question-answer and video pairs. To answer these fill-in-the-blank questions a model is expected to have basic understanding of scenes by detecting activities, objects, people and their interactions. Table 5 shows dataset statistics for existing VQA datasets.

2.1   MovieQA

MovieQA [36] is one of the unique benchmark datasets for VQA as it contains multiple sources of information. It comprises of video clips, subtitles, plots and scripts. It is aimed to perform automatic story comprehension from both text and video. The dataset includes 408 movies and 14,944 questions with semantic variations. [8,12,20,23] attempt to tackle the VQA task with MovieQA dataset [36]. The original paper [36] obtains an accuracy of ~35% which is gradually improved and reached to ~45% by [23]. A comparison of various models using MovieQA dataset is shown in Table 1.

2.2 YouTube2TextQA

YouTube2TextQA was one of the first dataset VQA datasets proposed by [48]. The data is taken from the YouTube2Text dataset. It comprises of 1987 videos and 122,708 automatically generated QA pairs. The dataset can address multiple-choice questions. Initially, an accuracy of 61.4% was obtained with r-ANL [5] method and it reached to 83.9% with one of the latest models L-GCN [21].

Table 1. Best test accuracies of models using MovieQA and Youtube2TextQA datasets

Datasets	YouTube2TextQA				MovieQA			
Model	r-ANL [5]	DLAN [9]	Fan et al. [13]	L-GCN [21]	DEMN [8]	MDAM [12]	PAMN [20]	AMN [23]
Accuracy	61.4	36.33	82.5	83.9	30	41.41	42.53	45.31

2.3 MSRVTT-QA and MSVD-QA

MSRVTT-QA and MSVD-QA were first used by [7] based on MSRVTT [45] and MSVD [49] video datasets. MSRVTT-QA consists of ~10K video clips and ~243K QA pairs. MSVD-QA is mainly used in video captioning experiments but due to its large data size it is also used for VQA. It has 1970 video clips and ~50.5K QA pairs. A comparison of various models using MSRVTT-QA and MSVD-QA datasets is shown in Table 2.

Table 2. Accuracies of models using MSRVTT-QA(Left) and MSVD-QA(Right)

Model	MSRVTT-QA						MSVD-QA					
	What	Who	How	When	Where	All	What	Who	How	When	Where	All
Xu et al. [7]	26.2	43	80.2	72.5	30	32.5	20.6	47.5	83.5	72.4	53.6	32
Fan et al. [13]	26.5	43.6	82.4	76	28.6	33	22.4	50.1	73	70.7	42.9	33.7
CAN [22]	26.7	43.7	83.7	75.3	35.2	33.2	21.1	47.9	84.1	74.1	57.1	32.4
Resnet+CDC(AA)[15]	27.3	44.2	86.3	73.9	34.4	34.1	–	–	–	–	–	–
QueST [28]	27.9	45.6	83	75.7	31.6	34.6	24.5	52.9	79.1	72.4	50	36.1
TSN [17]	27.9	46.1	84.1	77.8	37.6	35.4	25	51.3	83.8	78.4	59.1	36.7
Jin et al. [19]	29.5	45	83.2	74.7	42.4	35.4	24.2	49.5	83.8	74.1	53.6	35
Jiang et al. [31]	29.2	45.7	83.5	75.2	34	35.5	23.5	50.4	83	72.4	46.4	34.7
HCRN [29]	–	–	–	–	–	35.6	–	–	–	–	–	36.1
MHMAN [30]	28.7	47.1	85.1	77.1	35.2	35.6	23.3	50.7	84.1	72.4	53.6	34.6
Zhang et al. [15]	–	–	–	–	–	–	21.3	48.3	82.4	70.7	53.6	32.6

2.4 VideoQA

VideoQA dataset [4] consists of the questions in free-form natural language as opposed to previous fill-in-the-blank questions [42]. The dataset is made from the internet videos having user-curated descriptions. QA pairs are generated automatically from the descriptions by using a state-of-the-art question generation method. A collection of 18,100 videos and 175,076 candidate QA pairs make up this large-scale dataset for VQA.

2.5 Pororo-QA

In [8], authors construct a large-scale VQA dataset - PororoQA from cartoon video series. The dataset is made up of videos having simple but a coherent story structure as compared to videos in previous datasets. The dataset includes ~16K dialogue-scene pairs, ~27K descriptive sentences and ~9K multiple-choice questions. DEMN [8] achieved best accuracy of 68% at the time of release. A comparison of various models using PororoQA dataset is shown in Table 3.

2.6 TVQA+ and TVQA

TVQA+ dataset [41] is built upon one of the largest VQA datasets - TVQA [40]. TVQA [40] is a dataset built on natural video content, collected from 6 famous TV series. The videos have rich dynamic interactions and the QA pairs are formed by the people considering both the videos and the accompanying dialogues. The provision of temporal annotations by the TVQA dataset [40] is one of the key properties. However, this dataset does not consider the spatial annotations which are equally important as temporal. Hence, TVQA+ [41] is a spatio-temporally grounded VQA dataset made by adding bounding boxes to the essential regions of the videos from a subset of the TVQA dataset [40]. It comprises of ~29.4K multiple-choice questions annotated in both spatial and temporal domains. A comparison of various models using TVQA dataset is shown in Table 3.

2.7 ActivityNet-QA

[35] introduces ActivityNet-QA, a fully manually annotated large-scale dataset. It consists of ~58K QA pairs based on ~5.8K web videos taken from the popular ActivityNet dataset [50]. The dataset consists of open-ended questions. A comparison of various models using ActivityNet-QA dataset is shown in Table 3.

Table 3. Best test accuracies of models using PororoQA, TVQA and ActivityNetQA datasets

Datasets	PororoQA			TVQA			ActivityNetQA	
Model	DEMN [8]	MDAM [12]	Yang et al. [27]	PAMN [20]	AMN [23]	Yang et al. [27]	CAN [22]	MAR-VQA(a+m) [25]
Accuracy	68.0	48.9	53.79	66.77	70.70	73.57	35.4	34.6

2.8   TGIF-QA

[44] proposes a TGIF-QA dataset from the earlier (Tumblr GIF) TGIF dataset which was used for video captioning task. TGIF-QA dataset [44] contains ~165K QA pairs from ~72K animated GIFs taken from the TGIF dataset. The TGIF dataset was ideal for VQA task as it comprised of attractive animated GIFs in a concise format for the purpose of story telling. A comparison of various models using TGIF-QA dataset is shown in Table 4 (Lower the count value, better is the model performance).

Table 4. Models using TGIF-QA dataset

Model	Action	Trans	Frame	Count
MAR-VQA(a+m) [25]	71.1	71.9	56.6	5.35
PSAC [16]	70.4	76.9	55.7	4.27
STA(R) [18]	72.3	79	56.6	4.25
Jin et al. [19]	72.7	80.9	57.1	4.17
Co-Mem [10]	68.2	74.3	51.5	4.1
Jiang et al. [31]	75.4	81	55.1	4.06
Fan et al. [13]	73.9	77.8	53.8	4.02
L-GCN [21]	74.3	81.1	56.3	3.95
HCRN [29]	75	81.4	55.9	3.82

2.9   LifeQA

The LifeQA dataset from [33] considers the real life scenarios in its video clips unlike the movies or TV shows which contain the edited content. In real-life QA systems, this dataset will be useful because of its relevance to day-to-day lives. It comprises of 275 videos and 2328 multiple-choice questions.

2.10   DramaQA

[43] proposes DramaQA dataset which focuses upon character-centred representations with richly annotated 217,308 images taken from the TV drama "Another Miss Oh". The annotations consider the aspects of behaviour and emotions of the characters. In total, the dataset consists of ~23K video clips and ~16K QA pairs.

2.11   Social-IQ

Social-IQ dataset [38] also referred as Social Intelligence Queries dataset targets to analyze the unscripted social situations which comprise of natural interactions. The dataset is prepared extensively from the YouTube videos as they are rich in dynamic social situations. This dataset comprising of 1,250 social videos, 7,500 questions and a total of 52,500 answers including 30,000 correct and 22,500 incorrect answers is created with an aim to benchmark social intelligence in AI systems.

2.12   MarioQA

The main motivation behind creating MarioQA dataset [37] was to understand temporal relationships between video scenes to solve VQA problems. Before MarioQA [37], all the datasets required excessive reasoning through single frame inferences. The dataset was synthesized from super mario gameplay videos. It consists of ~92K QA pairs from video sets as large as 13 h.

2.13   EgoVQA

The EgoVQA dataset proposed by [34] presents a novel perspective of considering first-person videos as most of the previous VQA datasets [MSRVTT-QA [7], MSVD-QA [7], Youtube2TextQA [48]] focus on third-person video datasets. Answering questions about first-person videos is a unique and challenging task because the normal action recognition and localization techniques based on the keywords in questions will not be effective here. To address the same, 600 question-answer pairs with visual contents are taken across 5,000 frames from 16 first-person videos.

2.14   Tutorial-VQA

Tutorial-VQA [32] was prepared to address the need for a dataset comprising of non-factoid and multi-step answers. The technique for data acquisition was same as MovieQA [36] and VideoQA [4] as it was build from movie-scripts and news transcripts. The answers of MovieQA [36] had a shorter span than the answers collected in Tutorial-VQA [32]. In VideoQA dataset [4], questions are based on single entity in contrast to instructional nature of Tutorial-VQA [32]. It consists of 76 videos and 6195 QA pairs.

2.15   KnowIT-VQA

KnowIT VQA (knowledge informed temporal VQA) dataset [39] tries to resolve the limited reasoning capabilities of previous datasets by incorporating external knowledge. External knowledge will help reasoning beyond the visual and textual content present in the videos. The collected dataset comprises of videos annotated with knowledge-based questions and answers from the popular TV show - The Big Bang Theory. There are a total of ~12K video clips and ~24K QA pairs, making it one of the largest knowledge-based human-generated VQA dataset.

Table 5. Dataset statistics for existing video question answering datasets

Dataset	Videos	QA Pairs	QA Type	Source	QA Tasks
TutorialVQA [32]	76	6195	Manual	Screencast Tutorials	Open-ended
LifeQA [33]	275	2328	Manual	Daily life videos	Multiple-choice
EgoVQA [34]	520	600	Manual	Egocentric video studies	Open-ended and multiple-choice
ActivityNetQA [35]	5800	58,000	Manual	ActivityNet dataset	Open-ended
MovieQA [36]	6771	6462	Manual	Movies	Multiple-choice
PororoQA [8]	16,066	8913	Manual	Cartoon Videos	Multiple-choice
MarioQA [37]	–	187,757	Automatic	Gameplay videos	Multiple-choice
TGIF-QA [44]	71,741	165,165	Automatic and Manual	TGIF(internet animated GIFs)	Open-ended and multiple-choice
TVQA [40]	21,793	152,545	Manual	Sitcoms,medical and criminal drama	Multiple-choice
TVQA+ [41]	4198	29,383	Manual	Sitcoms,medical and criminal drama	Multiple-choice
Know-IT-VQA [39]	12,087	24,282	Manual	Big Bang Theory	Multiple-choice
Movie-FIB [42]	118,507	348,998	Automatic	LSMDC 2016	Open-ended
VideoQA [4]	18,100	175,076	Automatic	Online internet videos	Open-ended
DramaQA [43]	23,928	16,191	Automatic and Manual	TV Drama	Multiple-choice
Social-IQ [38]	1250	7500	Manual	YouTube Videos	Multiple-choice
MSRVTT-QA [7]	10,000	243,680	Automatic	MSRVTT(Web videos)	Open-ended
MSVD-QA [7]	1970	50,505	Automatic	MSVD(Web videos)	Open-ended
Youtube2TextQA [48]	1987	122,708	Automatic	Youtube2Text	Multiple-choice

3 Methods

VQA is a complex task which that combines the domain knowledge of both computer vision and natural language processing. The methods implemented for VQA tasks have been extended from the methods of Image-QA. VQA is more challenging compared to Image-QA because it expects spatial as well as temporal mapping and requires understanding of complex object interactions in the video scenes. One of the first works on VQA was proposed by [1]. They build a query-answering system based on a joint parsing graph from videos as well as texts. As an improvement over the model proposed by [1,3], proposes a flexible encoder-decoder architecture using RNN, which covers a wide range of temporal information. Leveraging the task of VQA, [4] proposes a model which is capable of generating QA pairs from learned video descriptions. [4] constructs a large-scale VQA dataset with ~18K videos and ~175K candidate QA pairs. In comparison to previous baseline methods of Image-QA, the model proposed in [4] proves to be efficient.

Following the aforementioned approaches, several methods have evolved in the literature to solve the complex task of VQA. The first sub-section presents the methods based on spatio-temporal approaches. The second sub-section deals with memory-based methods. Methods using attention mechanism are listed in the third sub-section. The fourth sub-section details the multimodal attention based approaches. In addition to these, various novel approaches for VQA are described in the later section. A summary of results from all the described models is abridged in Tables 1, 2, 3 and 4.

3.1   Spatio-Temporal Methods

Joint reasoning of spatial and temporal structures of a video is required to accurately tackle the problem of VQA. The spatial structure will give information about in-frame actions and temporal structure will analyze the sequence of actions taking place in the videos. Methods which work on the same motivation are described in detail in the below sections.

r-STAN (Zhao et al., 2017). The authors of [6] propose a hierarchical Spatio-temporal attentional encoder-decoder learning framework to address the problem of open-ended Video-QA. The proposed model jointly learns the representation of the sequential frames with targeted objects and performs multi-modal reasoning according to the text and video. [18] infers the long-range temporal structures from the videos to answer the open-ended questions.

STA(R) (Gao et al., 2019). A Structured Two-stream Attention network (STA) is developed by [18] with the help of a structured segment component and the encoded text features. The structured two-stream attention component focuses on the relevant visual instances from the video by neglecting the unimportant information like background. The fusion of video and question aware representations generate the inferred answer. Evaluation of the model is conducted on the famous TGIF-QA dataset.

L-GCN (Huang et al., 2020). Along with the extraction of spatio-temporal features, taking into account the information about object interactions in the videos helps to infer accurate answers for VQA. Considering this, [21] proposes a novel and one of the first graph model network called Location-Aware Graph Network. In this approach, an object is associated to each node by considering features like appearance and location. Inferences like category and temporal locations are obtained from the graph with the help of graph convolution. The final answer is then generated by merging the outputs of the graph and the question features. The proposed method [21] is evaluated on three datasets - YouTube2Text-QA [48], TGIF-QA [44], and MSVD-QA [7].

QueST (Jiang et al., 2020). Often more significance is given to spatial and temporal features derived from the videos but less to questions, therefore, accounting to this, the authors of [28] propose a novel Question-Guided Spatio-Temporal contextual attention network (QueST) method. In QueST [28], semantic features are divided into spatial and temporal parts, which are then used in the process of constructing contextual attention in spatial and temporal dimensions. With contextual attention, video features can be extracted from both spatio-temporal dimensions. Empirical results are obtained by evaluating the model on TGIF-QA [44], MSVD-QA [7], and MSRVTT-QA [7] datasets.

3.2 Memory-Based Methods

To answer questions for VQA, models need to keep track of the past and future frames as the answers would require inferences from multiple frames in time. Below are descriptions of some of the approaches which implemented memory-based models.

DEMN (Kim et al., 2017). A novel method based on video-story learning is developed by [8] which uses Deep Embedded Memory Network (DEMN) to reconstruct stories from a joint scene dialogue video stream. The model uses a latent embedding space of data. Long short term memory (LSTM) based attention model is used to store video stories for answering the questions based on specific words. The model consists of three modules: video story understanding, story selection and answer selection. Baseline models such as Bag-of-Words (BOW), Word2vec(W2V), LSTM, end to end memory networks, and SSCB are used to compare with the proposed model. In addition to the proposed model, authors also create PororoQA dataset which is the first VQA dataset having a coherent story line throughout the dataset.

Co-Mem (Gao et al., 2018). A novel method of motion appearance co-memory networks is developed by [10] which considers the following two observations. First, motion and appearance when correlated becomes an important feature to be considered for the VQA task. Second, different questions have different requirements of the number of frames to form answer representations. The proposed method uses two stream models that convert the videos into motion and appearance features. The extracted features are then passed to a temporal convolutional and deconvolutional neural network to generate the multi-level contextual facts. These facts are served as input to the co-memory networks which are built upon the concepts of Dynamic Memory Networks (DMN). Co-memory networks aid in joint modeling of the appearance and motion information with the help of a co-memory attention mechanism. The co-memory attention mechanism takes the appearance cues for generating motion attention and motion cues for generating appearance attention. The proposed model is evaluated on the TGIF-QA [44] dataset.

AHN (Zhao et al., 2018). To tackle the problem of long-form VQA, the authors of [11] create an adaptive hierarchical reinforced network (AHN). The encoder-decoder network comprises adaptive encoder network learning and a reinforced decoder network learning. In adaptive encoder learning, the attentional recurrent neural networks segment the video to obtain the semantic long-term dependencies.

It also aids the learning of joint representation of the segments and relevant video frames as per the question. The reinforced decoder network generates the open-ended answer by using the segment level LSTM networks. A new large scale long-form VQA dataset, created from the ActivityNet data [50], is also contributed by [11]. For the long-form VQA task, the proposed model is compared with state-of-the-art methods [STAN [6], MM+ [4] and ST-VQA [44]] on which the new model outperforms.

Fan et al., 2018. [13] proposes a novel heterogeneous memory component to integrate appearance and motion features and learn spatio-temporal attention simultaneously. A novel end-to-end VQA framework consisting of three components is proposed. The first component includes heterogeneous memory which extracts global contextual information from appearance and motion features. The second component of question memory enables understanding of complex semantics of questions and highlights queried subjects. The third component of multi-modal fusion layer performs multi-step reasoning by attending relevant textual and visual clues with self-attention. Finally, an attentional fusion of the multi-modal visual and textual representations helps infer the correct answer. [13] outperforms existing state-of-the-art models on the four benchmark datasets - TGIF-QA [44], MSVD-QA [7], MSRVTT-QA [7], and YouTube2Text-QA [48].

MHMAN (Yu et al., 2020). Humans give attention only to certain important information to answer questions from a video rather than remembering the whole content. Working on the same motivation, [30] proposes a framework comprising of two heterogeneous memory sub-networks. The top-guided memory network works at a coarse-grained level which extracts filtered-out information from the questions and videos at a shallow level. The bottom memory networks learn fine-grained attention by guidance from the top-guided memory network that enhances the quality of QA. Proposed MHMAN [30] model is evaluated on three datasets - ActivityNet-QA [35], MSRVTT-QA [7] and MSVD-QA [7]. Results demonstrate that MHMAN [30] outperforms the previous state-of-the-art models such as CoMem [10] and HME [13].

3.3   Attention-Based Methods

Attention mechanism plays an important role in VQA as it helps to extract answers efficiently from the videos by giving more substance to certain relevant words from the questions. The approaches mentioned below develop models by exploiting the different aspects of attention mechanism.

r-ANL (Ye et al., 2017). As an improvement to the earlier presented approach [3] which considered temporal dynamics, [5] also incorporates frame-level attention mechanism to improve the results. [5] proposes an attribute-augmented attention network that enables the joint frame-level attribute detection and unification of video representations. Experiments are conducted with both multiple-choice and open-ended questions on the YouTube2Text dataset [50]. The model is evaluated by extending existing Image-QA methods and it outperforms the previous Image-QA baseline methods.

Xu et al., 2017. In [7], authors exploit the appearance and motion information resided in the video with a novel attention mechanism. To develop an efficient VQA model, [7] proposes an end-to-end model which refines attention over the appearance and motion features of the videos using the questions. The weight representations and the contextual information of the videos are used to generate the answer. [7] also creates two datasets - MSVD-QA and MSRVTT-QA from the existing video-captioning datasets - MSVD [49] and MSRVTT [45].

DLAN (Zao et al., 2017). In [9], authors study the problem of VQA from the viewpoint of Hierarchical Dual-Level Attention Network (DLAN) learning. On the basis of frame-level and segment-level feature representations, object appearance and movement information from the video is obtained. The object appearance is obtained from frame-level representations using a 2D-ConvNet whereas movement information based on segment-level features is done using a 3D-ConvNet. Dual-level attention mechanism is used to learn the question-aware video representations with question-level and word-level attention mechanisms. The authors also construct large scale VQA datasets from existing datasets - VideoClip [51] and YouTube2Text [48].

Resnet+CDC (AA) (WenqiaoZhang et al., 2019). Inspired from previous contributions based on attention mechanisms in VQA systems, [15] studies the problem related to unidirectional attention mechanism which fails to yield a better mapping between modalities. [15] also works on the limitation of previous works that do not explore high-level semantics at augmented video-frame level. In the proposed model, each frame representation with context information is augmented by a feature extractor- (ResNet+C3D variant). In order to yield better joint representations of video and question, a novel alternating attention mechanism is introduced to attend frame regions and words in question in multi-turns. Furthermore, [15] proposes the first-ever architecture without unidirectional attention mechanism for VQA.

PSAC (Li et al., 2019). In contrast to previous works like [10,16], proposes Positional Self Attention with Co-attention (PSAC) architecture without recurrent neural networks for addressing the problem of VQA. Despite the success of various RNN based models, sequential nature of RNNs pose a problem as they are time-consuming and they cannot efficiently model long range dependencies. Thus, positional self attention is used to calculate the response at each position by traversing all positions of sequence and then adding representations of absolute positions. PSAC [16] exploits the global dependencies of temporal information of video and question, executing video and question encoding processes in parallel. In addition to positional attention, co-attention is used to attend relevant textual and visual features to guarantee accurate answer generation. Extensive experimentation on TGIF-QA [44] shows that it outperforms RNN-based models in terms of computation time and performance.

TSN (Yang et al., 2019). The authors of [17] work on the failure of existing methods that take into account the attentions on motion and appearance features separately. Based on the motivation to process appearance and motion features synchronously, [17] proposes a Question-Aware Tube Switch Network (TSN) for VQA. The model comprises of two modules. First, correspondence between appearance and motion is extracted with the help of a mix module at each time slice. This combination of appearance and motion representation helps achieve fine-grained temporal alignment. Second, to choose among the appearance or motion tube in the multi-hop reasoning process, a switch module is implemented. Extensive experiments for the proposed method [17] have been carried out on the two benchmark datasets - MSVD-QA [7] and MSRVTT-QA [7].

PAMN (Kim et al., 2019). [20] proposes the Progressive Attention Memory Network (PAMN) for movie story QA. PAMN [20] is designed to overcome 2 main challenges of movie story QA: finding out temporal parts for discovering relevant answers in case of movies that are longer in length and to fulfill the need to infer the answers of different questions on the basis of different modalities. PAMN [20] includes 3 main features: progressive attention mechanism which uses clues from video and question to find the relevant answer, dynamic modality fusion and belief correction answering scheme that improves the prediction on every answer. PAMN [20] performs experiments on two benchmark datasets: MovieQA [36] and TVQA [40].

CAN (Yu et al., 2019). Earlier works in the domain of VQA have achieved favorable performance on short-term videos. Those approaches fail to work equally well on the long-term videos. Considering this drawback, [22] proposes a two-stream fusion strategy called compositional attention networks (CAN) for the VQA tasks. The method comprises of an encoder and a decoder. The encoder uses a uniform sampling stream (USS) which extracts the global semantics of the entire video and the action pooling stream (APS) is responsible for capturing the long-term temporal dynamics of the video. The decoder utilizes a compositional attention module (CAN) to combine the two-stream features generated by encoder using attention mechanism. The model is evaluated on two-short term (MSRVTT-QA [7] and MSVD-QA [7]) and one long-term (ActivityNet-QA [35]) VQA datasets.

3.4   Multimodal Attention Based Methods

The task of VQA demands representations from multiple modalities such as videos and texts. Therefore, several approaches have worked upon the concept of multi-modal attention to fuse the extracted knowledge from both modalities and obtain accurate answers.

MDAM (Kim et al., 2019). A video story QA architecture is proposed by [12], in which authors improve on [8] by proposing a Multimodal Dual Attention Model (MDAM). The main idea behind [12] was to incorporate a dual attention method with late fusion. The proposed model uses self attention mechanism to learn the latent concepts from video frames and captions. For a question, the model uses the attention over latent concepts. After performing late fusion, multimodal fusion is done. The main objective of [12] was to avoid the risk of over-fitting by using multimodal fusion methods such as concatenation or multimodal bilinear pooling. Using this technique, MDAM [12] learns to infer joint level representations from an abstraction of the video content. Experiments are conducted on the PororoQA and MovieQA datasets.

Chao et al., 2019. One of the main challenges in VQA is that for a long video, the question may correspond to only a small segment of the video. Thus, it becomes inefficent to encode the whole video using an RNN. Therefore, authors of [14], propose a Question-Guided Video Representation module which summarizes the video frame-level features by employing an attention mechanism. Multimodal representations are then generated by fusing video summary and question information. The multimodal representations and dialogue-contexts are then passed with attention as an input in seq2seq model for answer generation. The authors conduct experiments using Audio-Visual Scene-Aware Dialog dataset (AVSD) [46] for evaluating their approach.

AMN (Yuan et al., 2020). In [23], authors present a framework Adversarial Multimodal Network (AMN) to better understand movie story QA. Most movie story QA methods proposed by [8,12,20] have 2 limitations in common. The models fail to learn coherent representations for multimodal videos and corresponding texts. In addition to that, they neglect the information that retains story cues. AMN [23] learns multimodal features by finding a coherent subspace for videos and the texts corresponding to those videos based on Generative Adversarial Networks (GANs). The model consists of a generator and a discriminator which constitute a couple of adversarial modules. The authors compare AMN [23] with other extended Image-QA baseline models and previous state-of-the-art methods for MovieQA [36] and TVQA [40] datasets.

VQA-HMAL (Zhao et al., 2020). [24] proposes a method for open-ended long-form VQA which works upon the limitations of the short-form VQA methods. Short-form VQA methods do not consider the semantic representation in the long-form video contents hence they are insufficient for modeling long-form videos. The authors of [24] implement a hierarchical attentional encoder network using adaptive video segmentation which jointly extracts the video representations with long-term dependencies based on the questions. In order to generate the natural language answer, a decoder network based on a Generative Adversarial Network (GAN) is used on the multimodal text-generation task. Another major contribution of this paper accounts for generating three large-scale datasets for VQA from the ActivityNet [50], MSVD [49], TGIF [51] datasets.

Jin et al., 2019. [19] proposes a novel multi-interaction network that uses two different types of interactions, i.e. multimodal and multi-level interaction. The interaction between the visual and textual information is termed as multimodal interaction. Multi-level interaction occurs inside the multimodal interaction. The proposed new attention mechanism can simultaneously capture the element-wise and the segment-wise sequence interactions. Moreover, fine-grained information is extracted using object-level interactions that aid in fetc.hing object motions and their interactions. The model is evaluated over the TGIF-QA, MSVD-QA and MSRVTT-QA datasets.

3.5   Miscellaneous Models

In addition to the spatio-temporal, memory-based, attention-based and multimodal attention based models, the following methods use different novel approaches to solve the problem of VQA.

MAR-VQA(a+m) (Zhuang et al., 2020). [25] takes into consideration multiple channels of information present in a video such as audio, motion and appearance. It differs from earlier works which either represented the question as a single semantic vector or directly fused the features of appearance and motion leading to loss of important information. [25] also incorporates external knowledge from Wikipedia which suggests attribute text related to the videos and aids the proposed model with support information. The model is experimented on two datasets - TGIF-QA and ActivityNet-QA.

MQL (Lei et al., 2020). A majority of previous works in VQA consider video-question pairs separately during training. However, there could be many questions addressed to a particular video and most of them have abundant semantic relations. In order to explore these semantic relations, [26] proposes Multi-Question Learning (MQL) that learns multiple questions jointly with their candidate answers for a particular video sequence. These joint learned representations of video and question can then be used to learn new questions. [26] introduces an efficient framework and training paradigm for MQL, where the relations between video and questions are modeled using attention network. This framework enables the co-training of multiple video-question pairs. MQL [26] is capable to perform better understanding and reasoning even in the case of a single question. Authors carry out extensive experiments on 2 popular datasets- TVQA and CPT-QA.

Yang et al., 2020. In natural language processing, BERT has outperformed the LSTM in several tasks. However, [27] presents one of the first works which explores BERT in the domain of computer vision. [27] proposes to use BERT representations to capture both the language and the visual information with the help of a pre-trained language-based transformer. Firstly, Faster-RCNN is used to extract the visual semantic information from video frames as visual concepts. Next, two independent flows are developed to process the visual concepts and the subtitles along with the questions and answers. In the individual flows, a fine-tuned BERT network is present to predict the right answer. The final prediction is obtained by jointly processing the output of the two independent flows. The proposed model is evaluated on TVQA [40] and Pororo [8] datasets.

HCRN (Le et al., 2020). [29] proposes a general-purpose reusable Conditional Relation Network (CRN) that acts as a foundational block for constructing more complex structures for reasoning and representation on video. CRN [29] takes an input array of tensorial objects with a conditioning feature and computes an output array of encoded objects. The model consists of a stack of these reusable units for capturing contextual information and diverse modalities. This structural design of the model bolsters multi-step and high-order relational reasoning. CRN [29] hierarchy constitutes 3 main layers for capturing clip motion as context, linguistic context and video motion as context. One important advantage of CRN over [10, 13, 16] is that it scales well on long-sized video simply with the addition of more layers in the structure. Performance evaluation methods demonstrate that CRN [29] achieves high performance accuracy as compared to various state-of-the-art models. CRN [29] is evaluated on 3 benchmark datasets- TGIF-QA [44], MSRVTT-QA [7] and MSVD-QA [7].

4   Discussion

Video Question Answering (VQA) is a complex and challenging task in the domain of Artificial Intelligence. Extensive literature is available for VQA and to the best of our knowledge no previous survey has been conducted for the same. In this survey, we studied the prominent datasets and models that have been researched in VQA. We also compare the results of all the models in Tables 1, 2, 3 and 4 by grouping them according to the datasets. Significant improvements can be witnessed in the past years in the field of VQA. Therefore, there lies a huge potential for research in this field for future tasks.

References

1. Tu, K., et al.: Joint video and text parsing for understanding events and answering queries. IEEE MultiMed. **21**(2), 42–70 (2014)
2. Yu, Y., et al.: End-to-end concept word detection for video captioning, retrieval, and question answering. In: Proceedings of the IEEE Conference on Computer Vision and Pattern Recognition (2017)
3. Zhu, L., et al.: Uncovering the temporal context for video question answering. Int. J. Comput. Vis. **124**(3), 409–421 (2017)
4. Zeng, K., et al.: Leveraging video descriptions to learn video question answering. In: Thirty-First AAAI Conference on Artificial Intelligence (2017)
5. Ye, Y., et al.: Video question answering via attribute-augmented attention network learning. In: Proceedings of the 40th International ACM SIGIR conference on Research and Development in Information Retrieval (2017)
6. Zhao, Z., et al.: Video question answering via hierarchical spatio-temporal attention networks. In: IJCAI (2017)
7. Xu, D., et al.: Video question answering via gradually refined attention over appearance and motion. In: Proceedings of the 25th ACM International Conference on Multimedia (2017)
8. Kim, K., et al.: Deepstory: Video story qa by deep embedded memory networks. arXiv preprint arXiv:1707.00836 (2017)
9. Zhao, Z., et al.: Video question answering via hierarchical dual-level attention network learning. In: Proceedings of the 25th ACM International Conference on Multimedia (2017)
10. Gao, J., et al.: Motion-appearance co-memory networks for video question answering. In: Proceedings of the IEEE Conference on Computer Vision and Pattern Recognition (2018)
11. Zhao, Z., et al.: Open-ended long-form video question answering via adaptive hierarchical reinforced networks. In: IJCAI (2018)
12. Kim, K., et al.: Multimodal dual attention memory for video story question answering. In: Proceedings of the European Conference on Computer Vision (ECCV) (2018)
13. Fan, C., et al.: Heterogeneous memory enhanced multimodal attention model for video question answering. In: Proceedings of the IEEE Conference on Computer Vision and Pattern Recognition (2019)
14. Chao, G., et al.: Learning question-guided video representation for multi-turn video question answering. arXiv:1907.13280 (2019)
15. Zhang, W., et al.: Frame augmented alternating attention network for video question answering. IEEE Trans. Multimed. **22**(4), 1032–1041 (2019)
16. Li, X., et al.: Beyond rnns: positional self-attention with co-attention for video question answering. In: Proceedings of the AAAI Conference on Artificial Intelligence, vol. 33 (2019)
17. Yang, T., et al.: Question-aware tube-switch network for video question answering. In: Proceedings of the 27th ACM International Conference on Multimedia (2019)
18. Gao, L., et al.: Structured two-stream attention network for video question answering. In: Proceedings of the AAAI Conference on Artificial Intelligence, vol. 33 (2019)
19. Jin, W., et al.: Multi-interaction network with object relation for video question answering. In: Proceedings of the 27th ACM International Conference on Multimedia (2019)

20. Kim, J., et al.: Progressive attention memory network for movie story question answering. In: Proceedings of the IEEE Conference on Computer Vision and Pattern Recognition (2019)
21. Huang, D., et al.: Location-aware graph convolutional networks for video question answering. In: AAAI (2020)
22. Yu, T., et al.: Compositional attention networks with two-stream fusion for video question answering. IEEE Trans. Image Process. **29**, 1204–1218 (2019)
23. Yuan, Z., et al.: Adversarial multimodal network for movie story question answering. IEEE Trans. Multimed. **21**(2), 42–70 (2020)
24. Zhao, Z., et al.: Open-ended video question answering via multi-modal conditional adversarial networks. IEEE Trans. Image Proces. **29**, 3859–3870 (2020)
25. Zhuang, Y., et al.: Multichannel attention refinement for video question answering. ACM Trans. Multimed. Comput. Commun. Appl. **16**(1), 1–23 (2020)
26. Lei, C., et al.: Multi-question learning for visual question answering. In: AAAI (2020)
27. Yang, Z., et al.: BERT representations for video question answering. In: The IEEE Winter Conference on Applications of Computer Vision (2020)
28. Jiang, Jianwen, et al. "Divide and Conquer: Question-Guided Spatio-Temporal Contextual Attention for Video Question Answering." AAAI. 2020
29. Le, T.M., et al.: Hierarchical conditional relation networks for video question answering. In: Proceedings of the IEEE/CVF Conference on Computer Vision and Pattern Recognition (2020)
30. Yu, T., et al.: Long-term video question answering via multimodal hierarchical memory attentive networks. IEEE Trans. Circuits Syst. Video Technol. (2020)
31. Jiang, P., Yahong, H.: Reasoning with heterogeneous graph alignment for video question answering. In: AAAI (2020)
32. Teney, D., et al.: Visual question answering: a tutorial. IEEE Signal Process. Magaz. **34**(6), 63–75 (2017)
33. Castro, S., et al.: LifeQA: a real-life dataset for video question answering. In: Proceedings of the 12th Language Resources and Evaluation Conference (2020)
34. Fan, C.: EgoVQA-an egocentric video question answering benchmark dataset. In: Proceedings of the IEEE International Conference on Computer Vision Workshops (2019)
35. Yu, Z., et al.: Activitynet-QA: A dataset for understanding complex web videos via question answering. In: Proceedings of the AAAI Conference on Artificial Intelligence, vol. 33 (2019)
36. Tapaswi, M., et al.: Movieqa: understanding stories in movies through question-answering. In: Proceedings of the IEEE Conference on Computer Vision and Pattern Recognition (2016)
37. Mun, J., et al.: Marioqa: answering questions by watching gameplay videos. In: Proceedings of the IEEE International Conference on Computer Vision (2017)
38. Zadeh, A., et al.: Social-iq: a question answering benchmark for artificial social intelligence. In: Proceedings of the IEEE Conference on Computer Vision and Pattern Recognition (2019)
39. Garcia, N., et al.: KnowIT VQA: answering knowledge-based questions about videos. arXiv preprint arXiv:1910.10706 (2019)
40. Lei, J., et al.: Tvqa: localized, compositional video question answering. arXiv preprint arXiv:1809.01696 (2018)
41. Lei, J., et al.: Tvqa+: spatio-temporal grounding for video question answering. arXiv preprint arXiv:1904.11574 (2019)

42. Maharaj, T., et al.: A dataset and exploration of models for understanding video data through fill-in-the-blank question-answering. In: Proceedings of the IEEE Conference on Computer Vision and Pattern Recognition (2017)
43. Choi, S., et al.: DramaQA: character-centered video story understanding with hierarchical QA. arXiv preprint arXiv:2005.03356 (2020)
44. Jang, Y., et al.: Tgif-qa: toward spatio-temporal reasoning in visual question answering. In: Proceedings of the IEEE Conference on Computer Vision and Pattern Recognition (2017)
45. Xu, J., et al.: Msr-vtt: a large video description dataset for bridging video and language. In: Proceedings of the IEEE Conference on Computer Vision and Pattern Recognition (2016)
46. Alamri, H., et al.: Audio visual scene-aware dialog (avsd) challenge at dstc7. arXiv preprint arXiv:1806.00525 (2018)
47. Rohrbach, A., et al.: A dataset for movie description. In: Proceedings of the IEEE Conference on Computer Vision and Pattern Recognition (2015)
48. Guadarrama, S., et al.: Youtube2text: recognizing and describing arbitrary activities using semantic hierarchies and zero-shot recognition. In: Proceedings of the IEEE International Conference On Computer Vision (2013)
49. Chen, D., William B.D.: Collecting highly parallel data for paraphrase evaluation. In: Proceedings of the 49th Annual Meeting of the Association for Computational Linguistics: Human Language Technologies (2011)
50. Caba, H.F., et al.: Activitynet: a large-scale video benchmark for human activity understanding. In: Proceedings of the IEEE Conference on Computer Vision and Pattern Recognition (2015)
51. Li, Y., et al.: TGIF: a new dataset and benchmark on animated GIF description. In: Proceedings of the IEEE Conference on Computer Vision and Pattern Recognition (2016)

IQ-VQA: Intelligent Visual Question Answering

Vatsal Goel$^{(\boxtimes)}$ ⃝, Mohit Chandak ⃝, Ashish Anand ⃝, and Prithwijit Guha ⃝

Indian Institute of Technology, Guwahati, India
{vatsal29,mohit,anand.ashish,pguha}@iitg.ac.in

Abstract. Despite tremendous progress in the field of Visual Question Answering, models today still tend to be inconsistent and brittle. Thus, we propose a model-independent cyclic framework which increases consistency and robustness of any VQA architecture. We train our models to answer the original question, generate an implication based on the answer and then learn to answer the generated implication correctly. As part of the cyclic framework, we propose a novel implication generator which generates implied questions from any question-answer pair. As a baseline for future works on consistency, we provide a new human-annotated VQA-Implications dataset. The dataset consists of 30k implications of 3 types - Logical Equivalence, Necessary Condition and Mutual Exclusion - made from the VQA validation dataset. We show that our framework improves consistency of VQA models by ~15% on the rule-based dataset, ~7% on VQA-Implications dataset and robustness by ~2%, without degrading their performance.

Keywords: VQA · Implications · Consistency

1 Introduction

Visual Question Answering [3] task requires an AI system to answer natural language questions on a contextual image. Ideally, this system should be equipped with the ability to extract useful information (with reference to the question) by looking at the image. To answer these questions correctly, the system should not only identify the color, size, or shape of objects, but may also require general knowledge and reasoning abilities which makes the task more challenging.

Previous works [6,12] have pointed out strong language priors present in the VQA dataset. This could result in false impression of good performance by

---

V. Goel and M. Chandak—Equal Contribution.

Electronic supplementary material The online version of this chapter (https://doi.org/10.1007/978-3-030-68790-8_28) contains supplementary material, which is available to authorized users.

A. Del Bimbo et al. (Eds.): ICPR 2020 Workshops, LNCS 12662, pp. 357–370, 2021.
https://doi.org/10.1007/978-3-030-68790-8_28

many state of the art models, without them actually understanding the image. For instance, answering any question starting with "What sport is" by "tennis" results in 41% accuracy. Moreover, citing the 'visual priming bias' present in the VQA dataset, questions starting with "Do you see a .." result in "yes" 87% of the time.

Original	How many sailboats are there?	1
Logeq	Is there 1 sailboat?	no
Mutex	Are there 2 sailboats?	yes
Nec	Are there any sailboats?	no
Rep	What is the number of sailboats?	2
Rep	How many sailboats can you see?	2
Rep	How many sailboats do you see?	2

(a) Input image   (b) Implications and Rephrasings answered incorrectly

Fig. 1. An example of inconsistent and brittle nature of VQA models. Even though the model [24] correctly answers the original question, it fails to answer any of the 3 generated implications and rephrasings correctly.

Many recent works [5, 16, 18] have shown that despite having high accuracy on questions present in the dataset, these models perform poorly when rephrased or implied questions are asked and hence are not intelligent enough to be deployed in the real world. Figure 1 shows the inconsistent and brittle nature of VQA models. Despite answering the original question correctly, the model fails to answer rephrased and implied questions related to the original question. This shows that models learn from language biases in the dataset to some extent, rather than correctly understanding the context of the image.

Throughout the paper, **Implications** are defined as questions Q_{imp} which can be answered by knowing the original question Q and answer A without the knowledge of the context i.e. image I. We categorize these implications into 3 types - logical equivalence, necessary condition and mutual exclusion - as introduced in [16]. Figure 2 shows these 3 categories for a QA pair. **Consistency** is the percentage of implications answered correctly, given that the original question is answered correctly. **Rephrasings** Q_R are linguistic variations on original question Q keeping the answer A exactly same, as introduced in [18]. **Robustness** is defined as the accuracy on rephrasings, calculated only on correctly answered original questions.

We believe that any model can be taught to better understand the content of the image by enforcing intelligence through consistency and robustness among the predicted answers. In this paper, we present and demonstrate a cyclic training scheme to solve the above mentioned problem of intelligence. Our framework is model independent and can be integrated with any VQA architecture. The framework consists of a generic VQA module and our implication generation module tailored especially for this task.

Our framework ensures intelligent behaviour of VQA models while answering different questions on the same image. This is achieved in two steps: Implication generator module introduces linguistic variations in the original question based on the answer predicted by the VQA model. Then, the model is again asked to answer this on-the-fly generated question so that it remains consistent with the previously predicted answer. Thus, the VQA architecture is collectively trained to answer questions and their implications correctly. We calculate the consistency of different state of the art models and show that our framework significantly improves consistency and robustness without harming the performance of the VQA model.

We observe that there is no benchmark for consistency, which perhaps is the reason for limited development in this area. Hence, to promote robust and consistent VQA models in the future we collect a human annotated dataset of around 30k questions on the original VQA v2.0 validation dataset.

In later sections, we demonstrate the quality of these generated questions. We provide a baseline of our implication generator module for future works to compare with. We also perform a comparative study of the attention maps of models trained with our framework to those of baselines. We provide a qualitative and quantitative analysis and observe significant improvement in the quality of these attention maps. This proves that by learning on these variations, our framework not only improves the consistency and robustness of any generic VQA model but also achieves a stronger multi modal understanding of vision and language.

To summarize, our main contributions in this paper are as follows -

- We propose a model independent cyclic framework which improves consistency and robustness of any given VQA architecture without degrading the architecture's original validation accuracy.
- We propose a novel implication generator module, which can generate implications $G : (Q, A) \longrightarrow Q_{imp}$, for any given question answer pair.
- For future evaluation of consistency, we provide a new VQA-Implication dataset. The dataset consists of ~30k questions containing implications of 3 types - Logical Equivalence, Necessary Condition and Mutual Exclusion.

2   Related Works

Ever since Visual Question Answering [3] was introduced, numerous models have been proposed to combine techniques from computer vision and natural language processing using techniques such as CNNs and LSTMs. Some of the best models using complex attention mechanisms include [2, 7, 24]. The current state of the art is LXMERT [20], which uses a transformer network for self and cross attention between vision and language modalities.

Original	How many people? 4
Logeq	Are there 4 people? yes
Mutex	Are there 5 people? no
Nec	Are there any people? yes

(a) Input image example   (b) Generated implication example

Fig. 2. An example of the rule-based implication dataset. Note that the implications can be answered without looking at the image.

Analysis of the VQA v1 dataset [1,6] showed the presence of language priors in the dataset. Models were reportedly exploiting these priors as a shortcut for answering questions instead of understanding the image. To tackle this problem, VQA 2.0 was released which created complementary pairs in order to counter these priors. More specifically, for every image, question and answer triplet (I, Q, A), a complimentary image I^c and answer A^c were created. However, investigations in [12] found that even after these ramifications priors continue to exist and exploited by VQA models.

Recent works [5,16,18] in VQA have introduced novel benchmarks such as robustness and consistency of models as a step towards limiting the false sense of progress in VQA with just accuracy and proposing models with better multimodal understanding.

Consistency: Inconsistency in QA models on the VQA and SQUAD dataset was first studied in [16]. They show how even the best VQA models are inconsistent in their answers. For example, given a question "How many birds are in the picture?", the model correctly answers "3". But upon asking "Are there 3 birds in the picture?", the same model incorrectly answers "No". This shows that models lack high level language and vision capabilities and could still be exploiting biases present in the dataset. They proposed evaluating consistency of these models and a simple data augmenting technique for improvement. However, augmentation limits the scope of implied questions to the added dataset. We in-turn propose a generative model based solution without this limitation.

More recent work in [5] tackles inconsistency among binary i.e. "yes/no" questions. They argue that despite answering original question correctly, VQA models performs poorly on logical composition of these questions. Another work [17], focuses on improving consistency of models on reasoning questions. Unlike [16], these works provide model based solution but they target only a specific category of questions such as reasoning or binary questions. Unlike these, we show that our approach works better on the entire VQA v2.0 dataset rather than a small subset of it.

Authors of [14] previously attempted to improve consistency on the entire VQA v2.0 dataset. However, their concept of Entailed questions, generated from the Visual Genome [8] dataset, is quite different to our Implications. We believe that if the model is able to answer a question correctly, it should also be able to answer its implications correctly, which implies consistent behavior. But in [14], given a question Q as "Has he worn sneaker?" and answer "yes", an entailed question Q' is "Where is this photo?" with answer "street". Clearly Q and Q' have no direct relation and as per our definition of consistency, answering Q and Q' correctly does not exhibit consistent behavior.

Robustness: To decrease strong language priors, a number of works [6,25] have introduced balanced datasets in context of robustness. The concept of robustness as a measure of performance on linguistic variations in the questions known as rephrasings was first introduced by [18]. However, they used a 'consensus score' metric whereas we use a metric similar to consistency for evaluation, to provide uniformity. To motivate future works in this field, they provide a VQA-Rephrasings dataset which we use to evaluate robustness of our models.

Question Generation: There has been a thorough study of Natural Language generation in NLP, such as [9,10,15,23]. [15] extracts keywords from knowledge graphs and then formulate question generation from these keywords as Seq2Seq translation problem. [9] tackles the question generation problem from Reinforcement Learning point of view. They consider generator as an actor trying to maximise BLEU score as it's reward function. [23] propose a Transformer based Seq2Seq pretraining model which beats the current state of the art in many summarization and question generation tasks. To the best of our knowledge, we are the first ones to propose an implication generator module to improve consistency of any VQA architecture.

Cyclic Framework: Cyclic training for singular modality has been used in the past for tasks such as motion tracking [19] and text-based question answering [21]. For multi-modal tasks such as VQA, cyclic training was first introduced by [18]. They used a Visual Question Generator (VQG) module to generate rephrasings of the original question and then trained their VQA module on those rephrasings in a cyclic manner. Similar to [18], our framework is also model-independent and can be used for any VQA architecture. However, their aim was to make VQA models more robust to linguistic variations through rephrasings. Our aim, through our approach, is to make the models more accurate to not just rephrasings like in [18], but also on implications.

3   Approach

We use the rule-based approach in [16] to generate implications on entire VQA v2.0 dataset, referred to as the rule-based implication dataset. This rule-based method is unable to create all 3 implications for every QA pair, especially on yes/no type questions. Due to these restrictions by this approach, the rule-based implication dataset contains implications from about 60% of the original dataset. Moreover, all generated implications are of 'yes/no' type, this serves as a strong prior for our implication generator module. Additional details about the rule-based implication dataset can be found in Sect. 4 (Fig. 3).

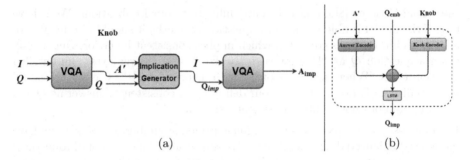

(a)   (b)

Fig. 3. Proposed Model Architecture (a) Abstract representation of our cyclic framework. Given an input image I and question Q, a VQA model predicts answer A'. Then our proposed Implication generator transforms the original question Q to Q_{imp} using A' and a control knob. This generated implication (and image) is passed to the VQA model to obtain answer A_{imp}. (b) Detailed architecture of our implication generator. The predicted answer A' and control knob are encoded to a latent space using respective encoders. They are then summed up along with question embedding and fed to a LSTM to generate implication Q_{imp}.

3.1   Implication Generator Module

The role of this module is to generate implications of a given QA pair. This can be formulated as a transformation $G : (Q, A) \longrightarrow Q_{imp}$ where Q_{imp} is the generated implication. In the VQA setting, this QA pair is provided by the VQA model. Any generic VQA model takes (Q, I) to predict A' where Q is the original question, I is the image and A' is the predicted answer. Our implication generator takes as input, the learned question encoding of the original question Q, the predicted answer scores A' and a control knob (as one hot vector) to select between the three implication categories.

The implication generation module consists of three linear encoders that transform question encoding obtained from VQA model, the predicted answer scores, and the knob to lower dimensional feature vectors. These three inputs are then added together, and passed through a single layered LSTM with hidden size of 1024. This LSTM is trained to generate implications and optimized by minimizing the negative log likelihood with corresponding ground truth implication from the rule-based implication dataset. One thing to note is that we use answers scores over the entire vocabulary instead of a particular answer label, which increases performance on questions with more than one possible correct answer. Also, this provides a distribution over the entire set of answers which is a slightly rich and dense signal to learn from.

The implication generator module - by generating implications - introduces stronger linguistic variations than rephrasings as proposed in [18]. Thus we believe that by learning on these implications, models trained with our approach should also perform better on rephrasings thus leading to improvement in robustness, in addition to consistency.

3.2 Knob Mechanism

Instead of using an implied answer selected randomly from (yes, no) as input to the implication generator module, we use a three way knob to switch between logical equivalence, necessary condition and mutual exclusion. This helps the model to have better control over the generated implications. In our training dataset, implications from two categories - logical equivalence and necessary condition have 'yes' as the correct answer. While training the implication generator using implied answer, we noted that model tends to generate necessary implications when provided 'yes' as the implied answer. We believe that generating a necessary condition is easier as compared to logical equivalence and without having any control signal, model might learn to generate necessary implications all the time. Hence, we provide this control signal in the form of a one hot vector between the three implication categories.

3.3 Cyclic Framework

To integrate our implication generator module with any VQA module, we use a cyclic framework. The confidence score over answers generated by the VQA module is used by the implication generator module. The implications are then passed as question to the VQA module, along with the image I to give implied answer A_{imp}. This enables the VQA module to learn on these implications and improve its consistency. Inspired by [18], we incorporate gating mechanism and late activation in our cyclic architecture. So, instead of passing all implied questions, we filter out undesirable implications which have cosine similarity less than threshold T_{sim} with the ground truth implication. Also, as part of the late activation scheme, we disable cyclic training before A_{iter}.

We use three loss functions in our architecture, namely VQA loss L_{vqa}, question loss L_Q and implication loss L_{imp}. L_{vqa} is the standard binary cross-entropy (BCE) loss between predicted answer A' and ground truth A^{gt}. L_Q is the negative log likelihood loss between generated implication Q_{imp} and ground truth implication Q_{imp}^{gt}. L_{imp} is also the BCE loss between A_{imp} and A_{imp}^{gt}. Combining the three losses with their respective weights, we get total loss L_{tot} as:

$$L_{tot} = L_{vqa}(A', A^{gt}) + \lambda_Q L_Q(Q_{imp}, Q_{imp}^{gt}) + \lambda_{imp} L_{imp}(A_{imp}, A_{imp}^{gt}) \quad (1)$$

Table 1. Consistency and robustness performance on rule-based validation, VQA-Implications and VQA-Rephrasings dataset. Consistency and robustness are defined as percentage of correctly answered implications and rephrasings respectively, generated only on correctly answered original questions. All the models trained with our approach outperform their respective baselines in all categories, keeping the validation accuracy almost same.

Method	Val acc	Consistency (rule-based)				Consistency (VQA-Imp)	Robustness
		Logeq	Nec	Mutex	Overall		
BUTD [2]	63.62	64.3	71.1	59.8	65.3	67.14	79.21
BUTD + IQ	63.57	**88.5**	**96.7**	**77.0**	**88.1**	**74.38**	**80.77**
BAN [7]	65.37	67.1	77.6	61.1	69.0	66.57	79.93
BAN + IQ	65.28	**89.3**	**97.9**	**79.8**	**89.6**	**74.61**	**81.62**
Pythia [24]	64.70	69.7	76.4	67.7	70.0	70.89	79.31
Pythia + IQ	65.60	**88.7**	**97.6**	**79.0**	**88.7**	**76.55**	**82.40**

4 Experiments Setup

4.1 Datasets

We use the VQA v2.0 dataset for training and evaluating our model's VQA performance. The VQA v2.0 training split consists of 443,757 questions on 82,783 images and the validation split contains 214,354 questions over 40,504 images.

To train and evaluate our implication generator module and consistency, we use the implication dataset made by the rule-based approach in [16]. This dataset consists of 531,091 implied questions in training split and 255,682 questions for the validation split.

We also evaluate our model's consistency performance on human annotated VQA-Implications dataset which consists of 30,963 questions. For this dataset, we randomly select 10,500 questions from the VQA v2.0 validation set and create 3 implications (logeq, nec and mutex) per question.

For robustness evaluation, we use the VQA-Rephrasings dataset provided by [18]. The dataset consists of 121,512 questions by making 3 rephrasings from 40,504 questions on the VQA v2.0 validation set.

4.2 VQA Models

In order to show the model independent behaviour of our proposed method, we evaluate intelligence of three VQA models: BUTD, BAN, Pythia. We use the open-source implementation of these models for training and evaluation. These models are trained with hyperparameters proposed in respective papers.

BUTD [2] uses bottom up attention mechanism from pretrained Faster-RCNN features on the images. Visual Genome [8] dataset is used to pretrain and extract top-K objects in the images during the preprocessing step. This model won the annual VQA challenge in 2017. For training BUTD, we used the fixed top-36 objects RCNN features for every image. Their model achieves 63.62% accuracy on the VQA v2.0 validation split.

BAN [7] uses bilinear model to reduce the computational cost of learning attention distributions, whereby different attention maps are built for each modality. Further, low-rank bilinear pooling extracts the joint representations for each pair of channels. BAN achieves 65.37% accuracy on the VQA v2.0 validation split.

Pythia [24] extracts image features from detectron also pretrained over visual genome. It also uses Resnet-152 features and ensembling over 30 models, but we didn't use these techniques in our study. Glove embeddings are used for question and its implications. Pythia was the winning entry of 2018 VQA challenge and achieves 64.70% accuracy on the VQA v2.0 validation split.

4.3 Implementation Details

For the gating mechanism and late activation, $T_{sim} = 0.9$ and $A_{iter} = 5500$ for Pythia and $A_{iter} = 10,000$ for BAN and BUTD. The LSTM hidden state size for implication generator module is 1024 and Glove embeddings are used of $dim = 300$. The weights for the losses are kept as $\lambda_Q = 0.5$ and $\lambda_{imp} = 1.5$. All models are trained on training split and evaluated on validation split of VQA v2.0 dataset.

5 Results and Analysis

5.1 Consistency Performance

As defined in Sect. 1, consistency of any VQA model is its ability to answer the implications of a question correctly, if it correctly answers the original question. Implications are generated on the correctly answered questions from validation VQA v2.0 dataset, and consistency score is calculated as the fraction of correct predictions to total implications. These generated implications are binary yes/no questions, and hence randomly answering them would give about 50% consistency score.

As seen in Table 1 All the 3 models achieve an average consistency score of ~70%. i.e. they fail 30% of the times on implications of correctly predicted questions. Intuitively, Nec-implication serves as the neccessary condition which the models should know in order to answer the question. For eg.: In order to answer "How many birds are there?", they should understand if "Are there any birds in the picture?" Consistency score of ~75% Nec-implication shows the lack of image understanding in these models. Using our approach, the three models achieve ~97% on Nec-implication.

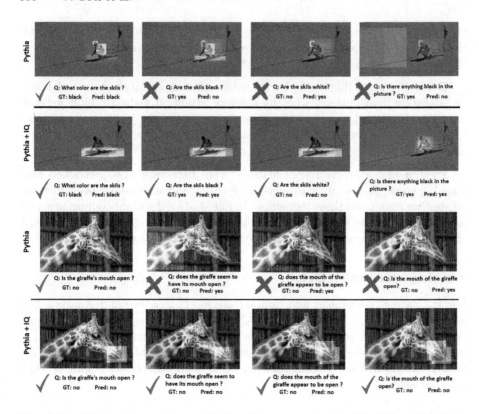

Fig. 4. Qualitative comparison of attention maps for Pythia [24] and Pythia + IQ. Top 2 rows represent implications and bottom 2 rows represent rephrasings. As evident from the figure, Pythia does not attend to relevant regions, whereas upon using our approach, the attention maps are much more focused on relevant regions.

5.2   Robustness Performance

We evaluate our framework's robustness performance on the VQA-Rephrasings dataset introduced in [18]. Robustness is evaluated only on correctly answered original questions. Note that just like the models in [18], we also do not train our models on the VQA-Rephrasings dataset. The results in Table 1 show that models trained using our approach are more robust compared to baseline. This is consistent with the hypotheses that our models learn to improve on a stronger linguistic variation than rephrasings by learning on implications and hence improvement in robustness is expected.

5.3   Attention Map Analysis

As a qualitative analysis, we compare the attention maps of [24] with our approach. As we can see in Fig. 4, the attention maps generated by our approach are

Table 2. Attention map analysis. Logeq (Rephrasing) denotes the mean Euclidean distance between attention maps of original question and Logeq (Rephrasing). Models trained with our approach produce better results highlighting stronger multi-modal understanding.

Method	Logeq $(\times 10^{-4})$	Rephrasing $(\times 10^{-4})$
BUTD [2]	31.72	15.51
BUTD + IQ (ours)	**26.73**	**13.88**
BAN [7]	8.09	5.03
BAN + IQ (ours)	**5.41**	**3.64**
Pythia [24]	11.41	5.40
Pythia + IQ (ours)	**5.83**	**3.37**

significantly better than those of [24] for both implications and rephrasings. Further qualitative comparisons of attention maps can be found in Supplementary material Sect. 1.

To ensure appropriate visual grounding, we believe that the model should look at same regions as original question for logically equivalent and rephrased questions. As a quantitative comparison, we compute the mean Euclidean distance between attention weights for logically equivalent (Logeq) and rephrased questions with their respective original question. Table 2 shows that models trained with our approach tend to focus on same regions to answer the original question, its rephrasing and its logical equivalent counterpart. These analysis show that multi-modal understanding of vision and language is enhanced using our approach.

5.4 Data Augmentation

Since we are using an extra dataset (rule-based implications) in addition to VQA v2.0 to train our models, we also compare our models' consistency with models finetuned using data augmentation. Table 3 summarizes these results. Better performance of our models on the human annotated VQA-Implications dataset shows that models trained with our approach generalize better and hence would do better than data augmentation in the outside world.

5.5 Implication Generator Performance

We train our implication generator on the rule-based training dataset and evaluate our module on rule-based validation split. We use common question generator metrics such as BLEU [13], ROUGE-L [11], METEOR [4] and CIDEr [22] scores for evaluation. We also demonstrate the importance of using the knob mechanism instead of an implied answer as input to the module. Table 4 shows the results of the implication generator module. Further examples of generated implications can be found in Supplementary material Sect. 2.

Table 3. Consistency comparison of data augmentation vs our approach.
VQA-Imp denotes our VQA-Implications dataset and DA stands for models fine-tuned on rule-based training implications. Even though our models lack on rule-based dataset, they consistently outperform their respective baselines on the VQA-Implication dataset.

Method	Consistency (rule-based)	Consistency (VQA-Imp)
BUTD + DA	**93.1**	74.24
BUTD + IQ (ours)	88.1	**74.38**
BAN + DA	87.6	74.33
BAN + IQ (ours)	**89.6**	**74.61**
Pythia + DA	**89.7**	76.19
Pythia + IQ (ours)	88.7	**76.55**

Table 4. Implication generation performance on rule-based Implication validation dataset. Note that using the knob mechanism instead of an implied answer gives significant improvement.

Method	BLEU-1	BLEU-2	BLEU-3	BLEU-4	ROUGE-L	METEOR	CIDEr
Pythia + IQ	0.627	0.520	0.443	0.381	0.632	0.288	3.343
Pythia + IQ + Knob	**0.785**	**0.715**	**0.647**	**0.581**	**0.795**	**0.409**	**5.263**

6 Conclusion and Future Works

Our contributions in this paper are three fold. First, we propose a model-independent cyclic training scheme for improving consistency and robustness of VQA models without degrading their performance. Second, a novel implication generator module for making implications using the question answer pair and a knob mechanism. Third, a new annotated VQA-Implications dataset as an evaluation baseline for future works in consistency.

Our implication generator being trained on rule-based implications dataset, has its own limitations. Firstly, the implications are restricted to 3 types - Logical Equivalence, Necessary Condition and Mutual Exclusion and all implications are limited to 'yes/no' type. We believe that learning on implications not restricted to these limitations should lead to better performance. Furthermore, the rule-based implications come from a fixed distribution and are not as diverse as human annotated implications would be. This limitation can also be quantitatively seen by observing the difference between models' performance on rule-based and human annotated implications.

References

1. Agrawal, A., Batra, D., Parikh, D., Kembhavi, A.: Don't just assume; look and answer: overcoming priors for visual question answering. In: 2018 IEEE/CVF Conference on Computer Vision and Pattern Recognition, pp. 4971–4980 (2018)
2. Anderson, P., et al.: Bottom-up and top-down attention for image captioning and visual question answering. In: The IEEE Conference on Computer Vision and Pattern Recognition (CVPR), June 2018
3. Antol, S., et al.: VQA: visual question answering. In: International Conference on Computer Vision (ICCV) (2015)
4. Denkowski, M., Lavie, A.: Meteor universal: language specific translation evaluation for any target language. In: Proceedings of the EACL 2014 Workshop on Statistical Machine Translation (2014)
5. Gokhale, T., Banerjee, P., Baral, C., Yang, Y.: VQA-LOL: visual question answering under the lens of logic. In: Vedaldi, A., Bischof, H., Brox, T., Frahm, J.-M. (eds.) ECCV 2020. LNCS, vol. 12366, pp. 379–396. Springer, Cham (2020). https://doi.org/10.1007/978-3-030-58589-1_23
6. Goyal, Y., Khot, T., Summers-Stay, D., Batra, D., Parikh, D.: Making the V in VQA matter: elevating the role of image understanding in visual question answering. In: Conference on Computer Vision and Pattern Recognition (CVPR) (2017)
7. Kim, J.H., Jun, J., Zhang, B.T.: Bilinear attention networks. Adv. Neural Inf. Process. Syst. **31**, 1571–1581 (2018)
8. Krishna, R., et al.: Visual genome: connecting language and vision using crowdsourced dense image annotations. Int. J. Comput. Vision **123**(1), 32–73 (2017)
9. Kumar, V., Ramakrishnan, G., Li, Y.F.: Putting the horse before the cart: a generator-evaluator framework for question generation from text. In: Proceedings of the 23rd Conference on Computational Natural Language Learning (CoNLL), pp. 812–821 (2019)
10. Lan, Z., Chen, M., Goodman, S., Gimpel, K., Sharma, P., Soricut, R.: Albert: a lite bert for self-supervised learning of language representations. In: International Conference on Learning Representations (2020). https://openreview.net/forum?id=H1eA7AEtvS
11. Lin, C.Y.: ROUGE: a package for automatic evaluation of summaries. In: Text Summarization Branches Out, pp. 74–81. Association for Computational Linguistics, Barcelona, July 2004. https://www.aclweb.org/anthology/W04-1013
12. Manjunatha, V., Saini, N., Davis, L.S.: Explicit bias discovery in visual question answering models. In: The IEEE Conference on Computer Vision and Pattern Recognition (CVPR), June 2019
13. Papineni, K., Roukos, S., Ward, T., Zhu, W.J.: BLEU: a method for automatic evaluation of machine translation. In: Proceedings of the 40th Annual Meeting on Association for Computational Linguistics, ACL 2002, pp. 311–318. Association for Computational Linguistics (2002). https://doi.org/10.3115/1073083.1073135
14. Ray, A., Sikka, K., Divakaran, A., Lee, S., Burachas, G.: Sunny and dark outside?! Improving answer consistency in VQA through entailed question generation. In: Proceedings of the 2019 Conference on Empirical Methods in Natural Language Processing and the 9th International Joint Conference on Natural Language Processing (EMNLP-IJCNLP), pp. 5863–5868 (2019)

15. Reddy, S., Raghu, D., Khapra, M.M., Joshi, S.: Generating natural language question-answer pairs from a knowledge graph using a RNN based question generation model. In: Proceedings of the 15th Conference of the European Chapter of the Association for Computational Linguistics: Volume 1, Long Papers, pp. 376–385. Association for Computational Linguistics, Valencia, April 2017. https://www.aclweb.org/anthology/E17-1036

16. Ribeiro, M.T., Guestrin, C., Singh, S.: Are red roses red? Evaluating consistency of question-answering models. In: Proceedings of the 57th Annual Meeting of the Association for Computational Linguistics, pp. 6174–6184. Association for Computational Linguistics, Florence, July 2019. https://doi.org/10.18653/v1/P19-1621. https://www.aclweb.org/anthology/P19-1621

17. Selvaraju, R.R., et al.: SQuINTing at VQA models: introspecting VQA models with sub-questions. In: Proceedings of the IEEE/CVF Conference on Computer Vision and Pattern Recognition, pp. 10003–10011 (2020)

18. Shah, M., Chen, X., Rohrbach, M., Parikh, D.: Cycle-consistency for robust visual question answering. In: The IEEE Conference on Computer Vision and Pattern Recognition (CVPR) (2019)

19. Sundaram, N., Brox, T., Keutzer, K.: Dense point trajectories by GPU-accelerated large displacement optical flow. In: Daniilidis, K., Maragos, P., Paragios, N. (eds.) ECCV 2010, Part I. LNCS, vol. 6311, pp. 438–451. Springer, Heidelberg (2010). https://doi.org/10.1007/978-3-642-15549-9_32

20. Tan, H., Bansal, M.: LXMERT: learning cross-modality encoder representations from transformers. In: Proceedings of the 2019 Conference on Empirical Methods in Natural Language Processing (2019)

21. Tang, D., et al.: Learning to collaborate for question answering and asking. In: Proceedings of the 2018 Conference of the North American Chapter of the Association for Computational Linguistics: Human Language Technologies, Volume 1 (Long Papers), pp. 1564–1574. Association for Computational Linguistics, New Orleans, June 2018. https://doi.org/10.18653/v1/N18-1141. https://www.aclweb.org/anthology/N18-1141

22. Vedantam, R., Lawrence Zitnick, C., Parikh, D.: CIDEr: consensus-based image description evaluation. In: The IEEE Conference on Computer Vision and Pattern Recognition (CVPR), June 2015

23. Yan, Y., et al.: ProphetNet: predicting future N-gram for sequence-to-sequence pre-training (2020)

24. Jiang, Y., Natarajan, V., Chen, X., Rohrbach, M., Batra, D., Parikh, D.: Pythia v0.1: the winning entry to the VQA challenge 2018. arXiv preprint arXiv:1807.09956 (2018)

25. Zhang, P., Goyal, Y., Summers-Stay, D., Batra, D., Parikh, D.: Yin and Yang: balancing and answering binary visual questions. In: Proceedings of the IEEE Conference on Computer Vision and Pattern Recognition, pp. 5014–5022 (2016)

CVAUI 2020 - 4th Workshop on Computer Vision for Analysis of Underwater Imagery

Preface

The 4th Workshop on Computer Vision for the Analysis of Underwater Imagery (CVAUI 2021), was held on-line, in conjunction with the International Conference on Pattern Recognition (ICPR) on 10 January 2021. This workshop further consolidated the series that was started in Stockholm (CVAUI 2014), and continued in Cancun (CVAUI 2016) and Beijing (CVAUI 2018).

Monitoring marine and freshwater ecosystems is of critical importance in developing a better understanding of their complexity, including the effects of climate change and other anthropogenic influences. The collection of underwater video and imagery, whether from stationary or moving platforms, provides a non-invasive means of observing submarine ecosystems in situ, including the behaviour of organisms. Oceanographic data acquisition has been greatly facilitated by the establishment of cabled ocean observatories, whose co-located sensors support interdisciplinary studies and real-time observations. Scheduled recordings of underwater video data and static images are gathered with Internet-connected fixed and PTZ cameras, which observe a variety of biological processes. These cabled ocean observatories, such those operated by Ocean Networks Canada (www.oceannetworks.ca), offer a 24/7 presence, resulting in unprecedented volumes of visual data and a "big data" problem for automated analysis. Due to the properties of the environment itself, the analysis of underwater imagery imposes unique challenges which need to be tackled by the computer vision community in collaboration with biologists and ocean scientists.

This workshop provided a forum for researchers to share and discuss new methods and applications for underwater image analysis. We received 9 submissions, out of which 6 were accepted based on a thorough, single-blind peer review process. Most of the submitted papers were of high quality, so the acceptance rate reflects a self-selection process performed by the authors. We thank the members of Program Committee for lending their time and expertise to ensure the high quality of the accepted workshop contributions.

January 2021

Maia Hoeberechts
Alexandra Branzan Albu

Organization

Workshop Chairs

Maia Hoeberechts	Ocean Networks Canada
Alexandra Branzan Albu	University of Victoria, Canada

Program Committee

Jacopo Aguzzi	Institut de Ciéncies del Mar, Spain
Duane Edgington	Monterey Bay Aquarium Research Institute, USA
Bob Fisher	University of Edinburgh, UK
Hervé Glotin	Université de Toulon, France
Nina S. T. Hirata	University of São Paulo, Brazil
Anthony Hoogs	Kitware Inc., USA
Reinhard Koch	University of Kiel, Germany
Kevin Köser	Helmholtz Centre for Ocean Research, Germany
Tim Nattkemper	Bielefeld University, Germany
Timm Shoening	Helmholtz Centre for Ocean Research, Germany
Gaoang Wang	Zhejiang University, China

Organization

Deep Sea Robotic Imaging Simulator

Yifan Song$^{(\boxtimes)}$, David Nakath, Mengkun She, Furkan Elibol, and Kevin Köser

Oceanic Machine Vision, GEOMAR Helmholtz Centre for Ocean Research Kiel,
Kiel, Germany
{ysong,dnakath,mshe,felibol,kkoeser}@geomar.de
https://www.geomar.de/en/omv

Abstract. Nowadays underwater vision systems are being widely applied in ocean research. However, the largest portion of the ocean - the deep sea - still remains mostly unexplored. Only relatively few image sets have been taken from the deep sea due to the physical limitations caused by technical challenges and enormous costs. Deep sea images are very different from the images taken in shallow waters and this area did not get much attention from the community. The shortage of deep sea images and the corresponding ground truth data for evaluation and training is becoming a bottleneck for the development of underwater computer vision methods. Thus, this paper presents a physical model-based image simulation solution, which uses an in-air texture and depth information as inputs, to generate underwater image sequences taken by robots in deep ocean scenarios. Different from shallow water conditions, artificial illumination plays a vital role in deep sea image formation as it strongly affects the scene appearance. Our radiometric image formation model considers both attenuation and scattering effects with co-moving spotlights in the dark. By detailed analysis and evaluation of the underwater image formation model, we propose a 3D lookup table structure in combination with a novel rendering strategy to improve simulation performance. This enables us to integrate an interactive deep sea robotic vision simulation in the Unmanned Underwater Vehicles simulator. To inspire further deep sea vision research by the community, we release the source code of our deep sea image converter to the public (https://www.geomar.de/en/omv-research/robotic-imaging-simulator).

Keywords: Deep sea image simulation · Underwater image formation · UUV perception

1 Introduction

More than 70% of Earth's surface is covered by water, and more than 90% of it is deeper than 200 m, where nearly no natural light reaches. Due to physical obstacles, even nowadays, most of the deep sea is still unexplored. Deep sea exploration is however receiving increasing attention, as it is the largest living space on Earth, contains interesting resources and is the last uncharted area of our planet. Since humans cannot easily access this hostile environment, Unmanned Underwater Vehicles (UUVs) have been used for deep sea exploration for decades.

© Springer Nature Switzerland AG 2021
A. Del Bimbo et al. (Eds.): ICPR 2020 Workshops, LNCS 12662, pp. 375–389, 2021.
https://doi.org/10.1007/978-3-030-68790-8_29

With the rapid development of underwater robotic techniques, UUVs are able to autonomously reach and to measure even in several kilometer water depth nowadays, providing platforms for carrying various sensors to explore, measure and map the oceans.

Optical sensors, e.g. cameras, are able to record the seafloor as high resolution images which are advantageous for human interpretation. Consequently, many UUV platforms are equipped with camera systems for visual mapping of the seafloor due to the significant improvement of imaging capabilities during the last decades. However, underwater computer vision remains less investigated than on land because underwater images are suffering from several effects, such as attenuation and scattering, which significantly decrease the visibility and the image quality. In addition, since no natural light penetrates the deep ocean, artificial light sources are also needed. This non-homogeneous illumination on limited-size platforms causes anisotropic backscatter that can not be modeled by atmospheric fog models, as often done for shallow water in sunlight, and further degrades image quality. The above effects often make computer vision solutions struggle or fail in (deep) ocean applications.

The recent trend to employ machine learning methods for various vision tasks even increases the performance gap between underwater vision and approaches on land, since learning methods usually require a large amount of training data to achieve good performance. However, the lack of appropriate underwater (especially deep sea) images with ground truth data is a bottleneck for developing learning-based approaches in this field. Simulation of deep sea images, in particular with illumination, attenuation and scattering effects could be one way to obtain development or training material for UUV perception.

This paper therefore proposes a physical model-based deep sea underwater image simulator which uses in-air texture images and corresponding depth maps as inputs to simulate synthetic images with underwater optical effects. The simulator considers spotlights (with main direction and angular fall-off) and with arbitrary poses in the model for the special conditions in the deep sea. Several optimization strategies are introduced to improve the computational performance of the simulator, which enables us to integrate the deep sea camera simulation into common underwater robotic simulation platforms (e.g. the Gazebo-based UUV simulator [10]).

2   Related Work and Main Contributions

Light rays are attenuated and scattered while traversing underwater volumes, which can be formulated by corresponding radiometric physical models [14]. [7] and [13] decompose underwater image formation into three components: direct signal, forward-scattering and backscatter, which is known as the Jaffe-McGlamery model. [17] describes the underwater image formation for shallow water cases. Underwater image formation has been intensively studied in underwater image restoration that can be considered as the inverse problem of underwater image formation. The most widely applied model has been presented by

[6], which was initially used to recover the depth cues from atmospheric scattering images (e.g. in fog or haze in sunlight):

$$I = J \cdot e^{-\eta \cdot d} + B \cdot (1 - e^{-\eta \cdot d}). \tag{1}$$

In the above fog model, the image I is described as a weighted linear combination of object color J and background color B. Here, d is the distance between the camera and scene point, while η represents the attenuation coefficient.

Current underwater image simulators are mostly based on the fog model: [23] adds a color transmission map and presents a method to generate synthesized underwater images, given an "in-air" image and a depth map that encodes, for each pixel, the distance to the imaged 3D surface. In the literature, such pairs of color images (RGB) and depth maps (D) are also called RGB-D images, and we will use this notation also for the remainder of this paper. [11] proposes a generative adversarial network (GAN) - WaterGAN, which has been trained with shallow water images. It also requires in-air RGB-D images as the input to generate synthetic underwater images. The target function of the GAN discriminator is also based on the fog model.

However, the fog model is only valid in shallow water cases, where the scene has global homogeneous illumination from the sunlight. [1] addresses many weaknesses of this model, which introduces significant errors in both direct signal and backscatter components. Obviously, the fog model does not apply to deep sea scenarios where artificial light sources are required to illuminate the scene and the resultant light distribution is extremely inhomogeneous. The light originates from the artificial sources attached to the robot and interacts with the water body in front of the camera, leading to very different visual effects in the images, especially in the backscatter component (see Fig. 1). Hence, the underwater image formation model in deep sea requires additional knowledge about the light sources like corresponding poses and properties. [22] uses the recursive rendering equation adapted to underwater imagery considering point light sources in their model. [18] proposes an underwater renderer based on physical models for refraction, but not focusing on realistic light sources. Since backscatter is computed for each pixel for each image, the simulation is quite demanding and does not allow real-time performance. For image restoration rather than simulation, [5] considers a spotlight with Gaussian characteristics in the image formation model and applies it to restore the true color of underwater scenes. Consequently, there is no simulator available to the community that generates realistic deep sea image sequences at interactive frame rates.

A key use case for deep sea image simulation is integrating it into a UUV simulation platform, which enables developing, testing and coordinating performance of underwater robotic systems before risking expensive hardware in real applications. Current ray-tracing solutions are too heavy to integrate to real-time robotic simulation platforms. For instance, general robotic simulators provide the simulation of a normal camera and a depth sensor, which can jointly be extended to undewater cases. [16] developed a software tool called UWSim, for visualization and simulation of underwater robotic missions. This simulator

Fig. 1. Different artificial lighting configurations strongly affect the appearance of deep sea images, especially the backscatter pattern (light cones), which can not be modeled by the fog model. Images courtesy GEOMAR/CSSF/Schmidt Ocean Institute, JAGO Team GEOMAR, AUV Team GEOMAR.

includes a camera system to render the images as seen by underwater vehicles but without any water effect. [12] extended the open-source robotics simulator Gazebo to underwater scenarios, called UUV Simulator. This simulator uses so-called RGB-D sensor plugins to generate the depth and color images, and then converts them to underwater scenes by using the fog model (Eq. 1).

Another RGB-D based underwater renderer [4] applies trained convolutional neural networks to style transfer the image output from [12] and additionally add forward scattering and haze effect. However, their improvements still rely on the fog model and the haze addition just manually adds two bright spots, which lacks a physical interpretation. [3] integrated the ocean atmosphere radiative transfer (OSOA) model into their simulator SOFI and created look-up tables to compose the back scatter component. However, the OSOA model only describes the sunlight transformation at the ocean atmosphere interface, which is only suitable for shallow water scenarios.

The main contributions of this paper are: (1) A deep sea underwater image simulation solution based on the Jaffe-McGlamery model considering multiple spotlights (with angular characteristics) with corresponding poses and properties. (2) Analysis of the components in the deep sea image formation model and several optimizations to improve the simulator's performance in particular for rigid robotic configurations. (3) Integration of the deep sea imaging simulator into the UUV robotic simulator, which can be applied for underwater robotic development and rapid prototyping. (4) Open source renderer for facilitating the development and testing in underwater vision and robotics communities.

3   Deep Sea Image Formation Model

In the deep sea scenario, there is no sun light to illuminate the scene. Only artificial light sources, which are attached to the underwater vehicles, provide the illumination. This moving light source configuration makes the appearance of deep sea images strongly depend on the geometric relationships between the camera, light source and the object (see Fig. 2).

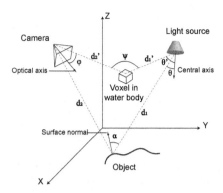

Fig. 2. Geometry components involved in the deep sea image formation model (modified from [18]).

3.1 Radiation of the Light Source

This paper considers spotlights, which are commonly used on the UUV platforms. This type of light source usually has the highest light emanation along its central axis and an intensity drop-off with increasing angle to the central axis. This angular characteristic can be formulated as radiation intensity distribution (RID) curve. Often the RID is approximated using a Gaussian function (see e.g. [5]). In our simulator it is also possible to directly use the sparse measurements as a lookup-table and interpolate the RID values (see Fig. 3). In the Gaussian model, the radiance along each light ray can be calculated as:

$$I_\theta(\lambda) = I_0(\lambda)e^{-\frac{1}{2}\frac{\theta^2}{\sigma^2}}. \tag{2}$$

Where $I_\theta(\lambda)$, $I_0(\lambda)$ are the relative light irradiance at angle θ and the maximum light irradiance along the central axis respectively. The dependency on the wavelength λ can be obtained from the color spectrum curve of the LED, which is often provided by the manufacturer or can be measured by a spectrophotometer.

3.2 Attenuation and Reflection

Light is attenuated when it travels through the water, where the loss of irradiance depends on the traveling distance and the water properties. Different wavelengths of light are absorbed with different strengths, which causes the radiometric changes in underwater images. This is because different types of water hold different water attenuation coefficients, resulting in variations of color shifts in images (e.g. coastal water images often appear more greenish, while the deep water images appear more blueish, see Fig. 4). [8] measured and classified Earth's waters into five typical oceanic spectra and nine typical coastal spectra. [2] shows how the corresponding attenuation curves vary between the different types and can serve as a first approximation for typical coefficients (and their expected variations). Due to the point

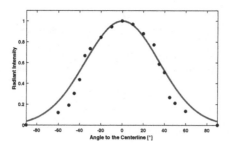

Fig. 3. Radiation characteristics of the light source used in this paper, blue dots: our underwater lab measurement, red line: its approximation by using a scaled Gaussian function ($\sigma = 35°$). (Color figure online)

source property of the spotlight, the Inverse Square Law must be applied in order to simulate the quadratic decay of the light irradiance along the distance from the point-source it originated from. When we combine the attenuation effect with the object reflection model, which assumes light is reflected equally in all directions on the object surface (Lambertian surface), the entire attenuation and reflection model can be formulated as:

$$E(\lambda) = J(\lambda) \cdot I_\theta(\lambda) \frac{e^{-\eta(\lambda)(d_1+d_2)}}{d_1^2} \cos \alpha. \tag{3}$$

Here, $E(\lambda)$ is the irradiance which arrives at the pixel of the image and $J(\lambda)$ is the object color. The attenuation parameter η indicates the strength of irradiance attenuation through the specific type of water on wavelength λ. d_1 and d_2 refer to the distance from light to object and from object to camera, respectively. α indicates the incident angle between the light ray from the light source and surface normal. In the multiple light sources case, the computation is a summation of camera viewing rays for all light sources. Note that the denominator only contains d_1 because with increasing d_2 each pixel will simply integrate the light from a larger surface area.

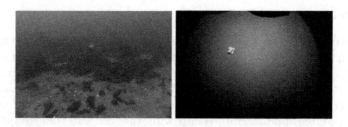

Fig. 4. Different types of water appear in different colors. Left: coastal water in Baltic Sea. Right: deep sea water in SE Pacific.

3.3   Scattering

The rendering of scattering in this paper is based on the Jaffe-McGlamery model, and is the most complex part of the involved physical models due to its accumulative character. In the Jaffe-McGlamery model, the scattering is partitioned into two parts: forward scattering and backscatter. Forward scattering usually describes the light which is scattered by a very small angle, resulting in unsharpness of the scene in the images. This paper approximates the forward scattering effect with a Gaussian filter $g(\bar{d})$ and the size of filter mask depends on the local scene depth $\bar{d}$. We neglect the forward scattering from light to the scene because the RID curve of the light is usually very smooth (e.g. modeled as a Gaussian function), where a small extra smoothing can be neglected. Backscatter refers to light rays which are interacting with ocean water and scattered backwards to the camera, this leads to a "veiling light" effect in the medium. This effect is happening along the whole light path. Following [13], the 3D field in front of the camera can be discretized by slicing it into several slabs with certain thicknesses, the irradiance on each slab is then accumulated in order to form up the backscatter component:

$$
\begin{cases}
E'(\lambda) = I'_\theta(\lambda)\dfrac{e^{-\eta(\lambda)(d'_1+d'_2)}}{d'^2_1} \\
E'_f(\lambda) = E'(\lambda) * g(\overline{d'_2}) \\
E_b(\lambda) = \sum_{i=1}^{N} \beta(\pi - \psi)[E'(\lambda) + E'_f(\lambda)]\Delta z_i \cos(\varphi).
\end{cases}
\tag{4}
$$

Equation 4 gives the computation of the backscatter component from each light source. Here i indicates the slab index and $E'(\lambda)$ denotes the direct irradiance reaching slab i. d'_1 and d'_2 represent the distances from slab voxel to light source and camera respectively. $E'_f(\lambda)$ denotes the forward scattering component of the slab which convolves $E'(\lambda)$ by the Gaussian filter $g(\overline{d'_2})$ and $*$ indicates the convolution operator. $\beta(\pi - \psi)$ refers to the Volume Scattering Function (VSF), where ψ is the angle between the light ray that hits the voxel and the light ray scattered from the voxel to the camera (see Fig. 2). The VSF model in this paper applies the measurements from [15] but can be adapted easily to other VSFs. Δz_i is the thickness of the slab and φ is the angle between the camera viewing ray and the central axis.

[5,7,18] also consider optics and electronics of the camera (e.g. vignetting, lens transmittance and sensor response) in their models. They are needed to simulate the image of a particular camera and could be added also to our simulator if needed. This is however out of scope for this contribution, where we focus rather on efficient rendering of realistic backscatter. As discussed in [19], underwater dome ports can be adjusted in a way to avoid refraction, which is why we also consider adding refraction as a non-mandatory step for underwater simulators (if needed it can be added using the methods proposed in [18,20]).

4   Implementation

This section shows the implementation of our deep sea robotic imaging simulator. The complete workflow is illustrated in Fig. 5.

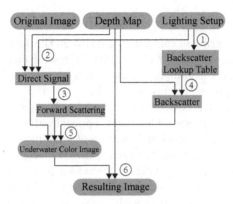

Fig. 5. Workflow.

1. Establish the 3D backscatter lookup table, each unit cell accumulates the backscatter elements along the viewing ray from the camera which is calculated by Eq. 4.
2. Compute the forward scattering component by smoothing the direct signal through a Gaussian filter.
3. Generate the direct signal component considering attenuation and object surface reflection according to Eq. 4.
4. Interpolate the backscatter component from the backscatter lookup table with respect to the depth value from the depth map.
5. Form up the underwater color image by combining the direct signal, forward scattering and the backscatter component.
6. Optionally, add refraction effect to the image.

Several optimization procedures are employed in order to improve the performance of the deep sea imaging simulator, as described in the following.

4.1   Optimizations for Rendering

In deep sea image simulation, one of the most computationally costly parts is the simulation of the backscatter component. Backscatter happens through the water body between the camera and the 3D scene, which is an accumulative phenomenon in the image. However, when the relative geometry between camera and light source is fixed, given the same water, backscatter remains constant in the 3D volume in front of the camera. For example, if there are no objects but only water in front of the camera, the image will be relatively constant and only contains the backscatter component. Once the object appears in the scene, the backscatter volume is cut depending on the depth between the object and the camera, the remaining part is accumulated to form up the image backscatter component.

To this end, we construct a 3D frustum of a pyramid for the camera's field of view and slice it into several volumetric slabs with certain thicknesses parallel to

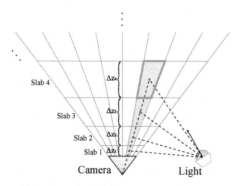

Fig. 6. Pre-rendered backscatter field, each unit cell in the slab (green) stores the accumulated backscatter component (yellow) along the camera viewing ray. (Color figure online)

Fig. 7. Rendering of backscatter component under the same setups ($d_{max} = 10$m, $N = 3$, single light which is at $(1\,\mathrm{m},\ 1\,\mathrm{m},\ 0\,\mathrm{m})$ in camera coordinate system and pointing parallel to the camera optical axis.) with different slab thickness sampling approaches. Left: by equal distance sampling, Right: by Eq. 5.

the image plane (see Fig. 6). Each slab is rasterized into unit cells according to the image size. We pre-compute the accumulative backscatter elements for each unit cell and store them in a 3D lookup table. Since the backscatter component of each pixel is an integration of all the illuminated slabs multiplied by the corresponding slab thickness along the viewing ray, the calculation of the backscatter for a pixel with depth D then is simplified by interpolating the value between the closest two unit cells along the viewing ray.

During the rendering of the slabs, we noticed that in practically relevant UUV camera-light configurations, the backscatter component appearance is dominated by the irradiance from the water volume close to the camera and scattering becomes smoother and eventually disappears in the far field. This depends strongly on the relative pose of the light source(s) and is different in each individual camera system but this is a fundamental difference to the shallow water cases, where also far away from the camera a lot of light from the sun is still available. Sample "scatter irradiance" patterns on slabs can be seen in Fig. 9.

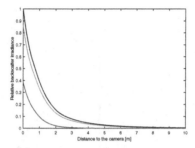

Fig. 8. Normalized backscattered irradiance along camera optical axis at different depth of slabs. Each curve describes the backscatter behavior of Jerlov water type II with the same light settings as Fig. 7. It can be seen that in this configuration almost no scattered light reaches the sensor from more than 8 m distance. This puts an upper limit on the extent of the lookup table for backscatter.

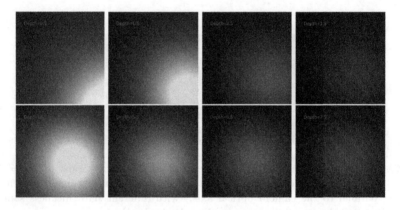

Fig. 9. Backscatter components of different slabs from 0.5 m to 7.5 m depth (Second row images' intensities are amplified 10 times).

In order to generate an accurate backscatter component with less number of slabs, we propose an adaptive slab thickness sampling function based on Taylor series expansion of the exponential function:

$$\Delta z_i = s \cdot \frac{N^{(i-1)}}{(i-1)!} \qquad (i = 1, 2, ..., N) \qquad (5)$$

where Δz_i indicates the slab thickness of slab index i. The scale factor $s = 2.2 \cdot d_{max}/e^N$, where d_{max} refers to the maximum depth of the scene field which is divided into number of slabs N. Here, e^N normalizes the Taylor series and $2.2 \cdot d_{max}$ ensures the slab thickness is monotonically increasing in $(1 < i < N)$ and $\sum_{i=1}^{N} \Delta z_i \approx d_{max}, (N > 3)$. This equation leads to denser slab samplings closer to the camera. As it is shown in Fig. 7, under the light setup described in its caption, the brightest spot should be at the bottom right corner of the image. The sampling of slab thickness by Eq. 5 gives a more plausible backscatter rendering result than the equal distance sampling approach.

The value of maximum depth of the scene d_{max} is also an important factor which affects the backscatter rendering quality and performance. In Fig. 8 we demonstrate the normalized backscattered irradiance of the voxels along the optical center axis in deep ocean water. This figure can be a good reference for finding d_{max} to simulate the underwater images under different conditions or settings.

4.2 Rendering Results

As it is shown in Fig. 10, (a) and (b) are the inputs from the RGB-D sensor plugin. The direct signal (c) and backscatter (d) components are computed respectively, then the simulated underwater color image (e) is constructed by the direct signal, the smoothed direct signal (forward scattering) and the backscatter. In the end, the refraction effect is added to the underwater color image in (f) by using the method from [20].

4.3 Integration in Robotic UUV Simulation Platform

Gazebo is an open-source robotics simulator. It utilizes one out of four different physics engines to simulate the mechanisms and dynamics of robots. Additionally, it provides the platform for hosting various sensor plugins. [12] proposes the UUV Simulator which is based on Gazebo and extends Gazebo to underwater

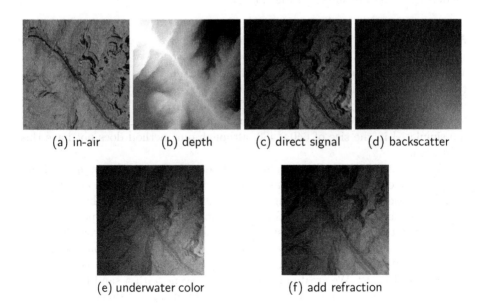

(a) in-air   (b) depth   (c) direct signal   (d) backscatter

(e) underwater color   (f) add refraction

Fig. 10. Deep sea image simulation results.

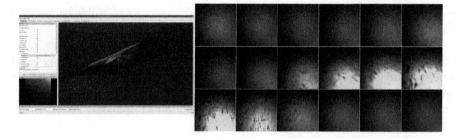

Fig. 11. Left: camera path overview in simulator. Right: rendered image sequence. Due to the physically correct model, already in the simulation we can see that some images will be overexposed with the settings chosen. Consequently, the exposure control algorithm of the robot can be adapted already after simulation without wasting precious mission time at sea.

scenarios. The UUV Simulator additionally takes into account the hydrodynamic and hydrostatic forces and moments for simulating vehicle dynamics in underwater environments. Several sensor plugins which are commonly deployed on UUVs are also available, including inter alia: inertial measurement unit (IMU), magnetometer, sonar, multi-beam echo sounders and camera modules. We integrate our deep sea camera simulator into the UUV Simulator camera plugin which provides in-air and depth images as the input and it is able to reach interactive speeds for 800×800 size of images using OpenMP without any GPU acceleration on a 16-core CPU consumer hardware. The workspace interface and sample rendering results are shown in Fig. 11.

5   Evaluation

We evaluate our deep sea image simulator by comparing with three state-of-the-art methods, which use in-air and depth images as the input to synthesize underwater images: UUV Simulator [12], WaterGAN [11] and UW_IMG_SIM [4]. Due to the image size limitation from WaterGAN, all the evaluated images are simulated in the size of 640×480, although our method does not have this limitation.

To render the realistic deep sea images close to the images shown in Fig. 1, we initialize the camera-light setups as: two artificial spotlights which are 1m away from the camera on the left and right sides, both tilt $45°$ towards the image center. The real image was taken in the Niua region (Tonga) in south pacific ocean, according to the map of global distribution of Jerlov water types from [9], water in this region belongs to type IB and the corresponding attenuation parameters are $(0.37, 0.044, 0.035)$ $[m^{-1}]$ for RGB channels. The simulation comparisons are given in Fig. 12. We create an in-air virtual scene with a sand texture, and simulate the corresponding underwater images by using the different methods. Since only our method considers the impact of lighting geometry configuration, the other methods are not able to add the shading effect on the texture image.

To fairly compare our approach to others, we first add the in-air shading in the texture image and feed it to the other simulators, even though this in-air shading with a specific light RID is not available in any standard renderers.

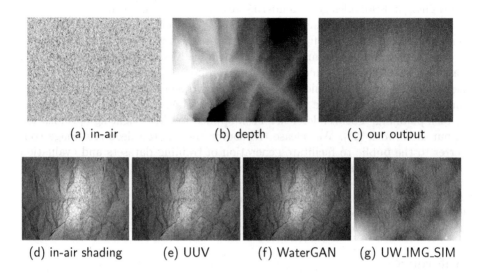

(a) in-air   (b) depth   (c) our output

(d) in-air shading   (e) UUV   (f) WaterGAN   (g) UW_IMG_SIM

Fig. 12. Outputs of different underwater image simulators for the same scene.

As it is shown in Fig. 12, the UUV Simulator is only able to render the attenuation effect based on the fog model without considering the impact of the light sources, the backscatter pattern caused by lighting is completely missing in their image. Their attenuation effect only considers the path from the scene points to the camera, which makes the rendered color also not conform to the deep sea scenario. The same problem also occurs in the WaterGAN results, due to the lack of deep sea images with depth maps and ground truth in-air images, the GAN is trained using the parameters given in the official repository[1] on the *Port Royal, Jamaica* underwater dataset[2]. Therefore the color and the backscatter pattern of the light source is highly correlated with the training data which does not fulfill the setup in this evaluation case. UW_IMG_SIM presents the backscatter pattern of the light source. However this effect is just adding the bright spots into the image without any physical interpretation, their direct signal component also has no dependence to the light source, which also is not realistic. Our proposed approach captures all discussed effects present in real images better than the other methods, it not only renders the color much closer to the real image, but also simulates attenuated shading on the topography and back scatter caused by the artificial light sources which is missing in other approaches.

---

[1] https://github.com/kskin/WaterGAN.
[2] https://github.com/kskin/data.

6   Conclusion

This paper presents a deep sea image simulation framework readily usable in current robotic simulation frameworks. It considers the effects caused by artificial spotlights, and provides good rendering results in deep sea scenarios at interactive framerates. Earlier underwater imaging simulation solutions are either not physically accurate, or far from real-time to be integrated into a robotic simulation platform. By detailed analysis of the deep sea image formation components, based on the Jaffe-McGlamery model, we propose several optimization strategies which enable us to achieve interactive performance and makes our deep sea imaging simulator fit to be integrated into the UUV simulator for prototyping or task planning. This renderer has been applied for AUV lighting optimization in our later work [21]. We release the source code of the deep sea image converter to the public to facilitate generation of training datasets and evaluation of underwater computer vision algorithms.

Acknowledgements. This publication has been funded by the German Research Foundation (Deutsche Forschungsgemeinschaft, DFG) Projektnummer 396311425, through the Emmy Noether Programme.

References

1. Akkaynak, D., Treibitz, T.: A revised underwater image formation model. In: Proceedings of the IEEE Conference on Computer Vision and Pattern Recognition, pp. 6723–6732 (2018)
2. Akkaynak, D., Treibitz, T., Shlesinger, T., Loya, Y., Tamir, R., Iluz, D.: What is the space of attenuation coefficients in underwater computer vision? In: 2017 IEEE Conference on Computer Vision and Pattern Recognition (CVPR), pp. 568–577. IEEE (2017)
3. Allais, A., et al.: Sofi: a 3D simulator for the generation of underwater optical images. In: OCEANS 2011, Spain, pp. 1–6. IEEE (2011)
4. Álvarez-Tuñón, O., Jardón, A., Balaguer, C.: Generation and processing of simulated underwater images for infrastructure visual inspection with UUVs. Sensors **19**(24), 5497 (2019)
5. Bryson, M., Johnson-Roberson, M., Pizarro, O., Williams, S.B.: True color correction of autonomous underwater vehicle imagery. J. Field Robot. **33**(6), 853–874 (2016)
6. Cozman, F., Krotkov, E.: Depth from scattering. In: Proceedings of IEEE Computer Society Conference on Computer Vision and Pattern Recognition, pp. 801–806. IEEE (1997)
7. Jaffe, J.S.: Computer modeling and the design of optimal underwater imaging systems. IEEE J. Oceanic Eng. **15**(2), 101–111 (1990)
8. Jerlov, N.: Irradiance optical classification. Opt. Oceanogr. 118–120 (1968)
9. Johnson, L.J.: The underwater optical channel. Dept. Eng., Univ. Warwick, Coventry, UK, Technical report (2012)
10. Koenig, N., Howard, A.: Design and use paradigms for gazebo, an open-source multi-robot simulator. In: 2004 IEEE/RSJ International Conference on Intelligent Robots and Systems (IROS) (IEEE Cat. No. 04CH37566), vol. 3, pp. 2149–2154. IEEE (2004)

11. Li, J., Skinner, K.A., Eustice, R.M., Johnson-Roberson, M.: WaterGAN: unsupervised generative network to enable real-time color correction of monocular underwater images. IEEE Robot. Autom. Lett. **3**(1), 387–394 (2017)
12. Manhães, M.M.M., Scherer, S.A., Voss, M., Douat, L.R., Rauschenbach, T.: UUV simulator: a gazebo-based package for underwater intervention and multi-robot simulation. In: OCEANS 2016 MTS/IEEE Monterey. IEEE, September 2016. https://doi.org/10.1109/oceans.2016.7761080
13. McGlamery, B.: A computer model for underwater camera systems. In: Ocean Optics VI, vol. 208, pp. 221–231. International Society for Optics and Photonics (1980)
14. Mobley, C.D.: Light and Water: Radiative Transfer in Natural Waters. Academic Press, San Diego (1994)
15. Petzold, T.J.: Volume scattering functions for selected ocean waters. Technical report, Scripps Institution of Oceanography La Jolla Ca Visibility Lab (1972)
16. Prats, M., Perez, J., Fernández, J.J., Sanz, P.J.: An open source tool for simulation and supervision of underwater intervention missions. In: 2012 IEEE/RSJ international conference on Intelligent Robots and Systems, pp. 2577–2582. IEEE (2012)
17. Schechner, Y.Y., Karpel, N.: Clear underwater vision. In: 2004 Proceedings of the 2004 IEEE Computer Society Conference on Computer Vision and Pattern Recognition, CVPR 2004, vol. 1, p. I. IEEE (2004)
18. Sedlazeck, A., Koch, R.: Simulating deep sea underwater images using physical models for light attenuation, scattering, and refraction (2011)
19. She, M., Song, Y., Mohrmann, J., Köser, K.: Adjustment and calibration of dome port camera systems for underwater vision. In: Fink, G.A., Frintrop, S., Jiang, X. (eds.) DAGM GCPR 2019. LNCS, vol. 11824, pp. 79–92. Springer, Cham (2019). https://doi.org/10.1007/978-3-030-33676-9_6
20. Song, Y., Köser, K., Kwasnitschka, T., Koch, R.: Iterative refinement for underwater 3d reconstruction: application to disposed underwater munitions in the baltic sea. ISPRS - International Archives of the Photogrammetry, Remote Sensing and Spatial Information Sciences XLII-2/W10, pp. 181–187 (2019)
21. Song, Y., Sticklus, J., Nakath, D., Wenzlaff, E., Koch, R., Köser, K.: Optimization of multi-led setups for underwater robotic vision systems. In: Proceedings of the Computer Vision for Automated Analysis of Underwater Imagery Workshop (CVAUI). Springer (2020)
22. Stephan, T., Beyerer, J.: Computer graphical model for underwater image simulation and restoration. In: 2014 ICPR Workshop on Computer Vision for Analysis of Underwater Imagery, pp. 73–79. IEEE (2014)
23. Ueda, T., Yamada, K., Tanaka, Y.: Underwater image synthesis from RGB-D images and its application to deep underwater image restoration. In: 2019 IEEE International Conference on Image Processing (ICIP), pp. 2115–2119. IEEE (2019)

Optimization of Multi-LED Setups for Underwater Robotic Vision Systems

Yifan Song[1(✉)], Jan Sticklus[1], David Nakath[1], Emanuel Wenzlaff[1],
Reinhard Koch[2], and Kevin Köser[1]

[1] GEOMAR Helmholtz Centre for Ocean Research Kiel, Kiel, Germany
{ysong,jsticklus,dnakath,ewenzlaff,kkoeser}@geomar.de
[2] Department of Computer Science, Kiel University, Kiel, Germany
rk@informatik.uni-kiel.de

Abstract. In deep water conditions, vision systems mounted on underwater robotic platforms require artificial light sources to illuminate the scene. The particular lighting configurations significantly influence the quality of the captured underwater images and can make their analysis much harder or easier. Nowadays, classical monolithic Xenon flashes are gradually being replaced by more flexible setups of multiple powerful LEDs. However, this raises the question of how to arrange these light sources, given different types of seawater and-depending-on different flying altitudes of the capture platforms. Hence, this paper presents a rendering based coarse-to-fine approach to optimize recent multi-light setups for underwater vehicles. It uses physical underwater light transport models and target ocean and mission parameters to simulate the underwater images as would be observed by a camera system with particular lighting setups. This paper proposes to systematically vary certain design parameters such as each LED's orientation and analyses the rendered image properties (such as illuminated image area and light uniformity) to find optimal light configurations. We report first results on a real, ongoing AUV light design process for deep sea mission conditions.

Keywords: Underwater imaging system · Illumination optimization

1   Introduction and Previous Work

Camera vision systems are being widely applied in underwater exploration tasks, because optical data provides high resolution and is directly understandable by humans. Additionally, optical underwater data is increasingly evaluated automatically. More details on the developments in underwater imaging are discussed in [6,8]. One important aspect is that more than half of Earth's surface is situated so deep in the ocean that no natural light reaches there. To investigate and map this uncharted terrain of the deep sea with camera systems, underwater vehicles (robots) have to bring their own lights to illuminate the scene. Ocean water strongly scatters visible light and largely absorbs the red end of the visible

© Springer Nature Switzerland AG 2021
A. Del Bimbo et al. (Eds.): ICPR 2020 Workshops, LNCS 12662, pp. 390–397, 2021.
https://doi.org/10.1007/978-3-030-68790-8_30

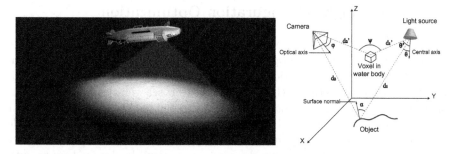

Fig. 1. Left: a robot carrying a camera and lighting system in the deep sea (sketch). Right: geometry components involved in the underwater image formation model [15].

spectrum, depending on the distance the light has to travel through the medium. Thus, the configuration of artificial light sources significantly affects the image quality of the underwater camera system: The lights have to be tuned well to the camera system used, to the desired working distance to the seafloor and to the water properties on specific locations, in order to capture images with ideal quality.

The fundamental physics of light transmission in water have been well studied [10] and the optical properties of different water types have been analyzed (e.g. in [7,11]). Nowadays, it is possible to simulate physical underwater lighting effects and to render realistic underwater images [5,12,15]. The state-of-the-art underwater image formation model by [5] decomposes the captured image into three components: (attenuated) direct signal, forward scattering and backscatter. Practical design principles and issues in underwater imaging systems are discussed in [1,2,5]. Since the underwater imaging quality heavily relies on the lighting conditions, a good relative lighting configuration to the camera is required, as already addressed by [4] and [5]. The properties of a survey path given a particular camera and light system to maximize new information is discussed in [14]. This is useful since underwater robotic missions, especially in the deep sea, are very expensive, so it is desirable to preplan as much as possible and also to adjust and optimize the required lighting configuration already on land. This holds in particular for autonomous underwater vehicles (AUVs, see sketch in Fig. 1) which execute their missions without any user interaction, and therefore also without possibility to tune or adapt the light to the situation on site.

While in early days, single light-source solutions could be built using expert knowledge (e.g. [5]) and possible later human visual inspection of the captured images to adjust the light configuration, this becomes more and more unpractical for more complex systems composed of several light sources. With the advent of systems consisting of many flexibly attachable LEDs, this process has become even more complicated. To alleviate this problem, this paper proposes to predict the deep sea images using an appropriate underwater image simulation and automatically analyses an objective function to find out the optimal multi-LEDs lighting configuration that can be adapted to the mission needs.

2   Multiple Light Configuration Optimization

This paper focuses on finding an optimal configuration for a multi-light design on a deep diving robot. The proposed approach allows optimizing various parameters such as position, orientation, power (or other LED properties such as radiation intensity distribution (RID)), practical deep sea vehicles have also other constraints, such as limited space, predefined mount points that should be far away from the camera, limited power, hydrodynamic properties of the vehicle and so on. This contribution therefore conceptually discusses only how to vary the LED orientations, but this can be seen as an example also for how to optimize other parameters. When the number of light sources is high, the restriction on orientation optimization also reduces the complexity for the grid search algorithm in Sect. 2.3.

2.1   Underwater Image Simulation

Since light setups significantly influence the appearance of the underwater image, the underwater image renderer should be able to simulate the direct signal and backscatter components properly according to different lighting configurations. The renderer should additionally consider the following physical features in order to generate realistic images: (1) Practical light sources are not isotropic point lights, but rather spotlights with a principal direction and an angular characteristic (RID). (2) The wavelength-dependent attenuation model should consider the complete traveling path from light source to object to camera. (3) The scattering component, especially the backscatter, should cope with different lighting configurations.

The renderer used in this paper is proposed by [15], which contains most of the requirements we listed above. We only need to add the radiometric effects caused by the real digital camera in our application. The radiometric model of a real digital camera consists of the optical effect caused by lens systems, and sensor characteristics when it converts analog light intensity to digital pixel values. The optical effect model was adapted from [5,12]. It depends on the transmittance of the camera lens T_l, f-number f_n and the angle between the light ray arriving on the camera lens and the optical center line of the camera ϕ. For the digital signal transformation part, the camera is assumed with a linear response function, additional three other factors $b(\lambda)$, $c(\lambda)$ and k are also considered in the model. $b(\lambda)$ represents the white balance parameters which changes the relative intensity for different color channels and $c(\lambda)$ is the relative sensitivity of the camera for RGB channels. k is the global scale factor that integrates many effects like the absolute irradiance of the light source, ISO and shutter speed of the camera, analog-to-digital (A/D) conversion and so on. The final pixel value I_{pixel} can be computed from the irradiance I (direct signal + scattering components) arriving on the lens of the camera by:

$$I_{pixel} = k \cdot b(\lambda) \cdot c(\lambda) \cdot \frac{cos^4\phi T_l^2}{4f_n} \cdot I. \tag{1}$$

In our work we focus on cameras behind spherical glass domes that can be considered as single-center of projection cameras [13].

2.2 Evaluation Factors

There is no gold standard approach to evaluate the quality of underwater photo illumination. The latter highly depends on the actual details that the photos should depict, what the observer is finally interested in or which analysis algorithm should be applied. Some people might look for subtle color differences of animals, while others would like to maximize the amount of light that falls into the image center. This paper targets generic mapping applications, where the goal is to use as much of the image as possible rather than optimizing for a particular color. Since the overlay of many light cones can produce unpleasant artificial patterns that could be misinterpreted as patterns in the scene, a desirable property of the illumination is smoothness.

This paper defines an energy function that contains several terms which quantify desired properties of the illumination pattern. The pattern is analyzed by rendering a piece of homogeneous seafloor from a predefined altitude and with predefined water properties using a target camera model. The resulting image is then inspected and represents a certain configuration (i.e. setting of illumination directions). Obviously, a high priority goal is to make as many pixels as possible well-lit. Thus, the first factor considered is E_{ratio}, which denotes the ratio of pixels under good illumination (in the moderate intensity range). Besides absolute intensity, the homogeneity of the illumination for each pixel is another goal to be pursued, i.e. high-frequency illumination patterns should be avoided. Two factors are computed to consider the homogeneity of the illumination, one is the local gradient $E_{gradient}$ and another is the entropy of the image $E_{entropy}$. Therefore, the final evaluation energy E as a weighted sum (with empirical weights w_i) is defined as:

$$E = w_1 E_{gradient} + w_2 E_{entropy} - w_3 E_{ratio}. \tag{2}$$

2.3 Optimization Algorithms

Since in a multi-LED system, the energy defined in the last section is nonconvex, this paper proposes a two-step approach for finding a good illumination configuration (see Fig. 2). First a grid is applied to search on the (naturally confined) parameter space (orientations of lights are only useful and physically realisable in a certain interval, it would also be adapted for optimization of light positions on a robot), which generates an image for certain LED tilt angles (with fixed intervals between angles). Each light can also be rotated orthogonal to the tilt, which adds more degrees of freedom to the entire setup. However, for mechanical reasons, the lights are typically mounted in groups, such that lights cannot be rotated independently in all directions. This reduces the dimension of the search space. For a discrete grid of all allowed combinations we calculate the energy values for the corresponding image (grid search, similar to Fig. 3 Left).

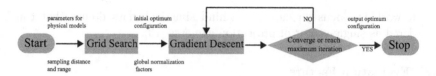

Fig. 2. Workflow of the optimization algorithms.

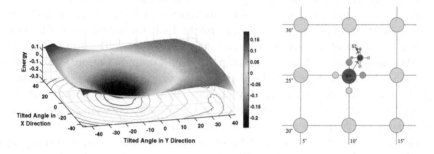

Fig. 3. Left: energy function results of a single light during grid search. (In this case, the camera is located 1 m away from the light source along Y-axis and they all initially point perpendicular to the seafloor with 4 m flying altitude.) Right: optimization strategy in gradient descent steps of a single light.

The grid point corresponding to the image with the lowest energy will then be used as the initial configuration in the next step, which is a local optimization.

For the local optimization in the second step, a variant of simple gradient descent is implemented (Fig. 3 Right): At the best position in parameter space found so far, we numerically compute the gradient of the energy with respect to the lighting configuration parameters, i.e. the energy for small delta steps of all parameters and assemble the gradient from the small differences computed. The interval of the previous grid search on the one hand and the mechanical precision reachable for fixing physical LEDs to a robot provide approximate maximum and minimum useful step size in parameter space. The algorithm then repeats taking a step in gradient direction until convergence (or maximum iterations reached). Many other downhill techniques can also be applied, to completely automatize the light configuration optimization.

3   Implementation and Test Results

The proposed solution is applied to optimize a real deep sea camera-lighting system on an AUV [9]. The system has been developed for flying at high altitudes (up to 10 m) and the goal of our optimization is to find optimal light configurations for a 4 m and 7 m altitude, respectively. The AUV is equipped with a DSLR camera inside a dome-port housing. Twenty-four (4 rows and 6 columns) LEDs are placed 1.9 m away from the camera (see Fig. 4). The images captured with

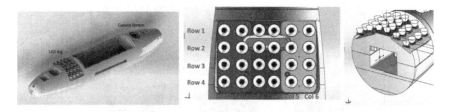

Fig. 4. LED layout on the AUV rig, the camera is located 1.9 m below the rig along the X-axis and its optical axis is along the Z-axis. The initial settings of all LEDs are perpendicular to the surface of the rig.

Fig. 5. Simulation results of the reference image. From left to right: direct signal, back scatter, complete rendering result of the virtual seafloor, the reference image. Since the old LED rig was manually setup in the cruise, the precise light configuration of the reference image is not available. Only rough measurements are used for our simulator and the light pattern in the rendered image will be slightly different from the real one.

the mentioned camera system are available at PANGAEA (https://doi.pangaea. de/10.1594/PANGAEA.881850), which will be used as the reference data in our simulation model.

Details about the parameter set for the renderer are provided in the following: (1) The peak wavelengths with respect to RGB channels are {650, 510, 440} [nm], and the following values for the channels all refer to these wavelengths. (2) Same LEDs as described in [15] are applied. The relative spectrum of the light source is {0.25, 0.35, 0.4} measured by a spectrometer. (3) The expected Jerlov water type in the area is II, with coefficients {0.37, 0.044, 0.035}[m^{-1}]. (4) The seafloor is simulated as a flat Lambertian surface with uniform brown albedo. (5) The camera in the AUV is a Canon EOS 6D with no significant vignetting. Intrinsic parameters are obtained by the standard calibration. White Balance parameters $b(\lambda)$ are embedded in the metadata of the reference image. For Relative Sensitivity $c(\lambda)$, the CIE-D50 Color Response of the Canon EOS 6D can be obtained from internet sources such as [3]. (6) Comparing the reference pixel value with the computed absolute light irradiance value (see Fig. 5) allows us to get the scale factor k in Eq. (1).

For different deep sea exploration tasks, images with certain AUV operating distance above the seafloor are simulated to find the respective optimal light configurations.

In the optimization, the sampling distance of tilt angle was set to 15° ∈ [−30°, 30°] in both x, y directions (see Fig. 4) of each LED. Assume LEDs on each row or column share the same tilt angle and the LED layout is

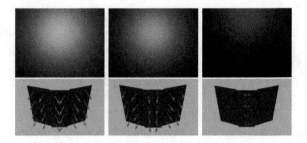

Fig. 6. Final optimization results and their light layout. First row: optimization result with 4 m altitude (left), 7 m altitude (middle) and a bad example of illumination with 7 m altitude (right). Second row: corresponding light layouts which are visualized by the Visualization Toolkit (VTK).

Fig. 7. Left: technical drawing of optimized LED configuration for 4 m flying attitude mission in the Central Pacific. Right: corresponding 3D printed LED rig bases.

symmetric along columns, which continuously reduce the number of grid search samples from 5^{48} to 5^7. The rendering of the scattering effect is by far the most time consuming part, but the lights are fixed relative by far from the camera. Hence, the direct component pattern at the seafloor is changing stronger than the scattering components. Consequently, we temporarily disabled the scattering in the grid search step and only evaluate on the direct component, the scattering will be rendered only during the gradient descent step. The empirical weights $\{0.2, 0.2, 0.6\}$ are applied in the evaluation function (Eq. (2)). Once the initial optimum from the grid search is obtained, the gradient descent was run and always converged before 50 iterations. The final optimal light configurations for 4 m and 7 m flying attitude are illustrated in Fig. 6.

With the optimized lighting configuration, we are able to design the new lighting rigs which can be quickly assembled and disassembled for different mission conditions during the next cruises (see Fig. 7).

4   Conclusion and Further Work

This paper integrates a deep sea imaging simulator into an optimization strategy that automatically synthesizes and analyses rendered images with different light configurations for an AUV. The optimization process then suggested optimal lighting configurations for particular flying altitudes of a robot to improve the image quality for different deep sea exploration tasks and to support the system designers in making decisions. During the grid search step, the rendering of images for each lighting configuration is independent of each other, which could be well parallelizable by using GPU hardware in future work.

Acknowledgements. This publication has been funded by the German Research Foundation (Deutsche Forschungsgemeinschaft, DFG) Projektnummer 396311425, through the Emmy Noether Programme.

References

1. Bonin-Font, F., Burguera, A., Oliver, G.: Imaging systems for advanced underwater vehicles. J. Maritime Res. **8**, 65–86 (2011)
2. Brutzman, D.P., Kanayama, Y., Zyda, M.J.: Integrated simulation for rapid development of autonomous underwater vehicles. In: Proceedings of the 1992 Symposium on Autonomous Underwater Vehicle Technology, pp. 3–10. IEEE (1992)
3. DxOMark: Canon EOS 6D Measurement. http://www.dxomark.com/Cameras/Canon/EOS-6D---Measurements (2013). Accessed Oct 2020
4. Hatchett, G.L.: Optimization of light sources for underwater illumination. In: Underwater Photo Optics I, vol. 7, pp. 150–156. International Society for Optics and Photonics (1966)
5. Jaffe, J.S.: Computer modeling and the design of optimal underwater imaging systems. IEEE J. Oceanic Eng. **15**(2), 101–111 (1990)
6. Jaffe, J.S.: Underwater optical imaging: the past, the present, and the prospects. IEEE J. Oceanic Eng. **40**(3), 683–700 (2014)
7. Jerlov, N.G.: Irradiance optical classification. In: Jerlov, N.G. (ed.) Optical Oceanography. Elsevier Oceanography Series, vol. 5, pp. 118–120. Elsevier (1968)
8. Kocak, D.M., Dalgleish, F.R., Caimi, F.M., Schechner, Y.Y.: A focus on recent developments and trends in underwater imaging. Marine Technol. Soc. J. **42**(1), 52–67 (2008)
9. Kwasnitschka, T., et al.: DeepSurveyCam–a deep ocean optical mapping system. Sensors **16**(2), 164 (2016)
10. Mobley, C.D.: Light and Water: Radiative Transfer in Natural Waters. Academic Press, San Diego (1994)
11. Petzold, T.J.: Volume scattering functions for selected ocean waters. Technical report, Scripps Institution of Oceanography La Jolla Ca Visibility Lab (1972)
12. Sedlazeck, A., Koch, R.: Simulating deep sea underwater images using physical models for light attenuation, scattering, and refraction (2011)
13. She, M., Song, Y., Mohrmann, J., Köser, K.: Adjustment and calibration of dome port camera systems for underwater vision. In: Fink, G.A., Frintrop, S., Jiang, X. (eds.) DAGM GCPR 2019. LNCS, vol. 11824, pp. 79–92. Springer, Cham (2019). https://doi.org/10.1007/978-3-030-33676-9_6
14. Sheinin, M., Schechner, Y.Y.: The next best underwater view. In: Proceedings of the IEEE Conference on Computer Vision and Pattern Recognition, pp. 3764–3773 (2016)
15. Song, Y., Nakath, D., She, M., Elibol, F., Köser, K.: Deep sea robotic imaging simulator. In: Proceedings of the Computer Vision for Automated Analysis of Underwater Imagery Workshop (CVAUI). Springer (2020)

Learning Visual Free Space Detection
for Deep-Diving Robots

Nikhitha Shivaswamy[1,2], Tom Kwasnitschka[1], and Kevin Köser[1(✉)]

[1] GEOMAR Helmholtz Centre for Ocean Research Kiel, Kiel, Germany
{nshivaswamy,tkwasnitschka,kkoeser}@geomar.de
[2] Department of Electrical Engineering and Information Technology,
Paderborn University, Paderborn, Germany

Abstract. Since the sunlight only penetrates a few hundred meters into the ocean, deep-diving robots have to bring their own light sources for imaging the deep sea, e.g., to inspect hydrothermal vent fields. Such co-moving light sources mounted not very far from a camera introduce uneven illumination and dynamic patterns on seafloor structures but also illuminate particles in the water column and create scattered light in the illuminated volume in front of the camera. In this scenario, a key challenge for forward-looking robots inspecting vertical structures in complex terrain is to identify free space (water) for navigation. At the same time, visual SLAM and 3D reconstruction algorithms should only map rigid structures, but not get distracted by apparent patterns in the water, which often resulted in very noisy maps or 3D models with many artefacts. Both challenges, free space detection, and clean mapping could benefit from pre-segmenting the images before maneuvering or 3D reconstruction. We derive a training scheme that exploits depth maps of a reconstructed 3D model of a black smoker field in 1400 m water depth, resulting in a carefully selected, ground-truthed data set of 1000 images. Using this set, we compare the advantages and drawbacks of a classical Markov Random Field-based segmentation solution (graph cut) and a deep learning-based scheme (U-Net) to finding free space in forward-looking cameras in the deep ocean.

Keywords: Water segmentation · Deep learning · Underwater robotics · Deep sea mapping

1 Introduction

The oceans cover a large fraction of our planet; most of Earth's surface even lies in the deep sea, with high pressures, no sunlight nor satellite coverage. Currently, camera-equipped robots are the key effective means to map this unknown area visually. While flat regions, e.g., in the Abyssal plains [5], can be mapped using fixed altitude paths between preprogrammed waypoints, navigation in more complex terrain like hydrothermal vent fields [2], requires either a skilled human operator, or automatic detection of free space and obstacles, or a combined

© Springer Nature Switzerland AG 2021
A. Del Bimbo et al. (Eds.): ICPR 2020 Workshops, LNCS 12662, pp. 398–413, 2021.
https://doi.org/10.1007/978-3-030-68790-8_31

Fig. 1. Black smoker image (left), depth map from 3D reconstruction (center) and free space mask (right) for local navigation or for stabilizing 3D reconstruction/SLAM. The depth map in the center image stores the distance to each point as a floating point number, e.g. the one marked red on the smoker. If we consider everything further away than 10 m as "free space", we obtain the mask to the right. Many raw images (with their masks) can then be used for training the appearance of free space, such that the system can later predict from one image alone where are potential obstacles. Photo from Falkor cruise "Virtual Vents" 03/2016: GEOMAR/CSSF/Schmidt Ocean Institute.

approach. Since underwater light propagation involves scattering and attenuation [10,11,15,16], and for deep-sea robotic missions, also requires to bring (co-moving) light sources, color, and lighting correction is a much more complex task in the deep sea than with homogeneous illumination (as is often the case in shallow water). Even when the navigation problem is solved, visual mapping systems [12] can struggle because of hallucinated structures in the smoke and can create noisy reconstructions with many outliers. Both, navigation and 3D reconstruction could benefit from pre-segmented images, i.e. if the computer could classify the image into regions with closeby structure and closeby empty regions (see Fig. 1). As we envision this as a first online preprocessing step, such a solution should be fast, and it should not require to solve the dense depth estimation problem for the frame at hand as this creates a chicken-and-egg problem, where mapping requires segmentation, but segmentation requires mapping.

In this manuscript, we, therefore, suggest training a deep neural network on deep-sea footage of black smokers, for which we have performed 3D reconstruction with manual guidance. Binary images obtained from the depth maps, encoding how far structures are away from the camera, will be used as masks for training. We compare a recent learning-based approach to a more classical energy-based optimization, where we can explicitly specify weights for smoothness and data term in a Markov-Random-Field type of model and a baseline color segmentation (Gaussian-mixture model).

2   Previous Work and Contributions

Underwater imaging has a long tradition, and the fundamental principles and physical models for light propagation and underwater photography have been proposed already several decades ago [10,11,15,16]. However, for multiple-light source equipped deep-diving robots, underwater 3D reconstruction and visual

SLAM are still challenging [13]. A critical problem for autonomous robots is to identify obstacles [3, 4, 6, 9, 14], where classically systems employ acoustic sensors such as forward-looking sonar (FLS) [3]. However, in optical mapping missions of very complex terrain, the system needs to be very close to the structures requiring close-range mapping [9]. For optical obstacle avoidance, systems based on single-image depth estimation have used the dark channel prior [4] originally proposed for haze removal in air [8] or learn a transmission map [6] or descattering [14]. We approach the problem from another direction since our additional goal is to mask the water and exclude "water areas" in forward-looking images from SLAM and 3D reconstruction. Such regions often create hallucinated correspondences and noise in 3D reconstruction [12]. Therefore we do not want to detect particular objects or structures, but we want to learn the appearance of "empty water" to exclude these areas from the 3D reconstruction. At the same time, this approach can be used as a free-space cue in navigation.

Essentially, this is a segmentation problem (see [23] for a recent survey) that has been addressed in the literature in different ways: In classical pixel-based segmentation, the image is segmented based on color similarity (e.g., using Gaussian Mixture models), which does not guarantee continuous areas. This drawback is overcome by Split-and-Merge approaches that subdivide the image in a coarse-to-fine approach and vice-versa. Other techniques use Markov Random Fields to model each pixel's probability of belonging to the foreground or background based on its color and impose smoothness by setting a high probability that neighboring pixels share the same label. Neighboring pixel similarity based on gray level, color, texture, or distance between the neighbors can influence the smoothness. Such segmentation systems are more costly to solve than pixel-wise, purely color-based systems but can be solved relatively efficiently by graph cut due to their binary nature (see [20] for an overview). Interactive-based graph cut (e.g., GrabCut [18]), is a variant of the problem, letting the user draw a bounding box around an object. As a representative of energy-based segmentation, we choose [18] in this paper, but we only use the energy formulation of [18], whereas the interactive part is only applied for manual labeling of complex training data.

Finally, neural network-based segmentation is quite different from the other techniques. Here, the segmentation is implemented in a network of several layers of neuron-inspired computation units that each weight many input signals to compute an output signal. The network weights are then trained offline by minimizing a loss function that penalizes wrong network predictions on the training data, and the actually learned functionality is encoded in the weights (see [7] for a recent overview). A key drawback is that these networks need a lot of labeled training data, which is difficult for underwater scenarios. However, the training set's size can be increased by applying image warps and intensity transformations to the original training images (augmentation). The U-Net [17] architecture is one instance of these learning approaches that will be inspected in more detail in the next section. Other recent approaches include SegNet [1] or HRNet [21]. After training, the trained model will segment previously unseen test images. The advantages of such networks are that they do not need user

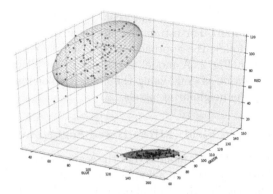

Fig. 2. GMMs in RGB color space: The BG (water) is rather in the blue-green area of the color space whereas the FG components contain more red. Only one component is shown, whereas in the experiments we used a mixture of five Gaussian components to represent the likelihoods for FG and for BG. (Color figure online)

interaction during the segmentation process and can work in parallel (e.g., on a GPU) to increase the computation speed. The downside is that they adapt only to the training data and act as black boxes, such that it is often hard to tell what exactly the network has learned. In this paper, we demonstrate how to prepare data and to solve the free space detection problem on deep sea images. We compare the advantages and drawbacks of two main approaches on a novel data set of 1000 labeled images of a vast hydrothermal vent field. In the next section, we will briefly discuss the layout of the different approaches as well as the baseline for free space detection.

3   Segmentation Approaches

We start by outlining a baseline pixel-wise classification approach using the Gaussian mixture model on pixel colors. Secondly, we describe an MRF-based approach derived from the GrabCut energy [18], and thirdly, we discuss segmentation method based on the U-Net architecture [17]. Lastly, we explain an approach to obtain ground truth since all the approaches mentioned above are supervised. In the following, we refer to water pixels as background (or BG) and black smoker or seafloor pixels as foreground (or FG).

3.1   Gaussian Mixture Models (GMM) in Color Space

Gaussian mixture models are a simple approach for learning FG and BG colors. We use a variant similar in spirit to [22] as a baseline method to compare the performance of the other approaches. For a set of training images, where we know a pixel is FG or BG, we compute the color space likelihoods in RGB space, i.e., the probability for the specific color of the pixel in case it belongs to FG or BG. These likelihoods are represented as a sum of several Gaussians (Gaussian Mixture Model). For estimating the means and covariance matrices of the Gaussians, we follow the approach of [18]. Sample clusters in RGB color space are displayed in Fig. 2. A model is trained using all pixels in all training images; the prediction is done using the trained color model for each pixel without considering the neighboring pixels.

3.2   Markov Random Field (MRF) Based Segmentation

A drawback of using the GMMs alone is that the segmentation is often very noisy. Introducing a smoothness term can help to overcome this drawback. MRFs allow us to specify the probability that neighboring pixels share the same label (FG or BG) and support a data term that considers how compatible is a pixel with the FG or BG color likelihood. We base the MRF-based segmentation on the energy optimization derived in GrabCut [18]. We train the color model similar to the previous approach and also include the smoothness term used in [18] to encourage continuous regions. Effectively, the colors are learned from the training images, but the structure is specified manually through the energy function.

3.3   Deep Learning Based Approach

Exemplarily, we discuss the U-Net architecture [17] for the free space segmentation. A basic U-net architecture consists of 23 convolutional layers in the encoder or contraction path and the decoder or expansion path (see [17] for more detailed information). In the Encoder, the size of the image gradually reduces while the depth gradually increases (double the number of feature channels). In the decoder, the size of the image gradually increases, and the depth gradually decreases (half the number of feature channels). However, the output layer has a 1×1 convolution that maps each 64-component feature vector to the number of classes, giving us the segmented mask. The two major observations for the U-net architecture are:

1. Pooling operators are supplemented by successive upsampling operators, which increase the size of the output mask. This upsampled output is combined with high-resolution features from the Encoder path to enable the convolutional layers to gather a more accurate output.
2. The size of the output mask is the same as the input image, but U-net does not demand a fixed size of the input image.

To better adapt to our setting, we modify U-Net as outlined below inheriting the basic U-Net architecture:

1. Increase the number of convolutional layers from two 3×3 to three 3×3 in every block to increase the feature extraction layers.
2. Increase the size of all convolutional kernels from 3×3 to 5×5 or 7×7 to explore the variation in the smoothness of the edges.

Also, we vary the input image size to match the aspect ratio of training images.

3.4 Efficiently Obtaining Masks for Training

We used three methods to label the training data, namely:

1. **Estimated depth maps:** Depth maps contain distance information of how far the surface of an object is from the camera's viewpoint. They can be obtained from 3D reconstruction, such as [12] by dense matching of images. For underwater, and in particular deep-sea reconstruction, generating 3D models and depth maps is still a challenging problem, but in those areas where we automatically obtain such depth maps, they automatically segment many images. In Fig. 1 the marked pixel's estimated depth is represented by the floating-point number in the grey value image. We set a threshold depth and consider everything further away to be the BG (water resp. free space). Similarly, all the pixels within the considered threshold would be considered FG/smoker.
2. **GrabCut:** As described above, depth maps are only available for some images. We also generate segmentation masks from applying the GrabCut algorithm on training images. GrabCut performs well on images containing prominent FG and BG color distributions. In the case of images with high color distortion, GrabCut does not yield accurate results. We manually verified all the automated labelings and corrected where necessary.
3. **Manual Labelling:** Images which are difficult to be labeled using GrabCut were manually labeled.

Overall we obtain a set of 1000 labeled images, which we subdivide into a subsets of training images with different sizes and an independent test set of 100 images that do not include any training images (or very similar images) (Fig. 3).

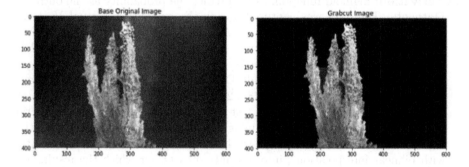

Fig. 3. Example of a poorly segmented image (water between arms of the smoker). When generating training data, such masks have to be reworked manually.

4   Implementation Details

4.1   Model Hyperparameters

Hyperparameters play a vital role in the model training process, and the challenge of tuning these values depends on each task and each dataset. The training of the U-Net has been performed using these settings:

1. **Learning rate (LR):** Very small LR can take plenty of epochs to find the local minima, whereas a large LR can skip the local minima. Hence, we use a callback function from Keras called "ReduceLROnPlateau" that monitors the dice coefficient value. LR is reduced if the coefficient value does not improve for a certain number of epochs.
2. **Batch size (BS):** Very large BS will require enormous memory, whereas a very small BS will increase the noise in the error calculations. Hence we increase the BS as we increase the dataset size.
3. **Number of Epochs:** We use a callback function from Keras called "EarlyStopping" to monitor the dice coefficient value. Training stops if there is no improvement in the coefficient value for a certain number of epochs.

4.2   Other Details

1. **Optimizer:** We use the Adam optimizer, which is a combination of RMSprop and the Stochastic Gradient Descent method with adaptive estimation of first-order and second-order moments.
2. **Loss:** We use Binary Cross-Entropy that calculates the probability distribution between the pixels belonging to foreground and background sets; Dice Loss [19] finds similarities between two pixels.
3. **Activation:** Sigmoid function, as we predict the probability as the output.
4. **Datasets:** The data was captured during the expedition Virtual Vents in 2016 in the southwest Pacific (Niua). A Remotely Operated Vehicle (ROV), which carried high-resolution survey cameras coupled with powerful LED flash, photographed a 500m diameter area hydrothermal vent field (images courtesy GEOMAR/CSSF/Schmidt Ocean Institute). 900 of such images, downsampled from resolution 3648×2432 have been used as a training set (see Fig. 4) and 100 images as the test set (see Fig. 5). We conducted several experiments to derive the best training results. In this process, we divide the big dataset of 1000 images, image size 448×320 (maintaining the aspect ratio with bilinear interpolation), into multiple subsets, for which the results are reported in the experimental section.
   (a) **Using depth maps vs. GrabCut results as segmentation masks:** We trained the model with 100images once using the depth maps and once with GrabCut generated masks to compare the accuracy obtained from these models.
   (b) **Dividing the big dataset into various small subsets:** We trained the model using 25, 50, 100, 150, 300, 600, 900image datasets to observe the learning curve as the dataset size increases.

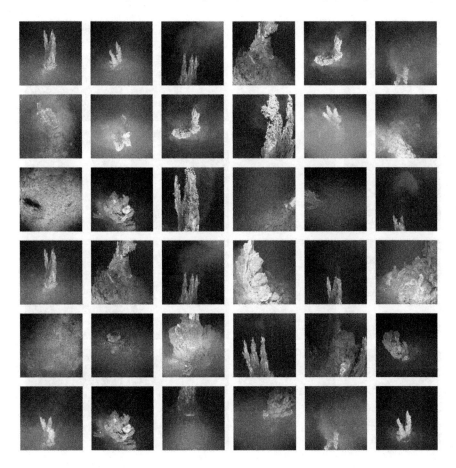

Fig. 4. Some training images (Images: GEOMAR/CSSF/Schmidt Ocean Institute)

(c) **Data Augmentation:** We trained the model with 25images, once with data augmentation, and once without augmentation to observe the importance of augmentation. Augmentation included increasing the brightness of the images, flipping, scaling, rotating the images, and finally adding Gaussian Noise.

(d) **Change Image size:** As mentioned earlier, we trained our model with image size 448×320 in the above experiments. To examine the image resolution interference in the training process, we trained the model with 256×256 images (resized with bilinear interpolation).

Fig. 5. Some test images (Images: GEOMAR/CSSF/Schmidt Ocean Institute).

(e) **Change filter size:** We trained the model with 900images once with convolution kernel size 3×3 and with 5×5 to find the best kernel size for our images. We also added two 3×3 kernels to the U-net architecture instead of using one 5×5 kernel.

5. **ML framework:** Tensorflow 2.x, Keras (python Deep Learning API)
6. **IDE:** Jupyter Notebook
7. **Computer vision libraries:** OpenCV and various python libraries.

5   Evaluation

5.1   Influence of CNN Parameters

1. **Data Augmentation vs. No Data Augmentation:** In Fig. 4, we observe the diversity of the training images. Most of the images are quite bright (illuminated by multiple high-intensity lights on the ROV, required due to the darkness at that depth). However, the training set also consists of images with lower light conditions. Hence, augmentation techniques to brighten the low light images and darken the bright images, along with flip, scale and rotate techniques, not only increases the amount and variety of the data set but also make our model robust to slight variations in the test set and finally prevents the model from overfitting. From the Table 1, we can infer that the Test dice coefficient value is higher for the model trained using augmentation than the model trained without using augmentation techniques.

Table 1. CNN experiment details and results

Dataset size	Image size	Batch size	Filter sizes	Augmentation	Validation Dice Coef	Test Dice Coef
25 images	448×320	2	3×3	Yes	0.8311	0.4058
25 images	448×320	2	3×3	No	0.7238	0.3369
50 images	448×320	2	3×3	Yes	0.8007	0.8552
100 images	448×320	4	3×3	Yes	0.7589	0.8908
150 images	448×320	4	3×3	Yes	0.8446	0.8965
300 images	448×320	4	3×3	Yes	0.8412	0.9163
300 images	448×320	4	5×5	Yes	0.8520	0.9249
300 images	448×320	4	7×7	Yes	0.8666	0.9266
600 images	448×320	6	3×3	Yes	0.7850	0.8907
600 images	448×320	6	two 3×3	Yes	0.8081	0.9088
900 images	448×320	8	3×3	Yes	0.7934	0.9295
900 images	256×256	8	3×3	Yes	0.8446	0.9162

2. **3×3 vs. 5×5 vs. 7×7 filter size:** From Table 1, we can infer that 7×7 filter size yielded us the best results, although smaller kernels (e.g. 3×3) extract detailed and complex features. Figure 6 depicts the prediction results of 3×3, 5×5 and 7×7 filter size trained models respectively. The model trained on a 7×7 filter has learned and extracted features along with edge information very close to the ground truth.
3. **One 5×5 Kernel vs. Two 3×3 Kernels:** Replacing a 5×5 convolution layer with two 3×3 convolution layers reduces the parameter count by sharing the weights and reuses the activations between the adjacent tiles. From Table 1, we can deduce that this setup performs better. Training 25images

using one 5×5 convolution takes 7 m 7 s whereas, training 25images using two 3×3 convolution takes 6m due to reduced parameters and operations.

4. **Dataset Sizes:** From Table 1, we can infer that as the diversity of training images increases, the model learns better features and segments the FG from the BG more accurately. There is a drastic improvement in the dice value from 25images trained model to 900images trained model. However, we observe only a subtle improvement from the 300 to 900images trained model. That could indicate that beyond 300images, no substantial extra information is added in our training set.

5. **Training Image Size 448,320 vs. 256,256:** A model trained on the slightly larger resolution images performs better than that of low resolution. The reason might simply be the reduced number of data points since the images are already quite small. Also, the rotation with the wrong aspect ratio produces less realistic training data than when rotating the correct image (rotation and stretch are not commutative).

Fig. 6. Filter size: Test image, ground truth, 3×3, 5×5, 7×7

5.2   Training Time and Prediction Time

This section provides numbers for how long it took on our Workstation (intel Xeon 3.5 GHz, 64 GB RAM) with Quadro K4000 GPU to train and predict. These times have not been optimized and are just reported to document the order of magnitude for the runtime.

Table 2. Computation time of the both methods

Segmentation Method	Number of Images	Training Time	Test Time 100 images
CNN	25	4 min 19 s	44 s
CNN	300	78 min 95 s	44 s
Graph cut	25	1 min 3 s	42.9 s
Graph cut	300	15 min 7 s	42.9 s

Training time is the main caveat of CNN. Our model, trained on 900images with input image size 448×320, takes four times more time to train than the

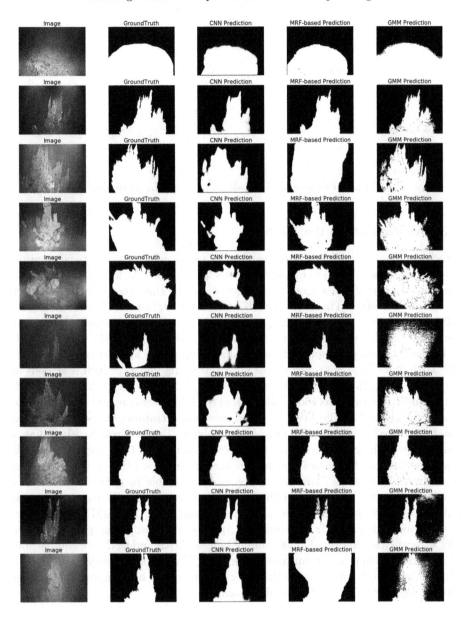

Fig. 7. Exemplary predictions of all the methods

MRF approach Table 2. That is clearly due to a high number of hidden layers in the network, and unlike the MRF, CNN is a feedforward learning approach where there are multiple forward and backward passes. However, this Table is shown more from a practitioner's point of view; the learning process of both methods is too distinct to compare the computation time in detail.

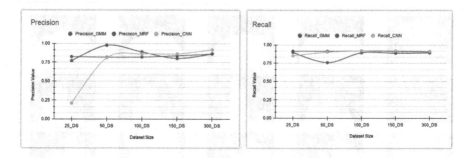

Fig. 8. Precision and Recall graph of all the methods.

5.3   Analysis of All the Methods

Precision reports the fraction of correct FG predictions, whereas recall reports how many of the actual FG pixels were detected. The graphs in Fig. 8 depict such precision and recall values, on the test set, of all the three methods with respect to the dataset sizes. The CNN model, with augmentation and input image size 448×320, trained on only 25images has the least precision, which shows that the model could not extract enough features and predicted BG pixels as FG. Nonetheless, both precision and recall values of the CNN are getting higher than the MRF and GMM approach values for 300 training images, which is evident in Fig. 7. However, the overall performance of the MRF (and even GMM) approach is not unsatisfactory. Precision and recall of the GMM are constant. Based on the graph cut algorithm, the MRF required more color training to counter-balance the hard-coded structures for neighboring pixels. The last row in Fig. 7 has a test image with unclear FG, as the smoker is quite far from the camera. We can observe that the neural network outperformed the other two methods in segmenting this image. The GMM approach is unable to decide FG and BG pixels close to the smoker, and the MRF result is completely inappropriate in this case. However, in the some of the predictions, MRF and GMM approaches have segmented sharp edges as both the approaches classify each pixel as BG or FG. Overall, for a low number of training images, MRF and CNN are comparable in performance and online runtime, while the CNN breaks for very low numbers of training data. It outperforms the other methods for a high number of training data at the cost of long offline training times.

5.4   SLAM/SfM Reconstruction

In Fig. 9, we compare an unconstrained 3D reconstruction (left) and a reconstruction where we guide the structure-from-motion system by masks obtained from U-Net segmentation. We observe that parts, where the smoke is close to the smoker appear cleaner, but this is only a subjective impression since we have no ground truth information about the 3D model. Nevertheless, the approach avoids mismatches in the water during the sparse 3D reconstruction phase.

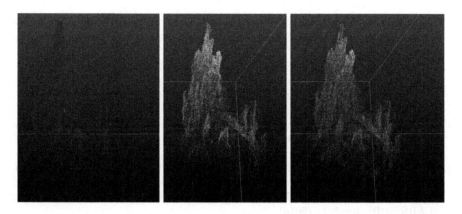

Fig. 9. 3D model: without using mask, using mask and comparison of both

6 Conclusion

We have shown how to prepare data in order to learn water versus seafloor structure characteristics and segment previously unseen images with high accuracy. Free space was segmented with good accuracy and can also support 3D reconstruction. Classical MRF-based segmentation achieves similar performance with the structural model hard-wired as compared to the fully learned approach, when the training set is rather small or even outperforms the neural network on our smallest training set size of 25images. Both are preferable to the local color-only approach because they provide continuous regions rather than isolated pixels. As expected, the neural network approach shows more asymmetry in the required time for training and application, but both systems would be fast enough to run at interactive rates on a real robot without the need to compute detailed depth maps in advance.

Acknowledgements. This work has been funded by the German Research Foundation (Deutsche Forschungsgemeinschaft, DFG) Projektnummer 396311425 (DEEP QUANTICAMS), through the Emmy Noether Programme. We would also like to thank the ROPOS team and the crew of RV Falkor, as well as Schmidt Ocean Institute, for supporting the cruise "Virtual Vents" to the Niua South Hydrothermal Vent Field.

References

1. Badrinarayanan, V., Kendall, A., Cipolla, R.: Segnet: A deep convolutional encoder-decoder architecture for image segmentation. IEEE Trans. Pattern Anal. Mach. Intell. **39**, 2481–2495 (2017)
2. Baker, E.T., German, C.R.: On the global distribution of mid-ocean ridge hydrothermal vent-fields. Am. Geophys. Union Geophys. Monograph **148**, 245–266 (2004)

3. Braginsky, B., Guterman, H.: Obstacle avoidance approaches for autonomous underwater vehicle: Simulation and experimental results. IEEE J. Oceanic Eng. **41**(4), 882–892 (2016)
4. Drews, P., Hernández, E., Elfes, A., Nascimento, E.R., Campos, M.: Real-time monocular obstacle avoidance using underwater dark channel prior. In: 2016 IEEE/RSJ International Conference on Intelligent Robots and Systems (IROS), pp. 4672–4677 (2016)
5. Erik Simon-Lledó, E., et al.: Biological effects 26 years after simulated deep-sea mining. Scientific Reports **8040**(9) (2019). https://doi.org/10.1038/s41598-019-44492-w
6. Gaya, J.O., Gonçalves, L.T., Duarte, A.C., Zanchetta, B., Drews, P., Botelho, S.S.C.: Vision-based obstacle avoidance using deep learning. In: 2016 XIII Latin American Robotics Symposium and IV Brazilian Robotics Symposium (LARS/SBR), pp. 7–12 (2016)
7. Goodfellow, I., Bengio, Y., Courville, A.: Deep Learning. MIT Press (2016). http://www.deeplearningbook.org
8. He, K., Sun, J., Tang, X.: Single image haze removal using dark channel prior. IEEE Trans. Pattern Anal. Mach. Intell. **33**(12), 2341–2353 (2011)
9. Hernández, J.D., et al.: Autonomous underwater navigation and optical mapping in unknown natural environments. Sensors **16**(8) (2016). https://www.mdpi.com/1424-8220/16/8/1174, https://doi.org/10.3390/s16081174
10. Jaffe, J.S.: Computer modeling and the design of optimal underwater imaging systems. IEEE J. Oceanic Eng. **15**(2), 101–111 (1990). https://doi.org/10.1109/48.50695
11. Jerlov, N.G.: Marine Optics. Elsevier Scientific Publishing Company (1976)
12. Jordt, A., Köser, K., Koch, R.: Refractive 3d reconstruction on underwater images. Methods in Oceanography **15–16**, 90–113 (2016). https://doi.org/10.1016/j.mio.2016.03.001, http://www.sciencedirect.com/science/article/pii/S2211122015300086
13. Köser, K., Frese, U.: Challenges in underwater visual navigation and SLAM. In: Kirchner, F., Straube, S., Kühn, D., Hoyer, N. (eds.) AI Technology for Underwater Robots. ISCASE, vol. 96, pp. 125–135. Springer, Cham (2020). https://doi.org/10.1007/978-3-030-30683-0_11
14. Li, Y., Lu, H., Li, J., Li, X., Li, Y., Serikawa, S.: Underwater image descattering and classification by deep neural network. Comput. Electr. Eng. **54**, 68–77 (2016). https://doi.org/10.1016/j.compeleceng.2016.08.008, http://www.sciencedirect.com/science/article/pii/S0045790616302075
15. McGlamery, B.L.: Computer analysis and simulation of underwater camera system performance. Technical report, Visibility Laboratory, Scripps Institution of Oceanography, University of California in San Diego (1975)
16. Mobley, C.D.: Light and Water: Radiative Transfer in Natural Waters. Academic Press, San Diego (1994)
17. Ronneberger, O., Fischer, P., Brox, T.: U-net: convolutional networks for biomedical image segmentation. In: Navab, N., Hornegger, J., Wells, W.M., Frangi, A.F. (eds.) Medical Image Computing and Computer-Assisted Intervention - MICCAI 2015, pp. 234–241. Springer, Cham (2015). https://doi.org/10.1007/978-3-319-24574-4_28
18. Rother, C., Kolmogorov, V., Blake, A.: "grabcut": interactive foreground extraction using iterated graph cuts. ACM Trans. Graph. **23**(3), 309–314 (2004). https://doi.org/10.1145/1015706.1015720

19. Tustison, N., Gee, J.: Introducing dice, Jaccard, and other label overlap measures to itk. Insight J. (2009)
20. Yi, F., Moon, I.: Image segmentation: a survey of graph-cut methods. In: International Conference on Systems and Informatics (ICSAI2012), pp. 1936–1941 (2012)
21. Yuan, Y., Chen, X., Wang, J.: Object-contextual representations for semantic segmentation. In: Vedaldi, A., Bischof, H., Brox, T., Frahm, J.-M. (eds.) ECCV 2020. LNCS, vol. 12351, pp. 173–190. Springer, Cham (2020). https://doi.org/10.1007/978-3-030-58539-6_11
22. Chuang, Y.-Y., Curless, B., Salesin, D.H., Szeliski, R.: A Bayesian approach to digital matting. In: Proceedings of the 2001 IEEE Computer Society Conference on Computer Vision and Pattern Recognition. CVPR 2001, vol. 2, p. II (2001)
23. Zaitoun, N.M., Aqel, M.J.: Survey on image segmentation techniques. Procedia Comput. Sci. **65**, 797–806 (2015)

Removal of Floating Particles from Underwater Images Using Image Transformation Networks

Lei Li[1], Takashi Komuro[1]($\boxtimes$), Koichiro Enomoto[2], and Masashi Toda[3]

[1] Saitama University, Saitama 338-8570, Japan
lilei@is.ics.saitama-u.ac.jp, komuro@mail.saitama-u.ac.jp
[2] The University of Shiga Prefecture, Hikone 522-8533, Japan
enomoto.k@e.usp.ac.jp
[3] Kumamoto University, Kumamoto 860-8555, Japan
toda@cc.kumamoto-u.ac.jp

Abstract. In this paper, we propose three methods for removing floating particles from underwater images. The first two methods are based on Generative Adversarial Networks (GANs). The first method uses CycleGAN which can be trained with an unpaired dataset, and the second method uses pix2pixHD that is trained with a paired dataset created by adding artificial particles to underwater images. The third method consists of two-step process – particle detection and image inpainting. For particle detection, an image segmentation neural network U-Net is trained by using underwater images added with artificial particles. Using the output of U-Net, the particle regions are repaired by an image inpainting network Partial Convolutions. The experimental results showed that the methods using GANs were able to remove floating particles, but the resolution became lower than that of the original images. On the other hand, the results of the method using U-Net and Partial Convolutions showed that it is capable of accurate detection and removal of floating particles without loss of resolution.

Keywords: Underwater imagery · Noise removal · Generative adversarial networks · Image segmentation · Image inpainting

1 Introduction

In fishing industry, resource surveys using underwater images are widely conducted. Since the result has a large influence on the preservation of fishery resource and fishing, it is required to improve the accuracy. However, underwater images are sometimes unclear due to floating particles such as plankton and seaweed and it is required to obtain clear images for accurate survey. To solve this problem, some methods of underwater image color correction [1, 2], underwater image enhancement [3, 4], underwater image restoration [5, 6], and underwater image dehazing [7, 8] have been proposed. Although these methods can improve the quality of underwater images under certain circumstances, they can hardly remove big and dense floating particles in underwater images. When images are captured in the water, various shapes and sizes of particles

© Springer Nature Switzerland AG 2021
A. Del Bimbo et al. (Eds.): ICPR 2020 Workshops, LNCS 12662, pp. 414–421, 2021.
https://doi.org/10.1007/978-3-030-68790-8_32

diffuse in front of the camera, which affects the quality of the images as shown in Fig. 1. Therefore, it is necessary to remove the floating particles in underwater images, which would be a great help for fisheries resource surveys.

Some studies have achieved good results in removing raindrops [9–11] and snow [12–14] in images. Unlike weather images, the appearance of floating particles changes in various ways. Therefore, the methods for the removal of raindrops and snow may not be effective for the removal of underwater floating particles.

To remove floating particles in underwater images, some methods using deep neural networks have been proposed. Jiang et al. [15] proposed a method using a generative adversarial network (GAN) for removing floating particles in underwater images. However, it synthesizes spot noisy images by using Photoshop to create training images. Moreover, the test images only contain a small amount of floating particles, and some of them remained in the removal results. Michal et al. [16] proposed a method using a fully convolutional 3D neural network to detect the positions of marine snow and then a median filter is applied to remove the marine snow. This method also used time-consuming manual annotation, and the test images also only contain a small amount of floating particles.

In this study, we propose three methods for removing floating particles from underwater images which do not require time-consuming annotation and that can remove a plenty of floating particles. The first two methods are based on Generative Adversarial Networks (GANs). The first method uses CycleGAN [17] which can be trained with an unpaired dataset. The second method uses pix2pixHD [18] that is trained with a paired dataset created by adding artificial particles to underwater images. Since the methods using GANs may cause a decreased resolution, we also propose the third method that consists of two-step process – particle detection and image inpainting. For particle detection, an image segmentation neural network U-Net [19] is trained by using a dataset of underwater images added with artificial particles. Using the output of U-Net, the particle regions are repaired by an image inpainting network Partial Convolutions [20]. We show the result of an experiment using these three methods and compare the quality of the output images.

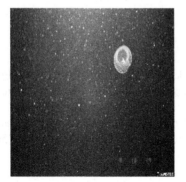

Fig. 1. Examples of underwater images with floating particles.

2 Method Description

CycleGAN with Unpaired Data. CycleGAN [17] is a deep neural network that can transform images from one style to another. CycleGAN does not require a dataset with one-to-one correspondence. Since it is difficult to prepare a paired dataset of underwater images with and without floating particles, the benefit of using CycleGAN is that we do not have to prepare a paired dataset.

pix2pixHD with Paired Data. pix2pixHD [18] has also the ability of image style transformation, but unlike CycleGAN, it requires a paired dataset before and after transformation. Therefore, we generate training images by adding artificial particles to underwater images without floating particles.

U-Net and Partial Convolutions. To remove a large number of floating particles and keep the resolution of input images, we propose the third method that combinates U-Net [19] and Partial Convolutions [20]. A graphical overview of this method is shown in Fig. 2. The first step of this method is to detect the regions of floating particles using U-Net, and the second step is to repair the particle regions using Partial Convolutions.

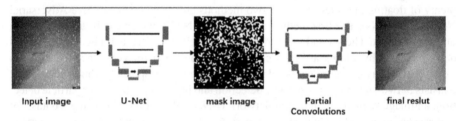

| Input image | U-Net | mask image | Partial Convolutions | final reslut |

Fig. 2. Method using U-net and Partial Convolutions.

3 Experimental Study

3.1 Dataset

All the underwater images in the dataset were downloaded from J-EDI (JAMSTEC e-library of deep-sea images) database. The dataset is divided into underwater images without and with floating particles. There are 1263 images without particles and 1273 underwater images with particles. Examples of images without and with particles are shown in Fig. 3 and Fig. 4, respectively. All images in the dataset were resized to 256 × 256 pixels and the aspect ratio were converted to 1: 1.

Fig. 3. Examples of underwater images without floating particles in J-EDI database.

Fig. 4. Examples of underwater images with floating particles in J-EDI database.

3.2 Training Details

CycleGAN. Since CycleGAN only needs an unpaired dataset for training, we used 1263 underwater images without floating particles and 1273 underwater images with floating particles in the J-EDI database. During the training process, we used the Adam optimizer with a batch size of 1. All networks were trained from scratch with a learning rate of 0.0002. We kept the same learning rate for the first 100 epochs and linearly decreased the rate to zero over the next 100 epochs.

pix2pixHD. Since pix2pixHD needs a paired dataset for training, we used 1263 underwater images without particles in the J-EDI database, and the same number of underwater images with particles that were created by adding artificial particles to the images without particles. We created an empty image, added with randomly generated patterns with different shapes, sizes and colors that are similar to those of natural floating particles, and then applied Gaussian filter on the image. By adding the created image to an underwater image, an image with artificial particles was obtained as shown in Fig. 5. Same as CycleGAN, we used the Adam optimizer with a batch size of 1. The networks were trained from scratch with a learning rate of 0.0002. We kept the same learning rate for the first 100 epochs and linearly decreased the rate to zero over the next 100 epochs.

U-Net for Particle Region Detection. U-Net can detect and segment the target region of an image. Since U-Net also requires a paired dataset, we created underwater images with artificial particles using the same way as above. However, the number of data was not enough for training U-Net, we applied data augmentation and the number was increased to 11367. Consequently, we trained the network using 11367 underwater images with artificial particles and the same number of corresponding mask images which were

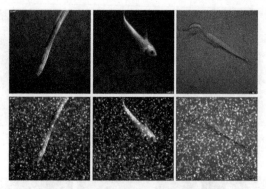

Fig. 5. Examples of underwater images in the paired dataset. Top row: underwater images without particles, Bottom row: Underwater image with artificial particles.

obtained by binarizing the particle images. For training U-net, we used the Adam optimizer with a batch size of 4. The networks were trained from scratch with a learning rate of 0.00005 and the training process lasted for 7 epochs.

Partial Convolutions for Image Inpainting. Partial Convolutions can repair the missing part in an image. We trained the network using 11367 underwater images without particles, and the same number of mask images, which were used for the training of U-Net. For training Partial Convolutions, we used the Adam optimizer with a batch size of 4. The networks were trained from scratch with a learning rate of 0.0004 and the training process lasted for 175 epochs.

3.3   Results

Results of Floating Particle Removal. The results using the three methods are show in Fig. 6. For the images with a small amount of floating particles, both CycleGAN and pix2pixHD were able to remove floating particles, but the outputs of pix2pixHD were better than CycleGAN. However, for the images with a large amount of floating particle, either pix2pixHD or CycleGAN were only able to remove a part of floating particles. Besides, all the removal results using pix2pixHD and CycleGAN had the problem of resolution degradation compared with the original input images. On the other hand, the results using U-Net and Partial Convolutions showed good removal performance of floating particles without loss of resolution.

Image Quality Assessment. We created a synthetic test image with artificial particles for assessing image quality of the three methods. The original image, the synthetic image with artificial particles, and the outputs of the three methods are shown in Fig. 7. SSIM (Structural Similarity Index Measure) and PSNR (Peak Signal-to-Noise Ratio), which are commonly-used full-reference image metrics, were employed to evaluate the quality of the output images. Higher values of PSNR and SSIM represent better removal performance. As shown in Table 1, the method using U-net + Partial Convolutions excellently removed floating particles and acquired higher scores of PSNR and SSIM metrics than the other two methods.

(a)   (b)   (c)   (d)

Fig. 6. Results of floating particles removal. (a) input underwater images with floating particles, (b) output of CycleGAN, (c) output of pix2pixHD, (d) output of U-Net + Partial Convolutions.

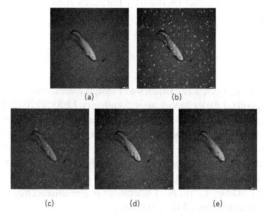

Fig. 7. Results using a synthetic image: (a) ground truth image, (b) synthetic test image with artificial particles, (c) output of CycleGAN, (d) output of pix2pixHD, (e) output of U-net + Partial Convolutions.

Table 1. PSNR/SSIM results for the three methods on a synthetic test image.

Method	PSNR [dB]	SSIM
Synthetic test image	24.540	0.740
CycleGAN	25.375	0.773
pix2pixHD	28.322	0.771
U-net + Partial Convolutions	32.157	0.858

4   Conclusions

In this paper, we proposed three methods for removing floating particles from underwater images. The first two methods are based on Generative Adversarial Networks (GANs), and the third method uses U-Net for particle detection and Partial Convolutions for image inpainting. Experimental results showed that the methods using GANs were basically able to remove floating particles, but the resolution of the output images became lower than that of the original images. However, the third proposed method using U-Net and Partial Convolutions was capable of accurate detection and removal of floating particles without loss of resolution.

Future work includes removal of floating particles from underwater video images using 3D neural networks.

References

1. Lee, H.S., Moon, S.W., Eom, I.K.: Underwater image enhancement using successive color correction and superpixel dark channel prior. Symmetry **12**(8), 1220 (2020)
2. Yeh, C.H., Huang, C.H., Lin, C.H.: Deep learning underwater image color correction and contrast enhancement based on hue preservation. In: 2019 IEEE Underwater Technology (UT), pp. 1–6 (2019)
3. Li, C., et al.: An underwater image enhancement benchmark dataset and beyond. IEEE Trans. Image Process. **29**, 4376–4389 (2019)
4. Zhang, S., Wang, T., Dong, J., Yu, H.: Underwater image enhancement via extended multi-scale Retinex. Neurocomputing, **245**, 1–9 (2017)
5. Zhang, M., Peng, J.: Underwater image restoration based on a new underwater image formation model. IEEE Access, 58634–58644 (2018)
6. Li, C., Guo, J., Chen, S., Tang, Y., Pang, Y., Wang, J.: Underwater image restoration based on minimum information loss principle and optical properties of underwater imaging. In: 2016 IEEE International Conference on Image Processing (ICIP), pp. 1993–1997 (2016)
7. Li, C.Y., Guo, J.C., Cong, R.M., Pang, Y.W., Wang, B.: Underwater image enhancement by dehazing with minimum information loss and histogram distribution prior. IEEE Trans. Image Process. **25**(12), 5664–5677 (2016)
8. Hou, G., Li, J., Wang, G., Pan, Z., Zhao, X.: Underwater image dehazing and denoising via curvature variation regularization. Multimedia Tools Appl. **79**(27), 20199–20219 (2020)
9. Hao, Z., You, S., Li, Y., Li, K., Lu, F.: Learning from synthetic photorealistic raindrop for single image raindrop removal. In: Proceedings of the IEEE International Conference on Computer Vision Workshops, p. 0 (2019)

10. Luo, W., Lai, J., Xie, X.: Weakly supervised learning for raindrop removal on a single image. IEEE Trans. Circuits Syst. Video Technol. (2020)
11. Lin, J., Dai, L.: X-net for single image raindrop removal. In: 2020 IEEE International Conference on Image Processing (ICIP), pp. 1003–1007 (2020)
12. Li, Z., et al.: Single image snow removal via composition generative adversarial networks. IEEE Access **7**, 25016–25025 (2019)
13. Huang, S.C., Jaw, D.W., Chen, B.H., Kuo, S.Y.: Single image snow removal using sparse representation and particle swarm optimizer. ACM Transa. Intell. Syst. Technol. (TIST) **11**(2), 1–15 (2020)
14. Patel, K.F., Tatariw, C., MacRae, J.D., Ohno, T., Nelson, S.J., Fernandez, I.J.: Soil carbon and nitrogen responses to snow removal and concrete frost in a northern coniferous forest. Can. J. Soil Sci. **98**(3), 436–447 (2018)
15. Jiang, Q., Chen, Y., Wang, G., Ji, T.: A novel deep neural network for noise removal from underwater image. Signal Process. Image Commun. **87**, 115921 (2020)
16. Koziarski, M., Cyganek, B.: Marine snow removal using a fully convolutional 3D neural network combined with an adaptive median filter. In: Zhang, Z., Suter, D., Tian, Y., Branzan Albu, A., Sidère, N., Jair Escalante, H. (eds.) ICPR 2018. LNCS, vol. 11188, pp. 16–25. Springer, Cham (2019). https://doi.org/10.1007/978-3-030-05792-3_2
17. Zhu, J.Y., Park, T., Isola, P., Efros, A.: Unpaired image-to-image translation using cycle-consistent adversarial networks. In: Proceedings of the IEEE International Conference on Computer Vision, pp. 2223–2232 (2018)
18. Wang, T.C., Liu, M.Y., Zhu, J.Y., Tao, A., Kautz, J., Catanzaro, B.: High-resolution image synthesis and semantic manipulation with conditional GANs. In: Proceedings of the IEEE Conference on Computer Vision and Pattern Recognition, pp. 8798–8807 (2018)
19. Ronneberger, O., Fischer, P., Brox, T.: U-net: convolutional networks for biomedical image segmentation. In: Navab, N., Hornegger, J., Wells, W.M., Frangi, A.F. (eds.) MICCAI 2015. LNCS, vol. 9351, pp. 234–241. Springer, Cham (2015). https://doi.org/10.1007/978-3-319-24574-4_28
20. Liu, G., Reda, F.A., Shih, K.J., Wang, T.-C., Tao, A., Catanzaro, B.: Image inpainting for irregular holes using partial convolutions. In: Ferrari, V., Hebert, M., Sminchisescu, C., Weiss, Y. (eds.) ECCV 2018. LNCS, vol. 11215, pp. 89–105. Springer, Cham (2018). https://doi.org/10.1007/978-3-030-01252-6_6

Video-Based Hierarchical Species Classification for Longline Fishing Monitoring

Jie Mei[1]([✉])[iD], Jenq-Neng Hwang[1][iD], Suzanne Romain[2], Craig Rose[2], Braden Moore[2], and Kelsey Magrane[2]

[1] University of Washington, Seattle, WA 98195, USA
{jiemei,hwang}@uw.edu
[2] EM Research and Development, National Oceanic and Atmospheric Administration (NOAA) Affiliate, Pacific States Marine Fisheries Commission, Seattle, WA 98115, USA
{suzanne.romain,craig.rose,braden.j.moore,kelsey.magrane}@noaa.gov
https://ipl-uw.github.io/

Abstract. The goal of electronic monitoring (EM) of longline fishing is to monitor the fish catching activities on fishing vessels, either for the regulatory compliance or catch counting. Hierarchical classification based on videos allows for inexpensive and efficient fish species identification of catches from longline fishing, where fishes are under severe deformation and self-occlusion during the catching process. More importantly, the flexibility of hierarchical classification mitigates the laborious efforts of human reviews by providing confidence scores in different hierarchical levels. Some related works either use cascaded models for hierarchical classification or make predictions per image or predict one overlapping hierarchical data structure of the dataset in advance. However, with a known non-overlapping hierarchical data structure provided by fisheries scientists, our method enforces the hierarchical data structure and introduces an efficient training and inference strategy for video-based fisheries data. Our experiments show that the proposed method outperforms the classic flat classification system significantly and our ablation study justifies our contributions in CNN model design, training strategy, and the video-based inference schemes for the hierarchical fish species classification task.

Keywords: Electronic monitoring · Hierarchical classification · Video-based classification · Longline fishing

1 Introduction

1.1 Electronic Monitoring (EM) of Fisheries

Automated imagery analysis techniques have drawn increasing attention in fisheries science and industry [1–3,7,8,14–16,20], because they are more scalable and deployable than conventional manual survey and monitoring approaches.

© Springer Nature Switzerland AG 2021
A. Del Bimbo et al. (Eds.): ICPR 2020 Workshops, LNCS 12662, pp. 422–433, 2021.
https://doi.org/10.1007/978-3-030-68790-8_33

One of the emerging fisheries monitoring methods is electronic monitoring (EM), which can effectively take advantage of the automated imagery analysis for fisheries activities [7]. The goal of EM is to monitor fish captures on fishing vessels either for catching counting or regulatory compliance. Fisheries managers need to assess the amount of fish caught by species and size to monitor catch quotas by vessel or fishery. Such data are also used in analyses to evaluate the status of fish stocks. Managers also need to detect the retention of specific fish species or sizes of particular species that are not allowed to be kept. Therefore, accurate detection, segmentation, length measurement, and species identification are critically needed in the EM systems.

1.2   Hierarchical Classification

Especially in the EM systems, a hierarchical classifier is more meaningful for the fisheries than a flat classifier with the standard softmax output layer. The hierarchical classifier can predict coarse-level groups and fine-level species at the same time. If the system predicts some images with high confidence in one coarse-level group but with low confidence in the corresponding fine-level species, then a hierarchical classifier stops predictions of those images at the correct coarse-level group and allows fisheries personnel to assign corresponding experts to review those images and get the correct fine-level labels.

To address the hierarchical classification needs, in this paper, we develop a video-based hierarchical species classification system for the longline fishing monitoring, where fish are caught are caught on hooks and viewed as they are pulled up from the sea and over the rail of the fishing vessel as shown in Fig. 1.

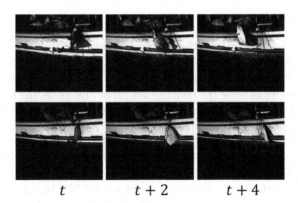

t $t + 2$ $t + 4$

Fig. 1. Longline Fishing: Each column is a sequence of an individual fish caught on a longline hook, as it is being pulled up from the sea and over the rail of the fishing vessel.

The proposed hierarchical prediction, which allows coarse-level prediction to be the final output if fine-level confidence score is too low, improves accuracy on tail-class species when training data follows a long-tail (imbalanced) distribution. Our contributions can be summarized as follows: 1) Our proposed CNN architecture enforces an effective hierarchical data structure. 2) An efficient training strategy. 3) Two robust video-based hierarchical inference schemes.

The remaining sections of this paper are organized as follows. In Sect. 2, overviews of the related works for flat classifiers with the standard softmax output layer and hierarchical classifiers are provided. Section 3 describes the proposed system in details. The experimental results, including the ablation study, are demonstrated and discussed in Sect. 4. Finally, Sect. 5 gives the conclusion of this work.

2   Related Work

2.1   Flat Classifiers

We use 'flat classifiers' to represent all deep learning classification systems with softmax as the final layer to normalize the outputs of all classes, without introducing any hierarchical level of prediction.

AlexNet [11] is the first CNN-based winner in 2012 ImageNet Large Scale Visual Recognition Challenge (ILSVRC), which introduces the 1000-way softmax layer for classifying the 1000 classes of objects. The subsequent ILSVRC winners, VGGNet [12], GoogLeNet [13], and ResNet [5] continue to use softmax as the final layer to achieve good performance. Until now, flat classifiers with softmax operations as the final layer are the dominant design structure for classification tasks.

2.2   Hierarchical Classifier

A hierarchical classifier means the system can output all confidence scores at different levels in the hierarchical data structure. One obvious advantage is that if the confidence score of a sample is too low at the fine level but very high at coarse level, then we can use the coarse-level prediction to be the final prediction. In contrast, flat classifiers have no alternative ways if the confidence score is too low at the final prediction.

Hand-crafted features are used in [6] for hierarchical fish species classification. Hierarchical medical image classification [10] and text classification [9] use cascaded flat classifiers to be their hierarchical classifiers, which use only one flat classifier for each level's prediction. They stack CNN-based models with flat classifiers without considering any hierarchical architecture design and increased computational complexity. HDCNN [18] introduces confidence-score multiplication operations to enforce hierarchical data structure but the model uses the same feature maps for both coarse level and fine level, resulting in learning an overlapping hierarchy of training data. B-CNN [19] uses different feature maps for different levels' predictions without enforcing any hierarchical data structure in the architecture. Deep RTC [17] adopts hierarchical classification to deal with long-tailed recognition, resulting in improved accuracy of tail classes. It adopts a simple confidence-score thresholding method, which is also adopted in our app-roach, to decide to output fine-level prediction or coarse-level prediction. But Deep RTC predicts an overapping hierarchical data structure in the first place, which is different with our situation.

3 Proposed Method

3.1 Hierarchical Dataset

The hierarchical dataset utilized for training our system is professionally labeled and provided by the Fisheries Monitoring and Analysis (FMA) Division, Alaska Fisheries Science Center (AFSC) of NOAA, researchers can contact AFSC directly about getting permission to access this dataset and the corresponding hierarchical data structure.

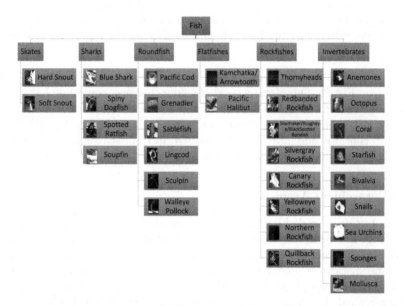

Fig. 2. Hierarchical Data Structure: The dataset, labeled and provided by NOAA fish-eries scientists, includes frames and corresponding labels which are bounding box loca-tion, start and end frames' IDs of each individual fish, coarse-level group ground truth, and fine-level species ground truth. The sample images shown here are randomly chosen from the dataset.

To construct the dataset used for our system, we use labels of bounding box location to crop objects from raw videos and use labeled start and end frames' IDs for each individual fish to divide raw videos into individual tracks (video clips). There are 6 coarse-level groups and 31 fine-level species in this hierarchical

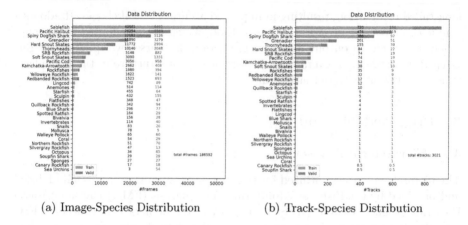

(a) Image-Species Distribution   (b) Track-Species Distribution

Fig. 3. Dataset Distribution: The left column black numbers in both figures are the number of images or tracks for training while the green numbers are for evaluation. There is a 0.5 in (b) because of only one track in the whole dataset, therefore we have to split it into 2 tracks and denote each as 0.5 track. 'SRB Rockfish' represents Shortraker/Rougheye/BlackSpotted Rockfish. Our dataset follows a long-tail (imbalanced) distribution.

dataset (see Fig. 2). Our dataset is challenging because some fine-level species are very similar to one another. The total number of frames is 186,592 (see Fig. 3(a)). The total number of video clips/tracks is 3,021 (see Fig. 3(b)). Each video clip contains one individual fish pulled up from sea surface to the fishing vessel during the longline fishing activities.

3.2   Hierarchical Architecture

Instead of using cascaded flat classifiers in our species identification in the longline fishing, as inspired by the success of Mask-RCNN [4], which feeds shared feature maps extracted from the backbone to different heads for object classification and instance segmentation at the same time, our proposed architecture is also an end-to-end training network including two parts: a backbone and several hierarchical classification heads (see Fig. 4). Inspired by B-CNN [19], we use ResNet101 as our backbone to extract shallow feature maps from images for 'Head-1' and shared deeper feature maps for the other 6 classification heads. We use Head-1 for coarse-level (6 groups) classification and Head-2 to Head-7 for fine-level (31 species in total) predictions.

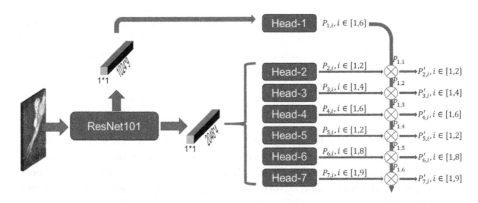

Fig. 4. Hierarchical Architecture: We call our 7 classification heads 'Hierarchical Heads'. Head-1 is for 6 groups in coarse-level and uses shallower feature maps extracted from the backbone for predictions, while the rest of 6 heads are for fine levels, i.e., Head-2 for 'Skates' group, Head-3 for 'Sharks' group, Head-4 for 'Roundfish' group, Head-5 for 'Flatfishes' group, Head-6 for 'Rockfishes' group, and Head-7 for 'Invertebrates' group, respectively. All fine-level heads use the shared deeper feature maps from the same backbone for predictions. Head-1 has two fully connected layers followed by a softmax layer. The rest 6 fine-level heads have one fully connected layer followed by a softmax layer.

Enforcing Hierarchical Data Structure. We use confidence-score multiplication operations to enforce the hierarchical data structure in our system. The final confidence score of one specific species is the product of the confidence score in the corresponding coarse group and the confidence score in that specific (fine) species as shown in the following equation, i.e.,

$$P'_{j,i} = P_{1,j-1} \cdot P_{j,i}, \quad j \in [2, 7], \tag{1}$$

where $P_{1,j-1}$ represents the confidence score in the $(j-1)^{th}$ group in coarse level, $P_{j,i}$ represents the confidence score in the ith species of the $(j-1)^{th}$ group, $P'_{j,i}$ represents the final confidence score in the ith species of the $(j-1)^{th}$ group. As a result, the final confidence score, $P'_{j,i}$, includes both scores from coarse level and fine level so that this CNN architecture enforces a hierarchical data structure when using the final confidence score product to calculate the training loss. Training loss is meaningful when using $P'_{j,i}$ because the final layers in all heads are softmax outputs, which satisfy the following equations:

$$\begin{cases} \sum_i P_{j,i} = 1, \\ \sum_{j=2}^{7} \sum_i P'_{j,i} = 1. \end{cases} \tag{2}$$

Efficient Training Strategy. During the image-based training, one input image has both labeled coarse-level ground truth as well as fine-level ground truth. For our architecture, there are two options experimented about how to use these two ground truths.

The first option is training the 'Head-1' and the fine-level head corresponding to the ground truth coarse-level group. Since the corresponding fine-level head is picked by coarse-level group ground truth, therefore the losses are only calculated based on these two heads.

$$Loss_1 = -\sum_i y_{1,i} \cdot log(P_{1,i}) - \sum_i y_{2,i} \cdot log(P'_{j,i}), \tag{3}$$

where the first summation is the cross entropy loss in the 'Head-1' and $y_{1,i}$ is coarse-level ground truth. The second summation is the cross entropy loss in the corresponding fine-level head using final predictions, $P'_{j,i}$, after confidence-score multiplication operations. Note that $y_{2,i}$ is the ground-truth among species within this fine-level head. This regular loss does not involve P'_{ji} from other heads, therefore, it does not fully enforce hierarchical data structure during training and only trains two heads each time.

The second option is training the 'Head-1' and all the other fine-level heads using final predictions P'_{ji} after confidence-score multiplication operations, resulting in a more efficient training strategy because it enables confidence-score multiplication operations to fully enforce hierarchical data structure during training and all heads can be trained simultaneously.

$$Loss_2 = -\sum_i y_{1,i} \cdot log(P_{1,i}) - \sum_{j=2}^{7} \sum_i y'_{2,i} \cdot log(P'_{j,i}), \tag{4}$$

where $y'_{2,i}$ denotes the ground-truth among 31 species. The reason why given one input image we can calculate cross entropy on all final predictions, $P'_{j,i}$, is that after the confidence-score multiplication operations, summation of these products is still 1.

Video-Based Inference Schemes. Although we use image-based training, where training loss is calculated on each individual input image, two video-based (track-based) inference methods are implemented and compared. Since for each input image frame, our system outputs confidence scores of 31 species, $P'_{j,i}$, therefore the first way is to pick the species with the maximum average confidence score of all frames in each track to be the prediction for each track.

The second way is to pick the species with maximum confidence score for every frame in each track, then uses majority vote to select one species as the prediction for that track. Finally, we calculate the average confidence scores among frames corresponding to the selected species. We report performance under these two video-based inference schemes and calculate their average confidence scores with image-based confidence scores in the following section. These two inference schemes can be summarized into the following equations:

$$\begin{cases} p_{1,i} = 1/T \cdot \sum_t P'_{1,i,t}, \\ p_{2,i} = 1/T \cdot \sum_t P'_{j,i,t}, \quad j \in [2,7], \end{cases} \tag{5}$$

where t is frame index. $P'_{j,i,t}$ is $P'_{j,i}$ at the tth frame. In the first way, T is the total number of frames in one video clip (a track from the start-frame to the end-frame of one catching), while in the second way, it is the total number of frames corresponding to the selected species in one video clip. As a result, $p_{1,i}$ is video-based average confidence scores in 6 groups and $p_{2,i}$ is video-based average confidence scores in 31 species.

4 Experiments and Discussion

4.1 Data Split

We use the video-based data split, i.e., each short video clip (a track) is associated with one individual fish and all frames from 80% of all tracks are used as training data for image-based training. All frames from the rest 20% tracks are the evaluation data (see Fig. 3). As a result, images for training and evaluation are totally from tracks of different individual fishes. All hyper-parameters like training epochs, learning rate, data augmentation, and so on are kept the same in the following different competing approaches.

4.2 Baseline

The dominant species classification architecture is extracting deep features using CNN followed by a flat classifier. As a result, for the baseline, we use ResNet101 as the backbone and two fully connected layers followed by a 31-way softmax layer as the flat classifier head, which is a classic deep learning classification architecture. During training, we only use fine-level ground truth to calculate the cross-entropy loss based on the flat classifier output confidence scores in 31 species without any coarse-level predictions. *From Table 1, we can see the accuracy of the baseline is far below our hierarchical method.*

4.3 Evaluation Methods

Using all frames from the rest 20% tracks for evaluation, we try the following evaluation methods, where we calculate both image-based accuracy as well as video-based (track-based) accuracy, denoted in the 'Unit' column in Table 1.

   We also calculate classification accuracy on the coarse level based on coarse-level ground truth, denoted as 'Level-1' in Table 1.

   Moreover, with confidence scores in the coarse level as well as the fine level, we can pick the species with maximum fine-level confidence score under the group with maximum coarse-level confidence score as the final prediction, which is denoted as 'Level-2 A' in Table 1. While, with final confidence scores in 31 species, we can directly pick the species with the maximum confidence score product of coarse and fine levels as the final prediction, which is denoted as 'Level-2 B' in Table 1. For these two metrics ('Level-2 A' and 'Level-2 B') in video-based schemes, we further use either maximum average confidence score (denoted as

Table 1. Comparison with Flat Classifier and Ablation Study: *'video'* denotes video-based inference by using average confidence score among 31 species to pick one predicted species for each track. *'video*'* denotes video-based inference through majority vote to pick one species for each track. Two numbers under 'Level-2 C' column following the accuracy value are total number of stopping at coarse-level and total number of proceeding to fine-level respectively.

Model	Unit	Level-1	Level-2 A	Level-2 B	Level-2 C
Baseline	img	-	-	78.3	-
Scheme-1	img	86.3	77.4	77.4	82.0 (8567, 27393)
	video*	93.2	86.5	86.6	93.2 (298, 319)
	video	93.4	86.5	86.8	93.4 (293, 324)
Scheme-2	img	88.4	79.9	80.0	84.6 (8660, 27300)
	video*	94.3	88.6	88.9	94.3 (329, 288)
	video	94.9	88.9	88.8	94.9 (328, 289)
Scheme-3	img	91.0	82.3	82.3	86.3 (5830, 30130)
	video*	96.3	90.6	90.3	96.3 (286, 331)
	video	**96.4**	**90.9**	**90.9**	**96.4** (293, 324)

'video') or majority vote (denoted as *'video*'*) to report the performance, as discussed in the 'Video-based Inference Schemes' in Sect. 3.2.

Finally, with these final confidence scores, $P'_{j,i}$, $j \in [2, 7]$, in 31 species, for image-based inference, we can also decide to go back to the coarse level if the maximum confidence score product is below a threshold and calculate the accuracy on the coarse level for that input, otherwise stay at the fine level and calculate the accuracy for that input. For video-based inference methods, we compare the average confidence score mentioned in 'Video-based Inference Scheme' section with the threshold. This metric, being able to stay at a coarse level, is denoted as 'Level-2 C' in Table 1. Theoretically, the ceiling limit of 'Level-2 C' is 'Level-1' if all samples stop at the coarse level. Therefore we use the greedy search to find a threshold for each scheme in Table 1 to make sure that after stopping at the coarse level, the overall video-based inference accuracy will not degrade. We fix these thresholds in image-based inference for every competing scheme.

From Table 1, we can see video-based inference is always better than image-based inference in all competing schemes. And these two video-based inference methods, average confidence and majority vote, are comparable with each other.

Scheme-3 is our full system demonstrated in Fig. 4, which includes confidence-score multiplication operations to enforce hierarchical data structure and uses the efficient training strategy ($Loss_2$). Scheme-2 only removes the efficient training strategy and uses $Loss_1$ instead. Scheme-1 removes confidence-score multiplication operations in the architecture but keeps 7 heads. It also removes the efficient training strategy and instead uses standard cross-entropy losses on 'Head-1' and the fine-level head corresponding to the ground truth coarse-level

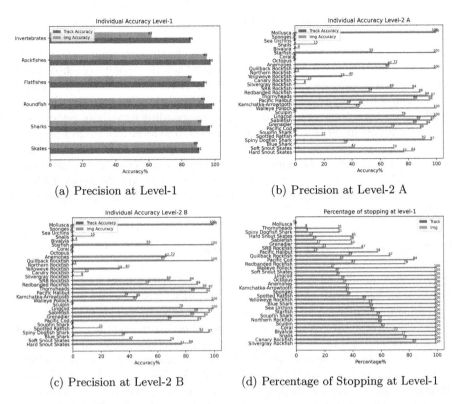

(a) Precision at Level-1

(b) Precision at Level-2 A

(c) Precision at Level-2 B

(d) Percentage of Stopping at Level-1

Fig. 5. Detailed Accuracy on Coarse Level and Fine Level of the complete proposed system: The orange bar is image-based inference and the blue bar is video-based inference. (d) is under 'Level-2 C' evaluation method. We can see most tail species stop at coarse-level prediction, which makes 5.5% improvement in overall video-based accuracy showed in Table 1 (Color figure online).

group. Scheme-1 shares the same architecture as B-CNN [19]. When evaluating under 'Level-2 B' and 'Level-2 C' for Scheme-1, we have to multiply the coarse-level confidence scores with the fine-level confidence score in advance to get the final confidence scores.

Detailed accuracy on the coarse level and fine level of Scheme-3 (our proposed complete system), based on the maximum average confidence score (denoted as '*video*' in Table 1) is in Fig. 5.

4.4 Ablation Study

Scheme-1 and Scheme-2 are experimented mainly for ablation study purposes. *Comparing Scheme-1 with Scheme-2 from Table 1, we can see confidence-score multiplication operations can effectively enforce hierarchical data structure and improve the performance even when Scheme-2 only trains two heads each time. Comparing Scheme-2 with Scheme-3, we can see our efficient training strategy*

(Loss₂) improves the performance by fully enforcing hierarchical data structure during training.

Under 'Level-2 C', the completing systems' final predictions can stop at a coarse level if the final confidence score is lower than the greedy-searched threshold mentioned in the previous section. We call 'Level-2 C' as hierarchical prediction, which is one big advantage of the hierarchical classifier over flat classifiers, which allows fisheries managers to assign corresponding experts to review those images in a certain group and get the correct fine-level labels. *Besides, from Fig. 5(d), we can see most tail-class species identification stop at a coarse level, resulting in a significantly higher overall accuracy in 'Level-2 C' over that of 'Level-2 B' in Table 1. Also, our full system, Scheme-3, has the greatest number of images or tracks proceeding to fine level and at the same time achieves the best performance.*

5   Conclusions and Future Work

We proposed an efficient hierarchical CNN classifier to enforce hierarchical data structure for fish species identification, combined with an efficient training strategy, and two video-based inference schemes. Our experiments show that the integrated use of these three main strategies indeed improves accuracy clearly. Additionally, hierarchical predictions allow images that cannot be confidently classified at the fine level to be confidently classified at a coarse level for experts future examination, which especially improve overall accuracy on tail-class species identification significantly by stopping at coarse-level predictions. Moreover, our method greatly outperforms the baseline method, a flat classifier. Future work will be devoted to adding more techniques like data sampling or additional training losses for tail-class species identification. It would be interesting to combine more strategies for long-tailed data with hierarchical classification.

References

1. Chuang, M.C., Hwang, J.N., Rose, C.S.: Aggregated segmentation of fish from conveyor belt videos. In: 2013 IEEE International Conference on Acoustics, Speech and Signal Processing, pp. 1807–1811. IEEE (2013)
2. Chuang, M.C., Hwang, J.N., Williams, K., Towler, R.: Tracking live fish from low-contrast and low-frame-rate stereo videos. IEEE Trans. Circuits Syst. Video Technol. **25**(1), 167–179 (2014)
3. Gupta, S., et al.: Trends in application of imaging technologies to inspection of fish and fish products (2011)
4. He, K., Gkioxari, G., Dollár, P., Girshick, R.B.: Mask R-CNN. CoRR abs/1703.06870 (2017). http://arxiv.org/abs/1703.06870
5. He, K., Zhang, X., Ren, S., Sun, J.: Deep residual learning for image recognition. In: Proceedings of the IEEE Conference on Computer Vision and Pattern Recognition, pp. 770–778 (2016)
6. Huang, P.X., Boom, B.J., Fisher, R.B.: Hierarchical classification with reject option for live fish recognition. Machine Vis. Appl. **26**(1), 89–102 (2014). https://doi.org/10.1007/s00138-014-0641-2

7. Huang, T.W., Hwang, J.N., Romain, S., Wallace, F.: Live tracking of rail-based fish catching on wild sea surface. In: 2016 ICPR 2nd Workshop on Computer Vision for Analysis of Underwater Imagery (CVAUI), pp. 25–30. IEEE (2016)

8. Huang, T.W., Hwang, J.N., Rose, C.S.: Chute based automated fish length measurement and water drop detection. In: 2016 IEEE International Conference on Acoustics, Speech and Signal Processing (ICASSP), pp. 1906–1910. IEEE (2016)

9. Kowsari, K., Brown, D.E., Heidarysafa, M., Meimandi, K.J., Gerber, M.S., Barnes, L.E.: Hdltex: Hierarchical deep learning for text classification. In: 2017 16th IEEE International Conference on Machine Learning and Applications (ICMLA), pp. 364–371. IEEE (2017)

10. Kowsari, K., et al.: Hmic: hierarchical medical image classification, a deep learning approach. Information **11**(6), 318 (2020)

11. Krizhevsky, A., Sutskever, I., Hinton, G.E.: Imagenet classification with deep convolutional neural networks. In: Advances in Neural Information Processing Systems, pp. 1097–1105 (2012)

12. Simonyan, K., Zisserman, A.: Very deep convolutional networks for large-scale image recognition. arXiv preprint arXiv:1409.1556 (2014)

13. Szegedy, C., et al.: Going deeper with convolutions. In: Proceedings of the IEEE Conference on Computer Vision and Pattern Recognition, pp. 1–9 (2015)

14. Wang, G., Hwang, J.N., Williams, K., Cutter, G.: Closed-loop tracking-by-detection for rov-based multiple fish tracking. In: 2016 ICPR 2nd Workshop on Computer Vision for Analysis of Underwater Imagery (CVAUI), pp. 7–12. IEEE (2016)

15. White, D.J., Svellingen, C., Strachan, N.J.: Automated measurement of species and length of fish by computer vision. Fisheries Res. **80**(2–3), 203–210 (2006)

16. Williams, K., Lauffenburger, N., Chuang, M.C., Hwang, J.N., Towler, R.: Automated measurements of fish within a trawl using stereo images from a camera-trawl device (camtrawl). Methods in Oceanography **17**, 138–152 (2016)

17. Wu, T.Y., Morgado, P., Wang, P., Ho, C.H., Vasconcelos, N.: Solving long-tailed recognition with deep realistic taxonomic classifier. arXiv preprint arXiv:2007.09898 (2020)

18. Yan, Z., et al.: Hd-cnn: hierarchical deep convolutional neural networks for large scale visual recognition. In: Proceedings of the IEEE International Conference on Computer Vision (ICCV), December 2015

19. Zhu, X., Bain, M.: B-cnn: branch convolutional neural network for hierarchical classification. arXiv preprint arXiv:1709.09890 (2017)

20. Zion, B.: The use of computer vision technologies in aquaculture-a review. Comput. Electron. Agriculture **88**, 125–132 (2012)

Robust Fish Enumeration by Multiple Object Tracking in Overhead Videos

Hung-Min Hsu$^{(\boxtimes)}$, Ziyi Xie, Jenq-Neng Hwang, and Andrew Berdahl

University of Washington, Seattle, WA 98195, USA
{hmhsu,ziyixie,hwang,berdahl}@uw.edu

Abstract. In this paper, we design a framework called detection-tracking fish counting (DTFC) for overhead videos of fish moving in shallow water. DTFC includes three steps, fish tracking, counting zone generation and fish counting. In this work, we adopt two deep learning framework FasterRCNN and DeepSORT as our detector and tracker, respectively. Conventional fish counting methods need to predefine the counting zone, and as a result, they become very inefficient. In DTFC, we use the tracking to refine the generated trajectories and automatically generate the counting zone. For the detection of small objects, we divide the overhead image into four sub-images and enlarge the sub-images for detector training. Our experimental results show that DTFC is very suitable for fish tracking and counting, and may provide a valuable resource for fisheries management.

Keywords: Fish enumeration · Fish tracking · Fish counting

1 Introduction

The Bristol Bay sockeye fishery is a vital economic, ecological and cultural resource. This fishery has been managed sustainably in part due to a lack of development but largely due to careful management. This management is based on daily counts of salmon that escape past the commercial fishery and migrate up rivers. In many of the rivers, in-season escapement counts are estimates, based on by-eye human counting for ten minutes of each hour. Such methods are imprecise as well as cost and labor intensive. Leveraging large technological leaps in deep learning neural nets and camera, hard drive and battery capabilities, we aim to develop a computer-vision system capable of accurately and efficiently estimating the salmon run sizes, using the sockeye salmon migration in the Wood and Kvichak River systems of Bristol Bay as case studies.

For marine scientists, conservationists and fisheries managers, it is important to accurately enumerate the real-time abundance of fish and closely monitor changes in population size. In contrast to laborious manual sampling, many

This publication is [partially] funded by the Joint Institute for the Study of the Atmosphere and Ocean (JISAO) under NOAA Cooperative Agreement NA15OAR4320063, Contribution No. TBD.

© Springer Nature Switzerland AG 2021
A. Del Bimbo et al. (Eds.): ICPR 2020 Workshops, LNCS 12662, pp. 434–442, 2021.
https://doi.org/10.1007/978-3-030-68790-8_34

computer-based automatic fish sampling solutions have been proposed in underwater video. However, there is no optimal solution for automatic fish tracking. Therefore, we plan to establish a fish tracking and counting system based on tracking by detection framework.

Tracking, especially under natural conditions, is a very challenging task. For example, due to unreliable target detection, severe occlusion, different directions of the same target, low video resolution, and changes in lighting conditions, it is difficult to track targets through many cameras. For data association in tracking, the appearance features of the same target may be different due to different lighting and viewing angles in different cameras. Because fish may continuously enter or leave the field of view (FOV), an offline trained detector is needed to detect live fish from video data. However, due to the difference in fish posture, the deformation of the fish body shape, and the color similarity between the fish and the background, the detection performance is greatly reduced, resulting in tracking and counting offset errors. To solve this problem, we need to develop a more robust approach for tracking and detection.

To overcome these challenges, we propose a hybrid solution called detection-tracking fish counting (DTFC) as shown in Fig. 1. DTFC includes DeepSORT [12] and FasterRCNN [7], which are two deep neural networks for tracking and detection. The fish detection system based on FasterRCNN is used for capturing static and clearly visible fish instances. In terms of tracking, we use DeepSORT to associate the detection results to generate the tracking results. We use Deep-SORT as a starting point, then refine our approach based on the provided fish videos. For example, sometimes the detection results are not reliable, we can use tracking results to interpolate the missing detection to improve the performance of detection. After tracking, we generate a counting zone so that the number of trajectories passing through the counting zone can be referred to as the count of fish for fish enumeration.

2   Related Work

Fish detection, tracking, and counting in the underwater videos have been studied for many years. Chen-Burger et al. [10] propose a texture and color analysis system based on histogram features for fish tracking. Chuang et al. [4] propose a tracking algorithm based on the deformable multiple kernels (DMK) and adopted color histogram, texture histogram, and the histogram of oriented gradients (HOG) as object features for tracking. Chuang et al. [3] propose an automatic fish segmentation algorithm to overcome the low-contrast issue in underwater video and modified the Viterbi algorithm for data association in low frame rate tracking. Wang et al. [11] proposed a closed-loop tracking by detection mechanism for the underwater videos from a remotely-operated vehicle (ROV). Pedersen et al. [6] present a system for 3D tracking of zebrafish. Romero-Ferrero et al. [9] use two convolutional networks for fish tracking. One is for classification to determine the type of image (crossing or individual), and the other one is for fish re-identification.

In terms of tracking, we need to use appearance features to associate the detected objects from the detector. Therefore, the appearance feature extractor plays an important role in fish tracking. Chuang et al. [2] propose a fully unsupervised feature learning technique and an error-resilient classifier for an underwater fish recognition system. Haurum et al. [1] use metric learning with Ensemble of Localized Features (ELF) and Local Maximal Occurrence (LOMO) for zebrafish re-identification. However, most of the research is focused on underwater videos or tracking the fish in a tank. In this work, we introduce a framework for fish enumeration in overhead videos, in which the camera looks down into shallow water.

3   Methodology

With the demands of fish resource management, fish counting has become a vital problem, which can be used to monitor fish migration and elevate the efficiency of fish resource management. Traditional fish counting problems focus on counting fishes in a predefined tracking zone and counting how much fishes passing through the tracking zone. They count fishes by movements of interest (MOI), which is pre-defined by humans. In this paper, a detection-tracking fish counting (DTFC) framework is proposed, which detects and tracks fishes in the region of interest (ROI), then counts those tracked trajectories by movements.

3.1   Video Collection

In this work, we pilot two automatic fish enumeration scenarios. The first is near the mouth of the Wood River in Bristol Bay, Alaska, where the Alaska Department of Fish and Game (ADFG) count sockeye salmon escapement (migration through fisheries) from a pair of bank-side towers. Sockeye salmon travel upstream exclusively near the river banks, so in principle, the majority of the fish are visible from the towers. ADFG employees count fish for ten minutes of each hour over the duration of the migration (several weeks) to estimate the total run size. To pilot an automated approach to this counting ADFG has provided sample video from their counting towers which we analyse here (Fig. 2). Once we have validated our approach, solar-powered cameras can be deployed on both banks (for complete enumeration) and on other river systems, potentially providing a high-accuracy, low-cost solution for sustainable salmon management.

Our second filming scenario is on the spawning (breeding) grounds, in small shallow tributaries of the Kvichak River in Bristol Bay, Alaska. Here salmon researchers perform regular salmon counts by walking the shores of several index streams as part of a long-term data set, used for population forecasts and studying the ecology of these fish. We collected drone-based overhead video alongside these traditional counts to explore the feasibility of an automated overhead video count (Fig. 3). This water is typically very shallow and clear and the fish are a highly conspicuous red color in these spawning habitats, so we expect high accuracy even from high altitudes, meaning the drone can efficiently cover large sections of habitat. Our approach has strong potential to yield a new tool for the continued long-term monitoring efforts of the these streams, allowing smaller crews to cover more streams more quickly.

3.2 Fish Tracking

From this video footage we will use bespoke computer vision algorithms to detect and count individual salmon traveling through the overhead video frame. Due to unreliable target detection, severe occlusion and changes in lighting conditions, tracking is a very challenging task. To overcome these challenges, we propose a hybrid solution DeepSORT and FasterRCNN which are two deep neural networks for tracking and detection, respectively. The fish detection system based on FasterRCNN is used for capturing static and clearly visible fish instances. We will then use DeepSORT to link detections into trajectories.

Before we illustrate our tracking approach, we provide the definition of the tracking problem first. The input of tracking is detection $\mathcal{D} = \{\mathcal{D}_1, \mathcal{D}_2, \cdots, \mathcal{D}_N\}$ of the camera which includes N frames in this video of the camera, where $\mathcal{D}_i = \{\mathcal{F}_i^1, \mathcal{F}_i^2, \cdots, \mathcal{F}_i^j\}$, where j is the index of the fish in the frame i. Thus, we can denote the trajectory set as $\mathcal{T} = \{\mathcal{T}_1, \mathcal{T}_2, \cdots, \mathcal{T}_V\}$. Each element $\mathcal{T}_i$ in $\mathcal{T}$ indicates a trajectory and V means the number of trajectory in this camera.

We adopt a FasterRCNN with Feature Pyramid Network (FPN) [5] as the fish detector for each video frame since the FPN can deal with different fish sizes. The fish detector is trained by one hundred frames of overhead images in each video, and infers on the rest of the frames in the corresponding video, then the conducting experiments of tracking are only based on these rest of frames in the corresponding video. We label the fish positions with bounding boxes. Moreover, the fish size in some videos is tiny because the altitude of the camera is very high. Therefore, we use a trick about dividing this kind of overhead image into four sub-images and enlarge the sub-images for detector training. After training the FasterRCNN, we apply the detector on the entire video sequences and obtain the bounding box and confident score for each detection, then we apply Deep-SORT as our tracking approach after obtaining these detections. DeepSORT is an online tracking algorithm which exploits the Kalman Filter to estimate the location and speed of objects from noisy detections. Moreover, DeepSORT takes advantage of the deep visual features for data association, which is much more representative than the traditional appearance feature. Fig. 1 shows the procedure of fish tracking. The detector FasterRCNN generates the bounding boxes of each fish as the detection results. Then, the Kalman filter is applied as the predictor to infer the bounding box of each fish in the next frame. After that, the matching scheme is based on Hungarian algorithm by considering the intersection of union (IOU), deep visual feature, speed, orientation, and location. Subsequently, we update the Kalman filter and predict the state of the tracks in the previous frame in the current frame again. The detail of the matching scheme is as follows: (1) calculate the cost matrix of tracks and detections based on the Mahalanobis distance of appearance feature. (2) perform feature

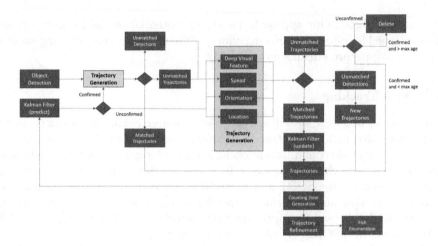

Fig. 1. Illustration for the proposed detection-tracking fish counting (DTFC) framework of fish enumeration.

matching and IOU matching. (3) get al.l matching pairs, unmatched tracks, and unmatched detections of the current frame. For each track that matches successfully, update it with its corresponding detection, and process unmatched tracks and detections.

3.3 Trajectory Refinement and Counting Zone Generation

In our DTFC framework, the first step is fish tracking, where we implement the DeepSORT. After using DeepSORT to generate the tracking results, we observe that there are some fish ID switches due to some isolated trajectories. Therefore, we propose trajectory refinement to deal with this issue and performing single camera ReID to reconnect the isolated trajectories. Then, the counting zone is required for the fish counting task so that we propose an unsupervised manner based on the MeanShift clustering algorithm to generate the counting zone. There are two steps of counting zone generation: 1) use the last positions of all the trajectories from the fish tracking results as counting zone nodes; 2) cluster the counting zone nodes into different groups by applying the MeanShift algorithm.

3.4 Fish Counting

According to the fish counting overhead videos, it has been observed that the main fish flow orientation occurred from the right counting region to the left counting region, since this moving orientation corresponds to the direction of migration of the fish (upstream). In this case, fish counting is an effortless task due to reliable tracking results. Thus, the only thing left is to define the fish-counting strategy. We use MeanShift algorithm to generate the exit zone as the counting zone. Once the central point of the bounding box of a fish passes through the exit zone, a fish counter is incremented. If the overhead video is a fish spawning video, which means there is no dominant fish flow based on the tracking results. In this case, we directly refer the number of the trajectories as the number of the fishes after filtering out the small trajectories. If the length of a trajectory is lower than a threshold which is shown in the Section 4, this trajectory will be removed.

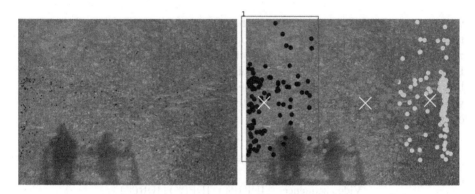

Fig. 2. Counting zone in the tower overhead image. Blue dots indicate the last positions of all the trajectories; red dots are the first positions of all the trajectories. The red bounding box is the generated fish counting zone from the MeanShift algorithm. (Color figure online)

4   Experiments

Our system is using fish tracking to achieve fish counting, therefore we conduct two experiments for this task. For fish tracking, we use IDF1, IDP, and IDR as evaluation metrics. IDF1 [8] is a widely used multiple object tracking (MOT) metric, and it can assess the precision and recall at the same time. IDF1 is used to calculate the ratio of correctly identified trajectories over the average number of ground truth and computed trajectories. In MOT, the identification precision (IDP) and the identification recall (IDR) are used to calculate the IDF1.

$$IDP = \frac{IDTP}{IDTP + IDFP}. \tag{1}$$

$$IDR = \frac{IDTP}{IDTP + IDFN}. \tag{2}$$

$$IDF1 = \frac{2IDTP}{2IDTP + IDFP + IDFN}. \tag{3}$$

IDP and IDR are obtained by the number of false negative ID (IDFN), true negative ID (IDTN) and true positive ID (IDTP). The definition of IDFN, IDTN and IDTP are as follows,

$$IDFN = \sum_{\tau} \sum_{t \in T_\tau} m(\tau, \gamma_m(\tau), t, \Delta), \tag{4}$$

$$IDFP = \sum_{\gamma} \sum_{t \in T_\gamma} m(\tau_m(\gamma), \gamma, t, \Delta), \tag{5}$$

$$IDTP = \sum_{\tau} len(\tau) - IDFN = \sum_{\gamma} len(\gamma) - IDFP, \tag{6}$$

where t is the frame index; τ and γ are the ground truth trajectory and computed trajectory, respectively; $\gamma_m(\tau)$ is the computed trajectory that best matches τ; $\tau_m(\gamma)$ indicates the ground truth trajectory that best matches γ; Δ means the IOU threshold for the computed bounding box and the ground truth bounding box (here we set $\Delta = 0.5$); $m(\cdot)$ is a boolean function for indicating the mismatch, which is set to 1 while there is a mismatch. In Table 1, we show the performance of the fish tracking.

Table 1. Quantitative results of fish tracking in the overhead image.

Video sequence	IDF1(%)	IDP(%)	IDR(%)
Tower	83.80	84.33	83.44
Spawn (large)	58.70	58.25	58.94
Spawn (medium)	48.19	48.05	48.21
Spawn (small)	83.58	89.30	77.67

In terms of fish counting, the performance can be estimated by the Counting Accuracy Rate (CAR) which is defined below:

$$CAR = 1 - \frac{|FC - NF|}{NF}, \tag{7}$$

where NF means the total number of individual fish; FC is the counted number of fish. For tower overhead video, we use the number of trajectory passing through the counting zone as the fish count. For large fish spawning video, we use the total number of trajectory as the fish count since the IDF1 is over 83%. In terms of medium and small fish spawning videos, we remove the trajectories in which the length are lower than a threshold (here we set the threshold as 14% of the total number of frames of the video) because the tracking performance is not sufficient for fish counting. There are many small trajectories due to the ID switching so that we need to filter out these small trajectories. Table 2 shows our experimental results of fish counting. Fig. 3 and Fig. 4 are the qualitative results of the fish tracking.

Table 2. Quantitative results of fish counting in the overhead image.

Video sequence	FC	NF	CAR(%)
Tower	115	123	93.49
Spawn (large)	28	24	83.33
Spawn (medium)	97	92	94.56
Spawn (small)	221	216	97.68

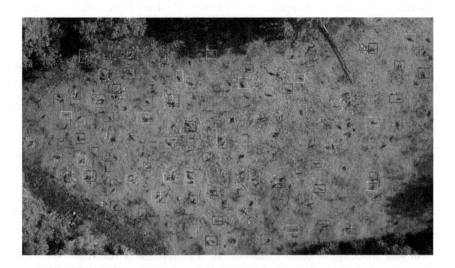

Fig. 3. Qualitative results of small fish tracking in the spawning overhead image.

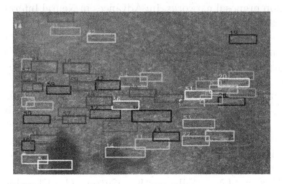

Fig. 4. Qualitative results of fish tracking in the tower overhead image.

5 Conclusion

In this paper, we propose DTFC - a new detection-tracking fish counting method designed for fish enumeration in overhead videos. We show that the introduced DTFC can estimate the number of fishes accurately. Moreover, we investigate the performance of fish tracking and fish counting in DTFC. We also propose an image processing trick for small fish detection. Our promising results show the capability of DTFC as a potential tool for automated fish enumeration, which will support a range of scientific- and management-focus projects.

References

1. Bruslund Haurum, J., Karpova, A., Pedersen, M., Hein Bengtson, S., Moeslund, T.B.: Re-identification of zebrafish using metric learning. In: Proceedings of the IEEE Winter Conference on Applications of Computer Vision Workshops, pp. 1–11 (2020)
2. Chuang, M.C., Hwang, J.N., Williams, K.: A feature learning and object recognition framework for underwater fish images. IEEE Trans. Image Process. **25**(4), 1862–1872 (2016)
3. Chuang, M.C., Hwang, J.N., Williams, K., Towler, R.: Tracking live fish from low-contrast and low-frame-rate stereo videos. IEEE Trans. Circ. Syst. Video Technol. **25**(1), 167–179 (2014)
4. Chuang, M.C., Hwang, J.N., Ye, J.H., Huang, S.C., Williams, K.: Underwater fish tracking for moving cameras based on deformable multiple kernels. IEEE Trans. Syst. Man Cybern. Syst. **47**(9), 2467–2477 (2016)
5. Lin, T.Y., Dollár, P., Girshick, R., He, K., Hariharan, B., Belongie, S.: Feature pyramid networks for object detection. In: Proceedings of the IEEE Conference on Computer Vision and Pattern Recognition, pp. 2117–2125 (2017)
6. Pedersen, M., Haurum, J.B., Bengtson, S.H., Moeslund, T.B.: 3d-zef: A 3d zebrafish tracking benchmark dataset. In: Proceedings of the IEEE/CVF Conference on Computer Vision and Pattern Recognition, pp. 2426–2436 (2020)
7. Ren, S., He, K., Girshick, R., Sun, J.: Faster r-cnn: Towards real-time object detection with region proposal networks. In: Advances in Neural Information Processing Systems, pp. 91–99 (2015)
8. Ristani, Ergys., Solera, Francesco., Zou, Roger., Cucchiara, Rita, Tomasi, Carlo: Performance measures and a data set for multi-target, multi-camera tracking. In: Hua, Gang, Jégou, Hervé (eds.) ECCV 2016. LNCS, vol. 9914, pp. 17–35. Springer, Cham (2016). https://doi.org/10.1007/978-3-319-48881-3_2
9. Romero-Ferrero, F., Bergomi, M.G., Hinz, R.C., Heras, F.J., de Polavieja, G.G.: Idtracker. ai: tracking all individuals in small or large collectives of unmarked animals. Nat. Methods **16**(2), 179–182 (2019)
10. Spampinato, C., Chen-Burger, Y.H., Nadarajan, G., Fisher, R.B.: Detecting, tracking and counting fish in low quality unconstrained underwater videos. VISAPP (2) **2008**(514–519), p. 1 (2008)
11. Wang, G., Hwang, J.N., Williams, K., Cutter, G.: Closed-loop tracking-by-detection for rov-based multiple fish tracking. In: 2016 ICPR 2nd Workshop on Computer Vision for Analysis of Underwater Imagery (CVAUI), pp. 7–12. IEEE (2016)
12. Wojke, N., Bewley, A., Paulus, D.: Simple online and realtime tracking with a deep association metric. In: 2017 IEEE International Conference on Image Processing (ICIP), pp. 3645–3649. IEEE (2017)

DEEPRETAIL 2020 - Workshop on Deep Understanding Shopper Behaviours and Interactions in Intelligent Retail Environments 2020

Workshop on Deep Understanding Shopper Behaviours and Interactions in Intelligent Retail Environments (Deep Retail)

Workshop Description

Deep Retail is a Workshop for researchers and companies working on understanding customer behaviors and trends, with Artificial Intelligence and Deep Learning methods. In fact, the use of artificial intelligence in retail environment covers several aspects of industry, such as demand forecasting, making pricing decisions, and the optimization of product placement. Many efforts have been devoted toward monitoring how shoppers move about in the retail space and interact with products. This challenge is still open due to several serious problems, which include occlusions, appearance changes and dynamic and complex backgrounds. Moreover, while the retail environment has several convenient features for computer vision (such as reasonable lighting), the huge number and assortment of products sold, and the aptitude of shopper movements mean that accurately understanding shopper behaviours is still challenging. The Workshop calls the attention on datasets collected in real retail environment, which are for example a store or a supermarket. Moreover, the aim is to build a network between research and industries and a roadmap for future deep learning retail applications.

The first Edition of Deep Understanding Shopper Behaviours and Interactions in Intelligent Retail Environments (Deep Retail) was held in Trento, Italy, in conjunction with the International Conference on Image Analysis and Processing (ICIAP 2019). The format of the workshop included a keynote followed by technical presentations. Besides, an Industrial session includes presentation from retailers and companies working in this field.

After an accurate and thorough peer-review process, we selected 9 papers for presentation at the workshop. The review process focused on the quality of the papers, their scientific novelty and applicability to an intelligent retail environment. The acceptance of the papers was the result of the reviewers' discussion and agreement. All the high-quality papers were accepted, and the acceptance rate was 62%. The accepted articles represent an interesting mix of techniques and datasets to solve recurrent as well as new problems when dealing with data coming from sensors and devices installed in retail. The workshop program was completed by the invited talk titled given by Francesco Marzoni Group Head of Analytics, Data & Integration at Nestlé and by the round table with companies leader in technological solution for retail.

November 2020   Emanuele Frontoni

Organization

Scientific Committee

Emanuele Frontoni	Department of Information Engineering, Università Politecnica delle Marche
Sebastiano Battiato	Dipartimento di Matematica ed Informatica, Università di Catania
Cosimo Distante	Institute of Applied Sciences and Intelligent Systems - ISASI CNR
Marina Paolanti	Department of Information Engineering, Università Politecnica delle Marche
Luigi Di Stefano	Dipartimento di Informatica - Scienza e Ingegneria, Università di Bologna
Giovanni Marina Farinella	Dipartimento di Matematica ed Informatica, Università di Catania
Annette Wolfrath	GFK Verein - Germany
Primo Zingaretti	Department of Information Engineering, Università Politecnica delle Marche
Alessandro Bruno	Faculty of Media & Communication, Bournemouth University - UK
Annalisa Milella	Institute of Intelligent Industrial Technologies and Systems for Advanced Manufacturing – CNR

Industrial Committee

Massimo De Benedictis	IPSOS – France, Italy
Luigi Caniglia	Acqua & Sapone – Italy
Rinaldo Rinaldi	Conad Adriatico – Italy
Fioravante Allegrino	Sogeda – Italy, Poland
Francesco Mammana	Huawei – China, Italy
Julian Oberndoerfer	ERA Europe - Germany
Stefan Shemann	Harald Wypior, GFK – Germany
Luca Di Camillo	Luxottica – US, Italy
Joe Baer	Zen Genius - US
Nicola Evoli	James Damian, Grottini – Italy, US
Vito Micunco	Software Design – Italy
Lorenzo Vorabbi	Datalogic – Italy
Patrizia Gabellini	Valerio Placidi, Grottini Lab – Italy

3D Vision-Based Shelf Monitoring System for Intelligent Retail

Annalisa Milella$^{(\boxtimes)}$ (ID), Roberto Marani(ID), Antonio Petitti(ID), Grazia Cicirelli(ID),
and Tiziana D'Orazio(ID)

Institute of Intelligent Industrial Technologies and Systems for Advanced
Manufacturing, National Research Council of Italy,
via G. Amendola 122/D, 70126 Bari, Italy
annalisa.milella@stiima.cnr.it

Abstract. This paper presents a 3D Vision-Based Shelf Monitoring system (3D-VSM) aimed at automatically estimating the On-Shelf Availability (OSA) of products in a retail store. The proposed solution exploits 3D data returned by a consumer-grade depth sensor to provide up-to-date information about product availability for customer purchase and eventually generate alerts on Out-Of-Stock (OOS) events, based on the comparison between a reference model of the shelf and its current status. The main advantage is that no a priori knowledge about the product characteristics is required, while the shelf reference model is automatically built, based on an initial training stage. The 3D-VSM system is integrated into an e-commerce application for electronic shopping and home delivery, developed in the context of the E-SHELF research project. Experimental tests carried out in a retail store show that the system is able to accurately estimate the on-shelf availability of products, overcoming time and labour problems of conventional audits.

Keywords: Intelligent retail · RGB-D sensors · 3D reconstruction and modelling

1 Introduction

Smart and automatic management of temporary store shelves is increasingly gaining interest among retailers, since frequent lack of products may lead to significant losses in sales. This event, conventionally known as Out-Of-Stock (OOS), is recognized to be a central concern for consumers, being the third most important issue for shoppers, after the desire for shorter lines at the cash register and more promotions [7,8]. If OOS conditions occur repeatedly, customer satisfaction is highly decreased with potentially negative effects for both retailers and manufacturers. Specifically, two different situations may occur:

- Shelf-OOSs (SOOSs): stockouts are due to inefficient shelf replenishment practices. This situation often occurs with Fast-Moving Consumer Goods (FMCG), which are depleted faster than their replenishment. The product is still available, but the shelf has not been refilled yet;

© Springer Nature Switzerland AG 2021
A. Del Bimbo et al. (Eds.): ICPR 2020 Workshops, LNCS 12662, pp. 447–459, 2021.
https://doi.org/10.1007/978-3-030-68790-8_35

Fig. 1. The 3D-VSM system looking down at a cheese countertop in a real retail environment.

- Store-OOSs: lack of products results from problems in the supply chain. The product is not available in the store warehouse and can not be refilled in a short time.

This work mainly focuses on SOOSs. In this case, most of the responsibility for lowering OOSs rests in the retail store. Higher checking frequency of shelves may be useful to reduce SOOSs. However, current human-based survey practices are labor-intensive and do not provide reliable assessment, especially for FMCG, fresh food and products with a due date. As a result, in the last decade, several sensor-based solutions have been developed to provide automated real-time shelf measurements [16], however research efforts are still required to reach higher reliability and efficiency.

In this research, a novel 3D Vision-based Shelf Monitoring (3D-VSM) approach using an RGB-D sensor, namely an Intel RealSense D435 camera, is proposed. The system is intended to continuously monitor the availability level of critical products such as fresh and perishable goods stored in countertop shelves, baskets or crates, for which conventional information logistic approaches are not applicable, although it can be extended to any type of products. The 3D-VSM module is integrated into an eco-sustainable electronic commerce platform, developed in the context of the E-SHELF research project funded by the Apulia Region, Italy [1].

The algorithm for OSA estimation was first presented by the authors in [12]. It uses depth data provided by the RGB-D sensor and exploits surface reconstruction and modelling techniques to build a model of the supporting shelf plane and estimate the product OSA, based on the comparison between the shelf reference model and its current status. Since each shelf can store multiple product categories, the estimation of the availability of a given product must be limited to specific Regions of Interest (ROIs), i.e. baskets or spaces, which contain a single product type. These ROIs are typically bounded by separators, whose positions can be altered by both shop assistants and customers while loading and grasping the goods. In this paper, a method using depth images to automatically detect and update the positions of shelf ROIs, allowing for simultaneous monitoring of multiple product types, is introduced. In addition, a demonstrator of the 3D-VSM system operating in a real retail environment is presented and its integration in the E-SHELF platform is discussed.

Experimental tests show that, despite the low quality of the sensor data, the proposed system is helpful in automatically extracting up-to-date information about the shelf status, overcoming time and cost problems related to traditional physical store audits.

The rest of the paper is organized as follows. Related work is reported in Sect. 2. The proposed visual shelf monitoring system and its integration in the E-SHELF framework are described in Sect. 3. Experimental results are presented in Sect. 4. Finally, conclusions are drawn in Sect. 5.

2   Related Work

Solutions based on sensor technologies and artificial intelligence have emerged in the last decade to automatically monitor on-shelf availability and detect OOS on a rapid basis, without having to conduct physical store audits. Pilot studies have shown that Radio-Frequency Identification (RFID) solutions are helpful to improve inventory tracking and reduce stockouts [4,9]. However, widespread use of RFID technology is often prevented by expensive and time-consuming item-level tagging. Smart shelves using weight sensors have been developed in [11,17] to detect weight changes and determine accordingly the number of products available on the shelf. However, weight sensors entail high installation costs. In addition, they can only determine the number of products stacked on the shelf without accounting for possible product misplacements, as they do not allow for product identification and tracking.

More recently, computer vision systems have been successfully demonstrated in smart retail applications. A comprehensive survey on computer vision-based

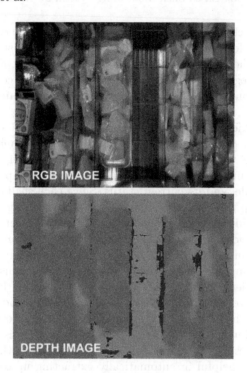

Fig. 2. RGB image (top) and corresponding depth map (bottom) acquired by the Intel RealSense D435 camera placed on the top of a horizontal shelf.

approaches for automatic identification of products in retail store is presented in [16]. They typically use feature-based or template-based matching algorithms for detection of misplaced products [13], verification of planogram compliance [5], and stock assessment [15]. To ensure higher robustness and accuracy under highly variable environmental conditions, deep learning-based approaches have been also developed [2,14], although at the cost of labor-intensive manual image annotations for training. A notable example is the Vispera Shelfsight system [2], which uses IoT cameras mounted on the aisles and state-of-the-art deep learning technology for out-of-stock detection and planogram compliance verification in real-time.

As an alternative to conventional 2D cameras, RGB-D sensors, i.e. sensors coupling a depth sensing device with an RGB camera to augment images with depth information, have been proposed in recent works. An example can be found in [10], where a top-view RGB-D camera is employed to monitor shoppers' behaviors. A Smart Shelf concept has been demonstrated by Intel [3] using Intel's RealSense depth cameras in combination with stock sensors, to automatically recognize products and display related information such as pricing, promotional offers or other appropriate content.

In this work, a novel framework using an Intel Real Sense D435 camera for 3D shelf monitoring is proposed. The main contribution relies on the use of depth information to not only provide alerts about OOSs, but also to accurately estimate the number of items actually available on the shelf, thus yielding a more precise and timely knowledge of the shelf status. In addition, the proposed system does not require any supervised training, nor a priori information on the shelf or product characteristics, whereas the only assumption is that the store items lay on a planar shelf surface whose model is automatically learnt via an initial calibration phase.

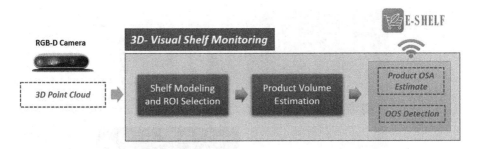

Fig. 3. Workflow of the 3D-VSM system: the output is transferred via WiFi connection to the E-SHELF platform [1].

3   3D Vision-Based Shelf Monitoring (3D-VSM)

A 3D Vision-Based Shelf Monitoring (3D-VSM) framework is proposed for automated product OSA estimation and early detection of OOS conditions. The system exploits a consumer-grade RGB-D sensor, namely an Intel RealSense D435 camera, to continuously survey critical products, such as perishable and fresh goods stored in countertop shelves, baskets or crates. A demonstrator of the system installed on the top of a horizontal cheese refrigerator is shown in Fig. 1. Typical sensor output is displayed in Fig. 2, including the RGB image and the corresponding depth map of the scene. It is worth noticing that the proposed approach is independent of the orientation in space of the shelf support surface and, therefore, also applies to inclined or vertical shelves. In particular, it can be used also with a forward-looking camera setup, where the shelf back panel would provide the reference shelf plane. Nevertheless, in such a configuration, the system would only be able to detect full stock ruptures, since forward products would occlude the backward portion of the shelf.

   The overall processing pipeline is shown in Fig. 3. The sensor provides a 3D reconstruction of the scene, which is computed based on an infrared (IR) stereo processing algorithm running on-board the camera and is used for OSA estimation, as follows. First, a reference model of the shelf in absence of products is built, using a plane fitting algorithm. Depth images of the shelf filled with products are also processed to extract different ROIs corresponding to different products on the shelf. Then, for each product in a given ROI, points sticking out of the shelf reference plane are retained as belonging to products and are used as input to a 2.5-dimensional occupancy grid algorithm to estimate the product volume. Such an estimate is considered as a measure of the product OSA.

The 3D-VSM module is integrated into an eco-sustainable e-commerce system, namely the E-SHELF platform [1] aimed at providing electronic commerce and home delivery services.

In the following, first, each sub-module of the 3D-VSM system is presented, then the integration in the E-SHELF platform is discussed.

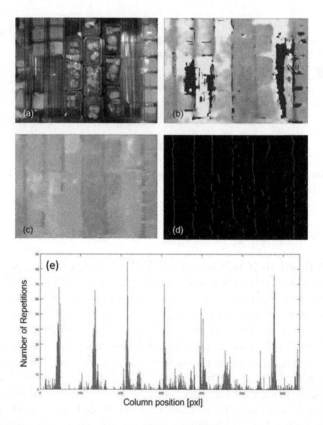

Fig. 4. Pipeline for the selection of the ROIs: (a) input RGB image and (b) corresponding depth map; (c) interpolated depth map; (d) gradient-thresholded binary image; (e) histogram of the detected vertical separators.

3.1   Shelf Modeling and ROI Selection

A planar model of the shelf in absence of products is preliminary built. Without loss of generality, a top-view camera configuration, with the camera looking downward at the shelf, is considered. To fit the plane model, the relations correlating multiple views of the shelf using a maximum likelihood estimation algorithm, i.e. the M-estimator SAmple Consensus (MSAC) [18], are computed.

Fig. 5. Result of the processing for the automatic selection of the ROIs.

Specifically, N subsequent point clouds spanning a few seconds of acquisition, are employed to solve a constrained nonlinear optimization problem aimed at minimizing the Maximum Likelihood Estimation (MLE) error, while filtering out measurement noise.

Typically, transparent glass or plastic mobile separators may be present on the shelf plane to separate different product categories. Therefore, a procedure to automatically detect the separators and define different Regions of Interest (ROIs), one for each product type, is needed to correctly estimate the availability level of different products simultaneously. To this end, the proposed pipeline takes advantage of the depth maps provided by the sensor under standard operative conditions, i.e. when the shelf is filled with products. With reference to Fig. 4 (a), the processing is aimed at detecting the vertical lines which separate the products (here cheese packs) by using the depth map of Fig. 4 (b). It is important to notice that this method can be applied for any direction of the separators, given a simple preprocessing of the image, able to rotate it and display the separators along the vertical direction. The detection of the ROIs is thus achieved as follows:

1. The depth map is linearly interpolated to fill missed acquisitions, typically due to direct reflections of light, carrying to local image saturation. The result of linear interpolation is shown in Fig. 4 (c).
2. A one-dimensional gradient is computed along the horizontal direction of the depth map. A threshold operation highlights the points corresponding to a significant discontinuity of the depth map, likely due to a separator. The threshold is set to the maximum thickness of the products of the shelf, known from the expected planogram of the shop. The resulting binary image is displayed in Fig. 4 (d).
3. The vertical histogram of the binary image is finally evaluated to group the column positions of the detected separators. An example of the result is shown in Fig. 4 (e). The histogram is then processed to detect the maximum occurrences of the column positions, but neglecting those local maxima which are too close to the most recurring ones.

The processing for the automatic selection of the ROIs produces the result in Fig. 5, where the vertical red lines separate the areas filled with products of the same kind.

3.2 Product OSA Estimation

In this work, the OSA level of a given product is defined as the percentage of the shelf volume occupied by product items with respect to the maximum available space [12]. In detail, product volume estimation is performed using a 2.5-dimensional occupancy grid approach, as follows. First, points whose distance from the shelf reference plane is higher than a threshold are segmented. Then, a squared 2D grid with size s is fitted to the product point cloud. Each cell $i = 1, 2, ..n$ of the grid is assigned a depth value H_i, corresponding to the average distance of all points of the cell from the shelf reference plane. The volume V_i of each cell is successively computed as:

$$V_i = s \times s \times H_i \tag{1}$$

The volume V_t occupied by the whole product at a certain observation time t is estimated as:

$$V_t = \sum_{i=1}^{n} V_i \tag{2}$$

Finally, the OSA level of the product is recovered as:

$$OSA_\% = \frac{V_t}{V_{max}} \times 100 \tag{3}$$

where V_{max} is the maximum product volume, i.e., the volume occupied by the product when the shelf is fully replenished. V_{max} can be either calculated based on the knowledge of the shelf geometry or directly estimated by the proposed volume estimation approach after complete shelf refill. An example of product volume estimation using the grid-based approach is shown in Fig. 6 (b) for a selected ROI on a cheese counter displayed in Fig. 6 (a).

Finally, if the maximum number of products N_{max} corresponding to V_{max} is known and under the assumption that all items of the product occupy a similar volume, the number of items per product at time t can also be easily recovered as:

$$N_t = OSA_\% \times N_{max} \tag{4}$$

3.3 Integration in the E-SHELF Platform

The 3D-VSM system was conceived and implemented in the context of the E-SHELF research project, which was aimed at developing an eco-sustainable electronic commerce and home delivery platform. The latter is characterized by interoperability and scalability and is capable of integrating Information and

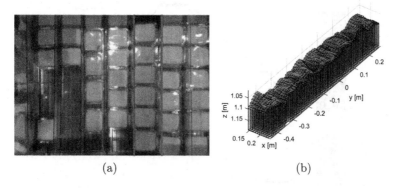

Fig. 6. Volume estimation for pre-packed Grated Parmesan cheese in a selected ROI of a cheese counter. (a) Color image with highlighted the shelf ROI; (b) product modelling using 2.5D occupancy grid for volume estimation. (Color figure online)

Communication Technology (ICT) devices present in the stores and on-board a fleet of transport vehicles by using a de-verticalizing middleware [1,6].

In the E-SHELF architecture, each store is equipped with the 3D-VSM system for online estimation of shelf stocks, in order to reduce the misalignment between the stock data and the actual product availability in the shop. The integration is performed based on the TCP/IP stack protocol. Specifically, the 3D-VSM system communicates to the E-SHELF server, every 15 min, all the information related to the current state of the monitored shelves, including:

- store identification number;
- shelf identification number;
- product identification number;
- product OSA.

A video showing the system demonstrator working online in a real retail environment can be found at the link https://youtu.be/R_M2gxwHJeI.

4 Experimental Results

Experimental tests were performed in a real retail store with the depth sensor looking downward at a cheese countertop as shown in Fig. 1. In each test, the shelf was gradually refilled starting from an OOS condition ($OSA_\% = 0$) or emptied after full replenishment ($OSA_\% = 100$). One every sixth frames was processed and the estimated OSA level was compared with the actual value obtained by manual inspection of the box.

Results obtained for four types of pre-packed cheese, namely (from top to bottom) Silano, Rodez, Scamorza Bianca and grated Parmesan, are reported in Fig. 7. Specifically, for each product type the first column shows a sample image, whereas the second column reports the result of the OSA estimation

456   A. Milella et al.

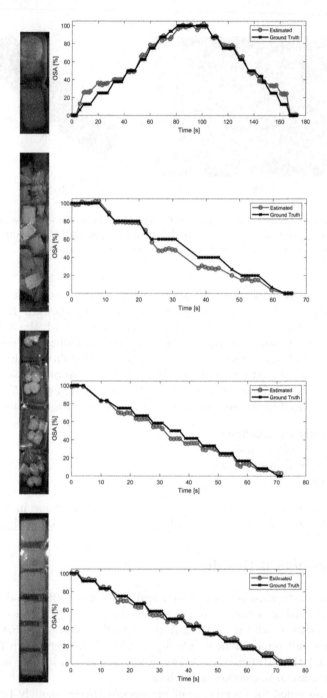

Fig. 7. OSA estimation for different pre-packed cheese varieties. From top to bottom: Silano, Rodez, Scamorza Bianca, Grated Parmesan. Left: color images. Right: 3D-VSM results (gray) compared with ground-truth (black). (Color figure online)

Table 1. OSA estimation for different cheese types: error analysis. Average error $\bar{E}$, standard deviation σ and r-squared coefficient are reported to compare estimated and ground-truth OSA.

Product type (N_{max})	$\bar{E}$[%]	σ[%]	R-squared
Silano (16)	3.991	4.228	0.981
Rodez (15)	5.007	4.591	0.981
Scamorza (12)	3.675	2.343	0.991
Grated parmesan (12)	2.453	1.506	0.992

algorithm (gray line) compared with the ground-truth (black line). The average discrepancy $\bar{E}$, the standard deviation σ and the r-squared coefficient were computed. Numerical results are reported in Table 1. A maximum average discrepancy between actual and estimated $OSA_\%$ of about 5.0 was obtained for the Rodez cheese, which for a maximum number of products $N_{max} = 15$ leads to an average error of 0.75 (i.e., less than 1 product) over the estimated number of available items. An r-squared coefficient higher than 0.98 was reached in all tests showing a good agreement between estimated and ground-truth OSA values.

5   Conclusion

A 3D Vision-Based Shelf Monitoring (3D-VSM) system is proposed, aimed at estimating online the on-shelf availability of products in a retail environment, using 3D data provided by an Intel Realsense D435. The system is intended to early detect out-of-stock events, as well as to provide continuously updated information on product availability for e-commerce apps and stock inventory purposes. Experimental results obtained in a real point of sale are presented, showing that the proposed framework is effective for online estimation of product on-shelf availability in a non-invasive and automatic way. In addition, the 3D-VSM system does not require *a priori* knowledge of shelf or product characteristics, whereas it only assumes that the shelf can be modelled as a planar surface whose parameters are learnt via an initial calibration.

Future work will deal with the development and integration of additional system functionalities, such as image segmentation and classification algorithms for automated product identification, detection of misplaced products or product occlusions and planogram compliance verification in general, also using RGB information. Further research objectives will address the integration of the proposed methods on-board mobile robots or smart shopping carts, so as to reduce the need for environment infrastructuring and increase the overall system flexibility.

Acknowledgements. The financial support of the grant Electronic Shopping & Home delivery of Edible goods with Low environmental Footprint (E-SHELF), POR Puglia FESR-FSE 2014–2020 (Grant Id. OSW3NO1) is gratefully acknowledged. The authors are thankful to Primo Prezzo (Carelli) for providing test facilities and valuable support in performing the experiments. The administrative and technical support by Vito Micunco (Software Design), Michele Attolico (CNR-STIIMA) and Giuseppe Bono (CNR-STIIMA) is also gratefully acknowledged.

References

1. https://www.eshelf.it/. Accessed 28 Sep 2020
2. https://vispera.co/products/shelfsight. Accessed 12 Nov 2020
3. https://www.intel.it/content/www/it/it/retail/digital-signage/intelligent-shelving.html. Accessed 28 September 2020
4. Bertolini, M., Gino, F., Vignali, G., Volpi, A.: Reducing out of stock, shrinkage and overstock through RFID in the fresh food supply chain: evidence from an Italian retail pilot. Int. J. RF Technol. Res. Appl. **4**, 107–125 (2013). https://doi.org/10.3233/RFT-120040
5. Frontoni, E., Mancini, A., Zingaretti, P.: Embedded vision sensor network for planogram maintenance in retail environments. Sensors (Switzerland) **15**(9), 21114–21133 (2015). https://doi.org/10.3390/s150921114
6. Grieco, L.A., et al.: E-shelf: Electronic shopping and home delivery of edible goods with low environmental footprint. In: Convegno Nazionale CINI sull'Intelligenza Artificiale, ITAL-IA 18–19 Marzo, 2019, Rome, Italy (2019)
7. Gruen, T., Corsten, D.: Rising to the challenge of out-of-stocks. ECR J. **2**, 45–58 (2002)
8. Gruen, T.W.: A comprehensive guide to retail out-of-stock reduction in the fast-moving consumer goods industry a research study conducted. https://www.semanticscholar.org/paper/A-Comprehensive-Guide-To-Retail-Out-of-Stock-In-the-Gruen/19280a78f21294dabcbe300c71791f0b7e9bf9bf. Accessed 28 Sep 2020
9. Hardgrave, B., Langford, S., Waller, M., Miller, R.: Measuring the impact of RFID on out of stocks at Wal-Mart. MIS Q. Executive **7**(4), 181–192 (2008)
10. Liciotti, D., Contigiani, M., Frontoni, E., Mancini, A., Zingaretti, P., Placidi, V.: Shopper analytics: a customer activity recognition system using a distributed RGB-D camera network. In: Distante, C., Battiato, S., Cavallaro, A. (eds.) VAAM 2014. LNCS, vol. 8811, pp. 146–157. Springer, Cham (2014). https://doi.org/10.1007/978-3-319-12811-5_11
11. Metzger, C., Meyer, J., Fleisch, E., Tröster, G.: Weight-sensitive foam to monitor product availability on retail shelves. In: LaMarca, A., Langheinrich, M., Truong, K.N. (eds.) Pervasive 2007. LNCS, vol. 4480, pp. 268–279. Springer, Heidelberg (2007). https://doi.org/10.1007/978-3-540-72037-9_16
12. Milella, A., Petitti, A., Marani, R., Cicirelli, G., D'orazio, T.: Towards intelligent retail: Automated on-shelf availability estimation using a depth camera. IEEE Access **8**, 19353–19363 (2020). https://ieeexplore.ieee.org/document/8963979
13. Moorthy, R., Behera, S., Verma, S., Bhargave, S., Ramanathan, P.: Applying image processing for detecting on-shelf availability and product positioning in retail stores. In: WCI 2015: Proceedings of the Third International Symposium on Women in Computing and Informatics, pp. 451–457 (2015). https://doi.org/10.1145/2791405.2791533

14. Paolanti, M., Sturari, M., Mancini, A., Zingaretti, P., Frontoni, E.: Mobile robot for retail surveying and inventory using visual and textual analysis of monocular pictures based on deep learning. In: 2017 European Conference on Mobile Robots (ECMR), pp. 1–6 (2017). https://doi.org/10.1109/ECMR.2017.8098666
15. Rosado, L., Gonçalves, J., Costa, J., Ribeiro, D., Soares, F.: Supervised learning for out-of-stock detection in panoramas of retail shelves. In: 2016 IEEE International Conference on Imaging Systems and Techniques (IST), pp. 406–411 (2016)
16. Santra, B., Mukherjee, D.P.: A comprehensive survey on computer vision based approaches for automatic identification of products in retail store. Image Vis. Comput. **86**, 45–63 (2019).https://doi.org/10.1016/j.imavis.2019.03.005, http://www.sciencedirect.com/science/article/pii/S0262885619300277
17. Sharma, N., Jain, P.: Real-time secure smart shelf management for supermarkets. Int. J. Adv. Res. Electron. Commun. Eng. (IJARECE) **5**, 2085–2088 (2016)
18. Torr, P.H.S., Zisserman, A.: MLESAC: a new robust estimator with application to estimating image geometry. Computer Vis. Image Underst. **78**, 2000 (2000)

Faithful Fit, Markerless, 3D Eyeglasses Virtual Try-On

Davide Marelli$^{(\boxtimes)}$, Simone Bianco , and Gianluigi Ciocca

DISCo - Department of Informatics, Systems and Communication,
University of Milano-Bicocca, Viale Sarca 336, 20126 Milan, Italy
{davide.marelli,simone.bianco,gianluigi.ciocca}@unimib.it

Abstract. Virtual try-on allows people to check the appearance of accessories, makeup, hairstyle, hair color, clothes, and potentially more on themselves. In this paper, we propose an eyewear virtual try-on experience that is performed on a 3D face reconstructed from an input image allowing the user to see the virtual face and eyeglasses from different viewpoints. The try-on process takes into account real face and glasses sizes to provide a realistic fit estimation; it is fully automated and only requires a face picture and selection of eyeglasses frames to test.

Keywords: 3D · Face reconstruction · Augmented reality · Virtual try-on

1 Introduction

Virtual try-on systems are a new trending solution that allows people to virtually check the appearance of accessories, makeup, hairstyle, hair color, clothes, and more using augmented reality. These systems are available in several forms such as web applications, mobile applications, or showcase displays for use in stores. The virtual try-on speeds up the process of trying several items and provides the ability to test products without the need to reach a physical store; it also allows us to try products that can be unavailable in the store. Another advantage is the possibility to easily verify the try-on results from multiple viewpoints also including usually non-possible ones. Sometimes is also possible to see the results after the try-on session; this comes handy especially when the user suffers from view problems and thus has difficulties in checking the appearance of glasses when not wearing lenses.

In this paper, we propose an eyeglasses and sunglasses virtual try-on web application that builds on 3D face reconstruction using a single input image allowing the user to see the virtual face and eyeglasses from different viewpoints. Figure 1 shows the underlying idea of how our virtual try-on application works. Real face and eyeglasses sizes are taken into account in order to provide a realistic fit estimation. The try-on process is fully automated and only requires a face picture and a selection of the glasses frames to test.

A. Del Bimbo et al. (Eds.): ICPR 2020 Workshops, LNCS 12662, pp. 460–471, 2021.
https://doi.org/10.1007/978-3-030-68790-8_36

Fig. 1. Proposed virtual try-on for glasses. The output is a 3D model that can be freely seen from different points of view.

The paper is organized as follows. A brief review of existing virtual try-on and 3D face reconstruction solutions is provided in Sect. 2. In Sect. 3 is described the proposed method for the try-on with realistic size fitting. Finally, in Sect. 4 we provide a review of the current state of the project, its limitations, and possible enhancements.

2 Related Work

In this paper, we focus on eyeglasses and sunglasses virtual try-on, and we briefly discuss the strengths and weaknesses of other available solutions. Since our try-on builds on top of 3D face reconstruction, we briefly present an overview of the available methods that require a single face image.

Virtual try-on due to the increasing demand by commercial companies, various virtual try-on applications have been developed in the last years. These solutions, usually available as mobile or web applications, aim to allow a potential customer to virtually try on himself some products sold by a store or manufacturer. At our knowledge, the arguably most popular virtual try-on solutions for face accessories and makeup are: Ditto's Virtual try-on [8] allows eyeglasses try-on from multiple viewpoints and takes dimensions into account using a credit card like object to perform size estimation; XLabz's Glassify [25] which renders eyeglasses and sunglasses on a single frontal face image, position and scale of the glasses can be adapted manually; Perfect Corp's YouCam Makeup [19] supports try-on of several items, including eyeglasses, some of the try-on features available work in real-time on the live camera video stream; Jeeliz [14] proposes a virtual try-on widget based on WebGL capable to perform eyeglasses try-on on a camera video feed in real-time or an image; Luxottica's Virtual Mirror [17] also provides eyewear try-on in real-time on a camera video feed.

Most of these solutions are available as standalone applications and frameworks that can be integrated with existing services and platforms. The majority of such applications offer integration with social services and store platforms to allow the user to share their virtual try-on sessions and buy the products. Key features of virtual try-on systems are summarized in Table 1 for ease of comparison.

3D face reconstruction over the years, many different approaches were proposed to solve the problem of 3D reconstruction, usually requiring specific hardware

Table 1. Comparison of eyeglasses virtual try-on applications.

Applications	Input	Output	3D glasses	Size fitting	Markerless
Ditto [8]	Video	Images	✓	✓	—
Glassify [25]	Image	Image	—	—	✓
YouCam [19]	Image	Image	—	—	✓
Jeeliz [14]	Video/image	Video/image	✓	—	✓
Virtual mirror [17]	Video	Video	✓	—	✓
Ours	Image	3D	✓	✓	✓

[1] or multiple images [3, 20, 22]. Our goal is to provide the user an easy to use application for virtual try-on; therefore, we want to simplify the acquisition and reconstruction process. For this reason, we chose to focus on 3D face reconstruction from a single image.

Common methods for face reconstruction using a single image are divided into two categories: *3DMM fitting-based* and *Shape regression-based*.

3D Morphable Model (3DMM) is a popular solution to the face reconstruction problem from a single view and it was introduced by [4]. Some common methods, such as [12], search for the correspondence of local features on the images and landmarks on the 3DMM. This correspondence is then used to regress the 3DMM coefficients that, applied to the model, generate a 3D face mesh similar to the one in the image. Newer methods, such as [24], use CNNs to regress the 3DMM coefficients. An advantage of this solution is the availability of the complete face 3D model under any circumstance, even the parts that are occluded in the input image are reconstructed using the geometry of the 3DMM model. The main limitation of these methods is that the reconstruction is often too similar to the original 3DMM model; characteristic traits usually defined by fine details of the face geometry are not preserved in the reconstruction process. A recent approach based on 3DMM models is proposed by [21] and claims to be able to obtain better results than previous works. The Ganfit method proposed by Gecer et al. [10] uses a different approach and generates a realistic face texture by means of Generative Adversarial Networks (GANs).

Shape regression based methods were developed to obtain more accurate reconstruction then the 3DMM based ones. A noteworthy method is the one proposed by [13], whose core idea is to straightforward map the input image pixels to full 3D face structure in a voxel space via volumetric CNN regression. As main advantage, the output face shape is not restricted to a face model space. Another approach is the one proposed by [9], similarly to [13] the face mesh is regressed directly from the input image but, in this case, without making use of a voxel space. As result the latter method is light-weight and faster to execute and also it usually leads to more precise and detailed reconstructions.

3 Proposed Method

The currently available solutions for face accessories virtual try-on present some limitations such as 2D only try-on or marker requirement for size estimation. With the proposed method, we want to define a solution that allows the user to try different glasses starting from a single image taken with a computer or mobile device camera providing him a realistic idea of how they will look on himself.

To accomplish this task we use a 3D face reconstruction process from a single image. After the 3D reconstruction, we estimate the face size and the fitting parameters for a database of eyeglasses models. The processing requires 0.65 s on average on a GTX 1080 GPU, running multiple steps in parallel. The result is displayed using a 3D rendering framework that also allows the user to rotate the reconstruction and test on it different glasses models, once completed both the reconstruction and fitting processes. An overview of the method's workflow is provided in Fig. 2. The following subsections provide further details.

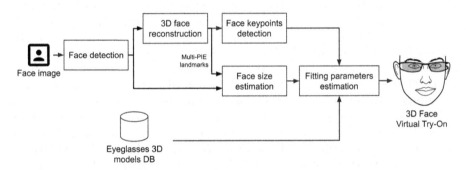

Fig. 2. 3D eyeglasses virtual try on workflow. The input face image is used for both the 3D face reconstruction and the face size estimation. The try-on fitting procedure makes also use of a database of 3D eyeglasses models with mesh, glasses sizes and keypoints.

3.1 3D Face Reconstruction

We base our face reconstruction over the PRNet [9] method. This CNN based method can regress a 3D face shape from a single RGB input image without requiring any particular restriction on face pose or occlusions. The face is detected using DLIB's CNN based face detector; in the case of multiple detections, the reconstruction and the subsequent try-on process run only on the biggest. If face detection fails, the application asks the user to repeat the acquisition. The bounding box around the face is then resized to 256×256 pixels and used as input of the CNN that regresses a 256×256 position map, where each element is a vertex of the reconstructed face. In addition, the method also recovers the face pose, and the 68 landmarks defined by the Multi-PIE landmark scheme [11] from the vertex map. This CNN is implemented using TensorFlow.

The vertex map is rotated to move the face in a frontal pose and scaled in a cube of size $1 \times 1 \times 1$ units in 3D world space, centered in the origin of a canonical right-handed global coordinate reference system. At the final processing stage, the 65k vertices face mesh is generated from the regressed vertex map and its texture image is obtained remapping the input image to the reconstructed mesh. If occlusions are present in the input image, the generated texture image will lack those portions. The average time required by this phase is 0.62 s of which: 0.06 s for face detection, 0.01 s for mesh reconstruction and the remaining for texture generation at resolution 512×512 pixels.

3.2 Face Size Estimation

One of the common limitations of the existing virtual try-on applications is the lack of face size estimation. When choosing between different eyeglasses frames the size of the glasses covers a key part in defining the final look, therefore is important to provide the user with a realistic try-on experience also in terms of size fitting. In some cases, the face size is estimated using known size markers such as a credit card placed on the forehead [8]. While this will provide accurate sizing it also requires the user to perform additional acquisitions. Here we propose the use of facial features to provide a rough estimation of the face size. In particular, we use the diameter of the iris to perform such estimation. The human iris diameter is closely related to the cornea diameter and according to Rüfer et al. [23], the average white-to-white horizontal diameter is 11.71 ± 0.42 mm. According to Kerrigan et al. [16], it is possible to obtain good iris segmentation on NIR eye images using off-the-shelf deep neural networks. Our images are RGB but we found out that those methods still work on them with some preprocessing. The eyes are identified in the input image using the landmarks provided by the face reconstruction method [9]. The Red channel is extracted from the image crops and 0.35 gamma correction is applied to clear the sclera and enhance the visibility of the iris and pupil. The result is fed into a DRN-D-22 network [26] to create the segmentation map of the iris (uses weights from Kerrigan et al. [16]). The neural network is implemented using PyTorch. The Hough Circle Transform is applied to the segmentation maps to fit a circle on the iris and find its radius in pixels. The size in millimeters of each pixel is finally computed as $11.71/2r$ where r is the iris' radius in pixels.

3.3 Glasses Try-On

To allow the user to virtually test different glasses frames on his face, we have defined a fitting procedure to generate the parameters required to render the face with the selected glasses frame. First of all, a database of available glasses frames includes the glasses meshes and some additional parameters. Some face keypoints are subsequently extracted and used to compute the correct glasses positioning parameters. The try-on process assumes that the 3D reconstructed face is always available in the front-view pose for the computation of the fitting parameters.

Glasses Model Parameters the database consists of the glasses mesh, a model name, a preview image, their width, and some fitting keypoints. The bridge location keypoint G_n and both the temples far endpoints G_l and G_r (where the glasses lean on the ears). See Fig. 3 for keypoints reference positions. The keypoints are manually defined in advance when a new glasses model is added to the available library in the virtual try-on application.

(a) Eyeglasses keypoints sample.

(b) Example of face keypoints.

(c) Example of eyeglasses fitting.

Fig. 3. Example of keypoints for glasses fitting. Nose keypoints in blue, left ear keypoints in green, right ear keypoints in red. (Color figure online)

Face Point Detection in a similar way, the fitting process requires some keypoints on the 3D face mesh. These include the face points that usually sustain the glasses, such as the nose top F_n and the ears top F_l and F_r, see Fig. 3 for reference. Those need to be computed for each reconstructed face as opposed to the glasses key points. The detection of such points is a complex task since it is not possible to assume that their position is always the same; each face has a different geometry hence different keypoints' placement. To detect the nose keypoint we select the vertex with the maximum Z coordinate value (uses the canonical right-handed coordinate reference system) as the nose tip. Then the positive direction of the Y-axis is iteratively searched for a point that decreases the Z value at least of a threshold; a minimum tolerance around the X coordinate is allowed. The search process runs until no new vertex with decreasing Z value that satisfies all the conditions exists at the maximum allowed distance. The final detected point is the nose keypoint. Since the reconstruction of the 3D vertex map is imprecise around ears location, the ears keypoint detection uses the vertex with the maximum X value as the right ear. The left ear is detected using the vertex that has the minimum X value among the reconstructed map. Similarly, the left ear keypoint is detected using the vertex that has the minimum X value among the reconstructed map. The two detected points are stored as the ears keypoints. The face point detection process is completed in 0.001 s on average.

Fitting Parameters Estimation the glasses fitting parameters are estimated by finding the correct transformation that aligns the glasses model on the reconstructed 3D face using the front view vertex map.

The glasses position is computed using the translation transformation that aligns the glasses bridge keypoint G_n with the face nose keypoint F_n as defined by Eq. 1. The scale factor is assumed to be the same for all the axes and is computed taking into account the face size estimation which is described in Sect. 3.2.

The correction needed to lay the glasses temples on the ears is computed using the projection of the keypoints on the 2D plane defined by the Z and Y axes, as shown in Fig. 4 and in Eq. 2. The two-dimensional unit vector $\hat{g}$ that represents the direction from the nose keypoint to the glasses left keypoint is computed. In a similar way is defined the unit vector $\hat{e}$ that represents the direction from nose keypoint to the left ear keypoint. These two vectors are then used to compute the angle of rotation α. A second angle β is computed in the same way but using the right face and glasses keypoints; the average of the angles is therefore used to build the X axis rotation transformation needed to lean the glasses on the ears.

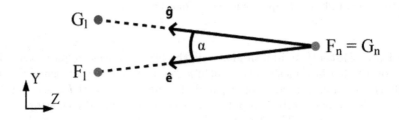

Fig. 4. Eyeglasses pitch angle between ear and temple keypoints.

$$t = F_n - G_n \tag{1}$$

$$
\hat{g} = \frac{G_l - G_n}{|G_l - G_n|}
$$
$$
\hat{e} = \frac{F_l - F_n}{|F_l - F_n|} \tag{2}
$$
$$
\alpha = arctan2(|\hat{g} \times \hat{e}|,\ \hat{g} \cdot \hat{e})
$$

Since the 3D reconstruction is directly stackable on the input image, it is possible to compute face size using the distance between the ear keypoints as the number of pixels between them multiplied by the size of each pixel in millimeters. Knowing the size difference between the coordinate system of the facial reconstruction and of the glasses mesh makes it possible to scale the latter keeping the real dimensions fit over the face. Figure 5 shows the fitting difference between *best fit* and *estimated sizes fit*, in the latter case the glasses are slightly thicker on the face. The best fit strategy does not take into account face and eyeglasses dimensions, the glasses are simply scaled to make the keypoints match the face

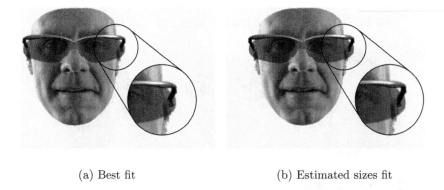

(a) Best fit   (b) Estimated sizes fit

Fig. 5. Examples of eyeglasses fitting using best fit and estimated sizes fit.

ones. Instead, the estimated size fit strategy uses real glasses dimensions and estimated face size (computed as explained above) to provide a realistic glasses fit, in this case face and glasses keypoints are only used to guide the eyeglasses placement and have no role in the eyeglasses scale definition.

In the case of a failure of the face size estimation, the scale factor is computed as the mean X-axis scale factor that best fits the temples on the ears (the *best fit* startegy). Finally, the angle of rotation that leans the glasses on the ears is computed as the mean of the rotation needed for both the ears. The lean angle for each ear is determined using the orientation difference between the 2D YZ projection of the vectors that connect the glasses' nose keypoint with the temple keypoint and with the ear keypoint. The fitting parameter estimation requires around 0.4 milliseconds for each eyeglasses model in the library.

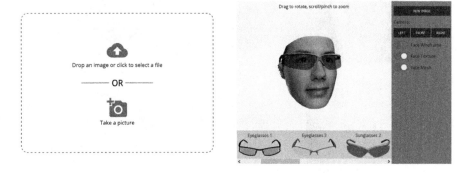

(a) Image upload interface   (b) Virtual try-on interface

Fig. 6. Screen captures of the virtual try-on web application user interface.

(a) Input image   (b) Output, front   (c) Output, side

Fig. 7. Examples of virtual try-on results on images from the FFHQ dataset [15]. Glasses models from CadNav [6] and Blend Swap [5].

User Interface the virtual try-on user interface is available as a web application that interact with the remote face reconstruction and fitting parameters estimation web service. Through a web browser, the user can upload a picture or take one using his device camera. The reconstructed face mesh is rendered in the browser using a 3D rendering framework; this allows the user to see the face from different angles and levels of zoom. The interface displays the list of available glasses frames; when the user selects a model, the application loads and render the relative mesh in the correct position using the estimated fitting parameters. Figure 6 shows the user interface that is composed of two main screens, the first is for face image upload or acquisition and the second one is to expose the 3D virtual try-on.

Some results of virtual try-on with different face pictures and eyeglasses models are visible in Fig. 7, the images used are taken from the FFHQ dataset [15] which provides high-quality image of human faces.

For the implementation of the web application we used *Vue.js*[1] as our core JavaScrip framework. The face picture can be acquired using a webcam or uploaded manually using an image file, file upload is handled by *Dropzone.js*[2]. Finally the 3D rendering framework is provided by *babylon.js*[3], while other JavaScript 3D rendering engines are publicly available (e.g. three.js[4]) we found out the babylon.js provides better support to some features used by this application such as Physically Based Rendering (PBR) Materials and lenses transparencies.

4   Conclusions and Future Works

We presented a web application for virtual try-on of glasses frames in 3D space that consider real measures to provide a realistic try-on experience. The try-on process only requires a picture portraying the user's face to perform size estimation using the iris diameter as a reference and does not require additional markers as other try-on solutions. However, some limitations are present: the generated mesh is not complete on the top and rear parts and does not include the hair; the texture is sometimes incomplete and generally of low quality. Furthermore, while the 3D reconstruction also works in the case of non-frontal pictures, the face size estimation provides correct values only on perfectly focused fronto-parallel pictures of high resolution. Another limitation is that while the 3D face reconstruction method still works with the presence of occlusions, the texture generation process does not take into account the presence of them. Since it does not really make sense to provide a try-on experience over a reconstruction that uses an incomplete or occluded texture an improvement will be to add a step of verification for presence of occlusions and in such case it will be better to ask the user to provide a new photo. Another possibility is to deal with occlusions

---

[1] https://vuejs.org/.

[2] https://www.dropzonejs.com/.

[3] https://www.babylonjs.com/.

[4] https://threejs.org/.

trough restoration techniques such as in [2,7]. Further work can be also done to provide a quantitative evaluation for both the face reconstruction and the eyeglasses fitting quality. Another useful advance could be an evaluation of the web application usability like the preliminary one done in [18].

Finally, a conversion of the web interface to a mobile application will provide better integration with user devices and enhanced ease of use; this will require to develop a new application for the user interface that will use the existing background web service for the process of face reconstruction and eyeglasses fitting.

References

1. 3dMD: 3dmdface system. https://3dmd.com/products/#!/face. Accessed 28 Sep 2020
2. Bianco, S., Ciocca, G., Guarnera, G.C., Scaggiante, A., Schettini, R.: Scoring recognizability of faces for security applications. In: Image Processing: Machine Vision Applications VII, vol. 9024, p. 90240L. SPIE (2014)
3. Bianco, S., Ciocca, G., Marelli, D.: Evaluating the performance of structure from motion pipelines. J. Imaging 4(8), 98 (2018)
4. Blanz, V., Vetter, T., et al.: A morphable model for the synthesis of 3d faces. Siggraph. 99, 187–194 (1999)
5. Blend Swap LLC: Blend swap website. https://www.blendswap.com. Accessed 11 Nov 2020
6. CadNav: Cadnav website. https://www.cadnav.com. Accessed 11 Nov 2020
7. Colombo, A., Cusano, C., Schettini, R.: Three-dimensional occlusion detection and restoration of partially occluded faces. J. Math. Imaging Vis. 40(1), 105–119 (2011)
8. DITTO Technologies: Ditto website. https://ditto.com. Accessed 28 Sep 2020
9. Feng, Y., Wu, F., Shao, X., Wang, Y., Zhou, X.: Joint 3d face reconstruction and dense alignment with position map regression network. In: Proceedings of the European Conference on Computer Vision (ECCV), pp. 534–551 (2018)
10. Gecer, B., Ploumpis, S., Kotsia, I., Zafeiriou, S.: Ganfit: Generative adversarial network fitting for high fidelity 3d face reconstruction. In: Proceedings of the IEEE Conference on Computer Vision and Pattern Recognition, pp. 1155–1164 (2019)
11. Gross, R., Matthews, I., Cohn, J., Kanade, T., Baker, S.: Multi-Pie. Image Vis. Comput. 28(5), 807–813 (2010)
12. Huber, P., Kopp, P., Christmas, W., Rätsch, M., Kittler, J.: Real-time 3d face fitting and texture fusion on in-the-wild videos. IEEE Signal Process. Lett. 24(4), 437–441 (2017)
13. Jackson, A.S., Bulat, A., Argyriou, V., Tzimiropoulos, G.: Large pose 3d face reconstruction from a single image via direct volumetric CNN regression. In: Proceedings of the IEEE International Conference on Computer Vision, pp. 1031–1039 (2017)
14. Jeeliz: Jeeliz virtual try-on. https://github.com/jeeliz/jeelizGlassesVTOWidget. Accessed 28 Sep 2020
15. Karras, T., Laine, S., Aila, T.: A style-based generator architecture for generative adversarial networks. In: Proceedings of the IEEE conference on computer vision and pattern recognition, pp. 4401–4410 (2019)

16. Kerrigan, D., Trokielewicz, M., Czajka, A., Bowyer, K.W.: Iris recognition with image segmentation employing retrained off-the-shelf deep neural networks. In: 2019 International Conference on Biometrics (ICB), pp. 1–7. IEEE (2019)
17. Luxottica Group: Virtual mirror. http://www.luxottica.com/en/virtual-mirror-technology-arrives-valentinocom. Accessed 28 Sep 2020
18. Marelli, D., Bianco, S., Ciocca, G.: A web application for glasses virtual try-on in 3d space. In: 2019 IEEE 23rd International Symposium on Consumer Technologies (ISCT), pp. 299–303. IEEE (2019)
19. Perfect Corp: Youcam makeup. https://www.perfectcorp.com/app/ymk. Accessed 28 Sep 2020
20. Piotraschke, M., Blanz, V.: Automated 3d face reconstruction from multiple images using quality measures. In: Proceedings of the IEEE Conference on Computer Vision and Pattern Recognition, pp. 3418–3427 (2016)
21. Ranjan, A., Bolkart, T., Sanyal, S., Black, M.J.: Generating 3d faces using convolutional mesh autoencoders. In: Proceedings of the European Conference on Computer Vision (ECCV), pp. 704–720 (2018)
22. Roth, J., Tong, Y., Liu, X.: Adaptive 3d face reconstruction from unconstrained photo collections. In: Proceedings of the IEEE Conference on Computer Vision and Pattern Recognition, pp. 4197–4206 (2016)
23. Rüfer, F., Schröder, A., Erb, C.: White-to-white corneal diameter: normal values in healthy humans obtained with the orbscan ii topography system. Cornea **24**(3), 259–261 (2005)
24. Tuan Tran, A., Hassner, T., Masi, I., Medioni, G.: Regressing robust and discriminative 3d morphable models with a very deep neural network. In: Proceedings of the IEEE Conference on Computer Vision and Pattern Recognition, pp. 5163–5172 (2017)
25. XLabz Technologies: Glassify ios application. https://apps.apple.com/it/app/glassify-tryon-virtual-glass/id1166851088. Accessed 28 Sep 2020
26. Yu, F., Koltun, V., Funkhouser, T.: Dilated residual networks. In: Proceedings of the IEEE conference on computer vision and pattern recognition, pp. 472–480 (2017)

Performance Assessment of Face Analysis Algorithms with Occluded Faces

Antonio Greco, Alessia Saggese$^{(\boxtimes)}$, Mario Vento, and Vincenzo Vigilante

Department of Information Engineering, Electrical Engineering and Applied
Mathematics, University of Salerno, Salerno, Italy
{agreco,asaggese,mvento,vvigilante}@unisa.it
http://mivia.unisa.it

Abstract. In retail environments, it is important to acquire information
about customers entering in a selling area, by counting them, but also by
understanding stable traits (such as gender, age, or ethnicity) and temporary feelings (such as the emotion). Anyway, in the last year, due to the
COVID-19 pandemic, it is becoming mandatory to wear a mask, covering
at least half of the face, thus making the above mentioned face analysis
tasks definitely more challenging. In this paper, we evaluate the drop in
the performance of these analytics when the face is partially covered by
a mask, in order to evaluate how existing face analysis applications can
perform with occluded faces. According to our knowledge, this is the first
time a similar analysis has been performed. Furthermore, we also propose two new datasets, designed as extensions with masked faces of the
widely adopted VGG-Face and RAF-DB datasets, that we make publicly available for benchmarking purposes. The analysis we conducted
demonstrates that, except for gender and ethnicity recognition whose
accuracy drop is quite limited (less than 10%), further investigations are
necessary for increasing the performance of methods for age estimation
(MAE drop between 4 and 10 years) and emotion recognition (accuracy
decrease between 45% and 55%).

Keywords: COVID-19 · Deep learning · Face analysis

1 Introduction

Analyzing the customers entering in the selling area and understanding their
feelings allows to improve the customer experience and then to enhance the
development of the business [17]. Among the different technologies, video analysis surely plays a crucial role, since it allows to know in real time the number
of people inside a store, the typical paths of the customers [6], their behaviours

---

This research was partially supported by the Italian MIUR within PRIN 2017 grants,
Projects Grant20172BH297 002CUP D44I17000200005 I-MALL, and by A.I. Tech -
www.aitech.vision.

A. Del Bimbo et al. (Eds.): ICPR 2020 Workshops, LNCS 12662, pp. 472–486, 2021.
https://doi.org/10.1007/978-3-030-68790-8_37

[2] but also the stable traits (such as gender, age and ethnicity) and the temporary feeling (emotion). In particular, stable traits and temporary feeling can be obtained by analyzing the faces of the customers. Due to COVID-19 pandemic, it is becoming mandatory in several countries all around the world to wear a mask on the face in indoor environments, both public and private.

The presence of occlusion on the face may have a severe impact on the effectiveness of face analysis tasks, such as gender, ethnicity, age or emotion, which are analyzed in literature from different points of view.

Gender recognition is among the most simple face analysis task [33], being a binary problem (male, female). In recent years, several methods adopting local [3] and global [4] features have been proposed, but the advent of deep learning moved the interest to fast and effective convolutional neural networks (CNNs) [20]; some of these methods have also been optimized for running in real-time on board of smart cameras [10] and social robots [14,36].

Ethnicity recognition [15] is probably the task less investigated, maybe due to the absence of standard, large and reliably annotated datasets. Therefore, the attention of the researchers is focused on the collection of new datasets and on the realization of extensive benchmarks with modern CNNs [18,27].

Age estimation is one of the most challenging face analysis tasks even for humans. Around 10 years ago, a comprehensive survey on the topic [16] outlined the main concepts related to the aging process and described the methods based on handcrafted features widely adopted for the problem at hand. However, a significant performance improvement has been experienced after the deep learning revolution [9]. To this aim, in the last two years various surveys reporting methods based on CNNs [9,21] or comparisons between handcrafted and automatically learned features [5] have been proposed.

Emotion recognition is perhaps the most fashionable face analysis task in this period, especially for its possible usage in social robotics [19] or business intelligent applications. Encouraged by various recent challenges on this topic [29], various research groups are moving towards the design of novel methods based on CNNs for image-based [30] or multi-modal expression recognition [11].

Anyway, there is still a limited available literature related to the evaluation of the performance of the above mentioned face analysis tasks in presence of occlusions, and in particular in presence of masks. For several years, the scientific community has mainly focused on the detection of the presence of a mask, or on the detection of face masked, for security reasons. Indeed, the presence of a face covered by a mask, for instance of a person entering a bank, has to be detected since it can be associated to a criminal acts. In [41] the authors use Gabor filters for identifying facial features, and then define geometric heuristics for the detection of the mask. In [7] a CNN-based cascade framework has been designed in order to detect masked faces and distinguish them from non masked faces.

The presence of a mask on the face has been also investigated for face recognition tasks. Indeed, differently from the above mentioned face analysis tasks, several methods have been proposed with the aim to increase the robustness of face recognition methods in presence of a mask. In [13] and [12] the authors define a novel face recognition algorithm, able to also work with faces covered by a mask. The auhors propose a detector specially devised for face masked, namely a Multi-Task Cascaded Convolutional Neural Network (MTCNN); then, a feature vector is extracted by means of Google FaceNet [1] embedding model, and a SVM classifier is employed for classification. In [39] the authors propose MaskNet, a convolutional module able to adaptively generate different feature map masks for different occluded face images.

With the only exception of face recognition tasks, according to our knowledge, there are not any contribution for evaluating the drop of performance in other face analysis tasks, such as gender, age, ethnicity and emotion recognition. Starting from this assumption and given the recent interest in this topic, in this paper we provide a two-fold contribution: 1. we propose two new datasets, namely VGGFace2-M and RAF-DB-M, obtained by adding a mask on the faces contained in the original VGGFace2 and RAF-DB datasets, respectively. 2. we perform an experimental analysis to evaluate the performance gap of face analysis convolutional neural networks in presence of mask, considering different deep neural networks, typically used for face analysis tasks, and three different datasets, namely the available LFW+M and our VGGFace2-M and RAF-DB-M. In the whole, more than 3.3 million of face images have been experimentally evaluated in our experiment.

The paper is organized as follows. The architectures used in our benchmark are detailed in Sect. 2; the datasets used, together with the procedure adopted for generated the novel VGGFace2-M dataset are reported in Sect.3. The adopted protocol, together with the achieved results, is reported in Sect. 4. A discussion is also done in Sect. 5. Finally, some conclusions are drawn in Sect. 6.

2 The Architectures

2.1 Gender Recognition

For analyzing the performance drop of gender recognition algorithms in presence of masks, we partially reproduced the benchmark proposed in [20]. It includes various versions of MobileNet, ShuffleNet, SqueezeNet and Xception, but we have reused only the most effective. In particular, we adopted the full versions (depth multiplier equal to 1) with input size 224×224 for MobileNet, ShuffleNet and SqueezeNet and 150×150 for Xception.

These CNNs have been trained from scratch through Xavier initialization for 100 epochs with the training set of VGG-Face2, starting from a learning rate of 0.005 halved every 20 epochs. Adam was chosen as optimizer, with a batch size equal to 64 and a dropout rate of 0.2, adopted for regularization purposes.

In addition, we extended the framework with other convolutional neural networks, namely VGG, SENet and DenseNet, detailedly described in Sect. 2.3. Also for these CNNs we used the same training procedure described above.

2.2   Ethnicity Recognition

The benchmark adopted as reference for ethnicity recognition is the one proposed in [18], which includes five different convolutional neural networks pre-trained on the VMER dataset: VGG-Face, VGG-16, MobileNet, ResNet-50 and ResNet-34. The CNNs trained with the procedure described in this paper demonstrated a good generalization capability on different datasets. However, the analysis of the class activation maps reported in this scientific work shows that the neural networks focus their attention in the region of nose and mouth for specific ethnicity categories; therefore, we expect a performance drop in presence of masks.

The CNNs have been trained by using the VMER dataset for 20 epochs by starting from the ImageNet weights, except for VGG-Face. The procedure was designed with a batch size equal to 64 and the use of Adam optimizer. The learning rate has been scheduled to start from 0.0005, with a decay of 0.5 every 6 epochs. An early stopping mechanism was implemented to freeze the training whether the accuracy on the validation set did not improve for 3 consecutive epochs.

2.3   Age Estimation

Since there are no standard benchmarks including modern convolutional neural networks for age estimation, we selected and trained 4 popular architectures: VGG, SENet, DenseNet and MobileNet.

VGG, defined in [38], is the family of CNNs most widely used for face analysis tasks. It adopts 3×3 filters to build larger filters and obtain a larger receptive field while reducing the number of weights. In this paper, we use the VGG-16 version, which consists of 13 convolutional and 3 fully connected layers and requires an input size equal to 224×224.

SENet, described in [24], is a modified version of the ResNet architecture [23], with the addition of various *Squeeze and Excitation* modules. This architectural change allows to weigh the channels of the feature maps, in order to use only the most relevant features for the classification. In this paper, we adopt the SENet-50 version, which consists of 1 convolutional layer, 16 shortcut modules and 1 fully connected layer, and requires an input size equal to 224×224.

DenseNet, introduced in [25], is a family of CNNs characterized by the fact that each layer is connected to all the others through dense blocks, in order to aggregate the features learned at different resolutions, normalized through tailored pooling operations in the transition layers. For our experiments, we adopted the DenseNet-121 version, with 224×224 input size.

MobileNet, proposed in [37], is a family of CNNs optimized for mobile and embedded devices. It reduces the necessary number of weights and operations with respect to other existing architectures through the use of depthwise convolutions followed by pointwise convolutions and bottleneck layers. In this paper, we use the MobileNet V2 full version, which requires an input size of 224×224 pixels.

For all the CNNs, we start from the weights pre-trained over ImageNet, to reduce the overfitting effect and the convergence time [9]. Of course, we transform them in regressors, by replacing the classification layer with a single neuron providing an output in the range [0, 100]. The training set adopted for this experiment is IMDB-Wiki, by extracting the face images through the a face detector based on SSD [32] and rescaling them to a resolution of 224 × 224 pixels. As suggested in [34], we subtract the average face computed over the VGG-Face dataset to center in average the input distribution around zero and achieve faster convergence. We perform data augmentation by randomly applying random crop, horizontal flip, rotation, skew, brightness and contrast to the original images. The training over IMDB-Wiki is performed for 70 epochs, by using a SGD optimizer and an initial learning rate of 0.005 (reduced of 20% every 20 epochs) for all the CNNs except for VGG, which requires a smaller value (0.00005) probably for the absence of batch normalization.

2.4   Emotion Recognition

According to the same considerations done for age estimation, we train our own CNNs for emotion recognition. The adopted CNN architectures are the same, used directly as classifiers; also, the faces are extracted and resized in the same way, but the training procedure is slightly different. Indeed, we divided the RAF-DB training set in two folds: 80% for the training and 20% for the validation. We performed the training for 220 epochs, by adopting a SGD optimizer and an initial learning rate equal to 0.002, halved every 40 epochs. The used loss function is the cross entropy, with a weight decay set to 0.005. The validation set has been used to obtain the optimal values of the parameters.

3   Dataset

Fig. 1. Examples of images from the LFW+M Dataset.

In our experimentation we use the following three datasets for testing: LFW+M, VGGFace2-M and RAF-DB-M.

LFW+M has been recently introduced by [40] as an extension of the well known LFW+ dataset [22,26]. It is composed by 15,699 unconstrained face images of about 8,000 subjects. In the original LFW+, for each face image, three MTurk workers were asked to provide their estimates of age, gender, and

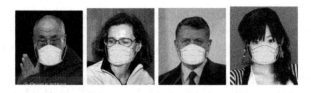

Fig. 2. Examples of images from the VGGFace2-M dataset.

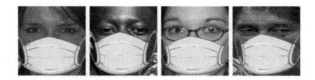

Fig. 3. Examples of images from the RAF-DB-M dataset.

ethnicity (White and No White classes). The apparent age is determined as the average of the three estimates, while gender and ethnicity are determined through the majority voting rule. Some images belonging to the dataset have been reported in Fig. 1. We use this dataset for evaluating the drop in the performance of the following three tasks: gender recognition, ethnicity recognition and age estimation.

We designed VGGFace2-M and RAF-DB-M in this paper as an extension of the original VGGFace2 and RAF-DB datasets. These datasets include the same face images, but covered by a synthetically added mask. In order to generate both VGGFace2-M and RAF-DB-M, we adopted the procedure proposed in [40]: the dlib face detector [28] has been applied and 68 facial landmarks have been identified on each face.

The position and the shape of the mask on the face is determined by using 4 points: the nose point, the chin bottom point, the chin right point and the chin left point. The left part of the mask is built by using the chin left point, the chin bottom point and the nose point, while the right part is obtained symmetrically with the chin right point. The two parts are then merged, by obtaining a mask whose height is equal to the distance between the nose point and the bottom chin point and whose width is equal to the sum of the widths of the two parts. Finally, the mask is rotated of the angle between the nose point and the extremity of one of the eyes.

In more details, **VGGFace2-M**, like the original VGGFace2 [8], contains over 3 millions of face images belonging to 9,131 different subjects, annotated with gender and identity. It is so far the largest publicly available dataset wih face images. Some samples of the dataset have been reported in Fig. 2. In our analysis, it has been used for evaluating the drop in the performance of the gender recognition task.

RAF-DB-M, like the original RAF-DB (Real-world Affective Faces Database) [31], is composed by about 15,339 face images, each one labelled by about 40 human annotators with the neutral class and the six basic emotions: anger, disgust, fear, happiness, sadness and surprise. The dataset is well known for its reliability, since the annotation procedure is very careful and accurate. It is already divided in training and test set. Thus, we used this for analyzing the drop in the emotion recognition task. Some examples are shown in Fig. 3.

Both the datasets will be made publicly available in the MIVIA Lab web site[1].

4   Experimental Analysis

In this Section we report and comment the results of the experiments carried out for the tasks of gender recognition, ethnicity recognition, age estimation and emotion recognition in presence of mask.

4.1   Gender Recognition Results

As mentioned in Sect. 2.1, the benchmark for gender recognition methods include MobileNet, ShuffleNet, SqueezeNet, Xception, VGG, SENet and DenseNet, which are applied over the test sets of VGGFace2, LFW+ and their masked versions VGGFace2-M and LFW+M. The results of these experiments are reported in Table 1 and 2, respectively. The tables show the accuracy achieved on the original dataset, the one obtained on the masked version and the relative performance drop. As expected, there is drop over both the datasets, but its entity is less than 10% in average, namely 6.59% for VGGFace2 and 5.12% for LFW+. This is surely an encouraging result, since the performance is more or less preserved despite the presence of the mask.

VGG is the best performing CNN over the original and masked datasets, achieving an accuracy of 96.1% on VGGFace2 and 98.7% on LFW+ and a performance of 92.1% on VGGFace2-M and 94.6% on LFW+M. This CNN architecture was already known for its excellent gender recognition accuracy and the drop of 4.2% over both the datasets also demonstrates a good resilience to mouth and nose occlusions.

Another CNN which proved a substantial robustness to mask occlusions is ShuffleNet, which achieves the second top accuracy over VGGFace2-M (91.0%) and LFW+M (94.2%) and the lowest relative performance drop over the latter (4.0%). Even MobileNet, obtaining an accuracy higher than 90% over both the masked datasets (90.2% and 93.6%) and a relative performance drop lower than 6% (5.8% and 4.8%), can be considered an excellent alternative, especially due to its additional advantages in terms of processing time. On the other hand, Xception achieves substantially lower accuracy over both the original and masked datasets.

The other CNN architectures have a less stable behavior. SqueezeNet achieves the fourth best performance over VGGFace2-M (89.9%), but the worst accuracy

---

[1] https://mivia.unisa.it/.

on LFW+M (88.9%), with a drop around $6 - 7\%$ over both the datasets. SENet and DenseNet obtain a remarkable accuracy over LFW+M (93.6% and 93.7%) and a drop less than 5% with respect to the original dataset, but a substantially lower accuracy (89.6% and 88.5%) and a drop over the average on VGGFace2-M.

Table 1. Gender recognition accuracy achieved by the considered convolutional neural networks over VGGFace2 and VGGFace2-M. The methods are ordered for increasing accuracy over VGGFace2-M, so that the best result is at the top.

Method	VGGFace2	VGGFace2-M	Drop (%)
VGG	96.1	92.1	4.2
ShuffleNet	95.8	91.0	5.0
MobileNet	95.8	90.2	5.8
SqueezeNet	95.7	89.9	6.1
SENet	96.0	89.6	6.7
DenseNet	95.7	88.5	7.5
Xception	94.3	84.1	10.8

Table 2. Gender recognition accuracy achieved by the considered convolutional neural networks over LFW+ and LFW+M. The methods are ordered for increasing accuracy over LFW+M, so that the best result is at the top.

Method	LFW+	LFW+M	Drop (%)
VGG	98.7	94.6	4.2
ShuffleNet	98.1	94.2	4.0
DenseNet	98.3	93.7	4.7
SENet	98.3	93.6	4.8
MobileNet	98.3	93.6	4.8
Xception	95.7	89.9	6.1
SqueezeNet	95.9	88.9	7.3

4.2 Ethnicity Recognition Results

The benchmark for ethnicity recognition includes the methods described in [18], namely VGG-Face, VGG-16, MobileNet, ResNet-50 and ResNet-34, applied over LFW+ and LFW+M. The results of this experiment are summarized in Table 3 and even in this case we report the accuracy achieved on the original dataset, the one obtained on the masked version and the relative performance drop.

We note on this face analysis task a higher regularity. Indeed, the gap between the best performing method, namely VGG-Face (97.8% over LFW+ and 92.8% over LFW+M), and the worst, i.e. ResNet-34 (96.1% over LFW+ and 91.7%

over LFW+M), is around $1 - 2\%$. The variance of the drop is quite small, with an average of about 4.8%. Also this result is positive, demonstrating a good capability of the CNNs to generalize even in presence of mask occlusions.

Table 3. Ethnicity recognition accuracy achieved by the considered convolutional neural networks over LFW+ and LFW+M. The methods are ordered for increasing accuracy over LFW+, so that the best result is at the top.

Method	LFW+	LFW+M	Drop (%)
VGG-Face	97.8	92.8	5.1
MobileNet	97.1	92.2	5.0
VGG-16	96.8	92.2	4.7
ResNet-50	96.4	91.9	4.6
ResNet-34	96.1	91.7	4.5

4.3 Age Estimation Results

The benchmark for age estimation includes modern CNN architectures such as VGG, SENet, DenseNet and MobileNet, applied over LFW+ and LFW+M. The results in terms of Mean Absolute Error (MAE) are reported in Table 4, by considering the value achieved over the original dataset, its masked version and the absolute performance drop.

The drop for age estimation is between 4 and 10 years, with a MAE of around 6 years over LFW+ and between 10 and 17 years over LFW+M. This result means that the age estimation error is quite high in presence of mask and, probably, this task requires further investigations.

VGG achieves the best performance even in this case, with a MAE of 6.2 over LFW and of 10.2 over LFW+M; the gap of 4 years is the lowest with respect to all the other CNN architectures. DenseNet obtains the second top performance over LFW+ (6.5), but the worst MAE over LFW+M (16.5), with a drop of 10 years. SENet achieve the second top MAE on LFW+M (12.4) with a drop 1.5 years higher than VGG (5.5), while MobileNet obtains a MAE over LFW+M

Table 4. Age estimation MAE achieved by the considered convolutional neural networks over LFW+ and LFW+M. The methods are ordered for increasing MAE over LFW+, so that the best result is at the top.

Method	LFW+	LFW+M	Drop
VGG	6.2	10.2	4.0
SENet	6.9	12.4	5.5
MobileNet	7.3	13.4	6.1
DenseNet	6.5	16.5	10.0

equal to 13.4 with a gap of 6.1 years with respect to the performance achieved on the original dataset (7.3).

4.4 Emotion Recognition Results

Also the benchmark for emotion recognition includes VGG, SENet, DenseNet and MobileNet, applied over RAF-DB and RAF-DB-M. The results of this experiment are summarized in Table 5. The performance are for this task extremely negative. Indeed, we notice a very high average drop of around 50%. All the CNNs achieve an accuracy ranging from 83% and 87% over the original dataset, while the highest performance over RAF-DB-M is 46.5%. Of course, this is not acceptable for real applications. The presence of mask seems to be decisive in a negative sense.

VGG and SENet achieve similar performance. VGG obtains 85.7% of accuracy over RAF-DB and the best 46.5% over RAF-DB-M, while SENet achieves 86.3% on RAF-DB and 46.2% over its masked version; the gap is around 46%. DenseNet, instead, starts from an accuracy of 84.1% over RAF-DB, but suffers a drop of around 55% over its masked version, achieving a performance of only 37.9%. MobileNet achieves the worst accuracy over RAF-DB (83.8%), but the gap of 48.3% on RAF-DB-M is smaller than DenseNet.

We can conclude that the systems nowadays available for emotion recognition can not be reliably applied when part of the face is occluded by a mask, since even the best performing CNNs are not able to achieve a satisfying performance.

Table 5. Emotion recognition accuracy achieved by the considered convolutional neural networks over RAF-DB and RAF-DB-M. The methods are ordered for increasing accuracy over RAF-DB, so that the best result is at the top.

Method	RAF-DB	RAF-DB-M	Drop (%)
VGG	85.7	46.5	45.7
SENet	86.3	46.2	46.5
MobileNet	83.8	40.4	48.3
DenseNet	84.1	37.9	54.9

5 Discussion

From the analysis of the results we can extrapolate two main insights.

The first evidence is that VGG seems to be the most robust architecture with respect to the occlusions caused by the presence of mask. While the literature already sealed the effectiveness of this CNN in face analysis tasks, its resilience to partial occlusions is a novel experimental finding. Therefore, we can conclude that VGG is the most effective solution for facial soft biometrics recognition, also in presence of mask.

The second insight got from the quantitative analysis is that gender and ethnicity recognition can be effectively carried out even in presence of mask, while age estimation and, especially, emotion recognition, suffer a significant accuracy drop. This means that the already existing CNN architectures, pre-trained over non masked faces, do not have sufficient features for determining the age and the emotion when most of the face is covered by a mask. This aspect can be further investigated with a qualitative analysis, focused on the region of eyes, namely the only part of the face usable in case of mask occlusion.

Figure 4 shows examples of males and females belonging to different ethnicity groups. As for gender recognition, the region of the eyes seems to contain features useful to distinguish men and women. Indeed, physiological studies demonstrate that testosterone and estrogen produce different facial traits in the areas of eyes and cheekbones, which are softer for females than for males due to the higher presence of subcutaneous fat [35]. This scientific evidence may explain the limited performance drop, which is only around 4% for VGG and about 10% in the worst case. Similar considerations can be done for ethnicity recognition, since the region of eyes include discriminant features for the problem at hand [15]. It is well known that the shape and the inclination of the eyes is a distinctive feature for east asians. Although the class activation maps and the activation maximization reported in [18] show a higher discriminant capability of features extracted from nose and mouth for african americans and caucasian latins, the performance gap of around 5% in presence of mask demonstrates that the region of the eyes is also useful for these ethnicity groups.

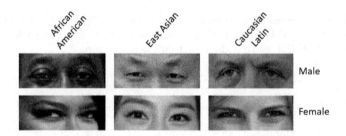

Fig. 4. Region of the eyes of males and females belonging to different ethnicities, namely african american, east asian and caucasian latin. The visible features allows to recognize these soft biometrics with a performance drop lower than 10%.

The task becomes harder when moving to age estimation, as evident from Fig. 5. A survey on the topic [16] states that the aging process mainly concerns a craniofacial growth from birth to adulthood, while skin changes such as wrinkles, eyelid bags and dropping cheeks from adulthood to old age. These shape and textural features are not easy to determine, even without occlusions, due to the reduced image quality in real environments and the possible presence of makeup in case of women. Therefore, we can expect that the CNNs may suffer a performance drop in case of strong occlusions of the face. Our experimental analysis shows that on masked faces the MAE is around 10 years in the best case, so confirming the negative effect of strong occlusions.

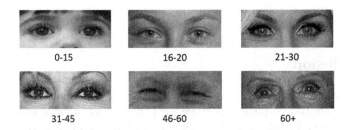

0-15   16-20   21-30

31-45   46-60   60+

Fig. 5. Region of the eyes for specific age groups. The visible features allows to estimate the age with a MAE between 4 and 10 years higher than when using the whole face.

As for emotion recognition, Fig. 6 is a clear proof that the region of the eyes is not sufficient for achieving accurate performance. For example, eyes widening is an action that can be related without substantial differences to a face expressing anger, fear or surprise; without also looking at the mouth, it is impossible to distinguish the three emotional states. A similar consideration applies to glowering, which can be associated to anger and disgust; the former requires the analysis of the whole expression of the face, while the latter requires a further verification of the nose (typically wrinkled) and of the mouth. Even sadness and fear can be confused by only looking at the eyes, especially if the person is crying. On the other hand, happiness seems to be easy to recognize even in presence of mask; however, this observation is true when the happy expression is very clear, otherwise a slight smile could easily be associated with a neutral face. All these qualitative insights justify the strong performance drop of more than 45% which make not sufficiently accurate (around 46% in the best case) the emotion recognition systems in presence of masks.

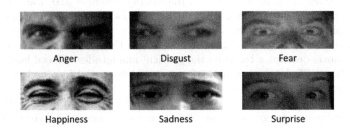

Anger   Disgust   Fear

Happiness   Sadness   Surprise

Fig. 6. Region of the eyes of people with the six basic expressions. The visible features are not sufficient to recognize the emotions, as demonstrated by the accuracy drop higher than 45%.

A possible common solution for increasing the performance of the CNNs for all the considered face analysis tasks may be the training through the region of eyes instead than the whole face. This training procedure may help the CNNs to maximize the discriminating power with the only part of the face visible in the presence of mask and to learn specific features extracted from the region of eyes. Such future direction can help to design and implement reliable solutions for face analysis in all those scenarios in which wearing a mask is mandatory.

6 Conclusions

In this paper we performed an extensive analysis to verify how reliable are the existing face analysis applications when the person wears a mask which occludes most of the face. To this aim, we designed and built two new datasets, namely VGGFace2-M and RAF-DB-M, based on the widely adopted VGGFace2 and RAFDB datasets, by automatically adding a mask on the faces of the persons. We used these datasets, together with the already existing LFW+M, to evaluate the accuracy of the most widely used networks for that specific tasks and to compute the performance drop in presence of mask. We obtained a quite impressive accuracy for the tasks of gender and ethnicity recognition, with $4-5\%$ of drop in the accuracy in the best case, thus confirming the reliability of the existing systems for these tasks even dealing with strong mouth and nose occlusions. Viceversa, this consideration can not be confirmed for the task of age estimation (10 years of MAE in the best case) and, especially, emotion recognition, whose 45% drop in the accuracy makes the systems currently available definitely unreliable and not usable for those contexts in which wearing a mask is still mandatory.

References

1. Google facenet scores almost 100% recognition. Biometric Technology Today 2015(4), 2–3 (2015). https://doi.org/10.1016/S0969-4765(15)30052-7
2. Acampora, G., Foggia, P., Saggese, A., Vento, M.: A hierarchical neuro-fuzzy architecture for human behavior analysis. Information Sciences **310**, 130 – 148 (2015). https://doi.org/10.1016/j.ins.2015.03.021, http://www.sciencedirect.com/science/article/pii/S0020025515001863
3. Azzopardi, G., Greco, A., Saggese, A., Vento, M.: Fast gender recognition in videos using a novel descriptor based on the gradient magnitudes of facial landmarks. In: 2017 14th IEEE International Conference on Advanced Video and Signal Based Surveillance (AVSS), pp. 1–6. IEEE (2017)
4. Azzopardi, G., Greco, A., Saggese, A., Vento, M.: Fusion of domain-specific and trainable features for gender recognition from face images. IEEE Access **6**, 24171–24183 (2018)
5. Bekhouche, S.E., Dornaika, F., Benlamoudi, A., Ouafi, A., Taleb-Ahmed, A.: A comparative study of human facial age estimation: handcrafted features vs. deep features. Multimedia Tools Appl. **79**(35), 26605–26622 (2020). https://doi.org/10.1007/s11042-020-09278-7
6. Brun, L., Saggese, A., Vento, M.: Dynamic scene understanding for behavior analysis based on string kernels. IEEE Trans. Circ. Syst. Video Technol. **24**(10), 1669–1681 (2014). https://doi.org/10.1109/TCSVT.2014.2302521

7. Bu, W., Xiao, J., Zhou, C., Yang, M., Peng, C.: A cascade framework for masked face detection. In: 2017 IEEE International Conference on Cybernetics and Intelligent Systems (CIS) and IEEE Conference on Robotics, Automation and Mechatronics (RAM), pp. 458–462 (2017)
8. Cao, Q., Shen, L., Xie, W., Parkhi, O.M., Zisserman, A.: Vggface2: a dataset for recognising faces across pose and age. In: IEEE Conference on FGR, pp. 67–74 (2018)
9. Carletti, V., Greco, A., Percannella, G., Vento, M.: Age from faces in the deep learning revolution. IEEE Trans. Pattern Anal. Mach. Intell. **42**(9), 2113–2132 (2020)
10. Carletti, V., Greco, A., Saggese, A., Vento, M.: An effective real time gender recognition system for smart cameras. J. Ambient Intell. Humanized Comput. **11**(6), 2407–2419 (2019). https://doi.org/10.1007/s12652-019-01267-5
11. Corneanu, C.A., Simón, M.O., Cohn, J.F., Guerrero, S.E.: Survey on rgb, 3d, thermal, and multimodal approaches for facial expression recognition: history, trends, and affect-related applications. IEEE Trans. Pattern Anal. Mach. Intell. **38**(8), 1548–1568 (2016)
12. Ejaz, M.S., Islam, M.R.: Masked face recognition using convolutional neural network. In: 2019 International Conference on Sustainable Technologies for Industry 4.0 (STI), pp. 1–6 (2019)
13. Ejaz, M.S., Islam, M.R., Sifatullah, M., Sarker, A.: Implementation of principal component analysis on masked and non-masked face recognition. In: 2019 1st International Conference on Advances in Science, Engineering and Robotics Technology (ICASERT), pp. 1–5 (2019)
14. Foggia, P., Greco, A., Percannella, G., Vento, M., Vigilante, V.: A system for gender recognition on mobile robots. In: Proceedings of the 2nd International Conference on Applications of Intelligent Systems, pp. 1–6 (2019)
15. Fu, S., He, H., Hou, Z.: Learning race from face: a survey. IEEE Trans. Pattern Anal. Mach. Intell. **36**(12), 2483–2509 (2014)
16. Fu, Y., Guo, G., Huang, T.S.: Age synthesis and estimation via faces: a survey. IEEE Trans. Pattern Anal. Mach. Intell. **32**(11), 1955–1976 (2010)
17. Greco, A., Saggese, A., Vento, M.: Digital signage by real-time gender recognition from face images. In: 2020 IEEE International Workshop on Metrology for Industry 4.0 IoT, pp. 309–313 (2020)
18. Greco, A., Percannella, G., Vento, M., Vigilante, V.: Benchmarking deep network architectures for ethnicity recognition using a new large face dataset. Mach. Vis. Appl. **31**(7), 1–13 (2020)
19. Greco, A., Roberto, A., Saggese, A., Vento, M., Vigilante, V.: Emotion analysis from faces for social robotics. In: 2019 IEEE International Conference on Systems, Man and Cybernetics (SMC), pp. 358–364. IEEE (2019)
20. Greco, A., Saggese, A., Vento, M., Vigilante, V.: A convolutional neural network for gender recognition optimizing the accuracy/speed tradeoff. IEEE Access **8**, 130771–130781 (2020)
21. Guo, G., Zhang, N.: A survey on deep learning based face recognition. Comput. Vis. Image Underst. **189**, 102805 (2019)
22. Han, H., Jain, A.K., Shan, S., Chen, X.: Heterogeneous face attribute estimation: a deep multi-task learning approach. IEEE Trans. Pattern Anal. Mach. Intell. **99**, 2597–2609 (2017). https://doi.org/10.1109/TPAMI.2017.2738004
23. He, K., Zhang, X., Ren, S., Sun, J.: Deep residual learning for image recognition. In: Proceedings of the IEEE Conference on Computer Vision and Pattern Recognition, pp. 770–778 (2016)

24. Hu, J., Shen, L., Sun, G.: Squeeze-and-excitation networks. In: Proceedings of the IEEE Conference on Computer Vision and Pattern Recognition, pp. 7132–7141 (2018)

25. Huang, G., Liu, Z., Van Der Maaten, L., Weinberger, K.Q.: Densely connected convolutional networks. In: Proceedings of the IEEE Conference on Computer Vision and Pattern Recognition, pp. 4700–4708 (2017)

26. Huang, G.B., Mattar, M., Berg, T., Learned-Miller, E.: Labeled faces in the wild: a database for studying face recognition in unconstrained environments. In: Workshop on faces in 'Real-Life' Images: Detection, Alignment and Recognition (2008)

27. Kärkkäinen, K., Joo, J.: Fairface: Face attribute dataset for balanced race, gender, and age. arXiv preprint arXiv:1908.04913 (2019)

28. Kazemi, V., Sullivan, J.: One millisecond face alignment with an ensemble of regression trees. In: Proceedings of the IEEE Conference on Computer Vision and Pattern Recognition, pp. 1867–1874 (2014)

29. Kollias, D., et al.: Deep affect prediction in-the-wild: Aff-wild database and challenge, deep architectures, and beyond. Int. J. Comput. Vis. **127**(6–7), 907–929 (2019)

30. Li, S., Deng, W.: Deep facial expression recognition: a survey. IEEE Trans. Affect. Comput. (2020)

31. Li, S., Deng, W., Du, J.: Reliable crowdsourcing and deep locality-preserving learning for expression recognition in the wild. In: 2017 IEEE Conference on Computer Vision and Pattern Recognition (CVPR), pp. 2584–2593. IEEE (2017)

32. Liu, W., et al.: SSD: single shot multibox detector. In: Leibe, B., Matas, J., Sebe, N., Welling, M. (eds.) ECCV 2016. LNCS, vol. 9905, pp. 21–37. Springer, Cham (2016). https://doi.org/10.1007/978-3-319-46448-0_2

33. Ng, C.-B., Tay, Y.-H., Goi, B.-M.: A review of facial gender recognition. Pattern Anal. Appl. **18**(4), 739–755 (2015). https://doi.org/10.1007/s10044-015-0499-6

34. Parkhi, O.M., Vedaldi, A., Zisserman, A., et al.: Deep face recognition. In: British Machine Vision Conference (BMVC) (2015)

35. Rhodes, G., Hickford, C., Jeffery, L.: Sex-typicality and attractiveness: are supermale and superfemale faces super-attractive? Br. J. Psychol. **91**(1), 125–140 (2000)

36. Saggese, A., Vento, M., Vigilante, V.: MIVIABot: a cognitive robot for smart museum. In: Vento, M., Percannella, G. (eds.) CAIP 2019. LNCS, vol. 11678, pp. 15–25. Springer, Cham (2019). https://doi.org/10.1007/978-3-030-29888-3_2

37. Sandler, M., Howard, A., Zhu, M., Zhmoginov, A., Chen, L.C.: Inverted residuals and linear bottlenecks: Mobile networks for classification, detection and segmentation. arXiv (2018)

38. Simonyan, K., Zisserman, A.: Very deep convolutional networks for large-scale image recognition. arXiv preprint arXiv:1409.1556 (2014)

39. Wan, W., Chen, J.: Occlusion robust face recognition based on mask learning. In: 2017 IEEE International Conference on Image Processing (ICIP), pp. 3795–3799 (2017)

40. Wang, Z., et al.: Masked face recognition dataset and application. arXiv preprint arXiv:2003.09093 (2020)

41. Wen, C.Y., Chiu, S.H., Tseng, Y.R., Lu, C.P.: The mask detection technology for occluded face analysis in the surveillance system. J. Forensic Sci. **50**, 593–601 (2005). https://doi.org/10.1520/JFS2004409

Who Is in the Crowd? Deep Face Analysis for Crowd Understanding

Simone Bianco⬤, Luigi Celona$^{(\boxtimes)}$⬤, and Raimondo Schettini⬤

Department of Informatics, Systems and Communication University
of Milano - Bicocca, viale Sarca, 336, 20126 Milan, Italy
{simone.bianco,luigi.celona,raimondo.schettini}@unimib.it

Abstract. Crowd understanding plays a vital role in management processes and continues to receive a growing interest in various public and commercial service sectors. Surveillance, entertainment, marketing, and social sciences are only a few of the fields that can benefit from the development of automatic systems for crowd understanding. Little attention has been paid to the study of who are the subjects that characterize the crowd. In this article, we present a crowd understanding system based on face analysis capable of providing statistics about people in the crowd in terms of demographic (i.e. gender and age group), affective state (i.e. eight emotions, valence and arousal), and fine-grained facial attributes.

Keywords: Crowd understanding · Face analysis · Affective state · Convolution neural network

1 Introduction

Crowd understanding has attracted much attention recently due to its wide range of application fields such as surveillance, entertainment, marketing, and social sciences. Much of the existing literature is devoted to crowd behavior analysis [3,11], crowd monitoring [10], and crowd segmentation [18]. Much of this advancement has been triggered by the creation of crowd datasets, as well as robust new features and templates for profiling intrinsic crowd properties. Most crowd understanding studies are scene-specific, meaning the crowd model is learned from a specific scene and thus poor in generalization to describe other scenes or contexts. In recent years, some studies have pointed the attention on the importance of providing an attribute-based description of the crowd as an alternative or complement to categorical characterization [13,14,19]. Indeed, attributes can express more information about a crowded scene. For example, in [13] the authors proposed a method for describing a video by answering "Who is in the crowd?", "Where is the crowd", and "Why is crowd here?". The 94 defined crowd-related attributes describe different scenarios, subjects, and events. While the above attributes can provide good crowd information, an extra level of detail can be achieved by analyzing the faces of people in the crowd.

© Springer Nature Switzerland AG 2021
A. Del Bimbo et al. (Eds.): ICPR 2020 Workshops, LNCS 12662, pp. 487–494, 2021.
https://doi.org/10.1007/978-3-030-68790-8_38

The human face is rich in information and conveys much of our age, emotions and lifestyle, so it is well suited to design an automatic crowd understanding system. Face analysis is extensively studied and applied in several domains. In the automotive field, for example it is useful to check the driver's state of stress or fatigue [5], in IoT applications it is used in smart TVs for children privacy prevention [7] or in smart mirrors to recognize the affective state of people [2]. Recently, many works about customer interest [8,15,17] have been presented.

Knowing who is in the crowd, what interests them, what they want, what they like or dislike is relevant to the management and planning processes in various sectors of public and commercial services. For example, in the world of marketing, customer knowledge is essential for retailers to know: who the customers are, what motivates them, what they want, need, love, or hate. On online websites, there are profiling systems analyzing customers' lifestyle, demographic, and activities to find out about their habits and needs. However, the number of automated systems for analyzing the crowd and its attitude are still limited or completely non-existent on site [6]. Knowing who customers are allows retailers to increase their satisfaction and improve the overall quality of their experience. At the same time, the design of crowd understanding systems makes it possible to implement Ambient Intelligence (AmI), namely: the set of technologies that permits the environment to unobtrusively interact and respond adequately to the human beings it contains [1]. These environments should be aware of people's needs, customizing requirements, and forecasting behaviors. The use of these technologies might serve to send tailored messages, push notifications, and ads to improve the "customer experience" [16]. Understanding who is in the crowd is also useful for safety reasons. Intuitively, the interactions between conference attendees differ from the interactions between museum visitors, and interested people have very different goals in both contexts.

Fig. 1. People standing in front of a shop window.

In this paper we propose a system capable of collecting statistics relating to the demographic composition, the emotional state and other characteristics of groups of people (see Fig. 1). Differently from existing methods it relies on the analysis of faces [13]. This crowd understanding system is suitable for different domains, from understanding customers in front of a shop window, to analyzing flows within a museum or other events.

The remainder of the paper is organized as follows: Section 2 describes the crowd comprehension system, in particular in Sect. 2.2 it is described which and how the characteristics of the faces of the crowd are estimated, and in Sect. 2.3 how these aspects are aggregated and displayed. A discussion of the obtained results are shown in Sect. 3. Finally in Sect. 4, conclusions and future works are outlined.

2   Crowd Understanding System

The proposed system is capable of gathering information about people in a crowd by conducting demographic and sentiment analysis on the faces detected in crowd images. This system is suitable for any event where visitor information is desired (e.g. exhibitions, shopping malls, concerts, shops). The Fig. 2 shows the process flow of the proposed system for crowd understanding. Live crowd image is captured through a camera, a face detector method is applied to localize all the faces in the image, then demographic and sentiment analysis is conducted on each facial image. Finally, the gathered information is aggregated into statistics and visualized in a dashboard.

Fig. 2. Crowd understanding system process flow.

2.1   Face Detection

Given an image of the crowd, a method for detecting faces is applied. The purpose of this part is to detect the position of faces within the image and locate 68 facial keypoints for each face. Among them, only the position of the eyes, nose, and mouth is required for face alignment. It is based on the *DLib* [9] face detector and landmark estimator.

2.2   Demographic and Sentiment Analysis of Facial Images

Each facial image is processed by the face analysis module to recognize several characteristics. It consists of two sub-modules: the first estimates face attributes, and the second recognizes emotions.

More in detail, the first module comprises the Multi-Task Convolutional Neural Network (MTL-CNN) proposed in [4]. By analyzing the facial image, this model can determine a wide variety of information regarding:

- demographic, i.e. gender ender and one among eight age groups (namely, 0–2, 4–6, 8–13, 15–20, 25–32, 38–43, 48–53, 60+ years);
- about 30 visual attributes (concerning: face shape, hairstyle, and beard style);
- perceptual attributes about image quality (e.g., blurry) and subject aesthetics (e.g., attractive);
- the presence of accessories (e.g., earrings, hat, lipstick, necklace, necktie, and eyeglasses).

The second module is based on a CNN designed for simultaneously estimating the affective state both in the discrete domain, i.e., in terms of eight discrete emotions, and the continuous domain, in terms of valence and arousal. To this end, we removed the last fully connected layer of a ResNet-50 architecture pretrained on ImageNet, and we put other two fully connected layers: the first outputs the logits for each emotion category; the second predicts two values corresponding to the scores of valence and arousal, respectively. We train our model on the AffectNet database [12], which consists of around 400,000 training images, and 5,000 validation images of faces annotated by only one human coder in terms of 8 discrete categories and valence-arousal scores.

Table 1 reports the entire list of characteristics the proposed system can recognize given an aligned facial image.

Table 1. Facial characteristics recognized by the proposed crowd understanding system.

Facial details	Parameters
Demographic	Gender, Age range
Affective state	8 emotions (Angry, Contempt, Disgusted, Happy, Neutral, Sad, Scared, Surprised) Valence Arousal
Other attributes	5 o'Clock Shadow, Arched Eyebrows, Bags Under Eyes, Bald, Bangs, Big Lips, Big Nose, Black Hair, Blond Hair, Brown Hair, Bushy Eyebrows, Chubby, Double Chin, Goatee, Gray Hair, Heavy Makeup, High Cheekbones, Mouth Slightly Open, Mustache, Narrow Eyes, No Beard, Oval Face, Pale Skin, Pointy Nose, Receding Hairline, Rosy Cheeks, Sideburns, Straight Hair, Wavy Hair
Perceptual quality	Attractive, Blurry
Accessories	Earring, Eyeglasses, Hat, Lipstick, Necklace, Necktie

2.3   Summarize and Display Statistics

The resulting characteristics for each face are aggregated and processed for the statistical analysis of the crowd. The following statistics are calculated in the proposed understanding system: (i) the number of present people; (ii) the demographic statistics of the people; (iii) the statistics on the affective state of the people; (iv) the statistics on the style of the people (hairstyle, makeup, beard style, and any accessories). Previous data are visualized in a dashboard where it is possible to choose whether to view statistics for the day, month, or the whole year. Figure 3 shows the tab corresponding to the dashboard containing the daily statistics of a shopping center. At the top of the dashboard, there is a line chart that shows the number of people accessed by different entrances. In the center, two modules show the demographic statistics using a pie chart for gender and one for age group; the affective state is shown as a percentage of people per emotion in a pie chart and surface diagram to show the distribution of people for the valence-arousal dimensions. At the bottom, a bar graph shows the percentage of individuals who have the various attributes recognized by the understanding system.

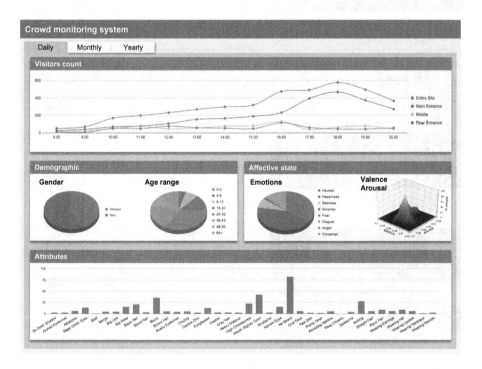

Fig. 3. The dashboard of the proposed crowd understanding system.

3   Discussion

In this section we show the results obtained by the proposed system in several contexts and we highlight the limitations of the current proposal. The image in Fig. 4a is a frontal shot of people at a concert: the faces of people are in a wide range of scales and poses. From the dashboard on the right, we highlight that the system has estimated that there is a high percentage of boys, that the majority of people are between 25 and 32 years old, and that they are happy or

(a)

(b)

(c)

Fig. 4. Results of our understanding system applied in a single instant in different contexts: (a) concert, (b) museum, and (c) shopping center.

neutral. As for the previous image, in Fig. 4b, the people visiting the Vatican museum are shot frontally. Statistics on this image show that it is a group of people, mostly men, of heterogeneous age with a majority in the age range of 38–43 years, and neutral facial expression. Finally, Fig. 4c captures people at a shopping center. The faces depicted are on the same scale, and only the variance of the head pose is high. The proposed system estimates that majority of people are women and there are people of different ages that look mainly happy and surprised.

On the basis of the qualitative results it is possible to note that one of the main problems of the proposed system is that only frontal or nearly frontal faces, sufficiently visible, are detected. This, albeit a limit, is reasonable because if the face is not totally or partially visible, analyzing its appearance is not feasible. The list of recognized attributes and estimated statistics may not be meaningful for all contexts and organizations.

4   Conclusions

In this article we present a crowd understanding system based on face analysis. Its purpose is to collect statistics on the demographic constitution of the crowd, on its affective state, and on other attributes. This system is applicable in various contexts where it is intended to collect information on people in a crowd for statistical purposes and can be used as a complement to existing crowd monitoring systems.

Based on the limitations previously highlighted, the proposed system could be improved by: (i) using a more effective face detector capable of detecting faces in more challenging poses and imaging conditions; (ii) including a face tracking algorithm to avoid recalculating statistics for the same individual in different frames; (iii) making the list of recognizable attributes customizable and also including for example the analysis of the upper body (e.g. outfit analysis).

References

1. Augusto, J.C.: Ambient intelligence: basic concepts and applications. In: Filipe, J., Shishkov, B., Helfert, M. (eds.) Software and Data Technologies, pp. 16–26. Springer, Berlin Heidelberg (2008)
2. Bianco, S., Celona, L., Napoletano, P.: Visual-based sentiment logging in magic smart mirrors. In: ICCE-Berlin, pp. 1–4. IEEE (2018)
3. Brunelli, R., Lanz, O., Santuari, A., Tobia, F.: Tracking visitors in a museum. PEACH-Intelligent Interfaces for Museum Visits, pp. 205–225. Springer, Berlin Berlin (2007)
4. Celona, L., Bianco, S., Schettini, R.: Fine-grained face annotation using deep multi-task cnn. MDPI Sens. **18**(8), 2666 (2018)
5. Celona, L., Mammana, L., Bianco, S., Schettini, R.: A multi-task cnn framework for driver face monitoring. In: ICCE-Berlin, pp. 1–4. IEEE (2018)
6. Chris, F.: Revealed: how facial recognition has invaded shops - and your privacy. The Guardian (2016). https://www.theguardian.com/cities/2016/mar/03/revealed-facial-recognition-software-infiltrating-cities-saks-toronto

7. Hung, P.C.K., Kanev, K., Huang, S.C., Iqbal, F., Fung, B.C.M.: Smart TV face monitoring for children privacy. J. Internet Technol. **19**(5), 1577–1583 (2018)
8. Kasiran, Z., Yahya, S.: Facial expression as an implicit customers' feedback and the challenges. In: CGIV, pp. 377–381 (2007)
9. King, D.E.: Dlib-ml: a machine learning toolkit. J. Mach. Learn. Res. **10**, 1755–1758 (2009)
10. Kulshrestha, T., Saxena, D., Niyogi, R., Cao, J.: Real-time crowd monitoring using seamless indoor-outdoor localization. IEEE Trans. Mobile Comput. **19**(3), 664–679 (2019)
11. Liu, W., Salzmann, M., Fua, P.: Context-aware crowd counting. In: CVPR, pp. 5099–5108. IEEE (2019)
12. Mollahosseini, A., Hasani, B., Mahoor, M.H.: Affectnet: a database for facial expression, valence, and arousal computing in the wild. IEEE Trans. Affect. Comput. **10**(1), 18–31 (2017)
13. Shao, J., Kang, K., Loy, C.C., Wang, X.: Deeply learned attributes for crowded scene understanding. In: CVPR, pp. 4657–4666. IEEE (2015)
14. Shao, J., Change Loy, C., Wang, X.: Scene-independent group profiling in crowd. In: CVPR, pp. 2219–2226 (2014)
15. Taha, A.E.M., Ali, A.: Monitoring a crowd's affective state: status quo and future outlook. IEEE Commun. Mag. **57**(4), 26–32 (2019)
16. Teixeira, T., Wedel, M., Pieters, R.: Emotion-induced engagement in internet video advertisements. J. Market. Res. **49**(2), 144–159 (2012)
17. Yolcu, G., Oztel, I., Kazan, S., Oz, C., Bunyak, F.: Deep learning-based face analysis system for monitoring customer interest. J. Ambient Intell. Humanized Comput. **11**(1), 237–248 (2019). https://doi.org/10.1007/s12652-019-01310-5
18. Zhang, C., Li, H., Wang, X., Yang, X.: Cross-scene crowd counting via deep convolutional neural networks. In: CVPR, pp. 833–841 (2015)
19. Zhou, B., Tang, X., Wang, X.: Measuring crowd collectiveness. In: CVPR, pp. 3049–3056 (2013)

A Saliency-Based Technique
for Advertisement Layout Optimisation
to Predict Customers' Behaviour

Alessandro Bruno[1]($\boxtimes$) (ID), Stéphane Lancette[1], Jinglu Zhang[1] (ID),
Morgan Moore[1], Ville P. Ward[2], and Jian Chang[1] (ID)

[1] National Centre for Computer Animation, Bournemouth University,
Poole BH12 5BB, UK
{abruno,slancette,zhangj,s5113911,jchang}@bournemouth.ac.uk
[2] Shoppar Ltd., Plexal, 14 East Bay Lane, Stratford London E20 3BS, UK
peter.ward@shopparapp.com
https://www.bournemouth.ac.uk/, https://shopparapp.co.uk/

Abstract. Customer retail environments represent an exciting and challenging context to develop and put in place cutting-edge computer vision techniques for more engaging customer experiences. Visual attention is one of the aspects that play such a critical role in the analysis of customers behaviour on advertising campaigns continuously displayed in shops and retail environments. In this paper, we approach the optimisation of advertisement layout content, aiming to grab the audience's visual attention more effectively. We propose a fully automatic method for the delivery of the most effective layout content configuration using saliency maps out of each possible set of images with a given grid layout. Visual Saliency deals with the identification of the most critical regions out of pictures from a perceptual viewpoint. We want to assess the feasibility of saliency maps as a tool for the optimisation of advertisements considering all possible permutations of images which compose the advertising campaign itself. We start by analysing advertising campaigns consisting of a given spatial layout and a certain number of images. We run a deep learning-based saliency model over all permutations. Noticeable differences among global and local saliency maps occur over different layout content out of the same images. The latter aspect suggests that each image gives its contribution to the global visual saliency because of its content and location within the given layout. On top of this consideration, we employ some advertising images to set up a graphical campaign with a given design. We extract relative variance values out the local saliency maps of all permutations. We hypothesise that the inverse of relative variance can be used as an Effectiveness Score (ES) to catch those layout content permutations showing the more balanced spatial distribution of salient pixel. A group of 20 participants have run some eye-tracking sessions over the same advertising layouts to validate the proposed method.

Keywords: Visual saliency · Retail environment · Layout optimisation · Computer vision · Deep learning

© Springer Nature Switzerland AG 2021
A. Del Bimbo et al. (Eds.): ICPR 2020 Workshops, LNCS 12662, pp. 495–507, 2021.
https://doi.org/10.1007/978-3-030-68790-8_39

1   Introduction

Over the last few years, it is observed an increasing demand for software tools for customer retail environments aiming for better understandings of customers' behaviours. As a matter of fact, computer vision-based algorithms have been widely adopted throughout heterogeneous application domains to automatise some context-aware tasks. Some retail companies invest in AI (Artificial Intelligence) and Computer Vision tools to strengthen their rank in a hugely competitive market. The analysis of visual attention processes during the customer experience represents quite a remarkable challenge to set up tasks which could turn out as stepping stones for retail companies. Digital screens are widely adopted in shops and retail environments to grab customers' attention over particular products and services. In this paper, we focus on the study of visual attention and, more particularly, visual saliency as a tool for the assessment and the optimisation of engaging advertisements for customer retail environments. From a computer vision perspective, much of progress has been made on the attempt of imitating and predicting the behaviour of HVS (Human Visual System) over the first seconds of observation of images [6,20]. Many techniques in the scientific literature approach the analysis of the visual attention focusing on the prediction of eye-movements over the first seconds of observation of a given scene. Visual Saliency is meant to decode spatial prediction of eye-movement returning the so-called saliency maps, that is, grey-scale maps encoding the probability each pixel might grab viewers' attention in the continuous range [0,1]. Most of the visual saliency approaches based on deep neural networks trained over publicly available datasets [3,4,7,8,16] allow to achieve high accuracy rates in the prediction of eye-movements in different scenarios. Several saliency-based techniques have been proposed in a wide range of contexts such as computer graphics, remote sensing, biomedical imaging [1,10,23] . Some applications such as image retargeting, image cropping and image quality assessment [2,5] employ visual saliency as a perceptually inspired means to accomplish tasks. In our work, we focus our efforts on how saliency maps can be used as a screening tool for the optimisation of the effectiveness of advertisements to predict customers' behaviour. The increasing interest towards computer vision techniques to make customer retail environments more engaging to potential customers reveals new application domains which need to consider visual attention processes as a leverage to improve the effectiveness of their advertisements. The main idea behind the proposed method is based on the relation between peaks of salient blobs in saliency maps and real eye-movements. If a saliency method reaches high benchmarks over different real fixation points datasets, it might occur the other way around either. We suppose that higher peaks in saliency maps should mostly correspond to the most beaten regions during observation. Furthermore, the sparser the peaks, the more likely viewers will look at areas all over the image. On the other side, the closer the saliency peaks are in a layout, the more likely some locally close regions in the image will grab viewers' attention. In our work, we conduct some experiments on a case study to validate the intuition above behind the proposed method. We rank all spatial permutation of a given advertisement layout with relative variance values of local saliency blobs. We validate the method using real feedback with a web-cam based eye-tracking, collecting eye-movements and fixation points in the first 10 s of observation of images. Our contributions are respectively an automatic saliency-based method to set up the most engaging advertisement content for a given layout and images, a validation session conducted with advertising images and a collection of real eye-movement to assess the robustness of the method. The remainder of the paper is organised as follows: Sect. 2 summarises the state-of-the-art techniques in this topic, Sect. 3 provides a detailed description of the proposed method, Sect. 4 shows the experimental results and Sect. 5 ends the manuscript with conclusions and future works.

2   Related Techniques

In this section, we provide the paper with a description of some articles in the scientific literature focused on the improvement of the customer experiences in retail environments. Researchers from different disciplines highlight interesting aspects on shopper behaviours related to visual attention, data collection, 3D mapping and reconstruction, trajectory detection and others which may literally represent a stepping stone for a new concept of customer retail environments.

The authors in [13] conducted some studies with wearable eye-trackers to detect the most important factors which play prominent roles in the final decision of buying a product. Visual attention is a key aspect in the act of looking longer or repeatedly at a product which will be more likely bought. Huddleston et al. [14] reviewed the progress of eye-tracking technology as a research tool in retail and retail marketing.

Khan [17] studied the impact of visual designs on customer perceptions of online assortments highlighting designs with simpler compositions to be the ones more liked by viewers. Paolanti et al. [21] tackled the topic of semantic store mapping using artificial intelligence models and a retail robot. They set up a system to build a 3D map of both the store and product locations. La Porta et al. [18] proposed a method based on a deep learning mobile application able to identify facial expressions and emotions of subjects from an image. On top of it, the lights of a Christmas tree run different special effects because of the emotions of the subjects around the tree.

Gabellini et al. [12] provided the scientific community with a large scale trajectory of shopper movements in stores employing a real-time locating system with Ultra-Wideband (UWB). Vaira et al. [25] put in place a system to avoid the so-called OOS (out of stock) problem in stores. The solution is based on two cameras, one with a depth sensor and a very high-resolution webcam recognition tasks.

The authors in [24] paired the utilisation of blue-tooth beacons' signal and sensor fusion approach with RGB-D camera to provide an accurate customer position detection and track shoppers movements. Liciotti et al. [19] proposed a software infrastructure consisting of computer vision algorithms and RGB-D

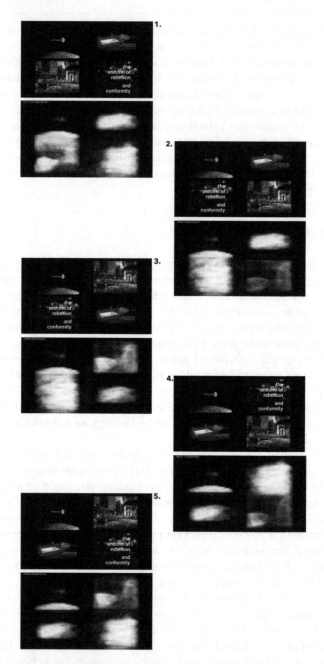

Fig. 1. Some layout content permutations and their saliency maps are shown above. Differences among salient blobs of corresponding images across permutations are noticeable.

cameras providing information related to the analysis of user-shelf interactions. Fuchs et al. [11] focused their efforts on the employment of edge device equipped with AI tools to help retailers improve the quality of their customers' experience. The next section is a detailed description of all steps of the proposed method.

3   Proposed Method

As briefly mentioned in Sect. 1, we propose a saliency-based method for optimising the effectiveness of advertisements in shops and customer retails. We detect salient pixels in images using a deep learning-based solution [26] trained over an object-oriented image and video dataset called DAVIS [22]. As a case of study, we consider a scenario where layout and images are provided to deliver an advertisement.

For the sake of clarity, we refer to the layout as the current combination of images within the given design configuration of images. In Fig. 1 an example with five permutations of a given $2 \cdot 2$ layout and four input images are shown. Each image represents a local region of the overall advertisement layout. The objective is to convey the most engaging advertisement possible with the given inputs (images and layout). We consider a generic M x N grid layout consisting of M x N images. We notice different spatial permutations of the same layout (see Fig. 1) showing different saliency maps. All five permutations in Fig. 1 show blobs whose saliency turns up differently because of the image location within the given layout. Saliency maps encode image pixels in the continuous range [0,1], then the most salient regions in images can be read as regions which may grab viewers' attention most likely. The aforementioned noticeable differences among saliency maps prompt us to further investigate visual saliency as leverage for predicting customers' behaviour in retail environments. A flow-chart in Fig. 2 describes the main steps of the proposed method. For a given M x N grid-based layout and M x N images composing the advertisement, all $P=(M$ x $N)!$ layout permutations are automatically generated. A saliency map out of each permutation is then extracted, summing up to P saliency maps. The most salient pixels are filtered in using a simple spatial threshold to detect the highest peaks of visual saliency in images.

We want to retrieve effectiveness scores out of each permutation saliency map, aiming for catching the one with the most well-balanced spatial distribution of salient pixels over all the images. If an image of the layout is much more salient than another, viewers will likely dwell on it for longer than the other images. The idea behind our algorithm is to employ scores which allow us to detect the layout content permutation that shows lower saliency variance among images. Due to different salient blobs over images in the same layout, we focus our efforts on the analysis of the variance of what we name 'local-saliency'. For the given layout made up of $M \cdot N$ images, we study the 'behaviour' of the overall layout saliency analysing the varying number of salient pixels of each of the image $M \cdot N$ images. In greater detail, we employ the inverse of the relative variance of local saliency maps as ES (Effectiveness Score). In Eq. 1 ES is the ratio between

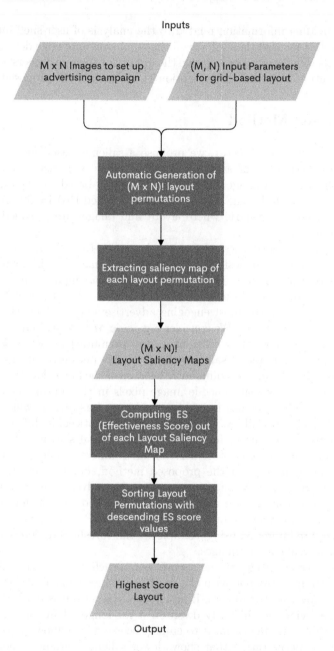

Fig. 2. Flow-chart of the proposed method.

the absolute mean and variance of $NMSP_k$ with $_k = 1, ..., (M \cdot N)$. $NMSP_k$ stands for Number of Most Salient Pixels of each image in the kth layout content permutation.

$$ES_{(i)} = \frac{\left|\mu(NMSP_k(Layout_{(i)}))\right|}{\sigma(NMSP_k(Layout_{(i)}))^2} \tag{1}$$

$$k = [1, ..., (M \cdot N)] \qquad i = \{1, ..., (M \cdot N)!\}$$

For a given layout with $M \cdot N$ images, $NMSP_h$ is the number of the most salient pixels in the local saliency map $LSM_{(h)}$ of the hth image (see Eq. 2).

$$NMSP_{(h)} = \sum_{i,j \, \epsilon \, Im} LSM_{(h)}(i,j) \geq th \tag{2}$$

Each *Layout* content permutation is the union of $M \cdot N$ images Im'_i as in the Eq. 3

$$Layout = \bigcup_{i'=1}^{M \cdot N} Im_{i'} \tag{3}$$

Each layout content permutation is then sorted along with its ES . The layout showing the highest score is the output of our proposed method. To confirm the robustness of our technique, we run through a validation session showing both best and worst score layout permutations to viewers and capture eye-movements over the first 10 s of observation of each permutation. To record eye-movements we employ Web Tool for eye-tracking called GazeRecorder [9].

4   Experimental Results

In this section, we show experimental sessions on five different advertising campaigns with a $2 \cdot 2$ layout. All images for our experiments are taken both from a publicly available dataset [15] at the link http://people.cs.pitt.edu/~kovashka/ads/ and over the Internet. For each layout, 24 permutations are computed. Their corresponding saliency maps are then extracted, and the 35% most salient pixels are filtered in with Eq. 2. In Fig. 3, we show all ES values of each layout content permutation of the four images in the first advertising campaign. ES values of each layout content permutation are quite different. The layout content permutation with the highest ES is highlighted with a red dot, while the black dot in the graph matches the lowest score layout content permutation. Highest and lowest score layout content permutations are the first and second layouts shown in Fig. 1. Local saliency maps are quite different between the two configurations. We run experimental sessions on 5 advertising campaigns to pick up the permutations with the highest and lowest ES. In Fig. 5, ES values are reported across the 5 advertising campaigns. In our case study, we consider 2 by 2 layouts which sum up to 24 possible permutations for each advertising campaign. The total number of layout content permutations is 120. As it is noticeable in the

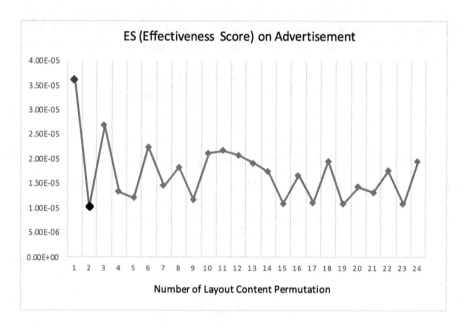

Fig. 3. The graph above shows all ES values for each layout content permutations in the advertising campaign 1. The red dot show the highest score while the black one shows the lowest score (Color figure online)

histogram chart in Fig. 5, gaps between the highest and lowest score over each campaign can be quite different. A plausible explanation is that the content of images which are part of the campaign consists of low, middle and high-level image features having a different impact on the saliency map. The experimental sessions shown so far concern the automatic detection of what we consider to be the most engaging layout content permutation for a given grid-based layout. The remainder of this section is focused on the validation sessions we conduct to assess the robustness of the algorithm behind the proposed method. In greater detail, we pick up all highest and lowest score layout permutations out of 5 advertising campaigns to run through webcam-based eye-tracking sessions. We used an off-the-shelf solution called GazeRecorder based on a web tool integrating a webcam calibration step before eye-tracking sessions. We employed GazeRecorder to record participants' eye-movements over the first 10 s of observation of layout permutations with highest and lowest ES. Twenty subjects took part in the validation sessions giving us the chance to extract statistically meaningful results out of the experiments. The participants in the eye-tracking sessions were equally distributed by gender. Their age was in the range of 25–40 years old. In our case study, we suppose that a digital screen in a retail environment displays an advertising campaign showing some objects or services the retail company or shop offer to customers. One of the objectives is to grab customers' attention to those objects in the advertisement itself. For this purpose, each of 20 participants

a.

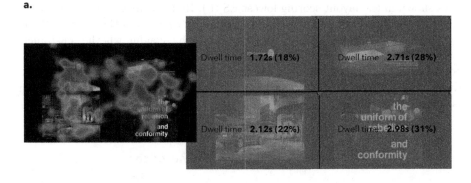

b.

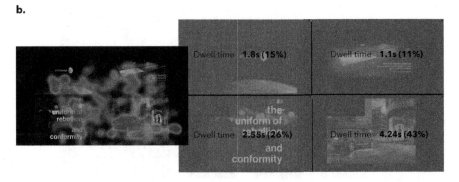

Fig. 4. Heat maps of real eye-movements and dwell times over each image in layouts are shown above. Best (a) and worst (b) layout content permutations show different amount of time spent by viewers over matching images between the two layouts.

is shown the layout content permutations with highest and lowest ES value as per the graph in Fig. 5. Each image is displayed for a time range of 10 s during which on-screen eye-movements of the viewer are recorded. GazeRecorder allows us to set up the validation sessions fine-tuning the time range and the order of layout contents to be shown. Furthermore, coordinates of eye-movements of all sessions are gathered and then displayed as heatmaps (see Fig. 4 overlaid with the input image. Other useful pieces of information are retrieved by manually drawing rectangular regions of interest on the layout content permutation. In Fig. 4, layout content permutations with highest (a) and lowest (b) ES are shown (left-hand side of the figure), the corresponding dwelling times over each image in the layouts are printed out in the right-hand side. Dwell times are meant to provide information on how long viewers spend their time focusing on one out of four pictures in the layout.

As noticeable in Fig. 4, the layout scoring the highest ES (a) show values of dwell times over each image which are more equally distributed than the ones shown in the layout scoring lowest ES (b). Red-car image observations take 2.12 s in the highest ES configuration while more than 4 s in the lowest ES configuration. We consider advertisements to be more engaging whether customer's attention gets caught over all products and services displayed with a sort of equalised distribution of dwell times. Due to the consideration above, the lower dwell time variance of the four images in the layout, the more the engagement is spatially distributed.

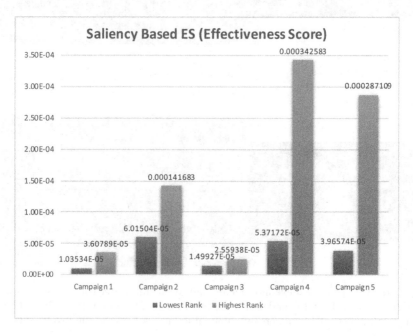

Fig. 5. The graph above shows highest and lowest ES values for each advertisement consisting of 4 images. Each advertisement layout is based on a 2 by 2 grid layout, summing up to a total of 24 permutations. Experiments are run over 5 advertisements, meaning 120 possible layout content permutations.

After describing experimental results related to validation session on the first graphical campaign, we show the variance of dwell times of each image for both highest and lowest ES layout permutation over all the 5 advertising campaigns (see Fig. 6. Highest score layout permutations have dwelling time variance values lower than the lowest score layout permutation. In the graph in Fig. 6, the lower dwell time variance values, the better layout configuration. The preliminary experimental sessions indicate the highest ES layout content permutations to grab viewers' attention over all the images of a given advertisement with more balanced dwell time distribution. The experiments on the automatic optimisation of the given advertisement layouts and images have been carried out on a 13-inch Mac-book Pro with 16 GB of RAM, 2.4 GHz Quad-Core Intel Core i5,

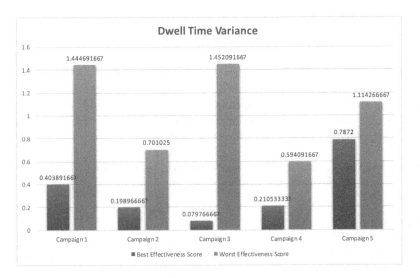

Fig. 6. This graph shows variance values of dwell time over each image in layouts with highest and lowest ES for a given advertising campaign. The first two histogram bars in the left-hand side of the graph are the dwell time variances of first campaign (see Fig. 4 a and b). We run experiments over 5 different campaigns.

Intel Iris Plus Graphics 655 1536 MB. The average running time of the method on the 2-by-2 grid layout is 40 s. The entire project has been developed with python version 3.8.0.

5   Conclusions and Future Works

In our work, we show how saliency maps as leverage for automatically optimising advertisement layout contents. We aim to provide retail environments with a lightweight, cost-effective and reliable software solutions to predict customers' eye-movements and dwell times on advertised products in a given graphical layout. Entirely meaningful changes in the level of engagement of an advertisement layout might occur because of changes in positions of its visual elements. We set up a new Effectiveness Score of a layout content as the ratio of absolute mean to the variance of the number of the most salient pixel of each image composing the advertisement itself. The score returns a measure of each image contribution to the overall spatial distribution of saliency in the layout. The final version of an advertisement is the combination of some ingredients such as layout and images. Our technique allows providing the most engaging layout content out of all possible permutations. Our method is validated by some preliminary experiments with real feedback of 20 subjects who underwent webcam-based eye-tracking sessions. Preliminary results show that the same layout permutations, which are best ranked with ES, are the same ones where visual attention is more equally distributed on all images in the advertisement itself. Our work suggests that visual saliency methods trained with artificial intelligence models can be used as a reliable tool to optimise and predict customers' behaviour when they cast their eyes over advertising layouts. Further investigations are necessary for the effectiveness of the method with finer grid layouts. Adding transformations such as scaling and rotation in the extraction of all possible permutations would probably make the functionalities appealing to a broader audience and application domains. We also aim for extending our work to other scenarios taking into account graphic elements with different priority scales.

Acknowledgment. This research was supported by Innovate UK. Smart Grants (39012) - Shoppar: Dynamically Optimised Digital Content.

References

1. Abouelaziz, I., Chetouani, A., El Hassouni, M., Latecki, L.J., Cherifi, H.: 3D visual saliency and convolutional neural network for blind mesh quality assessment. Neural Comput. Appl. **32**(21), 16589–16603 (2019). https://doi.org/10.1007/s00521-019-04521-1
2. Ardizzone, E., Bruno, A.: Image quality assessment by saliency maps. In: VISAPP (1), pp. 479–483 (2012)
3. Borji, A., Itti, L.: Cat 2000: a large scale fixation dataset for boosting saliency research. CVPR 2015 workshop on "Future of Datasets" (2015). arXiv preprint arXiv:1505.03581
4. Borji, A., Sihite, D.N., Itti, L.: Quantitative analysis of human-model agreement in visual saliency modeling: a comparative study. IEEE Trans. Image Process. **22**(1), 55–69 (2013)
5. Bruno, A., Gugliuzza, F., Ardizzone, E., Giunta, C.C., Pirrone, R.: Image content enhancement through salient regions segmentation for people with color vision deficiencies. I-Perception **10**(3), 2041669519841073 (2019)
6. Bruno, A., Gugliuzza, F., Pirrone, R., Ardizzone, E.: A multi-scale colour and keypoint density-based approach for visual saliency detection. IEEE Access **8**, 121330–121343 (2020)
7. Bylinskii, Z., et al.: Mit saliency benchmark. http://saliency.mit.edu/
8. Bylinskii, Z., Judd, T., Oliva, A., Torralba, A., Durand, F.: What do different evaluation metrics tell us about saliency models? arXiv preprint arXiv:1604.03605 (2016)
9. Deja, S.: Gazerecorder. https://api.gazerecorder.com/
10. Diao, W., Sun, X., Zheng, X., Dou, F., Wang, H., Fu, K.: Efficient saliency-based object detection in remote sensing images using deep belief networks. IEEE Geosci. Remote Sens. Lett. **13**(2), 137–141 (2016)
11. Fuchs, K., Grundmann, T., Fleisch, E.: Towards identification of packaged products via computer vision: convolutional neural networks for object detection and image classification in retail environments. In: Proceedings of the 9th International Conference on the Internet of Things, pp. 1–8 (2019)
12. Gabellini, P., D'Aloisio, M., Fabiani, M., Placidi, V.: A large scale trajectory dataset for shopper behaviour understanding. In: Cristani, M., Prati, A., Lanz, O., Messelodi, S., Sebe, N. (eds.) ICIAP 2019. LNCS, vol. 11808, pp. 285–295. Springer, Cham (2019). https://doi.org/10.1007/978-3-030-30754-7_29

13. Gidlöf, K., Anikin, A., Lingonblad, M., Wallin, A.: Looking is buying. how visual attention and choice are affected by consumer preferences and properties of the supermarket shelf. Appetite **116**, 29–38 (2017)
14. Huddleston, P.T., Behe, B.K., Driesener, C., Minahan, S.: Inside-outside: using eye-tracking to investigate search-choice processes in the retail environment. J. Retail. Consum. Serv. **43**, 85–93 (2018)
15. Hussain, Z., et al.: Automatic understanding of image and video advertisements. In: Proceedings of the IEEE Conference on Computer Vision and Pattern Recognition, pp. 1705–1715 (2017)
16. Judd, T., Durand, F., Torralba, A.: A benchmark of computational models of saliency to predict human fixations. In: MIT Technical Report (2012)
17. Kahn, B.E.: Using visual design to improve customer perceptions of online assortments. J. Retail. **93**(1), 29–42 (2017)
18. La Porta, S., Marconi, F., Lazzini, I.: Collecting retail data using a deep learning identification experience. In: Cristani, M., Prati, A., Lanz, O., Messelodi, S., Sebe, N. (eds.) ICIAP 2019. LNCS, vol. 11808, pp. 275–284. Springer, Cham (2019). https://doi.org/10.1007/978-3-030-30754-7_28
19. Liciotti, D., Frontoni, E., Mancini, A., Zingaretti, P.: Pervasive system for consumer behaviour analysis in retail environments. In: Nasrollahi, K. (ed.) FFER/VAAM -2016. LNCS, vol. 10165, pp. 12–23. Springer, Cham (2017). https://doi.org/10.1007/978-3-319-56687-0_2
20. Nguyen, T.V., Zhao, Q., Yan, S.: Attentive systems: a survey. Int. J. Comput. Vis. **126**(1), 86–110 (2018)
21. Paolanti, M., et al.: Semantic 3D object maps for everyday robotic retail inspection. In: Cristani, M., Prati, A., Lanz, O., Messelodi, S., Sebe, N. (eds.) ICIAP 2019. LNCS, vol. 11808, pp. 263–274. Springer, Cham (2019). https://doi.org/10.1007/978-3-030-30754-7_27
22. Perazzi, F., Pont-Tuset, J., McWilliams, B., Van Gool, L., Gross, M., Sorkine-Hornung, A.: A benchmark dataset and evaluation methodology for video object segmentation. In: Proceedings of the IEEE Conference on Computer Vision and Pattern Recognition, pp. 724–732 (2016)
23. Sran, P.K., Gupta, S., Singh, S.: Segmentation based image compression of brain magnetic resonance images using visual saliency. Biomed. Signal Process. Control **62**, 102089 (2020)
24. Sturari, M., et al.: Robust and affordable retail customer profiling by vision and radio beacon sensor fusion. Pattern Recogn. Lett. **81**, 30–40 (2016)
25. Vaira, R., Pietrini, R., Pierdicca, R., Zingaretti, P., Mancini, A., Frontoni, E.: An IOT edge-fog-cloud architecture for vision based pallet integrity. In: Cristani, M., Prati, A., Lanz, O., Messelodi, S., Sebe, N. (eds.) ICIAP 2019. LNCS, vol. 11808, pp. 296–306. Springer, Cham (2019). https://doi.org/10.1007/978-3-030-30754-7_30
26. Wang, W., Shen, J., Shao, L.: Video salient object detection via fully convolutional networks. IEEE Trans. Image Process. **27**(1), 38–49 (2017)

Data-Driven Knowledge Discovery in Retail: Evidences from the Vending Machine's Industry

Luca Marinelli[1], Marina Paolanti[2], Lorenzo Nardi[1(✉)], Patrizia Gabellini[3], Emanuele Frontoni[2], and Gian Luca Gregori[1]

[1] Dipartimento di Management, Università Politecnica delle Marche, Ancona, Italy
l.nardi@univpm.it
[2] Dipartimento di Ingegneria dell'Informazione, Università Politecnica delle Marche (DII), Ancona, Italy
[3] Grottini Lab S.R.L., Via Santa Maria in Potenza, 62017 Porto Recanati, Italy

Abstract. The purpose of this study is to investigate new forms of marketing data-driven knowledge discovery in the vending machine (VM) industry. Data of shopping activities understanding were gathered and analyzed by a system technology based on a RGBD camera. An RGBD camera, already tested in retail environments, is installed in top-view configuration on a VM to gather data that are processed and embedded within a knowledge discovery project. We adopted the knowledge discovery via data analytics framework (KDDA) that was applied to a real-world VM scenario. Real-world case tests, based on more than 17.000 shoppers measured in 4 different locations, were conducted. By using this method it was possible to verify the ability of this approach to generate new forms of marketing data-driven knowledge available on the VM industry. Main novelties are: i) the application of a KDDA project to the VM industry; ii) the use of RGBD data sources for the first time for a KDDA process; iii) the contribution to the practice by supporting retailers to carry out knowledge discovery processes more effectively; iv) the real-world testing process based on 4 locations and more than 17.000 shoppers.

Keywords: RGBD camera · Knowledge discovery via data analytics · KDDA · Big data analytics · Data-driven decision making

1 Introduction

In a world where everything is connected, big data is impacting significantly in many industries, contributing to the business practices improvement. According to an Accenture study, 79% of enterprise executives agree that companies which do not embrace Big Data will lose their competitive position and could face extinction [1]. From the Organizations perspective, one of the most relevant aspect deriving from these innovations is linked to the possibility of generating new forms of data-driven knowledge. This data-driven knowledge allows Organizations to have a better and more complete perspective of the business in which

A. Del Bimbo et al. (Eds.): ICPR 2020 Workshops, LNCS 12662, pp. 508–520, 2021.
https://doi.org/10.1007/978-3-030-68790-8_40

they operate. In fact, several studies show that the adoption of technologies and tools based on big data analytics is able to generate an actionable knowledge [6,29], that can be adopted in several business practices as a fundamental element in decision making processes. According to the McKinsey Analytics report, the sales and marketing functions are those in which the Big Data is providing the largest contributions [23]. At the same time, knowledge discovery models are also evolving. The traditional models described in literature were conceived in a pre-Big Data era [14,27], and are not able to sustain the enormous flow of data that can be generated by a plurality of sources and at a speed never seen before. While in many retail environments the adoption of big data analytics are consolidated practices [5,13,14], there is a lack of studies dealing with this topic in the vending machines (VM) industry. Therefore this study aims to provide a research contribution to fill this gap. In this work, a new system technology based on RGBD camera is proposed [17], which has been installed on real-world VM. Real-world case tests were conducted following a rigorous knowledge discovery process via data analytics (KDDA) [16], through which it was possible to verify the ability of this approach to generate new forms of marketing data-driven knowledge available for this industry. The system is based on cost-effective hardware and easy to install. The experimental results demonstrated that its performances are highly reliable especially in real world environment. Thanks to these actionable knowledge forms, it is possible to measure also the impact of new products, new offers, new promotion and new packaging. Main contributions of this paper are: i) the application of a KDDA project to the VM industry; ii) the use of RGBD data sources for the first time for a KDDA process; iii) the contribution to the practice by supporting retailers to carry out knowledge discovery processes more effectively; iv) the real-world testing process based on 4 locations and more than 17.000 shoppers. The paper is organized as follows: Sect. 3 describes the theoretical background and the research method. Section 4 presents a description of the system technology RGBD camera and its references in literature. Section 5 is the description of the tests following a Knowledge Discovery via Data Analytics framework and the discussion of the most relevant results, followed by conclusions in Sect. 6.

2   Big Data Analytics

One of the most widely accepted definition of big data appeared in 2011, when Mckinsey Global Institute in a research report defined Big Data as "huge datasets whose size is beyond the ability of typical database software tools to capture, store, manage, and analyze" [24].

The National Science Foundation underlines the complexity of this new data environment by defining Big Data as large, diverse, complex, longitudinal and/or distributed data sets generated from instruments, sensors, Internet transactions, email, video, click streams and/or digital sources available today and in the future [25].

Over the years Big Data have been described by examining the main aspects that characterize them. These characteristics, originally called the Three V's of big data [9, 22, 32], were subsequently expanded to the Five V's (5V's) [33] which are the following:

- *volume*: in terms of the size of the data sets that reaches at least the unit of measurement of the petabyte;
- *velocity*: referred to the flow of data that is produced very quickly by the various sources such as machines, business processes, networks, and human interaction with things;
- *variety*: big data can be both structured and unstructured data, since they come in multiple formats (sensor data, audio and video files, text and graphics);
- *veracity*: referred to the quality and reliability of the data despite the large volume of data involving certain and uncertain data;
- *value*: understood as the economic benefits deriving from the availability of big data.

Closely connected to the concept of Big Data is Big Data Analytics, [8] describes them as huge datasets requiring advanced and unique data storage, management, analysis and visualization technologies as well as statistical analysis. Big data analytics are seen as a "holistic approach to manage, process and analyze the" 5 Vs' data related dimensions in order to create actionable insights for sustained value delivery, measuring performance and establishing competitive advantages" [33]. Moreover, Big Data Analytics are described in literature as an important revolution [15], with emphasis on their ability to generate new forms of knowledge as well as to improve the decision-making processes within Organizations.

Several studies have confirmed the role of Big Data Analytics in supporting Organizations both in managing and solving problems, but also in making better decisions [9] through the production of actionable informations [28].

The use of Big Data Analytics has established itself in various fields of marketing. For example, the innovations introduced by the analytics may impact on the decision making processes, mostly in the areas of customer relationship management, marketing mix allocation, personalization, customer privacy, and security issues [34]. Thanks to Big Data Analytics, Organizations can now perform "predictive" analyzes [30] and "prescriptive" analyses. Organizations are therefore able to predict future trends or determine effect-cause relationships to be applied to business processes [26].

For example [5] have investigated the role of Big Data Analytics in retail. The authors have identified in the retailing field five possible dimensions in which big data can make a contribution in the "exploit the vast flows of information: customers, products, time, location, and channel". The authors have also described the new sources of data that are of interest to retailers defined as "related products locations in the store layout and on-line shop within an aisle". This work will be treating and discussing those sets of data. It is important to underline Big Data Analytics projects have never been managed in the VM industry so far.

3 Theoretical Background and Research Methodology

In this study, we discuss the role of the new technologies that through Big Data Analytics are able to generate actionable knowledge for the VM industry; at a conceptual level we refer to the evolutionary aspects of data-driven knowledge discovery processes.

The relationships between the knowledge discovery process and the data-driven approach were presented for the first time in a workshop in 1989 [11], where the term Knowledge Discovery in Databases (KKD) was coined to emphasize the concept that "knowledge" is the end product of a data-driven discovery process. Fayyad et al. (1996) define the Knowledge Discovery in Databases as "the non-trivial process of identifying valid, novel, potentially useful, and ultimately understandable patterns in data". From this definition clearly emerge the key concepts of KDD: the process, which involves the articulation in steps of this practice, and the patterns in data, that is what in this context is understood as "useful or interesting knowledge" that can be obtained working on large set of "real-world data".

One of the key component of data-driven knowledge discovery processes is data mining [4, 20, 21], defined as the process of searching and analyzing data in order to find implicit, but potentially useful, information [3, 12]. One of the most widespread model in data mining projects in business is the CRISP-DM (Cross-Industry Standard Process for Data Mining) [31], which is represented as a process made of six phases:

a) business understanding,
b) data understanding,
c) data preparation,
d) modeling,
e) evaluation,
f) deployment.

Due to the sudden changes in technology and the progress of data collection and analysis techniques such as Big Data Analytics, traditional data-driven knowledge discovery models are not able to guarantee the success of data mining projects in the business [14, 16, 27]. By reviewing the literature it is possible to notice that the most accepted models of knowledge discovery and data mining (KDDM) [10, 12, 21], were all developed before the advent of the Big Data Era.

In this scenario, Organizations are identifying new ways of the joint use of Information Technologies (IT) and advanced analytics techniques with the aim of improving the knowledge discovery process, making it faster, cheaper, more flexible and more reliable [16].

There is therefore a shift from traditional models of knowledge discovery and data mining (KDDM) to models that include the data analytics component that [30] defined as "the anaysis of data, using sophisticated quantitative methods, to produces insights that traditional approaches to business intelligence". [16] adopted the term KDDA to describe "the knowledge discovery process and practices in the analytic environment of an organization". In this work, in order to test the effectiveness of the RGBD configuration as a technology for a knowledge discovery project, we adopted as methodological framework the model proposed by [16], i.e. the Snail Shell KDDA process that has been validated in a real-world Big Data Analytics environments. The model is an iterative knowledge discovery process and consists in eight phases that will be addressed in the conduction of the following tests.

Problem formulation (PF) - The PF phase involves formulation of the business problems that the KDDA should address and its transformation towards an actionable analytic problem statement.

Business Understanding (BU) - The BU phase focuses on the business requirement elicitation that helps to translate high-level executive requirement into specific analytic needs.

Data understanding (DU) - This phase involves familiarizing with data from various sources that are relevant to solving the analytic problem.

Data preparation (DP) - Based on outputs from the PF, BU, and DU phases, an initial data integration requirement was created, including the way for processing each data element for modeling.

Modeling (MO) - This phase involves selecting applicable modeling techniques and building analytic models to provide most desirable outcomes for the stated analytic goal.

Evaluation (EV) - In this phase, 2 candidate models are evaluated against business objectives and business problems formulated in the PF phase.

Deployment (DE) - The strategy for DE was considered early in the KDDA process as part of the BU phase. The deployment plan was shared with all stakeholders to ensure that resources are available.

Maintenance (MA) - The MA phase includes all model management activities, including model selection, usage, retirement, and replacement.

The KDDA process model is represented in Fig. 1.

4   The System Technology

The system technology is based on an RGBD camera installed in the top-view configuration. The camera is positioned above the surface which is to be analysed. These kind of cameras are chosen because of their availability, reliability and affordability [19] and for their great value (both in accuracy and efficiency) in coping with severe occlusions among humans and complex background.

The camera captures depth and colour images, both with dimensions of 640×480 pixels, at a rate up to approximately 30 fps, and illuminates the scene/objects with structured light based on infrared patterns [18]. The RGBD adopted is an Asus Xtion Pro Live, since it allows acquiring colour and depth information in an inexpensive and fast way. The camera is able to detect the customer gestures when in front of a vending machine and a software is able to analyse the shoppers activities and report the results on a dashboard. The hearth of the technology is an innovative and cost-effective smart system able to understand the shoppers'

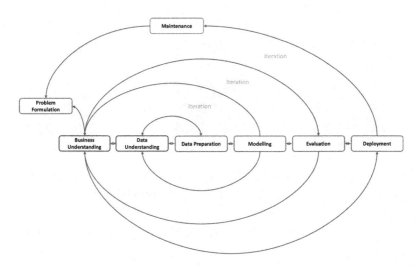

Fig. 1. The KDDA process model.

behaviour and, in particular, their interactions with the VM as previously done in [17]. Through the depth frames, the system detects shopper behaviour.

Data gathered from these sensors are used to evaluate the attraction (the level of attraction that the VM is creating on consumers), the attention (the time consumers spend in front of brand display) and the action (the number of consumers that stop in front of the VM and interact with merchandise).

This RGBD technology has the main purpose of measuring some relevant indexes, i.e. as for the most attractive area of the VM, a real vertical heat map shows how the shoppers interact with the available products reffig:vertical. Thanks to this information it is possible to measure the impact of new products, new offers, new promotion, new packaging and so on. It is also possible to evaluate the effect of the Out of Stock (OOS) situation in shoppers' preferences [2].

These information are extremely useful to understand the real interaction between shoppers and products. Moreover, it can be used to understand the traffic flow in front of the VM, as well as measure the real traffic of the shoppers in front of the VM (passing by, stop, avg time spent in front of the VM).

It is possible to evaluate the performance of category selected in terms of number of visitors per category, number of stops in category for a period greater than 5 s, number of interacting visitors, number of purchasers and several conversion ratio as: interaction/stop, average interactions per visitor, conversion ratio purchased/interaction and average purchases per interactor.

Table 1 is a list of definitions to better understand the metrics adopted and to define a common terminology baseline:

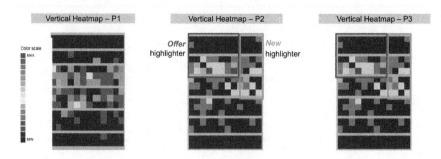

Fig. 2. Vertical Heatmap. Significantly higher level of positive interactions for the different executions. The vertical Heatmap shows the spatial distribution of interactions on the VM shelves of each category.

Table 1. Metrics

Main metrics	Definition
# total visitors	Customers counted and tracked by the RGBD camera
# Consumers passing by in front of VM	Number of different people passing by in front of VM
# Consumers stopping at VM	Number of different people stopping in front of VM
# Consumers interacting (select, chose, pay, grab)	Consumers interaction (select, chose, pay, grab)
Dwell time (VM area, specific VM)	Time each visitor spend in front of the VM (Time difference between VM exiting timestamp & VM entering timestamp)
Full Conversion Funnel	A page in the dashboard with funnel conversions (entering, passing, stopping) and interactions conversions (touch, bring back, buy)

5   Experimental Results

In this section, the eight phases followed in this work are described. Figure 3 depicted the KDDA process model designed for our application case.

The PF approach is applied to the problem of shopper behavior understanding in front of a VM with a strong focus on attraction, interaction and location intelligence. Actionable analytics are mainly expressed on attraction and interaction indexes, measuring the ratio between passing and stopping and passing and interacting.

In BU phase, we collected VM management and sales implicit knowledge by one-to-one interviews to understand the process of setting-up, evaluating and managing a VM in a specific location and location-related or target-related knowledge on specific needs (i.e. school vs. university, self-service store vs hospital, etc.).

In the current project for the DU, data coming from vending interactions, sell out and behavior data coming from RGBD data (described in Sect. 4) are involved the in the DU phase. Different tests are performed in the DU process in a lab environment to prove mining and effectiveness of the collected data from cameras and VM interfaces (i.e. touch screen monitor, cashier system, etc.).

In the DP phase, different dataset collected from 4 different location and 8 different VMs were collected and analyzed. Cameras data sources were proved to be suitable for consumer behavior modelling, while VM data sources (i.e. sell-out data and interface interactions) were designed for interaction conversion and interaction time modelling. 3 different iterations (2 weeks of data each) with different periods of the DU, DP, modeling, and BU were performed until an acceptable modeling dataset is produced.

The technique selection in the MO phase was performed in the 3 iterations described before and finally a multi-layer filter model with 3 filtering and modelling processes, going from the lower layer with raw data, to 2 different aggregation layers was performed. The probabilistic modeling phase is based on autocorrelation on time series. An outlier detection system is applied to the model to ensure proper data correction. Different models were tested during the next EV phase.

The objective of EV was to test and prove the effectiveness of the proposed analytics methodology and models to answer the main business questions derived from the BU phase. While analytic models are evaluated within the tool using objective KPIs measures such as accuracy (measured on a hourly time period against a ground truth manually evaluated), evaluation against business objectives is performed with focus groups with VM management teams.

The DE plan was the output of this phase and was confirmed as feasible with the following parameters: up to 50 different testing locations with more than 100 VMs monitored in real time; locations are statistical significant to test and prove performances.

A formal model MA process was established with assigned roles for business users and for data quality manager. A weekly call between the 2 different users ensure data publication after a data quality assessment.

The results of the experiments conducted for 12 weeks in 4 different locations and with different VM layout are reported. In particular, we analyse 4 different cases: University Lounge, Hospital Hall, Self-Service store, and Company Hall. Self Service Store VMs have interactive screens and are in the special location.

Table 2 reports the VM attraction level and time spent by a standard consumer in front of a VM. Consumers tend to interact only once and only limited percentage of consumers buy more than 1 product per time.

The average time spent at shelf by consumers is 14 sec. University consumers are investing more time for selection (Table 3). Furthermore, VM at university has also a cashless payment method.

Another evaluation is the placement of the vending. For example, in the hospital the best location for VM is close to EXIT (B). Passing by B are 34% more than passing by A (actual location). The different colors that are seen in

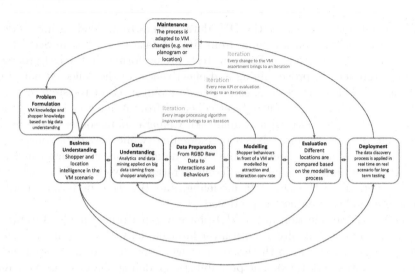

Fig. 3. The KDDA process model for shopper behaviour understanding in front of a vending machine with a strong focus on attraction, interaction and location intelligence.

this graphical representation have different meaning or depict different values for different data. In fact, the heat map shows the hot spots, dead areas, bottlenecks in performing A/B tests. Figure 4 depicts the horizontal heat map of the hospital.

Based on the results obtained it is inferred that shoppers who buy at a VM with Screens spent more time in front of the VM. Shoppers who buy at a VM with Screens buy more often (+12%) than the others. VM with Cashless payment systems seems to be preferred by the shoppers and also they spent more time even if not required on Screen Interaction. The design of A/B tests is to sell more attractive VM to your customers (my new machine is attracting +7% than the old one).

Information relevant for the operators are that shoppers are fast in their purchase decisions and they are not usually buying more than 1 product at the time, less than 14%. University shoppers are more interested to interact with the small screens, as learned from the Marketing tool. Instead, companys shoppers seems to be more attracted by cross merchandising proposal, attractive advertising, and design VM are more attractive to the shoppers. The more VM interaction is fast and easy the more shoppers buy multiple products. Company shoppers are more interested in the mix of the merchandising and it seems they prefer the offers not too large with no more than 5 different products. When VM are close to each other, shoppers are more interested in buying mixed merchandise.

The most attractive shelves of the VM are the ones on the top level. Shoppers are not usual buying more than 1 product at the time. Shoppers who buy at a VM with Screens spent more time in front of the VM (up to a threshold of 25 s). Shoppers who buy at a VM with Screens buy more often (+12%) than the others. Company's Shoppers seems to be more attract by cross merchandising proposal.

Table 2. VM attraction level and time spent by a standard consumer in front of a VM

	Model A University	Model B Hospital	Model C Self-Service	Model D Company
Visit the VM Area	N = 9432 100%	n = 4761 100%	n = 2140 100%	n = 679 100%
Stop at the VM (% stop vm/ visit area)	N = 1,562 17%	n = 1,103 23%	n = 2,103 98%	n = 553 81%
Interact with VM (% interact/ visit)	n = 1,325 84%	N = 977 86%	N = 1988 95%	n = 381 69%
Buying (% buying/ interact)	N = 1151 87%	N = 959 98%	N = 1976 99%	n = 381 100%
Multi Buying (% multiple buying/ interact)	N = 109 7,6%	N = 129 13%	N = 176 9%	n = 36 10%

Table 3. Average time spent at VM by consumers is 14 sec University consumers investing more time for selection

	Model A University	Model B Hospital	Model C Self-Service	Model D Company	Total
Spend 5–15 s to chose what to buy	62%	58%	71%	68%	66%
Spend over 15 s to chose what to buy	36%	32%	29%	31%	34%
AVG. Time Spent in front of the VM by positive interactors	18 s	12 s	13 s	13 s	14 s

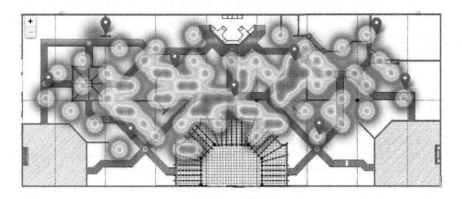

Fig. 4. Horizontal Heatmap VM Hospital

6 Conclusions

The results obtained from the tests show that the RGBD camera system can be adopted within a knowledge discovery process via data analytics (KDDA), through which it has been possible to obtain new forms of actionable knowledge. It is important to underline that this process was carried out in an industry in which Big Data Analytics projects have never been implemented so far. Thanks to the adoption of new systems of big data analytics, even VM companies that come from "traditional" industries from a marketing and sales point of view, are able to perform more accurate decision-making processes through [17]: i) the monitoring of vending machines, location, category, brand and performance of products; ii) the optimization of VM layouts, location or planogram; and iii) the evaluation of the performance of marketing and promotional activities. In this context, the VM industry is taken into consideration in a broader omni-channel retail environment. CB-Insights considers VM as one of the emerging channels for physical retailers who, having to identify new options that can dialogue with online channels, are considering VMs as an option for an innovative shopping experience [7]. However, this option can only be implemented if the VM is equipped with an analytics system capable of generating better data [5], for an omni-channel retail approach [27].

Further investigation will be devoted to creating connection between VM and smartphone, in this way we can have new detailed information about shoppers and react once they visit the VM, providing for example new offers based on their attitude (no sugar products), or providing a tailor made service. The system can react once the "frequent" shoppers is coming, detecting the presence through the proximity technology as bluetooth. Payment should be processed in moneyless way through paypal or applepay. The more the system can learn from shopper the more will be able to provide brand manufacturer with detailed input on what shoppers want. Another aspect can be seen in creating a never ending (available on demand when the shopper needs) communication line between VM and Shoppers through the social media or the shopper preferred option (sms, email, instant message). The Behavioural Science Program is based on creating the correct mix between data coming from different sources in order to provide real data mining to improve the shopping experience for shoppers and the business opportunities for VM industry.

References

1. Accenture: Big success with big data (2014)
2. Anupindi, R., Dada, M., Gupta, S.: Estimation of consumer demand with stock-out based substitution: an application to vending machine products. Mark. Sci. **17**(4), 406–423 (1998)
3. Berry, M.J., Linoff, G.: Data Mining Techniques: For Marketing, Sales, and Customer Support. John Wiley & Sons, Inc., Indianapolis (1997)
4. Bharara, S., Sabitha, A.S., Bansal, A.: A review on knowledge extraction for business operations using data mining. In: 2017 7th International Conference on Cloud Computing, Data Science & Engineering-Confluence, pp. 512–518. IEEE (2017)
5. Bradlow, E.T., Gangwar, M., Kopalle, P., Voleti, S.: The role of big data and predictive analytics in retailing. J. Retail. **93**(1), 79–95 (2017)
6. Bumblauskas, D., Nold, H., Bumblauskas, P., Igou, A.: Big data analytics: transforming data to action. Bus. Process Manage. J. **23**(3), 703–720 (2017)
7. CBINSIGHTS: 15 trends shaping tech in 2018 (2018)

8. Chen, H., Chiang, R.H., Storey, V.C.: Business intelligence and analytics: from big data to big impact. MIS Q. **36**, 1165–1188 (2012)
9. Cheng, Y., Chen, K., Sun, H., Zhang, Y., Tao, F.: Data and knowledge mining with big data towards smart production. J. Ind. Inf. Integr. **9**, 1–13 (2017)
10. Cios, K.J., Kurgan, L.A.: Trends in data mining and knowledge discovery. In: Pal, N.R., Jain, L.(eds.) Advanced Techniques in Knowledge Discovery and Data Mining, pp. 1–26. Springer, London (2005). https://doi.org/10.1007/1-84628-183-0_1
11. Fayyad, U.M., Piatetsky-Shapiro, G., Smyth, P., Uthurusamy, R.: Advances in knowledge discovery and data mining, vol. 21. AAAI press Menlo Park (1996)
12. Fayyad, U.M., Piatetsky-Shapiro, G., Smyth, P., et al.: Knowledge discovery and data mining: towards a unifying framework. In: KDD, vol. 96, pp. 82–88 (1996)
13. Fisher, M., Raman, A.: Using data and big data in retailing. Production and Operations Management (2017)
14. Griva, A., Bardaki, C., Pramatari, K., Papakiriakopoulos, D.: Retail business analytics: customer visit segmentation using market basket data. Expert Syst. Appl. **100**, 1–16 (2018)
15. John Walker, S.: Big data: a revolution that will transform how we live, work, and think (2014)
16. Li, Y., Thomas, M.A., Osei-Bryson, K.M.: A snail shell process model for knowledge discovery via data analytics. Decis. Support Syst. **91**, 1–12 (2016)
17. Liciotti, D., Contigiani, M., Frontoni, E., Mancini, A., Zingaretti, P., Placidi, V.: Shopper analytics: a customer activity recognition system using a distributed rgb-d camera network. In: Distante, C., Battiato, S., Cavallaro, A. (eds.) VAAM 2014. LNCS, vol. 8811, pp. 146–157. Springer, Cham (2014). https://doi.org/10.1007/978-3-319-12811-5_11
18. Liciotti, D., Paolanti, M., Frontoni, E., Mancini, A., Zingaretti, P.: Person re-identification dataset with rgb-d camera in a top-view configuration. In: Nasrollahi, K., et al. (eds.) FFER/VAAM -2016. LNCS, vol. 10165, pp. 1–11. Springer, Cham (2017). https://doi.org/10.1007/978-3-319-56687-0_1
19. Liciotti, D., Paolanti, M., Frontoni, E., Zingaretti, P.: People detection and tracking from an rgb-d camera in top-view configuration: review of challenges and applications. In: Battiato, S., Farinella, G.M., Leo, M., Gallo, G. (eds.) ICIAP 2017. LNCS, vol. 10590, pp. 207–218. Springer, Cham (2017). https://doi.org/10.1007/978-3-319-70742-6_20
20. Madni, H.A., Anwar, Z., Shah, M.A.: Data mining techniques and applications—a decade review. In: 2017 23rd International Conference on Automation and Computing (ICAC), pp. 1–7. IEEE (2017)
21. Mariscal, G., Marban, O., Fernandez, C.: A survey of data mining and knowledge discovery process models and methodologies. Knowl. Eng. Rev. **25**(2), 137–166 (2010)
22. McAfee, A., Brynjolfsson, E., Davenport, T.H., Patil, D., Barton, D.: Big data: the management revolution. Harvard Bus. Rev. **90**(10), 60–68 (2012)
23. McKinsey: Analytics comes of age (2018)
24. McKinsey, G., et al.: Big data: the next frontier for innovation, competition and productivity. McKinsey Global Institute (2011)
25. NSF: Directorate of computer and information science and engineering (cise). event bigdata webinar/core techniques and technologies for advancing big data science and engineering (bigdata). National science foundation (2014)
26. Philip, J.: An application of the dynamic knowledge creation model in big data. Technol. Soc. **54**, 120–127 (2018)

27. Pondel, M., Korczak, J.: A view on the methodology of analysis and exploration of marketing data. In: 2017 Federated Conference on Computer Science and Information Systems (FedCSIS), pp. 1135–1143. IEEE (2017)
28. Rajpurohit, A.: Big data for business managers—bridging the gap between potential and value. In: 2013 IEEE International Conference on Big Data, pp. 29–31. IEEE (2013)
29. Rothberg, H.N., Erickson, G.S.: Big data systems: knowledge transfer or intelligence insights? J. Knowl. Manage. **21**(1), 92–112 (2017)
30. Sallam, R., Cearley, D.: Advanced Analytics: Predictive, Collaborative and Pervasive. Gartner Group, Stamford (2012)
31. Shearer, C.: The crisp-dm model: the new blueprint for data mining. J. Data Warehouse. **5**(4), 13–22 (2000)
32. Sun, S., Zhu, S., Cheng, X., Byrd, T.: An examination of big data capabilities in creating business value. In: Proceedings of the 2015 Decision Sciences Institute Annual Meeting, Seattle, WA, November 21–24 (2015)
33. Wamba, S.F., Akter, S., Edwards, A., Chopin, G., Gnanzou, D.: How 'big data' can make big impact: findings from a systematic review and a longitudinal case study. Int. J. Prod. Econ. **165**, 234–246 (2015)
34. Wedel, M., Kannan, P.: Marketing analytics for data-rich environments. J. Mark. **80**(6), 97–121 (2016)

People Counting on Low Cost Embedded Hardware During the SARS-CoV-2 Pandemic

Giulia Pazzaglia[✉], Marco Mameli, Luca Rossi, Marina Paolanti,
Adriano Mancini, Primo Zingaretti, and Emanuele Frontoni

Dipartimento di Ingegneria dell'Informazione (DII),Università Politecnica delle
Marche, Via Brecce Bianche 12, 60131 Ancona, Italy
`giuliapazzaglia.94@gmail.com`

Abstract. Detecting and tracking people is a challenging task in a persistent crowded environment as retail, airport or station, for human behaviour analysis of security purposes. Especially during the global spread of SARS-CoV-2 virus that has become part of everyday life in every country, it is important to be able to manage the flows inside and outside buildings indoors. This article introduces an approach to detect and count people when they cross a virtual line. The methods used are based on deep learning and in particular on convolutional neural networks, specifically MobileNetV3 which is used for the detection task and MOSSE filter which is used for the tracking phase. The hardware system assembled for people counting is inexpensive, as it is formed by Raspberry Pi4 and a Picamera module v2. These devices have already been installed in some supermarkets and museums in the center of Italy, precisely in the area of the Marche region.

Keywords: People counter · Deep learning · Intelligent retail
environment

1 Introduction

Detecting and counting people is an important concept on image processing [17, 21]. Recently, many systems using a wireless sensor network and mobile nodes are applied to a lot of fields, such as visual surveillance inside supermarket or monitoring for protection of an ecosystem, or human-computer interaction. Furthermore, they can be used for for security teams to determine if there are people left after hours inside a building, or to measure the flow of people inside a train or bus [8,13].

Especially during the period of SARS-CoV-2 virus has attacked a large part of the world's population, the control of overcrowding in both outdoor and indoor environments is extremely useful. A solution could therefore be to enable a limited number of people at the entrance to the shopping center at a time: this means that the density of people inside is lower, also reducing possible contacts

© Springer Nature Switzerland AG 2021
A. Del Bimbo et al. (Eds.): ICPR 2020 Workshops, LNCS 12662, pp. 521–533, 2021.
https://doi.org/10.1007/978-3-030-68790-8_41

and queues at the checkout. It is therefore useful to create systems for the control of gates and queues, which help to automatically manage the density of people within an environment, by acting on traffic lights, automatic ticket machines, turnstiles and automatic doors to control the flow and prevent overcrowding. The management of the unidirectional flow of people in supermarkets, airports or museums is therefore essential [8, 22].

For these purposes, the most used methods are based on Deep Learning (DL) and in particular on Convolutional Neural Networks (CNNs). Unlike traditional techniques for extracting features by manual design, the CNN architectures can automatically extract features from raw data [19].

In this regard, this paper is focused precisely on the field of People Counting and in particular on the Line of Interest (LOI) approach as the video streams are captured from a camera with top-view perspective. As shown in the Framework in Fig. 2, MobileNetV3 [10] is used for the task of People Detection: it is a streamlined architecture that uses depthwise separable convolutions to construct lightweight deep convolutional neural networks and provides an efficient model for mobile and embedded vision applications. The hardware system consists of processing units that contain a Raspberry Pi4and a Picamera module v2. These devices have already been installed in some supermarkets and museums in Italy, precisely in the area of the Marche region, as shown in Fig. 3.

Fig. 1. Some installation of the device used for the acquisition of images and for people counting.

The main contributions of this paper with respect to the state-of-the-art approach are the following: i) solutions for real retail environments, useful and fundamental during the global pandemic SARS-CoV-2, with great variability of acquired data derived from a large experience, over 10.4 million shoppers having been observed in three years in different types of stores in different countries [20]; ii) people counting of contemporary shoppers; iii) CNNs to deal with videos from RGB cameras and iv) a new, dataset of shoppers acquired during the pandemic situation in the center of Italy.

The paper is organized as follows. Section 2 provides a description of the approaches for People Detection both in case of counting of the number of people who may be within a selected area, and the counting of the number of people entering and exiting crossing a line. Section 3 describes the materials and the methods used to reach the solution of this paper. In Sect. 4, there is an analysis of the results of the approach. Finally, in Sect. 5, we draw conclusions and discuss future directions for this field of research.

2   Related Works

In recent years, monitoring the access of people in a certain defined area or who pass through a certain entrance has become a crucial parameter both for the safety and for the management of a store, i.e. in the world of Retail. People detecting and counting problem can be studied by using two main approaches. The first approach is the line of interest (LOI), in which a camera is fixed on a virtual line and people crossing this line are counted [1,34]; the second approach is the region of interest (ROI) [25,29], in which people in a certain area are counted.

The first people counting approaches that appeared in the literature are designed to process the video stream produced by traditional RGB cameras [1,4,5]. Most of these methods employ a combination of a foreground detection method (frame differentiation or background subtraction) and tracing techniques (superimposed tracing or optical flow), while they differ in the way they refine the foreground detection phase with more or less sophisticated people detection phases. This approach tends to be error prone in case of passages of more than one person per time.

Recently, Deep Learning approaches have been used to overcome these problems. In particular, methods based on CNNs have been developed.

As for counting the number of people who are within an area, CNNs were used to estimate crowd density maps: the first work on this topic was carried out in 2015 by Wang et al. [30], where an end-to-end deep CNN regression model for counting people of images in extremely dense crowds was developed, and by Fu et al.[9], where they proposed to estimate crowd density by an optimized multi-stage CNN.

Unlike the two previous works, Zhang et al. [35] proposed a switchable training scheme with two related learning objectives, estimating density map and global count. For this purpose they presented a CNN based framework for cross-scene crowd counting. However, this method has a disadvantage that is the need to include perspective maps in both training and test scenes.

In 2016, Del Pizzo et al. [6] presented a vision based method for counting the number of persons who cross a virtual line. Here, the videos were acquired from a camera mounted in a zenithal position with respect to the counting line: the method analyzes the video stream, allowing to count the number of people crossing the virtual line and providing the crossing direction for each person.

In [27], Sheng et al. attempted to design an image representation which takes into consideration semantic attributes and spatial cues. In order to characterize the distribution of people number, they defined a semantic attributes at the pixel level and learn the semantic feature map via deep CNN.

In the paper [26], the authors proposed an end-to-end CNN architecture that takes as input a whole image and directly outputs the counting result. They first broadcast the image to a pre-trained CNN to achieve a high-level feature. Then the features are mapped to local counting numbers using recurrent network layers with memory cells.

In the work of Yao et al. [33], the model is based on CNN and long short term memory (LSTM): the authors put the images into a pretrained CNN to extract a set of high-level features. Then the features in adjacent regions are used to regress the local counts with a LSTM structure which takes the spatial information into consideration. The final global count is obtained by a sum of the local patches and apply this framework to a crowd counting dataset.

The authors in the paper [7], first leverage a Convolutional Neural Networks to estimate the density map for each frame and introduce a Locality-constrained Spatial Transformer (LST) module to relate the density maps between neighbouring frames and to estimate the density map of next frame with that of current frame.

Qi et al. [31] built a large-scale synthetic dataset with the help of a data collector and labeler, which can generate the synthetic crowd scenes and simultaneously annotate them without any labor costs. Furthermore, they proposed a crowd counting method via domain adaptation, which does not use any label of the real data, using a GAN-based method.

Crowd counting methods on single images can lead to incorrect or inconsistent counts for neighboring frames in the video crowd count. To overcome this problem, Liciotti et al. [14] introduced an approach to track and detect people in cases of heavy occlusions based on CNNs for semantic segmentation using top-view depth visual data. The purpose is the design of a novel U-Net architecture, U-Net3, that has been modified compared to the previous ones at the end of each layer. In particular, a batch normalization is added after the first ReLU activation function and after each max-pooling and up-sampling functions.

Lastly, in retail sector, the most recent work on the issue of people counting was carried out by Paolanti et al. in [20]. In this article, the authors have introduced a new VRAI Deep Learning Framework which uses 3 different types of CNN: VraiNet1 is used for counting the people passing by a certain selected area; VraiNet2 is used to perform the re-id with topview camera and finally VraiNet3 is used to measure the iteration that exists between the buyer and the shelf in a single RGB-D frame. This system is able to process data up to 10 frames per second, guaranteeing high performance even in the case of very short buyer-shelf iteration.

3   Materials and Methods

In this section, the overall idea is presented, as summarized in Fig. 2. In particular the following steps have been performed:

- data acquisition (Subsect. 3.1);
- transfer learning phase, with the process of image annotation;
- the third phase was that of people detection, where the MobileNetV3-SSD was used (Subsects. 3.2);
- the tracking phase with the use of MOSSE filter (Subsect. 3.3);
- the last phase is that of counting, through the line cross detection in which the networks produce the label of the input

Further details will be provided in the following Subsections as well as the metrics used for the evaluation of the performances of the network and the Hardware System used for the acquisition of images and for people counting.

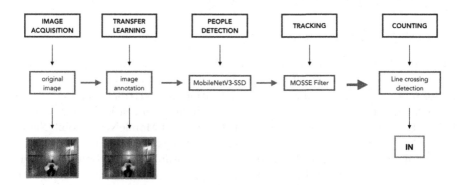

Fig. 2. Framework. The structure of the work carried out can be divided into 5 phases: the first phase can be identified by the data acquisition; the second is the transfer learning phase, with the process of image annotation; the third phase was that of people detection, where the MobileNetV3-SSD was used; the fourth is the tracking phase with the use of MOSSE filter; the last phase is that of counting, through the line cross detection in which the networks produce the label of the input.

3.1   Dataset

The dataset specifically collected for this work, was acquired in real retail installation for 1 class "top-view person". Using an RGB camera in top-view perspective, about 8000 image acquisitions were performed. The images have been subsequently manually annotated.

As shown in Fig. 3, the annotation of the images was carried out in three different moments: before the virtual red line, below the virtual red line and after the red line.

Fig. 3. Annotation of dataset. The red line is the virtual line of passage for people counting. (Color figure online)

To understand when a person crosses the virtual line, and then when the people counter starts working, detection centroid crossing is used: between two successive instants, a segment is drawn between the centroids of the respective detection; if this segment intersects the virtual line then a crossing has occurred, otherwise the crossing has not yet occurred.

3.2   Deep Learning Model

The network that has been chosen for the detection task is the MobileNet, which has proven to be the most effective in this kind of situation for its suitability to perform object detection, fine grain classification, face attributes and large-scale geo-localization [18]. These models have been built on increasingly more efficient building blocks. In particular there are three types of MobileNets: MobileNetV1, MobileNetV2, and MobileNetV3, that is the network used for this work.

MobileNetV1 [12] introduced depthwise separable convolutions as an efficient replacement for traditional convolution layers and as a way to reduce the number of parameters. Depthwise separable convolutions are defined by two separate layers: light weight depthwise convolution for spatial filtering and heavier 1 pointwise convolutions for feature generation.

MobileNetV2 [11] was built upon the ideas from MobileNet V1 but introduced two new features to the architecture: linear bottleneck layer and inverted residual structure. These two innovation was made in order to make even more efficient layer structures by leveraging the low rank nature of the problem.

In the latest version, called MobileNetV3 [10], the authors apply the squeeze and excite in the residual layer, and use different nonlinearity depending on the layer. MobileNetV3 is tuned to mobile phone CPUs through a combination of a Neural Architecture Search (NAS) [36] complemented by the NetAdapt algorithm [32]. NAS is the process of trying to make a model output a thread of modules that can be put together to form a model that gives the best accuracy possible by searching among all the possible combinations. The addition of squeeze and excitation module helps by giving un-equal weights to different

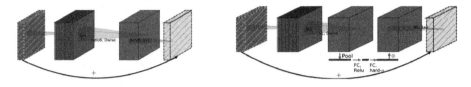

Fig. 4. Differences between MobileNetV2 and MobileNetV3

channels from the input when creating the output feature maps as supposed to equal weight that a normal CNN gives. Network Improvements have been made in two ways:

– **Layer removal.** In the last block, the 1×1 expansion layer taken from the Inverted Residual Unit from MobileNetV2 is moved past the pooling layer. Furthermore the expansion layer is moved behind a pooling layer, so the compression done by projection layer from the last layer from the previous block is not necessary.
– **Swish non-linearity.** It was introduced in [24] and it is used as a drop-in replacement for ReLU, that significantly improves the accuracy of neural networks. It is described by the equation

$$\text{swish } x = x \cdot \sigma(x)$$

However, as the sigmoid function is computationally expensive, the authors modify it with what is called hard swish or h-swish:

$$\text{h-swish}[x] = \frac{\text{ReLu } 6(x + 3)}{6}$$

There are two models of MobileNetV3: MobileNetV3- Large and Mobile NetV3-Small.These models are targeted at high and low resource use cases respectively. In fact, in this work, MobileNetV3 small was used.

Then, as regards the detection, SSD (Single Shot MultiBox Detector)[16] was used, an architecture with a single convolution network that learns to predict bounding box locations and classify these locations in one pass. The architecture is shown in Fig. 5.

Hence, SSD can be trained end-to-end. The SSD network consists of base architecture (MobileNetV3 in this case) followed by several convolution layers.

The whole network was pretrained on the COCO dataset [15], that is a large-scale object detection, segmentation, and captioning dataset containing 80 object categories and about 200.000 images labelled.

Finally, transfer learning was carried out to pass from the 80 classes of COCO dataset to 1 class "top-view person", not present in the pretrained dataset.

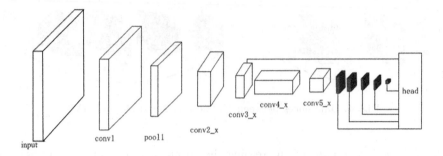

Fig. 5. SSD architecture

3.3   Mosse Tracking

For the tracking phase a filter has been adopted. Visual tracking can be defined as the process of detecting the position of an object in each frame of a video. In particular, Minimum Output Sum of Squared Error (MOSSE) is a filter introduced by Bolme et al. in [2]. MOSSE is a stable correlation filter which can be initialized on a single frame of a video. A tracker based upon MOSSE filter is robust to variations in lighting, scale, pose, and non-rigid deformations while operating at 669 frames per second.

Then, let $f = \{f_i\}$ a set of training input and $g = \{g_i\}$ a set of training output, that generally represents the ground truth. The filter H is found on the Fourier domain, therefore the Fast Fourier Transform (FFT) [23] of f_i and g_i is needed. Let be $F_i = \mathcal{F}(f_i)$ and $G_i = \mathcal{F}(g_i)$ the Fourier transform of f_i and g_i respectively. To find a filter H that maps the training input to the desired training output, MOSSE creates a filter that minimizes the sum of the squared errors among the filtered training inputs $(F_i \odot H^*)$ and the training output (G_i). Hence, the equation that describes the MOSSE filter is:

$$\min_{H^*} \sum_i |F_i \odot H^* - G_i|^2 \qquad (1)$$

3.4   Performance Evaluation

The metrics usually adopted for the performance evaluation of a machine learning or a deep learning model are **Precision**, also called *positive predictive value* and **Recall**, also known as *sensitivity*, described by the equations:

$$Precision = \frac{TP}{TP + FP} \qquad (2)$$

$$Recall = \frac{TP}{TP + FN} \qquad (3)$$

where

- TP is True Positive,
- FP is False Positive,
- TN is True Negative,
- FN is False Negative.

as described in [3].

Taken a range from 1 to n in the ranking it is possible to calculate the **Average Precision** (AP) and the **Average Recall** (AR) by evaluating Precision and Recall of each placement in the same interval.

$$\text{AP} = \sum_{i=1}^{n} \frac{\text{Precision}_i}{n} \tag{4}$$

$$\text{AR} = \sum_{i=1}^{n} \frac{\text{Recall}_i}{n} \tag{5}$$

By averaging the AP over all object classes, the **Mean Average Precision** (mAP) is found:

$$\text{mAP} = \frac{1}{N} \sum_{i=1}^{N} \text{AP}_i \tag{6}$$

where N is the total number of classes. In the case in which the class is unique, as in the case of this work, mAP and AP coincide [28].

3.5 Hardware Setup

This section shows the characteristics of the hardware system used for the acquisition of images and for people counting.

The hardware system consists of processing units that contain a Raspberry Pi4 and a Picamera module v2 (Fig. 6).

As can also be found on the net, these types of models are not expensive. The image processing takes place directly on the Raspberry with inference times equal to 0.11 s for Raspberry, using the MOSSE Tracking described in Sect. 3.3, and taking a frame every 3 in order to process 20 fps. These processing units are installed with a top-view perspective near the doors of the premises to which it is applied. In particular, one of these units has the function of *master*, i.e. through a websocket it collects from the other units, which are called *slaves*, messages indicating the crossings, with the relative direction (in or out). The master and slave units all have the same hardware, what changes is the software configuration.

Based on the information provided by the slaves, the master will calculate the net number of people who are currently inside the local. Once a threshold, set directly via a web interface by the store manager, has been exceeded, the master will activate the traffic light display system: this could be either a web page on a display with "forward" or "stop" written, or a real and own led traffic light activated by the raspberry GPIO pins.

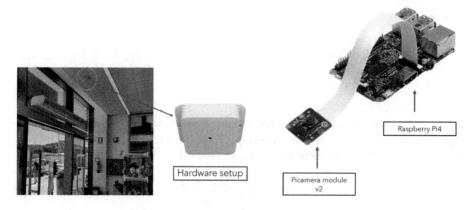

Fig. 6. Hardware setup

4 Results and Discussion

In this section, some preliminary results of the application are shown. For the training of the MobileNetv3, ReLU6 was used as an activation function and l2-regularized as a regularizer. The losses used for the different tasks were:

- weighted sigmoid focal as classification loss with parameters $\alpha = 0.75$ e $\gamma = 0.2$;
- weighted smooth l1 as localization loss with $\delta = 1.0$;
- sigmoid function as final loss of the model.

Furthermore, a batch size $= 128$ was used out of a total of steps equal to 20000. Finally, the momentum with parameter $\mu = 0.9$ and adaptive learning rate was used as an optimizer starting from a base of 0.05 and applying the cosine decay learning rate. The results obtained are shown in the graphs in the Fig. 7.

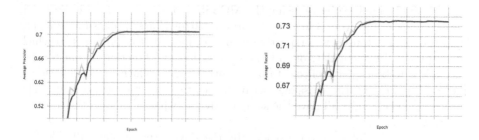

Fig. 7. Average precision and average recall for 20000 steps.

From these graphs, it is possible to notice that both the average precision and the average recall reach a maximum of 0.7 and 0.73 respectively.

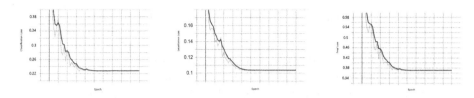

Fig. 8. Classificstion loss, localization loss and final loss for 20000 steps.

Comparing with the results inside of Howard's MobileNetV3 paper[10], it can be observed that the largest value obtained per class was 67.4. Therefore the results of this paper are better than those previously obtained. There is also the fact that in [10], the training was done in all COCO classes: therefore we can say that for the purpose of this work the MobileNetV3-Small performs well, but it could be improved.

5 Conclusions and Future Works

Deep Learning approaches, and in particular CNNs, have been used to overcome the problem of counting people that crossing a virtual line and the problem of counting people who are within a certain area of interest. Unlike traditional techniques for extracting features by manual design, the CNN architectures can automatically extract features from raw data. For this purpose, in this paper MobileNetV3 has been chosen for the task of People Detection and a MOSSE filter has been adopted for the tracking phase. The results of MobileNetV3 for the detection are promising. In fact, unlike the previous 0.67% of AP, the 0.7% of AP was reached.

Moreover, the tool developed for the acquisition of images and for people counting is inexpensive, as it is formed by Rasp-berry Pi4 and a Picamera module v2. These devices are already installed in real environments, such as some museums and supermarkets in the Marche, a region of central Italy.

It is also possible to observe that these kind of devices can be of great help for crowd management, given the large extent of the SARS-CoV-2 virus. In fact, they can be used in many other applications: at the entrance of banks, airports, train stations, buses and public places, both outdoors and indoors, in general.

Further investigations will be devoted to the improvement of the performance of the network, by increasing the dataset. Moreover, other networks will be tested and we will plan to test this tool in other environments.

References

1. Barandiaran, J., Murguia, B., Boto, F.: Real-time people counting using multiple lines. In: 2008 9th International Workshop on Image Analysis for Multimedia Interactive Services, pp. 159–162. IEEE (2008)

2. Bolme, D.S., Beveridge, J.R., Draper, B.A., Lui, Y.M.: Visual object tracking using adaptive correlation filters. In: 2010 IEEE Computer Society Conference on Computer Vision and Pattern Recognition, pp. 2544–2550. IEEE (2010)
3. Brismar, J., Jacobsson, B.: Definition of terms used to judge the efficacy of diagnostic tests: a graphic approach. AJR. Am. J. Roentgenol. **155**(3), 621–623 (1990)
4. Chen, C.H., Chen, T.Y., Wang, D.J., Chen, T.J.: A cost-effective people-counter for a crowd of moving people based on two-stage segmentation. J. Inf. Hiding Multimedia Sig. Process. **3**(1), 12–25 (2012)
5. Chen, T.H., Chen, T.Y., Chen, Z.X.: An intelligent people-flow counting method for passing through a gate. In: 2006 IEEE Conference on Robotics, Automation and Mechatronics, pp. 1–6. IEEE (2006)
6. Del Pizzo, L., Foggia, P., Greco, A., Percannella, G., Vento, M.: Counting people by rgb or depth overhead cameras. Pattern Recogn. Lett. **81**, 41–50 (2016)
7. Fang, Y., Zhan, B., Cai, W., Gao, S., Hu, B.: Locality-constrained spatial transformer network for video crowd counting. In: 2019 IEEE International Conference on Multimedia and Expo (ICME), pp. 814–819. IEEE (2019)
8. Ferracuti, N., Norscini, C., Frontoni, E., Gabellini, P., Paolanti, M., Placidi, V.: A business application of rtls technology in intelligent retail environment: defining the shopper's preferred path and its segmentation. J. Retail. Consum. Serv. **47**, 184–194 (2019)
9. Fu, M., Xu, P., Li, X., Liu, Q., Ye, M., Zhu, C.: Fast crowd density estimation with convolutional neural networks. Eng. Appl. Artif. Intell. **43**, 81–88 (2015)
10. Howard, A., et al.: Searching for mobilenetv3. In: Proceedings of the IEEE International Conference on Computer Vision, pp. 1314–1324 (2019)
11. Howard, A., Zhmoginov, A., Chen, L.C., Sandler, M., Zhu, M.: Inverted residuals and linear bottlenecks: Mobile networks for classification, detection and segmentation (2018)
12. Howard, A.G., et al.: Mobilenets: Efficient convolutional neural networks for mobile vision applications. arXiv preprint arXiv:1704.04861 (2017)
13. Liciotti, D., Paolanti, M., Frontoni, E., Zingaretti, P.: People detection and tracking from an rgb-d camera in top-view configuration: review of challenges and applications. In: Battiato, S., Farinella, G.M., Leo, M., Gallo, G. (eds.) ICIAP 2017. LNCS, vol. 10590, pp. 207–218. Springer, Cham (2017). https://doi.org/10.1007/978-3-319-70742-6_20
14. Liciotti, D., Paolanti, M., Pietrini, R., Frontoni, E., Zingaretti, P.: Convolutional networks for semantic heads segmentation using top-view depth data in crowded environment. In: 2018 24th International Conference on Pattern Recognition (ICPR), pp. 1384–1389. IEEE (2018)
15. Lin, T.-Y., et al.: Microsoft COCO: common objects in context. In: Fleet, D., Pajdla, T., Schiele, B., Tuytelaars, T. (eds.) ECCV 2014. LNCS, vol. 8693, pp. 740–755. Springer, Cham (2014). https://doi.org/10.1007/978-3-319-10602-1_48
16. Liu, W., et al.: SSD: single shot multibox detector. In: Leibe, B., Matas, J., Sebe, N., Welling, M. (eds.) ECCV 2016. LNCS, vol. 9905, pp. 21–37. Springer, Cham (2016). https://doi.org/10.1007/978-3-319-46448-0_2
17. Martini, M., Paolanti, M., Frontoni, E.: Open-world person re-identification with rgbd camera in top-view configuration for retail applications. IEEE Access **8**, 67756–67765 (2020)
18. Michele, A., Colin, V., Santika, D.D.: Mobilenet convolutional neural networks and support vector machines for palmprint recognition. Procedia Comput. Sci. **157**, 110–117 (2019)

19. Paolanti, M., Frontoni, E.: Multidisciplinary pattern recognition applications: a review. Comput. Sci. Rev. **37**, 100276 (2020)
20. Paolanti, M., Pietrini, R., Mancini, A., Frontoni, E., Zingaretti, P.: Deep understanding of shopper behaviours and interactions using rgb-d vision. Mach. Vis. Appl. **31**(7), 1–21 (2020)
21. Paolanti, M., et al.: Person re-identification with rgb-d camera in top-view configuration through multiple nearest neighbor classifiers and neighborhood component features selection. Sensors **18**(10), 3471 (2018)
22. Pierdicca, R., Paolanti, M., Frontoni, E.: etourism: Ict and its role for tourism management. J. Hospitality Tourism Technol. **10**, 90–106 (2019)
23. Press, W.H., Teukolsky, S.A., Vetterling, W.T., Flannery, B.P.: Numerical recipes 3rd edition: The art of scientific computing. Cambridge University Press (2007)
24. Ramachandran, P., Zoph, B., Le, Q.V.: Searching for activation functions. arXiv preprint arXiv:1710.05941 (2017)
25. Ryan, D., Denman, S., Fookes, C., Sridharan, S.: Crowd counting using multiple local features. In: 2009 Digital Image Computing: Techniques and Applications, pp. 81–88. IEEE (2009)
26. Shang, C., Ai, H., Bai, B.: End-to-end crowd counting via joint learning local and global count. In: 2016 IEEE International Conference on Image Processing (ICIP), pp. 1215–1219. IEEE (2016)
27. Sheng, B., Shen, C., Lin, G., Li, J., Yang, W., Sun, C.: Crowd counting via weighted vlad on a dense attribute feature map. IEEE Trans. Circ. Syst. Video Technol. **28**(8), 1788–1797 (2016)
28. Tan, R.: Breaking down mean average precision (map). Online, Mar (2019)
29. Vicente, A.G., Munoz, I.B., Molina, P.J., Galilea, J.L.L.: Embedded vision modules for tracking and counting people. IEEE Trans. Instrum. Meas. **58**(9), 3004–3011 (2009)
30. Wang, C., Zhang, H., Yang, L., Liu, S., Cao, X.: Deep people counting in extremely dense crowds. In: Proceedings of the 23rd ACM International Conference on Multimedia, pp. 1299–1302 (2015)
31. Wang, Q., Gao, J., Lin, W., Yuan, Y.: Learning from synthetic data for crowd counting in the wild. In: Proceedings of the IEEE Conference on Computer Vision and Pattern Recognition, pp. 8198–8207 (2019)
32. Yang, T.J., et al.: Netadapt: platform-aware neural network adaptation for mobile applications. In: Proceedings of the European Conference on Computer Vision (ECCV), pp. 285–300 (2018)
33. Yao, H., Han, K., Wan, W., Hou, L.: Deep spatial regression model for image crowd counting. arXiv preprint arXiv:1710.09757 (2017)
34. Yu, S., Chen, X., Sun, W., Xie, D.: A robust method for detecting and counting people. In: 2008 International Conference on Audio, Language and Image Processing, pp. 1545–1549. IEEE (2008)
35. Zhang, C., Li, H., Wang, X., Yang, X.: Cross-scene crowd counting via deep convolutional neural networks. In: Proceedings of the IEEE Conference on Computer Vision and Pattern Recognition, pp. 833–841 (2015)
36. Zoph, B., Le, Q.V.: Neural architecture search with reinforcement learning. arXiv preprint arXiv:1611.01578 (2016)

Shoppers Detection Analysis in an Intelligent Retail Environment

Laura Della Sciucca, Davide Manco, Marco Contigiani, Rocco Pietrini[✉],
Luigi Di Bello, and Valerio Placidi

Grottini Lab S.R.L., Via Santa Maria in Potenza, 62017 Porto Recanati, Italy
laura.dellasciucca@gmail.com, davide.manco@grottinilab.com,
marco.contigiani@grottinilab.com, rocco.pietrini@grottinilab.com,
luigi.dibello@grottinilab.com, valerio.placidi@grottinilab.com
http://www.grottinilab.com

Abstract. Mainly in the last years the analysis of the behaviour of shoppers inside a store is becoming a very attracting issue. Thanks to the use of technologies and artificial intelligence based approaches novel methods are applied to automatically evaluate the movement of shopper in the store, the interactions with the products on the shelves, the time spent inside the store, and more, in a passive mode without interviewing the consumers and preserving their privacy. The aim of this paper is to propose a method based on a Support Vector Machine classifier that classifies the interactions of the shopper with products solving the problem of unclassifiable interactions, when carts, baskets, or other objects are temporarily placed in front of the shelves. The implemented system is also able to solve the overcrowding problem that emerges when several shoppers are close to each other, and entering the analysis area of the camera together, were detected as a single person and not as distinct persons.

Keywords: Shopper behaviour · SVM · Neural network · Shelf interactions · Crowded environment

1 Introduction

Mainly in the last years in retail environment to understand the behaviour of shopper inside the store is becoming a more attractive challenge. In particular, to observe the behaviour of shoppers within different features of the store and shelf layouts, gives efficient perceptions to retailers and marketers to increase the shopping experience [3,10]. Since the customer satisfaction often corresponds to an increase of sales and profitability, the retailers intend to use innovative and non-invasive approaches to understand shoppers needs and satisfies them, also guaranteeing the customer privacy. Using business analysis, purchasing behavior models are studied in order to provide *ad hoc* services that adapt to the needs and preferences of different shoppers. The use of technology eased this purpose making the shopping experience involving and adaptable to the human presence [13,20]. For applications in retail environment interesting results have been

© Springer Nature Switzerland AG 2021
A. Del Bimbo et al. (Eds.): ICPR 2020 Workshops, LNCS 12662, pp. 534–546, 2021.
https://doi.org/10.1007/978-3-030-68790-8_42

obtained by the use computer vision and image processing techniques [15,16]. To monitor the shoppers behaviour inside a smart retail environment also integrated systems are implemented, in which data collected from different sensors installed in the store, are processed [14]. The latest technologies allow to study the shopper behaviour using innovative and non-invasive approaches [7]. So this continuous and implied detection allows to passively monitor the shoppers without affecting their purchases and interactions with products on the shelves. In [15] the analysis of consumer behavior towards the products on the shelves focuses on three different levels:

- LEVEL OF ATTRACTION of the individual shopper towards a store by evaluating his real interest, distinguishing the shoppers who enter it from those who are purely passing by;
- LEVEL OF ATTENTION of the shopper regarding a generic product based on the time he spends in front of the exhibition shelf;
- ACTION of the consumer towards the displayed product on the shelf or the type of interaction of the shopper with the product. The consumer's actions are themselves classifiable. In fact, we distinguish four different types of interaction:
  - *Positive*: when the shopper's hand is detected together with the product and the purchase of the article therefore follows;
  - *Negative*: when the consumer's hand is detected without the product. It occurs when the consumer is no longer willing to buy the product and puts it back on the shelf. In this case, therefore, the hand of the customer without the product is detected at the exit. It is therefore a negative interaction;
  - *Neutral*: when there is no interaction between the consumer and the shelf;
  - *Refill*: it occurs when the shelf is replenished, that is when in the same image the operator with the product in his hand and a box containing the same product that is being supplied are detected. Obviously this class has a higher priority over the others.

In this context is positioned this paper that has the aim to study the shopper interaction with the displayed products in the stores and in general the analysis of their behaviour in retail environments. The technologies are based on computer vision and artificial intelligence approaches. Consumer behavior is analyzed through the use of RGB-D cameras installed inside the store. Some of these are positioned at the entrance to the store, with the aim of counting and anonymously identifying individuals. In fact, the use of these cameras allows to preserve the privacy of shoppers, using image processing algorithms that avoid the faces detection. The remaining cameras are instead positioned above the individual shelves in order to detect and classify consumer behavior. The main problems occurred during the analysis of consumer behaviour have been:

- Unclassifiable interactions: occur at the instant in which carts, baskets, cartons or other objects are temporarily placed in front of the shelves and can become the cause of unclassifiable interactions with the products on the shelf.

It is a very frequent problem due to the fact that often customers inside a store are equipped with carts and baskets.

- Overcrowding: have emerged when several customers are close to each other, entering the analysis area of the camera together, were detected as a single person and not as two distinct persons. This problem arises in situations of greater crowding inside a store, or in the case in which several people simultaneously stop near a certain shelf, evaluating the purchase of the same product.

This paper proposes a solution to solve these two problems. As regards the problem relating to the presence of objects in front of the shelf, it was considered that it is often caused by carts or baskets that customers leave unattended in an area of the store for a variable period of time. It should be noted that the analysis of the area near the shelf is carried out through the use of a *model* generated when the software is started. By *model* we mean the acquisition of an image of the considered area, in order to allow the instant detection of the presence of new elements in the scene by comparing the original *model* with the current one. For this reason, the idea is to periodically update the base *model* saved when the software is started, as a solution to the problem of non-classifiable interactions. In doing so objects in front of the shelf and stopped for a certain period of time would become part of the *model* itself and any interactions of these objects with the shelf would not be considered valid.

Concerning the second problem, a solution based on *human detection* is proposed. The software has to be able to distinguish individual shoppers even in situations of highest presence. By analyzing *human detection* as a solution to the second problem, the possibility of solving also non-classifiable interactions with this technique was evaluated. In this way, the software would be able to distinguish shoppers from any other object, these would therefore not be detected and consequently also their possible interactions would not be considered valid. Therefore, since this solution is excellent for both problems analyzed, we decided to proceed in this direction. Subsequently, therefore, we will analyze the implementation of this technique and the evaluations regarding the improvements achieved.

2 Related Works

In this section we introduce some recent state of art applications which analyse the human behaviour in retail environments. The aim is to demonstrate as the introduction of technology in retail environments helps to obtain numerous and useful information concerning the shopper behaviour in a passive way according to [12] and [14]. The purpose is to evaluate how the shopper moves inside the store, how many interactions have with different items, how many times there are course deviations, to facilitate the task of retailers in improving the shopping experience and then in increasing sells.

Liu et al. [9] propose a system to automatically identify and track people in an indoor retail environment with the use of a single RGB-D camera. The aim of the paper is to present a novel approach that utilize a single RGB-D camera and that automatically detect and track people with different positions in dynamic stores. They develop a novel point ensemble image (PEI) representation using a transformation of RGB-D pixels to overcome segmentation problems. During the experimental phase they have used a purposely dedicated real-world clothing store dataset.

The work of Liciotti et al. [7] proposes a system able to overcome occlusion problems. Previously the authors have showed a literature review on the use of RGB-D camera to recognize and monitor people inside the store. Their aim is also to demonstrate that the use of a top-view configuration is the best method to monitor people mainly in crowded situations and if there are occlusions. Moreover, the advantage of this camera configuration preserve the privacy of the shoppers.

The work proposed by Dan et al. [2] have also used the cameras positioned on the ceiling of a store and also showed that the top-view configuration is robust even if the store is crowded and when the illumination changes. They use a morphological operator to process the image in order to solve problems related to optical noise and data loss. The silhouette of the shopper is then extracted through a human model of the depth image which is obtained after the processing phase. A high precision is showed through the experimental results.

The paper of Ravnik et al. [17] intends to propose a system that forecasts the shopper behaviour in a retail environment using machine learning methods on real-world digital signage viewership data, especially oriented towards the purchase decision process and the roles in purchasing situations. Moreover, they compare the performance of different machine learning algorithms, showing that the best performance are using the SVM classifier [1].

To determine the position of the shopping cart inside a store, RFID tags are used in the work of Li et al. [6]. Information about the movement inside the store are given by the tags and the knowledge about the most visited areas allows to create an optimization model.

Melià-Seguí and Pous [11] propose a system that studies the real-time interactions of the consumer with products based on a combination of 3 RFID sensors. The features extracted are used as input to supervised machine learning approaches, having encouraging results during the experimental phase.

The study conducted by [19] uses RFID technology to track shoppers inside the store. To define general shopping patterns, they consider three metrics: 1) the percentage of the store visited; 2) the number of products acquired; 3) the amount of time spent in the shopping trip. The results obtained during the experimental phase provided an important tool for retailers for the purpose of management and design of store and realization of marketing programs oriented to the improvement of the shopping experience.

Also in the work presented by [5] RFID technologies are implemented. The authors evaluate the displacement of the shopper inside the store using RFID tags positioned on the shopping carts. They also consider the shopping time a very important parameter and on the base of this they introduce three sets of clusters (short, medium, and long trips), identifying a total of 14 "canonical path types". The analysis only considers the shoppers movement inside the store without taking into consideration the purchasing methodology and the interactions with the products.

The RFID technology, named PathTracker, is also implemented in the paper of [18] where RFID tags are fixed under each shopping cart and using the traveling salesperson problem (TSP) [4] evaluate the shopping trajectory. In this work the TSP is defined as the shortest path that connects the entrance, all the products acquired by a shopper and the checkout. The behaviour of each shopper is compared with his TSP-path and the deviation are considered in order to evaluate the relations existing between the purchasing behaviour and the features of the path to purchase.

In the work of [12] to study the displacement of the shoppers inside a store, the duration of the traveling, and the traffic flow GPS technology is implemented. They implement a Bayesian belief network (BBN) that gives some indications of how the shopping path and the behaviour of each visitor may be interpreted departing from a particular observed outcome. For example the indications can concern the duration of shopping activity and the identification of the position of each activity.

The most recent paper of [3] uses an innovative RTLS system to study the behaviour of consumers inside the store. The system provides an extremely location precision and gives the possibility to monitor all the store zones, consider entirely the store map and collect data for a long and continuous period of observation, extending the observation for months and even years. They also propose a new method to elaborate a theory based on the preferred behaviour of the shoppers paying attention on the dependence of the state, since the choice made at the beginning of the path, influence the subsequent choice. This allow to study the key zones of the store and the consequent marketing activity to influence the behavior in the desired direction.

3 Materials and Methods

In this section the hardware and software tools used are described in detail. In particular, the characteristics of the cameras are reported, as well as the software which cameras is equipped and the dataset used for the training and the testing phase.

3.1 Setup and Configuration

The camera used is an RGB-D camera *Intel RealSense D435*, which is able to perceive the depth of the environment by analyzing in real time the information acquired by the sensors.

The cameras have been installed in the store in the target shelves, to ensure the greatest field of view and the detection of shoppers in front of them.

The camera installation in the company used to test the software was positioned above the testing shelf. An example of the image acquired by the testing camera is showed in Fig. 1.

Fig. 1. Image acquired by the testing camera

The *Shopper-Analytics-light* algorithm is used for the shopper analysis and gives in output two windows that show the stream of the camera, one in RGB and the other in RGB-D. Moreover when some interactions with products on the shelf occur, new windows are generated that depict the interaction itself. OpenCV (Open Source Computer Vision Library), a completely open source cross-platform library, has been chosen for our purpose.

3.2 Shoppers Dataset

The dataset used in our work has been collected by cameras installed above the shelves in the stores and has been splitted in training, validation and test set for the classification process.

The training set considers most of the total data available that are subdivided into two different classes:

– Positive: set of images in which at least a shopper is present that represents a detection element. Positive images are about 1500.
– Negative: set of images in which shelves, boxes and carts are present but never the shopper to have to take. In this work 300 negative images were necessary.

In addition, the same images rotated 90° in both directions have been added to the training set, so to have a greater dataset and increase the classification accuracy. Below are examples of positive (Figs. 2a and 2b) and negative (Figs. 2c and 2d) images.

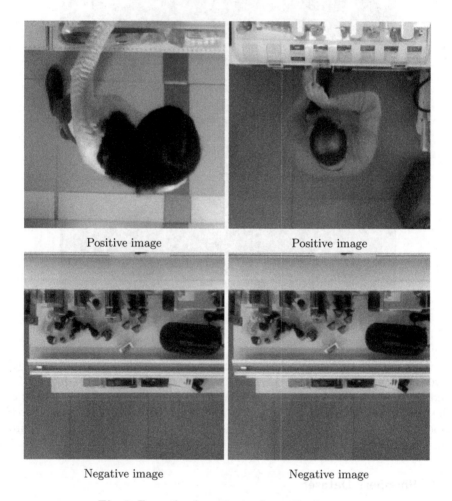

Positive image   Positive image

Negative image   Negative image

Fig. 2. Example of positive and negative images

However, he initial test was performed using TVHEADS (Top-View Heads) Dataset [8]. It consists of a dataset of people shot from above since cameras were in a top-view configuration. The dataset is made up of a total of 1815 depth images (16 bits) with a size of 320×240 pixels. The images acquire a crowded retail environment with a maximum of three people per square meter who can also be in contact with each other. There are also the same 8-bit images which highlight the contrast and brightness and consequently the silhouettes of people. These latest images are the ones we used. Before the actual use of the dataset, a phase of selecting the images to be used was necessary, as well as a phase of resizing the images in which there were several individuals in order to select a single shopper and to be able to use that image as a positive image. In Fig. 3 are examples of positive images belonging to the TVHEADS Dataset.

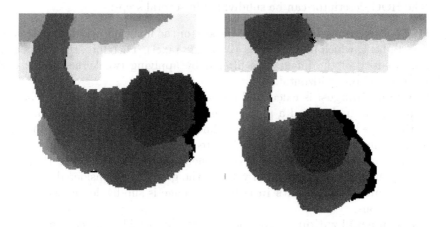

Fig. 3. Example of positive images of top-view heads dataset

3.3   Classification Model

The human detection algorithm is based on a combination of a Histogram of Oriented Gradient (HOG) features extractor and a Support-Vector Machine (SVM) [1] classifier.

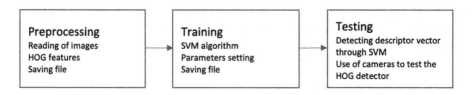

Fig. 4. Block diagram of the architecture

Figure 4 represents the block diagram of the implemented classifier training and testing.

During the preprocessing phase the training set images are subdivided in positive and negative images. Following the images have been resized so that all the images have the same dimension of 128×128 pixels. Moreover all the images have been rotated in both directions so to have a greater number of images and a more precise classification. The following phase is the HOG implementation that is a descriptor of features used in computer vision for object recognition. Its operation is based on the calculation of the direction of the gradient vector on different points of the considered image.

The HOG descriptor can be subdivided in several steps:

- **HOG-Preprocessing**: has input the size of the image on which to calculate the gradient, which in our case is 128×128 pixels. Therefore, the colors and the intensity of the images are filtered by applying two kernel filters: one vertical and one horizontal.
- **Gradient Images**: is calculated on each pixel of the image and is characterized by a module and a phase. Removes all irrelevant information, while at the same time highlighting the outlines of the figures present. In the case of color images, the module is calculated considering the maximum of the modules of the three chromatic components while for the phase the angle corresponding to the maximum value of the gradient is considered.
- **Histogram of Gradients in cells**: the image is subdivided in cells of 8×8 dimensions.
- **Block normalization**: in this phase spatially neighboring cells have been grouped into blocks of 16×16 dimension. For each block a normalization factor is extracted and used to correct the weight of each sub-cell. The normalization factor used is l2-norm [21].
- **Calculation of the HOG feature vector**: at the end of the calculation of the vectors of each single block of the image, a single large vector is created.
- **HOG visualization**: in this final phase, the HOG descriptor is visualized. In this way it is immediate to notice how the dominant direction of the vectors delineate the perimeter of the object to be recognized.

In this second phase, the images of the training and validation set are used to train the linear model of the SVM.

In this last phase the goal was to test the functioning of the previous phases. It was therefore necessary to consider a *test set* in order to check the correctness of the detection.

4 Results and Discussions

In this section, the results obtained during the experimental phase are in detail described. In this phase, also observing the results and the performance of the implemented software, the parameters are tuned in order to obtain the best performance. The output of the system is represented by images with the detection of the shopper for positive image or without detection for negative image. Figure 5 show an example of output obtained using the official dataset used in our study. It is interesting to note that even when here are more people close to each other, the algorithm detects distinctly individual persons.

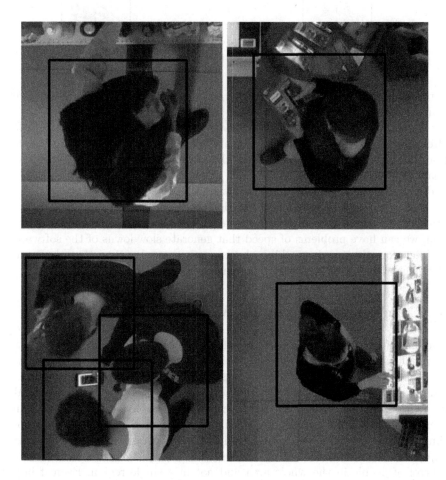

Fig. 5. People detection using our dataset

In addition to the dataset created in the company, we also present the results (Fig. 6) obtained using 8-bit depth images of the Top-View Heads Dataset.

During the testing phase of tracking is necessary to take into account the height of the camera and the brightness of the environment. In fact these two parameters vary considerably according to the point of the camera installation. Basing on these factors is necessary to modify the parameters by inserting the correct values which make the scene of the camera as similar as possible to that

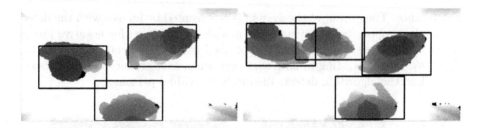

Fig. 6. People detection using TVHeads dataset

of the dataset used to train the model. For this reason the tracking phase is very delicate and requires a large number of test for each installation.

In this phase another aspect to consider is related to false positive. In fact often people are detected even if there are not, by generating noise in the detection. It is very important to manage this problem by improving the negative images that constitute the training set, making them more relevant to the case under analysis and inserting images of the shelf without shopper.

Finally it is necessary to find the right compromise between the quality of tracking and its speed. In fact if we are to achieve the absolute accuracy of detection, we can have problems of speed that generate slowdowns of the software.

Therefore the people tracking phase is more complex than the others, since it depends on aspects that vary from environment to environment.

5 Conclusions and Future Works

The aim of this paper is to introduce the human detection analysis as solution to problems related to the shopper behaviour in a store. In fact it allows to solve the issue of possible objects temporarily positioned in the scene and an interaction with the shelf or with products on display, creating an unclassifiable interaction. At the same time the overcrowding problem can be solved since there is a distinction among single humans even when two or more shoppers simultaneously enter in the scene acquired by the cameras. In the case of real applications, the idea is to simultaneously use multiple cameras so to have the control of people in the whole area and not in a single region. Even if in the training phase this method is very fast and efficient, it requires attention and precision during the tracking so that to obtain an instant detection and without noise.

This type of solution is at the same time very versatile. Considering the retail environment, it could in fact be useful to carry out the people counting at the entrance of any room or to analyze the position or movement of people in a specific area, for marketing surveys aimed at increasing the number of sales.

References

1. Cortes, C., Vapnik, V.: Support-vector networks. Mach. Learn. **20**(3), 273–297 (1995)
2. Dan, B., Kim, Y., Suryanto, Jung, J., Ko, S.: Robust people counting system based on sensor fusion. IEEE Trans. Consum. Electron. **58**(3), 1013–1021 (August 2012). https://doi.org/10.1109/TCE.2012.6311350
3. Ferracuti, N., Norscini, C., Frontoni, E., Gabellini, P., Paolanti, M., Placidi, V.: A business application of RTLS technology in intelligent retail environment: Defining the shopper's preferred path and its segmentation. J. Retail. Consum. Serv. **47**, 184–194 (2019)
4. Hahsler, M., Hornik, K.: TSP-infrastructure for the traveling salesperson problem. J. Stat. Softw. **23**(2), 1–21 (2007)
5. Larson, J.S., Bradlow, E.T., Fader, P.S.: An exploratory look at supermarket shopping paths. Int. J. Res. Mark. **22**(4), 395–414 (2005)
6. Li, H.B., Wang, W., Ding, H.W., Dong, J.: Mining paths and transactions data to improve allocating commodity shelves in supermarket. In: Proceedings of 2012 IEEE International Conference on Service Operations and Logistics, and Informatics, pp. 102–106. IEEE (2012)
7. Liciotti, D., Paolanti, M., Frontoni, E., Zingaretti, P.: People detection and tracking from an RGB-D camera in top-view configuration: review of challenges and applications. In: Battiato, S., Farinella, G.M., Leo, M., Gallo, G. (eds.) ICIAP 2017. LNCS, vol. 10590, pp. 207–218. Springer, Cham (2017). https://doi.org/10.1007/978-3-319-70742-6_20
8. Liciotti, D., Paolanti, M., Pietrini, R., Frontoni, E., Zingaretti, P.: Convolutional networks for semantic heads segmentation using top-view depth data in crowded environment. In: 2018 24th International Conference on Pattern Recognition (ICPR), pp. 1384–1389. IEEE (2018)
9. Liu, J., Liu, Y., Zhang, G., Zhu, P., Chen, Y.Q.: Detecting and tracking people in real time with RGB-D camera. Pattern Recogn. Lett. **53**, 16–23 (2015)
10. Martini, M., Paolanti, M., Frontoni, E.: Open-world person re-identification with RGBD camera in top-view configuration for retail applications. IEEE Access **8**, 67756–67765 (2020)
11. Melià-Seguí, J., Pous, R.: Human-object interaction reasoning using RFID-enabled smart shelf. In: 2014 International Conference on the Internet of Things (IOT), pp. 37–42. IEEE (2014)
12. Moiseeva, A., Timmermans, H.: Imputing relevant information from multi-day GPS tracers for retail planning and management using data fusion and context-sensitive learning. J. Retail. Consum. Serv. **17**(3), 189–199 (2010)
13. Paolanti, M., Frontoni, E.: Multidisciplinary pattern recognition applications: a review. Comput. Sci. Rev. **37**, 100276 (2020)
14. Paolanti, M., Liciotti, D., Pietrini, R., Mancini, A., Frontoni, E.: Modelling and forecasting customer navigation in intelligent retail environments. J. Intell. Robot. Syst. **91**(2), 165–180 (2018)
15. Paolanti, M., Pietrini, R., Mancini, A., Frontoni, E., Zingaretti, P.: Deep understanding of shopper behaviours and interactions using RGB-D vision. Mach. Vis. Appl. **31**(7), 1–21 (2020)
16. Paolanti, M., et al.: Person re-identification with RGB-D camera in top-view configuration through multiple nearest neighbor classifiers and neighborhood component features selection. Sensors **18**(10), 3471 (2018)

17. Ravnik, R., Solina, F., Zabkar, V.: Modelling in-store consumer behaviour using machine learning and digital signage audience measurement data. In: Distante, C., Battiato, S., Cavallaro, A. (eds.) VAAM 2014. LNCS, vol. 8811, pp. 123–133. Springer, Cham (2014). https://doi.org/10.1007/978-3-319-12811-5_9
18. Sorensen, H.: The science of shopping. Mark. Res. **15**(3), 30–30 (2003)
19. Sorensen, H., et al.: Fundamental patterns of in-store shopper behavior. J. Retail. Consum. Serv. **37**, 182–194 (2017)
20. Sturari, M., et al.: Robust and affordable retail customer profiling by vision and radio beacon sensor fusion. Pattern Recogn. Lett. **81**, 30–40 (2016)
21. Zhao, Y., Zhang, Y., Cheng, R., Wei, D., Li, G.: An enhanced histogram of oriented gradients for pedestrian detection. IEEE Intell. Transp. Syst. Mag. **7**(3), 29–38 (2015)

DLPR - Deep Learning for Pattern Recognition

Preface

Welcome to the 3rd ICPR International Workshop on Deep Learning for Pattern Recognition (DLLPR). This workshop was previously held in conjunction with 23rd International Conference on Pattern Recognition (ICPR 2016) and 24rd International Conference on Pattern Recognition (ICPR 2018). The 3rd workshop was originally to have taken place in Milan, Italy, September 2020. However, given the worldwide pandemic situation, DLPR is postponed until 11 January 2021 and converted into a fully virtual event.

Pattern recognition is one of the most important branches of artificial intelligence, which focuses on the description, measurement and classification of patterns involved in various data. In the past 60 years, great progress has been achieved in both the theories and applications of pattern recognition. A typical pattern recognition system is composed of pre-processing, feature extraction, classifier design and post-processing. Nowadays, we have entered a new era of big data, which offers both opportunities and challenges to the field of pattern recognition. We should seek new pattern recognition theories to be adaptive to big data. We should also push forward new pattern recognition applications benefited from big data.

Deep learning, which can be treated as the most significant breakthrough in the past 10 years in the field of pattern recognition and machine learning, has greatly affected the methodology of related fields like computer vision and achieved terrific progress in both academy and industry. It can be seen as a solution to change the whole pattern recognition system. It achieved an end-to-end pattern recognition, merging previous steps of pre-processing, feature extraction, classifier design and post-processing. It is expected that the development of deep learning theories and applications would further influence the field of pattern recognition. This workshop emphasizes the deep learning for pattern recognition and hopes to solicit original contributions, of leading researchers and practitioners from academia as well as industry, which address a wide range of theoretical and application issues in deep learning for pattern recognition.

We received 28 submissions in total. All paper submissions underwent a rigorous single-blind review process where the vast majority of papers received two or three reviews from the 15 members of the program committee, judging the originality of work, the relevance to deep learning for pattern recognition, the quality of the research or analysis, and the presentation. Of the 28 regular submissions received, 14 were accepted for presentation at the workshop (50%). The accepted regular papers are published in this proceedings volume in the Springer Lecture Notes in Computer Science series.

The final program comprises presentations by invited speaker and oral session. All 14 accepted papers are presented as oral. In addition to the contributed papers, the program includes one invited keynote presentation by distinguished member of the research community: Dr. Liang Zheng from Australian National University will speak about "Do We Really Need Ground Truths to Evaluate A Model?"

Last but not least, we would like to thank all the authors, program committee and additional reviewers, who made the workshop possible with their excellent work, rigorous and timely review process.

November 2020

Yongchao Xu
Meina Kan
Xiang Bai
Shiguang Shan
Jingdong Wang
Chunhua Shen
Gang Hua

Organization

General Chairs

Yongchao Xu	Huazhong University Science and Technology
Meina Kan	Institute of Computing Technology of Chinese Academy of Sciences
Xiang Bai	Huazhong University Science and Technology
Shiguang Shan	Institute of Computing Technology of Chinese Academy of Sciences
Jingdong Wang	Microsoft Research Asia
Chunhua Shen	University of Adelaide
Gang Hua	Wormpex AI Research Program Committee

Program Committee

Mingtao Fu	Huazhong University of Science and Technology, China
Zhenliang He	University of Chinese Academy of Sciences, China
Zhiwu Huang	ETH Zurich, Switzerland
Yong Li	Nanjing University of Science and Technology, China
Bingpeng Ma	University of Chinese Academy of Sciences, China
Yuhao Ma	University of Chinese Academy of Sciences, China
Tianyi Shi	Huazhong University of Science and Technology, China
Qian Tang	Wuhan University, China
Xijun Wang	University of Chinese Academy of Sciences, China
Zengmao Wang	Wuhan University, China
Yuxuan Xiong	Wuhan University, China
Chenfeng Xu	University of California, Berkeley
Guoping Xu	Wuhan Institute of Technology, China
Nan Xue	Wuhan University, China
Zhou Zhao	Laboratoire de Recherche et Développement de l'EPITA, France

Recurrent Graph Convolutional Network for Skeleton-Based Abnormal Driving Behavior Recognition

Shun Wang, Fang Zhou$^{(\boxtimes)}$, Song-Lu Chen, and Chun Yang

Department of Computer Science and Technology, School of Computer and
Communication Engineering, University of Science and Technology, Beijing, China
wangshunnn@gmail.com, zhoufang@ies.ustb.edu.cn, chenslvs7@gmail.com,
chunyang@ustb.edu.cn

Abstract. Abnormal driving behavior recognition is important in driving and traffic safety. Currently, skeleton-based action recognition has achieved significant improvement. However, how to effectively recognize abnormal driving behavior is still challenging in real applications, especially for subtle and similar behaviors. In this work, we propose a novel recurrent graph convolution network, which combines spatiotemporal graph convolutional networks and recurrent neural networks. First, we design a new spatial topological graph that includes the joints of the hands and face, which is advantageous to recognize subtle abnormal driving behaviors, such as yawning. Second, the proposed network can extract discriminative spatial and temporal representation features of the segmented skeleton sequences. Our method achieves an accuracy of 90.04% on the dataset collected by ourselves. Moreover, experiments on the Kinetics dataset verify the generalization ability of our method.

Keywords: Action recognition · Abnormal driving · Skeleton-based · Graph convolutional networks

1 Introduction

Abnormal driving behavior recognition is important in safe driving. The main factors influencing safe driving are not only from the vehicle exterior but also from the vehicle interior, especially the driver's behavior. Almost 90% of crashes in recent years are caused by abnormal driving behaviors (i.e., error, impairment, fatigue, and distraction) [4]. We find the following two difficulties in abnormal driving scenarios. First, some abnormal behaviors are very subtle, such as yawning, which is mainly reflected in the subtle changes in facial expressions. Second, there are some behaviors that are difficult to distinguish, such as drinking and smoking. In short, how to effectively recognize subtly similar behaviors is still a challenge.

© Springer Nature Switzerland AG 2021
A. Del Bimbo et al. (Eds.): ICPR 2020 Workshops, LNCS 12662, pp. 551–565, 2021.
https://doi.org/10.1007/978-3-030-68790-8_43

Abnormal driving behavior recognition can be seen as a branch of human action recognition. The mainstream deep learning network motion recognition method based on the RGB video model has performed well in video action recognition [27,30]. The methods based on RGB video models mainly utilize appearance and optical flow information in the video. However, methods based on the RGB video model are susceptible to various appearance and illumination.

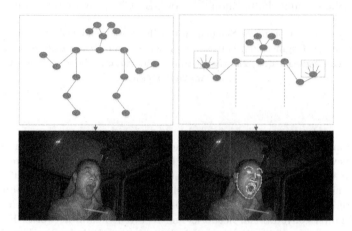

Fig. 1. The topological graph composed of 18 joints in the previous method (left) and the topological graph composed of 124 joints in our method (right).

Skeleton-based action recognition methods can focus more on the human body and be insensitive to the influence of various appearance and illumination [10]. Compared with networks based on RGB-model, graph neural networks can achieve better results because human skeleton data is more suitable for representation with graphs rather than pseudo-images [12,19,35]. Recently, many studies have begun to use graph convolutional networks to extract motion information in the human skeleton. A spatial temporal graph convolutional network has been proposed to model dynamic skeletons [34], but the spatial topological graph used in this method just contains 18 joints, so when it comes to recognizing subtle behaviors, such as the behavior "yawn", which is closely related to key points near the mouth and eyes, it fails to extract discriminative spatial semantic features. In view of this analysis, we designed the spatial topological graph of a total of 124 joints, which contains the 12 upper body joints, 70 facial joints, and 21 joints of each hand. More spatial semantic information can be extracted from the spatial topological graph composed of more joints to recognize subtle behaviors.

Moreover, the spatial temporal graph convolutional network is difficult to learn the temporal correlation between longer frames in the video. Some methods attempt to use the graph convolution network to extract the spatial semantic information of a single frame and then use the LSTM to extract the temporal semantic information contained in multiple frames [20,21]. However, these methods independently extract spatial and temporal features, so they cannot effectively represent the correlation between spatial and temporal information. The network we propose, which combines the advantages of graph convolutional network and LSTM, can learn the spatiotemporal correlation information through graph convolutional network, and then use LSTM to improve the temporal semantic representation of the skeleton sequence.

In this work, we propose a novel recurrent graph convolution network, which combines spatiotemporal graph convolutional networks and recurrent neural networks to address the challenge of recognizing subtle and similar abnormal driving behaviors. The architecture of our network is shown in Fig. 2. First, we process the video data and then split the skeleton sequence into several clips of the same length. Second, we use the GCN part to extract the spatiotemporal features of the segmented clips, and we design a spatial topological graph of 124 joints to replace the spatial topological graph of 18 joints used in the previous method, as shown in Fig. 3. Third, we use the LSTM part to explore deeper temporal features hidden between different clips. Finally, we use the fusion features formed by concatenating the output of our GCN part and the LSTM part to classify abnormal driving behaviors.

The main contributions can be summarized as follows:

- Aiming at the specific scenario of abnormal driving, we construct a new special topological graph, which is helpful to learn more spatial semantic information and recognize subtle abnormal driving behaviors.

- We combine our graph convolutional networks and LSTM networks to improve the temporal semantic representation of the skeleton sequence, helping to distinguish similar abnormal driving behaviors.

- We collect a video dataset for abnormal driving recognition. Extensive experiments on our dataset and Kinetics dataset verify the effectiveness of our proposed method.

2   Related Work

RGB-based Networks for Action Recognition. Conventional and mainstream human action recognition methods with deep learning networks are mainly based on RGB video models [2,7,11,22,26,27,30,36], which can be divided into three categories: CNN-RNN networks, Two-Stream networks and 3D-Convs networks. CNN-RNN networks is a tandem structure, which uses CNN and RNN to extract spatial and temporal features respectively [36]. Two-stream networks are a parallel structure, which uses CNN to explore spatial features in video frames as well as temporal features in optical flow respectively [22,30]. The 3D convolutional networks method uses 3D convolution to directly perform convolution operations on multiple frames of the video to obtain the spatiotemporal features [2,7,26,27]. However, methods based on the RGB video model are susceptible to various appearance and illumination.

Skeleton-based Networks for Action Recognition. Skeleton-based action recognition methods can focus on the human body, and external factors such as illumination conditions have less impact on this type of method. Moreover, human pose estimation has developed rapidly in recent years [1,24], which provides that the skeleton-based methods are feasible. Conventional skeleton-based methods mainly depends on hand-crafted features [9,28,29]. These methods usually model the coordinates of 3D joints based on manually selected features, such as normalizing the coordinates of 3D joints and use a covariance matrix to represent the characteristics of a certain length skeleton sequence [9]. With the development of deep learning, many methods have been proposed to learn the feature contained in skeleton data with convolutional neural networks (CNN) and recurrent neural networks (RNN) [5,6,12,13,15,19,23,31,35,37]. These methods usually convert skeleton sequence into pseudo image data that can be convolved by the CNN before the LSTM network [12]. Du et al. [5] divide human joints into five parts, then combine the different parts according to the simulation of the connection between adjacent body parts, and finally classifies the action by a hierarchical recurrent neural network. And some methods employ spatial attention and temporal attention mechanism to extract discriminative feature [23,31,35]. Now, many methods have attempted to utilize graph neural networks [14] to extract motion feature hidden in the human skeleton sequence [20,21,25,34]. Yan et al. [34] use a spatial-temporal graph convolutional network to model dynamic skeletons. However, the spatial topological graph used in these methods just contains 18 joints, so it fails to extract discriminative spatial semantic features of subtle behaviors. Si et al. [21] use graph neural networks and LSTM networks to extract spatial and temporal features respectively. Si et al. [20] add graph convolution operator and attention mechanism to LSTM. However, these methods lack the correlation between temporal and spatial information. Besides, unlike the previous dataset about driving [3,16–18,33], we collect a video dataset for abnormal driving recognition. In this work, we propose a novel network, which combines the advantages of graph convolutional networks and LSTM networks.

3   Approach

In order to accurately recognize abnormal driving behaviors, we propose a novel network. In this section, we describe the structure and components of our network in detail as follows.

3.1   Overall Architecture

As shown in Fig. 2, our network is mainly composed of two parts, the spatial temporal graph convolution network part (GCN Part), and the LSTM network part. Our input data is a skeleton sequence composed of human joints coordinates for each frame of the video. The skeleton data is shown in Fig. 1.

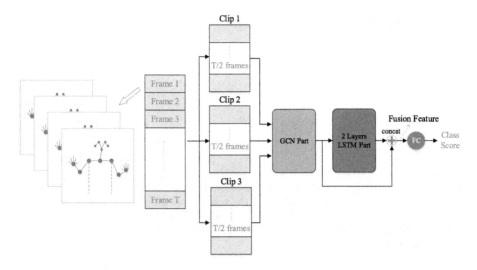

Fig. 2. Overall Architecture. First, we split the skeleton sequence into several clips of the same length. Second, we use the graph convolutional network part (GCN Part) to extract the spatiotemporal features of the segmented clips. Third, we use a 2 layers LSTM network (LSTM Part) to explore deeper temporal features hidden between different clips. Finally, we use the fusion features by concatenating the output of the GCN part and the LSTM part to classify abnormal driving behaviors.

Before feeding the data into the graph convolutional network part, we divide the complete skeleton sequence into several consecutive clips of the same length. And the adjacent fragments will contain overlapping parts, which helps us to efficiently multiplex the data at the same time during the graph convolution operation. The graph convolutional network part will perform convolution operations on the segmented fragments from the time and space dimensions. Then we feed the feature vectors of all fragments into a two-layer LSTM network. Finally, We will concatenate the feature vectors of all previous fragments with the output feature vectors of the LSTM to fusion features. The fusion features are used to calculate the classification score after the fully connected layer and softmax function. The entire model is trained end-to-end with the cross-entropy loss function.

$$Y = fc(concat(F_1, F_2, F_3, F_{LSTM}))$$ (1)

where F_1, F_2, F_3 are the 256 dimension feature of 3 fragments, extracted by spatial temporal graph convolutional network. F_{LSTM} is the output of the our LSTM network part. The classification score of the ith class F_{S_i} can be calculated as:

$$F_{Si} = \frac{e^{Yi}}{\sum_{n=1}^{C} e^{Yn}}, i = 1, 2, \ldots, C$$ (2)

where $F_S = (F_{S1}, F_{S2}, \ldots, F_{SC})$ is the predicted classification score of C classes.

3.2   Our Spatial Topological Graph

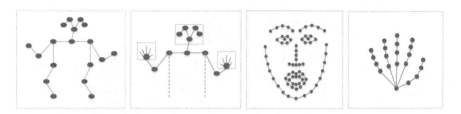

Fig. 3. The comparison of the spatial topological graph composed of 18 joints in the previous method (first from the left) and the topological graph composed of 124 joints in our method (second from the left). Specifically, the 124 joints are composed of 12 joints in the upper body, 70 joints in the face (third from the left) and 42 joints of hands (last from the left).

We infer that the topology graph with more joints can learn more spatial semantic information, which helps to distinguish subtle behaviors. As shown in Fig. 3, The spatial topology graph designed by the past method contains 18 joints(Fig. 3), lacking more joints of the hands and face. We construct the spatial topological graph of a total of 124 joints (Fig. 3). Specifically, the 124 joints are composed of 12 joints in the upper body, 70 joints in the face (Fig. 3) and 42 joints of hands (Fig. 3). Through the joints of the nose and the wrist, we can connect the face, hands and limbs to form the final spatial topology graph. In the actual driving scene, the lower body is not visible in the image, so we discard the joints data of the lower body. The temporal connection is the same as the previous methods, we connect the same joints in consecutive frames.

3.3   Our Graph Conventional Network Part

Graph convolutional network can perform convolution operation on the graph structure data and it can extract features hidden in the graph structure data. For skeletal data, we can construct a graph that can apply graph convolutional networks. We can simulate the natural connection of human joints to build a graph. For a single video frame at time t, we have a graph G_t as:

$$G_t = \{V_t, E_S\} \tag{3}$$

$$V_t = \{n_{t1}, n_{t2}, \ldots, n_{tJ}\} \tag{4}$$

$$E_S = \{n_{ti}n_{tj}|(i,j) \in H\} \tag{5}$$

where V_t is the nodes set of a single frame, and E_S is the edge set of a single frame, H represents the spatial connection relationship of J nodes in a single frame. We can build a neighbor nodes set $D(n_{ti})$ of a node n_{ti} as:

$$D(n_{ti}) = \{n_{tj}|d(n_{ti}, n_{tj}) \leq 1\} \tag{6}$$

where $d(n_{ti}, n_{tj}) \leq 1$ means the minimum joining distance is less than or equal to 1, which denote that two nodes are directly adjacent. In this work, we adopt the spatial temporal graph convolution model proposed in ST-GCN [34], which uses the labeling function $l_i : D(n_{ti}) \rightarrow \{0, 1, \ldots, K-1\}$ to partition the neighbor set $D(n_{ti})$ of node n_{ti} into a fixed number of K subsets, and each subset has a numeric label $k \in \{0, 1, \ldots, K-1\}$. The graph convolution on node n_{ti} in the spatial dimension can be described as:

$$f_{out}(n_{ti}) = \sum_{n_{tj} \in D(n_{ti})} \frac{1}{Z_{ti}(n_{tj})} f_{in}(n_{tj}) \cdot W(l_{ti}(n_{tj})) \tag{7}$$

where $f(n_{tj})$ is the feature of node n_{tj}. W is a weight function that can be indexed by the label l_{ti} from K weights. $Z_{ti}(n_{tj})$ is the cardinality of the corresponding subset, which is used to balance the contributions of different subsets. However the implementation of graph convolution in the spatial dimension is not as straightforward as image convolution. Kipf et al. [14] propose the computation of graph convolution as:

$$X_{out} = \Lambda^{-\frac{1}{2}}(A + I)\Lambda^{-\frac{1}{2}} X_{in} W \tag{8}$$

where A is the adjacency matrix and I is the identity matrix, $\Lambda = \sum_j (A^{ii} + I^{ii})$ is a degree matrix, and W is the weight matrix. X is the matrix representing node features. With spatial partition and labeling function l_i, we can replace the adjacency matrix $(A + I)$ with $\sum_k A_k$, and Eq. 7 can be transformed as

$$f_{out} = \sum_{k=0}^{K-1} \Lambda^{-\frac{1}{2}}(A_k \otimes M)\Lambda^{-\frac{1}{2}} f_{in} W_k \tag{9}$$

where A_k is the adjacency matrix in the spatial configuration of the label k. M is a learnable weight matrix that represents the importance of different nodes. $\otimes$ represents the element-wise product between two matrixes. For the temporal dimension, ST-GCN also extends temporal edges in consecutive frames as $E_T = \{n_{ti}n_{(t+1)i}\}$, and the adjacency matrix is extended to the time dimension. Equation 9 can also be used to extract features in both the spatial and temporal dimensions.

Every skeleton sequence data input into the graph convolutional network is a tensor of $[C, T_a : T_b, J]$, and C represents the coordinate dimension of the joints, and the coordinates of the joints can be expressed as the abscissa, the ordinate and the confidence score $(x, y, score)$, $T_a : T_b$ represents the segment of consecutive frames from a^{th} frame to b^{th} frame in the original sequence, J

represents the number of joints in the spatial topological graph. The output $\{F_1, F_2, F_3\}$ of the 3 input fragments in the graph network can be expressed as:

$$F_1 = f([C, 0 : T\frac{1}{2}, J])$$

$$F_2 = f([C, T\frac{1}{4} : T\frac{3}{4}, J]) \tag{10}$$

$$F_3 = f([C, T\frac{1}{2} : T, J])$$

where f means the spatial temporal graph convolution operators of our GCN part.

3.4   LSTM Network Part

With the development of deep learning, LSTM [8] has been proven to have advantages on models based on sequence data. The spatial temporal graph convolutional network can perform convolution operations in both the spatial and temporal dimensions. However, it will be difficult to learn the temporal relation between longer frames. Therefore, we add LSTM behind the graph network to make up for the shortcomings of our GCN part in temporal feature extraction and improve the temporal semantic representation of the skeleton sequence.

The input of the LSTM X_{in} is stacked by the 256 dimension feature vector F_1, F_2, F_3, extracted from the 3 segments in the previous network. We use a two-layer LSTM network to explore the timing characteristics of different fragments. The functions of the LSTM unit are defined as:

$$
\begin{aligned}
i_t &= \sigma(W_{xi} \cdot X_t + W_{hi} \cdot H_{t-1} + b_i) \\
f_t &= \sigma(W_{xf} \cdot X_t + W_{hf} \cdot H_{t-1} + b_f) \\
o_t &= \sigma(W_{xo} \cdot X_t + W_{ho} \cdot H_{t-1} + b_o) \\
u_t &= tanh(W_{xi} \cdot X_t + W_{hi} \cdot H_{t-1} + b_i) \\
C_t &= \sigma(f_t * C_{t-1}) + i_t * u_t \\
h_t &= o_t * tanh(C_t)
\end{aligned}
\tag{11}
$$

We use the short memory hidden output of the last time step as output F_{LSTM} we need. Then we concatenate the output of our GCN part F_1, F_2, F_3 and output of our LSTM part F_{LSTM}, and use SoftMax classification to class the concatenate feature.

4   Experiments

Our method is to recognize the abnormal driving behavior of the driver, but the current public dataset is mostly about general behavior recognition, which cannot be used as a benchmark for judging the performance of our method. Therefore, we collect a dataset dedicated to the identification of abnormal driving behaviors. We conduct experiments on our dataset and the public behavior recognition dataset Kinetics to prove the effectiveness of our method.

4.1   Our Driving Dataset

Fig. 4. Raw infrared images in our driving dataset. Five labels: "drink", "eye", "phone", "smoke", "yawn".

We collect a video dataset for abnormal driving recognition, which includes about 4850 short videos of 5 abnormal behaviors. This dataset includes five labels: "Drink", "Eye", "Phone", "Smoke", and "Yawn". As shown in Fig. 4, these five categories are common drivers' abnormal driving behaviors. Our videos are all infrared because infrared videos can reduce the impact of different lighting conditions. Each video lasts about 4 s and the frame rate is 10 frames per second. The resolution of each video is 1280×720 or 720×1280. The training set includes 4600 videos, of which there are approximately 1000 videos in each category. The test set consists of 250 videos, with 50 videos in each category. To obtain the skeleton data of these videos, we adopt the advanced pose estimation algorithm OpenPose [1], which can detect the joints of the limbs and face and hands. The coordinates of these joints are a tuple of (x, y, score) consisting of the abscissa, ordinate, and confidence score. As shown in Fig. 3, we select the coordinate information of a total of 124 joints, including 12 upper body joints, 70 facial joints, and 21 joints for each hand. To ensure the real-time performance of the pose estimation algorithm, we obtain the skeleton data in a way that guarantees the speed and some accuracy.

4.2   Kinetics Dataset

Kinetics [11] is the Deepmind Kinetics human action dataset. The human behavior dataset contains approximately 300,000 short videos clipped from YouTube. These videos cover up to 400 human action categories, including daily behaviors, motor behaviors, and some complex multiplayer behaviors, etc. Each clip video in the dataset lasts about 10 s. The Kinetics dataset only provides raw video clips, without skeleton data. Yan et al. [34] adjust all videos to a resolution of 340×256 and convert the frame rate to 30 FPS, then use the pose estimation algorithm OpenPose to obtain skeleton data of the Kinetics dataset. The coordinates of these joints are a tuple of (x, y, score) consisting of the abscissa, ordinate, and confidence score. Besides, this skeleton data set only contains skeleton information of the 18 joints of the body. We use this dataset to prove the generalization performance of our method and we evaluate recognition performance by calculating the top-1 and top-5 classification accuracy.

4.3   Implementation

We implement our network based on MMSkeleton[1] with 4 NVIDIA TITAN Xp GPUs. We use the spatial temporal graph convolution operators and spatial configuration partitioning in ST-GCN [34], which has 9 ST-GCN blocks, and every ST-GCN block is the combination of one spatial graph convolution layer, one temporal graph convolution layer and a dropout layer with the drop rate set as 0.5 to avoid overfitting. After each spatial and temporal graph convolution layer is a batch normalization layer and ReLU layer. The LSTM network part is a two-layer LSTM network with an intermediate hidden layer output of 512-dimension vectors. We use cross-entropy as the loss function to backpropagate gradients. The entire network is trained in an end-to-end manner.

When experimenting on our driving dataset, we uniformly process the input original skeleton sequence to 40 frames. Then we divide it into three overlapping clips with a length of 20 frames for training and testing. We adopt the SGD optimization algorithm as the optimization strategy, where the batch size is set to 32. The momentum set to 0.9 and weight decay is set to 0.0001. The initial learning rate is set as 0.1 and is divided by 10 at the 5_{th}, 10_{th} and 15_{th} epoch respectively. The training proc ess is ended at the 20_{th} epoch. For better comparison with ST-GCN [34], we train ST-GCN [34] on our dataset with the SGD optimization algorithm and the learning rate is set as 0.1 and is divided by 10 at the 5_{th}, 10_{th} and 15_{th} epoch, which is same as training of our network.

In addition, when training 3D-Resnet [7] and R(2+1)D [27] on our dataset, we choose the default training Settings. We train 3D-Resnet [7] with the pre-training model on the *Kinetics* dataset. We adopt the SGD optimization algorithm and the batch size is set to 128. The momentum set to 0.9 and weight decay is set to 0.0001, the training process is ended at the 200_{th} epoch. Especially, we use Resnxt-101 [32] as the backbone network of 3D-Resnet. We train R(2 + 1)D [27] with SGD optimization algorithm. The initial learning rate is set as 0.01 and is divided by 10 every 10 epochs, the training process is ended at the 45_{th} epoch.

When experimenting on the *Kinetics* dataset, we randomly choose 150 consecutive frames from the input skeleton sequence and then divide it into several overlapping clips. We adopt the SGD optimization algorithm as the optimization strategy, where the batch size is set to 256. The momentum set to 0.9 and weight decay is set to 0.0001. The initial learning rate is set as 0.1 and is divided by 10 at the 20_{th}, 30_{th} and 40_{th} epoch respectively. The training process is ended at the 50_{th} epoch.

---

[1] https://github.com/open-mmlab/mmskeleton..

4.4 Comparative Experiments

Table 1. Comparative experiments on our driving dataset.

Method		Top-1(%)
RGB-Based	3D-Resnet [7]	69.7
	R(2+1)D [27]	75.2
Skeleton-Based	ST-GCN [34]	75.6
	Ours	**90.4**

Table 2. Comparative experiments on the Kinetics dataset.

Method		Top-1(%)	Top-5(%)
RGB-Based	RGB [11]	57.0	77.3
	Optical Flow [11]	49.5	71.9
Skeleton-Based	Feature Enc. [6]	14.9	25.8
	Deep LSTM [12]	16.4	35.3
	Temporal Conv. [13]	20.3	40.0
	ST-GCN [34]	30.7	52.8
	Ours (with 2 clips)	29.6	52.5
	Ours (with 4 clips)	29.9	51.9
	Ours (with 3 clips)	**31.5**	**53.7**

The experimental results on our driving dataset are shown in Table 1. We train 3D-Resnet [7] and R(2 + 1)D [27] and ST-GCN [34] networks on our training set and calculated top-1 accuracy on the test set. We train ST-GCN with 18 joints. Experimental results show that our method achieves the best performance on driving dataset of 90.4%. We infer that the RGB-based methods are not as effective as expected because the infrared video contains less pixel information.

The experimental results on the Kinetics dataset are shown in Table 2. Experimental results on Kinetics datasets demonstrate that the performance of our method is a little better than that of previous methods. In addition, we compared the performance of our network when it was segmented into different numbers of clips, and the results show that our method with 3 clips performs best. We analyze the reasons for the lack of significant improvement in performance from two aspects. On the one hand, we only use the 18 joints of the body for training and testing. On the other hand, because a considerable part of the actions in the Kinetics dataset is not closely related to the skeleton, this is exactly why the performance of the skeleton-based method is significantly lower than that of the RGB-based method.

Table 3. Ablation study on our dataset about different topological graph.

Method	Top-1(%)
ST-GCN [34] (with default 18-graph)	75.6
ST-GCN [34] (with our 124-graph)	86.8
Ours (with default 18-graph)	76.0
Ours (with our 124-graph)	**90.4**

Table 4. Ablation study about different models of our networks on our driving dataset.

Model	Top-1(%)
Ours GCN Part (with 3 clips)	80.4
Ours LSTM Part (with 3 clips)	83.2
Fusion (with 3 clips)	**90.4**
Fusion (with 4 clips)	84.4

4.5   Ablation Study and Confusion Matrix

As shown in Table 3, we compare the results of our method with ST-GCN [34]. *18-graph* represents the topological graph of the default apstial topological graph of 18 joints in ST-GCN [34] , and *124-graph* represents the spatial topological graph of 124 joints we proposed in this work. When using the *124-graph* instead of default *18-graph*, the recognition results of ST-GCN [34] increased by 11.3% points. As shown in Fig. 5(top-left) and Fig. 5(top-right), it can be seen that the recognition results of all categories generally improve, and recognition accuracy of subtle behavior "Yawn" increased from 74% to 96%. But the two similar movements "drink" and "smoke" are still easily misidentified. From Fig. 5(bottom-left) and Fig. 5(bottom-right), it can be seen that when using our network, the recognition result improved by 3.6% to reach the best accuracy of 90.4%, and the recognition rate of two similar actions "drink" and "smoke" achieves the accuracy over 80%.

As shown in Table 4, we conduct ablation experiments on our dataset to evaluate the effectiveness of different modules in our network. We divide the 40-frame sequence into 3 overlapping 20-frames sequence clips, where we only use the output of the graph convolution network to classify. The result is 80.4%. Using only the output of the LSTM network portion to classify, the result is 83.2%. Fusion means that we use the fusion features by concatenating the output of our GCN Part and our LSTM Part for final classification, and the result achieves the best accuracy of 90.4%. Besides, we also experiment with the result of dividing into 4 overlapping 15-frame sequence clips, but the accuracy of 84.4% is not as good as dividing into 3 clips. We infer that 3 clips can well simulate the three stages of the beginning, proceeding, and ending of the behavior, so compared to dividing into 2 or 4 clips, 3 clips can get better experimental results.

	Drink	Eye	Phone	Smoke	Yawn
Drink	0.7	0	0.04	0.18	0.08
Eye	0	0.88	0.02	0	0.1
Phone	0.02	0	0.88	0	0.1
Smoke	0.08	0.02	0.2	0.58	0.12
Yawn	0.02	0	0.06	0.18	0.74

	Drink	Eye	Phone	Smoke	Yawn
Drink	0.84	0	0.02	0.12	0.02
Eye	0	0.96	0.02	0.02	0
Phone	0	0	0.96	0.02	0.02
Smoke	0.18	0.1	0.02	0.62	0.08
Yawn	0.02	0.02	0	0	0.96

	Drink	Eye	Phone	Smoke	Yawn
Drink	0.66	0.02	0.1	0.06	0.14
Eye	0	0.96	0	0	0.04
Phone	0	0	0.94	0	0.06
Smoke	0.02	0.04	0.22	0.5	0.22
Yawn	0.02	0.06	0.08	0.12	0.72

	Drink	Eye	Phone	Smoke	Yawn
Drink	0.82	0.02	0.02	0.14	0
Eye	0	0.98	0	0	0.02
Phone	0.02	0	0.94	0.02	0.06
Smoke	0.1	0.04	0.04	0.8	0.02
Yawn	0	0	0.02	0	0.98

Fig. 5. Confusion matrix for ST-GCN with default 18-graph(top-left) and our 124-graph(top-right) on our driving dataset, confusion matrix for our method with default 18-graph(bottom-left) and our 124-graph(bottom-right) on our driving dataset.

5 Conclusion

In this paper, we propose a novel recurrent graph convolution network that combines spatial temporal graph convolutional networks and LSTM networks for abnormal driving behavior recognition. Experiments show that through our spatial topological graph, we can better recognize subtle abnormal driving behaviors. Moreover, through our LSTM part, we can further improve the ability of our proposed network to distinguish similar behaviors. We collect a video dataset for abnormal driving recognition, and our method achieves the best accuracy of 90.04% on our driving dataset. Experiments on the Kinetics dataset show that our method has good generalization ability.

References

1. Cao, Z., Simon, T., Wei, S.E., Sheikh, Y.: Realtime multi-person 2D pose estimation using part affinity fields. In: Proceedings of the IEEE Conference on Computer Vision and Pattern Recognition, pp. 7291–7299 (2017)
2. Carreira, J., Zisserman, A.: Quo vadis, action recognition? a new model and the kinetics dataset. In: proceedings of the IEEE Conference on Computer Vision and Pattern Recognition, pp. 6299–6308 (2017)
3. Craye, C., Karray, F.: Driver distraction detection and recognition using RGB-D sensor. arXiv preprint arXiv:1502.00250 (2015)

4. Dingus, T.A., et al.: Driver crash risk factors and prevalence evaluation using naturalistic driving data. Proc. Natl. Acad. Sci. **113**(10), 2636–2641 (2016)
5. Du, Y., Wang, W., Wang, L.: Hierarchical recurrent neural network for skeleton based action recognition. In: Proceedings of the IEEE Conference on Computer Vision and Pattern Recognition, pp. 1110–1118 (2015)
6. Fernando, B., Gavves, E., Oramas, J.M., Ghodrati, A., Tuytelaars, T.: Modeling video evolution for action recognition. In: Proceedings of the IEEE Conference on Computer Vision and Pattern Recognition, pp. 5378–5387 (2015)
7. Hara, K., Kataoka, H., Satoh, Y.: Can spatiotemporal 3D CNNS retrace the history of 2D CNNS and imagenet? In: Proceedings of the IEEE conference on Computer Vision and Pattern Recognition, pp. 6546–6555 (2018)
8. Hochreiter, S., Schmidhuber, J.: Long short-term memory. Neural Comput. **9**(8), 1735–1780 (1997)
9. Hussein, M.E., Torki, M., Gowayyed, M.A., El-Saban, M.: Human action recognition using a temporal hierarchy of covariance descriptors on 3D joint locations. In: Twenty-Third International Joint Conference on Artificial Intelligence (2013)
10. Johansson, G.: Visual perception of biological motion and a model for its analysis. Percept. Psychophysics **14**(2), 201–211 (1973)
11. Kay, W., et al.: The kinetics human action video dataset. arXiv preprint arXiv:1705.06950 (2017)
12. Ke, Q., Bennamoun, M., An, S., Sohel, F., Boussaid, F.: A new representation of skeleton sequences for 3D action recognition. In: Proceedings of the IEEE Conference on Computer Vision and Pattern Recognition, pp. 3288–3297 (2017)
13. Kim, T.S., Reiter, A.: Interpretable 3D human action analysis with temporal convolutional networks. In: 2017 IEEE conference on computer vision and pattern recognition workshops (CVPRW), pp. 1623–1631. IEEE (2017)
14. Kipf, T.N., Welling, M.: Semi-supervised classification with graph convolutional networks. arXiv preprint arXiv:1609.02907 (2016)
15. Liu, J., Shahroudy, A., Xu, D., Wang, G.: Spatio-temporal LSTM with trust gates for 3D human action recognition. In: Leibe, B., Matas, J., Sebe, N., Welling, M. (eds.) ECCV 2016. LNCS, vol. 9907, pp. 816–833. Springer, Cham (2016). https://doi.org/10.1007/978-3-319-46487-9_50
16. Liu, T., Yang, Y., Huang, G.B., Yeo, Y.K., Lin, Z.: Driver distraction detection using semi-supervised machine learning. IEEE Trans. Intell. Transp. Syst. **17**(4), 1108–1120 (2015)
17. Martin, M., Popp, J., Anneken, M., Voit, M., Stiefelhagen, R.: Body pose and context information for driver secondary task detection. In: 2018 IEEE Intelligent Vehicles Symposium (IV), pp. 2015–2021. IEEE (2018)
18. Martin, M., et al.: Drive&act: a multi-modal dataset for fine-grained driver behavior recognition in autonomous vehicles. In: Proceedings of the IEEE International Conference on Computer Vision, pp. 2801–2810 (2019)
19. Shahroudy, A., Liu, J., Ng, T.T., Wang, G.: NTU RGB+ D: A large scale dataset for 3D human activity analysis. In: Proceedings of the IEEE Conference on Computer Vision and Pattern Recognition, pp. 1010–1019 (2016)
20. Si, C., Chen, W., Wang, W., Wang, L., Tan, T.: An attention enhanced graph convolutional lstm network for skeleton-based action recognition. In: Proceedings of the IEEE Conference on Computer Vision and Pattern Recognition, pp. 1227–1236 (2019)
21. Si, C., Jing, Y., Wang, W., Wang, L., Tan, T.: Skeleton-based action recognition with spatial reasoning and temporal stack learning. In: Proceedings of the European Conference on Computer Vision (ECCV), pp. 103–118 (2018)

22. Simonyan, K., Zisserman, A.: Two-stream convolutional networks for action recognition in videos. In: Advances in Neural Information Processing Systems, pp. 568–576 (2014)
23. Song, S., Lan, C., Xing, J., Zeng, W., Liu, J.: An end-to-end spatio-temporal attention model for human action recognition from skeleton data. In: Thirty-first AAAI Conference on Artificial Intelligence (2017)
24. Sun, K., Xiao, B., Liu, D., Wang, J.: Deep high-resolution representation learning for human pose estimation. In: Proceedings of the IEEE Conference on Computer Vision and Pattern Recognition, pp. 5693–5703 (2019)
25. Thakkar, K., Narayanan, P.: Part-based graph convolutional network for action recognition. arXiv preprint arXiv:1809.04983 (2018)
26. Tran, D., Bourdev, L., Fergus, R., Torresani, L., Paluri, M.: Learning spatiotemporal features with 3D convolutional networks. In: Proceedings of the IEEE International Conference on Computer Vision, pp. 4489–4497 (2015)
27. Tran, D., Wang, H., Torresani, L., Ray, J., LeCun, Y., Paluri, M.: A closer look at spatiotemporal convolutions for action recognition. In: Proceedings of the IEEE Conference on Computer Vision and Pattern Recognition, pp. 6450–6459 (2018)
28. Vemulapalli, R., Arrate, F., Chellappa, R.: Human action recognition by representing 3D skeletons as points in a lie group. In: Proceedings of the IEEE Conference on Computer Vision and Pattern Recognition, pp. 588–595 (2014)
29. Wang, J., Liu, Z., Wu, Y., Yuan, J.: Mining actionlet ensemble for action recognition with depth cameras. In: 2012 IEEE Conference on Computer Vision and Pattern Recognition, pp. 1290–1297. IEEE (2012)
30. Wang, L., et al.: Temporal segment networks: towards good practices for deep action recognition. In: Leibe, B., Matas, J., Sebe, N., Welling, M. (eds.) ECCV 2016. LNCS, vol. 9912, pp. 20–36. Springer, Cham (2016). https://doi.org/10.1007/978-3-319-46484-8_2
31. Xie, C., et al.: Memory attention networks for skeleton-based action recognition. arXiv preprint arXiv:1804.08254 (2018)
32. Xie, S., Girshick, R., Dollár, P., Tu, Z., He, K.: Aggregated residual transformations for deep neural networks. In: Proceedings of the IEEE conference on computer vision and pattern recognition, pp. 1492–1500 (2017)
33. Yan, C., Coenen, F., Zhang, B.: Driving posture recognition by convolutional neural networks. IET Comput. Vis. 10(2), 103–114 (2016)
34. Yan, S., Xiong, Y., Lin, D.: Spatial temporal graph convolutional networks for skeleton-based action recognition. In: Thirty-second AAAI Conference on Artificial Intelligence (2018)
35. Yang, Z., Li, Y., Yang, J., Luo, J.: Action recognition with spatio-temporal visual attention on skeleton image sequences. IEEE Trans. Circ. Syst. Video Technol. 29(8), 2405–2415 (2018)
36. Yue-Hei Ng, J., Hausknecht, M., Vijayanarasimhan, S., Vinyals, O., Monga, R., Toderici, G.: Beyond short snippets: deep networks for video classification. In: Proceedings of the IEEE Conference on Computer Vision and Pattern Recognition, pp. 4694–4702 (2015)
37. Zhang, P., Lan, C., Xing, J., Zeng, W., Xue, J., Zheng, N.: View adaptive neural networks for high performance skeleton-based human action recognition. IEEE Trans. Pattern Anal. Mach. Intell. 41(8), 1963–1978 (2019)

Supervised Autoencoder Variants for End to End Anomaly Detection

Max Lübbering[1]([✉]), Michael Gebauer[2], Rajkumar Ramamurthy[1], Rafet Sifa[1], and Christian Bauckhage[1]

[1] Fraunhofer IAIS, Schloss, Sankt Augustin, Germany
`max.luebbering@iais.fraunhofer.de`
[2] PriceWaterhouseCoopers GmbH, Berlin, Germany

Abstract. Despite the success of deep learning in various domains such as natural language processing, speech recognition, and computer vision, learning from a limited amount of samples and generalizing to unseen data still pose challenges. Notably, in the tasks of outlier detection and imbalanced dataset classification, the label of interest is either scarce or its distribution is skewed, causing aggravated generalization problems. In this work, we pursue the direction of multi-task learning, specifically the idea of using *supervised autoencoders* (SAE), which allows us to combine unsupervised and supervised objectives in an end to end fashion. We extend this approach by introducing an adversarial supervised objective to enrich the representations which are learned for the classification task. We conduct thorough experiments on a broad range of tasks, including outlier detection, novelty detection, and imbalanced classification, and study the efficacy of our method against standard baselines using autoencoders. Our work empirically shows that the SAE methods outperform *one class autoencoders, adversarially trained autoencoders* and *multi layer perceptrons* in terms of AUPR score comparison. Additionally, our analysis of the obtained representations suggests that the adversarial reconstruction loss functions enforce the encodings to separate into class-specific clusters, which was not observed for non-adversarial reconstruction loss functions.

Keywords: Multi task learning · Autoencoders · Outlier detection · Anomaly detection · Imbalanced datasets

1 Introduction

Deep learning methods have become ubiquitous in many real-world applications, such as understanding natural language, recognizing speech and processing of vision images. Despite their success, learning from a limited amount of training data and generalizing to unseen new data is an actively studied research area [29,40]. In addition to common strategies to avoid overfitting, such as weight decay, drop out and early stopping, an alternative direction has emerged in the form of *multi task learning* (MTL) [7]. The idea is to incorporate additional

© Springer Nature Switzerland AG 2021
A. Del Bimbo et al. (Eds.): ICPR 2020 Workshops, LNCS 12662, pp. 566–581, 2021.
https://doi.org/10.1007/978-3-030-68790-8_44

tasks into the learning process to improve the generalization of the model and to learn from a reduced number of samples. These additional tasks mostly focus on including unsupervised objectives [45,52] to learn intermediate layers of a deep network.

However, the problem of generalization still persists in imbalanced datasets and is not studied widely. In this setting, the distribution of categories is skewed with some group of data (majority class samples) appearing more frequently than others (minority class samples). This is generally seen in medical diagnosis [38], fraud detection [41,42], and image classification [24,30], where detection of minority samples is more crucial than majority samples. Due to the limited availability of training samples for these rare events, it has significant negative effects on the generalization performance of classifiers [5,38].

Handling imbalanced datasets have been studied under two main categories: Firstly, *data-level methods*, which focus on balancing the skewed data distribution by under-sampling of majority samples [6] or by over-sampling of minority samples [10]. Secondly, *algorithm-level methods*, that focus on adapting the learning algorithm directly to learn from minority samples by varying the decision threshold [32] and assigning higher costs [31] for minority samples.

Additionally, anomaly detection methods [26,35,48] using autoencoders or support vector machines have been used to learn a model from one class of samples and use reconstruction errors to classify majority and minority samples. When looking at these one-class solutions, it is clear that the imbalanced dataset problems are very similar to outlier detection problems. This similarity arises because of two main reasons. First, outliers are usually a minority as they occur rarely. Second, they are produced by a different underlying mechanism compared to inliers, which also applies to minority samples in an imbalanced classification setting [12,21,57].

Although one distinguishing factor is that outlier detection is generally seen as either unsupervised or semi-supervised problem, it is treated as a supervised problem when labeled data indicating if a data point is "normal" and "abnormal" is available. This setting corresponds to *supervised outlier detection* or *classification based anomaly detection*, which one of its subproblems is dataset imbalance [1,9].

In this work, we focus on imbalanced dataset problems, as well as, supervised outlier and novelty detection problems. In particular, we are interested in deep learning methods using *supervised autoencoders* (SAE) [34]. We pursue the direction of multi-task learning by training a SAE to jointly classify and reconstruct a sample. Typically, in this setting, while minimizing the auxiliary reconstruction loss, either all the samples (vanilla AEs) or just the majority samples (one-class AEs) are used during training. Both approaches become ineffective when majority and minority samples overlap in the feature space, which prevents the classifier from accurate discrimination. To address this limitation, we propose an auxiliary task based on the adversarial style of training autoencoders introduced in [37]. The main idea is instead of training the autoencoders to minimize only the reconstruction loss for majority samples, they can also be trained to maximize the loss for minorities, thereby enriching the features for classification.

Therefore, our main contributions are:

1. We introduce three novel autoencoder based end to end approaches by adaption of the vanilla SAE architecture. Namely, a) *adversarially trained supervised autoencoders* (ASAE), which incorporates an adversarial loss function instead of the mean squared error loss, b) *supervised autoencoders with reconstruction loss* (SAER) which forwards the reconstruction loss to the classification layers as a predictive feature and c) *adversarial supervised autoencoders with reconstruction* (ASAER) which is a combination of ASAE and SAER.
2. The proper functioning of all SAE variants is thoroughly verified by empirically analyzing the reconstruction loss distribution and clustering of autoencoders' encoding.
3. Additionally, we show the superiority of our methods compared to state-of-the-art autoencoder methods and a vanilla MLP on seven different datasets for imbalanced classification, novelty and anomaly detection problems.

2 Related Work

Deep learning for solving outlier detection problems come in a large variety. Approaches range from *autoencoders* (AE) over *variational autoencoders* (VAE) to *generative adversarial networks* (GAN). Some studies compare the fundamental approaches with other classes of neural network architectures [44,47,50,51]. Yet, these studies mostly lack incorporating recent developments, making the superiority of a single architecture (AE, VAE, or GAN) highly debatable.

Using AEs for outlier detection was first done in [22] with replicator networks. In later works, they were also used in a one-class fashion [8,13] or as an ensemble [11,46], which turned out to be computationally very expensive. *Robust deep autoencoders* were using ideas from *robust principal components analysis* and optimize the resulting loss function with *alternating direction method of multipliers* (ADMM) [56]. ADMM has slow convergence and mostly leads to modest accuracy in practice [4]. This limits the potential of Robust deep autoencoders.

Recently, SAE as an auxiliary-task model was investigated in [34]. For a plain MLP architecture an autoencoder is added, which reconstructs the given image. Hence, the model not only tries to minimize the classification error, but also the reconstruction error. The additional task introduced with the AE functions as a regularizer and prevents the model from overfitting [34,36]. The term *supervised autoencoders* was also coined in [17], where the authors take ideas from *denoising autoencoders*, in our work we do not relate to this approach, since we are interested in having an end-to-end approach with good generalization performance.

Since our approach uses adversarial training, we want to stress that our learning procedure is also not related to GANs or adversarial autoencoders. GAN-based outlier detection often involves additional training steps like in AnoGAN or OCGAN and measures outlierness by a residual loss, discriminative loss or feature mapping. In our case, outliers are detected via a subsequent classification layer, which takes the latent space and the reconstruction loss of the autoencoder as features. Architecturally our approach does not require a generator and discriminator for a min-max game, which is the foundation of the GAN architecture [2,44,47,55]. Furthermore, we also do not employ a prior distribution on the latent state like in adversarial autoencoders. In our case, the term originated from adversarial examples introduced in [19] to enhance model robustness.

3   Background

Throughout this paper, we follow the notation mentioned below. Majority and minority samples are denoted by m^+ and m^-, respectively. A sample set is defined by $\mathbb{X} = \{\mathbf{x}_1, \ldots, \mathbf{x}_n\}$ with corresponding targets $Y = \{y_1, \ldots, y_n\}$ where each target $y_i \in \{m^+, m^-\}$.

A neural network with weights $\boldsymbol{w}$ is represented by the function $f_w : \mathbf{x}_i \mapsto y_i'$, that maps a sample $\mathbf{x}$ to predictions y' or in the special case of an AE to the reconstruction $\mathbf{x}'$.

3.1   Autoencoders for Outlier and Novelty Detection

An autoencoder (AE) is a neural network, that reconstructs a given input vector by using dimensionality reduction. Hence, to encode the informative features, the encoder $f : \mathbb{R}^m \mapsto \mathbb{R}^k$ maps a given input $\mathbf{x}$ to a hidden vector $\mathbf{h}$ within a lower dimensional latent space. A decoder takes data points from the latent space and maps them back to the original space $g : \mathbb{R}^k \mapsto \mathbb{R}^m$ to form a reconstruction $\mathbf{x}'$. The loss of an AE is defined as the error between the original input $\mathbf{x}$ and its reconstruction $\mathbf{x}'$. For the linear case an AE can be defined as:

$$\mathbf{h} = f(\mathbf{x})$$
$$\mathbf{x}' = g(\mathbf{h})$$
$$f(\mathbf{x}) = W_f \mathbf{x} + b_f$$
$$g(\mathbf{h}) = W_g \mathbf{h} + b_g$$

The parameters involved are the weights and biases of the encoder $W_f, \mathbf{b}_f$ and the decoder $W_g, \mathbf{b}_g$. To train the AE, the parameters are updated to minimize the Mean Squared Error (MSE) defined as

$$L_{mse}(\mathbf{x}, \mathbf{x}') = \frac{1}{m} \sum_{i=1}^{m} (x_i - x_i')^2. \tag{1}$$

The classification of minorities is based on the assumption, that reconstruction errors for these data points are higher than for the majority class [22].

Therefore, a threshold t on the reconstruction error is found through brute force line search to detect minorities with a satisfactory performance.

$$y(\mathbf{x}, \mathbf{x}') = \begin{cases} m^-, & L_r(\mathbf{x}, \mathbf{x}') \geq t \\ m^+, & \text{otherwise.} \end{cases} \qquad (2)$$

Unfortunately this assumption may not hold, when majorities and minorities correlate in the feature space.

3.2 Supervised Autoencoders

SAEs were introduced by Le et al. in [33] to improve the generalization performance of neural networks by adding an unsupervised auxiliary task to the supervised learning task. From an architectural point of view, the neural network model is a combination of an AE performing the auxiliary task by reconstructing the inputs and a neural network performing the classification task, as shown in Fig. 1. Notably, both networks share the encoding layers. The overall architecture is trained in an end to end fashion by minimizing a combined loss function L_t, that is composed of both the reconstruction loss L_r for sample $\mathbf{x}$, $\mathbf{x}'$ and classification loss L_c for target y and prediction y':

$$L_t(\mathbf{x}, \mathbf{x}', y, y') = \lambda L_c(y, y') + L_r(\mathbf{x}, \mathbf{x}'). \qquad (3)$$

Notably, this loss function does not introduce any scaling of L_c and L_r, thus making the effect of the weighting hyperparameter λ dependent on the scale of the inputs. While the authors state that $\lambda = 0.01$ provided the best results, it is not interpretable whether this factor equalizes both losses or favors one of them.

4 Adversarial Supervised Autoencoders

Building upon SAEs, we introduce three novel variants specifically targeted at tackling imbalanced dataset, outlier detection and novelty detection problems. While the authors of SAE show that their architecture is less prone to overfitting and yields better generalization performance on the classification tasks, we investigate whether SAEs and our proposed variants can also achieve robust results on the settings mentioned above.

On top of that, our work builds upon the idea of exploiting the reconstruction loss as a highly predictive feature for anomaly detection, which was already investigated by [37] as part of their *adversarially trained autoencoder* (ATA) architecture. This architecture is trained in an adversarial supervised fashion, which minimizes the reconstruction loss for majorities and maximizes it for minorities, as formalized by:

$$L_{r\_adv}(\mathbf{x}, \mathbf{x}', y) =$$
$$\begin{cases} 0, & L_{mse} > t \wedge y \in m^- \\ L_{mse}(\mathbf{x}, \mathbf{x}'), & y \in m^+ \\ -\alpha L_{mse}(\mathbf{x}, \mathbf{x}'), & \text{otherwise.} \end{cases} \qquad (4)$$

Given the loss function L_{r_adv}, ATA learns to reconstruct majority samples having class label m^-. The loss of minority samples is maximized up to a chosen threshold t. The maximization intensity is controlled by outlier weighting factor $\alpha \in \mathbb{R}^+$. When the reconstruction loss exceeds this threshold t, the loss is zeroed out, thus zeroing out the gradients as well, resulting in no further learning w.r.t. the given sample. It has been shown by [37], that loss negation is equivalent to flipping the gradient.

By pushing the loss distribution of minority samples towards threshold t and the majority loss distribution towards 0, this adversarial training style enforces the high predictability of the reconstruction loss for anomalies. For estimating the decision boundary, they performed a brute force line search over the reconstruction loss space to find a suitable decision threshold. The authors empirically showed that their approach is robust in the three classification tasks, when compared to *one class autoencoders* (OCA) and *multi layer perceptrons* (MLP).

In this work, we combine the merits of both approaches by proposing two different extension to SAE, whose combinations result in three new architectures, as shown in Table 1.

In contrast to the loss function of the original SAE, as defined in Eq. (3), we improve the loss functions in two ways. First, we equalize the importance of each loss term L_c and L_r at the beginning of the training by unit scaling them with the scaling factors s_c and s_r, respectively. Secondly, we sum up a linear combination of both loss terms, instead of penalizing a single term individually. With these two changes, as defined in Eq. (5), the influence of each loss term can be investigated empirically.

$$L_t(\mathbf{x}, \mathbf{x}', y, y') = \frac{\lambda}{s_c} L_c(y, y') + \frac{1-\lambda}{s_r} L_r(\mathbf{x}, \mathbf{x}') \tag{5}$$

The first SAE extension replaces the reconstruction loss L_r within Eq. (5) by the adversarial reconstruction loss function L_{r_adv}, previously defined in Eq. (4). As shown in Table 1, models making use of the adversarial reconstruction loss function have a leading A in their name.

By our second SAE extension, the reconstruction loss L_r is passed to the MLP along with the encoding, as indicated by the dashed arrow in Fig. 1. As already shown for ATA and OCA, the reconstruction loss is a highly predictive feature and therefore reasonable to be included in the input of the MLP. In Table 1, the presence of this property is indicated by a trailing R in the respective model names.

5   Experiments and Results

In this section, the experiment setup and the model performances on seven different datasets are discussed. The datasets have been carefully chosen to represent all three tasks of imbalanced classification, supervised outlier and novelty detection. In outlier and novelty detection, AUROC metric is a commonly used metric for two main reasons; First, due to the integral over all possible thresholds, this

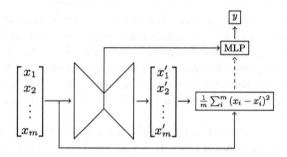

Fig. 1. Conceptual design of supervised autoencoder (SAE) and supervised autoencoder with readout loss (SAER). In both variants the autoencoder maps the input sample **x** to the reconstruction **x′**. In the SAE case, the MLP performs inference solely on the encoding of **x**, whereas in the case of SAER, the reconstruction loss is also passed to the MLP, as indicated by the dashed line.

Table 1. Summarization of the different model variations. The models are distinguished on an architectural level based on whether the reconstruction loss is passed to the MLP or not, as defined in the third column. Additionally, there are two different training styles by adaptation of the reconstruction loss function L_r in Eq. (5): Mean squared error L_{mse} and adversarial reconstruction loss L_{r_adv}.

Model	L_r	L_r to MLP
SAE	L_{mse}	No
SAER	L_{mse}	Yes
ASAE	L_{r_adv}	No
ASAER	L_{r_adv}	Yes

metric assesses the model at every possible threshold. Second, the metric can be interpreted as the probability of a random minority sample being ranked higher than a random majority sample [16]. However, as pointed out by [23], AUROC provides misleading scores when the class distributions are highly skewed, as AUROC does not take the classes' base rates into account.

To overcome this limitation of AUROC, we instead consider the AUPR score, which is also a threshold independent metric, but in contrast takes different base rates [14] into account. Furthermore, it can be interpreted as the proportion of samples to be predicted as minorities over the set of all predictions exceeding a randomly selected threshold [3]. It is generally applied to problems dealing with "finding a needle in a haystack" and therefore, it is more relevant for outlier and imbalanced dataset problems.

5.1   Datasets

For benchmarking our methods on three different classification tasks namely imbalanced classification, supervised outlier and novelty detection, we carefully

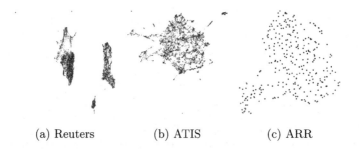

(a) Reuters   (b) ATIS   (c) ARR

Fig. 2. Imbalanced datasets: visualization of the Reuters, ATIS and arrhythmia dataset after projecting the samples onto a two dimensional plane using UMAP for dimensionality reduction. Minorities samples are highlighted in red, majorities in blue. (Color figure online)

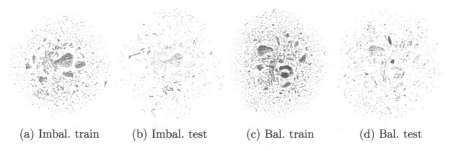

(a) Imbal. train   (b) Imbal. test   (c) Bal. train   (d) Bal. test

Fig. 3. KDD outlier and novelty datasets: visualization of the train and test splits after reducing the dimensionality to 2D using UMAP. Outlier samples are highlighted in red, inliers in blue and novelties in green. Figure a) and Fig. b) show the clustering of the imbalanced variant of datasets, while Fig. c) and Fig. d) show the clustering for the balanced datasets. Note that, as per definition, novelties only appear in the test set, whereas outliers appear in both. (Color figure online)

derived seven datasets from four widely accepted datasets. To test the models on imbalanced classification, we consider the Reuters[1], Arrhythmia[2] and ATIS[3] datasets, whereas for novelty and outlier detection, the models were benchmarked on four datasets derived from the KDD[4] dataset. Since we focus on binary classification settings, the datasets' class labels were binarized to have only two labels, resulting in the minority and majority class distributions depicted in Table 2. As part of the pre-processing, the textual data are vectorized using Glove word embeddings [43], whereas categorical features and continuous features were one-hot encoded and z-transformed, respectively.

---

[1] http://www.daviddlewis.com/resources/testcollections/reuters21578/.
[2] http://archive.ics.uci.edu/ml/datasets/Arrhythmia.
[3] http://www.ai.sri.com/natural-language/projects/arpa-sls/atis.html.
[4] https://www.unb.ca/cic/datasets/nsl.html.

Table 2. Datasets and their subdatasets: Majority and minority frequencies for train, validation and test split. A subclass share of all means that *all* subclasses are shared between the dataset splits, whereas a subclass share of *none*, indicates that none are shared between train/val and test split, which is an important requirement for novelty detection.

Dataset	Type	Majority classes	Minority classes	Subclass share	Train $\#m^+$	$\#m^-$	Val $\#m^+$	$\#m^-$	Test $\#m^+$	$\#m^-$
KDD	Outlier imbal.	NORMAL	U2R, R2L	all	47122	745	20221	302	9711	2236
	Outlier bal.	NORMAL	U2R, R2L, DOS, PROBE	all	47122	41059	20221	17571	9711	9083
	Novelty imbal.	NORMAL	U2R, R2L	none	47122	745	20221	302	9711	716
	Novelty bal.	NORMAL	U2R, R2L, DOS, PROBE	none	47122	41059	20221	17571	9711	3750
Reuters	Imbal.	CRUDE, TRADE INTEREST, MONEY-FX MONEY-SUPPLY	EARN, ACQ	all	3086	746	1349	293	1779	393
ATIS	Imbal.	FLIGHT	QUANT, AIRFARE, ABBR GSERVICE, REST, APORT ALINE, CITY, F.NO, F.TIME G.FARE, F.AIRFARE DIST, AIRCRAFT, CAPA	all	3173	1101	423	149	424	162
ARR	Imbal.	NORMAL	OTHERS (15 classes)	all	122	31	62	15	61	15

Reuters Dataset. This NLP dataset contains documents published by the news outlet Reuters. A single document is assigned to at least one of the pre-defined 90 classes. Combined with the high class imbalance in this dataset, this is a standard benchmarking dataset for multi-label document classification as well as outlier detection [27,28,53]. However, since multi-label classification is out of scope of this paper, all documents having multiple labels were filtered out. Therefore, this dataset is representative for imbalanced classification.

ATIS Dataset. This dataset contains transcribed queries that passengers requested against the air travel information system (ATIS) to receive flight related information. The resulting *ATIS Spoken Language Systems Pilot Corpus* was labeled with a total of 17 classes. The prevalent class frequency imbalance makes this dataset a suitable candidate for imbalanced classification problems.

ARR Dataset. This dataset contains samples on heart arrhythmia categorized into 18 classes. Due to its imbalanced nature and small size, this dataset is generally considered a difficult dataset for imbalanced classification.

KDD Dataset. This dataset consists of benign network communication samples, as well as, four types of network intrusions. Due to the low prevalence of those intrusions and their diversity, this is a common dataset to benchmark outlier and novelty detection problem [15,18,20,25]. Notably, the original KDD dataset had various inherent issues, including redundant features and information leakage. For this reason, we consider an improved version [49] that is devoid of these issues. The intrusion types contain several subtypes and these subtypes are either shared by all splits or are only present in either the train/validation sets or the test set. Therefore, four datasets are derived with a combination of outlier/novelty and balanced/imbalanced as shown in Table 2.

UMAP Visualization. To further support the different nature of these datasets, we projected the samples' features onto a two dimensional plane using

UMAP, an unsupervised manifold learning based algorithm for nonlinear dimensionality reduction [39]. Figure 2 shows the results for the datasets Reuters, ATIS and ARR representative for the imbalanced classification problem. The Reuters dataset expresses well separated clusters, on the other hand, for ATIS the minority and majority the clusters majorly overlap. The ARR dataset forms a single coherent cluster, with no separation between minority and majority samples, making this dataset the most difficult among the three imbalanced datasets. Similarly, the clusters of the four KDD based outlier and novelty datasets are displayed in Fig. 3. Since as per definition, novelties only appear in the test set, the UMAP model was trained on the training split first and subsequently applied to the train and test split separately. Especially for the balanced case, two insights can be deducted: a) the outliers and inliers form proper clusters on both splits and b) novelties do not cluster. Both insights support the definition of outlier and novelties, thus making these datasets valid representatives for the respective classification problem.

5.2   Results

To achieve a fair comparison, we keep the model complexity of each method roughly the same. The MLP has a single hidden layer of size 50 with sigmoid activations and an output layer with one neuron. The autoencoder of OCA, ATA and SAE models has a single hidden layer of size 50, also with sigmoid activations. Additionally, SAE methods have a single linear layer for classification. Furthermore, the three baselines, such as ATA, MLP and OCA are chosen based on our previous work of adversarially trained autoencoders [37]. It is noteworthy, that due to the decoder in the ATA, OCA and SAE methods, these methods have more parameters when compared to MLP. However, as shown in [37], tied weights between encoder and decoder did not have any significant effect on the performance of these models. Therefore, these additional parameters do not bring much improvements. We use Adadelta [54] as an optimizer for our models. Additionally, to account for overfitting and exploding gradients, we regularize with a weight decay of 1×10^{-4}. With respect to the training objective, the MLP minimizes the samples' binary cross entropy. OCA and ATA methods optimize the majorities w.r.t. the mean squared error. SAE methods optimize the multi task loss function as a linear combination of binary cross entropy for classification loss and mean squared error for reconstruction loss. We performed a formal grid search for each model on each dataset w.r.t. learning rate, outlier weighting factor α, loss term weighting factor λ and minority reconstruction loss threshold t for the applicable models. For instance, though balanced sampling does not apply to OCA, it has been included in the grid search for all SAE methods, ATA and the MLP. Then, for every dataset and every method, we select the best model based on the highest AUPR validation score.

Table 3 summarizes the test performance of the selected best models on each dataset. Overall, it can be seen that the SAE method and its variants are superior to the baselines of MLP, OCA and adversarial ATA. The SAE variants beat the MLP approaches in 21 out of 28 cases and the OCA in 22 out of 28 cases in

Table 3. Performance comparison of vanilla SAE, ASAE, SAER and ASAER to the baselines MLP, OCA and ATA based on AUPR scores throughout the seven datasets (ATIS, REU, ARR, KDD OI, KDD OB, KDD NI and KDD NB). For each dataset the score of the best model has been highlighted in bold face. It is observed that SAE approaches generally outperform the baselines.

	Imbalanced			Outlier		Novelty	
	ATIS	REU	ARR	KDD OI	KDD OB	KDD NI	KDD NB
MLP	97.34%	99.10%	78.02%	77.27%	93.26%	14.48%	75.95%
OCA	75.09%	49.16%	50.91%	65.63%	92.51%	20.35%	**76.37%**
ATA	96.09%	99.32%	82.39%	74.86%	**95.35%**	20.66%	69.06%
SAE	**98.51%**	99.31%	80.97%	**83.24%**	94.73%	21.00%	61.44%
ASAE	98.10%	99.37%	79.53%	80.11%	93.50%	15.67%	56.23%
SAER	98.02%	99.29%	73.52%	80.53%	93.21%	19.78%	75.31%
ASAER	97.81%	**99.43%**	**83.04%**	77.87%	90.42%	**21.34%**	66.46%

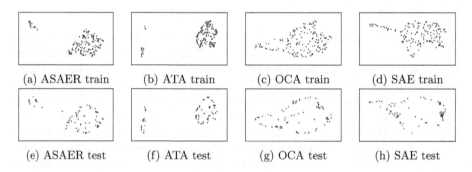

(a) ASAER train	(b) ATA train	(c) OCA train	(d) SAE train
(e) ASAER test	(f) ATA test	(g) OCA test	(h) SAE test

Fig. 4. Scatter plots of the encodings of minority (red points) and majority samples (blue points) generated by the best ASAER, ATA, OCA and SAE models on the ARR dataset split by train and test (Color figure online)

terms of AUPR score. On top of that, at least one of SAE approaches exceeds all baseline approaches in 5 out of 7 datasets, with SAE's and ASAER's performance significantly standing out. In conclusion, by direct AUPR score comparison, SAE and its variants are superior throughout the imbalanced classification, outlier, and novelty detection problem.

Next, we consider the representations learned by these best models and visualize them using UMAP clusters, as seen in Fig. 4. We can observe that for the adversarial approaches of ASAER and ATA, the resulting clusters are well separated, with each cluster mostly exclusively containing either inlier or outlier data points. However, for the non-adversarial approaches such as OCA and SAE, there is only a single cluster without any clear separation. This is a valuable insight, clearly showing that adversarial approaches force the autoencoder to learn better representations. Further, we visualize the reconstruction loss distributions of

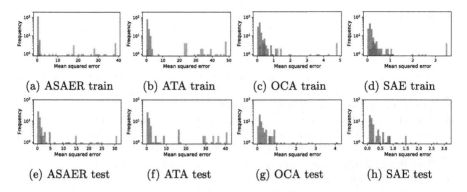

Fig. 5. Histograms of reconstruction loss distributions of minority (red bars) and majority samples (blue bars) for the best ASAER, ATA, OCA and SAE models on the ARR dataset split by train and test. Note y-axis is log scaled. (Color figure online)

outlier and inlier data points using a histogram as seen in Fig. 5. Here also, the loss distributions corresponding to inlier and outlier data points are strongly separated for adversarial methods; on the other hand, for non-adversarial methods, they overlap. This strong separation is due to the classification and reconstruction term within the adversarial loss functions of ASAER, as defined in Eq. (5); both the objectives are aligned with the core objective of separating majorities from minorities, resulting in better representations. In contrast, the SAE objective aims to minimize the reconstruction losses regardless of the sample's class. In other words, this auxiliary task of unsupervised learning is not aligned to the classification task. Our work, therefore, suggests choosing the auxiliary task (adversarial loss minimization) that is similar to the primary task.

6   Conclusion

In this work, we developed three novel end to end variants based on supervised autoencoders which can be used for imbalanced classification, outlier detection, and novelty detection. Having evaluated our approaches on a broad spectrum of datasets against competitive baselines such as MLP, OCA, and ATA methods, we show that SAE and its variants are superior in terms of AUPR scores. Additionally, the representations obtained from adversarial approaches show well-separated clusters suggesting for further investigation. Our work indicates that auxiliary tasks, when chosen relevant to the classification task, helps to obtain encodings useful for accurate classification. Finally, we would like to explore whether the reconstruction loss alone can be passed as a feature to the MLP without the encodings. This setup would automatically force the MLP to separate majority and minority samples based on reconstruction loss which may also render the adversarial objective redundant while classification loss only might be sufficient. We would pursue this approach as future work; thereby, we can reduce the multi-task learning problem into a simple classification problem with this informed architecture.

References

1. Aggarwal, C.C.: Outlier Analysis (2017)
2. Akcay, S., Atapour-Abarghouei, A., Breckon, T.P.: GANomaly: semi-supervised anomaly detection via adversarial training. In: Jawahar, C.V., Li, H., Mori, G., Schindler, K. (eds.) ACCV 2018. LNCS, vol. 11363, pp. 622–637. Springer, Cham (2019). https://doi.org/10.1007/978-3-030-20893-6_39
3. Boyd, K., Eng, K.H., Page, C.D.: Area under the precision-recall curve: point estimates and confidence intervals. In: Blockeel, H., Kersting, K., Nijssen, S., Železný, F. (eds.) ECML PKDD 2013. LNCS (LNAI), vol. 8190, pp. 451–466. Springer, Heidelberg (2013). https://doi.org/10.1007/978-3-642-40994-3_29
4. Boyd, S., Parikh, N., Chu, E., Peleato, B., Eckstein, J.: Distributed optimization and statistical learning via the alternating direction method of multipliers. Found. Trends® Mach. Learn. **3**(1) (2011)
5. Buda, M., Maki, A., Mazurowski, M.A.: A systematic study of the class imbalance problem in CNNs. Neural Netw. **106**, 249–259 (2018)
6. Cardie, C., Howe, N.: Improving minority class prediction using case-specific feature weights (1997)
7. Caruana, R.: Multitask learning. Mach. Learn. **28**(1), 41–75 (1997)
8. Chalapathy, R., Menon, A.K., Chawla, S.: Anomaly detection using one-class neural networks. arXiv preprint arXiv:1802.06360 (2018)
9. Chandola, V., Banerjee, A., Kumar, V.: Anomaly detection: a survey. ACM Comput. Surv. **41**(3), 1–58 (2009)
10. Chawla, N.V., Bowyer, K.W., Hall, L.O., Kegelmeyer, W.P.: SMOTE: synthetic minority over-sampling technique. J. Artif. Intell. Res. **16**, 321321–357 (2002)
11. Chen, J., Sathe, S., Aggarwal, C., Turaga, D.: Outlier detection with autoencoder ensembles. In: Proceedings of the SIAM International Conference on Data Mining (2017)
12. Chong, Y.S., Tay, Y.H.: Abnormal event detection in videos using spatiotemporal autoencoder. In: Proceedings of the International Symposium on Neural Networks (2017)
13. Dau, H.A., Ciesielski, V., Song, A.: Anomaly detection using replicator neural networks trained on examples of one class. In: Dick, G., Tang, K. (eds.) SEAL 2014. LNCS, vol. 8886, pp. 311–322. Springer, Cham (2014). https://doi.org/10.1007/978-3-319-13563-2_27
14. Davis, J., Goadrich, M.: The relationship between precision-recall and ROC curves. In: Proceedings of the 23rd International Conference on Machine Learning, pp. 233–240 (2006)
15. Divekar, A., Parekh, M., Savla, V., Mishra, R., Shirole, M.: Benchmarking datasets for anomaly-based network intrusion detection: KDD CUP 99 alternatives. In: Proceedings of 3rd International Conference on Computing, Communication and Security (ICCCS) (2018)
16. Fawcett, T.: An introduction to ROC analysis. Pattern Recogn. Lett. **27**(8), 861–874 (2006)
17. Gao, S., Zhang, Y., Jia, K., Lu, J., Zhang, Y.: Single sample face recognition via learning deep supervised autoencoders. IEEE Trans. Inf. Forensics Secur. **10**(10), 2108–2118 (2015)
18. Gogoi, P., Borah, B., Bhattacharyya, D., Kalita, J.: Outlier identification using symmetric neighborhoods. Procedia Technol. **6**, 239–246 (2012)

19. Goodfellow, I.J., Shlens, J., Szegedy, C.: Explaining and harnessing Adversarial Examples. arXiv preprint arXiv:1412.6572 (2014)

20. Hautamaki, V., Karkkainen, I., Franti, P.: Outlier detection using k-nearest neighbour graph. In: Proceedings of the 17th International Conference on Pattern Recognition, vol. 3 (2004)

21. Hawkins, D.M.: Identification of Outliers. Springer, Dordrecht (1980). https://doi.org/10.1007/978-94-015-3994-4

22. Hawkins, S., He, H., Williams, G., Baxter, R.: Outlier detection using replicator neural networks. In: Kambayashi, Y., Winiwarter, W., Arikawa, M. (eds.) DaWaK 2002. LNCS, vol. 2454, pp. 170–180. Springer, Heidelberg (2002). https://doi.org/10.1007/3-540-46145-0_17

23. Hendrycks, D., Gimpel, K.: A baseline for detecting misclassified and out-of-distribution examples in neural networks. In: Proceedings of International Conference on Learning Representations (2017)

24. Huang, C., Li, Y., Change Loy, C., Tang, X.: Learning deep representation for imbalanced classification. In: Proceedings of Conference on Computer Vision and Pattern Recognition (2016)

25. Ishii, Y., Takanashi, M.: Low-cost unsupervised outlier detection by autoencoders with robust estimation. J. Inf. Process. **27**, 335–339 (2019)

26. Japkowicz, N., Myers, C., Gluck, M., et al.: A novelty detection approach to classification. In: Proceedings of the International Joint Conference on Artificial Intelligence (1995)

27. Joachims, T.: Text categorization with Support Vector Machines: learning with many relevant features. In: Nédellec, C., Rouveirol, C. (eds.) ECML 1998. LNCS, vol. 1398, pp. 137–142. Springer, Heidelberg (1998). https://doi.org/10.1007/BFb0026683

28. Kannan, R., Woo, H., Aggarwal, C.C., Park, H.: Outlier detection for text data. In: Proceedings of the International Conference on Data Mining (2017)

29. Kawaguchi, K., Kaelbling, L.P., Bengio, Y.: Generalization in deep learning. arXiv preprint arXiv:1710.05468 (2017)

30. Kubat, M., Holte, R.C., Matwin, S.: Machine learning for the detection of oil spills in satellite radar images. Machine Learn. **30**, 195–215 (1998)

31. Kukar, M., Kononenko, I., et al.: Cost-sensitive learning with neural networks. In: Proceedings of European Conference on Artificial Intelligence (1998)

32. Lawrence, S., Burns, I., Back, A., Tsoi, A.C., Giles, C.L.: Neural network classification and prior class probabilities. In: Neural Networks: Tricks of the Trade (1998)

33. Le, L., Patterson, A., White, M.: Supervised autoencoders: improving generalization performance with unsupervised regularizers. In: Proceedings of Neural Information Processing Systems

34. Le, L., Patterson, A., White, M.: Supervised autoencoders: improving generalization performance with unsupervised regularizers. In: Advances in Neural Information Processing Systems (2018)

35. Lee, H., Cho, S.: The novelty detection approach for different degrees of class imbalance. In: King, I., Wang, J., Chan, L.-W., Wang, D.L. (eds.) ICONIP 2006. LNCS, vol. 4233, pp. 21–30. Springer, Heidelberg (2006). https://doi.org/10.1007/11893257_3

36. Liu, T., Tao, D., Song, M., Maybank, S.J.: Algorithm-dependent generalization bounds for multi-task learning. IEEE Trans. Pattern Analy. Mach. Intell. **39**(2), 227–241 (2016)

37. Lübbering, M., Ramamurthy, R., Gebauer, M., Bell, T., Sifa, R., Bauckhage, C.: From imbalanced classification to supervised outlier detection problems: adversarially trained auto encoders. In: Farkaš, I., Masulli, P., Wermter, S. (eds.) ICANN 2020. LNCS, vol. 12396, pp. 27–38. Springer, Cham (2020). https://doi.org/10.1007/978-3-030-61609-0_3

38. Mazurowski, M.A., Habas, P.A., Zurada, J.M., Lo, J.Y., Baker, J.A., Tourassi, G.D.: Training neural network classifiers for medical decision making: the effects of imbalanced datasets on classification performance. Neural Netw. **21**, 427–436 (2008)

39. McInnes, L., Healy, J., Melville, J.: UMAP: uniform manifold approximation and projection for dimension reduction. arXiv preprint arXiv:1802.03426 (2018)

40. Neyshabur, B., Bhojanapalli, S., McAllester, D., Srebro, N.: Exploring generalization in deep learning. In: Proceedings of the Advances in Neural Information Processing Systems, pp. 5947–5956 (2017)

41. Olszewski, D.: A probabilistic approach to fraud detection in telecommunications. Knowl.-Based Syst. **26**, 246–258 (2012)

42. Panigrahi, S., Kundu, A., Sural, S., Majumdar, A.K.: Credit card fraud detection: a fusion approach using Dempster-Shafer theory and Bayesian learning. Inf. Fusion **10**(4), 354–363 (2009)

43. Pennington, J., Socher, R., Manning, C.D.: GloVe: global vectors for word representation. In: Proceedings of Conference on Empirical Methods in Natural Language Processing (EMNLP) (2014)

44. Perera, P., Nallapati, R., Xiang, B.: OCGAN: one-class novelty detection using GANs with constrained latent representations. In: Proceedings of the IEEE Conference on Computer Vision and Pattern Recognition, pp. 2898–2906 (2019)

45. Ranzato, M., Szummer, M.: Semi-supervised learning of compact document representations with deep networks. In: Proceedings of International Conference on Machine learning (2008)

46. Sarvari, H., Domeniconi, C., Prenkaj, B., Stilo, G.: Unsupervised boosting-based autoencoder ensembles for outlier detection. arXiv preprint arXiv:1910.09754 (2019)

47. Schlegl, T., Seeböck, P., Waldstein, S.M., Schmidt-Erfurth, U., Langs, G.: Unsupervised anomaly detection with generative adversarial networks to guide marker discovery. In: Niethammer, M., et al. (eds.) IPMI 2017. LNCS, vol. 10265, pp. 146–157. Springer, Cham (2017). https://doi.org/10.1007/978-3-319-59050-9_12

48. Scholkopf, B., Smola, A.J.: Learning with Kernels: Support Vector Machines, Optimization, and Beyond, Regularization (2001)

49. Tavallaee, M., Bagheri, E., Lu, W., Ghorbani, A.A.: A detailed Analysis of the KDD CUP 99 Data Set. In: Proceedings of IEEE Symposium on Computational Intelligence for Security and Defense Applications (2009)

50. Vu, H.S., Ueta, D., Hashimoto, K., Maeno, K., Pranata, S., Shen, S.M.: Anomaly detection with adversarial dual autoencoders. arXiv preprint arXiv:1902.06924 (2019)

51. Wang, X., Du, Y., Lin, S., Cui, P., Yang, Y.: Self-adversarial variational autoencoder with gaussian anomaly prior distribution for anomaly detection. CoRR, abs/1903.00904 (2019)

52. Weston, J., Ratle, F., Mobahi, H., Collobert, R.: Deep learning via semi-supervised embedding. In: Montavon, G., Orr, G.B., Müller, K.-R. (eds.) Neural Networks: Tricks of the Trade. LNCS, vol. 7700, pp. 639–655. Springer, Heidelberg (2012). https://doi.org/10.1007/978-3-642-35289-8_34

53. Yang, Y., Liu, X.: A re-examination of text categorization methods. In: Proceedings of International Conference on Research and development in Information Retrieval (1999)
54. Zeiler, M.D.: ADADELTA: an adaptive learning rate method. arXiv preprint arXiv:1212.5701 (2012)
55. Zenati, H., Foo, C.S., Lecouat, B., Manek, G., Chandrasekhar, V.R.: Efficient GAN-based anomaly detection. arXiv preprint arXiv:1802.06222 (2018)
56. Zhou, C., Paffenroth, R.C.: Anomaly detection with robust deep autoencoders. In: Proceedings of the International Conference on Knowledge Discovery and Data Mining (2017)
57. Zou, J., Zhang, J., Jiang, P.: Credit Card Fraud Detection Using Autoencoder Neural Network. arXiv preprint arXiv:1908.11553 (2019)

Fuzzy-Based Pseudo Segmentation Approach for Handwritten Word Recognition Using a Sequence to Sequence Model with Attention

Rajdeep Bhattacharya[1], Samir Malakar[2], Friedhelm Schwenker[3]([✉]), and Ram Sarkar[1]

[1] Department of Computer Science and Engineering, Jadavpur University, Kolkata, India
`rajdeep.cse17@gmail.com`, `raamsarkar@gmail.com`
[2] Department of Computer Science, Ashutosh College, Kolkata, India
`malakarsamir@gmail.com`
[3] Institute of Neural Information Processing, University of Ulm, Ulm, Germany
`friedhelm.schwenker@uni-ulm.de`

Abstract. Sequence to sequence models have shown significant progress in the field of handwriting recognition. The recent trend has been that the input to these models is fed from a convolutional neural network (CNN) that acts as a generic feature extractor for the handwritten text images. The input to the CNN is usually either a sequence of patches extracted from the text image or the output of a segmentation algorithm applied on the text image to break it up into individual characters. However, patching is unable to convey proper information about character boundaries in the image, and the segmentation-based approach often suffers from over and under segmentation. To this end, we propose a fuzzy-based pseudo segmentation approach for handwritten word recognition using a sequence to sequence model. We use a fuzzy triangular function that generates column wise weights based on the distance of the nearest data pixel from the top of the text image. Thus the probable segmentation regions are assigned higher weights than the other regions. These weights are superimposed on the original text image and this modified input is fed patch wise to the CNN. The features extracted are then encoded as a first part of a sequence to sequence model and then decoded to obtain the sequence of characters in the input. An attention mechanism is used to ensure that the decoder focuses on the appropriate section of the features outputted by the encoder while generating each character. Our method is tested upon the IAM word database after the words in it undergo skew and slant correction. The architecture of the model is optimized based on exhaustive numerical simulations on the IAM database and it shows promising results.

Keywords: English word · Handwritten word recognition · IAM dataset · LSTM · Pseudo segmentation · Sequence to sequence model

1 Introduction

Recognition of textual contents of handwritten documents, prepared in offline mode, is still an open research problem as researchers need to deal with various challenges, mainly

© Springer Nature Switzerland AG 2021
A. Del Bimbo et al. (Eds.): ICPR 2020 Workshops, LNCS 12662, pp. 582–596, 2021.
https://doi.org/10.1007/978-3-030-68790-8_45

in terms of variations in writing style of individuals. The cursive nature of handwriting coupled with the low quality of images makes the problem of handwriting recognition a difficult task. Some of the notable applications of handwriting recognition, mentioned in [1], are postal address interpretation, bank cheque processing and form processing [2, 3]. Thus, an efficient recognition system for offline handwritten documents is an absolute necessity in the present technologically advanced society. In this work, we have primarily focused on the recognition of isolated handwritten word images.

Over the years, various researchers have come up with a variety of methods for the recognition of handwritten words. The word recognition methods mostly apply either a holistic approach or a classical approach. The techniques that follow the holistic approach consider the word images as an individual component whereas the methods following the classical approach first segment the word images into individual characters and then recognize the characters separately. Holistic approaches [1] become the obvious choice for limited, small and fixed size lexica, while for finding a solution for large lexica, researchers rely on the classical approach. Later, after the advent of Hidden Markov models (HMMs) and neural network based HMMs (NN-HMMs), a new era of word recognition approaches, known as segmentation-free approaches, began. Here, the entire word image is passed as the input to the system similar to holistic approach. In this context, we would like to mention that the works reported in [4] have provided segmented characters and achieved better recognition results than their counterparts. In [5], the authors have used a sliding window approach to obtain a set of feature vectors followed by modelling them with continuous density HMMs and achieved good results.

The deep learning based methods [6], which have already proven to be very successful for recognition of handwritten samples like isolated digit, character and word recognition, improvised over the traditional handcrafted feature based approaches. The use of feed forward neural networks, recurrent neural networks (RNNs) and variants of the same (e.g., long short term memory (LSTM) and bidirectional LSTM (BLSTM)) outperformed many traditional methods by a significant margin. However, these methods initially provided column wise features like NN-HMM based methods. To overcome the need of large training cost in RNNs, patch based approach was introduced as an alternative solution. However, relying on the outcome of the work in [4], one can segment the word images into components along the local valley regions, as shown in Fig. 1. However, not all local valley regions may represent absolute character boundaries e.g., valleys marked with red boxes in Fig. 1. To deal with this, in the present scope of the work, we apply a fuzzy membership function and transform the word images into pseudo segmented word images where columns belonging to absolute valley regions get higher fuzzy membership values than the rest. Next, these transformed word images are fed to an LSTM based sequence to sequence model. This transformation not only improves the recognition accuracy but also truncates the overall training cost on a fixed dataset.

2 Related Work

The backbone of our offline handwritten English word recognition method is a sequence to sequence model where we provide fuzzy triangular membership based transformed word images as input. In this section, we first describe some state-of-the-art work

Fig. 1. Representation of local valley regions of a word image. Regions marked with light blue color show absolute character boundary regions whereas the others do not.

that follow a similar approach and then some fuzzy membership function based word segmentation work.

2.1 Segmentation Free Word Recognition Models

Two handwritten word recognition models have been proposed by Menasri et al. [7]. The first model uses an RNN model whereas, the other one is a classifier combination based model having seven classifiers that is composed of one grapheme based Multi-Layer Perceptron (MLP) with HMM, two variants of sliding window based Gaussian mixture model (GMM) based HMM, and four variants of the current RNN based models. Graves et al. [8] use a language invariant method based on multidimensional RNNs with connectionist temporal classification using raw pixel data for word recognition. In another work, Bluche et al. [9] use the voting scheme called Recognizer Output Voting Error Reduction (ROVER) for combining four models: two BLSTM based and two deep MLP to recognize word images written in Roman script. A system proposed by Doetsch et al. [10] uses an LSTM with an additional parameter that controls the shape of the squashing functions in the gating units. A similar approach has been followed by Sueiras et al. [11] where an attention based method is used that concentrates on the history of the recognized character i.e. more weightage is provided on past history. The work proposed by Almazan et al. [12] encodes the input word image as Fisher vectors (FV), i.e. as an aggregation of the gradients of a GMM and Scale Invariant Feature Transformation (SIFT). The backbone of the n-gram based model, proposed by Poznanski and Wolf [13], is similar to the work reported in [12]. In this work, for a handwritten word image, it first uses a CNN model to estimate its n-gram frequency profile and then canonical correlation analysis based method is employed to match the estimated profile to the true profiles of all the words present in a large dictionary. Bluche et al. [14] use a multidimensional LSTM model with attention for end-to-end paragraph recognition written in Roman script.

Recently, Majid and Smith [15] propose a character spotting based recognition model for handwritten Bangla word images where they make use of faster R-CNN proposal that detects individual characters using C-Net and associated diacritics using D-Net separately. Märgner and Abed [16] use a hierarchy of multi-dimensional basic RNN and basic LSTM for recognition of Arabic word images. Stahlberg and Vogel [17] rely on a fully connected deep neural network (DNN) along with pixel and segment based features. Gui et al. [18] use an adaptive context-aware reinforced agent based model with low computation overheads for word recognition. Recently, Wu et al. [19] propose a handwritten word recognition method by combining a position embedding with residual network (ResNet) and BLSTM network. In the first step, ResNet is employed to extract features from the input image and in the later step, the output position embedding have been fed to BLSTM as indices of the character sequence corresponding to a word.

2.2 Fuzzy-Based Word Segmentation

Successful use of three fuzzy membership functions viz., triangular, trapezoidal and bell-shaped is found in literature of Matra based script character segmentation. Matra is a line over words found in many languages like Bangla, Hindi, etc. and serves the purpose of grouping the characters of a word together. The said fuzzy membership functions have been used either as Matra pixel generator or as segmentation pixels generator. The first instance of such type of work is found in [20], where the authors use fuzzy triangular membership function for marking some data pixels as Matra pixels as well as segmentation pixels. Later, Sarkar et al. [21] propose a two-stage mechanism to segment handwritten Bangla word images where in the first stage, the connected components (CCs) of a word image are classified as 'segment further' and 'do not segment' class. The CCs of 'segment further' class are then segmented into constituent characters using bell shaped fuzzy membership function. However, recently Malakar et al. [22] have shown experimentally that triangular fuzzy membership function is more useful for detecting segmentation points than the others.

2.3 Motivation

Relying on the outcome of the work reported in [22] and analysing the nature of valley regions in handwritten English word images, it could be said that the use of triangular fuzzy membership function would be a better choice for detecting pseudo-segmentation regions. In addition, literature study shows that segmentation-free approach is the state-of-the-art approach for handwritten word recognition. Therefore, we follow the same trend here to design our recognition model. However, the modification we have made is by using the image transformation wherein each column of the image carries implicit information about it being a segmentation line for character boundaries.

3 Proposed Method

In this work, we use a CNN based model, called LeNet [6], as a generic feature extractor followed by a sequence to sequence encoder-decoder model based on an LSTM architecture with an attention model. The input to the CNN, though, is a combination of pseudo segmentation and patch-wise approaches. In this approach, we assign a weight by using a fuzzy triangular membership function to identify the valleys in an image. These weights are superimposed on the original image to generate a new image, which is then fed patch-wise into the CNN. The overall process is shown in Fig. 2 and each module is described in the following subsections.

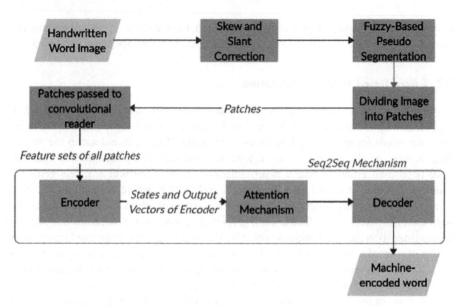

Fig. 2. Flowchart representing the overall word recognition process

3.1 Skew and Slant Correction

Patches of each word image are extracted by vertically segmenting it into multiple parts. However, skewed and slanted word images would create problems (see Fig. 3(a–b)). It would also create problems for fuzzy membership based image transformation. So, we have applied the method proposed by Bera et al. [23] for skew and slant correction. In this method, first busy zone of the word images are extracted and then a hypothetical ellipse is estimated. The angles of major and minor axes are then used there for estimating skew and slant angles respectively. One example of a skew and slant corrected image has been shown in Fig. 4 along with the original image.

(a)   (b)

Fig. 3. Illustration of the problem that arises during patch generation using vertical fragments on slanted word image. Red patches contain parts of two characters which may occur in general during such patch generation on non-slanted word images. However, the green patches contain three character parts which occur due to slanted or skewed word images.

3.2 Fuzzy-Based Pseudo Segmentation

In this section, we describe how we perform fuzzy-based pseudo segmentation. Here, segmentation information of characters is passed only in cases of cursive handwriting

(a)   (b)

Fig. 4. Represents (a) original image (b) skew and slant corrected image

where the characters are joined to each other and not in cases where the characters are already separated. The basic reasoning behind doing so is that we want to provide details of implicit segmentation columns to the recognizer. Mostly, the segmentation columns are where there are valleys in the image as they are the points where one character ends and another begins (as illustrated by the blue boxes in Fig. 1). We pass this information in terms of probability of each column being a valley column along with the image to the next step. We note that in valley columns, the data pixels are typically much lower than in other columns.

We use a modified fuzzy triangular function for assigning the weights to the columns representing the possible columns being selected as a segmentation point. A complete fuzzy triangular function is described in Eq. 1 and the corresponding graphical representation of the same is shown in Fig. 5(a).

$$\mu(x) = \begin{cases} \frac{x-a}{m-a}, & \text{if } a < x \leq m \\ \frac{b-x}{b-m}, & \text{if } m < x < b \\ 0, & \text{otherwise} \end{cases} \tag{1}$$

Here a, m, and b are constants and $a < m < b$. In this work, we use a modified form of this wherein we use only a part of the triangular function for mapping. We consider the part only for $a < x \leq m$ and set the value for the rest of the domain to be 0. The graphical representation of our function is shown in Fig. 5(b).

We first binarize the image using the Otsu binarization technique [24]. Next, we estimate the minimal bounding box of the entire word image containing all the data pixels of it. Let, the height and width of the minimal bounding box be h and w respectively. After this, we traverse the image column wise from left to right. For each column, we list the position of the nearest data pixel (say, $x[i]$ for column $i \in [1, w]$) from the top of the image. It is to be noted that if a column does not contain any data pixel then we set the value of $x[i] = h$. A few nearest data pixels for some columns upon left to right traversal of a word image are shown using arrows in Fig. 6. The red arrows represent valley columns while the blue arrows represent non-valley columns. $x[i]$ is then mapped to a value between 0 and 1 using a fuzzy membership function. This function is obtained by substituting the variables a and m in Fig. 5(b) by 0 and h respectively, and so Eq. (1) can be expressed by Eq. (2).

$$\mu[i] = \begin{cases} \frac{x[i]}{h}, & \text{if } x[i] \in [1, h] \\ 0, & \text{otherwise} \end{cases} \tag{2}$$

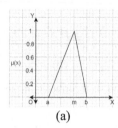
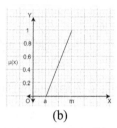

(a)   (b)

Fig. 5. Graphical representation of fuzzy triangular membership function (a) actual, and (b) used in the present work

In Eq. (2), $\mu[i]$ is the fuzzy membership value of ith column. We then calculate the mean (m) of these values and retain only those values which are above m, else we discard those values. Mathematically,

$$m = \frac{\sum_{i=1}^{w} \mu[i]}{w} \tag{3}$$

$$F[i] = \begin{cases} \mu[i], & if \; \mu[i] \geq m \\ 0, & if \; \mu[i] < m \end{cases} \tag{4}$$

Fig. 6. Pictorial Representation of column wise traversal along with minimum distance from the top of the image for each column. Red arrows denote valleys whereas blue arrows denote non-valleys

To apply these estimated fuzzy membership values to the input to the CNN, we multiply each column of the image with the corresponding fuzzy membership value for that column subtracted from one. This subtraction is done to ensure that the probable segmentation regions are darker than other areas. It can be represented mathematically as follows:

$$I_{ji}^{output} = (1 - F[i]) * I_{ji}^{original}, for \; 0 < j \leq h, 0 < i \leq w \tag{5}$$

An example of an original image and modified image is shown in Fig. 7.

3.3 Basic Recognition Architecture

In this section, we give an overview of our architecture for the word prediction model. Our model consists of: a convolutional reader, a sequence to sequence LSTM with an

(a)   (b)

Fig. 7. (a) Image obtained after skew and slant correction (b) Image obtained after fuzzy-based pseudo segmentation of (a)

attention model. A summary of our architecture is shown in Fig. 8. First of all, the modified image after pseudo segmentation is obtained and split up into patches. These patches are fed into the CNN, which follows LeNet-5 architecture, which generates a feature vector from each patch. These features are fed into an encoder, that is a part of a sequence to sequence model, which is obtained by the stacking of two BLSTMs and extracts the relationships between the feature sets fed into it. Thus, the encoder is used to understand the relationships among the letters of the word in the image and encode it in the form of a set of output features.

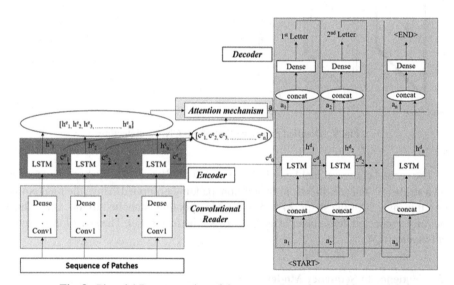

Fig. 8. Pictorial Representation of the proposed word recognition model

Next, the attention mechanism receives the output of the encoder and combines it with the corresponding state vectors of the encoder to generate its output. Its main utility is to help the decoder to focus on the correct part of the encoder output features while generating the output for that corresponding letter. The decoder, then, receives the output of the attention mechanism as input. The decoder, at each step, considers three things: output of the attention mechanism, current state vector and previous letter of the output. Using these, the decoder, at each step, generates each letter of the word as output. These letters come together to form the predicted word.

3.4 CNN Architecture

As mentioned earlier, our CNN architecture is inspired by the LeNet-5 architecture (see Fig. 9), where the input is fed patch wise. Each patch has a fixed width, which is 10, the choice of which is described in the next section. Thus, the patches are of size 48×10 and this will be the input size for the CNN. For the first convolutional layer, we consider 20 filters, each of shape 5×5. The input is padded such that the output for each filter is of same shape as the input. Thus, output from the first convolutional layer will have the shape $48 \times 10 \times 20$. Max pooling is applied to this with a stride of 2×2 after which the output is of shape $24 \times 5 \times 20$.

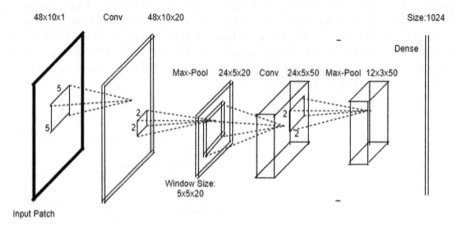

Fig. 9. Detailed representation of our CNN architecture

The second convolutional layer is composed of 50 filters, each of shape $5 \times 5 \times 20$ and a stride of 1 in each direction. In addition, padding is applied. Thus, the final output from this layer is of the shape $24 \times 5 \times 50$. Max pooling is applied to this after which one fully connected dense layer is used to extract the feature set. The size of this feature set is set to 1024.

3.5 Sequence to Sequence Model

In this section, we define the sequence to sequence model that we use here. A sequence to sequence model consists of an encoder and a decoder. In literature, the encoder and decoder are usually composed of RNNs. Here, we use LSTM which is a special type of RNN better than conventional RNNs in terms of handling long term dependencies.

The encoder consists of 2 layers of BLSTMs stacked one on top of the other. A conventional LSTM contains three gates: input, forget, and output. The input gate learns the information to update the state vector, forget gate learns the part of the input from the previous state which is not essential for the context vector and can be discarded, and output gate learns what information will be used to produce the final output. The sequence length for each layer is the number of patches that each image contains. The state vectors as well as outputs from this layer are passed on to the attention mechanism which is inspired from the work by Chorowski et al. [25]. It enables the decoder to focus on the important part of the encoder output while generating the output for the corresponding part. For example, if the output for the nth character is being determined, the attention mechanism enables the decoder to focus on the output of the encoder for this n^{th} character only. The attention mechanism used is represented by Eqs. (6) and (7).

$$e_{t,l} = w^T \tanh(Wc_{t-1}^e + Vh_l^e + b) \tag{6}$$

$$a_{t,l} = \frac{\exp(e_{t,l})}{\sum_{t=1}^{L} \exp(e_{t,l})} \tag{7}$$

Here, w and b represent weights and bias respectively, and W and V are trainable matrices. c_t^e is the state of the encoder at step t and h_l^e is the output of the encoder. L runs from the first output letter to the last letter. The decoder is also an LSTM network, but utilizes the attention mechanism. For each step, decoder uses the output of the attention model concatenated by the probability distribution of the previous letter, the state vector and the output vector. Initial letter of the decoder is initialized by the character <START> and the decoder stops operation when the character <END> appears at the output. Probability distribution for each output character is produced by applying Softmax activation over the output sequence. In the end, combining the outputs of each step, the entire word is reconstructed.

The main contribution of our method is the passing of implicit segmentation information in a weighted manner to the sequence-to-sequence model while preserving the originality of the word image at large by not segmenting it into individual characters. It is thus made easier for the model to identify the demarcation between characters while training and thus, identify the characters individually, and hence the word. The model is an attention-based sequence to sequence model which has been exhaustively optimized as described in Sect. 4. The two components together thus constitute a novel approach which gives competitive results as illustrated in Sect. 4.

4   Results

4.1   Dataset and Evaluation Metrics Used

For evaluation, we use the IAM offline handwriting database, prepared by Marti et al. [25]. The database has standard partition into train, test and validation sets with each containing 55891, 8895 and 25920 isolated word images respectively. However, we consider only the words having correct segmentation in the database and hence we obtain train, validation and test partitions containing 47952, 7558 and 20306 words respectively. As evaluation metrics, we use Character Error Rate (CER) and Word Error Rate (WER) [14]. CER is the Levenstein distance between the predicted word and the real word, whereas WER is simply the percentage of words wrongly recognized.

4.2   Optimizing the Parameters of the Word Recognition Model

After having pre-processed the data, all images are reshaped to 192x48. We perform certain experiments to justify our choices of pruning of fuzzy membership values, patch size, stride, and the CNN architecture (LeNet-5) used as feature extractor. In the first case, described in the Sect. 3.2, we test the system with pruning of fuzzy membership values by mean cut and without any such pruning. The results are illustrated in Table 1 using the test lexicon. This experiment has been performed using LeNet-5 based feature extraction and using half of the entire train set. We note that the results are better when pruning is taken into consideration. In the second case, we have also tested by varying the patch size (5 to 25 with multiple of 5) and patch size 10 provides the best performance. Similarly, we test for the best stride and 2 is found to be the best. As CNN model, we have performed experiments with both the VGG-16 [26] model and the LeNet-5 model using ReLU as activation function. The comparison of the best VGG and LeNet-5 models are summarized in Table 1. It clearly shows that LeNet-5 performs better than VGG.

Table 1. Effect of mean filtering and CNN architecture on WER and CER

Performance	Pruning of fuzzy membership value		CNN architecture used	
	No	Yes	LeNet-5	VGG
WER	24.2	18.2	23.6	82.3
CER	14.2	9.1	16.5	67.4

For the encoder, we have tested using both LSTM and BLSTM, and found the results better with BLSTM. We have also tested the performance of our system by varying the number of BLSTM layers for encoding. The comparison results are shown in Table 2. Hence, we are able to see that two BLSTM layers stacked over each other give the best results and so, we choose the same for our model. For the decoder, we use a single LSTM layer. Next, we illustrate the significance of fuzzy-based pseudo segmentation by comparing the results with and without using it on all 3 lexicons using the full training and test sets. The results of this are summarized in Table 2. As can be seen from Table 2, both WER and CER are better when applying fuzzy-based pseudo segmentation than without applying the same. Hence, we note that fuzzy-based pseudo segmentation helps in improving the overall system performance.

For the CNN, the filters are initialized randomly using a normal distribution and biases are initialized to a tensor of appropriate size with all values as 0.1. For the LSTM, the weights and projections are initialized randomly from a uniform distribution. Tensorflow library is used for the implementation and an Nvidia Quadro M4000 GPU is used to run the code. A dropout of 0.5 is used for the LSTM cells in the encoder with tanh as the activation function. The learning rate is initialized to 0.001 which has an exponential decay of 0.95 at each step and Adam optimizer is used for stochastic optimization. At each epoch, the order of training data is randomized. The stopping criteria is 10 epochs without any improvement in WER and CER.

Table 2. Comparison of WER and CER for one LSTM layer and two LSTM layers in Encoder and comparison of WER and CER without and with fuzzy-based pseudo segmentation using all 3 types of lexicons

Varying number of LSTM layers			Importance of performing fuzzy-based pseudo segmentation				
#LSTM Layer	WER	CER	Lexicon Used	Without fuzzy-based pseudo segmentation		With fuzzy-based pseudo segmentation	
One	20.6	11.2	No lexicon	WER	34.6	WER	31.3
				CER	14.5	CER	13.2
Two	18.2	9.1	IAM standard lexicon	WER	19.9	WER	18.6
				CER	10.9	CER	9.7
Three	21.5	12.3	Test lexicon	WER	18.8	WER	18.2
				CER	10.3	CER	9.1

4.3 Comparisons

To obtain the results, we use 3 approaches as described below.

Use of No Lexicon: In this section, we use no standard lexicon and consider the output produced by our model directly as the final output. The results obtained by us and the methods by others are shown in Table 3. Bluche et al. [9] used an MLP with HMM while in [14], an MDLSTM model is used with attention. We note that the WER is better for us than all other cases and the CER is competitive as well.

Table 3. Comparison of results of our method with other methods using no standard lexicon and IAM standard lexicon

No lexicon			IAM standard lexicon		
Method	WER	CER	Methods	WER	CER
Our method	31.3	13.2	**Our method**	18.6	9.7
Bluche [9] MLP	54.2	15.6	Bluche [9] MLP	25.5	8.0
Bluche [14]	–	12.6	Bluche [9] RNN	16.7	5.3

Use of IAM Entire Word Set Lexicon: Here, we use the lexicon comprising all the words in the IAM dataset. From the predicted words of our model, we find the closest words in terms of Levenstein distance from the lexicon set. The results are illustrated in Table 3. From these results, we note that our WER is competitive.

Use of Test Set Lexicon: Here, we use the lexicon comprising all the words in the test set where we find the closest words in terms of Levenstein distance. This helps us to realize the number of errors occurring when compared to the second case because of words that are in the standard lexicon and closer to the predicted word than the actual word in terms of Levenstein distance. The results are illustrated in Table 4. This table clearly indicates that our method performs better than the ones compared to.

Table 4. Comparison of the methods with test set lexicon

Method	WER	CER
Our Method	18.2	9.1
Almazan et al. [12]	20.1	11.2
Bluche et al. [27]	20.5	–

4.4 Error Case Analysis

Errors can occur in cases where the segmentation points appear in non-valley regions. For example, consider the Fig. 10(a). Here, the original word is "Gaitskell", and t and s are joined to each other but as their meeting point is quite high, the distance of first data pixel is negligible as compared to the maximum distance from the top. Hence, the fuzzy weight is less and not able to cross the threshold and so we cannot pass details about its segmentation. Further, in cases where the characters themselves contain valleys, they will be darkened out and wrong segmentation information will be passed. For example, in Fig. 10(b), the alphabet V contains a valley and hence wrong segmentation information will be passed. This scenario leads to the generation of erroneous results.

(a)   (b)

Fig. 10. Word images with erroneous output (a) first kind and (b) second kind

5   Conclusion and Future Work

In this paper, we have proposed a fuzzy-based approach for passing implicit segmentation information along with an image followed by a CNN and a sequence to sequence model for offline handwriting recognition. The main contribution of our work is the pseudo segmentation part where we use a fuzzy triangular function to modify the input image such that appropriate segmentation information can be conveyed to the network. This makes it easier for the network to identify the individual characters from the word image and hence reconstruct the word while avoiding problems of explicit segmentation such as under segmentation and over segmentation, but also improving over the conventional patch wise approach, which cannot convey any information about character boundaries. On comparing with other methods, we find that our method provides competitive results and outperforms other methods in most cases. In future, we plan to use this for recognizing handwritten words of other languages. Besides, we aim to remove the errors in pseudo segmentation by considering neighbouring regions while assigning column weights. Presently, we are using column wise data pixel distance for fuzzy-based pseudo segmentation, however, in future, other parameters like column-wise data pixels count and background pixels count and ratio of data and non-data pixels from the top can be used to facilitate the process. Also, in future, we plan to use patch based fuzzy weights with other metrics apart from mean such as standard deviation, instead of column wise fuzzy weights, for pseudo segmentation.

References

1. Madhvanath, S., Govindaraju, V., Member, S.: The role of holistic paradigms in handwritten word recognition. IEEE Trans. Pattern Anal. Mach. Intell. **23**, 149–164 (2001)
2. Ghosh, S., Bhattacharya, R., Majhi, S., Bhowmik, S., Malakar, S., Sarkar, R.: Textual content retrieval from filled-in form images. In: Sundaram, S., Harit, G. (eds.) DAR 2018. CCIS, vol. 1020, pp. 27–37. Springer, Singapore (2019). https://doi.org/10.1007/978-981-13-9361-7_3
3. Bhattacharya, R., Malakar, S., Ghosh, S., Bhowmik, S., Sarkar, R.: Understanding contents of filled-in Bangla form images. Multimed. Tools Appl. 1–42 (2020)
4. Roy, P.P., Bhunia, A.K., Das, A., Dey, P., Pal, U.: HMM-based Indic handwritten word recognition using zone segmentation. Pattern Recognit. **60**, 1057–1075 (2016). https://doi.org/10.1016/j.patcog.2016.04.012
5. Bunke, H., Bengio, S., Vinciarelli, A.: Offline recognition of unconstrained handwritten texts using HMMs and statistical language models. IEEE Trans. Pattern Anal. Mach. Intell. **26**, 709–720 (2004)
6. LeCun, Y., Bottou, L., Bengio, Y., Haffner, P.: Gradient-based learning applied to document recognition. Proc. IEEE **86**, 2278–2324 (1998)
7. Menasri, F., Louradour, J., Bianne-Bernard, A.-L., Kermorvant, C.: The A2iA French handwriting recognition system at the Rimes-ICDAR2011 competition. In: Document Recognition and Retrieval XIX, p. 82970Y. International Society for Optics and Photonics (2012)
8. Graves, A., Schmidhuber, J.: Offline handwriting recognition with multidimensional recurrent neural networks. In: Advances in Neural Information Processing Systems, pp. 545–552 (2009)
9. Bluche, T., Ney, H., Kermorvant, C.: A comparison of sequence-trained deep neural networks and recurrent neural networks optical modeling for handwriting recognition. In: Besacier, L., Dediu, A.-H., Martín-Vide, C. (eds.) SLSP 2014. LNCS (LNAI), vol. 8791, pp. 199–210. Springer, Cham (2014). https://doi.org/10.1007/978-3-319-11397-5_15
10. Doetsch, P., Kozielski, M., Ney, H.: Fast and robust training of recurrent neural networks for offline handwriting recognition. In: 2014 14th International Conference on Frontiers in Handwriting Recognition, pp. 279–284. IEEE (2014)
11. Sueiras, J., Ruiz, V., Sanchez, A., Velez, J.F.: Offline continuous handwriting recognition using sequence to sequence neural networks. Neurocomputing **289**, 119–128 (2018)
12. Almazán, J., Gordo, A., Fornés, A., Valveny, E.: Word spotting and recognition with embedded attributes. IEEE Trans. Pattern Anal. Mach. Intell. **36**, 2552–2566 (2014)
13. Poznanski, A., Wolf, L.: Cnn-n-gram for handwriting word recognition. In: Proceedings of the IEEE Conference on Computer Vision and Pattern Recognition, pp. 2305–2314 (2016)

14. Bluche, T., Louradour, J., Messina, R.: Scan, attend and read: End-to-end handwritten para-graph recognition with mdlstm attention. In: 2017 14th IAPR International Conference on Document Analysis and Recognition (ICDAR), pp. 1050–1055. IEEE (2017)

15. Majid, N., Smith, E.H.B.: Segmentation-free bangla offline handwriting recognition using sequential detection of characters and diacritics with a Faster R-CNN. In: 2019 International Conference on Document Analysis and Recognition (ICDAR), pp. 228–233. IEEE (2019)

16. Märgner, V., El Abed, H.: ICDAR 2009 Arabic handwriting recognition competition. In: 2009 10th International Conference on Document Analysis and Recognition, pp. 1383–1387. IEEE (2009)

17. Stahlberg, F., Vogel, S.: The QCRI recognition system for handwritten Arabic. In: Murino, V., Puppo, E. (eds.) ICIAP 2015. LNCS, vol. 9280, pp. 276–286. Springer, Cham (2015). https://doi.org/10.1007/978-3-319-23234-8_26

18. Gui, L., Liang, X., Chang, X., Hauptmann, A.G.: Adaptive context-aware reinforced agent for handwritten text recognition. In: BMVC, p. 207 (2018)

19. Wu, X., Chen, Q., You, J., Xiao, Y.: Unconstrained offline handwritten word recognition by position embedding integrated resnets model. IEEE Signal Process. Lett. **26**, 597–601 (2019)

20. Basu, S., Sarkar, R., Das, N., Kundu, M., Nasipuri, M., Basu, D.K.: A fuzzy technique for segmentation of handwritten Bangla word images. In: Proceedings-International Conference on Computing: Theory and Applications, ICCTA 2007, pp. 427–432. IEEE (2007). https://doi.org/10.1109/ICCTA.2007.7

21. Sarkar, R., Das, N., Basu, S., Kundu, M., Nasipuri, M., Basu, D.K.: A two-stage app-roach for segmentation of handwritten Bangla word images. In: Proceedings of International Conference on Frontiers in Handwriting Recognitions, pp. 403–408. Citeseer (2008)

22. Malakar, S., Sarkar, R., Basu, S., Kundu, M., Nasipuri, M.: An image database of handwritten Bangla words with automatic benchmarking facilities for character segmentation algorithms. Neural Comput. Appl. (2020)

23. Bera, S.K., et al.: A one-pass approach for slope and slant estimation of tri-script handwritten words. J. Intell. Syst. (2018). https://doi.org/10.1515/jisys-2018-0105

24. Otsu, N.: A threshold selection method from gray-level histograms. IEEE Trans. Syst. Man. Cybern. **9**, 62–66 (1979)

25. Chorowski, J.K., Bahdanau, D., Serdyuk, D., Cho, K., Bengio, Y.: Attention-based models for speech recognition. In: Advances in Neural Information Processing Systems, pp. 577–585 (2015)

26. Simonyan, K., Zisserman, A.: Very deep convolutional networks for large-scale image recognition. arXiv Prepr. arXiv1409.1556 (2014)

27. Bluche, T., Ney, H., Kermorvant, C.: Tandem HMM with convolutional neural network for handwritten word recognition. In: 2013 IEEE International Conference on Acoustics, Speech and Signal Processing, pp. 2390–2394. IEEE (2013)

Bifurcated Autoencoder for Segmentation of COVID-19 Infected Regions in CT Images

Parham Yazdekhasty[1], Ali Zindari[1], Zahra Nabizadeh-ShahreBabak[1],
Roshanak Roshandel[2], Pejman Khadivi[2(✉)], Nader Karimi[1], and Shadrokh Samavi[1,3]

[1] Department of Electrical and Computer Engineering, Isfahan University of Technology,
Isfahan, Iran
[2] Computer Science Department, Seattle University, Seattle, USA
khadivip@seattleu.edu
[3] Department of Electrical and Computer Engineering, McMaster University, Hamilton, Canada

Abstract. The new coronavirus infection has shocked the world since early 2020 with its aggressive outbreak. Rapid detection of the disease saves lives, and relying on medical imaging (Computed Tomography and X-ray) to detect infected lungs has shown to be effective. Deep learning and convolutional neural networks have been used for image analysis in this context. However, accurate identification of infected regions has proven challenging for two main reasons. Firstly, the characteristics of infected areas differ in different images. Secondly, insufficient training data makes it challenging to train various machine learning algorithms, including deep-learning models. This paper proposes an approach to segment lung regions infected by COVID-19 to help cardiologists diagnose the disease more accurately, faster, and more manageable. We propose a bifurcated 2-D model for two types of segmentation. This model uses a shared encoder and a bifurcated connection to two separate decoders. One decoder is for segmentation of the healthy region of the lungs, while the other is for the segmentation of the infected regions. Experiments on publically available images show that the bifurcated structure segments infected regions of the lungs better than state of the art.

Keywords: COVID-19 · Segmentation · Coronavirus · Multi-task learning

1 Introduction

Over the past few months, COVID-19 has spread worldwide, infecting over 30 million people and causing nearly a million people loss of life [1]. The pressure on the healthcare system to both detect and treat patients has been of great concern worldwide. Medical images are used to diagnose the disease and evaluate the severity of the infection, placing a significant burden on radiologists.

Researchers have been working on machine learning approaches to classify and segment lung images to aid COVID-19 diagnosis [2]. However, the accuracy and speed of the detection process have been of great concern.

Our work has mainly focused on computed tomography (CT) images from the lungs. We will review several segmentation methods used to partition infected lung regions from

© Springer Nature Switzerland AG 2021
A. Del Bimbo et al. (Eds.): ICPR 2020 Workshops, LNCS 12662, pp. 597–607, 2021.
https://doi.org/10.1007/978-3-030-68790-8_46

healthy regions in the following. Accurate segmentation often serves as a preliminary step for further image processing and analysis [3].

One of the classical medical image segmentation approaches is Support Vector Machine (SVM), where a supervised machine learning method classifies each image pixel. In [4], each CT image lungs' nodules region is segmented by using SVM. After introducing Convolutional Neural Networks (CNNs) by [5], a new horizon opened for medical image processing, especially for medical image segmentation. A few years later, Ronneberger invented U-Net architecture [6], which has become the foundation for many advances over the last few years. The challenges with the segmentation of lung lesions in CT images predate the Coronavirus pandemic. For example, in [7], the authors combined the powerful nonlinear feature extraction ability of CNNs and the accurate boundary delineation ability of Active Contour Models (ACMs) for segmenting lesion regions in the brain, liver, and lung. Amyar et al. [8] proposed a Multi-Task Learning (MTL) approach for segmentation and COVID-19 CT images classification. Deploying MTL can solve the problem of lacking data and could lead to a model with better generalization. The authors in [8] employ a shared encoder and two parallel networks to classify infected patients and segment infected regions.

Another approach for segmenting 3D-CT images is to use a 3D U-Net [9], the same method used in [10]. It is a regular U-Net, but its 2D layers are replaced with 3D convolutional layers, which gives the model the ability to process a 3D CT image. Nevertheless, it is always a good idea to customize U-Net's base architecture for more complex tasks. In [11], Tongxue Zhou et al. proposed a UNet-based architecture, which in each step, simple 2D convolutions are followed by a "res_dil" block – a residual architecture that is inspired from *ResNet* [12]. Also, in [11], They applied a new loss function to improve small Region of Interest (ROI) segmentation performance. They also took advantage of the *attention mechanism,* which assigns weights to emphasize important information. In their work, channel-wise and space-wise methods were used to avoid the model from distraction by non-important parts. Instead of max pooling, which is usually used for reducing the resolution of feature maps, authors in [11] have used convolution with different stride.

In some tasks, such as the lesion segmentation task, extracting textures can be helpful. In [13], this fact is used to improve the output of the segmentation task. As a preprocessing method, they segment the lungs region from CT Images. They then used similar architecture to predict the segmentation map for infected regions using the last part's output and the original image as inputs. Finding the mask for the lungs region will lead to higher segmentation accuracy.

In the early days of the COVID-19, authors of [14] focused on supporting the healthcare providers for diagnosing infections with deep learning. They used MTL to overcome challenges associated with small datasets. In this case, the primary contribution was a classification approach, but they first extracted the COVID-19 infected regions. For better prediction, the segmentation task was done by cascading two encoder-decoder based models. The first one is responsible for finding a mask for the lungs, and the second one uses the output of the previous model to segment infected regions.

In [15], each convolutional layer was replaced with a ResNet block, which is a residual block. The authors believe that the original UNet is not "deep" enough for this problem; therefore, they made the architecture deeper by adding another deep section. These ideas improved the accuracy of prediction by about 10% compared to the original UNet.

Laradji et al. [16] proposed a weakly-supervised method for segmentation. Using this technique reduces the cost of collecting data for training and the time an expert should spend. Given that the flipped image segmentation map is the same as flipping the original image segmentation map, the authors of [16] use flipped images to augment their training dataset. However, we believe that the model used in [16] could perform better if the lungs segmentation-maps were used.

In this paper, we propose a new bifurcated architecture for segmenting infected regions of COVID-19. Our architecture consists of an encoder and two decoders. Our approach relies on MTL because we have a small dataset of images for this task. Our proposed architecture learns how to segment the lungs and infected regions simultaneously.

The structure of this paper is as follows. In Sect. 2, the proposed method is explained. The metrics that are used for evaluation are described in Sect. 3. Then in Sect. 4, the experimental results are reported. In the final section, we conclude the article.

2   Proposed Method

Segmenting infected regions in CT images can help speed up the detection of COVID-19 infections. Due to the different sizes of infected regions and small datasets, achieving accurate results is difficult. In this paper, a new architecture for segmenting COVID-19 infected regions is proposed. This architecture is based on U-Net [6], which is previously used for segmenting medical images. It consists of four blocks: a *decoder*, a bifurcated structure with *two parallel encoders*, and a final *merging encoder*. The block diagram of our models is shown in Fig. 1. In the following, the details of the proposed architecture are explained.

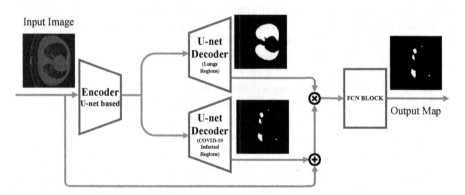

Fig. 1. The proposed bifurcated architecture

2.1 Architecture

One technique that we consider in our architecture is Multi-task learning (MTL) [17]. This technique is a learning algorithm to train a network for multiple related tasks. The basic idea is that multiple related tasks can have shared features. In such a case, the model learns which features are common and which features are distinct for each class. Hence, the problem of overfitting is avoided.

In MTL, deep learning is typically performed using either hard parameter sharing or soft parameter sharing. In hard parameter sharing, hidden layers are shared among all tasks while keeping several task-specific output layers. On the other hand, in soft parameter sharing, each task has its network architecture (model and parameters), and there is no shared layer between these tasks. In soft sharing, the distance between network parameters is regularized to make similar parameters for different tasks.

In this paper, we use MTL to increase the segmentation accuracy of infected regions. However, contrary to other works, both tasks in our work are segmentation, one for the lungs region, and one for infected regions. The encoder is shared for both tasks, but each task has a separate decoder. Since our architecture shares the encoder for the bifurcated branches, parameters are shared, and the complexity is reduced. In the following, the structure of each block that is shown in Fig. 1 is explained.

2.1.1 The Encoder

As a regular UNet architecture, our network begins with an encoder responsible for extracting useful feature maps to be used by the decoder. Our approach extracts both high-level and low-level features. We use a modified inception block [18] in the encoder's front layers when the image resolution is still high. As the feature maps go through the pooling layers, their sizes and resolutions are reduced. An inception block is shown in Fig. 2.

The inception block applies filters with different size kernels to the same input, rather than applying a stack of same-size kernel convolutional layers. The use of filters with multiple sizes enables us to extract different feature maps with various scales. Intuitively, the infected regions' sizes are different, so these feature maps help the model have better predictions by having different receptive fields.

There are four branches in our inception block:

a) Two 3×3 convolutional layers without dilation
b) Two 5×5 convolutional layers with dilation rate of "*d_rate*"
c) Two 9×9 convolution layers with dilation rate of "*d_rate*"
d) Max-pooling 2D with a size of 7×7 and a stride of 1×1.

Given that there are three inception blocks in our proposed architecture. In this inception block, two convolutional blocks are dilated type. The advantage of this type of convolution is having an extended field of view while the computational cost is not increased. For this purpose, a parameter determines the distance of neighbors from the center of the kernel. In our work, this parameter is called "d_rate." We determine the value for "*d_rate*" based on the corresponding inception block position in the network.

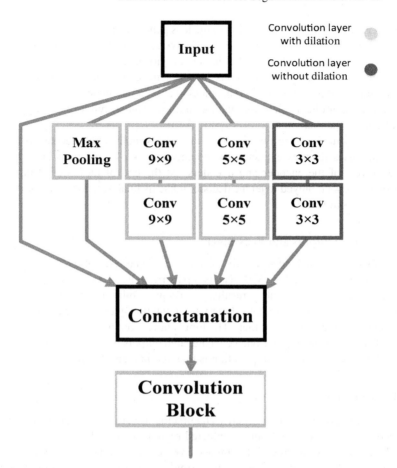

Fig. 2. Proposed modified inception block

For the inception blocks located toward each decoder's end, smaller "*d_rate*" values are used to allow for more conservative predictions.

2.1.2 Segmentation Decoders

The bifurcated part of the architecture is responsible for constructing desirable outputs, which have the same size as the input image. The decoder on the upper branch (Fig. 1) is responsible for the lung region's segmentation. The output of this upper decoder is a binary mask that specifies where the lungs are located. We will use the binary lungs mask as a region of interest (ROI) in our architecture's next stage. The decoder in the lower bifurcated branch is used to segment the infected regions of the lungs. This decoder's primary duty is to generate a probability map for the next stage to extract the COVID-19 infected regions.

Another reason for using the bifurcation structure is implementing the *deep supervision* technique. The deep supervision method is proposed in [19] for preventing the vanishing gradient descent. In addition to the loss generated at the output, deep supervision uses auxiliary loss from different network layers. We deployed the deep supervision method to train the network to predict infected regions. In our architecture, before the final output, the auxiliary loss function is defined at the end of the lower decoder. At this point, we have two loss functions in this block that help the model to learn features and avoid vanishing gradient. The loss function and the ground truth of the lower decoder are the same as the final output.

The architectures of the two decoders are very similar. Each decoder is a U-Net decoder that uses features extracted by the encoder. Skip connections are not shown between layers of the encoder and decoders. At the end of these decoders, another inception block is employed to let the decoder make its final decisions using multi-scale kernels in convolutions.

2.1.3   Merging Outputs and FCN Block

The Fully Convolutional Networks (FCN) block is the last part of our architecture that produces the final output. The FCN gets its inputs from the two segmentation decoders. To achieve accurate segmentation results, we use probability maps before binarization instead of using the infected regions' binary segmentation map. We then concatenate the input image and the probability map. The lung's binary output from the upper decoder is multiplied by the above-mentioned concatenated outputs for removing the misleading information. The output of multiplication is then fed into an FCN layer.

2.2   Loss Functions

Our loss function consists of three parts. The first loss function is from the lung decoder. The second loss function is from the infected region decoder. The third loss function is for the FCN block. We found that binary cross-entropy is the best loss function for classification and segmentation tasks. The formula of binary cross-entropy is as follow:

$$L = \frac{-1}{N} \sum_{i=1}^{N} y_i log(p(y_i)) + (1 - y_i).log(1 - p(y_i))$$

where N is the number of pixels, y_i is the label of the i-th pixel, and $p(y_i)$ is the predicted label.

Experimental results show that the effect of each of the three parts is different. Therefore, we use a weighted binary cross-entropy loss function. We use a weight of 0.5 for the lungs segmentation (W_{lung}), 1 for the auxiliary output (W_{aux}), and 2 for the final output (W_{fin}).

$$Loss = W_{lung}L_{lung} + W_{aux}L_{aux} + W_{fin}L_{fin}$$

3   Experimental Results

To evaluate our architecture and compare it with other state-of-the-art algorithms, we have used various standard metrics shown in Table 1.

Table 1. The metrics for evaluating

Metric	Formula
Sensitivity	$\frac{TP}{TP+FN}$
Specificity	$\frac{TN}{TN+FP}$
IoU	$\frac{TP}{TP+FP+FN}$
Dice Score	$\frac{2*TP}{2*TP+FP+FN}$
PPV	$\frac{TP}{TP+FP}$

We have used the standard definition for these metrics based on true/false negative/positive rates as in Table 1. Specifically, we use *sensitivity* because we are interested in evaluating how robust and trustworthy is our model in detecting infected regions. Intersection over Union (IoU) and Dice Score are commonly used in evaluating segmentation tasks to calculate the overlap between the ground truth and the region predicted by the algorithm. For emphasizing the positive samples, we used PPV (Positive Predicted Value).

We use a publicly available dataset [20] used for similar tasks by [10] to evaluate our work. This dataset contains 20 labeled COVID-19 CT scans. Left lung, right lung, and infections are labeled by a radiologist and verified by an experienced radiologist. Each instance of the dataset is a 3D-volume consisting of multiple slices. Since our dataset's images have different sizes, we resized each slice to 256×256 pixels to train the model. Also, the 3D dataset was converted into 2D slices to address computation limitations. The 3D CT images in the dataset have different slices and require a network capable of accepting 3D tensors. By using the 2D dataset, a simpler network can be used. The 2D dataset contains 3520 slices. About 52% (1841) of slices correspond to infected regions of the lungs, while the rest are images of healthy regions. We then selected 16 random training sets and four test sets. We selected 12% of training slices randomly as our validation set due to the slice level processing. We experimented with a five-fold and ran the model three times on each fold. The average number of slices in different folds for the training set is 2816, and the test set is 704. The average result is reported in this section.

As mentioned earlier, the last layer of our model's output is the class probability value for each pixel. A threshold should be selected to convert the probability into a binary value and generate the segmented mask. We use our validation data and experiment with different thresholds to obtain a suitable marker for selecting this threshold. We then apply this threshold to our test data. To evaluate our model's performance, we compare our results with [16], which uses weakly supervised learning, and [21], which uses a 3D U-Net model for solving this problem.

Table 2 shows the results of various metrics on five folds. The result of each fold is computed as the average of three runs. The last column reports the average and standard deviation of 5 the folds.

Table 2. Average metrics on five folds. Each fold is trained three times

	Fold 1	Fold 2	Fold 3	Fold 4	Fold 5	Average ± *std*
IOU	0.719	0.746	0.594	0.699	0.752	0.702 ± 0.057
Dice	0.778	0.812	0.669	0.762	0.819	0.763 ± 0.053
PPV	0.831	0.827	0.764	0.826	0.852	0.82 ± 0.029
Sensitivity	0.779	0.821	0.633	0.755	0.81	0.759 ± 0.067
Specificity	0.998	0.997	0.997	0.999	0.998	0.99 ± 0.0007

Table 3 compares our results with those of [21]. In [21], the authors also divided the dataset into five folds. The results of each fold for our model and the [21] model are shown in Table 3. Our model performs better on Dice score and sensitivity metrics while slightly lagging in specificity. The dice and sensitivity outperform the [21] due to different view fields using the inception block, which helps the model detect infection regions with different sizes accurately.

Table 3. Comparing our method with [21]

	Folds	Dice	Sensitivity	Specificity
[21]	Fold 1	0.556	0.447	0.999
	Fold 2	0.801	0.875	0.999
	Fold 3	0.829	0.796	0.999
	Fold 4	0.853	0.836	0.999
	Fold 5	0.765	0.697	0.999
	Average	0.761	0.730	**0.999**
Ours	Fold 1	0.778	0.779	0.998
	Fold 2	0.812	0.821	0.997
	Fold 3	0.669	0.633	0.997
	Fold 4	0.762	0.755	0.999
	Fold 5	0.819	0.81	0.998
	Average	**0.768**	**0.759**	0.997

Table 4 compares our results with those of [16]. Except for the sensitivity, our results outperform the model in [16].

Our proposed bifurcated method allows the model to concentrate on using features specifically located in the lungs region. Otherwise, the network could have been misled by features outside the lungs. Also, using the inception block made it possible to have different scales of features, which helped achieve accurate segmentation of infected regions with different sizes. In Fig. 3, some samples of the dataset and the results of our

Table 4. Comparing our method with [16]

	Dice	Sensitivity	Specificity	IOU	PPV
[16]	0.75	**0.86**	0.97	0.59	0.66
Ours	**0.768**	0.759	**0.997**	**0.702**	**0.82**

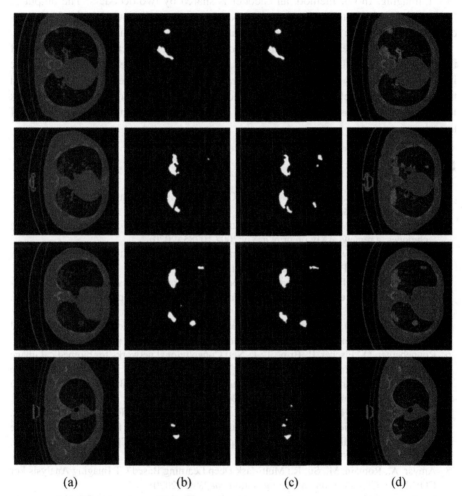

(a)	(b)	(c)	(d)

Fig. 3 (a) input images, (b) predicted mask, (c) ground truth, (d) predicted mask superimposed on the input image

models are shown. In this figure, the first column is the input image, the second column is the segmented results of our model, the third column is the ground truth, and the last one is the combined input image with segmented results. Since the infected regions have different sizes, our model could accurately segment the large regions, but some small regions are missed.

4   Conclusion

In this paper, we proposed a model for segmenting COVID-19 infected lung regions in CT imaging. In our method, an encoder is shared by two decoders. The output of the encoder is bifurcated and is fed two separate decoders. The two decoders' output is combined and used to predict the infected lungs areas' final segmentation map.

For more accurate and precise predictions, the model extracts the lungs' mask of the given input to avoid the image's useless regions. By using the inception blocks, the infected regions are segmented more accurately. Also, using deep supervision helped the model to learn better. The proposed method could be categorized as a multi-task learning system. By using these techniques in our model, the results surpassed state of the art.

References

1. Deng-Ping, F., et al.: Inf-Net: Automatic COVID-19 Lung Infection Segmentation from CT Images. IEEE Trans. Med. Imag. (2020)
2. Shoeibi, A., et al.: Automated detection and forecasting of covid-19 using deep learning techniques: a review. arXiv preprint arXiv:2007.10785 (2020)
3. Sharma, N., Aggarwal, L.M.: Automated medical image segmentation techniques. J. Med. Phys. Assoc. Med. Phys. India **35**(1), 3 (2010)
4. Keshani, M., Azimifar, Z., Tajeripour, F., Boostani, R.: Lung nodule segmentation and recognition using SVM classifier and active contour modeling: a complete intelligent system. Comput. Biol. Med. **43**(4), 287–300 (2013)
5. Krizhevsky, A., Ilya, S., Geoffrey, H.: Imagenet classification with deep convolutional neural networks. In: Advances in Neural Information Processing Systems, pp. 1097–1105 (2012)
6. Ronneberger, O., Fischer, P., Brox, T.: U-Net: Convolutional networks for biomedical image segmentation. In: Navab, N., Hornegger, J., Wells, W.M., Frangi, A.F. (eds.) Medical Image Computing and Computer-Assisted Intervention, vol. 9351, pp. 234–241. Springer, Cham (2015). https://doi.org/10.1007/978-3-319-24574-4_28
7. Hatamizadeh, A., Hoogi, A., Sengupta, D., Lu, W., Wilcox, B., Rubin, D., Terzopoulos, D.: Deep active lesion segmentation. In: Suk, H., Liu, M., Yan, P., Lian, C. (eds.) Machine Learning in Medical Imaging, vol. 11861, pp. 98–105. Springer, Cham (2019). https://doi.org/10.1007/978-3-030-32692-0_12
8. Amyar, A., Romain, M., Su, R.: Multi-task Deep Learning Based CT Imaging Analysis For COVID-19: Classification and Segmentation. *medRxiv* (2020)
9. Çiçek, Ö., Abdulkadir, A., Lienkamp, S.S., Brox, T., Ronneberger, O.: 3D U-Net: learning dense volumetric segmentation from sparse annotation. In: Ourselin, S., Joskowicz, L., Sabuncu, M.R., Unal, G., Wells, W. (eds.) Medical Image Computing and Computer-Assisted Intervention, vol. 9901, pp. 424–432. Springer, Cham (2016). https://doi.org/10.1007/978-3-319-46723-8_49
10. Ma, J., et al.: Towards efficient covid-19 ct annotation: a benchmark for lung and infection segmentation. arXiv preprint arXiv:2004.12537 (2020)
11. Zhou, T., Stéphane, C., Su, R.: An automatic COVID-19 CT segmentation network using spatial and channel attention mechanism.arXiv preprint arXiv:2004.06673 (2020)

12. He, K., Xiangyu, Z., Shaoqing, R., Jian, S.: Deep residual learning for image recognition. In: Proceedings of the IEEE Conference on Computer Vision and Pattern Recognition, pp. 770–778 (2016)
13. Elharrouss, O., Nandhini, S., Somaya, A.: An encoder-decoder-based method for COVID-19 lung infection segmentation. arXiv preprint arXiv:2007.00861 (2020)
14. Jin, S., et al.: AI-assisted CT imaging analysis for COVID-19 screening: Building and deploying a medical AI system in four weeks. *medRxiv* (2020)
15. Chen, X., Lina, Y., Yu, Z.: Residual attention U-Net for automated multi-class segmentation of COVID-19 chest CT images. arXiv preprint arXiv:2004.05645 (2020)
16. Laradji, I., Pau, R., Oscar, M., Keegan, L., Marco, L., Lironne, K., William, P., David, V., Derek, N.: A weakly supervised consistency-based learning method for COVID-19 segmentation in CT Images. arXiv preprint arXiv:2007.02180 (2020)
17. Caruana, R.: Multi task learning. Mach. Learn. **28**(1), 41–75 (1997)
18. Szegedy, Christian, Wei Liu, Yangqing Jia, Pierre Sermanet, Scott Reed, Dragomir Anguelov, Dumitru Erhan, Vincent Vanhoucke, and Andrew Rabinovich. "Going deeper with convolutions." In: Proceedings of the IEEE Conference on Computer Vision and Pattern Recognition, pp. 1–9 (2015)
19. Wang, L., Chen-Yu, L., Zhuowen, T., Svetlana, L.: Training deeper convolutional networks with deep supervision. arXiv preprint arXiv:1505.02496 (2015)
20. https://zenodo.org/record/3757476#.XqhRp_lS-5D
21. Müller, D., Iñaki, S.R., Frank, K.: Automated chest CT Image segmentation of COVID-19 lung infection based on 3D U-Net. arXiv preprint arXiv:2007.04774 (2020)

DeepPBM: Deep Probabilistic Background Model Estimation from Video Sequences

Rezaei Behnaz, Farnoosh Amirreza, and Sarah Ostadabbas[(✉)]

Augmented Cognition Lab, Electrical and Computer Engineering Department,
Northeastern University, Boston, MA 02115, USA
{brezaei,afarnoosh,ostadabbas}@ece.neu.edu
https://web.northeastern.edu/ostadabbas/

Abstract. This paper presents a novel unsupervised probabilistic model estimation of visual background in video sequences using a variational autoencoder framework. Due to the redundant nature of the backgrounds in surveillance videos, visual information of the background can be compressed into a low-dimensional subspace in the encoder part of the variational autoencoder, while the highly variant information of its moving foreground gets filtered throughout its encoding-decoding process. Our deep probabilistic background model (DeepPBM) estimation approach is enabled by the power of deep neural networks in learning compressed representations of video frames and reconstructing them back to the original domain. We evaluated the performance of our DeepPBM in background subtraction on 9 surveillance videos from the background model challenge (BMC2012) dataset, and compared that with a standard subspace learning technique, robust principle component analysis (RPCA), which similarly estimates a deterministic low dimensional representation of the background in videos and is widely used for this application. Our method outperforms RPCA on BMC 2012 dataset with 23% in average in F-measure score, emphasizing that background subtraction using the trained model can be done in more than 10 times faster (The source code is available at: https://github.com/ostadabbas/DeepPBM).

Keywords: Background subtraction · Probabilistic modeling · Unsupervised learning · Variational autoencoder

1 Introduction

Detection of moving objects or change detection in recorded videos can be seen as the process of separating the foreground from background. This process is a central component in every video surveillance, security, and traffic monitoring system. A huge body of research exists in the background vs. foreground separation topic since the introduction of simple yet effective mixture of Gaussian

---

R. Behnaz and F. Amirreza—Equal contribution.

© Springer Nature Switzerland AG 2021
A. Del Bimbo et al. (Eds.): ICPR 2020 Workshops, LNCS 12662, pp. 608–621, 2021.
https://doi.org/10.1007/978-3-030-68790-8_47

(MoG) model by Stauffer et al. [24]. Yet, development of an efficient background subtraction (BS) process for robust moving object detection that addresses the key challenges in dynamic backgrounds is not completely resolved. A competent BS algorithm should be fast and robust to the dynamic nature of the background. Furthermore, it should be implemented in an unsupervised manner to be able to generalize to the new scenes. Although several state-of-the-art algorithms have been proposed for adaptive background representation [1,4,22,23], a universal method that can address different BS challenges present in long-term videos is still missing. Recently, deep learning approaches based around using convolutional neural networks (CNNs) have shown promising results in BS problems in different challenging conditions [2,16,18,21,25,28,30]. However, all of these methods are supervised and were trained on ground truth video frames of benchmark datasets and tested on the same types of videos. In addition, non of these BS algorithms has been evaluated on long-term videos to demonstrate their adaptation performance in real-world applications, in which they are at least several hours in length.

To provide an unsupervised, generalizable and computationally efficient solution to the problem of background subtraction, we introduce a novel deep probabilistic background model (DeepPBM) estimation approach, which capitalizes on the power and flexibility of deep neural networks in approximating complex functions. Our approach is centered around two following hypotheses: (1) background in videos recorded by an stationary camera lies on a low-dimensional subspace represented by a series of latent variables, and (2) there is a Gaussian distribution model for the latent subspace of the background embedded by a non-linear mapping of the video frames. An important property of our DeepPBM approach is its generative modeling of the background, which can be used for creating synthetic backgrounds of the specific scene with different illuminations, shadings, and waving by variations in its latent variables. These synthetic backgrounds may be used for training purposes in deep learning models. The proposed DeepPBM shows high performance in background subtraction in the majority of the scenes in the BMC2012 dataset [26]. DeepPBM is also observed to have an acceptable performance in adapting the background model in long-term videos in this dataset performing orders of magnitude faster than its non-deep counterpart robust principle component analysis (RPCA).

The rest of the paper is structured as follows. Section 1.1 presents an overview of the recent work in the topic of background/foreground segmentation. Section 2 describes the details of the DeepPBM algorithm and the configuration of the employed network. Section 3 demonstrates the performance of the DeepPBM in BS on the videos from BMC2012 dataset and compares the results with other state-of-the-art method according to the objective BS metrics and also visual perception evaluations. Finally, in Sect. 4 we conclude the paper and discuss the possible future directions.

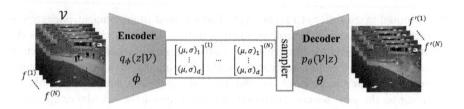

Fig. 1. Schematic of a variational autoencoder (VAE) for background subtraction used in our DeepPBM.

1.1 Overview of Background Subtraction Techniques

BS is usually achieved by creating a background model from video frame sequences, choosing a strategy to update this background model, and then subtracting each frame from this background model. The performance of the BS algorithm depends on how well each of theses steps can be implemented. After MoG presented in [24], various follow-up works in probabilistic background modeling have been proposed to improve the performance of the MoG approach through employing different learning methods and adaptation modifications [9,12,14,29]. All these methods nonetheless suffer from the noise in the initial frames as well as inflexibility to the the sudden changes in the background throughout the video.

In parallel, a significant amount of research effort has been dedicated to the modeling of the video backgrounds as a low-dimensional subspace in the original high-dimensional space of the video frames. Considering this assumption, the problem of BS has been formulated as an optimization problem in different works, in which an observation video matrix is decomposed into a low rank matrix forming background sequence and an additive part representing the moving object as the foreground [8,13,31]. Robust principle component analysis (RPCA) was the first problem formulation for this matrix decomposition [8]. Inspired by RPCA, various problem formulations for matrix decomposition have been proposed including robust subspace tracking (RST) [13], robust matrix completion (RMC) [17,19,20], and robust low rank minimization (RLRM) [31]. Although these algorithms work visually well in modeling the background and its gradual changes over time, they are based on an optimization problem with heavy structural properties that requires to be solved by computationally expensive iterations, which makes them impractical for online video inspection applications.

Recently, there has been several efforts to employ the capability of deep neural networks (DNNs) in performing BS [2,7,16,18,21,28,30]. However, all of these approaches are performed in a supervised manner, and therefore require manual foreground mask extraction from a subset of video frames for their learning phase. In [28], for the purpose of obtaining ground truth foreground segmentation, authors fed a small subset of manually annotated frames from each video as training data to a multi-resolution CNN with a cascaded architecture.

This choice of network architecture allowed them to handle larger foregrounds as well as to capture spatial coherence among neighboring pixels. The authors utilized this trained network to segment the rest of the frames, and reported the foreground mask detection results that were within the error margin of a human annotator. However, as they have noted, the selection of frames for CNN training needed to be done with caution especially for videos with sparse activities. In [16] authors proposed a triplet CNN with weak supervision for a multistage background feature embedding using an encoder-decoder structure. In [7], authors proposed a scene-specific BS algorithm by addressing the problem during the subtraction phase with the use of CNNs. A fixed background model was initially extracted by temporal median filtering of a short period of the scene. Afterwards, a CNN was trained on a subset of frames by feeding image patches from both video frames and the background image, and comparing the network output with the corresponding ground truth mask. The ground truth masks were obtained either from human expert labeling or alternatively generated with another existing BS algorithm, for which the performance was upper bounded by the mask generator. After the network was trained, it was used to generate binary foreground masks for the rest of the frames. In their approach, the CNN needed to be retrained for every new scene, and for this it required to have ground truth masks of a subset of frames in each video. In the work of [2], the authors took a similar approach in the subtraction phase of the BS task, however, they employed a different strategy for generating the background model. Their background model was generated/updated by applying temporal median filtering on a background pixel library built based on binary foreground masks obtained from another method [23]. This background pixel library has an adaptive memory length over the video sequence, which was dynamically tuned based on the ratio of moving pixels in each frame estimated from a flux tensor [27]. After preparing the background images, a CNN was trained with triplets of matching patches from background images, video frames, and ground truth masks on a random subset of frames from different scenes. This network is specifically trained to perform subtraction between original video frames and their corresponding estimated background images, therefore, its overall performance is highly dependant on the quality of the generated background images, and can degrade when the background model is challenged. Furthermore, this method still needs ground truth masks of a subset of frames for scenes that are used in the training process. Although the network did not need to be retrained for each new scene, its performance degraded if the scene was not included in the training set. BScGAN introduced [3] benefited from conditional generative adversarial networks (cGAN)in order to estimate a deep background model for the task of BS. By giving the original image, background and ground-truth foreground mask, generative networks generates a fake foreground mask for each given input and discriminator learns to distinguish between fake and real foreground mask examples under the condition of the input image. Lastly, BSPVGAN in [30] builds a BS model by using GANs to classify the pixels into foreground and background given the background images. Finally it uses parallel vision theory to improve the background subtraction results in complex scenes. However, one common problem of GANs is model collapse in which they fail estimating multimodal distributions. As we mentioned before, despite the high performance of above deep learning methods in foreground/ background segmentation, these methods are supervised and highly dependant on the quality of the extracted background model that they use for BS. Moreover none of them are tested on long videos to show their adaptation quality in real applications that need long-term video inspection.

Most recently there has been a few efforts in deep learning area for unsupervised modeling of the background in videos using the power of the auto-encoders methods. In the work of [11] authors employ denoising auto-encoders to generate a set of robust characteristics for each region or patch of the image, which will be the input of a probabilistic model to determine if that region is background or foreground. In this method the compact deep features are learned separately from the probabilistic classification step applied for classifying each patch into foreground or background. In a very recent attempt at [25] an unified method based on the GAN and image inpainting was presented for the BS. In this framework unsupervised visual features were learned based on the GAN followed by a semantic inpainting network for texture optimization. The whole pipeline of the BS was constructed of five steps including: motion mask extraction and object masking by computing the optical flow; context estimation using joint adversarial discriminator and context auto-encoder network; texture network for texture optimization; modified potion blending for refining the final background model; foreground detection by thresholding the difference image resulted from original image and background model. In compared to aforementioned deep unsupervised methods our model learns the probability distribution of the low-dimensional latent features learned by the deep neural network in an unified end-to-end manner. We refer the readers to [6] for a comprehensive review of DNN approaches on BS.

2   Proposed DeepPBM Estimation Approach

Variational autoencoders (VAEs) have emerged as one of the most popular approaches in unsupervised learning of complicated distributions that underlie models or generate data [10,15]. VAEs are compelling since they can be set up in the framework of deep learning (DL), and therefore benefit from the ongoing advances in this field. In the context of DL, a VAE consists of an encoder and a decoder. Illustrated in Fig. 1, encoder learns an efficient representation of its input data and projects that into a stochastic lower dimensional space, determined by latent variables. The decoder tries to recover the original data, given the probabilistic latent variables from the encoder. The entire network is trained by comparing the original input data with its reconstructed output [10]. We further discuss the mathematical details of each part in Sect. 2.1

From an information theoretic perspective, the compression of the high-dimensional input to a low-dimensional space as done in the encoder part of VAE, and then decompressing it back to the original space leads to the loss of high variant information (in our case moving objects), which is measured and used to learn the network. This lossy low-dimensional representation of the input data is a desired attribute that can be utilized in the context of BS in surveillance videos. This attribute follows similar principles employed in low-rank subspace learning approaches for unsupervised BS. Further, it can benefit from the power and flexibility of DL in learning a more effective low-dimensional space. Moreover, using DL allows us to transfer the computational cost of solving the subspace learning from the evaluation to the training process of DL, which could entirely be performed offline. Following aforementioned significance, the main idea behind our proposed DeepPBM is using VAE built on top of a DNN for the purpose of unsupervised BS considering the low-dimensional representation attribute of VAE along with the compression capacity of background images. Let's denote the encoder $q_\theta(z|x)$. We note that the lower-dimensional space is stochastic: the encoder outputs parameters to $q_\theta(z|x)$, which is a Gaussian probability density. We can sample from this distribution to get noisy values of the representations z. The decoder is another neural net. Its input is the representation z, it outputs the parameters to the probability distribution of the data, and has weights and biases ϕ. The decoder is denoted by $p_\phi(x|z)$. Information is lost because it goes from a smaller to a larger dimensionality. We measure this loss using the reconstruction log-likelihood $\log p_\phi(x|z)$. This measure tells us how effectively the decoder has learned to reconstruct an input image x given its latent representation z. The loss function of the variational autoencoder is the negative log-likelihood with a regularizer. Because there are no global representations that are shared by all datapoints, we can decompose the loss function into only terms that depend on a single datapoint. The total loss is then:

$$l_i(\theta, \phi) = -E_{z \sim q_\theta(z|x_i)}[\log p_\phi(x_i|z)] + KL(q_\theta(z|x_i)||p(z)) \qquad (1)$$

The first term is the reconstruction loss, or expected negative log-likelihood of the i-th datapoint. The expectation is taken with respect to the encoder's distribution over the representations. This term encourages the decoder to learn to reconstruct the data. The second term is the Kullback-Leibler divergence between the encoder's distribution $q_\theta(z|x)$. This divergence measures how much close q is to p. In the variational autoencoder, p is specified as a standard Normal distribution with mean zero and variance one, or $p(z) = \mathcal{N}(0, 1)$. If the encoder outputs representations z that are different than those from a standard normal distribution, it will receive a penalty in the loss. This regularizer term means 'keep the representations z of each digit sufficiently diverse'. We train the variational autoencoder using gradient descent to optimize the loss with respect to the parameters of the encoder and decoder θ and ϕ.

2.1 Probabilistic Modeling of the Background in Videos

Considering that video frames $f^{(i)} \in \mathcal{V}, i \in \{1, \ldots, N\}$, each of size $w \times h$ pixels, are generated from d underlying probabilistic latent variables vectorized in $z \in \mathbb{R}^d$ in which $d \ll w \times h$, the vector z is interpreted as the compressed representation of the video. A VAE considers the joint probability of the input video, $\mathcal{V}$, and its representation, z, to define the underlying generative model as $p_\theta(\mathcal{V}, z) = p_\theta(\mathcal{V}|z)p(z)$, where $p(z) = \mathcal{N}(0, I)$ is the standard Gaussian prior for latent variables z, and $p_\theta(\mathcal{V}|z)$ is the decoder part of a VAE that is parameterized by a DNN with parameters θ. In the encoder part of the VAE, the posterior distribution $p(z|\mathcal{V})$ is approximated with a variational posterior $q_\phi(z|\mathcal{V})$ with parameters ϕ. Each dimension of the latent space in this variational posterior is modeled independently with a Gaussian mean and variance for each video frame, as $q_\phi(z|f) = \prod_{k=1}^{d} \mathcal{N}(z_k|\mu_k^f, \sigma_k^{f^2})$, where μ^f, and σ^{f^2} are outputs of the encoder, $q_\phi(z|f)$, which is also parameterized by a DNN with parameters ϕ. We assume that observations $x^{(i)} \in \mathcal{X}, i = 1, \ldots, N$ are generated by combining K underlying factors $f = (f_1, \ldots, f_K)$. These observations are modelled using a real-valued latent/code vector $z \in \mathbb{R}^d$, interpreted as the representation of the data. The generative model is defined by the standard Gaussian prior $p(z) = \mathcal{N}(0, I)$, intentionally chosen to be a factorised distribution, and the decoder $p_\theta(x|z)$ parameterized by a neural network. The distribution of representations for the entire data set is then given by:

$$q(z) = \mathbb{E}_{p_{data}(x)}[q(z|x)] = \frac{1}{N} \sum_{i=1}^{N} q(z|x^{(i)}) \tag{2}$$

A disentangled representation would have each z_j correspond to precisely one underlying factor f_k. The efforts in making this variational posterior as close as possible to the true posterior distribution results in maximization of the evidence lower bound (ELBO) [5,15], such that the final VAE objective for the entire video becomes:

$$ELBO_\mathcal{V}(\theta, \phi) = \frac{1}{N} \sum_{i=1}^{N} \left[\mathbb{E}_{q_\phi(z|f^{(i)})} \left[\log p_\theta(f^{(i)}|z) \right] - KL\big(q_\phi(z|f^{(i)})||p(z)\big) \right] \tag{3}$$

The first term in Eq. (3) (expected likelihood term) can be interpreted as the negative reconstruction error, which encourages the decoder to learn to reconstruct the original input, and the second term is the Kullback-Leibler (KL) divergence between prior and variational posterior distribution of latent variables, which acts a regularizer to penalize the model complexity. The expectation is taken with respect to the encoder's distribution over the representations. This term encourages the decoder to learn to reconstruct the data.

For our purpose of BS, we used an l_1-norm loss function for reconstruction error of the VAE in order to capture the sparsity of the foreground assumed in the majority of low rank subspace factorization studies used in background/foreground separation. The KL term can also be calculated analytically in the case of Gaussian distributions. Therefore, the total loss function for our proposed DeepPBM becomes:

$$Loss(f, f', \mu^f, \sigma^{f^2}) = \sum_{i=1}^{N} |f^{(i)} - f'^{(i)}| - \frac{1}{2} \sum_{i=1}^{N} \left(1 + \log \sigma^{f^{(i)}2} - \mu^{f^{(i)}2} - \sigma^{f^{(i)}2}\right)$$

(4)

where f' is the reconstructed version of the input video frame, f, produced by the decoder [10, 15].

Table 1. DeepPBM network architecture.

Layer #	Encoder
	Input: $w \times h \times 3$ RGB image
1	4×4 conv, 32 Relu, stride 2, BatchNorm
2	4×4 conv, 64 Relu, stride 2, BatchNorm
3	4×4 conv, 128 Relu, stride 2, BatchNorm
4	4×4 conv, 128 Relu, stride 2, BatchNorm
	Intermediate output: $128 \times w' \times h'$ patch
5	FC 2400 ReLU, Dropout 0.3
6	FC $2 \times d$
	Output: $\mu_z, \sigma_z^2 \in \mathbb{R}^d$
	Decoder
	Input: $z \in \mathbb{R}^d$
1	FC 2400 ReLU
2	FC $128 \times w' \times h'$ ReLU, Dropout 0.3
3	4×4 deconv, 128 Relu, stride 2, BatchNorm
4	4×4 deconv, 64 Relu, stride 2, BatchNorm
5	4×4 deconv, 32 Relu, stride 2, BatchNorm
6	4×4 deconv, 3 Sigmoid, stride 2
	Output: $w \times h \times 3$ RGB image

2.2   DeepPBM Architecture and Training

The encoder and decoder parts of the VAE in the DeepPBM are both implemented using a CNN architecture specified in Table 1. The encoder takes the video frames as input and outputs the mean and variance of their underlying low dimensional latent variables distributions. The decoder takes samples drawn from latent distributions as input and output the recovered version of the original input. The network is trained by minimizing the error defined in Eq. (4). We trained the VAE using the gradient descent to optimize this loss with respect to the parameters of the encoder and decoder, θ and ϕ, respectively. The input video data is trained in batches of size 140 for 200 epochs.

Table 2. Benchmark metrics and execution time for the BS task of our DeepPBM compared to RPCA evaluated on the 6 short videos of BMC2012 dataset. For the fair comparison we ran the trained model on the CPU mode.

Algorithm	F-measure	Recall	Precision	Run Time
Big trucks – 1498 frames				
RPCA	0.68	0.6	0.80	18 min
DeepPBM ($d = 30$)	0.86	0.85	0.88	2.8 min
Wandering students – 795 frames				
RPCA	0.87	0.84	0.90	6.2 min
DeepPBM ($d = 20$)	0.94	0.92	0.95	1.1 min
Rabbit in the night – 1896 frames				
RPCA	0.60	0.59	0.61	28 min
DeepPBM ($d = 35$)	0.90	0.94	0.87	2.7 min
Beware of the trains – 1065 frames				
RPCA	0.68	0.61	0.78	11.5 min
DeepPBM ($d = 30$)	0.81	0.83	0.78	1.5 min
Train in the tunnel – 1726 frames				
RPCA	0.63	0.60	0.81	14.2 min
DeepPBM ($d = 30$)	0.70	0.70	0.71	2.4 min
Traffic during windy day – 793 frames				
RPCA	0.54	0.50	0.58	8.4 min
DeepPBM ($d = 1$)	0.76	0.74	0.79	1.1 min
Average over all the videos				
RPCA	0.67	0.62	0.75	14.4 min
DeepPBM	**0.83**	**0.83**	**0.83**	**1.9 min**

3   Performance Assessment

We evaluated the performance of our proposed algorithm, DeepPBM, in BS on
the BMC2012 benchmark dataset [26]. This benchmark contains 9 real world
surveillance videos along with encrypted ground-truth masks of the foreground
for a subset of frames in each video. This dataset focuses on outdoor situations with various weather and illumination conditions such as wind, sun, or
rain. Therefore, makes it suitable for performance evaluation of BS methods in
challenging conditions. We used the short videos in this dataset to compare the
estimation quality of DeepPBM against RPCA. We then used the long videos
to examine how our DeepPBM adapts to changes in the background model over
a long period of time. Please note that due to the shortage of the memory and
processing units required for running RPCA, we could not apply RPCA for the
long videos. The evaluation metrics are computed by the software that is provided with the dataset, based on the encrypted ground-truth masks. In order
to extract the masks of the moving objects in short videos (with less than 2000
frames), we first trained the DeepPBM network using all of the video frames as
explained in Sect. 2.2. The dimension of the latent variables, d, needs to be tuned
based on the dynamics/complexity of the background model in each video. For
videos with dynamic background (e.g. in windy, rainy or snowy conditions), a
larger d should be selected in order to capture variations in the background,
however, for videos with monotonic background (with slight or no changes in
background along video frames) a smaller d should be selected to prevent network from learning foreground. After the network was trained, we fed the same
frames to the network to estimate the background image for each individual
frame. Finally, we used the estimated background of each frame to find the
mask of the moving objects by thresholding the difference between the original
input frame and the estimated background. The quantitative results of the performance of DeepPBM in BS compared to the RPCA is reported in Table 2. As
it is observed, DeepPBM outperforms RPCA in all of the short videos by 23% in
F-measure. Further it performs more than 10 times faster than RPCA. Figure 2
illustrates sample results of applying DeepPBM and RPCA on short videos of
BMC2012 dataset. As seen, DeepPBM is quite successful in detecting moving
objects in these scenes, and generates acceptable masks of the foreground, while
RPCA fails to detect accurate foreground masks.

For the long videos, we used the first 20% of the video frames for training of
the DeepPBM, and then used this trained network to extract background images
for all of the frames. Table 3 shows the quantitative performance of DeepPBM
for the long videos which gives an average F-measure score of 0.69. Figure 3
illustrates how the network adapts to the changes in the background that happen

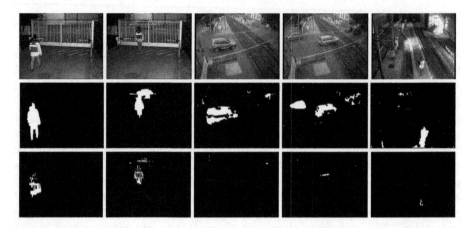

Fig. 2. Extracted masks from estimated background model for some of the frames selected from sample videos in the BMC2012 dataset. First row shows the original frames, second row shows the corresponding masks resulted from DeepPBM, and third row shows the masks resulted from the RPCA method.

Fig. 3. Adaptation of the background model estimated by DeepPBM to the changes of the scene for one of the long videos in BMC2012 dataset. First row shows the consecutive frames of the original video, second row shows the corresponding background model.

over a long period of time. The car in this sample scene is initially included as part of the background model in the first two frames, since it has been stationary for a long period, however, the network begins to detect that as foreground in the next two frames as soon as the car starts to move.

Table 3. Benchmark metrics for the BS task of our DeepPBM on 3 long videos of BMC2012 dataset.

Algorithm	F-measure	Recall	Precision	Run time
Video_001 – 22 min				
DeepPBM ($d = 30$)	0.73	0.76	0.71	4.69 min
Video_005 – 78 min				
DeepPBM ($d = 30$)	0.71	0.73	0.62	16.67 min
Video_009 – 72 min				
DeepPBM ($d = 20$)	0.63	0.70	0.68	15.34 min
Average over all the videos				
DeepPBM	0.69	0.73	0.67	12.23 min

4   Conclusion

In this paper, we presented our DeepPBM method using the framework of VAE for detecting the moving objects in videos recorded by stationary cameras. We evaluated the performance of our model in the task of background subtraction, and showed how well it adapts to the changes of the background in long-term monitoring on the BMC2012 dataset. According to the reported results, DeepPBM outperformed RPCA known as one of the standard and well-performed subspace learning methods for background modeling in both time efficiency and modeling performance. Note that our approach estimates a generative low-dimensional model of the background and task of the BS is performed by simply thresholding the difference between this model and the original input frame. One of the important directions in our future work will be performing selective background updates via adapting the background model to the pixels that were detected as background by the network, as opposed to the current network fine-tuning paradigm after specific time intervals.

References

1. Allebosch, G., Deboeverie, F., Veelaert, P., Philips, W.: Efic: edge based foreground background segmentation and interior classification for dynamic camera viewpoints. In: International Conference on Advanced Concepts for Intelligent Vision Systems, pp. 130–141 (2015)
2. Babaee, M., Dinh, D.T., Rigoll, G.: A deep convolutional neural network for video sequence background subtraction. Pattern Recogn. **76**, 635–649 (2018)
3. Bakkay, M.C., Rashwan, H.A., Salmane, H., Khoudour, L., Puigtt, D., Ruichek, Y.: Bscgan: deep background subtraction with conditional generative adversarial networks. In: 2018 25th IEEE International Conference on Image Processing (ICIP), pp. 4018–4022. IEEE (2018)

4. Bianco, S., Ciocca, G., Schettini, R.: How far can you get by combining change detection algorithms? In: Battiato, S., Gallo, G., Schettini, R., Stanco, F. (eds.) ICIAP 2017. LNCS, vol. 10484, pp. 96–107. Springer, Cham (2017). https://doi.org/10.1007/978-3-319-68560-1_9

5. Blei, D.M., Kucukelbir, A., McAuliffe, J.D.: Variational Inference: A Review for Statisticians. ArXiv e-prints (January 2016)

6. Bouwmans, T., Javed, S., Sultana, M., Jung, S.K.: Deep neural network concepts for background subtraction: a systematic review and comparative evaluation. Neural Netw. **117**, 8–66 (2019)

7. Braham, M., Van Droogenbroeck, M.: Deep background subtraction with scene-specific convolutional neural networks. In: IEEE International Conference on Systems, Signals and Image Processing (IWSSIP), Bratislava 23–25 May 2016, pp. 1–4 (2016)

8. Candès, E.J., Li, X., Ma, Y., Wright, J.: Robust principal component analysis? J. ACM (JACM) **58**(3), 11 (2011)

9. Chen, Y.T., Chen, C.S., Huang, C.R., Hung, Y.P.: Efficient hierarchical method for background subtraction. Pattern Recogn. **40**(10), 2706–2715 (2007)

10. Doersch, C.: Tutorial on Variational Autoencoders. ArXiv e-prints (June 2016)

11. García-González, J., Ortiz-de-Lazcano-Lobato, J.M., Luque-Baena, R.M., Molina-Cabello, M.A., López-Rubio, E.: Background modeling for video sequences by stacked denoising autoencoders. In: Herrera, F. (ed.) CAEPIA 2018. LNCS (LNAI), vol. 11160, pp. 341–350. Springer, Cham (2018). https://doi.org/10.1007/978-3-030-00374-6_32

12. Haines, T.S., Xiang, T.: Background subtraction with dirichletprocess mixture models. IEEE Trans. Pattern Anal. Mach. Intell. **36**(4), 670–683 (2014)

13. He, J., Balzano, L., Szlam, A.: Incremental gradient on the grassmannian for online foreground and background separation in subsampled video. In: 2012 IEEE Conference on Computer Vision and Pattern Recognition (CVPR), pp. 1568–1575 (2012)

14. KaewTraKulPong, P., Bowden, R.: An improved adaptive background mixture model for real-time tracking with shadow detection. In: Remagnino, P., Jones, G.A., Paragios, N., Regazzoni, C.S., (eds.) Video-based surveillance systems, pp. 135–144. Springer, Boston (2002) https://doi.org/10.1007/978-1-4615-0913-4_11

15. Kingma, D.P., Welling, M.: Auto-Encoding Variational Bayes. ArXiv e-prints (December 2013)

16. Lim, L.A., Keles, H.Y.: Foreground segmentation using convolutional neural networks for multiscale feature encoding. Pattern Recogn. Lett. **112**, 256–262 (2018)

17. Mansour, H., Vetro, A.: Video background subtraction using semi-supervised robust matrix completion. In: 2014 IEEE International Conference on Acoustics, Speech and Signal Processing (ICASSP), pp. 6528–6532 (May 2014)

18. Mondéjar-Guerra, V., Rouco, J., Novo, J., Ortega, M.: An end-to-end deep learning approach for simultaneous background modeling and subtraction. In: British Machine Vision Conference (BMVC), Cardiff (2019)

19. Rezaei, B., Ostadabbas, S.: Moving object detection through robust matrix completion augmented with objectness. IEEE J. Sel. Top. Sign. Proces. **12**(6), 1313–1323 (2018). https://doi.org/10.1109/JSTSP.2018.2869111

20. Rezaei, B., Ostadabbas, S.: Background subtraction via fast robust matrix completion. In: Proceedings of the IEEE International Conference on Computer Vision, pp. 1871–1879 (2017)

21. Sakkos, D., Liu, H., Han, J., Shao, L.: End-to-end video background subtraction with 3d convolutional neural networks. Multimedia Tools Appl. **77**(17), 23023–23041 (2017). https://doi.org/10.1007/s11042-017-5460-9

22. St-Charles, P.L., Bilodeau, G.A., Bergevin, R.: A self-adjusting approach to change detection based on background word consensus. In: 2015 IEEE Winter Conference on Applications of Computer Vision, pp. 990–997 (January 2015)
23. St-Charles, P.L., Bilodeau, G.A., Bergevin, R.: Subsense: a universal change detection method with local adaptive sensitivity. IEEE Trans. Image Process. **24**(1), 359–373 (2015)
24. Stauffer, C., Grimson, W.E.L.: Adaptive background mixture models for real-time tracking. In: 1999 IEEE Computer Society Conference on Computer Vision and Pattern Recognition, vol. 2, pp. 246–252 (1999)
25. Sultana, M., Mahmood, A., Javed, S., Jung, S.K.: Unsupervised deep context prediction for background estimation and foreground segmentation. Mach. Vis. Appl. **30**(3), 375–395 (2018). https://doi.org/10.1007/s00138-018-0993-0
26. Vacavant, A., Chateau, T., Wilhelm, A., Lequièvre: a benchmark dataset for outdoor foreground/background extraction. In: Asian Conference on Computer Vision, pp. 291–300 (2012)
27. Wang, R., Bunyak, F., Seetharaman, G., Palaniappan, K.: Static and moving object detection using flux tensor with split gaussian models. In: Proceedings of the IEEE Conference on Computer Vision and Pattern Recognition Workshops, pp. 414–418 (2014)
28. Wang, Y., Luo, Z., Jodoin, P.M.: Interactive deep learning method for segmenting moving objects. Pattern Recogn. Lett. **96**, 66–75 (2017)
29. Yong, X.: Improved gaussian mixture model in video motion detection. J. Multimedia **8**(5), 527 (2013)
30. Zheng, W., Wang, K., Wang, F.Y.: A novel background subtraction algorithm based on parallel vision and Bayesian GANs. Neurocomputing **394**, 178–200 (2019)
31. Zhou, X., Yang, C., Yu, W.: Moving object detection by detecting contiguous outliers in the low-rank representation. IEEE Trans. Pattern Anal. Mach. Intell. **35**(3), 597–610 (2013)

Tracker Evaluation for Small Object Tracking

Chang Liu[1], Chunlei Liu[1], Linlin Yang[2], and Baochang Zhang[1,3(✉)]

[1] School of Automation Science and Electrical Engineering, Beihang University, Beijing 100191, People's Republic of China
{lc1503,liuchunlei,bczhang}@buaa.edu.cn
[2] University of Bonn, Bonn, Germany
yangl@cs.uni-bonn.de
[3] Shenzhen Academy of Aerospace Technology, Shenzhen, China

Abstract. The small object problem becomes an increasingly important task because of its wide application. There are three significant challenges for small objects: 1) small objects have extremely vague and variable appearances, 2) due to the low resolution of the input images, their characteristic expression information is inadequate and, therefore, is prone to be absent after downsampling and 3) they draft drastically in the images when lens shake violently. Even though small object detection has been extensively studied, small object tracking is still in its infancy. To further explore small object tracking, we evaluate six latest trackers on OTB100 (normal object dataset) and small90 (small object dataset). According to our observation, we draw three instructive conclusions for the follow-up research of small object tracking. Firstly, due to the weak characteristics of small objects, existing trackers perform worse on small objects than on normal objects. Secondly, based on the results of ATOM, SPSTracker, DIMP, SiamFC and SiamMask, the trackers' performance on small objects is positively correlated with that on normal objects. Thirdly, trackers tend to perform better on small object datasets when they can handle drift, occlusion and out-of-view.

Keywords: Small object · Tracking · Evaluation · Feature

1 Introduction

Visual tracking has widespread applications in military, transportation, and other fields. Given one single object's size and position in an initial frame, the goal of visual tracking is to continually predict the object's size and position in subsequent frames. However, in some situations such as tracking in aerial images, objects might be small and difficult to be tracked. The significant challenges for small objects are: 1) small objects have extremely vague and variable appearances, 2) due to the low resolution of the input image, their characteristic expression information is inadequate and, therefore, is prone to be absent after

© Springer Nature Switzerland AG 2021
A. Del Bimbo et al. (Eds.): ICPR 2020 Workshops, LNCS 12662, pp. 622–629, 2021.
https://doi.org/10.1007/978-3-030-68790-8_48

downsample and 3) they drift drastically in the images when lens shake violently. Therefore, in order to explore more characteristic of small objects, we evaluate the performance of several latest trackers on small objects. Note that we define small objects as the objects whose sizes are less than 1% of image size in this paper.

In the field of detection, recent researches on small objects have drawn much attention. Inspired by Faster RCNN, feature pyramid networks (FPN) [9] use different layers of Region of Proposals Network (RPN) to get more information about small objects. In contrast, Multi-scale deconvolutional single shot detector (MDSSD) [10] adopts skip connections to increase the context information, which can enhance the feature maps at different scales. However, as for tracking, small objects are still tricky to handle. Existing tracking methods can be roughly divided into two classes: correlation filter methods and deep learning methods. The correlation filter algorithms, such as KCF [5] and C-COT [4], which have a fast operation speed, are to establish correlation filters to maximize the correlation response with the object. The deep learning methods, which use neural networks to extract features, benefit from their informative feature maps. However, the disadvantages are also obvious. Their training speed is relatively slower than correlation filter methods. The representative trackers of the deep learning method are SiamFC [1], SiamRPN [7], SiamRPN++ [8]. It is natural to raise a question: How the existing trackers perform on small object datasets? Regarding this, we evaluate six recent trackers on the small90 dataset [11]. We summarize our contributions as follows: 1) We evaluate existing well-known trackers on small90 for their performance on small objects. 2) We observe and analyze the performance of each tracker. We find that the performance of each tracker on small90 is lower than that of OTB100, which proves that small object tracking is more difficult than normal object tracking. Furthermore, we find that in most trackers, the performance on small90 is positively correlated with their performance on OTB100, which shows the ability to track small objects is positively related to the ability to track normal objects. In terms of the characteristics of the tracker, we find that if a tracker can solve problems of drift, occlusion and out-of-view well, it tends to perform better on small object datasets.

2   Related Works

In recent years, deep learning-based visual tracking methods progress rapidly. For example, on OTB100 [15], SiamFC [1] gets 0.785 in precision and 0.585 in success score, while the following work SiamRPN++ [8] boosts the precision and success score up to 0.905 and 0.695, respectively. The improvement is up to 10% both in precision score and success score. We classify recent deep learning-based methods into two categories, correlation filter-based methods and siamese-based methods, and introduce them separately.

Correlation filter-based deep learning trackers, such as HCF [12], benefiting from the powerful characterization capabilities of the depth features, directly extract the deep and the shallow features of the VGG network for object tracking. However, the resolution of each channel in HCF is the same size. Instead of using such a single resolution, C-COT [4] proposes a convolution operator in the continuous spatial domain, which can fuse multi-resolution feature maps and therefore improve the performance. Furthermore, to speed up C-COT's tracking speed, Danelljan et al. [4] introduces a Gaussian mixture model, which converts the frame update into components update, increases the diversity of samples, and reduces both the model parameters and the update frequency. Further, ATOM [3] first achieves high performance by using two components, an online object classification module and an offline object estimation module. The object classification module is used for rough positioning, and the object module is used for precise positioning. Based on ATOM tracker, Sub-Peak Response Suppression tracker (SPSTracker) [6] and DIMP [2] are proposed. SPSTracker with Peak Response Pooling (PRP) and Boundary Response Truncation (BRT) aggregates multiple sub-peaks on the tracking response map into a single enforced peak. Meanwhile, to obtain sufficient distinction between background and target, DIMP, an end-to-end tracking architecture for target model prediction, is derived from a discriminative learning loss and solved by a dedicated optimization process.

Unlike correlation filter-based methods, Bertinetto et al. [1] first uses Siamese architectures as the backbone and proposes SiamFC, which has become the dominant tracking framework. Inspired by SiamFC, more efficient tracking methods, such as SiamMask [13], SiamRPN [7] and SiamRPN++ [8] have been developed. Li et al.involves region proposal network into Siamese-based tracking methods. It abandons the traditional multi-scale testing and online tracking, which makes the tracking speed in a high fast. However, most Siamese networks are limited to ResNet-50 as a feature extractor and perform worse with deeper networks. To solve this, SiamRPN++ is proposed to distribute the positive samples evenly within a certain range, which alleviates the impact of destroying the strict translation invariance and can adopt deep networks in the tracking algorithm. On the other hand, with the observation that Siamese networks with only horizontal rectangular anchors fail to handle the object rotation, SiamMask tracker [13], which embeds a segmentation module into the Siamese network, is proposed. It integrates visual object tracking (VOT) with video object segmentation (VOS) and therefore can simultaneously estimate the anchors and the masks. Benefit from the predicted mask, SiamMask outperforms other state-of-the-art methods.

As for the tracker evaluation, the tracking benchmarks and the corresponding metrics have been improved gradually. A milestone OTB50 [14] is first released with 50 videos with full annotations. The proposed benchmark also introduces a novel metric, the area under curve (AUC) of success plot. Subsequently, OTB50 is further expanded to 100 video sequences and renamed as OTB100 [15]. Another famous tracking dataset is VOT, which is a competition that started in 2013. It uses both accuracy and robustness to evaluate the model. Recently, a small object tracking dataset small90 [11] is released. It includes 90 sequences with the same label format and evaluation index as OTB100, and all objects in this dataset with less than 1% of the whole image.

3 Experiments

In this section, we first introduce the datasets we use and then test the trackers introduced in previous section. At last, we analyze the evaluation results and draw conclusions.

3.1 Evaluation Dataset

In this paper, we use OTB100 [15], which contains 100 videos with full annotations and small90 [11], which includes 90 groups of small objects with full annotations.

To rank the object tracking algorithms, we use the precision plot of the central pixel distance curve and the area under curve (AUC) of each success plot. The precision plot shows the percentage of frames whose tracking results are within a certain distance, determined by a given threshold. The success plot shows the ratio of successful frames when the threshold changes from 0 to 1. Here the successful frame indicates that its overlap is greater than the given threshold. Those two metrics portray different aspects of the results.

3.2 Implementation Details

All the experiments are carried out with Pytorch on an Intel i9-9900X 3.5 GHz CPU and a single Nvidia GTX 2080ti GPU with 11 GB memory. All pretrained models used in our experiments are downloaded from their official github.

3.3 Experiments Results

We now conduct experiments on OTB100 and small90 to see the performance of each tracker. The experimental results are shown in Fig. 1, Fig. 2, Fig. 3 and Fig. 4. In Fig. 1 left and Fig. 2 left, we show the precision rate. The precision score of SiamRPN++ and SiamFC are reduced by 7.07% and 2.42% while the precision score of DIMP, ATOM, SPSTracker and SiamMask are increased by 3.36%, 6.19%, 3.92%, 6.67%, respectively. As shown in Fig. 1 right and Fig. 2 right, the success scores of trackers have different degrees of decline from OTB100 to small90. SiamRPN++, SiamFC, DIMP, ATOM, SPSTrackerand SiamMask are reduced by 12.09%, 13.50%, 2.62%, 1.60%, 1.34%, 0.15% respectively. Comparing all results, we observe that SiamRPN++ and SiamFC have the most significant percentage reduction in precision score and success score, indicating that SiamRPN++ and SiamFC have the worst adaptability to small object tracking. In contrast, SiamMask has the best adaptability to small object tracking with the largest percentage increase in precision score and the smallest percentage reduction in success score.

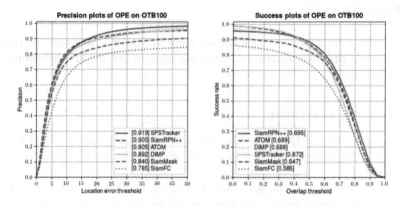

Fig. 1. Precision and success rate plots on OTB100.

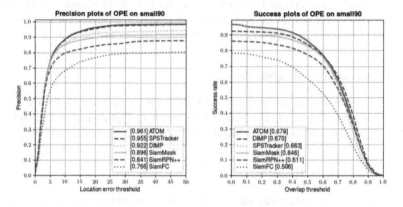

Fig. 2. Precision and success rate plots on small90.

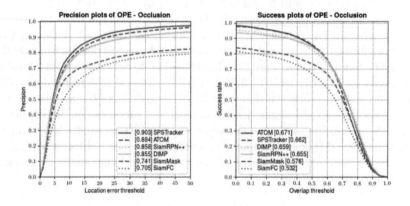

Fig. 3. Precision and success rate plots of occlusion on OTB100.

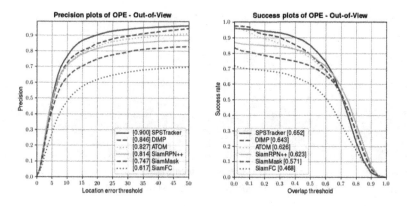

Fig. 4. Precision and success rate plots of out-of-view on OTB100.

3.4 Experiments Analysis

1) Analysis of Success and Precision Rate Plots: Comparing the performance of the six trackers on OTB100 and small90, we find that the rank of all the trackers on precision and success is consistent. Comparing all trackers on OTB100 and small90 longitudinally, we find that all trackers except SiamRPN++ and SiamFC achieve lower success scores but higher precision scores on small90 than OTB100. Since the horizontal coordinate of precision is a fixed pixel value, the relative size of small objects is larger than that of normal objects. In this case, small objects are easy to achieve higher precision than normal objects. But this doesn't mean the trackers work better on small objects. Because the performance mainly depends on the success with the IOU size as the horizontal coordinate. According to the success plots, we observe that all the tracking scores on small90 are lower than OTB100, proving that small object datasets are more difficult to track.

2) Object Drift: SiamRPN++ performs well on OTB100 but unsatisfactory on small90. To figure out the reason for its poor performance on small90, we show the visualization process on its worst performing sequences in Fig. 5. In these sequences, we find that when the object has a large drift, SiamRPN++ will easily lose the object, which results in the poor performance of SiamRPN++ on small90. Due to small objects are more prone to drift when lens are shaken, we need to strengthen the tracker's ability to deal with drift, so that we can improve the tracker's performance on small objects.

3) Occlusion and Out-of-View: Another issue which is worthy considering is the category of the videos. In most categories, SiamRPN++ performs best, but there is a clear underline in occlusion and out-of-view which is shown in Fig. 3 and Fig. 4. Coincidentally, the best performing trackers on these two categories, ATOM, DIMP and SPSTracker, also perform best on small90. Considering the relationship between these two categories and the perspective of characteristics

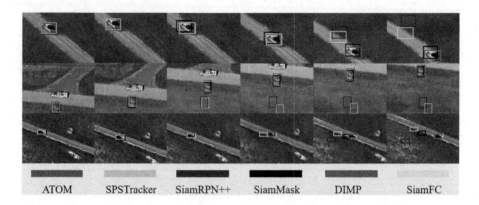

Fig. 5. Tracking visualization when drift occurs.

of small objects, we find that they all represent the situation of weakened characteristics. According to this result, we infer that when an tracker performs better on occlusion and out-of-view, it tends to perform better on small object datasets.

Our Findings: According to our analysis, we summarize our findings below: (1) The score of the small object dataset on success is lower than that of the normal object dataset, which shows that the small object is more difficult to be tracked, and this problem is worthy of being studied in the tracking field. (2) The performance of most trackers on small objects is positively correlated with the performance on normal objects. (3) When lens shake, small objects tend to have greater drift, which makes tracking more difficult. (4) Trackers that perform better on occlusion and out-of-view issues are more likely to perform better on small object dataset.

4   Conclusion

In this paper, we have evaluated six latest trackers on the small90 and OTB100 datasets. We found that the performance of most trackers on small object datasets is positively correlated with its performance on the normal object datasets. In addition, due to the area ratio of the small objects, drift phenomenon of small objects is more likely to occur. If a tracker can better cope with the drift phenomenon, then it will perform better on small object dataset. For the weak characteristics of small objects, we found that if a tracker performs better on occlusion and out-of-view, it could perform better on small object datasets. These findings provide new aspects for follow-up research on small object tracking. Small object tracking is still a significant challenge in computer vision and needs more research to improve the performance.

Acknowledgements. The work is supported by Shenzhen Science and Technology Program KQTD2016112515134654. Baochang Zhang is also with Shenzhen Academy of Aerospace Technology, Shenzhen, China.

References

1. Bertinetto, L., Valmadre, J., Henriques, J.F., Vedaldi, A., Torr, P.H.S.: Fully-convolutional Siamese networks for object tracking. In: Hua, G., Jégou, H. (eds.) ECCV 2016. LNCS, vol. 9914, pp. 850–865. Springer, Cham (2016). https://doi.org/10.1007/978-3-319-48881-3_56

2. Bhat, G., Danelljan, M., Gool, L.V., Timofte, R.: Learning discriminative model prediction for tracking. In: ICCV, pp. 6181–6190 (2019)

3. Danelljan, M., Bhat, G., Khan, F.S., Felsberg, M.: Atom: accurate tracking by overlap maximization. In: CVPR, pp. 4660–4669 (2019)

4. Danelljan, M., Robinson, A., Shahbaz Khan, F., Felsberg, M.: Beyond correlation filters: learning continuous convolution operators for visual tracking. In: Leibe, B., Matas, J., Sebe, N., Welling, M. (eds.) ECCV 2016. LNCS, vol. 9909, pp. 472–488. Springer, Cham (2016). https://doi.org/10.1007/978-3-319-46454-1_29

5. Henriques, J.F., Caseiro, R., Martins, P., Batista, J.: High-speed tracking with kernelized correlation filters. TPAMI 37(3), 583–596 (2014)

6. Hu, Q., Zhou, L., Wang, X., Mao, Y., Zhang, J., Ye, Q.: Spstracker: sub-peak suppression of response map for robust object tracking. arXiv:1912.00597 (2019)

7. Li, B., Yan, J., Wu, W., Zhu, Z., Hu, X.: High performance visual tracking with Siamese region proposal network. In: CVPR, pp. 8971–8980 (2018)

8. Li, B., Wu, W., Wang, Q., Zhang, F., Xing, J., Yan, J.: Siamrpn++: evolution of Siamese visual tracking with very deep networks. In: CVPR, pp. 4282–4291 (2019)

9. Lin, T.Y., Dollár, P., Girshick, R., He, K., Hariharan, B., Belongie, S.: Feature pyramid networks for object detection. In: CVPR, pp. 2117–2125 (2017)

10. Lisha, C., Lv, P., Xiaoheng, J., Zhimin, G., Bing, Z., Mingliang, X., et al.: Mdssd: Multi-scale deconvolutional single shot detector for small objects. SCIENCE CHINA Information Sciences (2018)

11. Liu, C., Ding, W., Yang, J., Murino, V., Zhang, B., Han, J., Guo, G.: Aggregation signature for small object tracking. TIP 29, 1738–1747 (2019)

12. Ma, C., Huang, J., Yang, X., Yang, M.: Hierarchical convolutional features for visual tracking. In: ICCV, pp. 3074–3082 (2015)

13. Wang, Q., Zhang, L., Bertinetto, L., Hu, W., Torr, P.H.: Fast online object tracking and segmentation: a unifying approach. In: CVPR, pp. 1328–1338 (2019)

14. Wu, Y., Lim, J., Yang, M.H.: Online object tracking: a benchmark. In: CVPR, pp. 2411–2418 (2013)

15. Wu, Y., Lim, J., Yang, M.H.: Object tracking benchmark. TPAMI 37(9), 1834–1848 (2015)

DepthOBJ: A Synthetic Dataset for 3D Mesh Model Retrieval

Francesco Carrabino⬤ and Lauro Snidaro$^{(\boxtimes)}$⬤

Department of Mathematics, Computer Science and Physics, University of Udine,
33100 Udine, Italy
lauro.snidaro@uniud.it

Abstract. In this work, we aim to obtain category-based 3D mesh models in .OBJ format through a Deep Neural Network from a single depth map. We introduce DepthOBJ, a synthetic dataset consisting of 3D mesh models divided in 54 categories and 19440 depth maps from 9 different angles in .PNG format. Recognizing the category in which the object depicted in the depth map belongs via a Convolutional Neural Network, we are able to produce the corresponding 3D mesh model in DepthOBJ using PyTorch3D.

Keywords: Dataset · 3D mesh retrieval · Deep learning

1 Introduction

The problem behind object recognition, classification and the consequent retrieval of a 3D shape, has been approached in many different ways and by many different researchers. New technologies and more computational power have ignited, in the last twenty years, the interest in this field and have lead to outstanding results and applications.

Approaches seem to vary between two different factors: the input (video, multi-view images, single-view) and the output type (voxels, point-cloud, mesh). The choice of an input and output format determines the type of approach needed and determines the complexity of the operation. Some of the approaches don't even require AI to obtain a precise 3D shape.

While a wealth of datasets for coloured RGB images are available on the web, very few and small datasets dedicated to depth maps are actually available for researchers and practitioners. Even if the capabilities of RGB images are astounding, the need to explore depth maps, images describing the depth of a scene or an object, is rapidly growing. Moreover, given the fact that this kind of images can be compressed into a single channel, being able to pair them with their corresponding RGB counterparts, opens up to RGB-D images, a powerful yet not much explored data format. Being able to retrieve a mesh model out of a depth map could lead to better quality meshes.

In this study, we focus on the retrieval of a 3D mesh (precisely a mesh in .OBJ format) from a single-view depth map, an image in which the pixels are

© Springer Nature Switzerland AG 2021
A. Del Bimbo et al. (Eds.): ICPR 2020 Workshops, LNCS 12662, pp. 630–643, 2021.
https://doi.org/10.1007/978-3-030-68790-8_49

the description of the distance of the object from the camera, using Deep Neural Networks able to categorize through the categories of our own dataset, called DepthOBJ, a collection of 19440 depth maps divided in 54 categories and in which each of these categories is represented by an .OBJ mesh.

The reason behind choosing to load category-representative 3D mesh models is to use them as a base for future feature-based deformations. The idea is to implement the method described in [25], in which an ellipsoid mesh is deformed extracting feature maps from different layers of the Neural Network. The problem they encountered was that an ellipsoid mesh can be deformed up to a certain point and struggles to achieve much more complex structures, such as holes and details.

Using depth maps' classification, we want to facilitate retrieving the closest mesh possible, so that deformations could be much more precise and effective, resulting in the generation of high-quality final meshes, an approach that could resemble the one proposed in [28]. In combination with ShapeNet, our dataset could represent a useful resource for future studies on mesh generation from RGB-D data.

1.1 Main Contributions

The main contributions of this paper are:

- To our knowledge, the vast majority of studies focus on RGB images. We focus on depth maps, images that can be used also as single-channel (for example, in RGB-D images), that are much more difficult to categorize and utilize, but that are a precise description of the shape in a three-dimensional space and that are free from elements of disturb in the retrieval of the shape, such as textures, colors and heterogeneous light conditions. One of the main limitations of this image format is the amount of data available to researchers and the necessity of a ToF sensor in a real environment: we solve both problems by introducing the new dataset and its related method of creation.
- The proposal of a completely new dataset[1] and its related method of creation and expansion. The dataset's taxonomy is easy to navigate and is made so that the dataset can be easily imported into projects through the use of Data Generators.
- Differentiating from other studies in which the focus is to generate 3D mesh models, we select a pre-existent mesh representative of the category of the image and use that model as a clean base for future deformations and feature-based transformations.
- We employed state-of-the-art Convolutional Neural Network (CNN) architectures and we try to demonstrate their viability in the task of categorization of depth maps, showing that is not compulsory for researchers to build specific architectures and networks.

---

[1] The dataset is available on GitHub at the link: https://github.com/francesco-carrabino/depthOBJ.

1.2 Organization of the Paper

Firstly, we will briefly analyze related works and the most recent researches on 3D object retrieval, depth maps, Deep Neural Networks and output types. We will then describe our new dataset, DepthOBJ, and give insight on the methodology and the criteria behind it. Then, we will describe our implementation of the dataset in Jupyter Notebooks, from which we conducted various experiments, showing the results and coming to some conclusions regarding our dataset and our methodology.

2 Related Works

In the last decade, 3D shape retrieval has been a topic of interest and many researchers have focused their efforts trying to achieve human-like perception of three-dimensional space from a bi-dimensional image with Neural Networks.

2.1 Single-View and Multi-view Approaches

Studies that focused on 3D object retrieval had to start from a deep knowledge of matrix space and surface reconstruction from a single image or multiple images [6,7]. Combined with new technologies and new devices, such as ToF sensors, particularly used thanks to the Microsoft Kinect, studies like KinectFusion [8] achieved great results.

Single-view approaches have become, throughout the years, more and more important for their vast applications and the scalability to multi-view approaches [3,17,25,26].

2.2 Depth Maps

Depth images, or depth maps, have played an important role since the first experiments [12], being a 2D representation of depth, therefore of the Z axis. Working with this kind of images carries some limitations, such as the absence of color and texture information, objects that are reflected or behind glasses are not shown and the loss of some details.

While these limitations are concrete and cannot be taken lightly, depth maps are an incredibly powerful tool in the representation of the shape in a three-dimensional space and, therefore, perfect for tasks such as ours. In fact, some studies have actually focused on the retrieval of depth information from RGB images [9] and have tried a similar approach as ours for completing 3D shapes with depth images [18].

2.3 Deep Neural Networks

Deep Neural Networks and 3D Neural Networks are the main tools behind detection, silhouette understanding and 3D object generation. Without these tools, single-view approaches would not be so viable thanks to the deep pattern recognition and the fast implementation of huge amounts of data. Some studies have underlined the necessity of depth images' specific CNNs [22], while our goal is to ease the use of existing architectures, trained on RGB images, by building and augmenting a dataset made only of depth images.

In this paper, moreover, we experiment with the deepest yet more efficient architectures, by exploiting sparsely-connected networks that are able to reach the same depth of fully-connected networks, while being much lighter and don't require the same amount of space and computational power [23].

2.4 Output Types

The output type in which the 3D object is retrieved, generated or transformed, is one of the key factors in determining the efficiency, the applicability and the quality of the final application. We have identified three main output types that have been used in literature:

- *Voxels:* they are a representation of the pixel in a three-dimensional space. The higher the resolution, the finest the shape. Voxels are a pretty good tool for applications in which the quality of the surface isn't the main focus and there is a need for fast volumetric calculations. While they have a vast application in military and industrial systems, they do not represent the best solution for achieving the finest 3D object.
- *Point-Cloud:* being the lightest and easiest to implement in an application, since the nature of Point-Cloud is a collection of points in a three-dimensional space, this output type has been used in many applications and researches. The main limitation to this format is the absence of a *mesh*, that defines the outer shape of an object by connecting point via faces and edges. The lack of more complex data prevents point-clouds from being reliable in the description of the shape.
- *Mesh:* heavily used in 3D applications and in many industries, meshes are a collection of vertices, edges and faces. This definition highlights the main advantage of this format: each vertex is connected to another in a structure that can vary in resolution, but that maintains the same structure by precisely describing the surface and only the surface.

While voxels and point-clouds have been used as output formats [3,15,27], they have also been used as starting points to create 3D meshes, by utilizing some techniques such as *marching cubes* [16] and the *Poisson reconstruction* [11], also in studies from which our research inherits a similar method and approach with depth maps [18].

Many other studies have utilized meshes as output formats [9,10,25,26], obtaining amazing results. Recently, Facebook has developed a library for 3D data, called PyTorch3D, that provides many useful tools and eases the way of handling meshes. Though this library, we then visualize and prepare for future deformations the category-based meshes located in our dataset.

3 Datasets

Approaching the task of 3D shape retrieval from depth maps, we stumbled across the problem of the narrowness of datasets currently available on the web for this type of images. Most of them are very limited in size and are manually obtained through ToF cameras or are community-based, with people recording and capturing real objects in real environment, which is a very time-consuming task and requires all the contributors to have ToF cameras.

This is why we developed depthOBJ, a synthetic dataset completely generated in a virtual environment with 3D models in .OBJ format.

3.1 ShapeNetCore V2

To collect the most 3D mesh models in .OBJ format, ShapeNetCore V2[2] was the easiest and most organized way we could possibly go. Thanks to its clear and ready-to-use organization, we manually collected 40 models for each of the 54 out of 55 categories and prepared them in folders where we could use our script in Python for batch-rendering, that can be run on Blender, an open-source 3D software that supports scripting. Not only ShapeNetCore was incredibly useful for the collection of models, operation that could have been much more costly in time if we had to manually collect them from the web, but also the models in ShapeNet come normalized and positioned on the origin, so that we could use the same camera for each rendering.

3.2 Methodology

Once set the folders in a simple taxonomy—basically a main folder containing a sub-folder for each category of objects—to render out the dataset, we used a script for batch-rendering, that renders all the models in the main folder with the selected angle of the camera. The script reads all the sub-folders in a set folder, looks for all the .OBJ models in them and renders them. In Blender, we made so that the rendering would only output the depth channel and we didn't include any type of background to make sure that the object, and only the object, would have been rendered. The reason why the object is not put in a context is twofold: on one hand, we just wanted to focus on the object, on the other hand, this could come as an helpful tool to extract a silhouette in this kind of operations. The images already come with normalized values, so that the data can be just loaded and is ready to be fed to Neural Networks.

3.3 Criteria of Choice

The criteria of choice in the selection of the models followed these principles: variety, homogeneity, shape; while the criteria behind the selection of the representative mesh followed: generality, size and topology.

---

[2] ShapeNet website: https://www.shapenet.org/.

Fig. 1. A depth map example for each of the 54 categories in DepthOBJ.

Noticeable in the dataset is the taxonomy, which is built to be readable by ImageDataGenerator, a real-time data augmentation function of Keras, but also very easily implementable by other Data Loaders or Generators available in other frameworks such as PyTorch.

3.4   Evaluation of the Dataset

Before proceeding, we believe it is necessary to make a deep evaluation of the quality of the dataset, of its limitations, how its nature will influence the performances of the CNNs' architectures and how the dataset could be optimized to produce the best results.

(a) Dishwasher   (b) Washer

Fig. 2. A couple of depth maps belonging to two completely different categories, yet very similar in shape and details.

As we can see from the images in Fig. 2, quite a few categories from the dataset represent objects very much different in use and category, yet very similar in shape and details. That means that the possibility of error during the prediction between the two categories is very high and that the overall accuracy of the model could be disrupted by these similarities. The solution to this kind of problem could be to group some of the most similar categories into super-categories, yet in this case having one representative mesh for a super-category would limit the ultimate goal of this paper, which is to lessen the need of deformation (by nature imprecise and approximate) by using as base the closest representative mesh.

4 Proposed Method

We created Jupyter Notebooks on Google Colaboratory [1], one for the Model and one for the Visualizer with the integration of PyTorch3D. We used Tensor-Flow2 and Keras to load the data, augment it and load the CNNs' architectures for the experiments.

4.1 Real-Time Augmentation

Data Generators remove the need to build huge datasets, basically allowing users to work with smaller amounts of data, and enhance the generalization's capability of the model. For our system, since our depth maps are pretty different already and are seen from many angles, we haven't applied drastic modifications since we already worked with non-fixed angles. We applied small rotations, horizontal flips, width and height shifts. Working on Colaboratory, we applied a batch size of 16 and worked with a 80:20 training/validation split.

4.2 Model

We evaluated the performances of various state-of-the-art networks on our dataset.

The instantiated model comes with pre-trained weights from ImageNet [4].

We used different input shapes for different architecture, generally between 224 × 224 and 299 × 299 pixels. Each model was then compiled, utilizing the *categorical_crossentropy* loss function and the Adam optimizer with learning rate of 0.0002. As performance metric, we utilized only the *accuracy* parameter for the purposes of our study.

4.3 Training

Instantiated and compiled the model, we then proceeded to train it on the augmented data generated by the ImageDataGenerator. Models have been trained for 30 epochs, using a batch-size of 16 and training for about 8 h total per model on Google Colaboratory.

4.4 Visualizer

In the Visualizer Notebook, we use the model trained in the previous Notebook to make predictions on new data, that the model has never seen. From that prediction, we then take the relative 3D mesh in the dataset and load it into the system with PyTorch3D functions. Predicted, selected and loaded the category-based mesh, we then use the structure of the Meshes class of PyTorch3D to only select the vertices and visualize them as a point-cloud.

One of the best features offered by PyTorch3D is the ease of handling meshes (in .OBJ or .PLY format), through the introduction of a Meshes class, from which it is possible to instantiate an object with the following attributes: a list of vertices, a list of faces and a list of auxiliary data. In fact, the *load_obj* function is called as assignment onto three variables, usually named *verts, faces, aux*, that are then used as parameters for the Meshes constructor. Thanks to this object, we now have full control of the mesh translated as lists of coordinates and data, that we can use in many different ways. PyTorch3D, moreover, implements the most important loss functions to apply during the deformation of the mesh, such as the Laplacian Smoothing [5] or the Chamfer Loss.

5 Experiments

In this section, we analyze the performances of various architectures on our dataset. In addition to comparing the results with previous depth maps classification works, we also performed an extensive evaluation of the number of parameters and the structure of the Networks. The metrics we focused on are the *val_loss* and the *val_accuracy* of the model's evaluation. The presented results show how our dataset can be used to perform state-of-the-art classification on depth maps.

5.1 Architectures

We firstly tested the dataset on a simple fully-connected network that we built, for an initial evaluation of the performances and to have some kind of metrics to compare to the state-of-the-art architectures. The basic CNN was composed of 4 blocks of Convolution 3×3, with successively 32, 64, 128 and 256 filters and a Pooling 2×2 layer, one Flatten layer, succeeded by a Dense layer made of 512 nodes, a Dropout set to 50% and the output Dense layer with 54 outputs, which are the total categories of the dataset. For each layer we used the ReLu activation function, as optimizer we used ADAM with the standard learning-rate and a *softmax* activation function for the output layer. The total number of trainable parameters were 29.907.254. Trained for 30 epochs, the results of the evaluation on the validation data were:

- *val_loss:* 1.8789
- *val_accuracy:* 0.6601

ResNet 50 V2. Switching to more complex and deeper networks, we first experimented with a ResNet50 V2, architecture divided in 5 phases in which each one contains a convolutional block and an identity block for the *skip-connections*. Each block, both convolutional and of identity, is formed by 3 Convolutional layers.

ResNet50 V2, on the images of ImageNet [4] has an accuracy of 75.6% and has a total of trainable parameters of 25.636.712. On our dataset, it has been trained for 30 epochs with 224×224 images with pre-trained weights and the results of the evaluation on the validation data were:

- *val_loss:* 0.1593
- *val_accuracy:* 0.9511

These results already outperform state-of-the-art studies, and highlight the potentialities of the dataset. The network has no problem classifying depth maps, but better results can be sought. These results, moreover, suggest that to an increasing number of parameters the *val_loss* value tends to decrease, but that at an increasing depth of the network the *val_accuracy* can get better.

Considering the results on ImageNet and our dataset, ResNet50 V2 can be considered viable for the goal of classifying depth maps.

ResNet 152 V2. ResNet 152 V2 is an extension of the ResNet50 and it is one of the deepest ResNet architectures, reaching 152 layers, keeping the same ResNet structure and increasing only the Convolutional and Identity blocks. ResNet 152 V2, on the images of ImageNet [4] has an accuracy of 77.8% and has a total of trainable parameters of 60.380.648. On our dataset, it has been trained for 30 epochs with 224×224 images with pre-trained weights and the results of the evaluation on the validation data were:

- *val_loss:* 0.1635
- *val_accuracy:* 0.9516

These results seem to prove that at an increasing numbers of parameters the *val_loss* decreases, with pre-trained weights on ImageNet. To the light of this results and after the evaluation of the dataset, we could come to the conclusion that there is no need of an excessive number of parameters, because the depth maps don't bring that much information as RGB images could, and that ResNet 152 V2 doesn't perform better than it's smaller version.

EfficientNet-B7. At the moment in which this paper is being written, EfficientNet-B7 has the best results in the classification of the images in ImageNet, with a stunning 84.4% of accuracy. EfficientNet [24] is a CNN architecture that is able to resize all three dimensions of a Neural Network, which are depth, width and resolution. Through this resize, EfficientNet is able to reach higher depths and resolutions, while keeping the computational request on the low. EfficientNet-B7 is able to work with images up to 600×600 and reaches 66 million trainable parameters. Even if the input resolution of this network corresponds to the actual resolution of the depth maps in our dataset, to keep reasonable training times and save computational power, we trained the model with 331×331 depth maps for 30 epochs, changing the batch_size to 8, with pre-trained weights on ImageNet. The results of the evaluation on the validation data were:

- *val_loss:* 0.1774
- *val_accuracy:* 0.9537

Again the accuracy didn't quite benefit from an excessive number of trainable parameters. EfficientNet-B7 could be re-evaluated with a more adequate hardware, to fully test it to its potential.

EfficientNet-B0. Considering the results of the B7 version and proved that an excessive number of trainable parameters lower the performances of the system, we have tested also EfficientNet-B0, architecture with much less trainable parameters (5.3 million) and that on ImageNet has a 76.3% of accuracy. On our dataset, it has been trained for 30 epochs with 224×224 images, with pre-trained weights on ImageNet, and the results of the evaluation on the validation data were:

- *val_loss:* 0.1700
- *val_accuracy:* 0.9475

These results disprove that the number of trainable parameters are of matters just in some architectures and at different depths, making some kind of architectures more suitable than others to certain kind of data. In fact, meanwhile on ImageNet the EfficientNet-B0's accuracy isn't one of the best, it performs excellently on our dataset.

Inception ResNet V2. Making a case for sparsely-connected Neural Network, Google presented in 2010 Inception V1, introducing the Inception layer, from which the architecture takes the name. This layer is a combination of three different Convolutional layers 1×1, 3×3 and 5×5, in which the output is the concatenation of each Convolutional layer's output. Therefore, the Inception layer allows the network to use filters of the right dimension every time.

Inception ResNet V2 is an architecture which combines the theory behind Inception and ResNet, using three different Inception layers and shortcuts for *skipping-connections* and reaching 50.6 million parameters. On our dataset, Inception ResNet V2 has been trained for 30 epochs with 299×299 images, with pre-trained weights on ImageNet, and the results of the evaluation on the validation data were:

- *val_loss:* 0.1084
- *val_accuracy:* 0.9673

Inception has a 80.3% accuracy on ImageNet. These results show that through skip-connections and adequate filter sizes, the number of parameters do not affect the accuracy and demonstrate the final goal of our paper, proving that these kind of architectures can be used on depth maps.

Table 1. Architectures and performances.

Architectures	Resolution	Param. (M)	val_loss	val_accuracy
Generic CNN	256×256	29.9	1.8789	0.6601
ResNet 50 V2	224×224	25.6	0.1593	0.9511
ResNet 152 V2	224×224	60.6	0.1635	0.9516
EfficientNet-B7	331×331	66	0.1774	0.9537
EfficientNet-B0	224×224	5.3	0.1700	0.9475
Inception ResNet	299×299	50.6	**0.1084**	**0.9673**

5.2 Results

In comparison to other works that focus on the classification of RGB-D images, taking specifically the results on the depth maps, we show how the experiments have produced state-of-the-art performances. Working on our dataset produces significantly better results than other works on datasets such as 2D3D[3], RGB-D Object [13], SUN RGB-D [21] and NYU Depth [20].

---

[3] 2D3D Dataset repository: https://github.com/alexsax/2D-3D-Semantics.

Table 2. Comparison between relevant works.

Researches	Dataset	Depth accuracy
Song et al. [22] (2017)	SUN RGB-D	0.658
Li et al. [14] (2019)	SUN RGB-D	0.562
	NYU Depth	0.677
Cai et al. [2] (2019)	NYU Depth & SUN RGB-D	0.682
Shao et al. [19] (2017)	RGB-D Object	0.755
	NYU Depth	0.565
	2D3D	**0.81**

6   Conclusions

In applications where an accurate description of surface and depth is necessary, depth maps are proven to be an important resource, since high-quality ToF sensors and millions of 3D models are available and ready to use. Retrieving a three-dimensional object from a bi-dimensional input is an argument of shapes versus colors, geometry versus detail.

In this paper, we show that a dataset entirely produced in a virtual environment, through the process of rendering of 3D mesh models, produces state-of-the-art results in combination with well-engineered Neural Networks. We further show that through the implementation of PyTorch3D, 3D mesh models can be imported and easily transformed.

Future developments could introduce the development of a specific CNN, built for the only purpose of depth maps classification. To better implement depth maps in most modern systems, using depth maps as depth channels in RGB-D images could be a viable expansion. Last but not least, implementing feature-based deformation to produce unique 3D mesh models is the natural extension of this research.

References

1. Bisong, E.: Google colaboratory. Building Machine Learning and Deep Learning Models on Google Cloud Platform, pp. 59–64. Apress, Berkeley, CA (2019). https://doi.org/10.1007/978-1-4842-4470-8_7
2. Cai, Z., Shao, L.: RGB-D scene classification via multi-modal feature learning. Cogn. Comput. **11**(6), 825–840 (2018). https://doi.org/10.1007/s12559-018-9580-y
3. Choy, C.B., Xu, D., Gwak, J.Y., Chen, K., Savarese, S.: 3D-R2N2: a unified approach for single and multi-view 3D object reconstruction. In: Leibe, B., Matas, J., Sebe, N., Welling, M. (eds.) ECCV 2016. LNCS, vol. 9912, pp. 628–644. Springer, Cham (2016). https://doi.org/10.1007/978-3-319-46484-8_38
4. Deng, J., Dong, W., Socher, R., Li, L.J., Li, K., Fei-Fei, L.: ImageNet: a large-scale hierarchical image database. In: 2009 IEEE Conference on Computer Vision and Pattern Recognition, pp. 248–255. IEEE (2009)

5. Field, D.A.: Laplacian smoothing and Delaunay triangulations. Commun. Appl. Numer. Methods **4**(6), 709–712 (1988)
6. Hartley, R., Zisserman, A.: Multiple View Geometry in Computer Vision. Cambridge University Press, Cambridge (2003)
7. Hoiem, D., Efros, A.A., Hebert, M.: Recovering surface layout from an image. Int. J. Comput. Vis. **75**(1), 151–172 (2007). https://doi.org/10.1007/s11263-006-0031-y
8. Izadi, S., et al.: KinectFusion: real-time 3D reconstruction and interaction using a moving depth camera. In: Proceedings of the 24th Annual ACM Symposium on User Interface Software and Technology, pp. 559–568 (2011)
9. Kaneko, M., Sakurada, K., Aizawa, K.: TriDepth: triangular patch-based deep depth prediction. In: Proceedings of the IEEE International Conference on Computer Vision Workshops (2019)
10. Kato, H., Ushiku, Y., Harada, T.: Neural 3D mesh renderer. In: Proceedings of the IEEE Conference on Computer Vision and Pattern Recognition, pp. 3907–3916 (2018)
11. Kazhdan, M., Bolitho, M., Hoppe, H.: Poisson surface reconstruction. In: Proceedings of the Fourth Eurographics Symposium on Geometry Processing, vol. 7 (2006)
12. Konrad, J., Wang, M., Ishwar, P.: 2D -to-3D image conversion by learning depth from examples. In: 2012 IEEE Computer Society Conference on Computer Vision and Pattern Recognition Workshops, pp. 16–22. IEEE (2012)
13. Lai, K., Bo, L., Ren, X., Fox, D.: A large-scale hierarchical multi-view RGB-D object dataset. In: 2011 IEEE International Conference on Robotics and Automation, pp. 1817–1824. IEEE (2011)
14. Li, Y., Zhang, Z., Cheng, Y., Wang, L., Tan, T.: MAPNet: multi-modal attentive pooling network for RGB-D indoor scene classification. Pattern Recogn. **90**, 436–449 (2019)
15. Lin, C.H., Kong, C., Lucey, S.: Learning efficient Point Cloud generation for dense 3D object reconstruction. In: AAAI (2018)
16. Lorensen, W.E., Cline, H.E.: Marching cubes: a high resolution 3D surface construction algorithm. ACM siggraph Comput. Graph. **21**(4), 163–169 (1987)
17. Pontes, J.K., Kong, C., Sridharan, S., Lucey, S., Eriksson, A., Fookes, C.: Image2Mesh: a learning framework for single image 3D reconstruction. In: Jawahar, C.V., Li, H., Mori, G., Schindler, K. (eds.) ACCV 2018. LNCS, vol. 11361, pp. 365–381. Springer, Cham (2019). https://doi.org/10.1007/978-3-030-20887-5_23
18. Rock, J., Gupta, T., Thorsen, J., Gwak, J., Shin, D., Hoiem, D.: Completing 3D object shape from one depth image. In: Proceedings of the IEEE Conference on Computer Vision and Pattern Recognition, pp. 2484–2493 (2015)
19. Shao, L., Cai, Z., Liu, L., Lu, K.: Performance evaluation of deep feature learning for RGB-D image/video classification. Inf. Sci. **385**, 266–283 (2017)
20. Silberman, N., Hoiem, D., Kohli, P., Fergus, R.: Indoor segmentation and support inference from RGBD images. In: Fitzgibbon, A., Lazebnik, S., Perona, P., Sato, Y., Schmid, C. (eds.) ECCV 2012. LNCS, vol. 7576, pp. 746–760. Springer, Heidelberg (2012). https://doi.org/10.1007/978-3-642-33715-4_54
21. Song, S., Lichtenberg, S.P., Xiao, J.: Sun RGB-D: a RGB-D scene understanding benchmark suite. In: Proceedings of the IEEE Conference on Computer Vision and Pattern Recognition, pp. 567–576 (2015)
22. Song, X., Herranz, L., Jiang, S.: Depth CNNs for RGB-D scene recognition: learning from scratch better than transferring from RGB-CNNs. In: Proceedings of the Thirty-First AAAI Conference on Artificial Intelligence, pp. 4271–4277 (2017)

23. Szegedy, C., et al.: Going deeper with convolutions. In: Proceedings of the IEEE Conference on Computer Vision and Pattern Recognition, pp. 1–9 (2015)
24. Tan, M., Le, Q.V.: EfficientNet: rethinking model scaling for Convolutional Neural Networks. arXiv preprint arXiv:1905.11946 (2019)
25. Wang, N., Zhang, Y., Li, Z., Fu, Y., Liu, W., Jiang, Y.G.: Pixel2Mesh: generating 3D mesh models from single RGB images. In: Proceedings of the European Conference on Computer Vision (ECCV), pp. 52–67 (2018)
26. Wen, C., Zhang, Y., Li, Z., Fu, Y.: Pixel2Mesh++: multi-view 3D mesh generation via deformation. In: Proceedings of the IEEE International Conference on Computer Vision, pp. 1042–1051 (2019)
27. Wu, Z., et al.: 3D ShapeNets: a deep representation for volumetric shapes. In: Proceedings of the IEEE Conference on Computer Vision and Pattern Recognition, pp. 1912–1920 (2015)
28. Xu, Q., Wang, W., Ceylan, D., Mech, R., Neumann, U.: DISN: deep implicit surface network for high-quality single-view 3D reconstruction. In: Advances in Neural Information Processing Systems, pp. 492–502 (2019)

GFTE: Graph-Based Financial Table Extraction

Yiren Li[1(✉)] , Zheng Huang[1], Junchi Yan[1], Yi Zhou[1], Fan Ye[2],
and Xianhui Liu[2]

[1] School of Electronic Information and Electrical Engineering, Shanghai Jiao Tong
University, Shanghai, China
{irene716,huang-zheng,yanjunch,izy_21th}@sjtu.edu.cn
[2] China Financial Fraud Research Center, Shanghai, China
yefan@xnai.edu.cn, lxianhui@tjzxcn.com

Abstract. Tabular data is a crucial form of information expression, which can organize data in a standard structure for easy information retrieval and comparison. However, in financial industry and many other fields, tables are often disclosed in unstructured digital files, e.g. Portable Document Format (PDF) and images, which are difficult to be extracted directly. In this paper, to facilitate deep learning based table extraction from unstructured digital files, we publish a standard Chinese dataset named FinTab, which contains more than 1,600 financial tables of diverse kinds and their corresponding structure representation in JSON. In addition, we propose a novel graph-based convolutional neural network model named GFTE as a baseline for future comparison. GFTE integrates image feature, position feature and textual feature together for precise edge prediction and reaches overall good results https://github.com/Irene323/GFTE.

Keywords: Deep learning · Document analysis · Document image processing

1 Introduction

In the information age, how to quickly obtain information and extract key information from massive and complex resources has become an important issue [1]. In the meantime, with the increase in the number of enterprises and the growing amount of disclosure of financial information, extracting key information has also become an essential means to improve the efficiency in financial information exchange process. In recent years, some new researches have also begun to focus on improving the efficiency and accuracy of information retrieval technology [2,3].

Table, as a form of structured data, is both simple and standardized. Hurst et al. [4] regard a table as a representation of a set of relations between organized hierarchical concepts or categories, while Long et al. [5] consider it as a superstructure imposed on a character-level grid. Due to its clear structure, table data

A. Del Bimbo et al. (Eds.): ICPR 2020 Workshops, LNCS 12662, pp. 644–658, 2021.
https://doi.org/10.1007/978-3-030-68790-8_50

can be quickly understood by users. Financial data, especially digital information, are often presented in tabular form. In a manner of speaking, table data, as key information in financial data, are increasingly valued by financial workers during financial data processing.

Table data in finance context have entirely different characteristics from ordinary table data in daily life or in academia:

1. The application of table data is widespread.
2. Complex table structure is difficult to be extracted.
   Tables in financial data have various sources and forms. Thus, the structure of financial table can be rather complicated. For example:
   - Some financial documents use various tables without complete table lines for elegant typography.
   - Many cells in financial table are merged to indicate values in different categories or in different stages.
   - There are often a lot of information in the cells of financial tables. For example, a cell may contain a large number that consists of multiple digits, or it may have many digits after the decimal point.
   - There are also cases when one cell holds a lot of text information. This may lead to the division of one single cell into two pages, especially when it is located at the end of one page.
   - One financial table may store hundreds of thousands of cells, which will occupy multiple consecutive pages.
   The results extracted from these complex tables often have data confusion or overlap.
3. Financial data demand high quality and accuracy.

Although table extraction is a common task in various domains, extracting tabular information manually is often a tedious and time-consuming process. We thus require automatic table extraction methods to avoid manual involvement. However, it is still difficult for the existing methods to accurately recover the structure of relatively complicated financial tables.

Figure 1 illustrates an intuitive example of the performance of different existing methods, i.e. Adobe Acrobat DC and Tabby [6]. Both of them fail to give the correct result. Meanwhile, it is not hard to notice that problems often occur at spanning cells, which very likely carry the information of table headers and are thus critical for table extraction and understanding. Therefore, the performance of table extraction methods are still hoped to be improved, especially by the complicated cases.

Based on these considerations, since the design of artificial intelligence algorithms relies on standard data and test benchmarks, we construct an open source financial benchmark dataset named FinTab. More specifically, sample collection, sample sorting and cleaning, benchmark data determining and baseline method test were completed. FinTab can be further used in financial context in terms of table extraction, key information extraction, image data identification, bill identification and other specific content. With a more comprehensive benchmark dataset, we hope to promote the emergence of more innovative technologies.

Year 2002	Return on equity		Earnings per share (yuan)	
	Fully diluted	Weighted average	Fully diluted	Weighted average
Income from main operation*	264%	287%	3.15	3.15

(a) The ground truth structure of part of a financial table. Each cell in the first two lines is marked with different colors.

Return on equity Earnings per share (yuan)				
Year 2002	Fully diluted	Weighted average	Fully diluted	Weighted average
Income from main operation*	264%	287%	3.15	3.15

(b) The recognized structure by Adobe Acrobat DC.

	Return on equity		Earnings per share (yuan)	
Year 2002				
	Fully diluted	Weighted average	Fully diluted	Weighted average
Income from main operation*	264%	287%	3.15	3.15

(c) The recognized structure using Tabby [6].

Fig. 1. An example of a table with spanned cells and the recovered table structures with the existing methods.

Further detailed information about our standard financial dataset will be introduced in Sect. 3.

Besides, this paper also proposes a novel table extraction method, named GFTE, with the help of graph convolutional network (GCN). GFTE can be used as a baseline, which regards the task of table structure recognition as an edge prediction problem based on graph. More specifically, we integrate image feature, textual feature and position feature together and feed them to a GCN to predict the relation between two nodes. Details about this baseline algorithm will be discussed in Sect. 4.

In general, the contributions of this work can be summarized as following:

1. A Chinese benchmark dataset FinTab of more than 1,600 tables of various difficulties, containing table location, structure identification and table interpretation information.
2. We propose a graph-based convolutional network model named GFTE as table extraction baseline. Extensive experiments demonstrate that our proposed model outperforms state-of-the-art baselines greatly.

2   Related Work

In this section, we will first familiarize the reader with some previous published datasets and some related contests, and then present a overview of table extraction technologies.

2.1 Previous Datasets

We introduce some existing public available datasets:

- The Marmot dataset [7] is composed of both Chinese and English pages. The Chinese pages are collected from over 120 e-Books in diverse fields of subject provided by Founder Apabi library, while the English pages are from Citeseer website. Derived from PDF, the dataset stores tree structure of all document layouts, where the leaves are characters, images and paths, and the root is the whole page. Internal nodes include textlines, paragraphs, tables etc.
- The UW3 dataset [8] is collected from 1,600 pages of skew-corrected English document and 120 of them contain at least one marked table zone. The UNLV dataset derives from 2,889 pages of scanned document images, in which 427 images include table.
- The ICDAR 2013 dataset [9] includes a total of 150 tables: 75 tables in 27 excerpts from the EU and 75 tables in 40 excerpts from the US Government, i.e. in total 67 PDF documents with 238 pages in English.
- This dataset for the ICDAR 2019 Competition on Table Detection and Recognition [10] is separated into training part and test part. The training dataset contains images of 600 modern documents and their bounding boxes of table region, as well as images of 600 archival documents, their table structures and bounding boxes of both table region and cell region. In the test dataset, images and table regions of 199 archival documents and 240 modern ones are offered. Besides, table structures and cell regions of 350 archival documents are also included.
- The PubTabNet dataset [11] contains more than 568 thousand images of tabular data annotated with the corresponding HTML representation of the tables. More specifically, table structure and characters are offered but the bounding boxes are missing.
- The SciTSR dataset [12] is a comprehensive dataset, which consists of 15,000 tables in PDF format, images of the table region, their corresponding structure labels and bounding boxes of each cell. It is split into 1,2000 for training and 3,000 for test. Meanwhile, a list of complicated tables, called SciTSR-COMP, is also provided.
- The TableBank [13] is an image-based table detection and recognition dataset. Since two tasks are involved, it is composed of two parts. For the table detection task, images of the pages and bounding boxes of tables region are included. For the table structure recognition task, images of the page and HTML tag sequence that represents the arrangement of rows and columns as well as the type of table cells are provided. However, textual content recognition is not the focus of this work, so textual content and its bounding boxes are not contained.

Table 1 gives more information for comparison.

2.2 Methods

Table extraction is considered as a part of table understanding [14], and conventionally consists of two steps [6]:

Table 1. Public datasets for table recognition

Name	Source	Content	Amount	Language
Marmot	e-Books and Citeseer website	Tree structure of all document layouts; bmp, xml	2,000	Chinese, English
UW3	Skew-corrected document	Images of page, manually edited bounding boxes of page frame, text and non-text zones, textlines, and words, type of each zone; png, xml	120	English
UNLV	Magazines, news papers, business letter, annual Report, etc.	Scanned images of page, bounding boxes for rows, columns and cells, OCR recognized words within the whole page; png, xml	427	English
ICDAR 2013	European Union and US Government websites	PDF documents, bounding boxes of table, textual content, structure labels, bounding boxes for each cell; pdf, xml	150	English
ICDAR 2019	Modern documents and archival ones with various formats	840 jpgs and xmls including bounding boxes of table in modern documents, 1149 jpgs and xmls including structure labels, bounding boxes of table and each cell in archival documents	About 2,000	English
PubTabNet	Scientific publications included in the PubMed Central Open Access Subset	Images of tabular data, textual content, table structure labels in HTML; png, json	More than 568,000	English
SciTSR	LaTeX source files	PDF, images, textual content, structure labels, bounding boxes for each cell	15,000	English
TableBank	Word and Latex documents on the internet such as official fillings, research papers, etc.	Table Detection: bounding boxes of table; Table Structure Recognition: table structure labels; jpg, HTML	417,234; 145,463	English
FinTab	Annual and semi-annual reports, debt financing, bond financing, collection of medium-term notes, short-term financing, prospectus	Textual ground truth, structure information and the unit of the table; pdf, json	1,685	Chinese

1. Table Detection. Namely, a certain part of the file is identified as a table in this step.
2. Table Structure Decomposition. This task aims to recover the table into its components as close to the original as possible. For example, the proper identification of header elements, the structure of columns and rows, correct allocation of data units, etc.

In the past two decades, a few methods and tools have been devised for table extraction. Some of them are discussed and compared in some recent surveys [15, 16].

There are generally three main categories in the existing approaches [17]: Predefined layout-based approaches, Heuristic-based approaches and Statistical or optimization-based approaches.

Predefined layout-based approaches design several templates for possible table structures. Certain parts of the document is identified as tables, if they correspond to certain templates. Shamilian [18] proposes a predefined layout-based table identification and segmentation algorithm as well as a graphical user interface (GUI) for defining new layouts. Nevertheless, it only works well in single-column cases. Mohemad et al. [19] present another predefined layout-based approach, which focuses on paragraph and tabular, then associated text using a combination of heuristic, rule-based and predefined indicators. However, a disadvantage of these approaches is that tables can only be classified into the previous defined layouts, while there are always limited types of templates defined in advance.

Heuristic-based approaches specify a set of rules to make decisions so as to detect tables which meet certain criteria. According to [16], heuristics-based approaches remain dominant in literature. [20] is the first relative research focusing on PDF table extraction, which uses a tool named pdf2hmtl to return text pieces and their absolute coordinates, and then utilizes them for table detection and decomposition. This technique achieves good results for lucid tables, but it is limited as it assumes all pages to be single column. Liu et al. [21] propose a set of medium-independent table metadata to facilitate the table indexing, searching, and exchanging, in order to extract the contents of tables and their metadata.

Statistical approaches make use of statistical measures obtained through offline training. The estimated parameters are then taken for practical table extraction. Different statistical models have been used, for example, probabilistic modelling [22], the Naive Bayes classifier [23, 24], decision trees [25, 26], Support Vector Machine [25, 27], Conditional Random Fields [27–29], graph neural network [12, 30, 31], attention module [32], etc. [33] uses a pair of deep learning models (Split and Merge models) to recover tables from images.

3   Dataset Collection

In general, there are currently following problems with the existing contests and standard datasets:

1. There are few competitions and standard datasets for extracting table information from financial documents.
2. The source for tabular information extraction lacks diversity.

In consideration of this, the benchmark dataset FinTab released this time
aims to make certain contribution in this field. In this dataset, we collect a total
of 19 PDF files with more than 1,600 tables. The specific document classification
is shown in Table 2. All documents add up to 3,329 pages, while 2,522 of them
contain tables.

Table 2. Document classification of our benchmark dataset FinTab

File type	Number of files
Annual and semi-annual reports	3
Debt financing	2
Bond financing	3
Collection of medium-term notes	2
Short-term financing	8
Prospectus	1

FinTab provides more comprehensive details of the table than any other
datasets introduced in Sect. 2. It is also worth noticing that FinTab has been
manually reviewed, which makes it much more reliable. We provide both characters and strings as textual ground truth. For structure ground truth of a table,
we present the detailed information of its cells and its table lines. More specifically, different kinds of ground truth of a table are stored in json files as shown
in Table 3.

Table 3. Ground truth provided in FinTab

Information type	Ground truth included
Table position	4 coordinates of the bounding box
Table line	2 coordinates of the start point and the end point, line color, fill color
Character	Textual content, font size, font name, font color, 4 coordinates of the bounding box
String	Textual content, font size, font color, 4 coordinates of the bounding box
Cell	Structure position in the table, 4 coordinates of the bounding box

To ensure that the types of forms are diverse, in addition to the basic forms of table, special cases with different difficulties are also included, e.g. semi-ruled table, cross-page table, table with merged cells, multi-line header table, etc. It is also worth mentioning that there are 119,021 cells in total, while the number of merged cells is 2,859, accounting for 2.4%. Detailed types and quantity distribution of tables are shown in Table 4.

Table 4. Document classification of our benchmark dataset FinTab

Table type	Number of tables	Percentage
Single-page table	523	62.5%
Double-page table	523	31%
Multi-page table	108	6.5%
Table with incomplete form line	140–150	8.3%-8.9%
Table with merged cells	583	34.59%
Total	1,685	100%

FinTab contains various types of tables. Here, we briefly introduce some of them in order of difficulty.

1. Basic single-page table. This is the most basic type of table, which takes up less than one page and does not include merged cells. It is worth mentioning that we offer not only textual ground truth and structure information, but also the unit of the table, because mostly financial table contains quite a few numbers.
2. Table with merged cells. In this case, the corresponding merged cells should be recovered.
3. Cross-page table. If the table appears to spread across pages, the cross-page table need to be merged into a single form. If the header of the two pages appears to be duplicated, only one needs to be remained. Page number and other useless information should also be removed. Another difficult situation to be noticed is that if a single cell is separated by two pages, it should be merged into one according to its semantics.
4. Table with incomplete form line. In this case, it is necessary to intelligently locate the dividing line according to the position, format, and meaning of the text.

4   Baseline Algorithm

In this paper, we also propose a novel graph-neural-network-based algorithm named GFTE to fulfill table structure recognition task, which can be used as a baseline. In this section, we introduce detailed procedure of this algorithm.

Figure 2 illustrates an overview of GFTE. Since our dataset is in Chinese, we give a translated version of the example in Table 5 for better understanding. To first train our model, the following steps are carried out on the training dataset:

a. Given a certain table, we load its ground truth, which consists of (1) image of the table region, (2) textual content, (3) text position and (4) structure labels.

b. Then, based on the ground truth (1)-(3) we construct an undirected graph $G = < V, R_C >$ on the cells.

c. After that, we use our GCN-based algorithm to predict adjacent relations, including both vertical relations (namely whether two nodes are in the same row) and horizontal relations (namely whether two nodes are in the same column).

d. By comparing the prediction with ground truth (4), i.e. the structure labels, we can calculate the loss and optimize the model.

After the model is trained to a satisfactory level, given an image of a certain table, we should be able to recognize the strings and their position in the image. Then, our GFTE model would predict the relationship between these strings and finally recover the structure of the table.

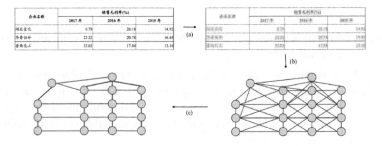

Fig. 2. Overview of our novel GCN-based algorithm.

Table 5. The translated version of the table we used for illustration in this paper.

Company name	Gross profit margin(%)		
	Year 2017	Year 2016	Year 2015
Hubei Yihua	0.79	20.18	14.92
Hualu Hengsheng	22.22	20.78	16.45
Luxi Chemical	33.03	17.84	13.16

In the next sub-section, we first introduce how we comprehend this table structure recognition problem.

Problem Interpretation. In a table recognition problem, it is quite natural to consider each character string in the table as a node. Then, the vertical or horizontal relation between a node and its neighbors can be understood as the feature of edges. More specifically, for a particular node, the vertical relation can be considered as "exist" only on the edges between this node and other nodes in the same column. Similarly, for this particular node, the horizontal relation only exists on edges between this node and other nodes in the same row.

If we use N to denote the set of nodes and E_C to denote the fully connected edges, then a table structure can be represented by a complete graph $G = < V, R_C >$, where R_C indicate a set of relations between E_C. More specifically, we have $R_C = E_C \times \{vertical, horizontal\}$.

Thus, we can interpret the problem as the following: given a set of nodes N and their feature, our aim is to predict the relations R_C between pairs of nodes as accurate as possible.

However, training on complete graphs is expensive. It is not only computationally intensive but also quite time-consuming. Meanwhile, it is not hard to notice that a table structure can be represented by far fewer edges, as long as a node is connected to its nearest neighbors including both vertical ones and horizontal ones. With the knowledge of node position, we are also capable of recovering the table structure from these relations.

Therefore in this paper, instead of training on the complete graph with R_C, which is of $O(|N|^2)$ complexity, we make use of the K-Nearest-Neighbors (KNN) method to construct R, which contains the relations between each node and its K nearest neighbors. With the help of KNN, we can reduce the complexity to $O(K * |N|)$.

GFTE. For each node, three types of information are included, i.e. the textual content, the absolute locations and the image, as shown in Fig. 3. We then make use of the structure relations to build the ground truth and the entire structure could be like Fig. 4. For higher accuracy, we train horizontal and vertical relations separately. For horizontal relations, we label each edge as 1: *in the same row* or 0: *not in the same row*. Similarly for vertical relations, we label each edge as 1: *in the same column* or 0: *not in the same column*.

Figure 5 gives the structure of our graph-based convolutional network GFTE. We first convert the absolute position into relative positions, which are further used to generate the graph. In the mean time, plain text is first embedded into a predefined feature space, then LSTM is used to obtain semantic feature. We concatenate the position feature and the text feature together and feed them to a two-layer graph convolutional network (GCN).

Meanwhile, we first dilate the image by a small kernel to make the table lines thicker. We also resize the image to 256×256 pixels in order to normalize the input. We then use a three-layer CNN to calculate the image feature. After that, using the relative position of the node, we can calculate a flow-field grid. By computing the output using input pixel locations from the grid, we can acquire the image feature of a certain node at a certain point.

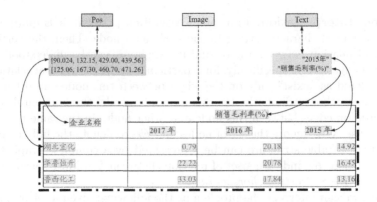

Fig. 3. An intuitive example of source data format.

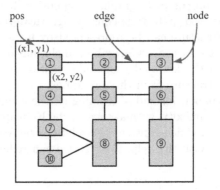

Fig. 4. Ground truth structure.

When these three different kinds of features are prepared, we pair two nodes on an edge of the generated graph. Namely we find two nodes of one edge and concatenate their three different kinds of features together. Finally, we use MLP to predict whether the two nodes are in same row or in same column.

5   Evaluation Results

In this section, we evaluate GFTE with prediction accuracy, as used in [31], for both vertical and horizontal relations. Our novel FinTab dataset is separated into train part and test part and is used to evaluate the performance of different GFTE model structures, meanwhile the SciTSR dataset is also used for validation.

Firstly, we train GFTE-pos. Namely we use the relative position and KNN algorithm to generate graph, and we train GFTE only with the position feature. Secondly, we train the network with the position feature as well as the text feature acquired by LSTM. This model is named GFTE-pos+text. Finally, our

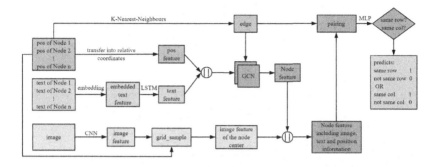

Fig. 5. The structure of our proposed GCN-based algorithm GFTE.

proposed GFTE is trained by further including the image feature with the help of grid sampling.

In Table 6, we give the performance of different models on FinTab dataset. As listed, the accuracy shows an overall upward trend when we concatenate more kinds of features. It improves distinctly when we include text feature, namely a rise of 10% by horizontal prediction and 5% by vertical prediction. Further including image feature seems to help improve the performance a little, but not too much.

Table 6. Accuracy results of different GFTE models on vertical and horizontal directions.

Network	Horizontal prediction	Vertical prediction
GFTE-pos	0.759836	0.842450
GFTE-pos+text	0.858675	0.903230
GFTE	0.861019	0.903031

Meanwhile, we notice a higher accuracy in vertical prediction than in horizontal prediction on FinTab. It is possibly caused by the uneven distribution of cells within a row of financial tables. Figure 6 gives some typical examples. In Fig. 6 (a), the nodes in the first 8 rows are distributed extremely far in horizontal direction. In Fig. 6 (b), when calculating K nearest neighbors for the first column, many vertical relations will be included, but very few horizontal relations, especially when K is small. These situations are rather rare in academic tables but not uncommon in financial reports.

In Table 7, we give the accuracy results of GFTE on different datasets, namely on the SciTSR test dataset and on our FinTab test dataset. It could be observed that our model reaches rather high accuracy on SciTSR validation dataset, which

名称（全名）	Development Principles Fund II L.P.	
成立时间	2008 年 3 月 14 日	
合伙人认缴资本	101,025,001.00 美元	
合伙人实际缴纳资本	92,589,110.00 美元	
注册地	开曼群岛	
主要生产经营地	开曼群岛	
合伙类型	有限合伙	
合伙期限	10 年	
合伙人姓名 或名称及认 缴资本金额	有限合伙人	Nederlandse Financierings-Maatschappij voor Ontwikkelingslanden N.V.	4,000 万美元
		DEG-Deutsche Investitions-und Entwicklungsgesellschaft MBH	2,000 万美元
		Société de Promotion et de Participation pour la Coopération Économique	2,000 万美元
		ITOCHU Finance (Asia) Limited	1,000 万美元
		ITOCHU Hong Kong Limited	1,000 万美元
		The Development Principles Group Limited	102.5 万美元
		Development Principles Fund II CIP Limited	1.0 万美元
	普通合伙人	Development Principles Fund II GP Ltd.	10,000 美元
执行事务合伙人	Development Principles Fund II GP Ltd.	
主营业务	投资	

(a) In this table, the last column is right-aligned. However, the bottom part gives all the partners' name and their capital subscriptions. The first 8 rows are thus far apart.

盈利能力指标	2015 年	2015 年备考	变动
营业收入	565,457.86	853,949.30	288,491.44
营业利润	75,589.69	77,115.31	1,525.62
利润总额	75,949.63	77,553.66	1,604.03
净利润	63,381.65	62,410.45	-971.20
主营业务毛利率	31.21%	30.36%	-0.85%
总资产报酬率	6.72%	5.19%	-1.53%

(b) In this profitability table, the first column is left-aligned for legibility and the other three columns are right-aligned because only then can the decimal points be aligned.

Fig. 6. Typical examples of unevenly distributed horizontal cells in financial tables.

implies that our algorithm works well as a baseline given enough training data. In addition, GFTE also achieves good results on FinTab test dataset, which suggests that GCN model also works well on more complex scenario.

Table 7. Accuracy results of both vertical and horizontal relations on validation dataset and test dataset.

Dataset	Horizontal prediction	Vertical prediction
SciTSR test dataset	0.954048	0.922423
FinTab test dataset	0.883612	0.918561

In conclusion, applying Graph Convolutional Network to a table extraction problem by integrating image feature, position feature and textual feature together is a novel solution. Since tables in financial context are much more difficult than ordinary tables to be exacted by existing methods, GFTE shows that integrating more types of table feature helps to improve the performance and is thus introduced and suggested as baseline method, which is hoped to be enlightening.

6 Conclusion

In this paper, we disclose a standard Chinese financial dataset from PDF files for table extraction benchmark test, which is diverse, sufficient and comprehensive. With this novel dataset, we hope more innovative and fine-designed algorithms of table extraction will emerge. Meanwhile, we propose a GCN-based algorithm GFTE as a baseline with a novel idea of integrating all possible types of ground truth together. We also discuss its performance and some possible difficulties by extracting tables from financial files in Chinese.

References

1. Etzioni, O., Fader, A., et al.: Open information extraction: the second generation. Proceedings of the Twenty-Second International Joint Conference on Artificial Intelligence, AAAI Press, Spain, 2011, pp. 3–10 (2011)
2. Baeza-Yates, R., Ribeiro-Neto, B.: Modern Information Retrieval. Addison-Wesley Longman Publishing Co., Inc., Boston, MA, USA (1999)
3. Manning, C.D., Raghavan, P., Schütze, H.: Introduction to Information Retrieval. Cambridge University Press New York, NY, USA (2010)
4. Hurst, M.: Towards a theory of tables. Int. J. Document Anal. Recogn. (IJDAR) 8(2–3), 123–131 (2006)
5. Long, V., Dale, R., Cassidy, S.: A model for detecting and merging vertically spanned table cells in plain text documents. In: International Conference on Document Analysis and Recognition, New York (2005)
6. Shigarov, A., Altaev, A., et al.: TabbyPDF: web-based system for PDF table extraction. In: 24th International Conference on Information and Software Technologies (2018)
7. Institute of Computer Science and Techonology of Peking University, Institute of Digital Publishing of Founder R&D Center, "Marmot Dataset," China (2011)
8. Shahab, A.: Table Ground Truth for the UW3 and UNLV datasets. In: German Research Center for Artificial Intelligence (DFKI) (2013)
9. Göbel, M., Hassan, T., Oro, E., Orsi, G.: ICDAR 2013 table competition. In: Proceedings of the 2013 12th International Conference on Document Analysis and Recognition, pp. 1449–1453 (2013)
10. Gao, L., Huang, Y., Dejean, H., Meunier, J.: ICDAR 2019 Competition on Table Detection and Recognition (cTDaR). In: International Conference on Document Analysis and Recognition (ICDAR), pp. 1510–1515 (2019)
11. Zhong, X., Tang, J., Yepes, A.J.: PubLayNet: largest dataset ever for document layout analysis. In: International Conference on Document Analysis and Recognition (ICDAR) (2019)
12. Chi, Z., Huang, H., Xu, H., Yu, H., Yin, W., Mao, X.: Complicated table structure recognition (2019). arXiv preprint arXiv:1908.04729
13. Li, M., Cui, L., Huang, S., Wei, F., Zhou, M., Li, Z.: TableBank: table benchmark for image-based table detection and recognition. In: The International Conference on Language Resources and Evaluation (2020)
14. Göbel, M., Hassan, T., Oro, E., Orsi, G., Rastan, R.: Table modelling, extraction and processing. In: Proceedings of the 2016 ACM Symposium on Document Engineering, pp. 1–2 (2016)
15. Coüasnon, B., Lemaitre, A.: Handbook of Document Image Processing and Recognition. Chap. Recognition of Tables and Forms, pp. 647–677 (2014)

16. Khusro, S., Latif, A., Ullah, I.: On methods and tools of table detection, extraction and annotation in PDF documents. J. Inf. Sci. **41**(1), 41–57 (2015)
17. Wang, Y.: Document analysis: table structure understanding and zone content classification. PhD thesis, Washington University (2002)
18. Shamilian, J.H., Baird, H.S., Wood, T.L.: A retargetable table reader. In: Proceedings of the Fourth International Conference on Document Analysis and Recognition. IEEE, New York (1997)
19. Mohemad, R., Hamdan, A.R., Othman, Z.A., Noor, N.M.: Automatic document structure analysis of structured PDF files. Int. J. New Comput. Architect. Appl. (IJNCAA) **1**(2), 404–411 (2011)
20. Yildiz, B., Kaiser, K., Miksch, S.: pdf2table: a method to extract table information from PDF files. In: IICAI (2005)
21. Liu, Y., Mitra, P., Giles, C.L., Bai, K.: Automatic extraction of table metadata from digital documents. In: Proceedings of the 6th ACM/IEEE-CS Joint Conference on Digital Libraries. ACM, New York (2006)
22. Wang, Y., Hu, J.: Detecting tables in HTML documents. In: Lopresti, D., Hu, J., Kashi, R. (eds.) DAS 2002. LNCS, vol. 2423, pp. 249–260. Springer, Heidelberg (2002). https://doi.org/10.1007/3-540-45869-7_29
23. Cohen, W.W., Hurst, M., Jensen, L.S.: A flexible learning system for wrapping tables and lists in HTML documents. In: Proceedings of the 11th International Conference on World Wide Web. ACM, New York (2002)
24. Oro, E., Ruffolo, M.: Xonto: an ontology-based system for semantic information extraction from pdf documents. In: ICTAI 2008: 20th IEEE International Conference on Tools with Artificial Intelligence. IEEE, New York (2008)
25. Wang, Y., Hu, J.: A machine learning based approach for table detection on the web. In: Proceedings of the 11th International Conference on World Wide Web. ACM, New York (2002)
26. Silva, A.C.: New metrics for evaluating performance in document analysis tasks - application to the table case. ICDAR (2007)
27. Liu, Y., Mitra, P., Giles, C.L.: Identifying table boundaries in digital documents via sparse line detection. In: Proceedings of the 17th ACM Conference on Information and Knowledge Management. ACM, New York (2008)
28. Pinto, D., McCallum, A., Wei, X., Croft, W.B.: Table extraction using conditional random fields. In: Proceedings of the 26th Annual International ACM SIGIR Conference on Research and Development in Informaion retrieval. ACM, New York (2003)
29. Wei, X., Croft, B., McCallum, A.: Table extraction for answer retrieval. Inf. Retrieval **9**(5), 589–611 (2006)
30. Qasim, S.R., Mahmood, H., Shafait, F.: Rethinking table recognition using graph neural networks. In: 2019, International Conference on Document Analysis and Recognition (ICDAR), pp. 142–147 (2019)
31. Pau, R., Anjan, D., Lutz, G., Alicia, F., Oriol, R., Josep, L.: Table detection in invoice documents by graph neural networks. In: 2019, International Conference on Document Analysis and Recognition (ICDAR), pp. 122–127 (2019)
32. Holeček, M., Hoskovec, A., Baudiš, P., Klinger, P.: Table understanding in structured documents. In: International Conference on Document Analysis and Recognition Workshops (ICDARW), pp. 158–164 (2019)
33. Chris, T., Morariu, V., Price, B., Cohen, S., Tony, M.: "Deep splitting and merging for table structure decomposition. In: 2019, International Conference on Document Analysis and Recognition (ICDAR), pp. 114–121 (2019)

Relative Attribute Classification with Deep-RankSVM

Sara Atito Ali Ahmed$^{(\boxtimes)}$ and Berrin Yanikoglu

Faculty of Engineering and Natural Sciences, Sabanci University,
34956 Istanbul, Turkey
{saraatito,berrin}@sabanciuniv.edu

Abstract. Relative attributes indicate the strength of a particular attribute between image pairs. We introduce a deep Siamese network with rank SVM loss function, called Deep-RankSVM, that can decide which one of a pair of images has a stronger presence of a specific attribute. The network is trained in an end-to-end fashion to jointly learn the visual features and the ranking function. The trained network for an attribute can predict the relative strength of that attribute in novel images.

We demonstrate the effectiveness of our approach against the state-of-the-art methods on four image benchmark datasets: LFW-10, PubFig, UTZap50K-2 and UTZap50K-lexi datasets. Deep-RankSVM surpasses state-of-art in terms of the average accuracy across attributes, on three of the four image benchmark datasets.

Keywords: Relative attributes · Rank SVM · Deep learning · Classification

1 Introduction

Identification and retrieval of images and videos with certain visual attributes are of interest in many real-world applications, such as image search/retrieval [6,7], video retrieval [3], image/video captioning [12,22], face verification [9], and zero-shot learning [2,4]. Visual attribute learning is studied in particular for *binary* attributes that indicate the presence or absence of a certain semantic attribute (smiling or not, has beard or not, etc.) [1,28].

Many attributes can not be considered as binary or categorical attributes (e.g. young). To address the issue of strength of an attribute, Parikh and Grauman [13] proposed to model *relative attributes*. The goal of relative attribute learning is to learn a function which predicts the relative strengths of a pair of images regarding a given attribute (e.g. which person is younger?). The network should be able to answer the comparisons, with more/less/equal of the presence of a specific attribute. Figure 1 shows image pairs for the attributes *mouth-open* and *sporty* from the LFW10 and UTZap50K-2 datasets, respectively.

© Springer Nature Switzerland AG 2021
A. Del Bimbo et al. (Eds.): ICPR 2020 Workshops, LNCS 12662, pp. 659–671, 2021.
https://doi.org/10.1007/978-3-030-68790-8_51

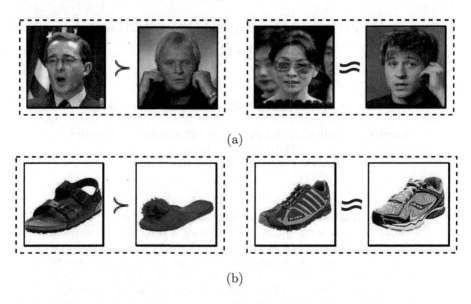

(a)

(b)

Fig. 1. Random samples of image pairs with relative attribute relations: (a) Images from the LFW10 dataset with respect to the *mouth-open* attribute. (b) Images from the UTZap50K-2 dataset with respect to the *sporty* attribute. Ordered pairs are given on the left and unordered pairs on the right.

Parikh and Grauman [13] formulated the solution using a set of image-pairs with human assessed relative ordering and a formulation similar to that of Support Vector Machines (SVMs). After the introduction of the problem, subsequent research approached the problem using either traditional features or deep learning approaches, as discussed in Sect. 2.

In this paper, we present a deep learning system that can compare the given two images in terms of their strength regarding a particular attribute. Specifically, we propose a convolutional Siamese network using rank SVM formulation for the relative attribute problem. The main contributions of our proposed model are summarized as follows:

- Proposing an end-to-end deep learning framework in which the network jointly learns the visual features and the rank SVM function, for relative attribute classification.
- Demonstrating the effectiveness of the proposed framework by comparing to our baseline [18], with improvements of 6%, 3%, 1%, and 2.65% in average ranking accuracy for LFW-10, PubFig, UTZap50K-2, and UTZap50K-lexi datasets, respectively.
- Surpassing the state-of-the-art results in LFW-10, PubFig, and UTZap50K-lexi datasets by about 2.0%, 0.2%, and 0.87% and obtaining slightly lower results (−0.15%) in UTZap50K-2 dataset.

The rest of this paper is organized as follows. A review of literature is presented in Sect. 2. Section 3 introduces our proposed method combining deep learning and the rank SVM objective function for learning relative attributes. Description of the employed datasets, implementation details, along with extensive experimental results are discussed in Sect. 4. Summary and future work are given in Sect. 5.

2 Related Works

The relative attribute learning problem has attracted significant attention, with researchers approaching it first using traditional approaches and later using deep learning approaches. Traditional and hand-crafted features are first used in [10, 13, 20, 23, 24]; however, more recently, deep convolutional neural networks are used to jointly learn the features and the ranking function in an end-to-end fashion [17, 18, 21, 26, 27].

2.1 Traditional Approaches

Parikh and Grauman [13] first proposed relative attributes where they used the GIST descriptor [11] and color histogram features, together with a constrained optimization formulation similar to that of SVMs. Our work is based on their rank SVM formulation; however unlike their use of traditional image features, we jointly learn visual features and the rank SVM function, in a deep convolutional network.

Later, Li et al. [10], introduced non-linearity by using the relative forest algorithm to capture accurate semantic relationships. More recently, Yu and Grauman [24], developed a Bayesian local learning strategy to infer when images are indistinguishable for a given attribute in a probabilistic learning manner.

2.2 Deep Learning Approaches

Hand-crafted feature representation may not be the best to capture the relevant visual features to describe relative attributes. As with many other visual recognition problems, deep learning approaches significantly outperform approaches that are based on hand-crafted features followed by shallow models.

Souri et al. [18] introduced RankNet, which is a convolutional neural network based on the architecture of VGG-16 [16]. A ranking layer is included to rank the strength of an attribute in the given pair of images based on the extracted features in an end-to-end fashion.

Using a similar approach, Yang et al. [21] proposed a DRA model which consists of five convolutional neural layers and five fully connected layers followed by a relative loss function.

Singh and Lee [17] trained a Siamese network based on AlexNet [8], with a pairwise ranking loss. The network consists of two branches, each branch consists of a localization module and a ranking module.

In Zhuang et al. [27], cross-image representation is considered via deep attentive cross-image representation learning (DACRL) model: an end-to-end convolutional neural network which takes a pair of images as input, and outputs a

Fig. 2. Proposed Deep-RankSVM architecture. The network for a given attribute takes a pair of images $(\mathbf{x_i}, \mathbf{x_j})$ as input and outputs 1 if $\mathbf{x_i}$ shows the given attribute more strongly compared to $\mathbf{x_j}$; 0 if they are comparable.

posterior probability that indicates the relative strengths of a specific attribute, based on cross-image representation learning.

Our work most resembles [18], except for the loss function. As our main contribution is embedding the SVM loss into the network, we compare our results to this system as the baseline, in addition to state-of-art approaches reported in literature [17,27].

3   Deep-RankSVM

We introduce Deep-RankSVM (DRSVM), a convolutional Siamese network trained with the rank SVM loss.

Following Parikh and Grauman's [13] notation, training images consist of a set of ordered image pairs $O_m = \{(\mathbf{x_i}, \mathbf{x_j})\}$ and a set of unordered image pairs $S_m = \{(\mathbf{x_i}, \mathbf{x_j})\}$, for every attribute m of a set of M attributes. An image pair $(\mathbf{x_i}, \mathbf{x_j})$ is ordered when the presence of attribute m in $\mathbf{x_i}$ is stronger than the presence of attribute m in $\mathbf{x_j}$ and unordered when $\mathbf{x_i}$ and $\mathbf{x_j}$ have similar presence strength of attribute m.

With these notations, we formulate the problem as learning the deep attribute representation $\mathbf{h}(\mathbf{x})$ of an image, for a specific attribute m, satisfying the following constraints:

$$\mathbf{w_m^T h}(\mathbf{x_i}) > \mathbf{w_m^T h}(\mathbf{x_j}); \ \forall (\mathbf{x_i}, \mathbf{x_j}) \in O_m$$
$$\mathbf{w_m^T h}(\mathbf{x_i}) = \mathbf{w_m^T h}(\mathbf{x_j}); \ \forall (\mathbf{x_i}, \mathbf{x_j}) \in S_m \tag{1}$$

We use the VGG-16 architecture [16] as the base of a Siamese network to jointly learn the deep attribute representation $\mathbf{h}(\mathbf{x})$ and the weights $\mathbf{w_m}$, to rank two input images for the given attribute m. The network is illustrated in Fig. 2.

As seen in this figure, the output of the two branches of the network are the 1,000-dimensional last fully connected layer of the VGG-16 architecture. In order to construct an end-to-end model, an additional layer is added to carry out the difference between the feature representations, $\mathbf{h}(\mathbf{x_i})$ and $\mathbf{h}(\mathbf{x_j})$. The output layer consists of a single node using the linear activation function, with weights that correspond to the $\mathbf{w_m^T}$ in Eq. 1. The target is set to +1 if $\mathbf{x_i}$ has the given attribute more strongly, 0 if the two images are similar regarding that attribute.

For the objective function, we use the rank SVM formulation for relative attributes proposed in [13]:

$$
\begin{aligned}
\min \quad & \frac{1}{2}\mathbf{w_m^T}\mathbf{w_m} + C_1\sum\xi_{ij}^2 + C_2\sum\gamma_{ij}^2 \\
\text{subject to} \quad & \mathbf{w_m^T}(\mathbf{h(x_i)} - \mathbf{h(x_j)}) \geq 1 - \xi_{ij} \qquad \forall(i,j) \in O_m \\
& |\mathbf{w_m^T}(\mathbf{h(x_i)} - \mathbf{h(x_j)})| \leq \gamma_{ij} \qquad \forall(i,j) \in S_m \\
& \xi_{ij} \geq 0, \gamma_{ij} \geq 0
\end{aligned}
\tag{2}
$$

where $\mathbf{w_m}$ is the trainable weights of the ranking layer; ξ_{ij} and γ_{ij} are the slack variables of the soft margin SVM formulation; and the constants C_1 and C_2, adjust the trade-off between maximizing the margin and satisfying the pairwise relative constraints. We use quadratic terms for the soft error, as done in [13].

We then combine the constraints on the slack variables ξ_{ij} and γ_{ij} and noting that the problem is one of minimization and simplifying, we obtain the corresponding unconstrained optimization problem (Eq. 4):

$$
\begin{aligned}
\xi_{ij} &\geq max(0, 1 - \mathbf{w_m^T}(\mathbf{h(x_i)} - \mathbf{h(x_j)})) \\
\gamma_{ij} &\geq max(0, |\mathbf{w_m^T}(\mathbf{h(x_i)} - \mathbf{h(x_j)})|)
\end{aligned}
\tag{3}
$$

$$
\begin{aligned}
\min_{\mathbf{w_m}} \quad & \frac{1}{2}\mathbf{w_m^T}\mathbf{w_m} + C_1 \sum_{(i,j)\in O_m} max\left(0, 1 - \mathbf{w_m^T}(\mathbf{h(x_i)} - \mathbf{h(x_j)})\right)^2 \\
& + C_2 \sum_{(i,j)\in S_m} (\mathbf{w_m^T}(\mathbf{h(x_i)} - \mathbf{h(x_j)}))^2
\end{aligned}
\tag{4}
$$

With this formulation, we are able to learn the deep features ($\mathbf{h(x_i)}$) and the rank SVM function ($\mathbf{w_m}$) in an end-to-end network.

4 Experimental Evaluation

We evaluate the effectiveness of our approach on the publicly available relative attributes datasets, described in Sect. 4.1. Our results are compared to the results of several systems that report accuracy as performance measurement, namely Rank SVM [13], FG-LP [23], spatial Extent [20], DeepSTN [17], DRA [21], and DACRL [27].

We consider the RankNet system proposed in [18] as our baseline: we use the same network (VGG-16 pre-trained on ILSVRC 2014) and the same data augmentation techniques, but instead of the cross-entropy loss used in RankNet, we use the rank SVM loss, as described in Sect. 3. In this way we aimed to evaluate the effectiveness of using the proposed formulation.

The network and implementation details are explained in Sect. 4.2. In Sect. 4.3, the performance of our proposed method is shown along with a comparison with the baseline and other state-of-the-art systems.

4.1 Datasets

Our proposed approach is evaluated on four different datasets from distinctive areas for comprehensive evaluation.

1. **LFW-10 Dataset** [15]: The dataset is a subset of the Labels Faces in the Wild (LFW) dataset [5]. It consists of $2,000$ images with 10 different face attributes (see Table 1). For each attribute, a random subset of 500 pairs of images have been annotated for training and testing sets.

2. **Public Figure Face Dataset** [13]: PubFig dataset consists of 772 images from 8 different identities with 11 semantic attributes (see Table 2). The ordering of the samples are annotated at the category level; in other words, all images with the same identity are ranked higher, equal, or lower than all images belonging to another identity with respect to a specific attribute. For instance, a person is said to have longer hair than another person, even if this may not be true in all of their photographs. This short-cut in annotation results in label inconsistencies.

3. **UTZap50K-2 Dataset** [23]: Large shoe dataset consists of $50,025$ images collected from Zappos.com. It consists of 4 shoe attributes: open, pointy, sporty, and comfort (see Table 3). After pruning out pairs with low confidence or agreement, the human-annotated examples consist of approximately $1,500$–$1,800$ training image-pairs for each attribute and in total $4,334$ image-pairs for testing.

4. **Zappos50K-lexi Dataset** [25]: It is based on UTZap50K dataset [23] with 10 additional fine-grained relative attributes: comfort, casual, simple, sporty, colorful, durable, supportive, bold, sleek, and open (see Table 4). The dataset consists of approximately $1,300$–$2,100$ image-pairs for each attribute.

In all of our experiments, we have used the provided training/testing split by the original publishers of the datasets. The evaluation is performed over the ordered pairs, excluding unordered pairs, in line with the literature. In other words, while the training sets consist of both ordered and unordered pairs, the test set only contains ordered pairs.

4.2 Implementation Details

We chose VGG-16 to be our base architecture, to have a fair comparison with our baseline [18] that also uses it.

The Siamese network contains two branches comprising the VGG-16 architecture up to the last layer, outputting $\mathbf{h}(\mathbf{x_i})$ and $\mathbf{h}(\mathbf{x_j})$ which are $1,000$ dimensional each, as illustrated in Fig. 2. The input to the model is two 224×224 RGB images.

We have used the pre-trained VGG-16 provided by Keras framework and fine-tuned all weights up to the output layer. The output layer weights $\mathbf{w_m}$ of the ranking layer is learned from scratch, after initialization using the Xavier method. We did not use a bias term, as suggested in [14].

A separate network for each attribute is trained. For LFW-10, UTZap50K-2, and UTZap50K-lexi datasets, we trained our model for 500, 200, and 200 epochs for each attribute, respectively. For training, stochastic gradient descent with RMSProp optimizer is used with a mini-batch size of 48 image-pairs. A unified learning rate is set to 10^{-5} for all of the layers. The training images are shuffled after every epoch. The constants C_1 and C_2 of the soft margin formulation are set to be equal as in [13], using a value of 0.1.

For PubFig dataset, the relative attributes are annotated at category-level, as explained in Sect. 4.1; hence one full epoch would contain a very large number of image-pairs. Therefore, we trained the model for 10,000 iterations where in every iteration a random selection of 48 image-pairs are chosen from the dataset, and ground-truth labels are assigned based on their categories.

Advanced data augmentation techniques have proven to improve performance in many studies specially for deep learning. However, to resemble our baseline and show the effectiveness of incorporating Rank SVM loss to the deep learning model, only simple on-the-fly data augmentation techniques are applied during training, namely rotation $[-15, 15]$, horizontal flipping, and random cropping.

4.3   Results

We compare the performance of Deep-RankSVM (DRSVM) with our baseline [18] and state-of-the-art, on four different datasets. The reported performance figures given in Tables 1, 2 and 3 are accuracies over ordered pairs, in line with the literature. In other words, while the training set comprises both ordered (O_m) and similar pairs (S_m), testing is evaluated over ordered pairs only.

Table 1 shows the results on the LFW-10 dataset where we outperform our baseline [18] by about 6% on average. We can attribute this to the use of the rank SVM loss as the loss function, as this is the main difference between our model and the baseline. Our results surpass the average accuracy by 1.2% points over the state-of-the-art [27], with best performance on 7 of the 10 attributes.

Table 2 shows the results on the PubFig dataset where our system improves over the baseline [18] by 3% points and obtains the best results on 6 out of 11 attributes compared to state-of-art [27]. The gain on this dataset is marginal (0.2%) which may be due to the category base annotation that may result in annotation inconsistencies, as discussed in Sect. 4.1.

Table 3 shows the results on the UTZap50K-2 dataset where we outperform the baseline by 1%, but slightly underperform the state-of-art [27] by -0.15%, while obtaining best results in 2 out of 4 attributes.

Finally, Table 4 shows the results on the UTZap50K-lexi dataset where we outperform the baseline by 2.65%. Furthermore, we improve the average accuracy by 0.87% over the state-of-the-art [27] and obtain the best results in 6 out of the 10 attributes.

Table 1. State-of-art results on LFW-10 dataset (in chronological order) compared with the results obtained in this work. Bold figures indicate the best results.

Attribute	[23]	[20]	[18] (baseline)	[17]	[27] (w/o attention)	[27]	DRSVM (proposed)
Bald Head	67.90	83.21	81.14	83.94	83.21	85.04	**90.75**
Dark Hair	73.60	88.13	88.92	92.58	91.99	92.58	**92.67**
Eyes Open	49.60	82.71	74.44	**90.23**	87.97	**90.23**	86.54
Good Look.	64.70	**72.76**	70.28	71.21	69.97	70.28	71.21
Masc. Look	70.10	93.68	98.08	96.55	97.70	**98.28**	95.05
Mouth Open	53.40	88.26	85.46	91.28	89.93	91.28	**92.67**
Smile	59.70	86.16	82.49	84.75	85.03	85.03	**88.64**
Teeth	53.50	86.46	82.77	89.85	88.00	89.23	**91.80**
Forehead	65.60	90.23	81.90	87.89	89.45	90.63	**90.84**
Young	66.20	75.05	76.33	80.81	74.84	76.55	**81.02**
Mean	62.43%	84.67%	82.18%	86.91%	85.81%	86.91%	**88.12%**

Table 2. Comparison of the state-of-the-art accuracies on the PubFig dataset.

Attribute	[23]	[18] (baseline)	[21]	[27] (w/o attention)	[27]	DRSVM (proposed)
Male	91.77	95.50	90.82	97.70	96.49	**97.74**
White	87.43	94.60	87.12	97.82	97.80	**98.61**
Young	91.87	94.33	91.49	97.10	**97.96**	96.32
Smile	87.00	95.36	92.68	97.03	**97.42**	96.14
Chubby	87.37	92.32	89.30	97.05	**97.22**	94.47
Forehead	94.00	97.28	94.39	98.30	98.05	**98.75**
Bushy Eyebrows	89.83	94.53	90.19	97.36	97.48	**98.68**
Narrow Eyes	91.40	93.19	90.60	**97.99**	96.91	97.28
Pointy Nose	89.07	94.24	91.03	97.26	97.74	**99.31**
Big Lips	90.43	93.62	90.35	94.36	96.83	**98.24**
Round Face	86.70	94.76	91.99	**98.04**	96.27	96.89
Mean	89.72%	94.42%	90.91%	97.27%	97.29%	**97.49%**

Table 3. Comparison of the state-of-the-art accuracies on UTZap50K-2 dataset.

Attribute	[13]	[23]	[18] (baseline)	[27] (w/o attention)	[27]	DRSVM (proposed)
Open	60.18	74.91	73.45	75.45	**75.66**	74.09
Pointy	59.56	63.74	68.20	69.80	70.65	**70.90**
Sporty	62.70	64.54	73.07	73.78	**73.87**	72.95
Comfort	64.04	62.51	70.31	68.54	69.56	**71.20**
Mean	61.62%	66.43%	71.26%	71.89%	**72.44%**	72.29%

Table 4. Comparison of the state-of-the-art accuracies on the UTZap50K-lexi dataset.

Attribute	[18] (baseline)	[27] (w/o attention)	[27]	DRSVM (proposed)
Comfort	90.48	**91.88**	**91.88**	91.59
Casual	90.43	94.44	91.36	**95.37**
Simple	90.40	89.93	90.16	**90.91**
Sporty	93.31	93.01	94.22	**96.57**
Colorful	95.43	**97.33**	95.81	95.95
Durable	90.47	92.65	92.33	**93.31**
Support	91.98	92.65	92.65	**94.98**
Bold	91.53	91.12	**92.56**	91.47
Sleek	86.31	89.24	**90.71**	89.30
Open	82.53	87.90	88.98	**89.99**
Mean	90.29%	92.02%	92.07%	**92.94%**

4.4 Discussion

The reported results in Tables 1, 2, 3 and 4 show that we outperformed our baseline [18] on all 4 datasets, by 6%, 3%, 1%, and 2.65% points, for LFW-10, PubFig, UTZap50K-lexi, and UTZap50K-2 respectively. This improvement over the baseline shows the effectiveness of using the rank SVM loss with the deep learning approach.

Furthermore, we surpassed the state-of-art on the LFW-10, PubFig, and UTZap50K-lexi datasets by 1.2%, 0.2%, and 0.87% points and obtained slightly lower results (-0.15%) on the UTZap50K-2 dataset.

We have employed the VGG-16 architecture as in our baseline [18] and in DACRL [27]. We expect that the performance will be even higher with a more advanced network (e.g. Inception-ResNet [19] or NasNetLarge [29]) and using heavy data augmentation.

Figure 3 shows random images x_i with their normalized predicted strength for the respective attribute. Specifically, we compute $\mathbf{w}_m^T \mathbf{h}(\mathbf{x_i})$ for each image x_i and attribute m and normalize by min-max normalization over all the images for that attribute. It is interesting to note that although our network is trained given only image-pairs, the network learns a global ranking for a given attribute.

The trained network is able to localize on the informative regions of the image related to a given attribute, without explicitly being taught to do so during training. We calculate the derivative of the output with respect to a given input and visualize the results of the last convolutional layer as shown in Fig. 4. The heat maps visualize the pixels in the images with the most contribution to the ranking prediction of the network.

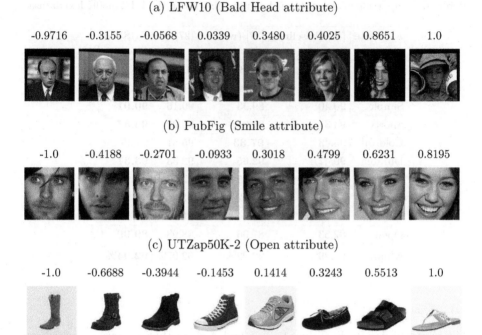

Fig. 3. Sample images ordered according to the output prediction of their associated attribute.

5   Summary and Future Work

In this paper, we propose the Deep-RankSVM (DRSVM) network for relative attribute learning, to jointly learn the features and the ranking function in an end-to-end fashion. While deep learning approach was used to learn relative attributes before [18], this is the first end-to-end deep formulation that combines deep learning with the rank SVM loss, to the best of our knowledge.

Our model is evaluated on four benchmarks, LFW-10, PubFig, UTZap50K-2 and UTZap50K-lexi and achieved state-of-the-art performance on LFW-10, PubFig, and UTZap50K-lexi datasets. These results shows the benefit of incorporating and jointly training the network with the Rank SVM loss function, for relative attributes. We believe that the performance can be further improved with a state-of-art network rather than the one used in this work (VGG-16), as well as using heavy data augmentation.

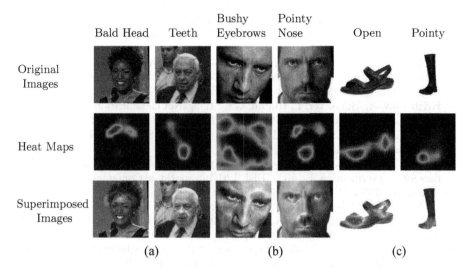

Fig. 4. Class Activation Maps showing the pixels with most contribution to the ranking prediction in (a) Bald Head and Teeth from LFW-10 dataset, (b) Bushy Eyebrows and Pointy Nose from PubFig dataset, and (c) Open and Pointy from UTZap50K-2 dataset.

Although the results show the ability of the network to localize on the informative regions in the image, adding a localization module similar to the one used in [27] can contribute to the performance, especially in the existence of some annotation inconsistencies, as in the case of PubFig dataset.

Acknowledgment. This work was supported by a grant from The Scientific and Technological Research Council of Turkey (TÜBİTAK) under project number 119E429.

References

1. Ahmed, S.A.A., Yanikoglu, B.: Within-network ensemble for face attributes classification. In: Ricci, E., Rota Bulò, S., Snoek, C., Lanz, O., Messelodi, S., Sebe, N. (eds.) ICIAP 2019. LNCS, vol. 11751, pp. 466–476. Springer, Cham (2019). https://doi.org/10.1007/978-3-030-30642-7_42
2. Bansal, A., Sikka, K., Sharma, G., Chellappa, R., Divakaran, A.: Zero-shot object detection. In: Proceedings of the European Conference on Computer Vision (ECCV), pp. 384–400 (2018)
3. Chen, L., Zhang, P., Li, B.: Instructive video retrieval based on hybrid ranking and attribute learning: a case study on surgical skill training. In: Proceedings of the 22nd ACM, pp. 1045–1048 (2014)
4. Fu, Y., Xiang, T., Jiang, Y.G., Xue, X., Sigal, L., Gong, S.: Recent advances in zero-shot recognition: toward data-efficient understanding of visual content. IEEE Sig. Process. Mag. **35**(1), 112–125 (2018)
5. Huang, G.B., Ramesh, M., Berg, T., Learned-Miller, E.: Labeled faces in the wild: a database for studying face recognition in unconstrained environments. Technical report 07–49, University of Massachusetts, Amherst, October 2007

6. Kovashka, A., Grauman, K.: Attributes for image retrieval. Visual Attributes. ACVPR, pp. 89–117. Springer, Cham (2017). https://doi.org/10.1007/978-3-319-50077-5_5
7. Kovashka, A., Parikh, D., Grauman, K.: WhittleSearch: interactive image search with relative attribute feedback. Int. J. Comput. Vis. **115**(2), 185–210 (2015). https://doi.org/10.1007/s11263-015-0814-0
8. Krizhevsky, A., Sutskever, I., Hinton, G.E.: ImageNet classification with deep convolutional neural networks. In: Pereira, F., Burges, C.J.C., Bottou, L., Weinberger, K.Q. (eds.) Advances in Neural Information Processing Systems, vol. 25, pp. 1097–1105. Curran Associates, Inc. (2012)
9. Kumar, N., Berg, A.C., Belhumeur, P.N., Nayar, S.K.: Attribute and simile classifiers for face verification. In: 2009 IEEE 12th International Conference on Computer Vision, pp. 365–372. IEEE (2009)
10. Li, S., Shan, S., Chen, X.: Relative forest for attribute prediction. In: Lee, K.M., Matsushita, Y., Rehg, J.M., Hu, Z. (eds.) ACCV 2012. LNCS, vol. 7724, pp. 316–327. Springer, Heidelberg (2013). https://doi.org/10.1007/978-3-642-37331-2_24
11. Oliva, A., Torralba, A.: Modeling the shape of the scene: a holistic representation of the spatial envelope. Int. J. Comput. Vis. **42**(3), 145–175 (2001). https://doi.org/10.1023/A:1011139631724
12. Pan, Y., Yao, T., Li, H., Mei, T.: Video captioning with transferred semantic attributes. In: Proceedings of the IEEE Conference on Computer Vision and Pattern Recognition, pp. 6504–6512 (2017)
13. Parikh, D., Grauman, K.: Relative attributes. In: International Conference on Computer Vision, pp. 503–510. IEEE (2011)
14. Ruff, L., et al.: Deep one-class classification. In: Dy, J., Krause, A. (eds.) Proceedings of the 35th International Conference on Machine Learning. Proceedings of Machine Learning Research, vol. 80, pp. 4393–4402. PMLR, Stockholmsmässan, Stockholm, 10–15 July 2018
15. Sandeep, R.N., Verma, Y., Jawahar, C.: Relative parts: distinctive parts for learning relative attributes. In: Proceedings of the IEEE Conference on Computer Vision and Pattern Recognition, pp. 3614–3621 (2014)
16. Simonyan, K., Zisserman, A.: Very deep convolutional networks for large-scale image recognition. arXiv preprint arXiv:1409.1556 (2014)
17. Singh, K.K., Lee, Y.J.: End-to-end localization and ranking for relative attributes. In: Leibe, B., Matas, J., Sebe, N., Welling, M. (eds.) ECCV 2016. LNCS, vol. 9910, pp. 753–769. Springer, Cham (2016). https://doi.org/10.1007/978-3-319-46466-4_45
18. Souri, Y., Noury, E., Adeli, E.: Deep relative attributes. In: Lai, S.-H., Lepetit, V., Nishino, K., Sato, Y. (eds.) ACCV 2016. LNCS, vol. 10115, pp. 118–133. Springer, Cham (2017). https://doi.org/10.1007/978-3-319-54193-8_8
19. Szegedy, C., Ioffe, S., Vanhoucke, V., Alemi, A.A.: Inception-v4, Inception-ResNet and the impact of residual connections on learning. In: Thirty-First AAAI Conference on Artificial Intelligence (2017)
20. Xiao, F., Lee, Y.J.: Discovering the spatial extent of relative attributes. In: Proceedings of the IEEE International Conference on Computer Vision, pp. 1458–1466 (2015)
21. Yang, X., Zhang, T., Xu, C., Yan, S., Hossain, M.S., Ghoneim, A.: Deep relative attributes. IEEE Trans. Multimed. **18**(9), 1832–1842 (2016)
22. Yao, T., Pan, Y., Li, Y., Qiu, Z., Mei, T.: Boosting image captioning with attributes. In: Proceedings of the IEEE International Conference on Computer Vision, pp. 4894–4902 (2017)

23. Yu, A., Grauman, K.: Fine-grained visual comparisons with local learning. In: Proceedings of the IEEE Conference on Computer Vision and Pattern Recognition, pp. 192–199 (2014)
24. Yu, A., Grauman, K.: Just noticeable differences in visual attributes. In: Proceedings of the IEEE International Conference on Computer Vision, pp. 2416–2424 (2015)
25. Yu, A., Grauman, K.: Semantic jitter: dense supervision for visual comparisons via synthetic images. In: Proceedings of the IEEE International Conference on Computer Vision, pp. 5570–5579 (2017)
26. Yu, A., Grauman, K.: Thinking outside the pool: active training image creation for relative attributes. In: Proceedings of the IEEE Conference on Computer Vision and Pattern Recognition, pp. 708–718 (2019)
27. Zhang, Z., Li, Y., Zhang, Z.: Relative attribute learning with deep attentive cross-image representation. In: Asian Conference on Machine Learning, pp. 879–892 (2018)
28. Zhuang, N., Yan, Y., Chen, S., Wang, H., Shen, C.: Multi-label learning based deep transfer neural network for facial attribute classification. Pattern Recogn. **80**, 225–240 (2018)
29. Zoph, B., Vasudevan, V., Shlens, J., Le, Q.V.: Learning transferable architectures for scalable image recognition. In: Proceedings of the IEEE Conference on Computer Vision and Pattern Recognition, pp. 8697–8710 (2018)

Adversarial Continuous Learning in Unsupervised Domain Adaptation

Youshan Zhang[✉] and Brian D. Davison

Computer Science and Engineering, Lehigh University, Bethlehem, PA, USA
{yoz217,bdd3}@lehigh.edu

Abstract. Domain adaptation has emerged as a crucial technique to address the problem of domain shift, which exists when applying an existing model to a new population of data. Adversarial learning has made impressive progress in learning a domain invariant representation via building bridges between two domains. However, existing adversarial learning methods tend to only employ a domain discriminator or generate adversarial examples that affect the original domain distribution. Moreover, little work has considered confident continuous learning using an existing source classifier for domain adaptation. In this paper, we develop adversarial continuous learning in a unified deep architecture. We also propose a novel correlated loss to minimize the discrepancy between the source and target domain. Our model increases robustness by incorporating high-confidence samples from the target domain. The transfer loss jointly considers the original source image and transfer examples in the target domain. Extensive experiments demonstrate significant improvements in classification accuracy over the state of the art.

Keywords: Adversarial learning · Unsupervised domain adaptation

1 Introduction

There is a high demand for automatic recognition of multimedia data. The availability of massive labeled training data is a prerequisite for creating machine learning models. Unfortunately, it is time-consuming and expensive to manually annotate data. Therefore, it is often necessary to transfer knowledge from an existing labeled domain to a new unlabeled domain. However, due to the phenomenon of data bias or domain shift [17], machine learning models do not generalize well from an existing domain to a novel unlabeled domain.

Domain adaptation has been a promising method to mitigate the domain shift problem. Existing domain adaptation methods assume that the feature distributions of the source and target domains are different, but share the same label space. These methods either aim to build a bridge between source and target domains [4,35] or identify the shared feature space between two domains [11,23].

Recently, deep neural network methods have achieved great success in domain adaptation problems. Adversarial learning shows its power within deep neural

© Springer Nature Switzerland AG 2021
A. Del Bimbo et al. (Eds.): ICPR 2020 Workshops, LNCS 12662, pp. 672–687, 2021.
https://doi.org/10.1007/978-3-030-68790-8_52

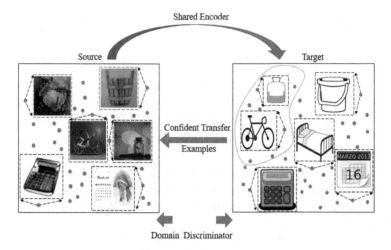

Fig. 1. The scheme of our proposed adversarial continuous learning in unsupervised domain adaptation (ACDA) model. It combines continuous learning and adversarial learning in a two-round classification framework. Initially, the shared encoder will learn a mapping from source images to target images and fool the domain discriminator (which attempts to distinguish examples from source versus target domains). In the second round, the shared encoder will be trained with a new training set which contains the original source images and confidence transfer examples from the target domain, resulting in an improved mapping. The yellow circle marks confident transfer examples.

networks to learn feature representations to minimize the discrepancy between the source and target domains [9,22]. These methods were inspired by the generative adversarial network (GAN) [5]. Adversarial learning also contains a feature extractor and a domain discriminator. The domain discriminator aims to distinguish the source domain from the target domain, while the feature extractor aims to learn domain-invariant representations to fool the domain discriminator [9,12,34]. The target domain risk is minimized via minimax optimization.

Although adversarial learning achieves remarkable results in domain adaptation, it still suffers from two challenges: (1) although the feature extractor is well trained on the source domain, its applicability to the target domain is lower; that is, the joint distributions of the two domains are not perfectly aligned; and, (2) generating proper transfer examples during training has not been well explored; such examples should enhance the positive transfer and alleviate the negative transfer, and these transfer examples should not affect the distributions of the original domains.

To address the above challenges, we take advantage of the source classifier, and generate transfer examples from the target domain, and then adversarially learn them during the second training. In this two-round paradigm, the feature extractor is not only trained with source data, but also the positive samples from the target domain. In addition, the transferred examples are adversarially learned during the training.

In this paper, we employed a two-round paradigm, and utilize five loss functions in one framework: classification loss, adversarial domain discrepancy loss, deep correlation loss, transfer loss and domain alignment loss. By aggregating these loss functions, our model can reduce the domain shift effect and thus enhance transferability from the source domain to the target domain. The scheme of our model is shown in Fig. 1.

Our principal contributions are three-fold:

1. We propose a novel method: adversarial continuous learning in unsupervised domain adaptation (ACDA). The proposed ACDA model adversarially learns high confidence examples from the target domain and confuses the domain discriminator;
2. We are the first to propose a deep correlation loss to help ensure that predictions are locally consistent (with those of nearby examples);
3. To better represent the learned features and train a robust classifier, we dynamically align both marginal and conditional distributions of source and target domains in a two-level domain alignment setting.

Extensive experiments on three highly competitive benchmark image datasets show that our ACDA model can significantly improve the classification accuracy over the state of the art.

2   Related Work

Domain adaptation has emerged as a prominent method to address the domain shift problem. Recently, adversarial learning models have been found to be a better mechanism for identifying invariant representations in domain adaptation.

Adversarial learning based models aim to define a domain confusion objective to identify the domains via a domain discriminator, and the feature extractor fools the discriminator [22,31]. The target domain risk is minimized via playing the minimax game. The domain-adversarial neural network (DANN) considers a minimax loss to integrate a gradient reversal layer to promote the discrimination of source and target domains [2]. The adversarial discriminative domain adaptation (ADDA) method uses an inverted label GAN loss to split the source and target domain, and features can be learned separately [22]. Zhang et al. [28] reweighed the target samples using the degree of confusion between source and target domains. The target samples are assigned by higher weights, which can confuse the domain discriminator. Miyato et al. incorporated virtual adversarial training (VAT) in semi-supervised contexts to smooth the output distributions as a regularization of deep networks [16]. Later, virtual adversarial domain adaptation (VADA) improved adversarial feature adaptation using VAT and harnessed the cluster assumption (decision boundaries cannot cross high-density data regions). It generated adversarial examples against the source classifier and adapted on the target domain [20]. Different from VADA methods, transferable adversarial training (TAT) adversarially generates transferable examples that fit the gap between source and target domain, yet these examples can affect original distributions of two domains [9].

3 Methods

3.1 Motivation

Existing domain adaptation theory shows that the risk in the target domain can be minimized by bounding the source risk and discrepancy between source and target domains (Theorem 1, from Ben-David et al. [1]). Inspired by GAN [5], adversarial learning [9,22] is designed to reduce the discrepancy between two domains. However, Tzeng et al.'s adversarial discriminative domain adaptation (ADDA) model does not generate any examples during training, and it produces hypotheses that require too large of an adaptability correction [22]. Although Liu et al.'s transferable adversarial training (TAT) model generates adversarial examples based on source classifier and domain discriminator, the generated examples are not real examples in either domain, and it still affects the distributions of both domains [9]. In contrast, we choose to identify high-confidence examples from the target domain and use them to help train the classifier. This not only confuses the domain discriminator, but also deceives the source classifier to push the decision boundary towards the target domain without changing the original data distributions.

Theorem 1 [1]. *Let θ be a hypothesis; $\epsilon_s(\theta)$ and $\epsilon_t(\theta)$ represent the source and target risk, respectively.*

$$\epsilon_t(\theta) \leq \epsilon_s(\theta) + d_{\mathcal{H}}(\mathcal{D}_{\mathcal{S}}, \mathcal{D}_{\mathcal{T}}) + \lambda \tag{1}$$

where $d_{\mathcal{H}}(\mathcal{D}_{\mathcal{S}}, \mathcal{D}_{\mathcal{T}})$ is the $\mathcal{H}$-divergence of source and target domains, λ is the adaptability to quantify the error in hypothesis space of two domains, which should be sufficiently small.

3.2 Problem and Notation

For unsupervised domain adaptation, given a source domain $\mathcal{D}_{\mathcal{S}} = \{\mathcal{X}_{\mathcal{S}i}, \mathcal{Y}_{\mathcal{S}i}\}_{i=1}^{n_s}$ of n_s labeled samples in C categories and a target domain $\mathcal{D}_{\mathcal{T}} = \{\mathcal{X}_{\mathcal{T}j}\}_{j=1}^{n_t}$ of n_t samples without any labels ($\mathcal{Y}_{\mathcal{T}}$ for evaluation only). Our ultimate goal is to learn a classifier f under a feature extractor G, that provides lower generalization error in target domain.

In this paper, we present our approach: adversarial continuous learning for domain adaptation (ACDA). It incorporates two paradigms: it selects high-confidence examples from the target domain for transferring to the training of the source classifier, and then adversarially trains those high-confidence transfer examples together with the original labeled source and unlabeled target domains, while the number of categories (C) is the same as during training.

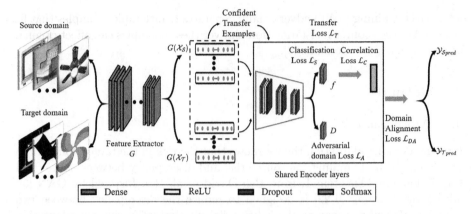

Fig. 2. The architecture of our proposed ACDA model. We first extract features from both source and target domains via G. In first round classification, the shared encoder layers are trained with examples from the labeled source domain and the unlabeled target domain. In the second round, the shared encoder layers are trained with additional high-confidence transfer examples with labels from the first round classifier in the target domain. The domain alignment loss reduces the difference of the marginal and conditional distributions between source and target domains. The red outline highlights the shared layers for both classifier f and domain discriminator D. $\mathcal{Y}_{S\,pred}$ and $\mathcal{Y}_{T\,pred}$ are the predicted labels of the source and target domains after performing domain distribution alignment.

3.3 Source Classifier

The task in the source domain is trained using the cross-entropy loss in Eq. 2:

$$\mathcal{L}_S(f(G(\mathcal{X}_S)), \mathcal{Y}_S) = -\frac{1}{n_s} \sum_{i=1}^{n_s} \sum_{c=1}^{C} \mathcal{Y}_{S\,ic} \log(f_c(G(\mathcal{X}_{Si}))), \qquad (2)$$

where $\mathcal{Y}_{S\,ic} \in [0,1]^C$ is the probability of each class c among true labels, and $f_c(G(\mathcal{X}_{Si}))$ is the predicted probability of each class.

3.4 Adversarial Domain Loss

Given the feature representation of feature extractor G, we can learn a discriminator D, which can distinguish between the two domains in Eq. 3.

$$\mathcal{L}_A(G(\mathcal{X}_S), G(\mathcal{X}_T)) = -\frac{1}{n_s} \sum_{i=1}^{n_s} \log(1 - D(G(\mathcal{X}_{Si}))) - \frac{1}{n_t} \sum_{j=1}^{n_t} \log(D(G(\mathcal{X}_{Tj})))$$

$$(3)$$

3.5 Deep Correlation Loss

Our features are extracted from a well-trained model using the last fully connected layer; we assume that two highly similar examples should belong to the same class:

$$\mathcal{Y}_m = \mathcal{Y}_n \text{ if } Sim(G(\mathcal{X}_m), G(\mathcal{X}_n)) > Sim(G(\mathcal{X}_m), G(\mathcal{X}_{n \neq m})), \tag{4}$$

where $\mathcal{X}_m$ and $\mathcal{X}_n$ are samples from the same domain. Sim is the cosine similarity. We then rank all similarity scores and calculate the top-K cosine similarity matrix for both source and target domains. Hence, we can compare correlated labels with the predicted labels from the source classifier. The loss is defined as:

$$\mathcal{L}_C(\mathcal{Y}_{pred}, \mathcal{Y}_{corr}) = \frac{1}{n_{s/t}} \sum_{i/j=1}^{n_{s/t}} ||\mathcal{Y}_{pred_{i/j}} - M(\mathcal{Y}_{pred}[\mathcal{Y}_{corr_{i/j}}])||, \tag{5}$$

where $\mathcal{Y}_{pred}$ is the prediction of either source domain or target domain from the first round source classifier, and $\mathcal{Y}_{corr}$ is the correlation label with the size of $n_{s/t} \times K$; it shows the top-K index, which is highly related to the instance that should be in the same class. $\mathcal{Y}_{pred}[\mathcal{Y}_{corr}]$ is the updated matrix for the predicted labels and $M(\cdot)$ selects the most frequent labels in the updated matrix as shown in Fig. 3. Therefore, the loss measures how different the predicted label is to its nearest neighbors, which are from the same domain. Notice that this loss can be applied in both source and target domains for inference since we are able to calculate the top-K correlated samples in each domain.

	$\mathcal{Y}_{pred}$	$\mathcal{Y}_{corr}$ (Top-3)	$\mathcal{Y}_{pred}[\mathcal{Y}_{corr}]$	$M(\mathcal{Y}_{pred}[\mathcal{Y}_{corr}])$
$\mathcal{X}_1$	0	2 3 4	1 0 0	0
$\mathcal{X}_2$	1	1 3 4	0 0 0	0
$\mathcal{X}_3$	0	1 2 4	0 1 0	0
$\mathcal{X}_4$	0	5 1 3	1 0 0	0
$\mathcal{X}_5$	1	4 1 3	0 0 0	0

Fig. 3. An example of top-3 correlation labels in updating predicted labels. Given five examples ($\mathcal{X}_1$ to $\mathcal{X}_5$), the prediction is the $\mathcal{Y}_{pred}$, which is from classifier f. The predictions of five examples are within two classes (0 and 1), and the true labels are all zeros. There are two incorrectly predicted results ($\mathcal{Y}_{pred_2}$ and $\mathcal{Y}_{pred_5}$). The top-3 $\mathcal{Y}_{corr}$ shows the top 3 instances that should be in the same class (e.g., $\mathcal{X}_1$ should have the same class as $\mathcal{X}_2$, $\mathcal{X}_3$ and $\mathcal{X}_4$.). The $M(\mathcal{Y}_{pred}[\mathcal{Y}_{corr}])$ changes the predicted labels and is the same as the truth label.

3.6 Continuous Learning

The purpose of continuous learning in unsupervised domain adaptation is to bring high-confidence transfer examples (positive samples) from the target domain to the source domain with a high probability threshold. The high probability ensures fewer negative samples are from the target domain, which causes negative transfer in the source classifier (f). In the first round prediction, we get $\mathcal{Y}'_{T_p}$ of $\mathcal{X}_T$ ($\mathcal{Y}'_{T_p}$ is the probability representation of prediction $\mathcal{Y}'_T$). The high-confidence transfer examples $\{\mathcal{X}'_{T_k}\}_{k=0}^{n_k}$, where $0 \leq n_k \leq n_t$, and n_k is determined by the hyper-parameter $\mathcal{P}$. Therefore, we have the following definition.

Definition: one sample in target domain is considered a high-confidence transfer example if and only if its prediction probability in the dominant class $\mathcal{Y}'_{T_{p_k}} \geq \mathcal{P}$. In the continuous learning setting, the new source domain consists of original source data plus high-confidence transfer examples from the target domain $\mathcal{X}'_S = \mathcal{X}_S + \mathcal{X}'_T$ with its new labels $\mathcal{Y}'_S = \mathcal{Y}_S + \mathcal{Y}'_T$.

Similar to training in the first round classification, we also train the new source domain via adversarial learning (in which high-confidence transfer examples effectively belong to both the target and new source domains); hence the transfer loss includes three new loss functions in the black box of Fig. 2.

$$
\mathcal{L}_T(\mathcal{X}'_S, \mathcal{Y}'_S, \mathcal{X}_T, \mathcal{Y}_{T\,corr}) = \mathcal{L}'_S(f(G(\mathcal{X}'_S)), \mathcal{Y}'_S) \\
+ \mathcal{L}_C(\mathcal{Y}'_{S\,pred}, \mathcal{Y}'_{S\,corr}) + \mathcal{L}_A(G(\mathcal{X}'_S), G(\mathcal{X}_T)),
\tag{6}
$$

where $\mathcal{L}'_S(f(G(\mathcal{X}'_S)), \mathcal{Y}'_S) = -\frac{1}{n'_s} \sum_{i=1}^{n'_s} \sum_{c=1}^{C} \mathcal{Y}'_{S\,ic} \log\,(f(G(\mathcal{X}'_{S_i})))$; $\mathcal{Y}'_{S\,pred}$ and $\mathcal{Y}'_{S\,corr}$ are predicted and correlated labels of the new source domain, respectively. The generated high-confidence transfer examples are together trained with original source data and new domain discrepancy is minimized in Eq. 7.

$$
\mathcal{L}_A(G(\mathcal{X}'_S), G(\mathcal{X}'_T)) = -\frac{1}{n'_s} \sum_{i=1}^{n'_s} \log(1 - D(G(\mathcal{X}'_{S_i}))) - \frac{1}{n_t} \sum_{j=1}^{n_t} \log(D(G(\mathcal{X}_{T_j})))
\tag{7}
$$

The discriminator also distinguishes between the new source domain and original target domain; the high-confidence transfer examples will further confuse the domain discriminator.

$$
\mathcal{L}_C(\mathcal{Y}'_{S\,pred}, \mathcal{Y}'_{S\,corr}) = \frac{1}{n'_s} \sum_{i=1}^{n'_s} ||\mathcal{Y}'_{S\,pred_i} - M(\mathcal{Y}'_{S\,pred}[\mathcal{Y}'_{S\,corr_i}])||
\tag{8}
$$

The deep correlation loss will also be minimized in the new source domain and original target domain. We adversarially generate the high-confidence transfer examples from the target domain, which not only maintains the distributions of two domains but also against both source classifier and domain discriminator. As aforementioned, we can minimize target domain risk via bounding the source risk and discrepancy between the source and target domains [1]. In the first round classification, we have a labeled source domain and unlabeled target domain. In the second round classification, we have a labeled new source domain (labels for high-confidence transfer examples are from the prediction of the source classifier), and unlabeled target domain. Hence, the source classifier trains with additional high-confidence transfer examples from the target domain; these examples will push the decision boundary of source classifier towards the target domain. Therefore, the target domain risk will be further reduced.

3.7   Shared Encoder Layers

The shared encoder layers begin with three repeated blocks and each block has a Dense layer, a "Relu" activation layer, and a Dropout layer. The numbers of units of the dense layer are 512, 128, and 64, respectively. The rate of the Dropout layer is 0.5. It ends with a Dense layer (the number of units is the number of classes in each dataset (10, 31, and 65 in our experiments)).

3.8 A Two-Level Dynamic Distribution Alignment

After the two-round classification, we can get the prediction of the target domain. However, we can further improve the predicted accuracy by employing a two-level dynamic distribution alignment, which can dynamically update the predicted labels in the target domain.

Manifold Embedded Distribution Alignment (MEDA), proposed by Wang et al. [25], aligns learned features from manifold learning. However, Zhang et al. showed that there are defects in estimating the geodesic of sub-source and sub-target domains [35]. We modified the domain alignment loss as follows:

$$
\begin{aligned}
\mathcal{L}_{\mathcal{D}\mathcal{A}}(\mathcal{D}_{\mathcal{S}}, \mathcal{D}_{\mathcal{T}}) = \arg\min(&\mathcal{L}_{\mathcal{G}}(f(G(\mathcal{X}_{\mathcal{S}})), \mathcal{Y}_{\mathcal{S}}) + \eta\|f\|^2 \\
&+ \lambda \overline{D_f}(\mathcal{D}_{\mathcal{S}}, \mathcal{D}_{\mathcal{T}}) + \rho R_f(\mathcal{D}_{\mathcal{S}}, \mathcal{D}_{\mathcal{T}}))
\end{aligned}
\tag{9}
$$

where f is the classifier from the shared encoder, $\mathcal{L}_{\mathcal{G}}$ is the sum of squares loss; $\|f\|^2$ is the squared norm of f; and the first two terms minimize the structure risk of shared encoder. $\overline{D_f}(\cdot, \cdot)$ represents the dynamic distribution alignment; $R_f(\cdot, \cdot)$ is a Laplacian regularization; η, λ, and ρ are regularization parameters. Specifically, $\overline{D_f}(\mathcal{D}_{\mathcal{S}}, \mathcal{D}_{\mathcal{T}}) = (1-\mu)D_f(P_{\mathcal{S}}, P_{\mathcal{T}}) + \mu\sum_{c=1}^{C} D_f^c(Q_{\mathcal{S}}, Q_{\mathcal{T}})$, where μ is an adaptive factor to balance the marginal distribution $(P_{\mathcal{S}}, P_{\mathcal{T}})$, and conditional distribution $(Q_{\mathcal{S}}, Q_{\mathcal{T}})$, and $c \in \{1, \cdots, C\}$ is the class indicator [25].

For overall training procedures, we first train the labeled source examples and unlabeled target examples using Eqs. 2, 3, and 5 in the first round classification. We then train the labeled new source domain and unlabeled target domain using Eq. 6 in the second round classification. Finally, we perform domain distribution alignment using Eq. 9.

3.9 The Overall Training Objective Function

The architecture of our proposed ACDA model is shown in Fig. 2. Taken together, our model minimizes the following objective function:

$$
\begin{aligned}
\mathcal{L}(\mathcal{X}_{\mathcal{S}}, \mathcal{Y}_{\mathcal{S}}, \mathcal{X}_{\mathcal{T}}, \mathcal{Y}_{\mathcal{T}corr}) = \arg\min \ &\alpha\mathcal{L}_{\mathcal{S}}(f(G(\mathcal{X}_{\mathcal{S}})), \mathcal{Y}_{\mathcal{S}}) + (1-\alpha)\mathcal{L}_{\mathcal{C}}(\mathcal{Y}_{pred}, \mathcal{Y}_{corr}) \\
&+ \beta\mathcal{L}_{\mathcal{A}}(G(\mathcal{X}_{\mathcal{S}}), G(\mathcal{X}_{\mathcal{T}})) + \mathcal{L}_{\mathcal{T}}(\mathcal{X}'_{\mathcal{S}}, \mathcal{Y}'_{\mathcal{S}}, \mathcal{X}_{\mathcal{T}}, \mathcal{Y}_{\mathcal{T}corr}) + \mathcal{L}_{\mathcal{D}\mathcal{A}}(\mathcal{D}_{\mathcal{S}}, \mathcal{D}_{\mathcal{T}})
\end{aligned}
\tag{10}
$$

where f is the classifier from the shared encoder module; $\mathcal{L}_{\mathcal{S}}$ is the classification loss, which is the typical cross-entropy loss in Eq. 2; $\mathcal{Y}_{pred}$ is the predicted label for the target domain and $\mathcal{Y}_{corr}$ is the correlated label, which shows the K most highly related instances ($\mathcal{Y}_{\mathcal{T}corr}$ is the correlated label for the target domain). The $\mathcal{L}_{\mathcal{C}}, \mathcal{L}_{\mathcal{T}}$ and $\mathcal{L}_{\mathcal{D}\mathcal{A}}$ represent the correlation loss, transfer loss, and the domain alignment loss, respectively. α and β are balance factors. $\{\mathcal{X}'_{\mathcal{S}}, \mathcal{Y}'_{\mathcal{S}}\}$ is the new source domain.

4   Experiments

In this section, we show how the ACDA model can enhance image recognition accuracy. Our model are evaluated using three public image datasets: Office + Caltech-10, Office-31 and Office-Home [19,25]. These datasets are widely used in many publications [4,6,25], and are the benchmarking data for evaluating the performance of domain adaptation algorithms by training in one domain and testing on another. In addition, we conduct ablation studies to investigate the impact of each component in the ACDA model[1].

4.1   Datasets

Office + Caltech-10 [4] consists of 2,533 images in four domains: Amazon (A), Webcam (W), DSLR (D) and Caltech (C) in ten classes. In experiments, C→A represents learning knowledge from domain C which is applied to domain A. There are twelve tasks in Office + Caltech-10 dataset. **Office-31** [19] is another benchmark dataset for domain adaptation, consisting of 4,110 images in 31 classes from three domains: Amazon (A), Webcam (W), and DSLR (D). We evaluate all methods across all six transfer tasks. **Office-Home** [24] contains 15,588 images from 65 categories. It has four domains: Art (Ar), Clipart (Cl), Product (Pr) and Real-World (Rw). There are also twelve tasks in this dataset.

4.2   Implementation Details

We extract features for the three datasets from a fine-tuned Resnet50 network [7]. We add one more Dense layer and the number of units is the same as the number of classes in each dataset. The 1,000 features are then extracted from the second-to-last fully connected layer [29,33]. Parameters in domain distribution alignment are $\eta = 0.1$, $\lambda = 10$, and $\rho = 10$, which are fixed based on previous research [25]. $\alpha = 0.5$, $\beta = 0.2$, $K = 3$, $\mathcal{P} = 0.9999$, batch size = 32 and number of iterations = 1000 are determined by performance on the source domain. We also compare our results with 19 state-of-the-art methods (including both traditional methods and deep neural networks).

4.3   Results

The performance on Office + Caltech-10, Office-Home and Office-31 are shown in Tables 1, 2, 3. For a fair comparison, we highlight in bold those methods that are re-implemented using our extracted features. Our ACDA model outperforms all state-of-the-art methods in terms of average accuracy (especially in the Office-Home dataset). It is compelling that our ACDA model substantially enhances the classification accuracy on difficult adaptation tasks (e.g., D→A task in the Office-31 dataset and the challenging Office-Home dataset (which has a larger

---

[1] Source code is available at https://github.com/YoushanZhang/Transfer-Learning/tree/main/Code/Deep/ACDA.

Table 1. Accuracy (%) on Office + Caltech-10 dataset

Task	C→A	C→W	C→D	A→C	A→W	A→D	W→C	W→A	W→D	D→C	D→A	D→W	Ave.
GFK [4]	94.6	94.9	95.5	92.6	90.5	94.3	93.5	95.7	**100**	93.6	95.9	98.6	95.0
GSM [35]	96.0	95.9	96.2	**94.6**	89.5	92.4	94.1	95.8	**100**	93.9	95.1	98.6	95.2
TJM [13]	94.7	86.8	86.6	83.6	82.7	76.4	88.2	90.9	98.7	87.4	92.5	98.3	88.9
JGSA [27]	95.1	97.6	96.8	93.9	94.2	96.2	95.5	95.9	**100**	94.0	96.3	99.3	96.2
ARTL [10]	**96.3**	94.9	96.2	93.9	98.3	97.5	94.7	**96.7**	**100**	**94.4**	96.2	99.7	96.6
MEDA [25]	**96.3**	98.3	96.2	**94.6**	99.0	**100**	**94.8**	96.6	**100**	93.6	96.0	99.3	97.0
DAN [11]	92.0	90.6	89.3	84.1	91.8	91.7	81.2	92.1	**100**	80.3	90.0	98.5	90.1
DDC [23]	91.9	85.4	88.8	85.0	86.1	89.0	78.0	83.8	**100**	79.0	87.1	97.7	86.1
DCORAL [21]	89.8	97.3	91.0	91.9	**100**	90.5	83.7	81.5	90.1	88.6	80.1	92.3	89.7
RTN [14]	93.7	96.9	94.2	88.1	95.2	95.5	86.6	92.5	**100**	84.6	93.8	99.2	93.4
MDDA [18]	93.6	95.2	93.4	89.1	95.7	96.6	86.5	94.8	**100**	84.7	94.7	99.4	93.6
ACDA	96.2	**100**	**100**	93.9	**100**	**100**	93.9	96.2	**100**	93.9	**96.7**	**100**	**97.6**

Table 2. Accuracy (%) on Office-Home dataset (A: Ar, C: Cl, R: Rw, P: Pr)

Task	A→C	A→P	A→R	C→A	C→P	C→R	P→A	P→C	P→R	R→A	R→C	R→P	Ave.
GFK [4]	40.0	66.5	71.8	56.8	66.4	65.1	58.1	43.0	74.1	65.3	44.9	76.3	60.7
GSM [35]	49.4	75.5	80.2	62.9	70.6	70.3	65.6	50.0	80.8	72.4	50.4	81.6	67.5
TJM [13]	50.0	60.1	61.3	42.9	60.0	57.3	38.6	34.7	63.3	48.7	38.0	67.2	51.8
JGSA [27]	45.8	73.7	74.5	52.3	70.2	71.4	58.8	47.3	74.2	60.4	48.4	76.8	62.8
ARTL [10]	56.5	80.4	81.4	68.7	81.6	81.7	70.1	56.2	83.3	72.8	58.2	85.6	73.0
MEDA [25]	49.1	75.6	79.1	66.7	77.2	75.8	68.2	50.4	79.9	71.9	53.2	82.0	69.1
DAN [11]	43.6	57.0	67.9	45.8	56.5	60.4	44.0	43.6	67.7	63.1	51.5	74.3	56.3
DANN [3]	45.6	59.3	70.1	47.0	58.5	60.9	46.1	43.7	68.5	63.2	51.8	76.8	57.6
JAN [15]	45.9	61.2	68.9	50.4	59.7	61.0	45.8	43.4	70.3	63.9	52.4	76.8	58.3
CDAN-RM [12]	49.2	64.8	72.9	53.8	62.4	62.9	49.8	48.8	71.5	65.8	56.4	79.2	61.5
CDAN-M [12]	50.6	65.9	73.4	55.7	62.7	64.2	51.8	49.1	74.5	68.2	56.9	80.7	62.8
TADA [26]	53.1	72.3	77.2	59.1	71.2	72.1	59.7	53.1	78.4	72.4	**60.0**	82.9	67.6
SymNets [34]	47.7	72.9	78.5	64.2	71.3	74.2	64.2	48.8	79.5	74.5	52.6	82.7	67.6
MDA [30]	54.8	81.2	82.3	71.9	82.9	81.4	71.1	53.8	82.8	75.5	55.3	86.2	73.3
ACDA	**58.4**	**83.1**	**84.4**	**74.6**	**84.1**	**83.5**	**74.4**	**58.7**	**85.2**	**77.2**	59.3	**87.5**	**75.9**

Table 3. Accuracy (%) on Office-31 dataset

Task	A→W	A→D	W→A	W→D	D→A	D→W	Ave.
GFK [4]	81.5	82.7	73.5	97.8	72.2	95.3	83.8
GSM [35]	85.9	84.1	75.5	97.2	73.6	95.6	85.3
TJM [13]	87.5	62.2	64.3	97.4	61.3	92.3	77.5
JGSA [27]	89.1	91.0	77.9	**100**	77.6	98.2	89.0
ARTL [10]	90.9	93.0	77.1	99.6	78.2	98.3	89.6
MEDA [25]	91.7	89.2	77.2	97.4	76.5	96.2	88.0
JAN [15]	85.4	84.7	70.0	99.8	68.6	97.4	84.3
TADA [26]	94.3	91.6	73.0	99.8	72.9	98.7	88.4
SymNets [34]	90.8	93.9	72.5	**100**	74.6	98.8	88.4
SSD [32]	92.7	91.6	77.5	99.4	77.8	98.0	89.4
CAN [8]	94.5	95.0	77.0	99.8	78.0	**99.1**	90.6
ACDA	**95.5**	**96.2**	**80.2**	99.2	**81.1**	98.4	**91.8**

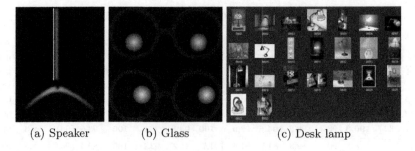

(a) Speaker   (b) Glass   (c) Desk lamp

Fig. 4. Example images to illustrate the importance of probability threshold $\mathcal{P}$. The task is A→W, (a) and (b) is from Clipart domain, and (c) is from Art domain in Office-Home dataset. When $\mathcal{P} = 0.9$, (a) and (b) are mistakenly treated as high-confidence examples in Desk lamp class, and these two examples will be excluded when $\mathcal{P} = 0.99$.

number of categories and different domains are visually dissimilar)). In addition, the improvement on the Office + Caltech-10 dataset is not large. This caused by the high state-of-the-art classification accuracy (more than 97%, it is hence difficult to make a significant improvement). However, our model still provides more than 0.6% (absolute) improvement over the best baseline method, which translates to a relative reduction in error of 20%. These experiments demonstrate the efficiency of the ACDA model in aligning both the marginal and conditional distributions of two domains.

4.4 Ablation Study

We first consider the effects of three different loss functions on classification accuracy ($\mathcal{L}_S$ and $\mathcal{L}_A$ are required for adversarial learning). Different combinations of loss functions are reported in Table 4, in which T represents transfer loss, C, correlation loss, and DA, distribution alignment loss. "ACDA−T/C/DA" is implemented without transfer loss, correlation loss, and distribution alignment loss. It is a simple model, which only reduces the source risk without minimizing the domain discrepancy using adversarial learning. "ACDA−T/C" directly aligns

Table 4. Ablation experiments on Office-31 dataset

Task	A→W	A→D	W→A	W→D	D→A	D→W	Ave.
ACDA−T/C/DA	88.2	90.6	75.3	97.8	74.4	94.8	86.9
ACDA−T/DA	85.8	91.6	75.6	98.0	76.3	97.4	87.5
ACDA−T/C	91.9	88.5	78.3	98.1	77.8	97.3	88.7
ACDA−C/DA	91.4	91.6	78.3	98.8	78.4	96.9	89.2
ACDA−T	94.5	94.0	77.7	99.0	78.8	97.1	90.2
ACDA−DA	94.3	94.2	79.5	99.6	79.5	97.7	90.8
ACDA−C	95.4	95.1	79.1	99.0	79.8	97.9	91.1
ACDA	**95.5**	**96.2**	**80.2**	**99.2**	**81.1**	**98.4**	**91.8**

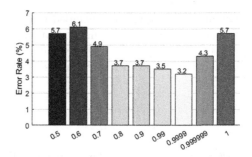

Fig. 5. Error rate of different probability thresholds $\mathcal{P}$ in Office + Caltech 10 dataset (x-axis is different $\mathcal{P}$).

Table 5. The number of high-confidence transfer examples under different $\mathcal{P}$ on Office + Caltech-10 dataset

$\mathcal{P}$	C→A	C→W	C→D	A→C	A→W	A→D	W→C	W→A	W→D	D→C	D→A	D→W
−	958	295	157	1123	295	157	1123	958	157	1123	958	295
0.5	956	295	157	1118	295	157	1112	950	157	1114	950	295
0.7	939	282	154	1076	278	150	969	894	157	1049	907	289
0.9	923	278	152	1056	272	147	804	675	156	900	685	273
0.9999	266	194	103	578	175	90	145	364	54	565	751	90
0.9999999	18	104	92	351	27	20	0	0	11	0	1	0
1	0	0	0	0	0	0	0	0	0	0	0	0

the joint distribution of the two domains. "ACDA−DA" reports results without performing the additional domain distribution alignment. We observe that with the increasing of the number of loss functions, the robustness of our model keeps improving. The usefulness of loss functions is ordered as $\mathcal{L}_{\mathcal{C}} < \mathcal{L}_{\mathcal{DA}} < \mathcal{L}_{\mathcal{T}}$. Therefore, the proposed continuous learning approach and correlation loss are effective in improving performance, and different loss functions are helpful and important in minimizing the target domain risk.

We further study the use of high-confidence transfer examples. To show the importance of selecting a high probability threshold $\mathcal{P}$, Fig. 4 shows two unusual examples, in which the task is from Art domain to Clipart domain. In our continuous learning setting, we bring high-confidence transfer examples into target domain (Clipart). Examples (a) and (b) are from the Clipart domain, and if $\mathcal{P} = 0.9$, these two will be treated as high-confidence transfer examples. However, these two are wrongly classified as desk lamps, while the original classes are speaker and glasses. When $\mathcal{P} = 0.9$, these two examples are included. Hence, $\mathcal{P} = 0.9999$ is a sufficient threshold to eliminate these negative examples.

We also show the performance of different $\mathcal{P}$ on classification error rate in Table 5 and Fig. 5. In Table 5, "−" lists the number of samples in the target domain (e.g., in C→A, the number of samples in A is reported). With the increasing of $\mathcal{P}$, the number of high confidence transfer examples is decreased, and error rate is first decreased and then increased. This phenomenon is induced by either the negative transfer examples are included in the training when $\mathcal{P}$ is small or fewer transfer examples are included in the training when $\mathcal{P}$ is too high. Therefore, $\mathcal{P} = 0.9999$ is the best hyper-parameter value for our model (hyperparameter $\mathcal{P}$ is tuned based on the top-3 correlated labels using the source domain data).

5 Discussion

There are two compelling advantages of the proposed ACDA model. First, we employed two-round classification and adversarially learned high-confidence transfer examples from the target domain. Secondly, our model surpasses the other models across all three datasets even using the same features, which reflects the value of our model on domain adaptation for image recognition. We also present the visualization of the W→A task in the Office-31 dataset to show the progress of our ACDA model. In Fig. 6(a), the feature representation of domain W of different methods is shown. For features of DAN, MEDA, JAN and Resnet50 model, the data distribution is not better aligned since data structures are mixing together. Although the features of the GSM model are slightly better

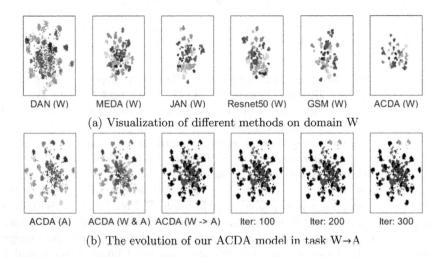

(a) Visualization of different methods on domain W

(b) The evolution of our ACDA model in task W→A

Fig. 6. t-SNE view of task W→A in Office-31 dataset. Different colors in (a) and first two images in (b) represent different categories. In the last four images of (b), black dots represent domain A, red dots represent domain W and green dots are high-confidence examples, which are adversarially learned in ACDA (Iter: iteration).

Table 6. Top-1 correlated label accuracy (%) on three benchmarking datasets.

Domain	Office + Caltech-10				Office-31			Office-Home			
	A	C	D	W	A	D	W	Ar	Cl	Pr	Rw
Correlated	95.7	96.6	100	100	88.4	99.0	99.2	76.7	76.0	93.3	85.0

than the first four methods, it is still worse than our features. Figure 6(b) shows the evolution of selecting high-confidence transfer examples (green color dots) using our ACDA model. As the number of iterations increases, the more high confidence transfer examples are adversarially generated from target domain (A), which will be included in source domain (W). The accuracy in the target can be improved via such an evolution.

To validate the assumption that "two highly correlated examples should belong to the same class" in Sect. 3.5, we further compare the classes and show the matching class accuracy using the top-1 correlated labels in Table 6. This demonstrates that the correlated labels are useful in updating predicted labels.

Unfortunately, performance is not always better than other methods, which is a function of the differences across specific domain adaptation tasks. Furthermore, we use the top-K highly correlated instances to construct the adjacency matrix (top-3 in our experiments); other K could be further explored.

6   Conclusion

We propose novel adversarial continuous learning in unsupervised domain adaptation (termed ACDA) to overcome limitations in generating proper transfer examples and aligning the joint distributions of two domains by minimizing five loss functions. The generated transfer examples can help to further learn the domain invariant of the two domains. As a component of our ACDA model, explicit domain-invariant features are learned through such a cross-domain training scheme. Experiments on three benchmark datasets show the robustness of our proposed ACDA model.

References

1. Ben-David, S., Blitzer, J., Crammer, K., Kulesza, A., Pereira, F., Vaughan, J.W.: A theory of learning from different domains. Machine Learn. **79**(1), 151–175 (2009). https://doi.org/10.1007/s10994-009-5152-4
2. Ganin, Y., et al.: Domain-adversarial training of neural networks. J. Mach. Learn. Res. **17**(1), 2030–2096 (2016)
3. Ghifary, M., Kleijn, W.B., Zhang, M.: Domain adaptive neural networks for object recognition. In: Pham, D.-N., Park, S.-B. (eds.) PRICAI 2014. LNCS (LNAI), vol. 8862, pp. 898–904. Springer, Cham (2014). https://doi.org/10.1007/978-3-319-13560-1_76
4. Gong, B., Shi, Y., Sha, F., Grauman, K.: Geodesic flow kernel for unsupervised domain adaptation. In: Proceedings of IEEE Conference on Computer Vision and Pattern Recognition (CVPR), pp. 2066–2073. IEEE (2012)
5. Goodfellow, I., et al.: Generative adversarial nets. In: Advances in Neural Information Processing Systems, pp. 2672–2680 (2014)
6. Gopalan, R., Li, R., Chellappa, R.: Domain adaptation for object recognition: an unsupervised approach. In: Proceedings of the IEEE International Conference on Computer Vision (ICCV), pp. 999–1006. IEEE (2011)

7. He, K., Zhang, X., Ren, S., Sun, J.: Deep residual learning for image recognition. In: Proceedings of the IEEE Conference on Computer Vision and Pattern Recognition (CVPR), pp. 770–778 (2016)
8. Kang, G., Jiang, L., Yang, Y., Hauptmann, A.G.: Contrastive adaptation network for unsupervised domain adaptation. In: Proceedings of the IEEE Conference on Computer Vision and Pattern Recognition, pp. 4893–4902 (2019)
9. Liu, H., Long, M., Wang, J., Jordan, M.: Transferable adversarial training: a general approach to adapting deep classifiers. In: International Conference on Machine Learning, pp. 4013–4022 (2019)
10. Long, J., Wang, M., Ding, G., Pan, S.J., Philip, S.Y.: Adaptation regularization: a general framework for transfer learning. IEEE Trans. Knowl. Data Eng. **26**(5), 1076–1089 (2013)
11. Long, M., Cao, Y., Wang, J., Jordan, M.I.: Learning transferable features with deep adaptation networks. arXiv preprint arXiv:1502.02791 (2015)
12. Long, M., Cao, Z., Wang, J., Jordan, M.I.: Conditional adversarial domain adaptation. In: Advances in Neural Information Processing Systems, pp. 1647–1657 (2018)
13. Long, M., Wang, J., Ding, G., Sun, J., Yu, P.S.: Transfer joint matching for unsupervised domain adaptation. In: Proceedings of the IEEE Conference on Computer Vision and Pattern Recognition, pp. 1410–1417 (2014)
14. Long, M., Zhu, H., Wang, J., Jordan, M.I.: Unsupervised domain adaptation with residual transfer networks. In: Advances in Neural Information Processing Systems, pp. 136–144 (2016)
15. Long, M., Zhu, H., Wang, J., Jordan, M.I.: Deep transfer learning with joint adaptation networks. In: Proceedings of the 34th International Conference on Machine Learning, vol. 70, pp. 2208–2217. JMLR.org (2017)
16. Miyato, T., Maeda, S., Koyama, M., Ishii, S.: Virtual adversarial training: a regularization method for supervised and semi-supervised learning. IEEE Trans. Pattern Anal. Mach. Intell. **41**(8), 1979–1993 (2018)
17. Pan, S.J., Yang, Q.: A survey on transfer learning. IEEE Trans. Knowl. Data Eng. **22**(10), 1345–1359 (2010)
18. Rahman, M.M., Fookes, C., Baktashmotlagh, M., Sridharan, S.: On Minimum Discrepancy Estimation for Deep Domain Adaptation. In: Singh, R., Vatsa, M., Patel, V.M., Ratha, N. (eds.) Domain Adaptation for Visual Understanding, pp. 81–94. Springer, Cham (2020). https://doi.org/10.1007/978-3-030-30671-7_6
19. K. Saenko, B. Kulis, M. Fritz, and T. Darrell. Adapting visual category models to new domains. In Proceedings of the European Conference on Computer Vision, pages 213–226. Springer, 2010
20. Shu, R., Bui, H.H., Narui, H., Ermon, S.: A dirt-t approach to unsupervised domain adaptation. arXiv preprint arXiv:1802.08735 (2018)
21. Sun, B., Saenko, K.: Deep CORAL: correlation alignment for deep domain adaptation. In: Hua, G., Jégou, H. (eds.) ECCV 2016. LNCS, vol. 9915, pp. 443–450. Springer, Cham (2016). https://doi.org/10.1007/978-3-319-49409-8_35
22. Tzeng, E., Hoffman, J., Saenko, K., Darrell, K.: Adversarial discriminative domain adaptation. In: Proceedings of the IEEE Conference on Computer Vision and Pattern Recognition, pp. 7167–7176 (2017)
23. Tzeng, E., Hoffman, J., Zhang, N., Saenko, K., Darrell, T.: Deep domain confusion: maximizing for domain invariance. arXiv preprint arXiv:1412.3474 (2014)
24. Venkateswara, H., Eusebio, J., Chakraborty, S., Panchanathan., S.: Deep hashing network for unsupervised domain adaptation. In: Proceedings of the IEEE Conference on Computer Vision and Pattern Recognition, pp. 5018–5027 (2017)

25. Wang, J., Feng, W., Chen, Y., Yu, H., Huang, M., Yu, P.S.: Visual domain adaptation with manifold embedded distribution alignment. In: Proceedings of the 26th ACM International Conference on Multimedia, MM 2018, pp. 402–410 (2018)
26. Wang, X., Li, L., Ye, W., Long, M., Wang, J.: Transferable attention for domain adaptation. Proc. AAAI Conf. Artif. Intell. **33**, 5345–5352 (2019)
27. Zhang, J., Li, W., Ogunbona, P.: Joint geometrical and statistical alignment for visual domain adaptation. In: Proceedings of the IEEE Conference on Computer Vision and Pattern Recognition, pp. 1859–1867 (2017)
28. Zhang, W., Ouyang, W., Li, W., Xu, D.: Collaborative and adversarial network for unsupervised domain adaptation. In: Proceedings of the IEEE Conference on Computer Vision and Pattern Recognition, pp. 3801–3809 (2018)
29. Zhang, Y., Allem, J.P., Unger, J.B., Cruz, T.B.: Automated identification of hookahs (waterpipes) on instagram: an application in feature extraction using convolutional neural network and support vector machine classification. J. Med. Internet Res. **20**(11), e10513 (2018)
30. Zhang, Y., Davison, B.D.: Modified distribution alignment for domain adaptation with pre-trained Inception ResNet. arXiv preprint arXiv:1904.02322 (2019)
31. Zhang, Y., Davison, B.D.: Adversarial consistent learning on partial domain adaptation of PlantCLEF 2020 challenge. In: CLEF working notes 2020, CLEF: Conference and Labs of the Evaluation Forum (2020)
32. Zhang, Y., Davison, B.D.: Domain adaptation for object recognition using subspace sampling demons. Multimedia Tools Appl. 1–20 (2020). https://doi.org/10.1007/s11042-020-09336-0
33. Zhang, Y., Davison, B.D.: Impact of ImageNet model selection on domain adaptation. In: Proceedings of the IEEE Winter Conference on Applications of Computer Vision Workshops, pp. 173–182 (2020)
34. Zhang, Y., Tang, H., Jia, K., Tan, M.: Domain-symmetric networks for adversarial domain adaptation. In: Proceedings of the IEEE Conference on Computer Vision and Pattern Recognition, pp. 5031–5040 (2019)
35. Zhang, Y., Xie, S., Davison, B.D.: Transductive learning via improved geodesic sampling. In: Proceedings of the 30th British Machine Vision Conference (2019)

A Survey of Deep Learning Based Fully Automatic Bone Age Assessment Algorithms

Yang Jia[1,2], Hanrong Du[1,2], Haijuan Wang[1,2], Weiguang Chen[1,2], Xiaohui Jin[3], Wei Qi[4], Bin Yang[3(✉)], and Qiujuan Zhang[4(✉)]

[1] School of Computer, Xi'an University of Posts and Telecommunications, Xi'an 710121, China
[2] Shanxi Key Laboratory of Network Data Intelligent Processing, Xi'an University of Posts and Telecommunications, Xi'an 710121, China
[3] Department of Radiology, Xi'an Honghui Hospital, Xi'an 710054, China
42044441@qq.com
[4] Department of Radiology, The Second Affiliated Hospital of Xi'an Jiaotong University, Xi'an 710004, China
Zhangqjlcx@aliyun.com

Abstract. Bone age assessment (BAA) is a technique for assessing the maturity of individual skeletal development, and it is the most accurate and objective method for assessing the deviation of individual development in clinical practice. It is used in the diagnosis of pediatric endocrine diseases, age determination of suspects in juvenile delinquency cases, height prediction, and athlete selection. Early computer-aided BAA mainly segments the skeleton of the hand and gives a final bone age by comparing the morphological descriptions of bones to standard atlas. In recent years, deep learning methods have developed rapidly, and a large number of end-to-end BAA methods based on deep learning have emerged, which have brought new development to the automatic BAA technology. This paper summarizes the technical basis, research status of automatic BAA. Firstly, the basic theory of medical BAA is introduced, then the commonly used traditional segmentation methods in BAA are analyzed. After that, we summarized the most popular methods of BAA based on deep learning, including the network model, data set and the assessment results. This survey also draws attention to a number of research challenges in the fully automatic BAA with deep learning.

Keywords: X-ray image · Bone age assessment (BAA) · Image segmentation · Regression · Deep learning

1 Introduction

Bone age is the abbreviation of "bone development maturity age". Bone age is more accurate than life age to reflect the maturity of the body. Bone age assessment (BAA) is a technology evaluating the maturity of individual bones by measuring the differences between bone age and time age according to the common characteristics of bone development of a specific population.

© Springer Nature Switzerland AG 2021
A. Del Bimbo et al. (Eds.): ICPR 2020 Workshops, LNCS 12662, pp. 688–702, 2021.
https://doi.org/10.1007/978-3-030-68790-8_53

Skeletal maturity can be shown in different parts of the human body, especially in the wrist and the features are universal and irreversible. When evaluating, the skeletal X-ray images are compared with the standard bone development map. If there are differences, further diagnostic evaluation will be needed. BAA is the most accurate and objective method for clinical evaluation of individual development. It plays an important role in the diagnosis of pediatric endocrine problems and children's growth disorders [1]. It is often used for screening symptoms such as endocrine disorders, growth retardation, congenital adrenal cortical hyperplasia, and it can also be used to evaluate the intervention effect of hormone use. In addition, bone age is an important index for identifying the real age of suspects in juvenile delinquency cases, the age of athletes in sports competitions [2], height prediction, and selection of athletes. X-ray films of the hands of the subjects are taken during the detection, and the films are read by doctors to evaluate the bone age.

Because of the tedious process, strong subjectivity and big discrepancy, the technique of BAA needs to be innovated and improved. Since the relevant evaluation standards in BAA have distinct and standardized corresponding bone age according to age, there are distinct descriptions, therefore, BAA is very suitable to be fully automated by machine learning method. And with the development of AI, image recognition, automatic BAA is becoming possible. The research can provide fundamental supports for the development of medical imaging intelligent diagnosis technology, and provide powerful auxiliary tools for orthopedics and pediatricians' scientific research and diagnosis, age determination in criminal investigation, and selection of athletes.

In the early stage, expert system was mainly used to estimate bone age through image processing and pattern recognition. In recent years, with the development of deep learning technology, end-to-end BAA based on deep neural network has emerged. This paper will review the latest research progress of BAA. The second part mainly introduces the basis of medical bone age evaluation; the third part mainly introduces the commonly used methods of X-ray hand bone segmentation, including threshold method, clustering method, active contour model method, level set, and optimization method; the fourth part reviews the related methods based on depth learning, including the basic algorithm framework, the network model and the calculation results of various models. Finally, we discusses the problems existing in the research.

2   Basic Medical Methods of BAA

With the development of bone age research and digital imaging technology, BAA has been widely used in the diagnosis of developmental diseases, height prediction, and true age test. There are mainly two categories of medical methods for BAA: G-P method [3] and TW scoring method [4]. G-P atlas describes the standard atlas of different age stages, lists the maturity indicators of wrist bones in detail, and indicates the bone age of each bone in the sample piece. At present, many hospitals in China are still using this method. However, it needs to consider various maturity indicators comprehensively in the whole comparison, so it is still very subjective.

TW series and Chinese bone age scoring method, CHN [4] method and RUS-CHN are all derived from TW scoring method [4]. The basic idea is to select the wrist bone for evaluation, as shown in Fig. 2. The epiphysis in TW score index is marked with boxes in Fig. 1(a). According to the law of development of skeletal indicators, the whole development process of each bone is divided into several parts and a certain score was given to each bone grade. A score of a grade was found by comparing with the score table of bone development grade. The grade of each bone was evaluated one by one. Then the total score of maturity of all the evaluated bones was calculated. Finally, the bone age was determined by the score and bone age comparison table. The scoring method makes BAA more accurate and effective. TW series is the most comprehensive and detailed method of introducing the legal basis of bone age research in China. It is also widely accepted all over the world and many countries have established their own bone age standards based on TW.

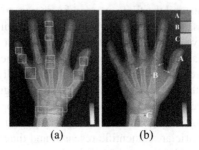

(a)   (b)

Fig. 1 Example of bones selected in TW scoring method. (a) Part of epiphysis in TW index; (b) red arrow marks the soft tissue area with great difficulty in segmentation, and green sampling points mark the gray distribution of bone (Color figure online)

3   Traditional Bone Extraction Method

An X-ray film of a hand can be divided into three regions: background, soft-tissue area and bone. The most challenging part is to remove the soft tissue area of the hand, and just keep the bone area for bone age assessment. Although it is generally believed that the gray value of bones in X-ray films has its own unique distribution range, due to the gray overlap of some pixels between bone region and soft tissue area, soft tissue area and background, especially in some epiphyseal regions, as shown with the red arrow in Fig. 2 (b), it is not easy to segment the complete hand bone from the soft tissue. Simu et al. [5] tested and compared 12 segmentation methods of X-ray films of hand bone with and without evolutionary algorithm, including threshold method, clustering method, level set method, etc. In [5], the author used Otsu threshold method and threshold method based on Tsallis entropy [6] to conduct segmentation experiments. The experimental results are shown in Fig. 2 (b, c). As shown in the original hand X-ray film in Fig. 1, the gray levels of sampling positions A, B and C are quite different, so just using threshold segmentation is difficult to meet the requirements of complete bone segmentation. The threshold method based on Tsallis entropy is not satisfactory in hand bone segmentation, but it can be used to segment the palm region. Clustering methods are also used commonly in bone segmentation, such as adaptive segmentation based on K-means clustering and Gibbs random field (KGRF) [5, 7], K-means with GLCM texture [8], adaptive regularized kernel fuzzy c-means clustering (ARKFCM) and biconvex fuzzy variational (BFV). The results are shown in Fig. 2 (d–i). In the clustering method, except that GLCM [8] can be used to analyze the palm region, other methods can not get a more complete segmentation result, which are unable not meet the needs of later BAA. Besides, active contour modeling, level set and optimization method are also used for bone segmentatiton, as shown in Fig. 2 (j–l).

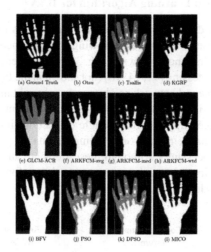

Fig. 2 Results of hand bone segmentation by 11 methods used in [5]

These segmentation methods are traditional segmentation methods. Because the gray level distributions of different parts of the hand bone are different, futher more, the gray levels of different tissues overlap, which means that the threshold method is not suitable for hand bone segmentation. The core idea of clustering method is to classify the homogeneous pixels into one group, but similar to the threshold method, it is also not suitable for hand bone segmentation; ASM and level set methods have high time consumption, and in many cases, users need to manually specify the initial point, and the workload is heavy and the operation is complex. In comparison, the results of hand bone segmentation based on optimization method are the best.

With the develop of deep learning techniques, there are also many deep learning based methods proposed these years. Details are as follows.

4   Deep Learning Based BAA

Medical image analysis based on deep learning has become a research hotspot in the field of medical and computer research, and it has also attracted the attention of industry. The North American Radiology Society (RSNA) has hosted the famous pediatric bone age machine learning competition Challenge [9], which has attracted a large number of interested institutions and individuals to participate in the competition, and it greatly promoted the development of automatic BAA technology. In recent years, our research group has also jointly carried out relevant research work with Xi'an Red Cross Society hospital and the Second Affiliated Hospital of Xi'an Jiaotong University, and published a series of research papers [10–13]. The general framework of automatic BAA is based on deep learning method. We reviewed the neural network models used in related studies and summarized the comparison of the BAA results.

4.1   Framework of Deep Learning Algorithm for BAA

There are two common deep learning algorithm frameworks of BAA, one is treating the hand as a whole object, another is following the TW theory, processing the key regions seperately. The first one is shown in Fig. 3, which is generally divided into three parts: input image preprocessing [11, 14–18] (not shown in the figure), feature extraction based on convolutional neural network and bone age value obtained by using regression network. The preprocessing part usually includes equalization of gray value of X-ray film of hand bone, image clipping, image angle correction (image angle standardization processing), and enhancement of original image data set [17, 19] (adding samples through deformation processing). Although these steps are auxiliary for segmentation and BAA, they play an important role in improving the performance of the algorithm. These methods are introduced below.

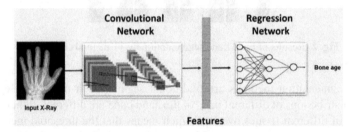

Fig. 3 Algorithm framework of BAA based on deep learning model [17]

The second strategy is extracting the key regions of a hand using region proposal network (RPN or Faster R-CNN) and evaluate each region seperately [20–23]. They integrate TW3 and CNN-based methods together while utilizing deep learning. With this strategy they can get both the maturity of all regions and the bone age, as shown in Fig. 4.

Equalization of Gray Value of X-Ray Film of Hand Bone. Hand bone images taken by different X-ray machines may be quite different. As shown in Fig. 5 in reference [38], the background of X-ray film of hand bone can be white or black. Firstly, the author sampled the image edge area to detect whether the background is white. If so, the background of all images is uniformly transformed to black (with image segmentation method), so as to reduce the impact of image difference. It is proved that removing the background is an important step for the later recognition [24].

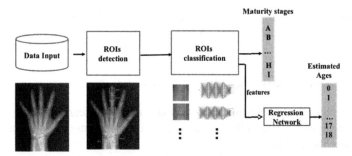

Fig. 4 Algorithm framework of BAA based on deep learning model with region detection [21]

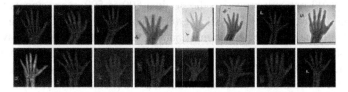

Fig. 5 X-ray of hand bone with different background and size [16]

In [14], the gray equalization method is used to suppress the region unrelated to the hand bone and enhance the hand bone region. In [13], the author uses GAN (generative counter neural network) to enhance the segmentation results to obtain better features for BAA. The experiment results show that using enhanced image as input can indeed improve the performance of the algorithm. Background suppression in preprocessing is beneficial to subsequent hand bone segmentation, and also plays an important role in later feature extraction.

Image Clipping and Transformation of X-Ray Film of Hand Bone. Many neural networks require the input to be square image. In the related literature, the original X-ray hand bone image has been expanded and cropped. The commonly used image sizes include 224 × 224 [14], 256 × 256 [25], 299 × 299 [11], and 512 × 512 [16]. Because most of the original images are 512 × 512 and 1024 × 1024, when the computing hardware conditions are satisfied, the image quality and computational complexity can be well balanced by selecting 512 × 512. We can also try 224 × 224 or 256 × 256 samples to accelerate the training.

In [26], in order to unify the position and direction of all hand bone images in X-ray film by rotating, the authors detected the center points of middle finger tip, thumb tip and palm head bone, and determine the position of the whole palm with these three points, and then carry out affine transformation (rotation and scaling) to realize angle correction of all hand bones.

In terms of image correction, there are mainly two ideas: the first is to process the image consistency and standardization through affine transformation, so as to reduce the difference of samples and the sample size to be learned [14]. And the second is to regard the hand bone as a non rigid object, increase the sample size through deformation and other processing, realize data enhancement, and cover more hand bone X-ray film samples [17]. These two completely different ideas are widely used in various literatures.

4.2 Deep Neural Network Model Used in BAA

In the research of BAA based on deep learning method, researchers have used some deep learning models, such as ResNet, alexnet, googlenet, densenet, etc. The following will introduce these models.

Inception V3+ DenseNet. Mark and Alexander won the first place in the rsna2017 bone age prediction challenge [27]. The specific network model is shown in Fig. 6. They used concept V3 network processes the original 500 × 500 image, inputs the gender information (0 for female, 1 for male) into the density layer of 32 neurons, connects the image information and gender information, and inputs the density layer of two 1000 neurons. Finally, the single output linear layer is used to obtain the final bone age prediction result. The MAE is about 4.3 months. Ren et al. [28] used regression CNN based on concept-v3 model for feature extraction and prediction of X-ray and attention map of hand bone [29], and the final Mae on RSNA data set was 5.2 months.

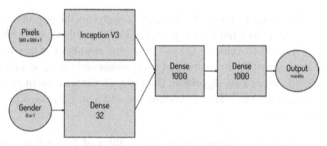

Fig. 6 Schematic diagram of network model used by Alexander et al. [9, 27] (The first in RSNA2017)

The most significant feature of GoogLeNet is that it uses the concept module to perform multiple convolution or pooling operations on the input image in parallel, so that all the output results are spliced into a deeper feature map. Concept V3 [30] network is a model proposed by Google in 2016. Although increasing the model size and computing cost can obtain better results, the author believes that higher computational efficiency and fewer parameters are conducive to the application of deep learning model in mobile terminal and big data scenarios. In this network, the convolution is decomposed and regularized to improve the computational efficiency. The core strategy is to decompose a large convolution kernel into several small convolution kernels. For example, the size of n × n convolution kernel is decomposed into 1 × N and N × 1 convolutions. When n is 3, the computational cost is reduced by 33%. In training, rmsprop optimizer [31] is used to replace SGD, batch normalization is used to prevent gradient dispersion, and label smoothing is used to prevent over fitting.

Synho et al. [16] of Massachusetts General Hospital and Harvard Medical School used googlenet for BAA. Firstly, the gray level distribution and size of the image are normalized, and then the hand bone part, other tissues, background area, marker text and other areas are sampled in block, and the hand bone region is segmented through CNN network. Then, googlenet is used to predict the bone age. The accuracy rate of the error less than one year old is 92.29%, and the time-consuming time is about 1.4–7.9 min. In this study, children aged 0–4 years are excluded, and their bone ages are given as "x years, 6 months". Because the bone age time resolution of the data set is half a year, the model may eventually introduce time errors.

Densenet [32] imports the output of each layer into all the following layers, ensuring the maximum information transmission between layers. This dense connection greatly reduces the amount of network parameters, and also reduces the phenomenon of over fitting. The network bypass strengthens the reuse of features and alleviates the problem of gradient disappearance to a certain extent. The output of densenet layer combines the outputs of all layers to ensure the integrity of information flow in the network.

In reference [13], our research group uses FC densenet to extract palm features, and uses QBC (query by) to extract palm features to get better input data for bone age evaluation. GAN is used to generate the enhanced palm area. At the same time, PTL (paced transfer learning) is proposed for top-down training. In the initial stage, only the random initialization of the top full connection layer is fine tuned. When the value of the loss function drops to the stable stage, the secondary high-level is fine tuned, and gradually downward until the model converges. The final MAE value is 5.991 months for men and 6.263 months for women. The overall framework of the algorithm is shown in Fig. 7:

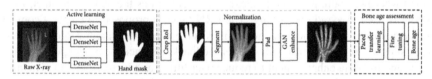

Fig. 7 Algorithm framework of BAA proposed in [13]

ResNet. ResNet [33] was proposed to solve the problem of neural network performance degradation when the network depth increases. Mutasa et al. [25] of Columbia University proposed the mabal model. By combining the concept model and the residual model, the 14 layer network was used to predict the bone age. The author tested the effects of different combinations of residual layer and inception layer. The results show that the former layer uses residual, and the later use inception has the best effect. The best MAE is about 0.637 years. Pan [9], who won the second place in the RSNA2017 bone age prediction challenge, used Imagenet and trained ResNet-50 to prevent over fitting and improve algorithm performance. The final MAD (median of absolute deviation) was about 4.4 months.

CaffeNet. Lee et al. [14] first cut and adjust the gray histogram of the original image, and then used the classic CNN model caffenet [34] after a series of convolution and pooling operations on 224 × 224 images, a 1000 dimensional feature vector is finally obtained. In the processing, the author uses the weight of the pre training model on Imagenet. The final average absolute error is about 6.4 months, and the maximum error is 24.4 months. However, the amount of data in this experiment is small, and the training data and test data are 400 and 200 respectively. The authors think that the scale of the dataset is too small and the performance may be improved by increasing the number of samples.

U-Net + VGG. Iglovikov et al. [26] used U-Net [35] to segment the hand, and detected the position of the opponent by key points, and then used VGG model to predict bone age. In particular, the author evaluated the effect of hand image, carpal region, metacarpal bone and phalangeal bone on bone age prediction. It was found that linear combination of these three regions had the best prediction effect, and the MAE was 7.52 months. Combining U-net segmentation and image feature extraction for BAA is a common operation. U-net regards image segmentation as a pixel classification problem. It extracts features through multiple convolutions, and then performs deconvolution to restore the position of the classified pixels to achieve image segmentation. VGG network is composed of 5 layers of convolution layer, 3 layers of full connection layer and softmax output layer. Max pooling is used between layers, and relu function is used for activation units of all hidden layers [36]. It is improved that the depth of VGG network and the use of small convolution kernel (3 × 3 convolution kernel) can improve the final classification and recognition effect of the network.

AlexNet. Alexnet [37] is an 8-layer deep neural network, including 5 layers of convolution layer and 3 layers of fully connected layer (excluding local response normalization layer (LRN) and pooling layer). In alexnet, relu is used as the activation function and dropout is used to avoid over fitting. The LRN layer is proposed to create a competition mechanism for local neurons, which makes the value with larger response become larger, suppress other neurons with smaller feedback, enhance the model generalization ability, and use data enhancement to reduce over fitting and increase model generalization ability. Hu tinghong et al. [38] used alexnet [37] as image recognition regression model to identify the bone age of left wrist joint of Uygur teenagers. When the error is 0.7 years old, the accuracy rate is about 70%. The author thinks that the accuracy rate of 0.7-year-old error can not be used in the practice of juvenile bone age identification in China. In the later stage, the sample scale will continue to be increased to further improve the accuracy rate, and gradually overcome the requirement of accuracy and the accuracy fluctuation.

BoNet (ad-hoc CNN). Spampinato et al. proposed Bonet [17] for bone age prediction, among which the key technologies include using the first layer convolution layer of overfeat [39] feature extraction network to encode low-level visual features, reduce the over fitting risk caused by the small number of X-ray films. They used the deformation layer to obtain the non-rigid deformation of hand bone, increase the number of samples. Bone age was obtained by the regression network. The author tested the convolution layer from 2 to 6 on the public data set. The best performance can be achieved by using five layers. The MAE is about 0.79 years. Bonet covers all the regions recommended by

TW method, as shown in Fig. 8. It is found that some areas are not necessary to predict bone age. This finding is worth of further excavation by doctors and researchers, and may have new findings.

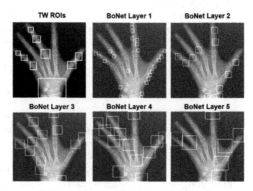

Fig. 8 Comparison between ROI region used in TW method and corresponding regions learned by BoNet

4.3 Results

In [15], the authors compared the performance of BAA by paediatric radiologists with or without artificial intelligence assistance, and found that artificial intelligence can improve radiologists' BAA performance by improving accuracy, reducing the variability and RMSE (Root Mean Square Error).

Table 1 is a summary of the bone age testing methods based on deep learning. The commonly used indicators in this paper are MAD and MAE (mean absolute error), both of which represent the average absolute error. Here, MAE is used uniformly.

Table 1. Summary of BAA methods based on deep learning methods

Reference	Method	Result (on testing set)	Dataset
1. Alexander et al. [9]	Inception V3 + DenseNet	MAE: 4.265 months	RSNA dataset Total data: 14,236 Training set: 12,611 Validation set: 1425 Test set: 200 The ratio of male to female is close to 1:1
2. Pan et al. [9]	Fine-tuned ResNet-50	MAE: 4.4 months	
3. Kitamura et al. [9]	Ice Module	MAE: 4.4 months	

(continued)

Table 1. (*continued*)

Reference	Method	Result (on testing set)	Dataset
4. Chen et al. [9]	U-Net segmentation and CNN based recognition	MAE: 4.5 months	
5. Ren et al. [28]	Regression CNN based on Inception-V3	MAE: 5.2 months	RSNA dataset SCH data set (Shanghai Children's Hospital): total data volume: 12,390 Training set: 9912 Validation set: 1239 Test set: 1239
6. Zhao et al. [13]	DenseNet + QBC + GAN + PTL	MAE: male 5.991 months, female 6.263 months	RSNA dataset
7. Mutasa et al. [25]	ResNet and inception	MAE: 6.432 months (0.536 years)	Total data: 20,581 (10,289 from the network, 8909 from Columbia University Hospital, 1383 from the public data set) Training set: 11,007 Verification set: 1105 Test set: 300
8. Lee et al. [14]	Caffenet Architecture	Maximum error: 24.4 months MAE: 6.4 months	RSNA dataset Training set: test set = 2:1
9. Iglovikov et al. [26] 10. Ronneberger et al. [35]	U-Net + VGG-style + Linear combination	MAE: 7.52 months	RSNA dataset
11. Hu et al. [38]	AlexNet (8 layers in total, 5 convolution layers and 3 full connection layers)	MAE: 8.4 months Accurace: male 71.2%, female 66.2%	DR images of left wrist joint of Uygur people, male 245, female 227 Training set: test set = 7:3
12. Spampinato et al. [17]	BoNet(ad-hoc CNN, 6 layers)	MAE: 9.48 months	Digital Hand Atlas Database System
13. Liu et al. [40]	NSCT-based Multi-Scale CNN (VGG16)	MAE: male 6.62 months, female 6.23 months	Total data: 1391 Male 700, female 691
14. Synho et al. [16]	CNN segmentation + Lenet BAA	Accuracy of MAE < 1 year: 92.29%	Male 4278, female 4047 Training set: verification set: test set = 70:15:15
15. Tajmir et al. [15]	LeNet-5	Accuracy of MAE < 1 year: 98.6%, RMSE 0.548 years	Total data: 8045 Training set: test set = 85:15
16. Liang et al.[20–23]	Faster RCNN + regression	MAE: 6.97 months	RMSE 9346, and MAPE 8128

5 Discussions

With the improvement of medical treatment level, parents pay much more attention to the growth and development of adolescents than before, and more and more parents take their kids to the hospital to evaluate their bone age to check their growth and actively take relevant interventions. In youth sports events, using BAA method to obtain the real age of adolescents is more common, and the needs of BAA in China is much greater. There is also a great demand for research and development of measurement methods and equipments. With the rapid development of artificial intelligence technology and the promotion of the market, some companies have launched relevant equipment and software, and the famous RSNA BAA competition has also attracted a large number of researchers to participate. At present, the software and algorithm of BAA based on the whole atlas are effective, but the time accuracy still needs to be further improved. The precise segmentation and more detailed assessment of bone age of each bone need to be studied. The specific problems and the possible future research directions are as follows:

(1) At present, the RSNA competition has provided more than 10,000 X-ray data sets of hand bones, which provides a great data support for research. However, in some studies, such as reference [15, 16], the accuracy of the dataset used by the authors is six months, which will inevitably affect the accuracy of the algorithm. Therefore, we should collect more accurate dataset in the later research. For example, improving the accuracy of X-ray films to months or days is conducive to train a more accurate BAA model. From the perspective of pediatrician's treatment of adolescent growth, BAA needs to have higher accuracy, which is more meaningful for short-term tracking of development and disease development. For example, it is more conducive for doctors to evaluate the effect of intervention such as growth hormone using, and timely adjust the treatment plan of patients.

(2) In BAA methods, the end-to-end deep learning methods input an X-ray film, and finally output the bone age. However, clinicians believe that the scoring method should give more accurate results, and they can see the development of each bone, rather than just give the final bone age results. If the software can automatically mark the bone development changes of a patient in time sequence, it is more valuable to analyze the patient's condition development and adjust the treatment plan. The research on the later bone age evaluation algorithm can be deeply involved in hand bone segmentation (multi-objective segmentation), hand bone age evaluation based on scoring method, minor difference analysis, and annotation of hand bone image.

(3) In hand bone segmentation, BFV, MICO, U-net and other methods are better, but the current segmentation are mainly single target segmentation, the whole palm area or hand bone area is regarded as the target to be segmented, and finally a binary image is obtained. In order to meet the needs of more accurate diagnosis in medicine, it is necessary to study the multi-objective segmentation method of hand bones, but the research is difficult, mainly due to the following reasons: (1) the number of segmentation targets is uncertain, the epiphysis of hand appears at different times during the development of children and the number of epiphyses needs to be segmented in X-ray films is different, which needs to be considered when using template method; (2) according to TW3 method, it is difficult to get

accurate segmentation results of all epiphysis and hand bones. A more robust multi-objective X-ray film segmentation method is needed. It can be considered to use GAN and densenet to carry out fine segmentation of hand bones.

(4) Some studies have analyzed the activation of different regions of the palm through deep neural network, and found that the contribution of different regions of hand bone to bone age is also different [13, 26, 28]. There is a certain possibility that only a small number of hand bones really play an important role in BAA, which can be confirmed by further deep learning analysis; with the improvement of living standards, the existing classic bone age standard may not be applicable to the current children, so we can extract features from new samples by using deep learning method and statistical analysis. The standard of bone age will be revised or a new bone age standard will be proposed.

(5) At present, there are some other methods for BAA, but the BAA based on hand bone X-ray film is still the simplest and most widely used method for most growth and development diagnosis and real age identification. It is also a method with the longest development history. It has more basic theoretical support and large-scale datasets, and is the easiest breakthrough in algorithm and technology research. However, other methods also have their own advantages, such as BAA for older adolescents (>18 years old), the use of ultrasound to avoid X-ray radiation, and so on.

6 Conclusion

In this paper, the development of automatic BAA algorithm in recent years is reviewed. The segmentation method, feature extraction method and end-to-end method based on deep learning model are analyzed. In addition, the current BAA system is investigated and the methods used are analyzed, doctors' expectations and possible implementation methods of BAA in the future are analyzed and prospected. The above contents provide methods and technical references for researchers and engineers in the field of medical image analysis and BAA. With the continuous improvement of the algorithm, high-performance parallel computing technology, higher quality and high-precision image sample set, the automatic BAA algorithm and system can get better development and provide more efficient technology and tools for medical research and disease diagnosis.

Acknowledgements. This research is supported by Foundation of Shaanxi Educational Committee (18JK0722), Key Research and Development Program of Shaanxi Province (2019GY-021), Open fund of Shaanxi Key Laboratory of Network Data Intelligent Processing (XUPT-KLND(201802, 201803)).

References

1. Martin, D.D., Wit, J.M., Hochberg, Z., et al.: The use of bone age in clinical practice – Part 1. Horm. Res. Paediatr. **76**(1), 1–9 (2011)
2. Yan, Y.: Research on ID identification and bone age testing of 2011 Double Happiness·New Star Cup National Children's table tennis players. Soochow University (2012)

3. Todd, T.W.: Atlas of skeletal maturation. J. Anat. **72**(Pt 4), 640 (1938)
4. Zhang, S.: The Skeletal Development Standards of Hand and Wrist for Chinese Children—China 05 and its application (2015)
5. Simu, S., Lal, S.: A study about evolutionary and non-evolutionary segmentation techniques on hand radiographs for bone age assessment. Biomed. Signal Process. Control **33**, 220–235 (2017)
6. de Albuquerque, M.P., Esquef, I.A., Mello, A.G.: Image thresholding using Tsallis entropy. Pattern Recogn. Lett. **25**(9), 1059–1065 (2004)
7. Pappas, T.N.: An adaptive clustering algorithm for image segmentation. IEEE Trans. Signal Process. **40**(4), 901–914 (1992)
8. Chai, H.Y., Wee, L.K., Swee, T.T., et al.: GLCM based adaptive crossed reconstructed (ACR) k-mean clustering hand bone segmentation. Book GLCM based adaptive crossed reconstructed (ACR) k-mean clustering hand bone segmentation, pp. 192–197 (2011)
9. Halabi, S.S., Prevedello, L.M., Kalpathy-Cramer, J., et al.: The RSNA pediatric bone age machine learning challenge. Radiology **290**(2), 498–503 (2018)
10. Wang, Y., Zhang, Q., Han, J., Jia, Y.: Application of Deep learning in Bone age assessment. In: IOP Conference Series: Earth and Environmental Scicence (EES), Guangzhou (2018)
11. Han, J., Jia, Y., Zhao, C., et al.: Automatic Bone age assessment combined with transfer learning and support vector regression. In: Proceedings of the 2018 9th International Conference on Information Technology in Medicine and Education (ITME) (2018)
12. Tang, Y., et al.: Assessment and analysis of wrist bone age in 190 adolescents in Xi'an. Shaanxi Med. J. **47**(12), 1661–1663 (2018)
13. Zhao, C., Han, J., Jia, Y., et al.: Versatile framework for medical image processing and analysis with application to automatic bone age assessment. J. Electr. Comput. Eng. (2018)
14. Lee, J.H., Kim, K.G.: Applying deep learning in medical images: the case of bone age estimation. Healthc. Inf. Res. **24**(1), 86 (2018)
15. Tajmir, S.H., Lee, H., Shailam, R., et al.: Artificial intelligence-assisted interpretation of bone age radiographs improves accuracy and decreases variability. Skeletal Radiol. **48**(2), 275–283 (2019)
16. Lee, H., Tajmir, S., Lee, J., et al.: Fully automated deep learning system for bone age assessment . J. Digit. Imaging **30**(4), 427–441 (2017)
17. Spampinato, C., Palazzo, S., Giordano, D., et al.: Deep learning for automated skeletal bone age assessment in X-ray images. Med. image Anal. **36**, 41–51 (2017)
18. Chen, X., Li, J., Zhang, Y., et al.: Automatic feature extraction in x-ray image based on deep learning approach for determination of bone age. Future Gener. Comput. Syst. (2020)
19. Jaderberg, M., Simonyan, K., Zisserman, A.: Spatial transformer networks. In: Proceedings of the Advances in Neural Information Processing Systems (2015)
20. Liang, B., Zhai, Y., Tong, C., et al.: A deep automated skeletal bone age assessment model via region-based convolutional neural network. Future Gener. Comput. Syst. **98**, 54–59 (2019)
21. Bui, T.D., Lee, J.J., Shin, J.: Incorporated region detection and classification using deep convolutional networks for bone age assessment. Artif. Intell. Med. **97**, 1–8 (2019)
22. Koitka, S., Kim, M.S., Qu, M., et al.: Mimicking the radiologists' workflow: Estimating pediatric hand bone age with stacked deep neural networks. Med. Image Anal. **64**, 101743 (2020)
23. Wibisono, A., Mursanto, P.: Multi region-based feature connected layer (RB-FCL) of deep learning models for bone age assessment. J. Big Data **7**(1), 1–17 (2020)
24. Wu, E., Kong, B., Wang, X., et al.: Residual attention based network for hand bone age assessment. In: Proceedings of the 2019 IEEE 16th International Symposium on Biomedical Imaging (ISBI 2019) (2019)
25. Mutasa, S., Chang, P.D., Ruzal-Shapiro, C., et al.: MABAL: a novel deep-learning architecture for machine-assisted bone age labeling. J. Digit. Imaging **31**(4), 513–519 (2018)

26. Iglovikov, V.I., Rakhlin, A., Kalinin, A.A., Shvets, A.A.: Paediatric Bone age assessment using deep convolutional neural networks. In: Stoyanov, D., et al. (eds.) DLMIA/ML-CDS -2018. LNCS, vol. 11045, pp. 300–308. Springer, Cham (2018). https://doi.org/10.1007/978-3-030-00889-5_34

27. Mark Cicero, A.B.: Machine learning and the future of radiology: how we won the 2017 RSNA ML challenge. In: 2017, Machine Learning and the Future of Radiology: How we won the 2017 RSNA ML Challenge (2017)

28. Ren, X., Li, T., Yang, X., et al.: Regression convolutional neural network for automated pediatric bone age assessment from hand radiograph. IEEE J. Biomed. Health Inf. **23**, 2030–2038 (2018)

29. Frangi, A.F., Niessen, W.J., Vincken, K.L., Viergever, M.A.: Multiscale vessel enhancement filtering. In: Wells, W.M., Colchester, A., Delp, S. (eds.) Medical Image Computing and Computer-Assisted Intervention—MICCAIx 1998. MICCAI 1998. Lecture Notes in Computer Science, vol. 1496. Springer, Heidelberg (1998). https://doi.org/10.1007/BFb005 6195

30. Szegedy, C., Vanhoucke, V., Ioffe, S., et al.: Rethinking the inception architecture for computer vision. In: Proceedings of the IEEE Conference on Computer Vision and Pattern Recognition (2016)

31. Tieleman, T., Hinton, G.: Divide the gradient by a running average of its recent magnitude. coursera: neural networks for machine learning. Technical report (2017)

32. Huang, G., Liu, Z., Van Der Maaten, L., et al.: Densely connected convolutional networks. In: Proceedings of the IEEE Conference on Computer Vision and Pattern Recognition (2017)

33. He, K., Zhang, X., Ren, S., et al.: Deep residual learning for image recognition. In: Proceedings of the IEEE Conference on Computer Vision and Pattern Recognition (2016)

34. Jia, Y., Shelhamer, E., Donahue, J., et al.: Caffe: convolutional architecture for fast feature embedding. In: Proceedings of the 22nd ACM International Conference on Multimedia. ACM (2014)

35. Ronneberger, O., Fischer, P., Brox, T.: U-net: convolutional networks for biomedical image segmentation. In: Navab, N., Hornegger, J., Wells, W.M., Frangi, A.F. (eds.) MICCAI 2015. LNCS, vol. 9351, pp. 234–241. Springer, Cham (2015). https://doi.org/10.1007/978-3-319-24574-4_28

36. Simonyan, K., Zisserman, A.: Very deep convolutional networks for large-scale image recognition. arXiv preprint arXiv:14091556 (2014)

37. Krizhevsky, A., Sutskever, I., Hinton, G.E.: ImageNet classification with deep convolutional neural networks. In: Proceedings of the Advances in Neural Information Processing Systems (2012)

38. Hu, T.-H., Liu, T.-A., et al.: Automated assessment for bone age of left wrist joint in Uyghur teenagers by deep learning. J. Forensic Med. **34**(1), 27–32 (2018)

39. Sermanet, P., Eigen, D., Zhang, X., et al.: OverFeat: Integrated recognition, localization and detection using convolutional networks. arXiv preprint arXiv:13126229 (2013)

40. Liu, Y., Zhang, C., Cheng, J., et al.: A multi-scale data fusion framework for bone age assessment with convolutional neural networks. Comput. Biol. Med. **108**, 161–173 (2019)

Unsupervised Real-World Super-resolution Using Variational Auto-encoder and Generative Adversarial Network

Kalpesh Prajapati[1], Vishal Chudasama[1], Heena Patel[1], Kishor Upla[1,2(✉)], Kiran Raja[2], Raghavendra Ramachandra[2], and Christoph Busch[2]

[1] Sardar Vallabhbhai National Institute of Technology (SVNIT), Surat, India
kalpesh.jp89@gmail.com, vishalchudasama2188@gmail.com,
hpatel1323@gmail.com, kishorupla@gmail.com
[2] Norwegian University of Science and Technology (NTNU), Gjøvik, Norway
{kiran.raja,raghavendra.ramachandra,christoph.busch}@ntnu.no

Abstract. Convolutional Neural Networks (CNNs) have shown promising results on Single Image Super-Resolution (SISR) task. A pair of Low-Resolution (LR) and High-Resolution (HR) images are typically used in the CNN models to train them to super-resolve LR images in a fully supervised manner. Owing to non-availability of true LR-HR pairs, the LR images are generally synthesized from HR data by applying synthetic degradation such as bicubic downsampling. Such networks under-perform when used on *real-world* data where degradation is different from the synthetically generated LR image. As obtaining true LR-HR pair is a tedious and resource (time and effort) consuming task, we propose a new approach and architecture to super-resolve the real-world LR images in an unsupervised manner by using a Generative Adversarial Network (GAN) framework with Variational Auto-Encoder (VAE). Along with a new network architecture, we also introduce a novel loss metric based on no-reference quality scores of SR images to improve the perceptual fidelity of the SR images. Through the experiments on NTIRE-2020 Real-World SR Challenge dataset, we demonstrate the superiority of the proposed approach over the other competing state-of-the-art methods.

Keywords: Unsupervised single image super-resolution · Generative Adversarial Network · Convolutional Neural Network · No-reference image quality assessment

1 Introduction

Many computer vision applications such as object detection, tracking, recognition, classification etc., demand and benefit from High-Resolution (HR) images as they comprise of fine-grain details of the scene being observed. However, acquiring HR images is limited by many factors such as cost which is directly

© Springer Nature Switzerland AG 2021
A. Del Bimbo et al. (Eds.): ICPR 2020 Workshops, LNCS 12662, pp. 703–718, 2021.
https://doi.org/10.1007/978-3-030-68790-8_54

proportional to sensor size, fabrication technique, etc. A reasonable alternative to overcome the requirement of HR images is to employ Super-Resolution (SR) techniques to generate HR image from its corresponding Low-Resolution (LR) observation. Given many real-world applications capture only single observation (i.e., single image), a number of approaches for the Single Image Super-Resolution (SISR) have been proposed in the recent years [4,5,9,12,13,27,30,32]. The SISR is a classical problem in computer vision which is inherently ill-posed in nature and presents challenges of complexity with higher up-sampling factors. Additionally, non-availability of exact ground-truth hinders the development of realistic quality assessment metrics to devise better approaches for SISR and leaves it as an open research problem in the community.

The remarkable progress in Graphical Processing Units (GPUs) and huge availability of annotated data has led the SISR to embrace deep learning approaches. The newer approaches based on deep learning have demonstrated superior performance in the SISR task over the traditional computer vision based methods [13,18,27,32]. Deep learning based SISR methods are designed to learn input-output relationship based on LR-HR pair of training data. As the abundance of such true LR-HR pair data is limited, most works have created training data by degrading the original HR image using known degradation like bicubic down-sampling for training SR networks. As the dataset comprises of such artificially degraded LR images and its corresponding HR images, many CNN networks have been developed in an efficient manner through supervised approaches [4,5,9,12,13,27,30,32]. Following the works of Dong et al. [4] who employed CNN based SR framework (*SRCNN*) in supervised manner for super-resolution task, a number of CNN based models have been proposed in the literature to improve the quality of SR images [5,9,12,13,27,30,32]. Despite these works providing a plausible solution, we note that they often generalize poorly for real-world data whose statistical distribution is different from synthetically generated LR images from corresponding HR images. As shown in Fig. 1, the statistical characteristics such as exposure, blur and sharpness of real and corresponding synthetic LR images from RealSR [2] and NTIRE-2020 Real-World SR Challenge validation [17] datasets do not correspond exactly despite being similar. The key reason for such an observation is the differing sensor noise and unknown degradation factors due to capture settings in real LR observation. Synthetically generated LR images thus limit the generalizability of the SISR approaches on real-world data where statistical properties are complex in relation. This can however be overcome by acquiring real-world LR-HR image pairs that requires dedicated cameras for capturing corresponding LR and HR images which increases the cost and also further demands expert human manpower to supervise minimum registration error [2] between acquired LR and HR image pair.

To mitigate this problem, an idea of unsupervised training of CNN network for SISR task was introduced by Lugmayr et al. [15]. In a similar direction, we propose a Generative Adversarial Network (GAN) based framework with Variational Auto-Encoder (VAE) to obtain the super-resolved version of the given LR observation in this work without using true LR-HR pairs. With suitable

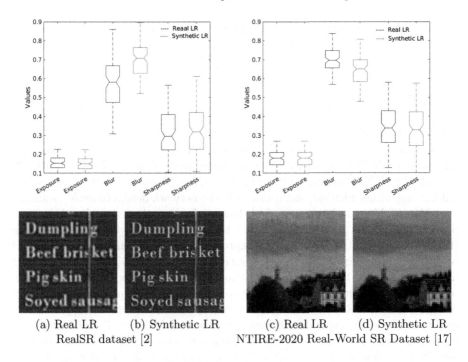

(a) Real LR (b) Synthetic LR (c) Real LR (d) Synthetic LR
RealSR dataset [2] NTIRE-2020 Real-World SR Dataset [17]

Fig. 1. Statistical characteristics for different parameters such as exposure, blur and sharpness between real and synthetic LR images obtained on RealSR dataset [2] and NTIRE-2020 Real-World SR Challenge dataset [17].

architectural changes, we also attain an upscaling factor of ×4 the original image size. Further, to generate high quality images with perceptual fidelity, we propose to employ a novel loss function based on no-reference quality metric which we refer as Quality Assessment (QA) loss. We assess the applicability of proposed network by evaluating it on NTIRE-2020 Real-world SR dataset [17] and report the set of results along with other competing state-of-the-art SISR methods. The key contributions of our proposed method are summarized as:

- A new unsupervised approach to super-resolve the real-world LR images is introduced for upscaling factor of ×4 the original LR image size by incorporating Variational Auto-Encoder (VAE) along with the discriminator network of GAN.
- A novel no-reference quality loss function referred as Quality Assessment (QA) loss is proposed to train the proposed network to enhance the perceptual fidelity of SR result.
- The proposed network is further trained in an unsupervised manner to generate SR images directly rather than learning degradation first and then training SR network in a supervised manner [15,29].

In the rest of the paper, Sect. 2 reviews the related works in the direction of Single Image Super-Resolution (SISR) using deep learning. The detailed description of the proposed method has been presented in Sect. 3 followed by experimental results with the comparison to other state-of-the-art methods in Sect. 4. Lastly, in Sect. 5 we draw the conclusions and list out future works.

2   Related Works

Dong et al. [4] attempted first to obtain SR of LR images using shallow CNN architecture with three layers which they referred as *SRCNN*. With suitable modifications in SRCNN [4] later on, similar works have been reported in FSRCNN [5] and VDSR [9]. Contrary to these networks, many other approaches use residual learning along with skip connections in the network design to avoid gradients vanishing problem and also to make it feasible for very deep neural architecture. Such modifications have been reported to obtain a better SR performance in [12,13]. An alternative approach to improve the SR performance was proposed by using dense connections in many recent networks [26,33]. Attention based model has further been proposed to allow the flexibility with sparse learning in order to select few features showing committed improvement for SR task. In similar lines, adversarial training has been reported to improve the perceptual quality of SR results using deep network in adversarial fashion [7]. Ledig et al. [11] employed such adversarial training and proposed a novel GAN based framework for SISR problem referred as *SRGAN* which obtained significant enhancement in terms of texture quality on the given LR observation. Inspired by SRGAN, many works such as SRFeat [18] and ESRGAN [27] have been recently reported to further enhance the perceptual quality in the SR images.

Despite the number of works, all of them suffer from common limitation of supervised training in which LR data is prepared through artificial degradation such as bicubic downsampling. The CNN model trained on such dataset often generalizes poorly for real-world data where degradation is significantly differ from that of bicubic downsampling (see Fig. 1). In order to learn the characteristics of the real-world data, the supervised deep networks require real LR-HR paired images. To this extent, Cai et al. [2] introduced RealSR dataset which was made publicly available through the NTIRE-2019 real-world SR challenge. Many representative SR works on RealSR dataset have been reported recently in [3,6,21,28].

In an alternative solution of creating true LR-HR pair data, training networks with multiple data degradation was proposed to model the realistic distribution of LR images [22,30]. Multiple data degradation for such a purpose have achieved better SR results [22,30]. Another alternative solution to real-world SR problem is to use unsupervised training in CNN network. In this direction, Lugmayr et al. [15] proposed a framework based on cycle consistency loss. Here, authors first learn real-world degradation to make LR-HR pair and then train the SR network in a supervised way. Similarly, Yuan et al. [29] use CycleGAN where one cycle is used to learn degradation distribution and other cycle is employed to learn HR image distribution. In order to further develop such novel idea of unsupervised learning for SISR task, AIM 2019 Challenge [16] (*in conjunction with ICCV*

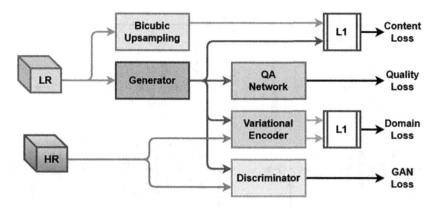

Fig. 2. The architecture framework of the proposed approach for super-resolution of single image.

2019) and NTIRE-2020 real-world SR Challenge [17] (*in conjunction with CVPR 2020*) were introduced. Motivated by our earlier work of unsupervised approach [19] to obtain SR images from LR images, we propose a new approach in this work as detailed in the next section.

3   Proposed Method

To super-resolve the noisy LR observation to a clean HR image, we propose a novel approach based on the combination of Generative Adversarial Network (GAN) and Variational Auto-Encoder (VAE). Figure 2 shows the overall architecture framework of the proposed approach. It consists of four deep networks: Generator (G), Discriminator (D), Variational Auto-Encoder (VAE) and Quality Assessment (QA) network. The generator network takes the LR observation as input and produces corresponding SR image. The SR outcome is provided to variational encoder network along with the *clean but not true HR image*. To discriminate the generated SR image from original clean image, we adapt encoding technique from VAE in which the encoding process follows the principle of multi-variate Gaussian distribution. Additionally, we also introduce a novel idea for improving quality of SR image based on human perception. A quality assessment deep network is specifically introduced to act on generated SR image and outputs quality score of that SR image. This measure is used as a loss function in the generator network. Further, one can see from the Fig. 2 that the LR image is upscaled using bicubic interpolation operator and then the upscaled image is compared with the generated SR image via L_1 based content loss function. Such loss helps the generator network to preserve the structure of the LR observation in the particular SR outcome.

Generator Network (G): The architecture of the generator network is depicted in Fig. 3. It can be divided into three different modules based on it's

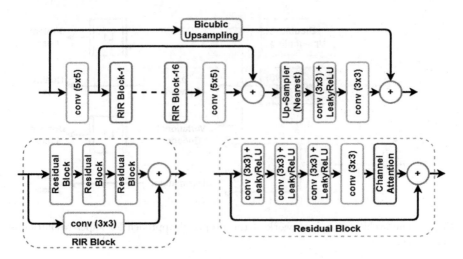

Fig. 3. The design architecture of the Generator (G) network.

functionality: Low Level Information Extraction (LLIE), High Level Information Extraction (HLIE) and SR reconstruction (SRRec) modules. First, the given LR observation (I_{LR}) is passed through LLIE module which consists of a convolutional layer with kernel size of 5 and channels of 32 to extract low level details (i.e., $I_{low-level}$). We employ a larger kernel to account for larger reception area to predict low level information. Mathematically, this can be expressed as,

$$I_{low-level} = f_{LLIE}(I_{LR}), \tag{1}$$

where, f_{LLIE} indicates the operation of the LLIE module.

The low level information extracted from LLIE module is passed further through the HLIE module to extract other details such as edges and fine structure from the LR observation. The HLIE module is designed using $m = 16$ number of Residual-In-Residual (RIR) blocks with one long skip connection. Each RIR block consists of three residual blocks with 1×1 convolutional layer in skip connection. The design of residual block contains 4 convolutional layers with kernel size of 3 and Channel Attention (CA) module. The function of CA module is to re-scale each channels separately based on statistical average of each channel which is inspired from [32]. However, it is notable from Fig. 3 that in residual block, we also use skip connection to incorporate the residual learning that helps the network to stabilize the deeper network training and to reduce the vanishing gradient problem. The output feature maps of the HLIE module can be represented as,

$$I_{high-level} = f_{HLIE}(I_{low-level}). \tag{2}$$

Here, the f_{HLIE} denotes the function of the HLIE module. The high-level information feature maps are fed into the SR Reconstruction (SRRec) module which consists of two up-sampling blocks and two convolutional layers that maps the feature maps to require number of channels for SR image (I_{SR}). Mathematically, this can be written as,

$$I_{SR} = f_{REC}(I_{high-level}),$$ (3)

where, f_{REC} indicates the reconstruction function of the SRRec module. The nearest neighbor approximation is used to perform 2× up-sampling in each up-sampling block. We employ the convolution layer with the kernel size of 3 and feature maps of 32. As suggested by authors in [9], a smaller filter size have small receptive field per layer which tends to capture smaller and complex features in an image. Additionally, such small filter size is also useful to reduce the number of parameters which makes the given algorithm computationally efficient. Similarly, one can increase the number of channels in each layer to extract more features with the cost of higher complexity in terms of trainable parameters. We choose 32 channels to optimize network performance with less number of training parameters.

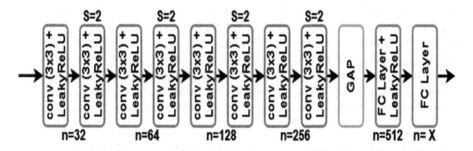

Fig. 4. The architecture design of discriminator and VAE networks. Here, n and s represents number of channels and stride respectively. The value of X is 1 for discriminator network and 256 (i.e., 128 values for mean and other 128 for variance value) for VAE network.

Variational Auto-encoder (VAE) and Discriminator (D) Networks: In the proposed method, we use Variational Auto-Encoder (VAE) along with Discriminator network to obtain domain and adversarial losses, respectively. The architecture of both these networks have been displayed in Fig. 4. They have been designed by following the guidelines suggested by Redford et al. [20]. These networks consist of eight convolutional layers alternatively with strided convolution. The number of channels are increased by factor of two after each strided convolution. However, instead of flattening layer, the Global Average Pooling (GAP) layer is used in which number of trainable parameters are reduced. Thus, computational complexity of these networks are decreased. Each convolutional layer is used with leaky ReLU activation with leaky constant of 0.2. In addition to generative adversarial network, we constrain the VAE network to generate values of means and variances that must follow roughly unit Gaussian distribution. To incorporate such constraint, we use KL divergence loss between generated latent vector of HR images and unit Gaussian distribution which can be expressed as [10],

$$L_{vae} = -1/2 \sum_i (1 + log(\sigma_i^2) - \mu_i^2 - \sigma_i^2), \qquad (4)$$

where, σ_i and μ_i are the ith value of variance and mean from VAE network respectively. In this work, dimension of σ and μ is 128 which makes total of 256 values from its output.

Quality Assessment (QA) Network: We further introduce a novel loss function based on quality as in training network to improve the perceptual reconstruction quality of SR images. We use deep network as shown in Fig. 5 inspired by VGG network [23]. Instead of single path of network, we use two paths to pass input to network and feature of these are subtracted to proceed further. Each VGG blocks consists of two convolutional layers in which second layer uses a strided value of 2 to reduce height and width of the features. The QA network also uses a Global Average Pooling (GAP) instead of flattening to reduce the number of trainable parameters. Deep network based image quality assessment frequently over-fits to training data because of limited number of images in the dataset. We overcame this problem by adding drop-out at fully connected layers. The QA network is trained on KADID-10K [14] dataset which consists of 10,050 images; it is largest dataset till date for quality assessment task. We split this dataset in the ratio of 70%-10%-20% for training-validating-testing purposes, respectively to account for generalizability of the testing data.

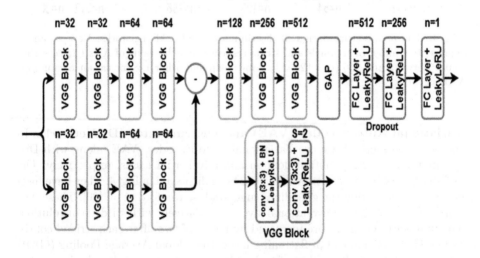

Fig. 5. The architecture of QA network.

3.1   Loss Functions

We use a combination of content, GAN, QA and domain losses with different weights to train the generator network as given below:

$$L_{gen} = \lambda_1 L_{content} + \lambda_2 L_{qa} + \lambda_3 L_{GAN}^G + \lambda_4 L_{domain}. \tag{5}$$

The values of λ_1, λ_2, λ_3 and λ_4 are set empirically to 1×10^{-1}, 5×10^{-6}, 1×10^{-3} and 1×10^{-4}, respectively. To preserve the content of LR image into generated SR image, content loss has been used which is a L_1 loss between bicubic up-sampled image and generated SR image. This can be expressed as:

$$L_{content} = \sum_{}^{N} \|G(I_{LR}) - B(I_{LR})\|_1 \tag{6}$$

where, N is the number of images in a batch during training process. G and B represent generator and bicubic upsampling operations, respectively. The proposed QA network estimates the quality of SR image in the range of 1–5 where greater value indicates better quality. As we need to minimize loss term, we use the predicted value as loss function and same is expressed as,

$$L_{qa} = \sum_{}^{N} \left(5 - Q(I_{SR})\right), \tag{7}$$

where, $Q(I_{SR})$ represents estimated quality score of SR image obtained from the proposed QA network. Further, the GAN loss has been used which is based on concept of relativistic average GAN [27]. Specifically, the standard discriminator is replaced with the relativistic average discriminator [8]. The standard discriminator in SRGAN can be expressed as $D(x) = f_{sig}(C(x))$, where f_{sig} is the sigmoid function and $C(x)$ is the non-transformed discriminator output. Then the relativistic average discriminator is formulated as $D_{Ra}(I_{HR}, I_{SR}) = f_{sig}(C(I_{HR}) - \sum^{N}[C(I_{SR})])$, where $\sum^{N}[\,\cdot\,]$ represents the operation of taking average for all fake data (super-resolved) in the mini-batch. Thus, the generator and discriminator losses are then defined as,

$$L_{GAN}^G = -\sum_{}^{N} \left[\log(1 - D_{Ra}(I_{HR}, I_{SR})) - \log(D_{Ra}(I_{SR}, I_{HR}))\right],$$

$$L_{GAN}^D = -\sum_{}^{N} \left[\log(D_{Ra}(I_{HR}, I_{SR})) - \log(1 - D_{Ra}(I_{SR}, I_{HR}))\right].$$

where L_{GAN}^D is the loss function used to train the discriminator network. For unsupervised domain translation (i.e. noisy to clean image), domain loss has been used which is based on concept of Variational Auto-Encoder [10] and same can be represented as,

$$L_{domain} = \sum_{}^{N} \left[\|\mu_{HR} - \mu_{SR}\|_1 + \|\sigma_{HR} - \sigma_{SR}\|_1\right], \tag{8}$$

where, $\mu.$ and $\sigma.$ are the mean and variance of image.

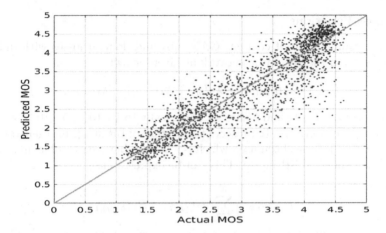

Fig. 6. Result of QA network: Actual MOS vs predicted MOS on KADID-10K testing dataset

4   Experimental Results

In order to see the effectiveness of the proposed method, we have conducted many experiments on three different datasets as described in the next section. All the experiments are performed on a computer with Intel Xeon(R) CPU with 128 GB RAM and NVIDIA Quadro P5000 GPU with 16 GB memory. In the following subsections, we elaborate the hyperparameter tuning as well as visual and quantitative evaluations of the proposed and other state-of-the-art methods.

4.1   Training Details and Hyper-parameter Tuning

To perform unsupervised SR experiment using the proposed approach, we use NTIRE-2020 Real-world SR Challenge (Track-1) dataset [17]. This dataset has been constructed with Flickr2k [13,25] and DIV2K [1] datasets where unknown noise artifacts have been added to prepare realistic source domain images. However, such degradation remains unknown to the model during training and hence it provides practical scenario of unsupervised learning with real-world images. A total of 2650 and 800 images from the above datasets are used for training of the proposed model. For the purpose of validation, an additional 100 pairs of LR-HR images have also been provided from DIV2K dataset [1]. The artificial degradation has been carried out enabling us to measure the performance in terms of PSNR or SSIM values. During training phase, the LR images are passed through different augmentations such as random rotation of $0°$ or $90°$, random horizontal flipping and random cropping operations. We train our model using Adam optimizer upto $200,000$ number of iterations with batch size of 32. We keep β_1 and β_2 values as 0.9 and 0.99, and set learning rate at 1×10^{-4} in this work. We decrease this learning rate by half after every $50,000$ iterations. The total number of trainable parameters of generator and discriminator

Table 1. The quantitative assessment of the proposed method without QA and without discriminator networks carried out on NTIRE 2020 Real-world SR Challenge validation dataset (Track-1).

Method	PSNR ↑	SSIM ↑	LPIPS ↓
w/o Discriminator	24.18	0.6027	0.572
w/o QA Network	24.11	0.5701	0.508
Proposed	**24.40**	**0.6487**	**0.458**

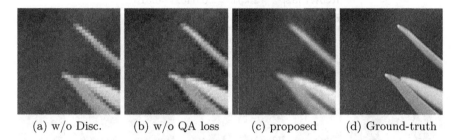

(a) w/o Disc.   (b) w/o QA loss   (c) proposed   (d) Ground-truth

Fig. 7. The SR results obtained using the proposed method (a) without discriminator network and (b) without QA network. (c) Proposed method and (d) ground-truth HR image. (better visualization in zoomed images)

result in $1.9M$ and $1.4M$, respectively. Similarly, training parameters of VAE and QA networks result in $1.4M$ and $5.35M$, respectively. We have in addition used QA network based loss in order to improve the perceptual quality of the SR images. As stated earlier, the proposed network has been trained on a KADID-10K [14] dataset. The training is carried upto $50,000$ number of iterations with 32 batch size using an Adam optimizer. Figure 6 shows fitting ability of the proposed trained QA network on the KADID-10K testing dataset. We employ Spearman's Rank Correlation Coefficient (SROCC) which is a non-parametric statistical measure to validate the performance of QA network by studying the strength of association between the two ranked variables [24] (ideal value of 1). We obtain the SROCC value of 0.89 that corresponds to close linearity of the QA network on KADID-10K testing dataset.

4.2 Ablation Study

We show the experimental justification for employing discriminator network in addition to VAE and QA networks to improve the quality of SR images in the proposed method in this section. The quantitative and qualitative results of this experiment carried out on NTIRE-2020 Real-World SR Challenge dataset [17] are depicted in Table 1 and Fig. 7, respectively. The quantitative assessments in terms of different distortion metrics such as PSNR & SSIM and in terms of perceptual measure such as LPIPS show that the proposed method with QA network in conjunction with VAE perform better (see Table 1) when compared

Table 2. The quantitative comparison of the proposed and other existing SR methods on NTIRE 2020 Real-world SR Challenge validation dataset (Track-1).

Method	PSNR ↑	SSIM ↑	LPIPS↓
ESRGAN [27]	19.04	0.2422	0.755
ZSSR [22]	25.13	0.6268	0.6160
EDSR [13]	25.35	0.6403	0.597
SRMD [30]	**25.52**	**0.6499**	0.5858
RCAN [32]	25.31	0.6402	0.576
Proposed	24.40	0.6487	**0.458**

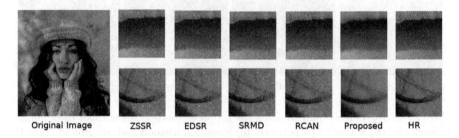

Original Image   ZSSR   EDSR   SRMD   RCAN   Proposed   HR

Fig. 8. The comparison of the SR results obtained using the proposed and other state-of-the-art methods on NTIRE-2020 Real world SR challenge validation dataset (Track-1) [17].

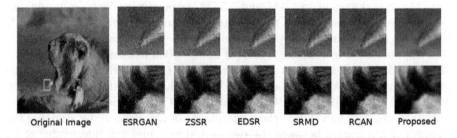

Original Image   ESRGAN   ZSSR   EDSR   SRMD   RCAN   Proposed

Fig. 9. The comparison of the SR results obtained using the proposed and other state-of-the-art methods on NTIRE-2020 Real world SR challenge testing dataset (Track-1, Real LR data) [17].

to the performance obtained using the proposed method without those modules. As shown in Fig. 7(a) and Fig. 7(b), the SR results obtained using the proposed network without the use of discriminator network and QA network is poorer than the proposed approach with both networks incorporated. We also illustrate the same in Fig. 7(c) to demonstrate the superiority of proposed approach and compare it with the respective ground-truth as indicated in Fig. 7(d). One can easily judge the improvement achieved in the perceptual quality from the proposed approach by looking at the Fig. 7.

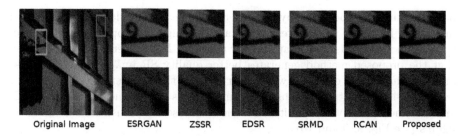

Fig. 10. The comparison of the SR results obtained using the proposed and other state-of-the-art methods on NTIRE-2020 Real world SR challenge testing dataset (Track-2, Smart Phone image) [17].

4.3 Quantitative Analysis

In order to compare the SR results obtained using the proposed method with other state-of-the-art methods, we estimate the values of PSNR and SSIM which are the standard measurements for super-resolution problem. However, these measurements can not account for quality based on human perception. We therefore use an additional LPIPS measurement [31] which is a deep network based full-reference perceptual quality assessment score. Lower LPIPS value indicates better visual quality. Table 2 provides the comparison of all the three metrics obtained on NTIRE-2020 Real-world SR Challenge validation dataset (Track-1) [17]. The proposed method obtains better SSIM and LPIPS values than that of other existing methods indicating the superiority of the proposed method in terms of quantitative evaluation.

4.4 Qualitative Analysis

In addition to quantitative comparison, in this section, we show the visual comparison to demonstrate the efficacy of the proposed method. We use the different state-of-the-art pre-trained networks such as ZSSR [22], EDSR [13], SRMD [30] and RCAN [32] and ESRGAN [27] to provide a comparison. We compare the SR results on an image of NTIRE-2020 Real-World SR challenge Track-1 validation dataset [17] in which original HR image is available as shown in Fig. 8. Looking at these SR results, one can observe that the amount of noise present in the SR image of the proposed method is reduced as compared to recently proposed competing methods.

The potential of the proposed method has also been verified on the testing datasets of NTIRE-2020 Real-World SR Challenge for Track-1 and Track-2, where original ground-truth images are not made available. In Track-1 dataset, the real-world LR images have been provided while in Track-2 the real-time data acquired using smart-phone camera are made available. We show the SR results obtained using different methods along with the proposed method on Track-1 dataset in Fig. 9. Similar to the earlier experiment, one can see the improvement of the performance from proposed method as compared to same with the other existing methods. Finally, the SR results on Track-2 dataset are depicted in Fig. 10 in which the SR results obtained using the proposed method preserves smooth regions than that of other methods. It can therefore be concluded that the proposed method generates better quality SR images with less noise artifacts than those obtained with other state-of-the-art methods. It is further supported by quantitative evaluation using various quality metrics (see Table 2) and obtained perceptual quality on various datasets as illustrated in different figures (see Fig. 8, 9 and 10).

5 Conclusion

In order to obtain an SR image of real-world LR data using deep learning approaches, generating true LR-HR pair dataset is tedious, costly and effort consuming process. We propose an alternative solution using unsupervised training of deep network to generate SR images without the use of true pair of LR-HR data. The proposed approach by incorporating Variational Auto-Encoder (VAE) along with the discriminator network of GAN can super-resolve the images upto a upscaling factor of ×4 of original LR image size. With a newly introduced deep network based on no-reference quality assessment as a loss function, the proposed approach also enhances the perceptual fidelity of SR result. Through the experiments on NTIRE-2020 Real-world SR challenge Track-1 and Track-2 datasets, this work has demonstrated consistently superior performance for super-resolution task both in terms of objective quality metrics and perceptual quality production. We intend to further explore alternative network architectures to mitigate the inadvertent blur and minor noise artifacts in the final SR image in future works in this direction.

Acknowledgment. This work is supported by ERCIM, who kindly enabled the internship of Kishor Upla at NTNU, Gjøvik. Authors are also thankful to Science and Engineering Research Board (SERB), a statutory body of Department of Science and Technology (DST), Government of India for providing support for this research work (ECR/2017/003268).

References

1. Agustsson, E., Timofte, R.: NTIRE 2017 challenge on single image super-resolution: dataset and study. In: 2017 IEEE Conference on CVPRW, pp. 1122–1131 (2017)
2. Cai, J., Zeng, H., Yong, H., Cao, Z., Zhang, L.: Toward real-world single image super-resolution: a new benchmark and a new model. In: ICCV, pp. 3086–3095, October 2019
3. Cheng, G., Matsune, A., Li, Q., Zhu, L., Zang, H., Zhan, S.: Encoder-decoder residual network for real super-resolution. In: CVPR Workshops, June 2019
4. Dong, C., Loy, C.C., He, K., Tang, X.: Image super-resolution using deep convolutional networks. IEEE Trans. PAMI **38**(2), 295–307 (2016)

5. Dong, C., Loy, C.C., Tang, X.: Accelerating the super-resolution convolutional neural network. In: Leibe, B., Matas, J., Sebe, N., Welling, M. (eds.) ECCV 2016. LNCS, vol. 9906, pp. 391–407. Springer, Cham (2016). https://doi.org/10.1007/978-3-319-46475-6_25

6. Du, C., Zewei, H., Anshun, S., et al.: Orientation-aware deep neural network for real image super-resolution. In: The IEEE Conference on CVPRW, June 2019

7. Goodfellow, I., Pouget-Abadie, J., Mirza, M., et al.: Generative adversarial nets. In: Ghahramani, Z., Welling, M., Cortes, C., Lawrence, N.D., Weinberger, K.Q. (eds.) Advances in Neural Information Processing Systems, vol. 27, pp. 2672–2680. Curran Associates, Inc. (2014)

8. Jolicoeur-Martineau, A.: The relativistic discriminator: a key element missing from standard GAN. arXiv preprint arXiv:1807.00734 (2018)

9. Kim, J., Lee, J.K., Lee, K.M.: Accurate image super-resolution using very deep convolutional networks. In: 2016 IEEE CVPR, pp. 1646–1654 (2016)

10. Kingma, D.P., Welling, M.: Auto-encoding variational bayes. arXiv preprint arXiv:1312.6114 (2013)

11. Ledig, C., Theis, L., Huszár, F., et al.: Photo-realistic single image super-resolution using a generative adversarial network. In: Proceedings IEEE Conference on CVPR, pp. 4681–4690 (2017)

12. Li, Y., Agustsson, E., Gu, S., Timofte, R., Van Gool, L.: CARN: convolutional anchored regression network for fast and accurate single image super-resolution. In: Leal-Taixé, L., Roth, S. (eds.) ECCV 2018. LNCS, vol. 11133, pp. 166–181. Springer, Cham (2019). https://doi.org/10.1007/978-3-030-11021-5_11

13. Lim, B., Son, S., Kim, H., Nah, S., Lee, K.M.: Enhanced deep residual networks for single image super-resolution. In: 2017 IEEE Conference on CVPRW, pp. 1132–1140 (2017)

14. Lin, H., Hosu, V., Saupe, D.: KADID-10K: a large-scale artificially distorted IQA database. In: 2019 10th Conference QoMEX, pp. 1–3. IEEE (2019)

15. Lugmayr, A., Danelljan, M., Timofte, R.: Unsupervised learning for real-world super-resolution. In: ICCV Workshops (2019)

16. Lugmayr, A., Danelljan, M., Timofte, R., et al.: Aim 2019 challenge on real-world image super-resolution: methods and results. In: ICCV Workshops (2019)

17. Lugmayr, A., Danelljan, M., Timofte, R., et al.: NTIRE 2020 challenge on real-world image super-resolution: methods and results. In: CVPRW (2020)

18. Park, S.-J., Son, H., Cho, S., Hong, K.-S., Lee, S.: SRFeat: single image super-resolution with feature discrimination. In: Ferrari, V., Hebert, M., Sminchisescu, C., Weiss, Y. (eds.) ECCV 2018. LNCS, vol. 11220, pp. 455–471. Springer, Cham (2018). https://doi.org/10.1007/978-3-030-01270-0_27

19. Prajapati, K., Chudasama, V., Patel, H., Upla, K., Ramachandra, R., Raja, K., Busch, C.: Unsupervised single image super-resolution network (USISResNet) for real-world data using generative adversarial network. In: 2020 IEEE CVPR-W, pp. 1904–1913 (2020)

20. Radford, A., Metz, L., Chintala, S.: Unsupervised representation learning with deep convolutional generative adversarial networks. arXiv preprint arXiv:1511.06434 (2015)

21. Shi, Y., Zhong, H., Yang, Z., Yang, X., Lin, L.: DDet: dual-path dynamic enhancement network for real-world image super-resolution. IEEE Signal Process. Lett. **27**, 481–485 (2020)

22. Shocher, A., Cohen, N., Irani, M.: Zero-shot super-resolution using deep internal learning. In: 2018 IEEE/CVF Conference on CVPR, pp. 3118–3126 (2018)

23. Simonyan, K., Zisserman, A.: Very deep convolutional networks for large-scale image recognition. arXiv preprint arXiv:1409.1556 (2014)

24. Spearman, C.: The proof and measurement of association between two things. In: Jenkins, J.J., Paterson, D.G. (eds.) Studies in individual differences: The search for intelligence, Appleton-Century-Crofts, pp. 45–58 (1961)

25. Timofte, R., Agustsson, E., Van Gool, L., Yang, M.H., Zhang, L.: NTIRE 2017 challenge on single image super-resolution: methods and results. In: Proceedings of IEEE Conference on CVPRW, pp. 114–125 (2017)

26. Tong, T., Li, G., Liu, X., Gao, Q.: Image super-resolution using dense skip connections. In: Proceedings of IEEE ICCV, pp. 4799–4807 (2017)

27. Wang, X., et al.: ESRGAN: enhanced super-resolution generative adversarial networks. In: Leal-Taixé, L., Roth, S. (eds.) ECCV 2018. LNCS, vol. 11133, pp. 63–79. Springer, Cham (2019). https://doi.org/10.1007/978-3-030-11021-5_5

28. Xu, X., Li, X.: Scan: spatial color attention networks for real single image super-resolution. In: The IEEE Conference on CVPRW, June 2019

29. Yuan, Y., Liu, S., Zhang, J., Zhang, Y., Dong, C., Lin, L.: Unsupervised image super-resolution using cycle-in-cycle generative adversarial networks. In: 2018 IEEE/CVF Conference on CVPRW, pp. 814–81409, June 2018

30. Zhang, K., Zuo, W., Zhang, L.: Learning a single convolutional super-resolution network for multiple degradations. In: Proceedings IEEE Conference on CVPR, pp. 3262–3271 (2018)

31. Zhang, R., Isola, P., Efros, A.A., Shechtman, E., Wang, O.: The unreasonable effectiveness of deep features as a perceptual metric. In: Proceedings of IEEE Conference on CVPR, pp. 586–595 (2018)

32. Zhang, Y., Li, K., Li, K., Wang, L., Zhong, B., Fu, Y.: Image super-resolution using very deep residual channel attention networks. In: Ferrari, V., Hebert, M., Sminchisescu, C., Weiss, Y. (eds.) ECCV 2018. LNCS, vol. 11211, pp. 294–310. Springer, Cham (2018). https://doi.org/10.1007/978-3-030-01234-2_18

33. Zhang, Y., Tian, Y., Kong, Y., Zhong, B., Fu, Y.: Residual dense network for image super-resolution. In: Proceedings of IEEE Conference on CVPR, pp. 2472–2481 (2018)

Training of Multiple and Mixed Tasks with a Single Network Using Feature Modulation

Mana Takeda, Gibran Benitez, and Keiji Yanai[✉]

The University of Electro-Communications,
1-5-1 Chofugaoka, Chofu-shi, Tokyo 182-8585, Japan
{takeda-m,gibran,yanai}@mm.inf.uec.ac.jp

Abstract. In recent years, multi-task learning (MTL) for image translation tasks has been actively explored. For MTL image translation, a network consisting of a shared encoder and multiple task-specific decoders is
commonly used. In this case, half parts of the network are task-specific,
which brings a significant increase in the number of parameters when
the number of tasks increases. Therefore, task-specific parts should be
as small as possible. In this paper, we propose a method for MTL image
translation using a single network with negligibly small task-specific
parts, in which we share not only the encoder part but also the decoder
part. In the proposed method, activation signals are adjusted for each
task using Feature-wise Linear Modulation (FiLM) which performs affine
transformation based on task conditional signals. In addition, we tried
to let a single network learn mixing of heterogeneous tasks such as a mix
of semantic segmentation and style transfer. With several experiments,
we demonstrate that a single network is able to learn heterogeneous
image translation tasks and their mixed tasks by following our proposed
method. In addition, despite its small model size, our network achieves
better performance than some of the latest baselines in most of the individual tasks.

1 Introduction

Deep Convolutional Neural Network (CNN) has achieved great success in various
image transformation tasks such as semantic segmentation, style transfer, and
coloring of grayscale photos. In general, each of these tasks is trained with a
single network independently, which is called as "Single Task Learning (STL)."
On the other hand, by employing Multi Task Learning (MTL), multiple tasks
can be trained with one network. Therefore, MTL is considered to be desirable
in actual applications that need to process multiple tasks with limited resources.
However, MTL models currently proposed for image translation tasks [19,23,29]
require task-specific parts, which are sometime about half portion of the whole
network, in addition to the parts shared by all the tasks. Typically, the MTL
image translation network consists of a shared encoder and multiple task-specific
decoders. In this case, the number of network parameters greatly increases as

© Springer Nature Switzerland AG 2021
A. Del Bimbo et al. (Eds.): ICPR 2020 Workshops, LNCS 12662, pp. 719–735, 2021.
https://doi.org/10.1007/978-3-030-68790-8_55

the number of tasks increases. Learning of multiple tasks with a single network consisting of a shared encoder and a shared decoder would have the advantage that the size of the network does not depend on the number of tasks involved. However, the distributions of the activations generally differ from task to task when multiple tasks are trained with a single network. It is necessary to adjust the distribution of the activations to the task-dependent distribution for each task. One of the techniques to enable adjusting the distribution of activations is Feature-wise Linear Modulation (FiLM) [5,32], which is a formalization of the conditional affine transformation. By applying the affine transformation based on the conditional signal to the network, the activation distribution can be adjusted for each task.

Furthermore, Dumoulin et al. [6], took advantage of the FiLM characteristics by proposing a multiple style transfer. They demonstrated that it was possible to mix multiple styles by providing a mixed conditional vector in addition to the conditional single style selection. Their results implied the possibility of extending mixing of multiple different styles to mixing of multiple different image transformation tasks. Therefore, in our work, we explore learning of mixing of multiple heterogeneous image translation tasks with a single network using FiLM, as well as learning of multiple tasks. We call this "mixed-task learning." Mixed-task learning is an unsolved problem in the existing MTL works. If mixed-task learning is possible with a single network, the trained network can process new tasks by combining existing tasks, which can be regarded as the first step to realize a general purpose image translation network. This is similar to arbitrary style transfer [10] which performs fast style transfer with any unknown styles by adaptive mixing of trained known styles. Note that in this paper, we assume two kinds of mixing of heterogeneous tasks: (1) sequential mixing and (2) mixing by region masking. "Sequential mixing" is mixing of multiple image translation tasks sequentially such as applying denoising first and applying style transfer next, while "mixing by region masking" is masking out the output image processed by one task using region masks estimated by the semantic segmentation task such as style transfer on only specific object region. Figure 1 shows some results on learning of both multiple and mixed tasks by our method.

The major contributions of this paper can be summarized as follows:

(1) We propose a single network capable of learning multiple heterogeneous image translation tasks with negligibly small task-specific parts, which is based on FiLM. (2) We enable mixed-task learning with the proposed network using synthesized mixed-task training samples. (3) We demonstrate the effectiveness of our proposed network on several experiments, which achieves better performance than the latest baselines (SGN [2] and Piggyback [25]), even with a smaller number of parameters.

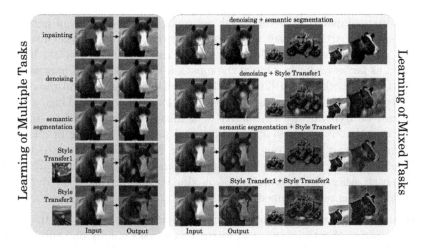

Fig. 1. The proposed network can learn multiple mixed tasks (Right) as well as multiple tasks (Left) on a single model with negligibly small task-specific parts.

2   Related Work

Multi-task Learning (MTL): In MTL, it has been shown that joint learning of multiple tasks leads to an improvement in accuracy [7,8,11,13,33,38]. However, feature sharing among multiple tasks can have a negative or positive effect depending on the combination of tasks (task interference) [23,34]. In our work, we use Feature-wise Linear Modulation (FiLM) [5,32] as a method to reduce task interference. By using FiLM, we apply a conditional transformation to activation signals, and realize a network which changes the operation dynamically for each task.

In MTL of image translation tasks, most of the networks have a shared encoder, and task-specific decoders or task-specific parts [3,21,28,37]. Cross-Stitch Network [29] includes a feed-forward network for each task, and uses cross-stitch units to share features between tasks. UberNet [23] proposes an image pyramid approach for processing images across multiple resolutions. At each resolution, an additional task-specific layer is formed on top of the shared network. Maninis et al. [26] proposes a network that modifies behavior based on task-specific features and attention. Task attention uses task-specific residual adapter branch and Squeeze-and-Excitation modulation. In addition, they also use adversarial training to force shared parts to be statistically indistinguishable across tasks. Strezoski et al. [35] proposes a task-specific binary mask which is applied to the activations of each channel, and the activations corresponding to each task were extracted. Bragman et al. [1] proposes the similar idea with stochastic assignments of convolution channels to tasks instead of binary assignments. Liu et al. [24] have proposed a multi-task attention network (MTAN) that can learn the attention of task-specific feature levels. MTAN consists of a single shared network including global feature sharing and a soft-attention module for each task. In the above-mentioned networks, it is necessary to train a network by replacing part of the network for each task. Therefore, more network parameters are required compared to a single task network, and the number of parameters of task-specific parts increases in proportion to the number of tasks involved.

No mixed-task learning with heterogeneous image translation tasks has been proposed as far as we know. One of the most similar work is Sym-parameterized Generative Network (SGN) [2]. In SGN, the distribution of multiple domains is learned by changing the weighted loss function dynamically. Thus, an input image can be dynamically translated to a mixed domain. SGN is similar to our work in that it uses a single network and performs learning of mixed loss functions. However, unlike our work, the tasks for which SGN mixed learning was applied were style transfer and domain transfer, both of which are similar tasks to each other. Heterogeneous task mixing such as mixing of inpainting and style transfer was not examined.

Continual Learning: Continuous learning [30] is a learning method that, when there is a series of n tasks $t_1, ..., t_n$, learns those tasks one by one and does not reduce the accuracy of the tasks learned in the past. However, catastrophic forgetting, which degrades the performance of $t_1, ..., t_{n-1}$ by learning t_n, is a major problem. Many studies have been proposed to address this problem. Piggyback [25] is the methods that transforms the output by applying a learned weight mask. Although in the original paper [25], Piggyback was applied to only image classification tasks, Matsumoto et al. [27] showed that Piggyback was able to be applied to image transformation tasks with an encoder-decoder network. Piggyback is similar to our work in that multiple tasks can be learned with a single network. However, it is not able to perform mixed-task learning.

Feature-wise Linear Modulation (FiLM): FiLM [5,32] is a formalization of the conditional affine transformation. By applying the affine transformation to the network based on the condition, the effect on the output of the network is learned. Specifically, as shown in Eq. (1), we learn the functions f_γ and f_β of an input conditional signal $\mathbf{c}$ that output the scaling coefficient γ_i and the shift coefficient β_i. The subscript i means the feature or feature map number. As shown in Eq. (2), γ_i and β_i regulate network activations.

$$\gamma_i = f_{\gamma,i}(\mathbf{c}) \quad , \quad \beta_i = f_{\beta,i}(\mathbf{c}) \tag{1}$$

$$FiLM(\mathbf{F}_i|\gamma_i, \beta_i) = \gamma_i \mathbf{F}_i + \beta_i \tag{2}$$

Various methods using FiLM have been proposed for image translation tasks. Dumoulin et al. [6] made it possible to combine multiple styles by applying FiLM to Fast Style Transfer [18]. AdaIN [15] is another study that applied FiLM to Style Transfer. The AdaIN module is widely used in various image translation tasks as well as style transfer because of the ability to manipulate network outputs using AdaIN parameters [16,20,31]. In our work, we propose using FiLM as a method for learning multiple tasks and mixed tasks using a single network. In the proposed method, we extend the work of Dumoulin et al. from multiple mixed style transfer to multiple mixed image transformation tasks. For FiLM generators, we use StyleGAN's Mapping Network [20] mechanism, which can generate AdaIN parameters from the conditional vectors with a sequence of fully-connected layers.

3   Proposed Method

In our work, Feature-wise Linear Modulation (FiLM) [5,32] is used for learning multiple image translation tasks using a single encoder-decoder network. Its effectiveness has been demonstrated in various image translation tasks. In other words, it can be said that FiLM has high versatility in image transformation tasks and is expected to be able to learn various image transformation tasks on a single network. Therefore, in our work, we propose learning of different kinds of image transformation tasks with a single network using FiLM, which is the first objective of this paper. The second objective is to accomplish mixed-task learning of multiple heterogeneous image translation tasks by a single network using FiLM, which is inspired by the work on mixing of multiple styles by Dumoulin et al. [6].

3.1   Task Conditional Vector

A task conditional vector is used for specifying tasks at the time of both training and inference of the network. At training time, the task conditional vector corresponding to the task to be learned by the network is given. Similarly, at inference time, the task conditional vector corresponding to the task to be executed is given. If the number of tasks is n, the task conditional vector $\mathbf{c}$ is defined as a n-dimensional vector $[c_1, ..., c_n]$. At inference time, c_i takes values in the range of 0.0 to 1.0, while c_i can be 0.0 or 1.0 at training time. Note that a zero conditional vector, $\mathbf{c} = [0.0, 0.0, 0.0]$, represents the identity transformation in which the output is the same as the input. Including the identity transformation in the training is needed to control the degree of the transformation, which can be explicitly defined by providing intermediate values between 0.0 and 1.0 in the task conditional vector.

3.2   FiLM-Based Network Architecture

We propose an architecture consisting of the FiLM generator and the FiLM network, as shown in Fig. 2. For the FiLM network, we adopted the Encoder-Decoder CNN with five Resblocks proposed by Johnson et al. [18]. This network is typically used in image translation tasks such as Pix2Pix [17] and Cycle-GAN [39]. As normalization layers, we used Instance Normalization (IN) [36] instead of Batch Normalization as used in the original network. We inserted FiLM layers after all the layers except for the last layer. The combination of IN and FiLM is equivalent to AdaIN [15]. The FiLM layer receives the FiLM parameters from the FiLM generator, and controls the operation of the network by affine transformation based on the FiLM parameters.

The FiLM generator is built with only Fully Connected (FC) layers, referring to the architecture of StyleGAN's mapping network [20]. The FC layers generate the FiLM parameters, γ_i and β_i, where i depicts an index of the FiLM layer. Although one FC layer is illustrated in Fig. 2, we compared one, three and five FCs for the FiLM generator in the preliminary experiments.

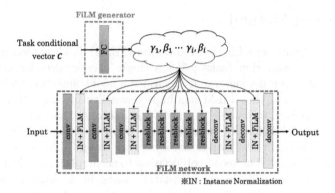

Fig. 2. Network Architecture. The FiLM network consists of three convolution layers, five residual blocks, and three deconvolution layers. FiLM layers are inserted after all the instance normalization layers including resblocks.

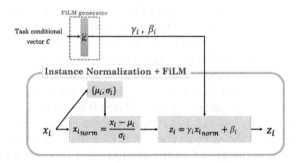

Fig. 3. Computation in a IN+FiLM layer. After normalizing the input features by Instance Normalization, affine transformation with FiLM parameters is applied. These parameters are obtained by the FiLM generator.

The computation of the FiLM layers is carried out based on Eq. 1 and Eq. 2 by using the FiLM parameters, γ_i and β_i, obtained by the FiLM generator after IN layers, as shown in Fig. 3. Here, γ represents scaling parameters, and β represents bias parameters.

As explained above, most part of the proposed network are shared with all the tasks. Only the number of input elements of the first FC layer in the FiLM generator depends on the number of tasks, which is the only task-specific part. Therefore, the number of the task-specific parameters is negligibly small.

3.3   Training of Mixed Tasks

We examine the possibility of a FiLM-based network for mixed-task learning in addition to learning of multiple tasks. We prepared three methods for mixed-task learning.

In Method 1, we train only single tasks individually. We expect the inference on mixed tasks is performed implicitly, in the same way as the FiLM-based multiple style transfer [6].

In Method 2, we train mixed tasks explicitly with the total loss functions of the multiple tasks we want to mix. We use a simple mixed loss, $L_{mixed} = L_{taskA} + L_{taskB}$. To calculate L_{mixed}, we calculate L_{taskA} and L_{taskB} with training samples of task A and B, respectively, and sum up both in a single function. For training, we add mixed tasks to a task set originally consisting of only single individual tasks. In the training loop, we select randomly one task from an extended task set at every mini-batch. Note that in Method 2 and 3, we use a mixed-task conditional vector in the form of $c = [1.0, 1.0]$ when a mixed task is selected.

In Method 3, we create new training samples for mixed tasks. Figure 4 shows how to create mixed-task samples. We prepare two ways to create mixed-task samples as shown in Fig. 4: (a) sequential mixing and (b) mixing by masking. Sequential mixing is a standard mixing way of multiple tasks, while mixing by masking is a combination of semantic segmentation with a different task. In (a) we need image translation networks in the second translation step, while we can use training ground truth (GT) samples instead of translation processing in the first step. In (b) we need the trained network of the fast style transfer if a mixed task contains style transfer, since no GT samples exist for this task. Therefore, in both ways, first, we train all the tasks except for semantic segmentation with a single FiLM-based network. Note that we can substitute a set of single-task networks for the multi-task network.

Next, we generate mixed-task samples. In (a), we pick up GT samples of the first task, and apply the trained model of the second task using the corresponding task conditional vectors. In the example of Fig. 4(a), the sequentially-translated image by Denoising and Style Transfer is generated. Thus, the synthesized image is used as a GT image (mixed GT) of the mixed-task of "Denoising + Style Transfer." We use another mixing way, (b) mixing by masking, for the case that mixed tasks contain semantic segmentation since we want to cut out the region belonging to specific objects as shown in Fig. 4(b). In (b), we pick up the segmentation mask from the training dataset of semantic segmentation, and the corresponding GT sample of the second task. If GT samples are not available in the second task, we apply the trained network to generate it. Next, we mask out the output of the second task with the segmentation mask. In the example of Fig. 4(b), we generate a mixed-task sample by combining Style Transfer and semantic segmentation on the horse image. To train mixed task samples, we use L2 loss. We compare these three methods in the experiments.

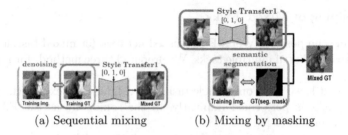

(a) Sequential mixing   (b) Mixing by masking

Fig. 4. Two ways to generate mixed-task training samples. (a) Denoising and Style Transfer are mixed sequentially. (b) The output of Style Transfer is masked out by the segmentation mask obtained Semantic Segmentation Task.

4   Experiments

We performed three kinds of experiments to verify the effectiveness of the proposed method: (1) learning of multiple different tasks, (2) learning of mixed tasks with the three training methods, and (3) comparison with the baselines.

In (1), we verified whether it is possible to learn multiple different image translation tasks with a single FiLM-based network. In (2), we verified whether the proposed method could learn mixed image transformation tasks with a single FiLM-based network. To do that, we qualitatively compared the three methods explained in Sect. 3.3. Finally, in (3), we quantitatively compared the proposed method with several baselines by evaluating their performance on all the tasks.

The dataset used for the experiments was Pascal VOC [9], a dataset containing general images such as people, vehicles, and animals with 20-class pixel-level annotation. In the experiment, out of 11,355 images on Pascal VOC 2011 for which Hariharan et al. [12] created the pixel-level annotation data, 8,498 were used as training data and 2,857 were used as test data.

As the network, we used the FiLM-based network explained in Sect. 3.2. In all the training, Adam [22] with a learning rate of 1e−4 was used as the optimization method, and the batch size were 32. To training multiple tasks with one network, one task is randomly selected from a task set at every minibatch. So we set the number of epochs as $300n$, where n is the total number of all the tasks to be learned. We did not use any weighting for loss functions, since several existing works [2,24] showed that no weighting or equal weighting was enough for training of multiple image translation tasks. Regarding the FiLM generator, we used one FC layer in the experiments, because we obtained good enough results with only one FC layer, which helped to keep the total model size small.

Table 1. The task list with the loss functions.

Task number	Task name	Loss function
Task 0	reconstruction (identity)	L2 loss
Task 1	inpainting	L2 loss
Task 2	denoising	L2 loss
Task 3	semantic segmentation	L2 loss + adversarial loss
Task 4	style transfer 1 (Gogh)	perceptual loss
Task 5	style transfer 2 (Munk)	perceptual loss

4.1 Task Sets for the Experiments

Table 1 shows six tasks and their loss functions used in the experiments. For all the tasks, both the inputs and the outputs are three-channel images.

Task 0 is the reconstruction of an input image which is equivalent to auto-encoder or identity transformation. In Task 1 (the inpainting task), we used training images masked with small square regions, as shown in the second column of Fig. 5, and in Task 2 (the denoising task) we used training images to which scratches were randomly added, as shown in the third column of Fig. 5.

Regarding Task 3 (the semantic segmentation task), the semantic segmentation network usually generates a semantic segmentation map (segmentation masks) with the number of channels equal to the number of classes, where a cross-entropy loss is used for training. However, in our work, to mix segmentation with a second task, the output of semantic segmentation is a three-channel image in which the region of the specified classes is cut out and the others are filled out. The target image used for learning of semantic segmentation was artificially created based on the annotation mask of the dataset. To train the semantic segmentation task, we used both L2 loss and adversarial loss instead of the conventional cross-entropy loss. In the experiment, 20 classes except for the background class conformed the set of the cut-out classes. In other words, the output of semantic segmentation is an image in which the background is filled with the background color. In this experiment, the RGB value of the background part of the semantic segmentation was set to (0,0,0) for the input image normalized by the ImageNet [4] mean and standard deviation.

Task 4 and 5 are Fast Neural Style Transfer with different styles using perceptual loss in the same way as Johnson et al. [18].

4.2 Experiment 1: Learning of Multiple Different Tasks

In Experiment 1, we qualitatively evaluate the proposed FiLM-based network to learn multiple different image translation tasks at the same time. We trained the proposed FiLM-based network with the six tasks shown in Table 1. To train the network and to get the results of all the tasks, we used one-hot vectors except for Task 0, which uses a zero vector. Figure 5 shows the results of all the six tasks.

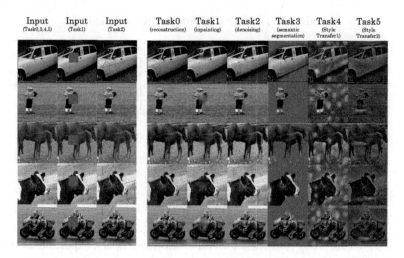

Input Input Input Task0 Task1 Task2 Task3 Task4 Task5
(Task0,3,4,5) (Task1) (Task2) (reconstruction) (inpainting) (denoising) (semantic segmentation) (Style Transfer1) (Style Transfer2)

Fig. 5. The results of all the six tasks generated by one trained FiLM-based network.

From the figure, it can be seen that most of the tasks are successfully executed by changing only the task conditional vector with a single trained network. No prominent task degradation and interference is not observed except for inpainting. Note that the result for inpainting (Task 1) is not perfect, partly because 8,498 training samples might be not enough for this task, which explicitly include images with square-masking. From these results, we conclude that multiple different image translation tasks can be learned with a single FiLM-based model.

4.3   Experiment 2: Learning of Mixed Tasks

In Experiment 2, we examine if it is possible to learn mixing of different image transformation tasks using the proposed method. As explained in Sect. 3.3, we prepared three methods: (1) training only individual tasks, (2) training mixed tasks using a compound loss function, and (3) training using synthesized mixed-task samples. For verification of mixed-task learning, four tasks were used: reconstruction, denoising, semantic segmentation, and Style Transfer 1, shown in Table 1. Then, experiments were performed on all the combinations of these patterns.

For Method 3, we used sequential mixing (Fig. 4(a)) for the pair of the denoising and the style transfer tasks, and mixing by masking (Fig. 4(b)) for the other two pairs. In the training time of Method 3, we repeat training of a FiLM-based network twice. First, we train individual single tasks with a single FiLM-based network, and after generating mixed samples, we train again both single tasks and mixed tasks with the FiLM-based network. To train mixed task samples, we use L2 loss between an output image and a synthesized mixed-task sample.

To train mixed tasks with Method 2 and 3, we used mixed-task conditional vectors shown in Table 2. Although the summed value of the conditional vector

Table 2. Task conditional vectors for mixed-task learning.

	Mixed tasks	Conditional vector	Mixing for Method 3
Mix 1	Denoising + Style Transefer 1	[1.0, 0.0, 1.0]	Sequential
Mix 2	Denoising + semantic segmentation	[1.0, 1.0, 0.0]	Masking
Mix 3	Semantic segmentation + Style Transfer1	[0.0, 1.0, 1.0]	Masking

is commonly set as 1.0, the degree of both task in the mixed task should be normal in case of sequential mix and mixing by masking. Therefore, we set 1.0 to the elements corresponding to both tasks, as shown in the table.

Figure 6 shows the experimental results. In case of Method 1, the results of Mix 2 and 3 look like weak combination of two tasks, while the result of Mix 1 remains many scratches, which means only style transfer without denoising was performed. In case of Method 2, the outputs are biased toward either output of the mixed task. For Mix 1 and 2, only denoising was carried out, while for Mix 3 only segmentation was performed. Therefore, we can conclude that with Method 1 in which only individual tasks are trained, and with Method 2 in which mixed-

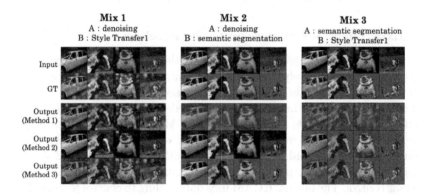

Fig. 6. The outputs of learning of the mixed tasks with Method 1, 2, and 3.

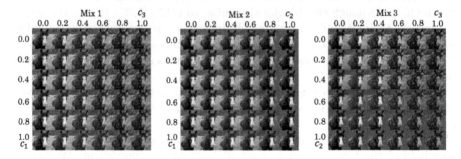

Fig. 7. The transitional results of mixed-task learning with Method 3 by changing the conditional weights by 0.2.

task was learned using the summed loss function, it is not possible to learn mixed tasks with the FiLM-based network. On the other hand, with Method 3, the results are almost the same as ground truth (GT), which means mixed-task learning succeeded. We found that mixed task learning of heterogeneous image translation tasks is possible when we use synthesized mixed-task training samples and L2 loss for training of mixed tasks.

In addition, we found an interesting characteristic of mixed-task learning with Method 3. Figure 7 shows the three kinds of mixed-task results of Method 3 by changing the conditional weights by 0.2. The elements of a task conditional vector corresponding to denoising, semantic segmentation, and Style Transfer1 is represented by c_1, c_2 and c_3, respectively. The figure shows that by performing mixed-task learning with mixed task conditions shown in Table 2 and synthesized mixed training samples, a natural transition from identity transformation to mixed-task image translation is possible by gradually changing the conditional vector at inference time, which covers both single task translations of the mixed tasks. This result indicates that the objective space between two tasks can be regarded as a linear combination space of the two tasks by using the proposed method, and the degree of the mixed task can be controlled over the 2D linear combination space among two target tasks during inference.

4.4   Experiment 3: Comparison to the Baselines

In Experiment 3, we compare the proposed method with the baselines both qualitatively and quantitatively. In this experiment, the basic network architecture of all the baselines was the same as the proposed method.

The first baseline is the Sym-parameterized Generative Network (SGN) [2], which adopts Conditional Channel Attention Module (CCAM) as an injection method of conditional signals. Inspired by SENet [14], CCAM controls feature channels based on Sigmoid attention by integrating both conditional signals and average-pooled feature map activations. The second baseline is the CCAM of SGN with bias control. The normal CCAM uses only scaling to control feature channels with Sigmoid function. Therefore, more flexible feature channel control is expected to be performed by adding bias control to CCAM. Note that no Sigmoid was used for the bias. Although this baseline is similar to FiLM [5,32], it uses the activation signals in addition to the task conditional vector to generate scaling and bias parameters. Besides it uses instance normalization as normalization layers and CCAM were inserted on only three parts of the network according to the SGN original paper [2].

The third baseline is Piggyback [25], which is a method for fixing the parameters of the base network learned first and learning the task-specific binary mask each time a new task is added. For the initialization of the real number mask, all parameters were initialized to 1e−2, as in the Piggyback's paper, and the threshold of the binary mask was set to 5e−3 in the experiment. We created two models, Piggyback1 and Piggyback2, which used different image translation tasks for learning the base network. For Piggyback1, inpainting was selected, and for Piggyback2, semantic segmentation was selected as the task used for learning

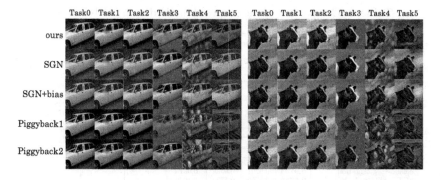

Fig. 8. The results of the proposed method and the baselines.

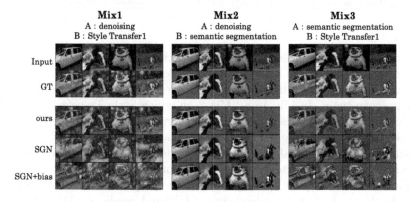

Fig. 9. Comparison of the results of mixed-task learning.

the base network. In the second and subsequent tasks, we learned a binary mask that selects effective weights for the newly added task from the base network. Since it is not possible to perform mixed-task learning with Piggyback, only the results of single tasks alone were compared.

First, we quantitatively evaluated the proposed method and the baselines. Figure 8 shows their outputs when learning of multiple heterogeneous image translation tasks. From the figure, it can be seen that in Piggyback1, much noise appears in the background part in the semantic segmentation. In Piggyback2, the outputs in the denoising still have a lots of scratches. In SGN and SGN+bias, in Style Transfer 1, the blue hue of the style was not transferred, although the style with yellow spots was transferred. Similarly, style transfer failed in Style Transfer2.

Figure 9 compares the output of the proposed method and the baseline for mixed-task learning. Following Method 3 explained in Sect. 3.3, we used synthesized mixed-task samples to train the networks of SGN and SGN+bias as well as the proposed network, since Method 1 and 2 did not work between heterogeneous image translation tasks in Experiment 2. From the figure, we can see that mix learning is possible at the two baselines as well as the proposed method. How-

Table 3. Comparison of the proposed method with the baselines.

		Ours	EncIN	SGN	SGN+bias	Piggyback1	Piggyback2	Single	SharedEnc
Task0	reconstruction	**0.1408**	0.1862	0.4182	0.4191	0.1598	0.1619	0.1222	0.1797
	(MSE↓, SSIM↑)	**0.9870**	0.9820	0.9579	0.9578	0.9854	0.9860	0.9892	0.9844
Task1	inpainting	**0.0932**	0.1041	0.4524	0.4518	0.1230	0.1535	0.0771	0.2375
	(MSE↓, SSIM↑))	0.9931	0.9920	0.9553	0.9554	**0.9939**	0.9885	0.9945	0.9806
Task2	denoising	**0.0839**	0.1038	0.4300	0.4309	0.1742	0.2022	0.0890	0.2045
	(MSE↓, SSIM↑)	**0.9936**	0.9916	0.8703	0.9572	0.9868	0.9830	0.9931	0.9825
Task3	semantic segmentation	**0.2112**	0.2298	0.5942	0.5892	0.3657	**0.2112**	0.2289	0.2516
	(MSE↓, SSIM↑, IoU↑)	0.9798	0.9775	0.9473	0.9500	0.9630	**0.9801**	0.9775	0.9753
		0.5907	0.5689	0.4299	0.4454	0.4030	**0.6014**	0.5639	0.5545
Task4	Style Transfer1 (ST1) (FID↓)	281.3	318.3	299.0	307.1	333.4	331.4	186.7	324.4
Task5	Style Transfer2 (ST2) (FID↓)	235.3	250.6	263.6	250.1	297.6	323.3	163.9	251.6
Mix1	denoising + ST1 (FID↓)	304.3	361.3	349.1	343.4	-	-	-	-
Mix2	denoising +	**0.2130**	0.2166	0.5699	0.5663	-	-	-	-
	semantic segmentation	**0.9794**	0.9791	0.9524	0.9519	-	-	-	-
	(MSE↓, SSIM↑, IoU↑)	**0.5755**	0.5752	0.4700	0.4625	-	-	-	-
Mix3	semantic segmentation + ST1	**313.0**	320.3	339.0	348.6	-	-	-	-
	(FID↓, IoU↑)	**0.5457**	0.5403	0.5011	0.4555	-	-	-	-
	Model size (num. of prams)	1,698,435	1,695,747	1,765,363	1,902,243	**1,688,835**	**1,688,835**	10,075,410	9,572,370

ever, compared to the proposed method, the two baselines showed more noise. In Mix 2 and Mix 3 in which semantic segmentation is included, the contrast of the cut-out part is increased, the output is dark overall, and noise in the background part is conspicuous. In Mix 1 and Mix 3, in which Style Transfer1 is included, in addition to the blue hue of the style not being transferred, white noise is seen in SGN, and orange noise is seen in SGN+bias. From Fig. 8 and Fig. 9, it can be said that the proposed method was superior to the baselines qualitatively, especially in the tasks of semantic segmentation and Style Transfer.

Furthermore, we quantitatively evaluate the proposed method and the baselines. In this evaluation, we added three additional baselines: (1) a set of single models trained with six tasks independently (Single), (2) a model consisting of a shared encoder and task-specific ResBlocks/decoders (SharedEnc), and (3) the modified FiLM-based network in which the encoder part uses no FiLM but standard INs (EncIn). Thus, (1) is the strong baseline, (2) is the standard MTL network for multi-task image translation, and (3) is a variant of the proposed FiLM-based network with a shared encoder. As evaluation indices, Frechet Inception Distance (FID) is used for tasks including Style Transfer, while Mean Square Error (MSE) and Structural Similarity (SSIM) are used for other tasks. For the tasks including semantic segmentation, Intersection over Union (IoU) is used as well.

Table 3 compares the performance of the proposed method and the baselines in the case of learning of both multiple single tasks and mixed tasks. From the table, we can see that despite its small model size, our method achieved the best evaluation scores in almost all the tasks compared to the MLT methods. Note that Piggyback1 initially started training on the single task of inpainting, while Piggyback2 started on the semantic segmentation. That is why both achieved the best score on each of these tasks, respectively. Besides, our proposal presents comparable results with the strong baseline (Single), even though its model size is the biggest (more than five times bigger). On the other hand, Piggyback has the smallest model size. However, it cannot learn mixed task.

Between the proposed method and SGN, the performance of most of the tasks was lower in SGN than in our proposed method in learning of multiple tasks and mixed tasks. One of the possible reasons is that CCAM are inserted on only three parts of the whole network, which is expected to make it hard to adopt the network to heterogeneous image translation tasks, such as style transfer and semantic segmentation. In addition, SGN uses Sigmoid functions to generate attention maps and the output values of attention maps are limited from 0.0 to 1.0, which is expected to restrict the adaptability of the network for various kinds of tasks. On the other hand, in our network, the combination of Instance Normalization and FiLM layers (IN+FiLM) without Sigmoid functions are inserted after all the convolutional layers except the last layer. Moreover, compared with EncIN, in which the encoder part uses no FiLM but standard non-conditional INs, the proposed network outperformed EncIN on all the tasks. Therefore, we can conclude that IN+FiLM should be used after all the convolutional layers except the last layer regardless of encoder and decoder parts.

5   Conclusions

In this work, we performed learning of multiple different image translation tasks and their mixed tasks with the single FiLM-based network. The experimental results showed that mixed-task learning using synthesized training images of mixed tasks is possible in addition to learning of multiple individual tasks. Furthermore, it was found that the objective space of those expressions could be complemented by changing the task conditional vector during inference, even though the intermediate representation of each task and the mutual tasks between multiple tasks were not learned. We also compared the proposed method with other baselines and showed its effectiveness.

In future work, we plan to add more tasks such as various kinds of image domain translation tasks and mix them. We aim to build a more practical network and task mixing by verifying learning of mixing of various tasks. In addition, we like to reduce task interference and improve accuracy by devising the network architecture. We will examine the effectiveness of the proposed network for image classification tasks as well. Extending the method to incremental learning is also one of the interesting topics.

References

1. Bragman, F.J., Tanno, R., Ourselin, S., Alexander, D.C., Cardoso, J.: Stochastic filter groups for multi-task CNNs: learning specialist and generalist convolution kernels. In: ICCV (2019)
2. Chang, S., Park, S., Yang, J., Kwak, N.: Sym-parameterized dynamic inference for mixed-domain image translation. In: ICCV (2019)
3. Chen, Z., Badrinarayanan, V., Lee, C., Rabinovich, A.: GradNorm: gradient normalization for adaptive loss balancing in deep multitask networks. In: ICML (2018)
4. Deng, J., Dong, W., Socher, R., Li, L.J., Li, K., Fei-Fei, L.: ImageNet: a large-scale hierarchical image database. In: CVPR (2009)

5. Dumoulin, V., Perez, E., Schucher, N., Strub, F., Vries, H.d., Courville, A., Bengio, Y.: Feature-wise transformations (2018). https://doi.org/10.23915/distill.00011. https://distill.pub/2018/feature-wise-transformations
6. Dumoulin, V., Shlens, J., Kudlur, M.: A learned representation for artistic style. In: ICLR (2017)
7. Dvornik, N., Shmelkov, K., Mairal, J., Schmid, C.: Blitznet: a real-time deep network for scene understanding. In: ICCV (2017)
8. Eigen, D., Fergus, R.: Predicting depth, surface normals and semantic labels with a common multi-scale convolutional architecture. In: ICCV (2015)
9. Everingham, M., Eslami, S.M.A., Van Gool, L., Williams, C.K.I., Winn, J., Zisserman, A.: The pascal visual object classes challenge: a retrospective. In: IJCV (2015)
10. Ghiasi, G., Lee, H., Kudlur, M., Dumoulin, V., Shlens, J.: Exploring the structure of a real-time, arbitrary neural artistic stylization network. In: BMVC (2017)
11. Girshick, R.: Fast R-CNN. In: ICCV (2015)
12. Hariharan, B., Arbelaez, P., Bourdev, L., Maji, S., Malik, J.: Semantic contours from inverse detectors. In: ICCV (2011)
13. He, K., Gkioxari, G., Dollar, P., Girshick, R.: Mask R-CNN. In: ICCV (2017)
14. Hu, J., Shen, L., Sun, G.: Squeeze-and-excitation networks. In: CVPR (2018)
15. Huang, X., Belongie, S.: Arbitrary style transfer in real-time with adaptive instance normalization. In: ICCV (2017)
16. Huang, X., Liu, M.-Y., Belongie, S., Kautz, J.: Multimodal unsupervised image-to-image translation. In: Ferrari, V., Hebert, M., Sminchisescu, C., Weiss, Y. (eds.) ECCV 2018. LNCS, vol. 11207, pp. 179–196. Springer, Cham (2018). https://doi.org/10.1007/978-3-030-01219-9_11
17. Isola, P., Zhu, J., Zhou, T., Efros, A.A.: Image-to-image translation with conditional adversarial networks. In: CVPR (2017)
18. Johnson, J., Alahi, A., Fei-Fei, L.: Perceptual losses for real-time style transfer and super-resolution. In: Leibe, B., Matas, J., Sebe, N., Welling, M. (eds.) ECCV 2016. LNCS, vol. 9906, pp. 694–711. Springer, Cham (2016). https://doi.org/10.1007/978-3-319-46475-6_43
19. Kaiser, L., et al.: One model to learn them all. arXiv:1706.05137 (2017)
20. Karras, T., Laine, S., Aila, T.: A style-based generator architecture for generative adversarial networks. In: CVPR (2019)
21. Kendall, A., Gal, Y., Cipolla, R.: Multi-task learning using uncertainty to weigh losses for scene geometry and semantics. In: CVPR (2018)
22. Kingma, D., Ba, J.: Adam: a method for stochastic optimization. In: ICLR (2015)
23. Kokkinos, I.: UberNet: training a universal convolutional neural network for low-, mid-, and high-level vision using diverse datasets and limited memory. In: ICCV (2017)
24. Liu, S., Johns, E., Davison, A.J.: End-to-end multi-task learning with attention. In: CVPR (2019)
25. Mallya, A., Davis, D., Lazebnik, S.: Piggyback: adapting a single network to multiple tasks by learning to mask weights. In: ECCV (2018)
26. Maninis, K.K., Radosavovic, I., Kokkinos, I.: Attentive single-tasking of multiple tasks. In: CVPR (2019)
27. Matsumoto, A., Yanai, K.: Continual learning of an image transformation network using task-dependent weight selection masks. In: ACPR (2019)

28. Guo, M., Haque, A., Huang, D.-A., Yeung, S., Fei-Fei, L.: Dynamic task prioriti-zation for multitask learning. In: Ferrari, V., Hebert, M., Sminchisescu, C., Weiss, Y. (eds.) ECCV 2018. LNCS, vol. 11220, pp. 282–299. Springer, Cham (2018). https://doi.org/10.1007/978-3-030-01270-0_17
29. Misra, I., Shrivastava, A., Gupta, A., Hebert, M.: Cross-stitch networks for multi-task learning. In: CVPR (2016)
30. Parisi, G.I., Kemker, R., Part, J.L., Kanan, C., Wermter, S.: Continual lifelong learning with neural networks: a review. arXiv:1802.07569 (2018)
31. Park, T., Liu, M.Y., Wang, T.C., Zhu, J.Y.: Semantic image synthesis with spatially-adaptive normalization. In: CVPR (2019)
32. Perez, E., Strub, F., De Vries, H., Dumoulin, V., Courville, A.: FiLM: visual rea-soning with a general conditioning layer. In: AAAI (2018)
33. Rosenfeld, A., Biparva, M., Tsotsos, J.K.: Priming neural networks. arXiv:1711.05918 (2017)
34. Standley, T., Zamir, A.R., Chen, D., Guibas, L.J., Malik, J., Savarese, S.: Which tasks should be learned together in multi-task learning? arXiv:1905.07553 (2019)
35. Strezoski, G., van Noord, N., Worring, M.: Many task learning with task routing. In: ICCV (2019)
36. Ulyanov, D., Vedaldi, A., Lempitsky, V.S.: Instance normalization: the missing ingredient for fast stylization. arXiv:1607.08022 (2016)
37. Vandenhende, S., Brabandere, B.D., Gool, L.V.: Branched multi-task networks: deciding what layers to share. arXiv:1904.02920 (2019)
38. Zhao, X., Li, H., Shen, X., Liang, X., Wu, Y.: A modulation module for multi-task learning with applications in image retrieval. In: Ferrari, V., Hebert, M., Sminchisescu, C., Weiss, Y. (eds.) ECCV 2018. LNCS, vol. 11205, pp. 415–432. Springer, Cham (2018). https://doi.org/10.1007/978-3-030-01246-5_25
39. Zhu, J.Y., Park, T., Isola, P., Efros, A.A.: Unpaired image-to-image translation using cycle-consistent adversarial networks. In: ICCV (2017)

Deep Image Clustering Using Self-learning Optimization in a Variational Auto-Encoder

Duc Hoa Tran[1(✉)], Michel Meunier[2], and Farida Cheriet[1]

[1] Department of Computer and Software Engineering, Polytechnique Montréal,
Montréal, Canada
{duc-hoa.tran,farida.cheriet}@polymtl.ca
[2] Department of Engineering Physics, Polytechnique Montréal, Montréal, Canada
michel.meunier@polymtl.ca

Abstract. Deep image clustering approaches typically use autoencoder architectures to learn compressed latent representations suitable for clustering tasks. However, they do not effectively regulate the latent space during training, leading to low performance and diminished applicability to different datasets. In this paper, we propose a deep clustering model combining maximum mean discrepancy (MMD) regularization and self-learning clustering optimization to mitigate this problem. Specifically, we first train the network to improve its image reconstruction ability by minimizing both reconstruction loss and MMD divergence from a target distribution. Then, the model gradually learns from its own high-confidence predictions to further optimize the latent distribution. We validate the network's performance on different benchmark image sets using standard clustering metrics, without changing network configuration or adjusting hyper-parameters between datasets. The proposed model provides top clustering performance across datasets while being more robust than state-of-the-art methods.

Keywords: Clustering · Self-learning · Variational Auto-Encoder

1 Introduction

Clustering is an unsupervised learning approach that is essential for image analysis tasks, especially image categorization and segmentation [6,24]. Conventional methods like K-means [18] and spectral clustering [1,20] have been applied to a wide range of applications. However, as they measure the similarity distance between points in shallow, high-dimensional feature spaces of raw pixels or gradient-based histograms, their clustering ability is largely limited to simple image datasets [22].

For their part, deep learning-based clustering approaches use deep neural networks to represent data as lower-dimension, hierarchical features such that conventional clustering techniques can be applied effectively [4]. For example, an

© Springer Nature Switzerland AG 2021
A. Del Bimbo et al. (Eds.): ICPR 2020 Workshops, LNCS 12662, pp. 736–749, 2021.
https://doi.org/10.1007/978-3-030-68790-8_56

encoder-decoder architecture can be used to produce a rich latent encoding of the input image with dramatically reduced dimensionality. Thus, a data grouping applying on the latent codes is more feasible than performing on the input images and can avoid the curse of dimensionality problem to which clustering algorithms are prone.

Typically, there are two major challenges when applying existing deep clustering algorithms to different image datasets: (1) maintaining high algorithm performance requires reconfiguring the network architecture or adjusting a large number of training hyper-parameters [5]; (2) complicated algorithms such as spectral clustering can achieve very high performance but dramatically increase model complexity and memory usage due to the need for extensive computations, e.g. computing the full graph Laplacian matrix [22]. Therefore, our aim is on designing an algorithm that has a limited number of hyper-parameters and computational complexity but still reaches top-level performance on several different image sets without the need for reconfiguring the network architecture.

In this paper, we present a deep clustering model based on a generative Variational Autoencoder (VAE) architecture, or MMD-VAE based **D**eep **E**mbedded **C**lustering, denoted as MMV-DEC, and devise training strategies to improve its clustering performance. Unlike conventional clustering algorithms or dimension reduction techniques, which use linear transformation, our method can perform complex non-linear transformations using a deep convolutional neural network (CNN). Our work differs from previous related works, especially another VAE-based approach in [11], in terms of architecture, optimization and performance. Firstly, instead of linear layers, we use convolutional layers to improve the feature extraction capability on image data. Secondly, inspired by a recent published work [25], we use the MMD divergence optimization approach instead of maximizing the Evidence Lower Bound Objective (ELBO) based on Kullback-Leibler (KL) divergence [12]. This helps to avoid the problem of vanishing mutual information between the input image and the embedded latent code. Finally, we integrate a self-learning optimization technique to improve the clustering quality. We demonstrate a significant improvement in performance and generalizability compared with state-of-the-art models by experiments on four different benchmark datasets.

2   Related Works

Deep learning-based clustering has been widely studied in recent years. An early work presented in [22] proposed to use a fully connected stacked autoencoder architecture, with a two-phase training strategy. In the first phase, the network learns a feature transformation via an image reconstruction task. Then in the second phase, a clustering objective is defined based on the network's own predictions to further optimize its parameters and the cluster centroids. Improvements to this architecture have been proposed, with the reconstruction task is maintained during the second phase to preserve the structure of the data generating distribution [9] and convolutional layers are used instead of fully

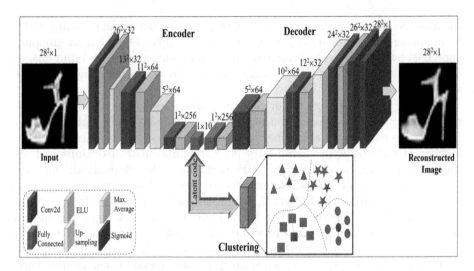

Fig. 1. Overview of our proposed model, which consists of encoder-decoder pathways and a clustering layer stacked on top of the latent layer. The dimensions shown above the layers are those of the feature maps generated at each layer.

connected layers [10]. A more effective approach was introduced in [5], based on jointly and simultaneously optimizing a clustering objective and the autoencoder parameters, without layer-wise pretraining of the autoencoder layers. To achieve noise-invariant predictions, the clustering loss objective is applied on the latent code of a denoising convolutional auto-encoder, whereas the reconstruction loss function is calculated between all the decoder layers and the clean encoder layers. Recently, the authors of [24] proposed to combine spectral clustering with a dual autoencoder, with one encoder pathway for clean input and the other for its noise-contaminated version. To learn more discriminative information from the inputs, they maximize mutual information calculated with a negative image sample randomly selected from the noisy batch. After training this autoencoder for initial latent representation, the latent representations are embedded into the eigenspace of their associated graph Laplacian matrix where clustering is performed. As an alternative training strategy, Joint Unsupervised Learning (JULE) [23] combines the feature representation of a CNN and agglomerative clustering in a recurrent manner. The algorithm starts with an initial over-clustering and alternates between two training steps: merging clusters based on the current network representation and updating network parameters using the current clustering result.

Variational Autoencoders (VAE) [12] and Generative Adversarial Networks (GAN) [7] also have an encoder-decoder network structure and are popular choices for modeling the process of data generation and synthesizing new images. Although VAEs and GANs yield better reconstruction performance in various applications, there is a limited number of published deep clustering algorithms based on VAE or GAN architectures. A clustering approach based on a VAE was introduced in [11]. To be more pertinent for clustering tasks, it models the data generation process by a Mixture of Gaussian prior instead of the original Gaussian prior. The model is optimized using the conventional method of maximizing the ELBO of the data log-likelihood as well as the re-parameterization trick. Another recent work exploits a generative model that performs latent space clustering in a GAN [19]. As the cluster structure is not held in the GAN latent space, they propose to use the mixture sampling of discrete and continuous latent variables and a set of optimization algorithms specialized for the discrete-continuous mixture.

3   Proposed Approach

Our deep clustering model, called MMV-DEC, is composed of two main parts: a VAE network based on MMD regularization, which we name MVAE and a clustering layer which is stacked on top of the latent layer of the MVAE to enable enhanced clustering (EC) optimization. The overall architecture is shown in Fig. 1.

3.1   MVAE

The proposed variational autoencoder network consists of two major components, an encoder and a decoder, each comprising a set of convolutional layers. The encoder extracts the input image features to produce the latent code, which has much lower dimension than the input image. This latent code is fed into the decoder layers to reconstruct the original image. The network functions as a generative model, with encoder g_θ and decoder f_ϕ being functions of the network parameter sets θ and ϕ:

$$X \xrightarrow{g_\theta} Z \xrightarrow{f_\phi} \hat{X}$$

We design the encoder network using three convolutional layers, with 32 kernels of size 3×3, 64 kernels of size 3×3 and 256 kernels of size 5×5, used in each layer, respectively. Each of these convolutional layers is followed by an Exponential Linear Unit (ELU) activation function. In addition, average pooling with kernel size 2×2 and stride of 2 pixels is applied on each of the activated feature maps, except for the third feature map. The pooling operation functions as a down-sampling layer to reduce the spatial dimension. After the third feature map, a fully connected layer is used to generate the latent vector of size 1×10 corresponding to the original grey-level image of size 28×28.

Conversely, the decoder component is designed to gradually increase the feature map dimensions, starting from the latent vector and ending with the output reconstructed image. The decoder pathway starts with one fully connected layer which is followed by four convolutional layers integrated with ELU activation functions. Each of the four convolutional layers uses a set of 64 kernels of size 5×5, 32 kernels of size 3×3, 16 kernels of size 3×3 and 1 kernel of size 3×3, respectively. In addition, appropriate padding is used in combination with bilinear up-sampling layers after each activated feature map such that the reconstructed images have the same size as the original images. The last layer uses a conventional sigmoid function to normalize the output values in the range of $[0,1]$ to produce the output image.

Typically, the cost function to train a VAE includes two terms: the reconstruction loss and the regularization loss. The main difference between a standard VAE and our MVAE network lies in the formulation of the regularization loss. Whereas conventional VAE optimization is based on minimizing the Kullback-Leibler divergence between the generated latent distribution and a prior distribution, we apply Maximum Mean Discrepancy (MMD) divergence [25] to optimize jointly with the reconstruction loss.

During training, each unlabeled image is fed into the network and the reconstruction loss, known as the binary cross entropy (BCE) function, is calculated. Supposing that in each training iteration, a batch of b images is processed, then the BCE loss between input images x_i of m pixels and their reconstructed counterparts $\hat{x}_i$ is measured element-wise by:

$$BCE = -\frac{1}{b \times m} \sum_{i=1}^{b} \sum_{j=1}^{m} [x_{ij} \ln \hat{x}_{ij} + (1 - x_{ij})(1 - \ln(1 - \hat{x}_{ij}))] \quad (1)$$

Simultaneously, the MMD divergence between the distribution of generated latent variables $q(z)$ and a target distribution $p(z')$ is computed by using the kernel embedding formula [8,25]:

$$MMD = E_{p(z'_i),p(z'_j)}[k(z'_i, z'_j)]$$
$$+ E_{q(z_i),q(z_j)}[k(z_i, z_j)]$$
$$- 2E_{p(z'_i),q(z_j)}[k(z'_i, z'_j)] \quad (2)$$

where

$$k(z_i, z_j) = e^{-\frac{||z_i - z_j||}{2\sigma^2}}$$

is a kernel to measure the similarity between two samples z_i, z_j in terms of Euclidean distance. Here, $q(z)$ is the distribution of latent variables generated by the encoder, while $p(z')$ is the prior distribution that we would like $q(z)$ to match. Intuitively, the MMD loss measures the difference between the average similarity of samples within each distribution and the average similarity of mixed samples from both distributions. When MMD reaches 0, the two distributions are matched. Finally, the model is trained to optimize the aggregate cost function, which is equal to the sum of the reconstruction and MMD losses:

$$L_1 = BCE + MMD \quad (3)$$

3.2   Enhanced Clustering Optimization (EC)

To further optimize the latent space for clustering purposes, we borrow the unsupervised self-learning optimization technique proposed in [22]. To achieve this, we connect an additional clustering layer to the latent variable layer, leaving the rest of the MVAE network intact. It is a fully connected layer that uses a t-distribution kernel [17] to measure the similarity between the latent code and the centroid of a target cluster. Specifically, the distance between a sample z_i and the centroid of a given cluster μ_j is calculated as:

$$q_{ij} = \frac{(1 + ||z_i - \mu_j||^2)^{-1}}{\sum_{j'}(1 + ||z_i - \mu_{j'}||^2)^{-1}} \tag{4}$$

This equation calculates the probability of a data point belonging to a cluster represented by its mean μ_j and is translated into the assignment of a class label to the input image. Note that standard K-means clustering is applied in the latent space to determine the initial clusters. Then, during the optimization process, the cluster centroids are updated as learnable parameters.

The self-learning optimization involves defining a target distribution p_{ij} and minimizing the Kullback-Leibler (KL) divergence between p_{ij} and the embedding distribution calculated in Eq. 4. We use a simple and effective empirical target distribution [22], defined as:

$$p_{ij} = \frac{q_{ij}^2 / \sum_i q_{ij}}{\sum_{j'}(q_{ij'}^2 / \sum_i q_{ij'})} \tag{5}$$

The KL divergence used to evaluate the matching between the target distribution and the clustering assignment of latent variables is computed by:

$$D_{KL}(P||Q) = \sum_i \sum_j p_{ij} \log \frac{p_{ij}}{q_{ij}} \tag{6}$$

This KL divergence is used as a regularization term and serves as a guidance criterion for refining the clusters. As p_{ij} is also a function of q_{ij}, optimizing this divergence is considered a self-training process. However, instead of using this single KL divergence as the cost function for optimization, we combine it with a reconstruction loss as suggested in [9] to preserve the local structure of the data distribution and avoid overfitting or getting stuck in local minima [5] during network optimization. Supposing that in each training iteration, a batch of b images is processed, then the element-wise mean squared error (MSE) between b input images x_i and their reconstructed images $\hat{x}_i$ is measured by:

$$MSE = \frac{1}{b \times m} \sum_{i=1}^{b} ||x_i - \hat{x}_i||_2^2 \tag{7}$$

So the final cost function for enhanced clustering optimization is formulated as the weighted sum of the KL divergence and reconstruction loss:

$$L_2 = MSE + \beta D_{KL} \tag{8}$$

where β is a hyper-parameter to adjust the weight of KL divergence and MSE is the reconstruction loss.

Algorithm 1: Training algorithm

Input: VAE network V, unlabeled data U

Stage 1: Training MVAE

repeat

> BCE, Z ← Calculate reconstruction loss, latent code;
> F ← Select target distribution ;
> MMD ← Calculate MMD divergence between Z and F;
> L1 ← Calculate the cost function by sum of BCE and MMD;
> M $\xleftarrow{L1}$ Train model by minimizing the cost function;

until *Maximum number of epochs*;

Finding initial clusters by K-means clustering in latent space ;

Stage 2: Enhanced Clustering Optimization

repeat

> MSE ← Calculate the reconstruction loss;
> Q ← Calculate cluster assignment probability;
> P ← Calculate target distribution;
> KL ← Calculate KL divergence between Q and P;
> L2 ← Calculate the cost function by weighted sum of MSE and KL divergence;
> M $\xleftarrow{L2}$ Update network parameters and clusters centroids;

until *Stopping condition*;

Output: Network weights W'; Cluster centroids μ_j and labels l

---

Fig. 2. Training algorithm for MMV-DEC.

3.3 Training

We use the training algorithm presented in Fig. 2 for all datasets, with the same network configuration and a fixed set of hyper-parameters. It is divided into two major stages: (1) training the MVAE network as a generative model and (2) training the network for enhanced clustering optimization. In the first stage, the training process uses standard back-propagation to update parameters, together with ADAM optimization at a fixed learning rate of 0.001. We select the Gaussian distribution $p(z') \sim \mathcal{N}(0, 0.5)$ as the target distribution for embedding variables. The maximum number of training epochs is set to 200 in order for the reconstructed images to achieve relatively good quality for all the tested datasets. Then, we use simple K-means clustering on the latent variables to find the initial cluster centroids, which are required for Eq. 4 in the first iteration of the clustering optimization stage.

In the second stage, we again use the ADAM optimizer with a fixed learning rate of 0.001 to update the network parameters, including the cluster centroids, after every training iteration. Similar to [9], we set $\beta = 0.1$ to balance the contributions of the loss terms in the L_2 cost function. The stopping condition is triggered when the difference in cluster assignments compared with the previous iteration is below a small threshold.

4 Experiments and Discussion

4.1 Datasets and Evaluation Metrics

To compare our performance results with recently published methods, we evaluated our proposed model on 4 reference image sets, namely MNIST [16], USPS [15], Fashion-MNIST [21], and Cifar-10 [13] as summarized in Table 1. For the Cifar-10 dataset which consists of color images, we adjust the dimension of the input tensor accordingly. Two standard unsupervised evaluation metrics for clustering performance were used: clustering Accuracy (ACC) and Normalized Mutual Information (NMI) [19,24]. The classification label for each input image was produced by applying the well known Hungarian algorithm [14] which maps the predicted clusters assignments to the groundtruth labels. To reduce measurement uncertainty, the performance measurements were averaged from 10 random trials.

Table 1. Image datasets for clustering evaluation

Dataset	# Images	# Classes	Dimension
MNIST	70,000	10	$28 \times 28 \times 1$
USPS	9,298	10	$16 \times 16 \times 1$
Fashion-MNIST	70,000	10	$28 \times 28 \times 1$
Cifar-10	60,000	10	$32 \times 32 \times 3$

4.2 Image Reconstruction

The MVAE embeds the input image into a low-dimensional latent code by the encoder layers and then reconstructs the original image by the decoder layers. By minimizing the integrated reconstruction loss and MMD regularization term, the network is supposed to learn the underlying data representation more effectively than a conventional autoencoder or VAE and thus produce better reconstruction quality. Examples of original and reconstructed images obtained by our model are illustrated in Fig. 3.

Even though the latent representation is significantly compressed compared with input dimensions, in the three datasets MNIST, USPS and Fashion-MNIST, the reconstructed images produced by the decoder (in the even rows) are visually very close to the original ones (in the odd rows), with some minor blurring. Therefore, the useful information to discriminate differing patterns is maintained in the latent code, which is a desired condition for clustering to be directly performed on it. On the other hand, the reconstructed images for Cifar-10 dataset are very blurry which implies the necessity of the preprocessing step and scaling up of the deep CNN architecture. For example, the authors of [2] used Sobel transformation to convert the color images into grayscale images and employed AlexNet or VGG networks for their feature extraction.

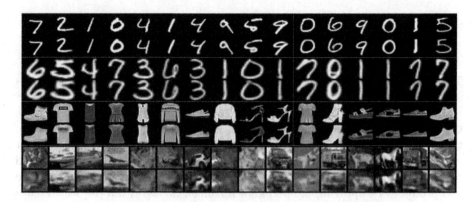

Fig. 3. Original and reconstructed images from the three different datasets: MNIST (rows 1 & 2), USPS (rows 3 & 4), Fashion-MNIST (rows 5 & 6) and Cifar-10 (rows 7 & 8).

4.3   Analysis of Training Strategies

We validated the effectiveness of applying the MMD regularization and enhanced clustering (EC) optimization strategies by comparing five variants of our model: (1) Convolutional autoencoder using our encoder-decoder architecture but trained with only the reconstruction loss (ConvAE); (2) Convolutional auto-encoder using enhanced clustering optimization (ConvAE+EC); (3) Convolutional autoencoder trained with both reconstruction and conventional KL divergence losses (ConvAE+KL div.); (4) Convolutional autoencoder trained with both reconstruction and MMD losses (ConvAE+MMD), which is the MVAE network; (5) MVAE network using enhanced clustering optimization (ConvAE +MMD+EC), which is our proposed MMV-DEC model.

As shown in Table 2, each training strategy of MMD loss and EC improves the clustering ACC and NMI results effectively on all four benchmarks, especially on the MNIST and USPS datasets. The application of both techniques consistently produces the highest performance, thus demonstrating the appropriateness of combining them in training. Although the improvement is not significant in the case of the Fashion-MNIST dataset and Cifar-10, the relatively high performance of the basic configuration (ConvAE) implies that the network architecture is well designed, laying the foundation for other optimization strategies. As can also be seen in the table, the application of conventional KL divergence loss during the network training is inferior to MMD loss optimization and even deteriorate the clustering performance of the autoencoder network in case of the Fashion-MNIST dataset.

Figure 4 provides a visual comparison of the latent space for the USPS dataset by applying the t-SNE visualization method [17] on the embedded code space Z. This visualization reveals that training with the combined reconstruction and MMD losses (MVAE network) produces more compact clusters that are

Table 2. Analysis of different training strategies

Strategies	MNIST		USPS		Fashion		Cifar-10	
	ACC	NMI	ACC	NMI	ACC	NMI	ACC	NMI
ConvAE	86.6	77.7	72.3	70.9	61.5	64.0	21.9	8.9
ConvAE+EC	95.2	91.2	78.7	82.7	62.1	65.8	23.3.0	9.5
ConvAE+KL div.	86.9	79.7	78.2	72.5	50.4	48.2	22.8	9.1
ConvAE+MMD	90.0	81.0	85.9	78.6	61.7	64.9	23.0	9.8
Our MMV-DEC	**96.8**	**93.3**	**96.4**	**91.2**	**62.9**	**66.2**	**24.1**	**10.4**

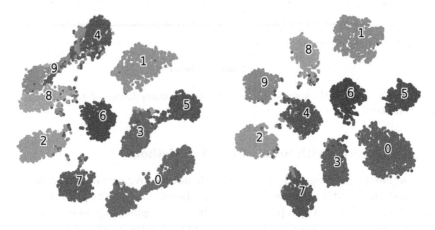

Fig. 4. Comparison of latent representations of USPS dataset produced by our model without (left) and with (right) MMD regularization. Colors represent true labels of the samples, and class numbers are positioned at cluster centroids. (Color figure online)

easier to discriminate, compared with using only the reconstruction loss (ConvAE network).

Figure 5 displays the improvement in clustering accuracy and NMI during Stage 2 of training (EC optimization) of our network for one trial example. We can see here that the clustering optimization technique is generally very fast and effective. In addition, the output performance of the Stage 1-trained MVAE plays a vital role in laying the basis for further clustering optimization. Indeed, the clustering optimization increases performance significantly on the MNIST (ACC: $\sim$ 7%, NMI: $\sim$ 12%) and USPS (ACC: $\sim$ 7%, NMI: $\sim$ 11%) datasets; in these cases, the Stage 1-trained MVAE provides high clustering capability to initiate Stage 2. For the Fashion-MNIST and Cifar-10 datasets, however, the improvement is more limited, as it is impacted by the quality of the previously trained MVAE.

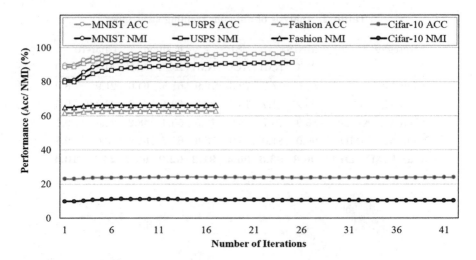

Fig. 5. ACC and NMI metrics during EC optimization (Stage 2 of training) with MVAE network.

4.4 Comparison with State of the Art Methods

We compared our proposed approach with both conventional clustering baselines and state-of-the-art deep clustering algorithms. Peformance results for these other methods were reported either in their original papers or compiled in recently published papers [19,24]. For the Cifar-10 dataset, we obtain the results by running the released codes of corresponding works and for those results that are not practical to obtain, we indicate by dash marks (−). As can be seen in Table 3, our proposed approach outperforms the conventional methods based on K-means or Spectral clustering (SC-LS) on all four datasets by a large margin. Furthermore, our model yields better performances than several other deep embedded clustering methods, including DEC, IDEC, DCEC, as well as a generative model based on the variational autoencoder (VaDE). Compared with the methods achieving the highest performances in the literature, namely JULE, DEPICT, Dual-AE and the GAN models (ClusterGAN, InfoGAN), our proposed MMV-DEC achieves higher overall clustering performance. In particular, it outperforms other methods on Cifar-10 dataset following both ACC and NMI criteria. It also achieves the highest accuracy (96.4%) on the USPS and the highest NMI (66.2%) on the Fashion-MNIST dataset. In general, our MMV-DEC secures at least 2nd best performance according to the two metrics on all benchmarks.

More importantly, the experiment results demonstrate the better generalization ability across datasets of our proposed framework. Note that none of the previous methods gain a top-two performance across all datasets and the latest state of the art methods could not provide consistent results. For example, although the USPS and MNIST datasets are similar, the difference in clustering accuracy is more than 10% in most of recent works [9,10,22,24] and it is even

Table 3. Comparison of different clustering algorithms on benchmark datasets based on NMI and ACC metrics. The top two performances are highlighted in each column.

Methods	MNIST		USPS		Fashion		Cifar-10	
	ACC	NMI	ACC	NMI	ACC	NMI	ACC	NMI
K-means [18]	53.2	50.0	66.8	60.1	47.4	51.2	19.8	7.6
SC-LS [1]	71.4	70.6	74.6	75.5	49.6	49.7	20.6	9.1
DEC [22]	86.3	83.4	76.2	76.7	51.8	54.6	21.6	8.4
JULE [23]	96.4	91.3	**95.0**	**91.3**	56.3	60.8	–	–
VaDE [11]	94.5	87.6	56.6	51.2	57.8	63.0	20.1	8.1
IDEC [9]	88.1	86.7	76.1	78.5	52.9	55.7	19.6	8.3
DCEC [10]	89.0	88.5	79.0	82.57	–	–	22.3	8.7
DEPICT [5]	96.5	91.7	89.9	90.6	39.2	39.2	22.8	9.6
InfoGAN [3]	89.0	86.0	–	–	61.0	59.0	–	–
ClusterGAN [19]	95.0	89.0	–	–	**63.0**	64.0	–	–
Dual-AE [24]	**97.8**	**94.1**	86.9	85.7	**66.2**	64.5	**23.9**	9.8
Our MMV-DEC	96.8	93.3	96.4	91.2	62.9	**66.2**	24.1	10.4

more than 35% for the similar approach based on conventional VAE network [11]. With our proposed method, this performance gap is non-remarkable and without the need of adjusting the training hyper-parameters.

5 Conclusion

In this paper, we present a new unsupervised deep clustering method that is based on a variational autoencoder architecture. The application of self-learning mechanism and MMD loss optimization consistently produces the highest effectiveness and generalization ability. The model also has the advantages of low computational complexity and few hyper-parameters to adjust. Experiments on four image benchmarks demonstrate that our proposed MMV-DEC model can reach state-of-the-art performance without requiring to reconfigure the network architecture or change the clustering hyper-parameters. Our further work will focus on more realistic images, where it is necessary to scale up the deep network architecture and apply preprocessing steps on the input images.

References

1. Cai, D., Chen, X.: Large scale spectral clustering via landmark-based sparse representation. IEEE Trans. Cybern. **45**(8), 1669–1680 (2015)
2. Caron, M., Bojanowski, P., Joulin, A., Douze, M.: Deep clustering for unsupervised learning of visual features. In: Ferrari, V., Hebert, M., Sminchisescu, C., Weiss, Y. (eds.) Computer Vision – ECCV 2018, Part XIV. LNCS, vol. 11218, pp. 139–156. Springer, Cham (2018). https://doi.org/10.1007/978-3-030-01264-9_9

3. Chen, X., Duan, Y., Houthooft, R., Schulman, J., Sutskever, I., Abbeel, P.: Info-GAN: interpretable representation learning by information maximizing generative adversarial nets. In: Advances in Neural Information Processing Systems 29: Annual Conference on Neural Information Processing Systems, pp. 2172–2180 (2016)
4. Dilokthanakul, N., et al.: Deep unsupervised clustering with gaussian mixture variational autoencoders. CoRR (2016)
5. Dizaji, K.G., Herandi, A., Deng, C., Cai, W., Huang, H.: Deep clustering via joint convolutional autoencoder embedding and relative entropy minimization. In: IEEE International Conference on Computer Vision, ICCV, pp. 5747–5756 (2017)
6. Forsyth, D.A., Ponce, J.: Computer Vision - A Modern Approach, 2nd edn. Pearson, Lewiston (2012)
7. Goodfellow, I.J., et al.: Generative adversarial nets. In: Advances in Neural Information Processing Systems 27: Annual Conference on Neural Information Processing Systems, pp. 2672–2680 (2014)
8. Gretton, A., Borgwardt, K.M., Rasch, M.J., Schölkopf, B., Smola, A.J.: A kernel method for the two-sample-problem. In: Schölkopf, B., Platt, J.C., Hofmann, T. (eds.) Advances in Neural Information Processing Systems 19, Proceedings of the Twentieth Annual Conference on Neural Information Processing Systems, pp. 513–520 (2006)
9. Guo, X., Gao, L., Liu, X., Yin, J.: Improved deep embedded clustering with local structure preservation. In: Proceedings of the Twenty-Sixth International Joint Conference on Artificial Intelligence, IJCAI, pp. 1753–1759 (2017)
10. Guo, X., Liu, X., Zhu, E., Yin, J.: Deep clustering with convolutional autoencoders. In: Liu, D., Xie, S., Li, Y., Zhao, D., El-Alfy, E.S. (eds.) ICONIP 2017. LNCS, vol. 10635, pp. 373–382. Springer, Cham (2017). https://doi.org/10.1007/978-3-319-70096-0_39
11. Jiang, Z., Zheng, Y., Tan, H., Tang, B., Zhou, H.: Variational deep embedding: an unsupervised and generative approach to clustering. In: Proceedings of the Twenty-Sixth International Joint Conference on Artificial Intelligence, IJCAI, pp. 1965–1972 (2017)
12. Kingma, D.P., Welling, M.: Auto-encoding variational Bayes. In: 2nd International Conference on Learning Representations, ICLR (2014)
13. Krizhevsky, A.: Learning multiple layers of features from tiny images. Technical report (2009)
14. Kuhn, H.W.: The Hungarian method for the assignment problem. In: Jünger, M., et al. (eds.) 50 Years of Integer Programming 1958-2008, pp. 29–47. Springer, Heidelberg (2010). https://doi.org/10.1007/978-3-540-68279-0_2
15. Le Cun, Y., et al.: Handwritten zip code recognition with multilayer networks. In: [1990] Proceedings. 10th International Conference on Pattern Recognition, vol. ii, pp. 35–40 (1990)
16. Lecun, Y., Bottou, L., Bengio, Y., Haffner, P.: Gradient-based learning applied to document recognition. In: Proceedings of the IEEE, pp. 2278–2324 (1998)
17. van der Maaten, L., Hinton, G.: Visualizing data using t-SNE. J. Machine Learn. Res. **9**, 2579–2605 (2008)
18. Macqueen, J.: Some methods for classification and analysis of multivariate observations. In: 5th Berkeley Symposium on Mathematical Statistics and Probability, pp. 281–297 (1967)
19. Mukherjee, S., Asnani, H., Lin, E., Kannan, S.: ClusterGAN: latent space clustering in generative adversarial networks. In: The Thirty-Third AAAI Conference on Artificial Intelligence, AAAI, pp. 4610–4617 (2019)

20. Shi, J., Malik, J.: Normalized cuts and image segmentation. IEEE Trans. Pattern Anal. Machine Intell. **22**(8), 888–905 (2000)
21. Xiao, H., Rasul, K., Vollgraf, R.: Fashion-MNIST: a novel image dataset for benchmarking machine learning algorithms. CoRR abs/1708.07747 (2017)
22. Xie, J., Girshick, R.B., Farhadi, A.: Unsupervised deep embedding for clustering analysis. In: Proceedings of the 33nd International Conference on Machine Learning, ICML, vol. 48, pp. 478–487 (2016)
23. Yang, J., Parikh, D., Batra, D.: Joint unsupervised learning of deep representations and image clusters. In: 2016 IEEE Conference on Computer Vision and Pattern Recognition, CVPR, pp. 5147–5156 (2016)
24. Yang, X., Deng, C., Zheng, F., Yan, J., Liu, W.: Deep spectral clustering using dual autoencoder network. In: IEEE Conference on Computer Vision and Pattern Recognition, CVPR, pp. 4066–4075 (2019)
25. Zhao, S., Song, J., Ermon, S.: InfoVAE: balancing learning and inference in variational autoencoders. In: The Thirty-Third AAAI Conference on Artificial Intelligence, AAAI, pp. 5885–5892 (2019)

Author Index

Printed in the United States
By Bookmasters